Lecture Notes in Computer Science 11603

Commenced Publication in 1973
Founding and Former Series Editors:
Gerhard Goos, Juris Hartmanis, and Jan van Leeuwen

More information about this series at http://www.springer.com/series/7412

Jan Lellmann · Martin Burger ·
Jan Modersitzki (Eds.)

Scale Space
and Variational Methods
in Computer Vision

7th International Conference, SSVM 2019
Hofgeismar, Germany, June 30 – July 4, 2019
Proceedings

 Springer

Editors
Jan Lellmann 🆔
University of Lübeck
Lübeck, Germany

Martin Burger
University of Erlangen-Nuremberg (FAU)
Erlangen, Germany

Jan Modersitzki
University of Lübeck
Lübeck, Germany

ISSN 0302-9743   ISSN 1611-3349 (electronic)
Lecture Notes in Computer Science
ISBN 978-3-030-22367-0   ISBN 978-3-030-22368-7 (eBook)
https://doi.org/10.1007/978-3-030-22368-7

LNCS Sublibrary: SL6 – Image Processing, Computer Vision, Pattern Recognition, and Graphics

This Springer imprint is published by the registered company Springer Nature Switzerland AG
The registered company address is: Gewerbestrasse 11, 6330 Cham, Switzerland

Dedicated to Mila Nikolova

Preface

Welcome to the proceedings of the 7th International Conference on Scale Space and Variational Methods in Computer Vision (SSVM) 2019.

This conference was held at Hofgeismar in the vicinity of Kassel, Germany. The conference continued the tradition of biannual events taking place in beautiful and more or less remote places, stimulating an optimal exchange of ideas between its participants. The conference series originates from a biannual series of conference on scale space—Utrecht (1997), Corfu (1999), Vancouver (2001), Isle of Skye (2003), Hofgeismar (2005)—and a Workshop on Variational, Geometric, and Level Set Methods (VLSM) in Vancouver (2001). SSVM successfully merged the communities and was continued under its current name at Ischia (2007), Voss (2009), Ein-Gedi (2011), Seggau (2013), Lège-Cap-Ferret (2015), and Kolding (2017).

The conference provides a platform for state-of-the-art scientific exchange on topics related to computer vision and image analysis, including diverse themes such as 3D vision, convex and non-convex modeling, image analysis, inpainting, inverse problems in imaging, level-set methods, manifold-valued data processing, optimization methods in imaging, PDEs in image processing, registration, restoration and reconstruction, scale-space methods, segmentation, and variational methods. The 44 contributions in this proceedings volume demonstrate a strong and lively community. All submissions underwent a rigorous peer-review process similar to that of a high-ranking journal in the field.

Following the successful tradition of previous SSVM conferences, four outstanding researchers were invited to give keynote presentations: Gitta Kutyniok (TU Berlin, Germany), Édouard Oudet (Université Grenoble Alpes, France), Alessandro Sarti (EHESS, France), and Julia Schnabel (King's College London, UK).

Finally, we would like to thank all those who contributed to the success of this conference. We thank the DAGM and DFG for financial support. We are very grateful to the organizers of SSVM 2017 for invaluable advice on organizing the event: François Lauze, Yiqiu Dong, and Anders Bjorholm Dahl. We also thank Joachim Weickert for continuous support and advice on the more difficult questions. A very special thank you goes to the busy bees in the background: the members of our three institutes, and in particular Kai Brehmer, Dana Knabe, Kerstin Sietas, and Thomas Vogt.

Most importantly, we would like to thank all authors and reviewers for their hard work. Without your help, and without your enthusiasm and commitment, an event like this would not have been possible!

April 2019

Jan Lellmann
Martin Burger
Jan Modersitzki

Tribute to Mila Nikolova (1962–2018)

Our colleague and friend Mila Nikolova, Research Director at Centre National de la Recherche Scientifique (CNRS), Microsoft Fellow, Simons Fellow, and IEEE Senior Member, passed away on June 20, 2018.

Mila was very active and influential in the international image processing community with an eye for interesting theoretical results that are useful in practice. She worked diligently on the Program Committee for several SSVM conferences. Mila's contributions to mathematics and signal processing were substantial and lasting. She established important models used for compressive sensing and studied variational models with non-smooth data-fidelity terms, demonstrating their ability to exactly match given data. Her two-phase approach to restore images corrupted by blur and impulse noise is still state of the art. Mila was awarded with the Michel Monpetit Prize from the French Academy of Science in 2010 "for the originality and depth of her research in mathematical image processing and in solving certain inverse problems."

She will remain on our minds through her important scientific work, her generosity, and her enthusiasm.

Gabriele Steidl

Organization

Organizing Committee and Editors

Jan Lellmann   University of Lübeck, Germany
Martin Burger   University of Erlangen-Nürnberg (FAU), Germany
Jan Modersitzki   University of Lübeck and Fraunhofer MEVIS, Germany

Scientific and Program Committee

Luis Alvarez   Universidad de las Palmas de Gran Canaria, Spain
Freddie Åström   Heidelberg University, Germany
Jean-François Aujol   University of Bordeaux, France
Benjamin Berkels   RWTH Aachen, Germany
Michael Breuß   Brandenburg Technical University, Germany
Freddy Bruckstein   Technion, Israel
Andrés Bruhn   University of Stuttgart, Germany
Antonin Chambolle   École Polytechnique, France
Agnès Desolneux   CMLA ENS Paris Saclay, France
Yiqiu Dong   DTU Compute, DTU, Denmark
Remco Duits   Eindhoven University of Technology, The Netherlands
Jean-Denis Durou   IRIT, France
Jalal Fadili   Normandy University, CNRS, France
Guy Gilboa   Technion, Israel
Martin Holler   École Polytechnique, France
Atsushi Imiya   IMIT Chiba University, Japan
Sung Ha Kang   Georgia Tech, USA
Francois Lauze   University of Copenhagen, Denmark
Tony Lindeberg   KTH Royal Institute of Technology, Sweden
Dirk Lorenz   TU Braunschweig, Germany
Russel Luke   University of Göttingen, Germany
Serena Morigi   University of Bologna, Italy
Nicolas Papadakis   CNRS/IMB, France
Pascal Peter   Saarland University, Germany
Stefania Petra   Heidelberg University, Germany
Gabriel Peyré   CNRS and ENS, France
Thomas Pock   Graz University of Technology, Austria
Yvain Quéau   Technical University Munich, Germany
Martin Rumpf   University of Bonn, Germany
Lars Ruthotto   Emory University, USA
Otmar Scherzer   University of Vienna, Austria
Bernhard Schmitzer   University of Lübeck, Germany
Christoph Schnörr   Heidelberg University, Germany

Carola Schoenlieb	Cambridge University, UK
Fiorella Sgallari	University of Bologna, Italy
Gabriele Steidl	TU Kaiserslautern, Germany
Xue-Cheng Tai	University of Bergen, Norway
Joachim Weickert	Saarland University, Germany
Martin Welk	UMIT Hall/Tyrol, Austria

Invited Speakers

Gitta Kutyniok	Einstein Chair and Head of the Applied Functional Analysis Group, Technical University of Berlin, Germany
Édouard Oudet	Université Grenoble Alpes, France
Alessandro Sarti	Research Director at the Center of Mathematics CAMS (CNRS-EHESS), France
Julia Schnabel	Chair of Computational Imaging at the School of Biomedical Engineering and Imaging Sciences, King's College London, UK

Sponsoring Institutions

DAGM, Deutsche Arbeitsgemeinschaft für Mustererkennung e.V.
DFG, Deutsche Forschungsgemeinschaft
MIC, Institute of Mathematics and Image Computing, University of Lübeck

Contents

3D Vision and Feature Analysis

The Fractional Harris-Laplace Feature Detector. 3
  Matthew Adams

Macrocanonical Models for Texture Synthesis 13
  *Valentin De Bortoli, Agnès Desolneux, Bruno Galerne,
  and Arthur Leclaire*

Finding Structure in Point Cloud Data with the Robust Isoperimetric Loss. . . 25
  Shay Deutsch, Iacopo Masi, and Stefano Soatto

Deep Eikonal Solvers 38
  Moshe Lichtenstein, Gautam Pai, and Ron Kimmel

A Splitting-Based Algorithm for Multi-view Stereopsis
of Textureless Objects. 51
  Jean Mélou, Yvain Quéau, Fabien Castan, and Jean-Denis Durou

Inpainting, Interpolation and Compression

Pseudodifferential Inpainting: The Missing Link Between
PDE- and RBF-Based Interpolation. 67
  Matthias Augustin, Joachim Weickert, and Sarah Andris

Towards PDE-Based Video Compression with Optimal Masks
and Optic Flow. 79
  Laurent Hoeltgen, Michael Breuß, and Georg Radow

Compressing Audio Signals with Inpainting-Based Sparsification 92
  Pascal Peter, Jan Contelly, and Joachim Weickert

Alternate Structural-Textural Video Inpainting for Spot Defects Correction
in Movies 104
  *Arthur Renaudeau, François Lauze, Fabien Pierre,
  Jean-François Aujol, and Jean-Denis Durou*

Inverse Problems in Imaging

Iterative Sampled Methods for Massive and Separable Nonlinear
Inverse Problems. 119
  Julianne Chung, Matthias Chung, and J. Tanner Slagel

Refitting Solutions Promoted by ℓ_{12} Sparse Analysis Regularizations
with Block Penalties 131
  Charles-Alban Deledalle, Nicolas Papadakis, Joseph Salmon,
  and Samuel Vaiter

An Iteration Method for X-Ray CT Reconstruction
from Variable-Truncation Projection Data 144
  Limei Huo, Shousheng Luo, Yiqiu Dong, Xue-Cheng Tai,
  and Yang Wang

A New Iterative Method for CT Reconstruction with Uncertain
View Angles 156
  Nicolai André Brogaard Riis and Yiqiu Dong

Optimization Methods in Imaging

Time Discrete Geodesics in Deep Feature Spaces for Image Morphing 171
  Alexander Effland, Erich Kobler, Thomas Pock, and Martin Rumpf

Minimal Lipschitz Extensions for Vector-Valued Functions
on Finite Graphs.. 183
  Johannes Hertrich, Miroslav Bačák, Sebastian Neumayer,
  and Gabriele Steidl

PDEs and Level-Set Methods

The Convex-Hull-Stripping Median Approximates Affine
Curvature Motion 199
  Martin Welk and Michael Breuß

Total Variation and Mean Curvature PDEs on the Space of Positions
and Orientations 211
  Remco Duits, Etienne St-Onge, Jim Portegies, and Bart Smets

A Variational Convex Hull Algorithm 224
  Lingfeng Li, Shousheng Luo, Xue-Cheng Tai, and Jiang Yang

PDE Evolutions for M-Smoothers: From Common Myths
to Robust Numerics.. 236
  Martin Welk and Joachim Weickert

Registration and Reconstruction

Variational Registration of Multiple Images with the SVD Based SqN
Distance Measure 251
  Kai Brehmer, Hari Om Aggrawal, Stefan Heldmann,
  and Jan Modersitzki

Multi-tasking to Correct: Motion-Compensated MRI via Joint
Reconstruction and Registration   263
    *Veronica Corona, Angelica I. Aviles-Rivero, Noémie Debroux,
    Martin Graves, Carole Le Guyader, Carola-Bibiane Schönlieb,
    and Guy Williams*

Variational Image Registration for Inhomogeneous-Resolution Pairs   275
    Kento Hosoya and Atsushi Imiya

Scale-Space Methods

Computing Nonlinear Eigenfunctions via Gradient Flow Extinction.   291
    Leon Bungert, Martin Burger, and Daniel Tenbrinck

Sparsification Scale-Spaces.   303
    Marcelo Cárdenas, Pascal Peter, and Joachim Weickert

Stable Explicit p-Laplacian Flows Based on Nonlinear
Eigenvalue Analysis   315
    Ido Cohen, Adi Falik, and Guy Gilboa

Provably Scale-Covariant Networks from Oriented Quasi Quadrature
Measures in Cascade.   328
    Tony Lindeberg

A Fast Multi-layer Approximation to Semi-discrete Optimal Transport.   341
    Arthur Leclaire and Julien Rabin

Segmentation and Labeling

Global Similarity with Additive Smoothness for Spectral Segmentation   357
    Vedrana Andersen Dahl and Anders Bjorholm Dahl

Segmentation of 2D and 3D Objects with Intrinsically Similarity Invariant
Shape Regularisers   369
    Jacob Daniel Kirstejn Hansen and François Lauze

Lattice Metric Space Application to Grain Defect Detection   381
    Yuchen He and Sung Ha Kang

Learning Adaptive Regularization for Image Labeling Using
Geometric Assignment.   393
    *Ruben Hühnerbein, Fabrizio Savarino, Stefania Petra,
    and Christoph Schnörr*

Direct MRI Segmentation from k-Space Data by Iterative
Potts Minimization   406
    Lukas Kiefer, Stefania Petra, Martin Storath, and Andreas Weinmann

A Balanced Phase Field Model for Active Contours   419
    Jozsef Molnar, Ervin Tasnadi, and Peter Horvath

Unsupervised Labeling by Geometric and Spatially
Regularized Self-assignment   432
    Matthias Zisler, Artjom Zern, Stefania Petra, and Christoph Schnörr

Variational Methods

Aorta Centerline Smoothing and Registration Using Variational Models   447
    Luis Alvarez, Daniel Santana-Cedrés, Pablo G. Tahoces,
    and José M. Carreira

A Connection Between Image Processing and Artificial Neural Networks
Layers Through a Geometric Model of Visual Perception   459
    Thomas Batard, Eduard Ramon Maldonado, Gabriele Steidl,
    and Marcelo Bertalmío

A Cortical-Inspired Model for Orientation-Dependent Contrast Perception:
A Link with Wilson-Cowan Equations   472
    Marcelo Bertalmío, Luca Calatroni, Valentina Franceschi,
    Benedetta Franceschiello, and Dario Prandi

A Total Variation Based Regularizer Promoting
Piecewise-Lipschitz Reconstructions   485
    Martin Burger, Yury Korolev, Carola-Bibiane Schönlieb,
    and Christiane Stollenwerk

A Non-convex Nonseparable Approach to Single-Molecule
Localization Microscopy   498
    Raymond H. Chan, Damiana Lazzaro, Serena Morigi,
    and Fiorella Sgallari

Preservation of Piecewise Constancy under TV Regularization
with Rectilinear Anisotropy   510
    Clemens Kirisits, Otmar Scherzer, and Eric Setterqvist

Total Directional Variation for Video Denoising   522
    Simone Parisotto and Carola-Bibiane Schönlieb

Joint CNN and Variational Model for Fully-Automatic
Image Colorization   535
    Thomas Mouzon, Fabien Pierre, and Marie-Odile Berger

A Variational Perspective on the Assignment Flow   547
  Fabrizio Savarino and Christoph Schnörr

Functional Liftings of Vectorial Variational Problems
with Laplacian Regularization.   559
  Thomas Vogt and Jan Lellmann

Author Index   573

3D Vision and Feature Analysis

The Fractional Harris-Laplace
Feature Detector

Matthew Adams$^{(\boxtimes)}$ (iD)

University of Calgary, 2500 University Drive NW, Calgary, Alberta T2N 1N4, Canada
Matthew.Adams@ucalgary.ca
http://contacts.ucalgary.ca/info/math/profiles/1-7240647

Abstract. Fractional calculus is an extension of integer-order differentiation and integration which explains many natural physical processes. New applications of the fractional calculus are in constant development. The current paper introduces fractional differentiation to feature detection in digital images. The Harris-Laplace feature detector is adapted to use the non-local properties of the fractional derivative to include more information about image pixel perturbations when quantifying features. Using fractional derivatives in the Harris-Laplace detector leads to higher repeatability when detecting features in grayscale images. Applications of this development are suggested.

Keywords: Feature detection · Image processing · Fractional calculus

1 Introduction

Soon after the inception of the differential and integral calculus, questions began circulating regarding the extension of integer-order derivatives and integrals to non-integer orders [9]. The extension of the calculus to non-integer orders resulted in new definitions of the 'fractional' derivative and integral. Several definitions of the fractional derivative (or integral) exist [10]. This paper considers only the Grünwald-Letnikov fractional derivative.

The topic of fractional calculus has been studied with renewed vigor since the work of Oldham and Spanier in 1974 [9,10]. Since that time, several applications have been found in control theory and differential equations [1]. More recently, the fractional calculus has been applied to problems within the field of digital image processing [5,7,12].

This paper introduces the fractional calculus to the Harris-Laplace feature detector [8], a refinement of the classic Harris detector described in [4]. It has been shown that fractional derivatives have greater immunity to random image noise than integer derivatives [3,5,12]. While SIFT [6] is used in many applications for its accuracy, methods based on the work of Harris and Stephens still enjoy wide popularity due to their ease of implementation, accuracy in highly textured images [8], and their inclusion in the open source image processing and computer vision library OpenCV [2]. The adapted feature detector is called

© Springer Nature Switzerland AG 2019
J. Lellmann et al. (Eds.): SSVM 2019, LNCS 11603, pp. 3–12, 2019.
https://doi.org/10.1007/978-3-030-22368-7_1

the 'Fractional Harris-Laplace' feature detector. In this paper, it is determined that the Fractional Harris-Laplace feature detector leads to an improvement in the repeatability of the Harris-Laplace detector, and in the number of matched features.

2   Fractional Differentiation

Since the inception of the differential and integral calculus, the fractional calculus has been a companion subject to mathematical analysis [9]. The aim of this subject is to give definitions and properties for linear operators of the form

$$D^\alpha := \frac{d^\alpha}{dx^\alpha},$$

where α is not an integer. Indeed, the term 'fractional' is a misnomer, since α need not take on rational values. In the fractional calculus, we may consider any $\alpha \in \mathbb{R}$, though the geometric and physical intuition behind non-integer derivative and integral orders becomes quite difficult. At present, the most important property of a fractional derivative is its non-locality, which can clearly be seen in the Grünwald-Letnikov (GL) definition of the fractional derivative.

Definition 1. *For a fractional differentiable function f, the backward Grünwald-Letnikov derivative of order α, between the limits 0 and x, is defined as*

$$D^\alpha f(x) = \lim_{N \to \infty} \left(\frac{N}{x} \right)^\alpha \frac{1}{\Gamma(-\alpha)} \sum_{k=0}^{N-1} \frac{\Gamma(k-\alpha)}{\Gamma(k+1)} f\left(x - \frac{kx}{N} \right), \qquad (1)$$

where $\Gamma(\cdot)$ is the gamma function.

2.1   The CRONE Operator

A signal-processing approach was undertaken by Mathieu, Melchior, Oustaloup, and Ceyral [7] to define an edge-detecting filter based on the GL fractional derivative. The resulting operator was called 'Contour Robuste d'Ordre Non Entier' (Robust Edges of Non-integer Order) or CRONE. This operator is defined as follows.

$$D^\alpha f(x) = h^{-\alpha} \sum_{k=0}^{\infty} b_k [f(x - kh) - f(x + kh)], \qquad (2)$$

where h is the step size, the b_k coefficients are defined as

$$b_k = \frac{(-\alpha)_k}{\Gamma(k+1)}, \qquad (3)$$

and we have used the Pochhammer symbol $(a)_k = a(a+1)\cdots(a+k-1)$.

In practice, the summation may only extend as far as the size of the array of function values. Thus, the sum in (2) is truncated at the upper limit of N, where N is the size of the array to be differentiated.

The CRONE operator agrees with the definition of the GL derivative at the endpoints of an array of function values. However, away from the endpoints of the array to be differentiated, the CRONE operator represents a weighted sum of both the forward and backward GL derivatives. The analysis in [7] shows that the use of the CRONE operator results in a differentiation filter that features improved immunity to random noise and increased selectivity over integer-order differentiation filters.

To compute using the CRONE operator, the coefficients b_k are placed into an array of the form

$$B = [b_N, b_{N-1}, \dots, b_1, 0, -b_1, \dots, -b_{N-1}, -b_N].$$ (4)

The array B is then convolved with the array of function values. Thus, if F represents the array of function values of size N, the CRONE algorithm may be written as

$$[D^\alpha f(x)]_{CRONE} = h^{-\alpha}(F * B).$$ (5)

The CRONE algorithm has computational complexity $\mathcal{O}(N^2)$.

3   The Harris-Laplace Detector

In the original Harris corner detector, the detection of salient features centered on the analysis of the following matrix.

$$A = w * \begin{bmatrix} I_x^2 & I_x I_y \\ I_x I_y & I_y^2 \end{bmatrix},$$ (6)

The matrix A will hereafter be referred to as the *stucture tensor*. The original authors of [4] presented an 'inspired formulation' for determining the extent to which a group of pixels represented a corner. The corner response function was given as

$$R = \det(A) - k \cdot \mathrm{Tr}(A),$$ (7)

where $k = 0.04$. The specific selection criteria for the value of k was not presented in the original article. The explicit reasoning behind the formulation of the corner response function was similarly avoided, however the authors mentioned that its purpose was to avoid the calculation of eigenvalues. This corner response function has the following interpretation:

$$\begin{cases} R > 0, & \text{corner detected.} \\ R < 0, & \text{edge detected.} \\ R \approx 0, & \text{flat region detected.} \end{cases}$$ (8)

A direct descendant of the Harris detector, the Harris-Laplace (HL) detector in [8] uses the Harris detector across various image scales to detect features, and then uses a Laplacian-of-Gaussian (LoG) kernel to match features across scales. To accomplish this task, the HL algorithm iterates through a range of image scales $\sigma_n = \xi^n \sigma_0$. The scale factor ξ represents the change in scale between two images and is set at 1.4, and σ_0 is the scale of the original image. In order to obtain the scale change for a given image, the image is smoothed with a Gaussian filter with scale parameter σ_n. Non-maximal suppression is used at each successive scale, and non-maximal features within an 8-neighborhood of a given feature are discarded. The structure tensor used in the classic Harris detector was adapted for use across scales by introducing the differentiation scale σ_D and the integration scale σ_I. The differentiation scale represents the size of the Gaussian filter used in calculating the local derivatives, and the integration scale σ_I is the smoothing weight, also a Gaussian, from (6). The scale-adapted stucture tensor is given by the expression

$$A = \sigma_D^2 \, G(\sigma_I) * \begin{bmatrix} \tilde{I}_x^2 & \tilde{I}_x \tilde{I}_y \\ \tilde{I}_y \tilde{I}_y & \tilde{I}_y^2 \end{bmatrix}, \tag{9}$$

where $\tilde{I}_j$ represents the derivative in the j-direction computed with the Gaussian filter $G(\sigma_D)$, and $G(\sigma_I)$ represents the local smoothing filter [14]. Since the derivative of a Gaussian will have the reciprocal of the scale parameter as a factor, the presence of the scale factor σ_D^2 ensures that the stucture tensor is properly scaled after differentiation. The cornerness measure from (7) was used with $k = 0.05$.

After computing features across image scales σ_n, the LoG is used to detect maxima in feature pixels over scales. The LoG function detects blob-like structures, and is approximated by the DoG function in the SIFT algorithm. It not only detects the scale at which the detected features are found, but locates them as well. The magnitude of the LoG function is given as

$$|\mathrm{LoG}(\sigma_n)| = \sigma_n^2 |L_{xx}(\sigma_n) + L_{yy}(\sigma_n)|. \tag{10}$$

When the LoG function attains its maxima, a match between the detected feature and the underlying object in the image is indicated [8]. This occurs when the LoG kernel has the same size as a given local image structure. In this case, these local image structures are dilated image features.

4   The Fractional Harris-Laplace Detector

The fractional derivative with order $\alpha \in (0, 1)$ is now introduced to the Harris-Laplace detector. The only major change from the original HL detector is the stucture tensor A in (9). With the fractional derivative of order α in the j-direction represented as $_\alpha I_j$, the stucture tensor is now written as

$$\tilde{A} = G(\sigma_I) * \begin{bmatrix} _\alpha I_x^2 & _\alpha I_x \, _\alpha I_x \\ _\alpha I_x \, _\alpha I_x & _\alpha I_y^2 \end{bmatrix}. \tag{11}$$

Since the fractional derivative is non-local in nature, the derivatives are no longer computed with the derivative of a local Gaussian kernel, as in (9). The same LoG function is computed in the fractional Harris-Laplace (FHL) detector as in the original HL detector, and the same cornerness function (7) is used with $k = 0.05$. To ensure fast computation of the fractional derivative and high selectivity with immunity to noise, the CRONE algorithm is used. The derivative order of $\alpha = \frac{1}{2}$ has been chosen, since this order of derivative is located exactly 'halfway' between the identity operator and the classical first derivative. Similar to the Harris-Laplace detector, scale invariance is achieved by iterating over the scales σ_n to detect features at each scale. In comparing the fractional Harris-Laplace detector to its namesake, the scale factor was chosen as $\xi = 1.2$. The LoG function is then used to determine the 'characteristic' scale at which the feature is best represented.

While the CRONE operator does not give an exact representation of a fractional derivative at a given image pixel, it does has convenient odd symmetry which improves its selectivity. Numerical implementations of the Riesz potential, such as those explored in [11], may be used to obtain a true centred-difference fractional derivative in future iterations of this project.

The introduction of the fractional derivative results in a change of perspective when interpreting the stucture tensor. In the Harris and HL detectors, the stucture tensor is a measure of the local perturbation in the neighborhoods of image pixels in two orthogonal directions. However, with the fractional stucture tensor defined in (11), the perturbation in orthogonal directions is now *non-local*. The original Harris-Laplace detector used derivative filters computed with local Gaussians of radius σ_D. However, when using the fractional derivative, the change in a given pixel in either the x- or y-direction now depends on a weighted sum of all other pixels that enter the sum before the pixel. For fractional derivatives in the x-direction, these weighted non-local neighbors lie both to the left and the right of a given pixel. In the y-direction, they lie above and below the pixel. The non-locality of the fractional derivative provides more information about an image pixel than an integer derivative. Thus, in areas with high image detail extending in orthogonal directions, the FHL will have more likelihood of detecting features. In areas of low detail, features may still be detected based on the texture of the surrounding pixels. These features must be evaluated for reliability as in the next section.

5   Results

The FHL detector was evaluated using the repeatability criterion established by Schmid, Mohr, and Bauckhage [13]. Repeatability is measured by the percentage of features matched between two images, with one image transformed by a known homography, within the boundaries of the larger image. To measure the repeatability of a feature detector, let x_0 be a feature detected in the original image I_0, and x_1 be a feature detected in the transformed image I_1. Furthermore, let H be the homography between the images that defines the common region of both images. The set of features in the common region of both images is

$$\{x_0 \colon Hx_0 \in I_1\} \cap \{x_1 \colon H^{-1}x_1 \in I_0\}.$$

The homography H represents a translation, rotation, and scale change between the two images in the image sequence. Since the features are detected at the various scales σ_n, we expect the ratio of the characteristic scales determined by the LoG function for the features in the two images to agree with the scale change in the homography H.

According to Schmid, Mohr, and Bauckhage, the set of features that match within a neighborhood of radius ε is given by

$$R(\varepsilon) = \{(x_0, x_1) : |Hx_0 - x_1| < \varepsilon\}. \tag{12}$$

Now let n_0 and n_1 be the number of features detected in the images I_0 and I_1, respectively. The repeatability rate [13] is defined as

$$r(\varepsilon) = \frac{|R(\epsilon)|}{\min(n_0, n_1)}. \tag{13}$$

Two real image sequences were considered: Boats1, Boats2, which are two different transformation sequences of a scene at a marina. The images used are part of the Oxford University affine covariant features image library.

5.1   Boats1 Sequence

The first image sequence features a scene of boats moored at a marina. The original image is transformed by a variation in focal length (zoom) and a rotation. To perform the transformation of detected features, the homography H is approximated by

$$H = \begin{bmatrix} 0.4274\cos(8°) & -0.4274\sin(8°) & 266.35003 \\ 0.4274\sin(8°) & 0.4274\cos(8°) & 174.60201 \\ 0 & 0 & 1 \end{bmatrix}. \tag{14}$$

First, features are detected in the original image. Next, features are detected in the transformed image. The original image features are transformed by the homography H to their position in the transformed image. Finally, features are matched using the repeatability criterion. The transformation and detected features are shown in Fig. 1. Various values for the localization error ε were considered, and a plot of the repeatability of both the HL and FHL detectors is included in Fig. 2. The highest value for ε was set at 5, as feature matches for errors higher than this could be due to anomalies [13].

5.2   Boats2 Sequence

The images in the Boats2 sequence contain the same boats and marina as in the Boats1 sequence. In this sequence, the transformation between images consists of a more dramatic rotation. The same procedure for the Boats1 sequence is followed for the Boats2 sequence, with the transformation and detected features shown in Fig. 3. The homography between the images is approximately

a)   b)   c)

Fig. 1. FHL results for the Boats1 sequence: (a) FHL features detected in original image; (b) FHL features detected in transformed image; (c) FHL features transformed from the original image to the transformed image with the homography H.

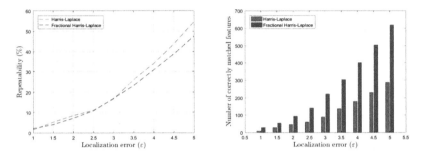

Fig. 2. Repeatability as a function of localization error ε for the Boats1 image sequence (left image); plot of the overall number of matches as a function of localization error (right image).

$$H = \begin{bmatrix} 0.5327\cos(80°) & 0.5327\sin(80°) & 205.87932 \\ -0.5327\sin(80°) & 0.5327\cos(80°) & 534.54522 \\ 0 & 0 & 1 \end{bmatrix}. \tag{15}$$

The repeatability for the HL and FHL detectors was again calculated for the Boats2 sequence, and the results are given in Fig. 4.

6   Discussion

Introducing the fractional derivative to the Harris-Laplace feature detector has led to a scale-invariant feature detector with improved repeatability and a higher number of matched features. The best results were for image sequences with large transformational changes between images. For the Boats1 sequence, which utilized a small rotation and moderate scale change, the FHL showed higher repeatability rates than the HL detector only when the localization error was 1 or 3 pixels. At all other localization errors, the HL showed higher repeatability. However, when the one-sided GL derivative was used in the FHL detector, the

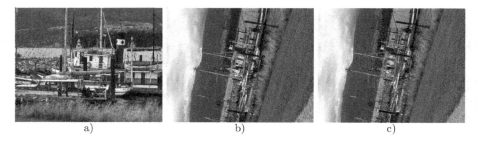

Fig. 3. FHL results for the Boats2 sequence: (a) FHL features detected in original image; (b) FHL features detected in transformed image; (c) FHL features transformed from the original image to the transformed image with the homography H.

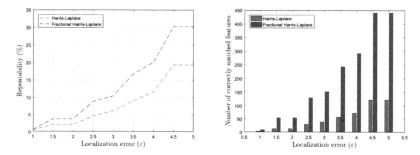

Fig. 4. Repeatability as a function of localization error ε for the Boats2 image sequence (left image); plot of the overall number of matches as a function of localization error (right image).

FHL had higher repeatability at every localization error. At the lowest localization error of 1, the FHL had repeatability of 2.082%, while the HL had repeatability of 1.527%. When the localization error was 3, the FHL showed repeatability of 17.039% and the HL had repeatability of 16.794%. A second measure, the number of correct matches, reported in Fig. 2, indicates that the FHL matched a higher number of features across images for all localization errors.

For the Boats2 sequence (see Fig. 4), at the highest localization error of 5, the FHL had repeatability of 30.282% and the HL had repeatability of 19.334%. It is noteworthy that the Boats2 sequence involved a more significant rotation between images than the Boat1 sequence, giving evidence that the FHL is more effective in image sequences with large changes between images. The repeatability was consistently higher across all localization errors. The number of correctly matched features was again higher for the FHL than the HL across all localization errors.

The FHL has the distinct tendency to detect more features as a result of its non-local property. The inclusion of non-local points in the computation of the derivative leads to larger values in the stucture tensor, and hence larger eigenvalues in the spectrum of the stucture tensor. Therefore, more candidate feature points are detected. This explains the FHL having consistently higher numbers of feature matches between images in the sequence. If there are more features detected, then there are more features to potentially match. It is noteworthy that the FHL matched more points even at the lowest localization error of 1. This indicates that the matches were not merely the result of anomalies introduced by the matching algorithm. The matching algorithm dilated each feature to a radius of the localization error, and then the dilated feature was compared to the LoG function. Therefore, if a single LoG function at a given pixel overlapped with multiple dilated features (given the density of the detected features, this is possible), false matches would be reported. The high number of matched features across the range of localization errors indicates that the FHL has outperformed the HL in terms of reliability.

7   Future Directions

The current paper adapted the Harris-Laplace (HL) feature detector to use fractional derivatives in the stucture tensor. The focus has been on grayscale images, and the extension to color images is a task to be completed. The Harris-Laplace detector assumes a translation-rotation-scale change model of motion between frames of the image sequence. However, the rotations between images are strictly along a single axis. Affine transformations, which allow for rotations along any of the three coordinate axes, are included in the generalized Harris-Affine detector [8]. The introduction of the fractional derivative to the larger stucture tensor of the Harris-Affine detector could improve the repeatability of the detector as it has improved the Harris-Laplace detector. Possible uses for the FHL detector exist in object tracking and image registration, due to the high number of correctly matched features.

The symmetry of the CRONE operator was used to improve the selectivity and immunity to noise of the FHL detector. However, since the CRONE operator represents a fractional derivative strictly at the image boundary, a true fractional derivative should be used in future implementations. There is a possibility to use the centred difference approximations to Riesz potentials for symmetric fractional derivatives [11] in the future, once the computational issues cited by the author of [11] have been addressed.

References

1. Baleanu, D., Diethelm, K., Scalas, E., Trujillo, J.J.: Fractional Calculus: Models and Numerical Methods. World Scientific, Singapore (2012)
2. Bradski, G.: The OpenCV library. Dr. Dobb's J. Softw. Tools 25, 120–126 (2000)
3. ElAraby, W.S., Median, A.H., Ashour, M.A., Farag, I., Nassef, M.: Fractional canny edge detection for biomedical applications. In: 28th International Conference on Microelectronics (ICM), pp. 265–268. IEEE (2016)
4. Harris, C., Stephens, M.: A combined corner and edge detector. In: Fourth Alvey Vision Conference, vol. 15, pp. 147–151 (1988)

5. Liu, Y.: Remote sensing image enhancement based on fractional differential. In: International Conference on Computational and Information Sciences (ICCIS), pp. 881–884. IEEE (2010)
6. Lowe, D.G.: Distinctive image features from scale-invariant keypoints. Int. J. Comput. Vis. **60**(2), 91–110 (2004)
7. Mathieu, B., Melchior, P., Oustaloup, A., Ceyral, C.: Fractional differentiation for edge detection. Signal Process. **83**(11), 2421–2432 (2003)
8. Mikolajczyk, K., Schmid, C.: Scale & affine invariant interest point detectors. Int. J. Comput. Vis. **60**(1), 63–86 (2004)
9. Oldham, K., Spanier, J.: The Fractional Calculus: Theory and Applications of Differentiation and Integration to Arbitrary Order, vol. 111. Academic Press Inc., New York (1974)
10. Ortigueira, M.D., Machado, J.T.: What is a fractional derivative? J. Comput. Phys. **293**, 4–13 (2015)
11. Ortigueira, M.D.: Riesz potential operators and inverses via fractional centred derivatives. Int. J. Math. Math. Sci. **2006**, 12 (2006)
12. Pu, Y., Wang, W., Zhou, J., Wang, Y., Jia, H.: Fractional differential approach to detecting textural features of digital image and its fractional differential filter implementation. Sci. China Ser. F: Inf. Sci. **51**(9), 1319–1339 (2008)
13. Schmid, C., Mohr, R., Bauckhage, C.: Evaluation of interest point detectors. Int. J. Comput. Vis. **37**(2), 151–172 (2000)
14. Weickert, J.: Anisotropic Diffusion in Image Processing, vol. 1. Teubner, Stuttgart (1998)

Macrocanonical Models for Texture Synthesis

Valentin De Bortoli[1(✉)], Agnès Desolneux[1], Bruno Galerne[2],
and Arthur Leclaire[3]

[1] Centre de mathématiques et de leurs applications, CNRS, ENS Paris-Saclay,
Université Paris-Saclay, 94235 Cachan Cedex, France
`valentin.de-bortoli@cmla.ens-cachan.fr`
[2] Institut Denis Poisson, Université d'Orléans, Université de Tours, CNRS,
Tours, France
[3] Univ. Bordeaux, IMB, Bordeaux INP, CNRS, UMR 5251, 33400 Talence, France

Abstract. In this article we consider macrocanonical models for texture synthesis. In these models samples are generated given an input texture image and a set of features which should be matched in expectation. It is known that if the images are quantized, macrocanonical models are given by Gibbs measures, using the maximum entropy principle. We study conditions under which this result extends to real-valued images. If these conditions hold, finding a macrocanonical model amounts to minimizing a convex function and sampling from an associated Gibbs measure. We analyze an algorithm which alternates between sampling and minimizing. We present experiments with neural network features and study the drawbacks and advantages of using this sampling scheme.

Keywords: Texture synthesis · Gibbs measure ·
Monte carlo methods · Langevin algorithms · Neural networks

1 Introduction

In image processing a texture can be defined as an image which contains repetitive patterns but also randomness in the pattern placement or in the pattern itself. This vague and unformal definition covers a large class of images such as the ones of terrain, plants, minerals, fur and skin. Exemplar-based texture synthesis aims at synthesizing new images of arbitrary size which have the same perceptual characteristics as a given input texture. It is a challenging task to give a mathematical framework which is not too restrictive, thus describing many texture images, and not too broad, so that computations are numerically feasible. In the literature two classes of exemplar-based texture synthesis algorithms have been considered: the parametric and the non-parametric texture algorithms. Non-parametric texture methods do not rely on an explicit image model in order to produce outputs. For instance copy-paste algorithms such as [6] fill the output image with sub-images from the input. Another example is

J. Lellmann et al. (Eds.): SSVM 2019, LNCS 11603, pp. 13–24, 2019.
https://doi.org/10.1007/978-3-030-22368-7_2

given by [8] in which the authors apply optimal transport tools in a multiscale patch space.

In this work we focus on parametric exemplar-based texture synthesis algorithms. In contrast to the non-parametric approach they provide an explicit image model. Output textures are produced by sampling from this image model. In order to derive such a model perceptual features have to be carefully selected along with a corresponding sampling algorithm. There have been huge progress in both directions during the last twenty years.

First, it should be noted that textures which do not exhibit long-range correlations and are well described by their first and second-order statistics can be modeled with Gaussian random fields [9,25]. These models can be understood as maximum entropy distributions given a mean and a covariance matrix. Their simplicity allows for fast sampling as well as good mathematical understanding of the model. However, this simplicity also restricts the class of textures which can be described. Indeed, given more structured inputs, these algorithms do not yield satisfactory visual results. It was already noted by Gagalowicz [7] that first and second-order statistics are not sufficient to synthesize real-world textures images. In [3] the authors remark that multiscale features capture perceptual characteristics. Following this idea, algorithms based on steerable pyramids [13], wavelet coefficients [20] or wavelet coefficients combined with geometrical properties [19] provide good visual results for a large class of textures. Using Convolutional Neural Networks (CNN), and especially the VGG model [22], Gatys et al. in [10] obtain striking visual results using Gram matrices computed on the layers of the neural network. All these models are called microcanonical textures according to Bruna and Mallat [2], in the sense that they approximately match statistical constraints almost surely (a.s.). Indeed, the previously introduced algorithms start from a noisy input containing all the randomness of the process, then use a (deterministic) gradient descent (or any other optimization algorithm) in order to fit constraints.

On the other hand, models relying on constraints in expectation have been considered in [26]. They correspond to macrocanonical textures according to [2]. They have the advantage to be described by exponential distributions and thus, since their distribution can be made explicit up to some parameters, standard statistical tools can be used for mathematical analysis. However, as noted in [2] they often rely on Monte Carlo algorithms which can be slow to converge. Zhu et al. [26] consider a bank of linear and non-linear filters in order to build an exponential model. Texture images are supposed to be quantized and a Gibbs sampler on each pixel is used in order to update the image. In [17] the authors propose to use first-order statistics computed on CNN outputs. They also suggest to use a Langevin algorithm in order to update the whole image at each iteration. It has also been remarked in [23] that specific Generative Adversarial Networks (GAN) [15] which produce satisfying outputs from a perceptual point of view but lack mathematical understanding can be embedded in an expectation constraint model using the Maximum Mean Discrepancy principle [12].

Our contribution is both theoretical and experimental. After recalling the definition of microcanonical models in Sect. 2.1 we give precise conditions under which macrocanonical models, *i.e.* maximum entropy models, can be written as exponential distributions in Sect. 2.2. In Sect. 2.3, we examine how these conditions translate into a neural network model . Assuming that the maximum entropy principle is satisfied we then turn to the search of the parameters in such a model. The algorithm we consider, which was already introduced without theoretical proof of convergence in [17], relies on the combination of a gradient descent dynamic, see Sect. 3.1, and a discretized Langevin dynamic, see Sect. 3.2. Using new results on these Stochastic Optimization with Unadjusted Kernel (SOUK) algorithms [4] convergence results hold for the algorithm introduced in [17], see Sect. 3.3. We then provide experiments and after assessing the empirical convergence of our algorithm in Sect. 4.1 we investigate our choice of models in Sect. 4.2. We draw the conclusions and limitations of our work in Sect. 5.

2   Maximum Entropy Models

2.1   Microcanonical Models

Let x_0 be a given input texture. For ease of exposition we consider that $x_0 \in \mathbb{R}^d$, with $d \in \mathbb{N}$, but our results extend to images and color images. We aim at sampling x from a probability distribution satisfying $f(x) \approx f(x_0)$, where $f : \mathbb{R}^d \to \mathbb{R}^p$ are some statistics computed over the images. However if such a probability distribution exists it is not necessarily unique. In order for the problem to be well-posed we introduce a reference function $J : \mathbb{R}^d \to (0, +\infty)$ such that $\int_{\mathbb{R}^d} J(x)\mathrm{d}\lambda(x) < +\infty$ and we associate to J a probability distribution Π_J such that $\Pi_J(A) = Z_J^{-1} \int_A J(x)\mathrm{d}\lambda(x)$ with $Z_J = \int_{\mathbb{R}^d} J(y)\mathrm{d}\lambda(y)$ for any $A \in \mathscr{B}(\mathbb{R}^d)$, the Borel sets of $\mathbb{R}^d$. Let $\mathscr{P}$ be the set of probability distributions over $\mathscr{B}(\mathbb{R}^d)$. If $\Pi \in \mathscr{P}$ is absolutely continuous with respect to the Lebesgue measure λ we denote by $\frac{\mathrm{d}\Pi}{\mathrm{d}\lambda}$ the probability density function of Π. We introduce the J-entropy, see [14], $H_J : \mathscr{P} \to [-\infty, +\infty)$ such that for any $\Pi \in \mathscr{P}$

$$H_J(\Pi) = \begin{cases} -\int_{\mathbb{R}^d} \log\left[\frac{\mathrm{d}\Pi}{\mathrm{d}\lambda}(x)J(x)^{-1}\right]\frac{\mathrm{d}\Pi}{\mathrm{d}\lambda}(x)\mathrm{d}\lambda(x) & \text{if } \frac{\mathrm{d}\Pi}{\mathrm{d}\lambda} \text{ exists ;} \\ -\infty & \text{otherwise.} \end{cases}$$

The quantity H_J is closely related to the Kullback-Leibler divergence between Π and Π_J. We recall that, if Π is absolutely continuous with respect to λ, we have $\mathrm{KL}(\Pi|\Pi_J) = \int_{\mathbb{R}^d} \log\left[\frac{\mathrm{d}\Pi}{\mathrm{d}\lambda}(x)\frac{\mathrm{d}\Pi_J}{\mathrm{d}\lambda}(x)^{-1}\right]\frac{\mathrm{d}\Pi}{\mathrm{d}\lambda}(x)\mathrm{d}\lambda(x)$, and $+\infty$ otherwise. Since $\frac{\mathrm{d}\Pi_J}{\mathrm{d}\lambda}(x) = Z_J^{-1}J(x)$ we obtain that for any $\Pi \in \mathscr{P}$, $H_J(\Pi) = -\mathrm{KL}(\Pi|\Pi_J) + \log(Z_J)$. The following definition gives a texture model for which statistical constraints are met a.s.

Definition 1. *The probability distribution function $\widetilde{\Pi} \in \mathscr{P}$ is a* microcanonical model *associated with the exemplar texture $x_0 \in \mathbb{R}^d$, statistics $f : \mathbb{R}^d \to \mathbb{R}^p$ and reference J if*

$$H_J(\widetilde{\Pi}) = \max\{H_J(\Pi), \ \Pi \in \mathscr{P}, \ f(X) = f(x_0) \ a.s. \ if \ X \sim \Pi\} . \quad (1)$$

Most algorithms which aim at finding a microcanonical model apply a gradient descent algorithm on the function $x \mapsto \|f(x) - f(x_0)\|^2$ starting from an initial white noise. The intuition behind this optimization procedure is that the entropy information is contained in the initialization and the constraints are met asymptotically. There exists few theoretical work on the subject with the remarkable exception of [2] in which the authors prove that under technical assumptions the limit distribution has its support on the set of constrained images, *i.e.* the constraints are met asymptotically, and provide a lower bound on its entropy.

2.2   Macrocanonical Models

Instead of considering a.s. constraints as in (1) we can consider statistical constraints in expectation. This model was introduced by Jaynes in [14] and formalized in the context of image processing by Zhu et al. in [26].

Definition 2. *The probability distribution function $\widetilde{\Pi} \in \mathscr{P}$ is a macrocanonical model associated with the exemplar texture $x_0 \in \mathbb{R}^d$, statistics $f : \mathbb{R}^d \to \mathbb{R}^p$ and reference J if*

$$H_J(\widetilde{\Pi}) = \max \{H_J(\Pi), \ \Pi \in \mathscr{P}, \ \Pi(f) = f(x_0)\} \ , \tag{2}$$

where $\Pi(f) = \mathbb{E}_\Pi(f)$.

Macrocanonical models can be seen as a relaxation of the microcanonical ones. A link between macrocanonical models and microcanonical models is highlighted by Bruna and Mallat in [2]. They show that for some statistics, macrocanonical and microcanonical models have the same limit when the size of the image goes to infinity. This transition of paradigm has important consequences from a statistical point of view. First, the constraints in (2) require only the knowledge of the expectation of f under a probability distribution Π. Secondly, in Theorem 1 we will show that the macrocanonical model can be written as a Gibbs measure, *i.e.* $\frac{d\widetilde{\Pi}}{d\lambda}(x) \propto \exp(-\langle \tilde{\theta}, f(x) - f(x_0)\rangle)J(x)$ for some $\tilde{\theta} \in \mathbb{R}^p$. Given $\theta \in \mathbb{R}^p$, when it is defined we denote by Π_θ the probability distribution defined by

$$Z(\theta) = \int_{\mathbb{R}^d} e^{-\langle \theta, f(x) - f(x_0)\rangle} J(x) d\lambda(x) \quad \text{and} \quad \frac{d\Pi_\theta}{d\lambda}(x) = \frac{e^{-\langle \theta, f(x) - f(x_0)\rangle}}{Z(\theta)} J(x) \ .$$

Theorem 1 (Maximum entropy principle). *Assume that for any $\theta \in \mathbb{R}^p$ we have*

$$\int_{\mathbb{R}^d} e^{\|\theta\|\|f(x)\|} J(x) d\lambda(x) < +\infty \quad \text{and} \quad \lambda \left(\{x \in \mathbb{R}^d, \ \langle \theta, f(x)\rangle < \langle \theta, f(x_0)\rangle\}\right) > 0 \ . \tag{3}$$

Then there exists $\tilde{\theta} \in \mathbb{R}^p$ such that $\Pi_{\tilde{\theta}}$ is a macrocanonical model associated with the exemplar texture $x_0 \in \mathbb{R}^d$, statistics f and reference J. In addition, we have

$$\tilde{\theta} \in \arg\min \left\{ \log \left[\int_{\mathbb{R}^d} \exp(-\langle \theta, f(x) - f(x_0)\rangle)J(x) d\lambda(x) \right], \theta \in \mathbb{R}^p \right\} \ . \tag{4}$$

Proof Without loss of generality we assume that $f(x_0) = 0$. First we show that there exists $\tilde{\theta} \in \mathbb{R}^p$ such that $\Pi_{\tilde{\theta}}$ is well-defined and $\Pi_{\tilde{\theta}}(f) = f(x_0)$. The first condition in (3) implies that $Z(\theta) = \int_{\mathbb{R}^d} \exp(-\langle \theta, f(x) \rangle) J(x) d\lambda(x)$ is defined for all $\theta \in \mathbb{R}^p$. Let $\theta_0 \in \mathbb{R}^p$ we have for any $\theta \in \mathrm{B}(\theta_0, 1)$, the unit ball centered on θ_0, and $i \in \{1, \ldots, p\}$, using that for any $t \in \mathbb{R}$, $t \leqslant e^t$ and the Cauchy-Schwarz inequality,

$$
\int_{\mathbb{R}^d} \left| \frac{\partial}{\partial \theta_i} \left[\exp(-\langle \theta, f(x) \rangle) \right] J(x) \right| d\lambda(x) \leqslant \int_{\mathbb{R}^d} \|f(x)\| \exp(-\langle \theta, f(x) \rangle) J(x) d\lambda(x)
$$

$$
\leqslant \int_{\mathbb{R}^d} \exp((\|\theta_0\| + 2)\|f(x)\|) J(x) d\lambda(x) < +\infty .
$$

Therefore $\theta \mapsto \log(Z)(\theta)$ is differentiable and we obtain that for any $\theta \in \mathbb{R}^p$, $\nabla \log(Z)(\theta) = -\mathbb{E}_{\Pi_\theta}(f) = -\Pi_\theta(f)$. In a similar fashion we obtain that $\log(Z) \in \mathcal{C}^2(\mathbb{R}^p, \mathbb{R})$ and we have $\frac{\partial^2 \log(Z)}{\partial \theta_i \partial \theta_j}(\theta) = \Pi_\theta(f_i f_j) - \Pi_\theta(f_i)\Pi_\theta(f_j)$. The Hessian of $\log(Z)$ evaluated at θ is the covariance matrix of $f(X)$ where $X \sim \Pi_\theta$ and thus is non-negative which implies that $\log(Z)$ is convex. We also have for any $\theta \in \mathbb{R}^p$ and $t > 0$

$$
\log(Z)(t\theta) = \log \left[\int_{\mathbb{R}^d} \exp(-t\langle \theta, f(x) \rangle) J(x) d\lambda(x) \right]
$$

$$
\geqslant \log \left[\int_{\langle \theta, f(x) \rangle < 0} \exp(-t\langle \theta, f(x) \rangle) J(x) d\lambda(x) \right] \xrightarrow[t \to +\infty]{} +\infty , \quad (5)
$$

where we use the first condition in (3) and the monotone convergence theorem. Therefore $\log(Z)$ is coercive along each direction of $\mathbb{R}^p$. Let us show that $\log(Z)$ is coercive, *i.e.* for any $M > 0$, there exists $R > 0$ such that for all $\|\theta\| \geqslant R$, $\log(Z)(\theta) \geqslant M$. Suppose that $\log(Z)$ is not coercive then there exists a sequence $(\theta_n)_{n \in \mathbb{N}}$ such that $\lim_{n \to +\infty} \|\theta_n\| = +\infty$ and $\log(Z)(\theta_n)_{n \in \mathbb{N}}$ is upper-bounded by some constant $M \geqslant 0$. We can suppose that $\theta_n \neq 0$. Upon extracting a subsequence we assume that $(\theta_n/\|\theta_n\|)_{n \in \mathbb{N}}$ admits some limit $\theta^\star \in \mathbb{R}^p$ with $\|\theta^\star\| = 1$. Let $M_0 = \max[M, \log(Z)(0)]$, we have the following inequality for all $t > 0$

$$
\log(Z)(t\theta^\star) \leqslant \inf_{n \in \mathbb{N}} \left[|\log(Z)(t\theta^\star) - \log(Z)(t\theta_n/\|\theta_n\|)| + \log(Z)(t\theta_n/\|\theta_n\|) \right] \leqslant M_0 ,
$$

where we used the continuity of $\log(Z)$ and the fact that for n large enough $t < \|\theta_n\|$ and therefore by convexity $\log(Z)(t\theta_n/\|\theta_n\|) \leqslant t/\|\theta_n\| \log(Z)(0) + (1 - t/\|\theta_n\|) \log(Z)(\theta_n) \leqslant M_0$. Hence for all $t > 0$, $\log(Z)(t\theta^\star)$ is bounded which is in contradiction with (5). We obtain that $\log(Z)$ is continuous, convex, coercive and defined over $\mathbb{R}^p$. This ensures us that there exists $\tilde{\theta}$ such that $\log(Z)(\tilde{\theta})$ is minimal and therefore $\nabla_\theta \log(Z)(\tilde{\theta}) = -\Pi_{\tilde{\theta}}(f) = 0$. Note that we have

$$
H_J(\Pi_{\tilde{\theta}}) = \int_{\mathbb{R}^d} \langle \tilde{\theta}, f(x) \rangle \frac{d\Pi_{\tilde{\theta}}}{d\lambda}(x) d\lambda(x) + \log(Z)(\tilde{\theta}) = \log(Z)((\tilde{\theta}) .
$$

Now let $\Pi \in \mathscr{P}$ such that $\Pi(f) = 0$. If Π is not absolutely continuous with respect to the Lebesgue measure, then $H_J(\Pi) = -\infty < H_J(\Pi_{\hat{\theta}})$. Otherwise if Π is absolutely continuous with respect to the Lebesgue measure we have the following inequality

$$H_J(\Pi) = -\int_{\mathbb{R}^d} \log\left[\frac{d\Pi}{d\lambda}(x)J(x)^{-1}\right]\frac{d\Pi}{d\lambda}(x)d\lambda(x)$$

$$= -\int_{\mathbb{R}^d} \log\left[\frac{d\Pi}{d\lambda}(x)\left(\frac{d\Pi_{\tilde{\theta}}}{d\lambda}(x)\right)^{-1}\right]\frac{d\Pi}{d\lambda}(x) - \log\left[\frac{d\Pi_{\tilde{\theta}}}{d\lambda}(x)J(x)^{-1}\right]\frac{d\Pi}{d\lambda}(x)d\lambda(x)$$

$$= -\mathrm{KL}\left(\Pi|\Pi_{\hat{\theta}}\right) + \log(Z)(\tilde{\theta}) \leqslant \log(Z)(\tilde{\theta}) = H_J(\Pi_{\hat{\theta}}) ,$$

which concludes the proof.

Theorem 1 gives a method for finding the optimal parameters $\theta \in \mathbb{R}^p$ by solving the convex problem (4). We address this issue in Sect. 3.

2.3 Some Feature Examples

In the framework of exemplar-based texture synthesis, f is defined as the spatial statistics of some image feature. For instance, let $\mathscr{F} : \mathbb{R}^d \to \prod_{i=1}^{p}\mathbb{R}^{d_i}$ be a measurable mapping lifting the image $x \in \mathbb{R}^d$ in a higher-dimensional space $\prod_{i=1}^{p}\mathbb{R}^{d_i}$. Classical examples include wavelet transforms, power transforms or neural network features. Let $(\mathscr{F}_i)_{i=1,\ldots,p}$ such that for any $x \in \mathbb{R}^d$, $\mathscr{F}(x) = (\mathscr{F}_1(x),\ldots,\mathscr{F}_p(x))$ and $\mathscr{F}_i : \mathbb{R}^d \to \mathbb{R}^{d_i}$. Then the statistics f can be defined for any $x \in \mathbb{R}^d$ as follows

$$f(x) = \left(d_1^{-1}\sum_{k=1}^{d_1}\mathscr{F}_1(x)(k),\ldots,d_p^{-1}\sum_{k=1}^{d_p}\mathscr{F}_p(x)(k)\right) . \tag{6}$$

Note that this formulation includes histograms of bank of filters [20], wavelet coefficients [19] and scattering coefficients [2]. The model defined by such statistics is stationary, *i.e.* translation invariant, since we perform a spatial summation. In the following we focus on first-order features, which will be used in Sect. 4.1 to assess the convergence of our sampling algorithm, and neural network features, extending the work of [17].

Neural Network Features. We denote by $\mathcal{A}_{n_2,n_1}(\mathbb{R})$ the vector space of the affine operators from $\mathbb{R}^{n_1}$ to $\mathbb{R}^{n_2}$. Let $(A_j)_{j\in\{1,\ldots,M\}} \in \prod_{j=1}^{M}\mathcal{A}_{n_{j+1},n_j}(\mathbb{R})$, where we let $(n_j)_{j\in\{1,\ldots,M+1\}} \in \mathbb{N}^{M+1}$, with $M \in \mathbb{N}$ and $n_1 = d$. Let $\varphi : \mathbb{R} \to \mathbb{R}$. We define for any $j \in \{1,\ldots,M\}$, the j-th layer feature $\mathscr{G}_j : \mathbb{R}^d \to \mathbb{R}^{n_j}$ for any $x \in \mathbb{R}^d$ by

$$\mathscr{G}_j(x) = \left(\varphi \circ A_j \circ \varphi \circ A_{j-1} \circ \cdots \circ \varphi \circ A_1\right)(x) ,$$

where φ is applied on each component of the vectors. Let $p \in \{1,\ldots,M\}$ and $(j_i)_{i\in\{1,\ldots,p\}} \in \{1,\ldots,M\}^p$ then we can define $\mathscr{F}$ as in (6) by

$$f(x) = \left(n_{j_1}^{-1}\sum_{k=1}^{n_{j_1}}\mathscr{G}_{j_1}(x)(k),\ldots,n_{j_p}^{-1}\sum_{k=1}^{n_{j_p}}\mathscr{G}_{j_p}(x)(k)\right) .$$

Assuming that φ, the non-linear unit, is $\mathcal{C}^1(\mathbb{R})$ we obtain that f is $\mathcal{C}^1(\mathbb{R}^d, \mathbb{R}^p)$. The next proposition gives conditions under which Theorem 1 is satisfied. We denote by df the Jacobian of f.

Proposition 1 (Differentiable neural network maximum entropy). *Let $x_0 \in \mathbb{R}^d$ and assume that $df(x_0)$ has rank $\min(d, p) = p$. In addition, assume that there exists $C \geqslant 0$ such that for any $x \in \mathbb{R}$, $|\varphi(x)| \leqslant C(1 + |x|)$. Then the conclusions of Theorem 1 hold for any $J(x) = \exp(-\varepsilon \|x\|^2)$ with $\varepsilon > 0$.*

Proof The integrability condition is trivially checked since f is sub-linear using that $|\varphi(x)| \leqslant C(1 + |x|)$. Turning to the proof of the second condition, since $f \in \mathcal{C}^1(\mathbb{R}^d, \mathbb{R}^p)$ and $df(x_0)$ is surjective we can assert the existence of an open set U as well as $\Phi \in \mathcal{C}^1(\mathsf{U}, \mathbb{R}^d)$ with $f(x_0) \in \mathsf{U}$ such that for any $y \in \mathsf{U}$, $f(\Phi(y)) = y$. Now consider $\theta \in \mathbb{R}^p$. If $\theta \in f(x_0)^{\perp}$ then for $\varepsilon > 0$ small enough $f(x_0) - \varepsilon\theta \in \mathsf{U}$ and we obtain that $\langle \theta, f(\Phi(f(x_0) - \varepsilon\theta)) \rangle = -\varepsilon \|\theta\|^2 < 0$. If $\theta \notin f(x_0)^{\perp}$ then there exists $\varepsilon > 0$ small enough such that $[f(x_0)(1 - \varepsilon), f(x_0)(1 + \varepsilon)] \subset \mathsf{U}$. Then for any $\alpha \in (-\varepsilon, \varepsilon)$ we get that $\langle \theta, f(\Phi((1 + \alpha)f(x_0))) \rangle - \langle \theta, f(x_0) \rangle = \alpha \langle \theta, f(x_0) \rangle$. By choosing $\alpha > 0$, respectively $\alpha < 0$, if $\langle \theta, f(x_0) \rangle > 0$, respectively $\langle \theta, f(x_0) \rangle < 0$, we obtain that for any $\theta \in \mathbb{R}^p$, there exists $x \in \mathbb{R}^d$ such that $\langle \theta, f(x) \rangle < \langle \theta, f(x_0) \rangle$. We conclude using the continuity of f.

3   Minimization and Sampling Algorithm

3.1   Maximizing the Entropy

In order to find $\tilde{\theta}$ such that $\Pi_{\tilde{\theta}}$ is the macrocanonical model associated with the exemplar texture x_0, statistics f and reference J we perform a gradient descent on $\log(Z)$. Let $\theta_0 \in \mathbb{R}^p$ be some initial parameters. We define the sequence $(\theta_n)_{n \in \mathbb{N}}$ for any $n \in \mathbb{N}$ by

$$\theta_{n+1} = P_\Theta \left[\theta_n - \delta_{n+1} \nabla \log(Z)(\theta_n) \right] = P_\Theta \left[\theta_n + \delta_{n+1} \left(\Pi_{\theta_n}(f) - f(x_0) \right) \right] , \quad (7)$$

where $(\delta_n)_{n \in \mathbb{N}}$ is a sequence of step sizes with $\delta_n \geqslant 0$ for any $n \in \mathbb{N}$ and P_Θ is the projection over Θ. The introduction of the projection operator P_Θ is a technical condition in order to guarantee the convergence of the algorithm in Sect. 3.3. Implementing the algorithm associated to (7) requires the knowledge of the moments of the statistics f for any Gibbs measure Π_θ with $\theta \in \mathbb{R}^p$. The more complex the texture model is the more difficult it is to compute the expectation of the statistics. This expectation can be written as an integral and techniques such as the ones presented in [18] could be used. We choose to approximate this expectation using a Monte Carlo strategy. Assuming that $(X_n)_{n \in \mathbb{N}}$ are samples from Π_θ, we have that $n^{-1} \sum_{k=1}^n f(X_k)$ is an unbiased estimator of $\Pi_\theta(f)$.

3.2   Sampling from Gibbs Measures

We now turn to the problem of sampling from Π_θ. Unfortunately, most of the time there is no easy way to produce samples from Π_θ. Nonetheless, using the

ergodicity properties of specific Markov Chains we can still come up with esti-
mators of $\Pi_\theta(f)$. Indeed, if $(X_n)_{n\in\mathbb{N}}$ is a homogeneous Markov chain with kernel
K and invariant probability measure Π_θ, $i.e.$ $\Pi_\theta K = \Pi_\theta$ we obtain under suitable conditions over f and K that $\lim_{n\to+\infty} \mathbb{E}\left[n^{-1}\sum_{k=1}^n f(X_k)\right] = \Pi_\theta(f)$. This
leads us to consider the following Langevin dynamic for all $n \in \mathbb{N}$

$$X_{n+1} = X_n - \gamma_{n+1}\sum_{i=1}^p \theta_i \nabla f_i(X_n) + \sqrt{2\gamma_{n+1}}Z_{n+1} \quad \text{and} \quad X_0 \in \mathbb{R}^d , \quad (8)$$

where $(Z_n)_{n\in\mathbb{N}^*}$ is a collection of independent d-dimensional zero mean Gaussian
random variables with covariance matrix identity and $(\gamma_n)_{n\in\mathbb{N}}$ is a sequence of
step sizes with $\gamma_n \geqslant 0$. The, possibly inhomogeneous, Markov Chain $(X_n)_{n\in\mathbb{N}}$
is associated with the sequence of kernels $(R_{\gamma_n})_{n\in\mathbb{N}}$ with $R_{\gamma_n}(x,\cdot) = \mathcal{N}(x - \gamma_n\sum_{i=1}^p \theta_i\nabla f_i(x), 2\gamma_n\,\mathrm{Id})$. Note that (8) is the Euler-Maruyama discretization
of the continuous dynamic $\mathrm{d}X_t = -\sum_{i=1}^p \theta_i\nabla f_i(X_t)\mathrm{d}t + \sqrt{2}\mathrm{d}B_t$ where $(B_t)_{t\geqslant 0}$
is a d-dimensional Brownian motion.

3.3   Combining Dynamics

We now combine the gradient dynamic and the Langevin dynamic. This algorithm is referred as Stochastic Optimization with Unadjusted Langevin (SOUL)
algorithm in [4] and is defined for all $n \in \mathbb{N}$ and $k \in \{0, m_n - 1\}$ by the following
recursion

$$X_{k+1}^n = X_k^n - \gamma_{n+1}\sum_{i=1}^p \theta_i\nabla f_i(X_k^n) + \sqrt{2\gamma_{n+1}}Z_{k+1}^n , X_0^n = X_{m_{n-1}}^{n-1}, n \geqslant 1 ,$$

$$\theta_{n+1} = P_\Theta\left[\theta_n + \delta_{n+1}m_n^{-1}\sum_{k=1}^{m_n}(f(X_k^n) - f(x_0))\right] ,$$

with $X_0^0 \in \mathbb{R}^d$, $\theta_0 \in \mathbb{R}^p$, $(\delta_n)_{n\in\mathbb{N}}$, $(\gamma_n)_{n\in\mathbb{N}}$ real positive sequences of step sizes
and $(m_n)_{n\in\mathbb{N}} \in \mathbb{N}^{\mathbb{N}}$, the number of Langevin iterations. In [4] the authors study
the convergence of these combined dynamics.

4   Experiments

In this section, we present experiments conducted with neural network features.
Texture synthesis with these features has been first done in [10] in which the
authors compute Gram matrices, $i.e.$ second-order information, on different layers of a network. The underlying model is microcanonical. In [17] the authors
consider a macrocanonical model with convolutional neural network features corresponding to the mean of filters at a given layer. In our model we consider the
features described in Sect. 2.3 where the linear units and rectifier units are given

by the VGG-19 model [22] which contains 16 convolutional layers. We consider the following settings and notations:

- Trained (**T**) or Gaussian (**G**): if option T is selected the weights used in the VGG convolutional units are given by a classical pretraining for the classification task on the ImageNet dataset [5]. If option G is selected we replace the pretrained weights with Gaussian random variables such that the weights of each channel and each layer have same mean and same standard deviation.
- Shallow (**3**), Mid (**6**), Deep (**8**): in our experiments we consider different settings regarding the number of layers and, more importantly, the influence of their depth. In the Shallow (3) setting we consider the linear layers number 3, 4 and 5. In the Mid (6) setting we consider the linear layers number 3, 4, 5, 6, 7 and 11. In the Deep (8) setting we consider the linear layers number 3, 4, 5, 6, 7, 11, 12 and 14.

4.1   Empirical Convergence of the Sampling Algorithm

We assert the experimental convergence of the SOUL algorithm in Fig. 1. We choose $\delta_n = \mathcal{O}(n^{-1})$, $\gamma_n = \mathcal{O}(n^{-1})$, $m_n = 1$, $\Theta = \mathbb{R}^p$ and $\varepsilon = 0$, *i.e.* $J = 1$. The algorithm is robust for these fixed parameters for a large number of images. Note that even if this case is not covered by the theoretical results of [4], the convergence is improved using these rates. The drawbacks of not using parameter projection ($\Theta = \mathbb{R}^p$) or image regularization ($\varepsilon = 0$) is that the algorithm may diverge for some images, see Fig. 2.

Interestingly, while neural network features capture perceptual details of the texture input they fail to restore low frequency components such as the original color histogram. In order to alleviate this problem, we perform a histogram matching of the output image, as in [11]. In the next section we investigate the advantages of using CNN channel outputs as features.

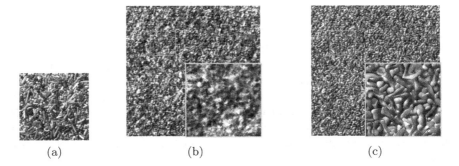

(a)   (b)   (c)

Fig. 1. *Empirical convergence.* In (a) we present a 512×512 objective texture. In (b) we show the initialization of our algorithm, a 1024×1024 Gaussian random fields with same mean and covariance as a zero-padded version of (a). In (c) we present the result of our algorithm after 5000 iterations with (T–8). In the bottom-right corner of each image (b) and (c) we present a $\times 3$ magnification of some details of the images.

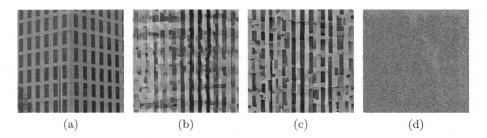

(a)   (b)   (c)   (d)

Fig. 2. *Depth influence.* The image in (a) is the original texture. In (b), (c) and (d) we consider the outputs of the sampling algorithms (T–3), (T–6) and (T–8) algorithms. Note that more geometrical structure is retrieved in (c) than in (b) and that the model has diverged in (d).

4.2   Neural Network Features

Number of Layers. We start by investigating the influence of the number of layers in the model by running the algorithm for different layer configurations. If too many layers are considered the algorithm diverges. However, if the number of layers considered in the model is reduced we observe different behaviors. This is illustrated in Fig. 2 where the objective image exhibits strong midrange structure information.

Model Choice. In all previous experiments the weights considered in the CNN architecture are pretrained on a classification task as in [10] and [17]. It is natural to ask if such a pretraining is necessary. In accordance with the results obtained by Gatys et al. [10] we find that a model with no pretraining does not produce perceptually satisfying texture samples, see Fig. 3. Note that in [24] the authors obtain good results with random convolutional layers and a microcanonical approach.

4.3   Comparison with State-of-the Art Methods

To conclude this experimental study we provide a comparison of our results with state-of-art texture synthesis methods in Fig. 4. Regarding regular textures our model misses certain geometrical constraints, which are encoded by the Gram matrices in [10] for instance. However, our model relies only on 2k features, using (T–8), whereas Gatys et al. use at least 10k parameters. One way to impose the lost geometrical constraints could be to project the spectrum of the outputs at each step of the algorithm as it was done by Liu et al. [16] in a microcanonical model.

5   Perspectives

There still exists a gap between the theoretical analysis of those algorithms which relies on control theory tools [2], stochastic optimization techniques [1] or general

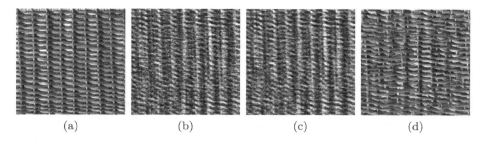

(a)   (b)   (c)   (d)

Fig. 3. *Noisy weights.* In (a) we present the input image, in (b) the initialization of the algorithm and in (c), respectively (d), the output of the algorithm (G–8), respectively (T–8) after 5000 iterations. Note that no spatial structure is retrieved in (c) which is close to its Gaussian initialization.

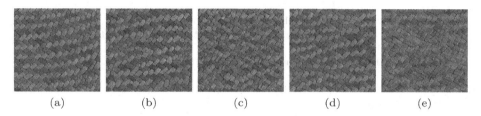

(a)   (b)   (c)   (d)   (e)

Fig. 4. *Comparison.* The input image is shown in (a). In (b) we show the output of the algorithm in [10], in (c) the output of the DeepFRAME method [17] and in (d) we present a result obtained with GAN texture synthesis algorithm [15]. Our result (e) after 5000 iterations of (T–8) is comparable but lacks spatial organization. Images (a)–(d) extracted from [21].

state space Markov chain results and the experimental study. Indeed the class of functions which is handled by these theoretical results is constrained (regularity assumptions, drift conditions...) and has yet to be extended to more general CNN features. In addition, they scale badly with the dimension of the data which is high in our image processing context. In a future work we wish to extend our theoretical understanding of SOUL algorithms applied to macrocanonical models and draw parallels with the microcanonical results obtained in [2].

References

1. Atchadé, Y.F., Fort, G., Moulines, E.: On perturbed proximal gradient algorithms. J. Mach. Learn. Res. **18**, 310–342 (2017). Paper No. 10, 33
2. Bruna, J., Mallat, S.: Multiscale sparse microcanonical models. arXiv e-prints, arXiv:1801.02013, January 2018
3. Cano, D., Minh, T.: Texture synthesis using hierarchical linear transforms. Signal Process. **15**(2), 131–148 (1988)
4. De Bortoli, V., Durmus, A., Pereyra, M., Fernandez Vidal, A.: Stochastic optimization with unadjusted kernel: the SOUK algorithm (2019, preprint)

5. Deng, J., Dong, W., Socher, R., Li, L., Li, K., Li, F.: Imagenet: a large-scale hierarchical image database. In: CVPR, pp. 248–255 (2009)
6. Efros, A.A., Leung, T.K.: Texture synthesis by non-parametric sampling. In: ICCV, pp. 1033–1038 (1999)
7. Gagalowicz, A., Ma, S.D.: Model driven synthesis of natural textures for 3-D scenes. Comput. Graph. **10**(2), 161–170 (1986)
8. Galerne, B., Leclaire, A., Rabin, J.: A texture synthesis model based on semi-discrete optimal transport in patch space. SIIMS **11**(4), 2456–2493 (2018)
9. Galerne, B., Gousseau, Y., Morel, J.: Random phase textures: theory and synthesis. IEEE Trans. Image Process. **20**(1), 257–267 (2011)
10. Gatys, L.A., Ecker, A.S., Bethge, M.: Texture synthesis using convolutional neural networks. In: NIPS, pp. 262–270 (2015)
11. Gatys, L.A., Ecker, A.S., Bethge, M., Hertzmann, A., Shechtman, E.: Controlling perceptual factors in neural style transfer. In: CVPR, pp. 3730–3738 (2017)
12. Gretton, A., Borgwardt, K.M., Rasch, M.J., Schölkopf, B., Smola, A.J.: A kernel method for the two-sample-problem. In: NIPS, pp. 513–520 (2006)
13. Heeger, D.J., Bergen, J.R.: Pyramid-based texture analysis/synthesis. In: ICIP, pp. 648–651 (1995)
14. Jaynes, E.T.: Information theory and statistical mechanics. Phys. Rev. **106**, 620–630 (1957)
15. Jetchev, N., Bergmann, U., Vollgraf, R.: Texture synthesis with spatial generative adversarial networks. CoRR (2016)
16. Liu, G., Gousseau, Y., Xia, G.: Texture synthesis through convolutional neural networks and spectrum constraints. In: ICPR, pp. 3234–3239 (2016)
17. Lu, Y., Zhu, S., Wu, Y.N.: Learning FRAME models using CNN filters. In: AAAI, pp. 1902–1910 (2016)
18. Ogden, H.E.: A sequential reduction method for inference in generalized linear mixed models. Electron. J. Stat. **9**(1), 135–152 (2015)
19. Peyré, G.: Texture synthesis with grouplets. IEEE Trans. Pattern Anal. Mach. Intell. **32**(4), 733–746 (2010)
20. Portilla, J., Simoncelli, E.P.: A parametric texture model based on joint statistics of complex wavelet coefficients. IJCV **40**(1), 49–70 (2000)
21. Raad, L., Davy, A., Desolneux, A., Morel, J.: A survey of exemplar-based texture synthesis. Ann. Math. Sci. Appl. **3**, 89–148 (2018)
22. Simonyan, K., Zisserman, A.: Very deep convolutional networks for large-scale image recognition. CoRR (2014)
23. Ulyanov, D., Lebedev, V., Vedaldi, A., Lempitsky, V.S.: Texture networks: Feed-forward synthesis of textures and stylized images. In: ICML, pp. 1349–1357 (2016)
24. Ustyuzhaninov, I., Brendel, W., Gatys, L.A., Bethge, M.: Texture synthesis using shallow convolutional networks with random filters. CoRR (2016)
25. van Wijk, J.J.: Spot noise texture synthesis for data visualization. In: SIGGRAPH, pp. 309–318 (1991)
26. Zhu, S.C., Wu, Y.N., Mumford, D.: Filters, random fields and maximum entropy (FRAME): towards a unified theory for texture modeling. IJCV **27**(2), 107–126 (1998)

Finding Structure in Point Cloud Data with the Robust Isoperimetric Loss

Shay Deutsch[1(✉)], Iacopo Masi[2], and Stefano Soatto[1]

[1] University of California, Los Angeles, USA
shaydeu@math.ucla.edu, soatto@cs.ucla.edu
[2] Information Sciences Institute, University of Southern California,
Marina Del Rey, CA, US
iacopo@isi.edu

Abstract. We present a new regularization method to find structure in point clouds corrupted by outliers. The method organizes points into a graph structure, and uses isoperimetric inequalities to craft a loss function that is minimized alternatingly to identify outliers, and annihilate their effect. It can operate in the presence of large amounts of outliers, and inlier noise. The approach is applicable to both low-dimensional point clouds, such as those obtained from stereo or structured light, as well as high-dimensional ones.

Keywords: Point Cloud Denoising · Isoperimetric inequalities

1 Introduction

Many problems in vision boil down to finding needles in haystacks: From 3D reconstruction in structure-from-motion – where putative correspondences contain a small percentage of correct matches – to visual recognition and unsupervised learning, where most of the features computed from the images are uninformative. Accordingly, there has been considerable interest in manifold learning, where the goal is finding relatively low-dimensional structures in high-dimensional embedding spaces. A critical component is to decide which point is "close" to the low-dimensional structure, without explicit knowledge of the latter, and which should be discarded (outliers). The discipline of (classical) Robust Statistics is devoted to this problem: How to simultaneously fit a relatively simple model, and decide which data belong to it – without wasting resources to model the latter.

We propose a method to jointly handle outliers and inlier noise by alternating two steps: First, an outlier estimation criterion is computed; then, a new regularization method is applied to annihilate them, while simultaneously regularizing inlier data, thus providing a unified treatment of both outliers and inlier noise.

© Springer Nature Switzerland AG 2019
J. Lellmann et al. (Eds.): SSVM 2019, LNCS 11603, pp. 25–37, 2019.
https://doi.org/10.1007/978-3-030-22368-7_3

1.1   Previous Work and Challenges

Robust statistical inference has been critical to the evolution of structure-from-motion (SFM), where RANSAC [10] has been credited with making early attempts practical. Tensor Voting (TV) [15] has also been applied broadly from perceptual organization to manifold learning [4–6,8].

More recent applications focus on unsupervised learning, and methods including Robust PCA [1] and its many extensions [14] can be effective when the percentage of outliers is small and the data lies close to a linear space. Similarly, sparse subspace clustering [9] uses the sparsest representation via l_1 minimization to define the affinity matrix of an undirected graph, resulting in state-of-the-art results in motion segmentation, mostly assuming linear structures with distant outliers. Other related work include [18] which proposed a scale estimator of inliers for heavily corrupted multiple structure data. Here, we focus on robust estimation of manifolds with complex topology corrupted by a high percentage of outliers, with the goal of estimating the underlying structure. Recent work [3] using Isoperimetric loss (IPL) as a regularizer was shown effective in the presence of noise, but did not handle outliers. Applying regularization to outliers may distort the manifold and alter its topological structure. shu

The rest of the paper is organized as follows: in Sect. 2 we describe the notation and main assumptions. In Sect. 3 we describe our proposed framework and Sect. 4 presents the experimental results. Discussion is provided in Sect. 5.

2   Preliminaries and Main Assumptions

Given a set of N points, $\mathbf{x} = \{\mathbf{x}_i\}_{i=1}^{N}, \mathbf{x}_i \in \mathbb{R}^D$ which consists of an unknown number N_I of inliers and N_o outliers, where the inliers $\{\mathbf{x}_i\}_{i=1}^{N_I}, \mathbf{x}_i \in \mathbb{R}^D$ live close to a low dimensional manifold M. The inliers are sampled from a complex manifold M or union of a sub-manifolds M_i and may include intersections or other types of singularities. For each $\mathbf{x}_i \in M$, let $T_{\mathbf{x}_i}M$ denote it tangent space, and let O_i be the subspace which corresponds to the local tangent space estimate of $T_{\mathbf{x}_i}M$.

An undirected, weighted graph $\mathcal{G} = (\mathcal{V}, \mathbf{W})$ is constructed over $\mathbf{x}$, where $\mathcal{V}$ corresponds to the nodes and $\mathbf{W}$ to the set of edges on the graph. In this work, the adjacency matrix $\mathbf{W} = (w_{ij})$ which consists of the weights w_{ij} between node i and node j, is constructed based on local tangent space similarity which is filtered using the Gaussian kernel function as follows:

$$w_{ij} = \begin{cases} \tilde{w}_{ij}F(O_i, O_j) & \text{if } \mathbf{x}_j \in \text{kNN}(\mathbf{x}_i) \\ 0 & \text{else} \end{cases} \tag{1}$$

where

$$\tilde{w}_{ij} = \exp\left(\frac{-||\mathbf{x}_i - \mathbf{x}_j||_2^2}{2\sigma_d^2}\right) \tag{2}$$

kNN($\mathbf{x}_i$) denotes the k nearest Euclidean neighbors of $\mathbf{x}_i$, ($|| \cdot ||_2^2$) denotes the l_2 norm, and σ_d^2 is the parameter of a radial basis function (RBF). The local tangent space affinity function $F(O_i, O_j)$ is chosen based on the maximal principal angle between the sub-spaces O_i, O_j as follows [11]:

$$F(O_i, O_j) = \min_{u \in O_i} \max_{\tilde{v} \in O_j} \langle u, \tilde{v} \rangle \tag{3}$$

The local tangent space can be estimated using Local PCA [19] or Tensor Voting. The degree of a vertex $i \in \mathcal{V}$ is defined as $d(i) = \sum_{(i,j) \in V} w_{ij}$. $\mathbf{L}$ denotes the combinatorial graph Laplacian, defined as $\mathbf{L} = \mathbf{D} - \mathbf{W}$, with $\mathbf{D}$ the diagonal degree matrix with entries $d_{ii} = d(i)$. The eigenvalues and eigenvectors of $\mathbf{L}$ are denoted by $\lambda_1, \ldots, \lambda_N$ and $\phi_1, \ldots, \phi_N$, respectively. Note that the eigenvalues and the corresponding eigenvectors carry with them a notion of frequency such that increasing eigenvalues corresponds to increasing the variation with respect to the intrinsic structure of the graph.

3   Proposed Approach

Our framework for a joint inlier and outlier regularization alternates between two main steps:

i **Outlier estimation** using the current estimate of the local geometric structure and a function $\mathbf{f} : \mathcal{V} \to \mathbb{R}$ indicating the likelihood of being an inlier; partition the set into inliers $\mathbf{S}$, and outliers $\mathbf{S}^c$.
ii **Inlier regularization** based on the current partition of $\{\mathbf{x}_i\}_{i=1}^N$ into a set of candidate inliers $\mathbf{S}$, and outliers $\mathbf{S}^c$ using the graph $\mathcal{G} = (\mathcal{V}, \mathcal{W})$.

Our scheme aims to identify outliers to perform an adaptive regularization that attenuates inlier noise from the data while annihilating outliers. The identification of outliers is incorporated in the regularization scheme, that is adapted to the type of noise detected in a local neighborhood at each point. For example, if a point is detected as an inlier and also lies in the vicinity of inliers, then this point and its neighbors will be smoothed and projected on a smooth sub-manifold. On the other hand, the position of a detected outlier would not be changed much. Therefore our process will amplify confidence of inliers. This strategy is different than typical outlier denoising, where outliers are either prematurely discarded, or inlier noise is not addressed explicitly.

The resulting approach is an automatic method for outliers classification which seeks to control the amount of regularization applied to the input points, in a way which is aware of the type of noise present in the data, i.e inlier or outlier noise. Next, we describe the two steps in more detail.

3.1   Outlier Estimation

In the first step we estimate outliers from the local geometric structure by estimating the proximity to the local tangent space, which will serve as an indicator of whether a point is an inlier or an outlier. Under this model we assume that if

$||\mathbf{x_i} - \tilde{\mathbf{x}}_i|| < \tau$ for some threshold τ then $\tilde{\mathbf{x}}_i$ is likely to be an inlier, otherwise if $||\mathbf{x_i} - \tilde{\mathbf{x}}_i|| > \tau$ then it is likely to be an outlier. In practice τ is unknown and may vary in different parts of the manifold. To estimate the proximity to the local tangent space, we employ the Tensor Voting operator [15] acting on a point's local neighborhood $B(\mathbf{x}_i, \sigma)$ (a ball of radius σ around $\mathbf{x}_i$), according to

$$\mathbf{T}_{ij} = \exp^{\frac{-||\mathbf{x}_i - \mathbf{x}_j||^2}{\sigma}} \left(\mathbf{I} - \frac{(\mathbf{x}_i - \mathbf{x}_j)(\mathbf{x}_i - \mathbf{x}_j)^T}{||\mathbf{x}_i - \mathbf{x}_j||^2} \right) \tag{4}$$

and the local information accumulated at each point is

$$\mathbf{T}_i = \sum_{\mathbf{x}_j \in B(\mathbf{x}_i, \sigma)} \mathbf{T}_{ij} \tag{5}$$

where $\mathbf{I}$ is the $D \times D$ identity matrix, and $\mathbf{T}_i$ is a semi-positive matrix. Note that (5) is used in the Tensor Voting framework [15] as a first step to detect outliers. For each $\mathbf{T}_i$ from the eigensystem decomposition we obtain a set of eigenvalues $\{\alpha_{ik}\}_{k=1}^{D}, \alpha_{i1} \geq \alpha_{i2}..., \alpha_{iD}$ and its corresponding eigenvectors $\{\hat{e}_{ik}\}_{k=1}^{D}$. Without loss of generality, let $\mathbf{x}_i = 0 \in \mathbb{R}^3$, and define $\tilde{\mathbf{x}}_j \doteq \mathbf{x}_j - \mathbf{x}_i$ (centered datum), and $\bar{\mathbf{x}}_j = \frac{\tilde{\mathbf{x}}_j}{||\tilde{\mathbf{x}}_j||}$ (scaled datum), $\hat{\mathbf{x}}_j$ is the skew-symmetric span of $\tilde{\mathbf{x}}_j^{\perp}$ (the 3×3 skew-symmetric matrix that spans the orthogonal complement of $\tilde{\mathbf{x}}_j$), and $-(\hat{\mathbf{x}}_j)^2 = \mathbf{I} - \bar{\mathbf{x}}_j\bar{\mathbf{x}}_j^T$ is the symmetric span of $\bar{\mathbf{x}}_j^{\perp}$ (the 3×3 symmetric matrix that spans the orthogonal complement of $\bar{\mathbf{x}}_j$). The weighted average of these spans is a 3×3 matrix that has full rank unless all the vectors span the same direction, and its spectrum can be used to measure the "dispersion" of the vector ensemble relative to some direction, captured by the eigenvector corresponding to the largest eigenvalue. Specifically, note that if we define

$$\mathbf{x}'_j \doteq e^{-||\tilde{\mathbf{x}}_j||} \frac{\tilde{\mathbf{x}}_j}{||\tilde{\mathbf{x}}_j||}$$

the spectrum of $\tilde{\mathbf{T}}_i = \sum_j -(\hat{\mathbf{x}}'_j)^2$ is identical to the spectrum of $\mathbf{T}_i^{\perp} = \sum_j \mathbf{x}'_j {\mathbf{x}'_j}^T$. The eigenvectors to the latter are the principal directions, whereas the former are the spans of their orthogonal complements. The span of $\mathbf{T}_i$ is different in that the weight multiplies the identity matrix, so the spectrum is rescaled unless the weights are normalized. However, it captures the same qualitative properties of the spectrum. We use α_{i1} ($\{\alpha_{ik}\}$ are the eigenvalues of the tensor $\mathbf{T}_i$ that represents local geometric structure saliency) to construct the graph signal $\mathbf{f}$ that serves as an initial estimate for outliers/inliers classification as follows:

$$\mathbf{f}(\mathbf{T}_i) = \frac{1}{1 + \exp(-(\beta_i))} \tag{6}$$

where $\beta_i = \log(\alpha_{i1})$, α_{i1} is the largest eigenvalue of $\mathbf{T}_i$. Note that if $\alpha_i \to 0$ then we have that $\mathbf{f}(\mathbf{T}_i) \to 0$, while if $\alpha_i \approx k$ where k was the k nearest neighbors voted for $\tilde{\mathbf{x}}_i$ then $\mathbf{f}(\mathbf{T}_i) \to 1$. By construction, the suggested function $\mathbf{f}$ in (6) incorporates both density and local tangent space saliency, assigning a scalar value $0 \leq \mathbf{f}(\mathbf{T}_i) \leq 1$ for each point. Let the following partition be induced by the values of $\mathbf{f}(\mathbf{T}_i)$:

$$\begin{cases} i \in \mathbf{S} \text{ if } \mathbf{f}(\mathbf{T}_i) \geq 1/2 \\ i \in \mathbf{S}^c \text{ if } \mathbf{f}(\mathbf{T}_i) < 1/2 \end{cases} \tag{7}$$

3.2   Step 2: Adaptive Regularization Using the Isoperimetric Loss (IPL)

Our key idea is to use the isoperimetric inequality in a new adaptive regularization scheme, which enjoys a number of desirable properties for the suggested framework:

(i) It can be utilized for general weighted graphs in an adaptive way, depending to the type of noise estimated at each point.

(ii) The IPL can be easily employed directly in the graph spectral domain in a way which provides both vertex-spectral localization (i.e. it is a local graph transform which also makes use of band limited spectral information).

(iii) It has a fast implementation which does not require expensive global computations.

In this section we describe in more detail our graph smoothing based on the isoperimetric loss (IPL).

Our baseline gives us a graph $\mathcal{G}$ with noisy weights w_{ij} and the noisy manifold coordinates $\{\mathbf{x}_i\}_{i=1}^{N_I}$ (each $\mathbf{x}_i$ is an attribute, as a vector to the corresponding vertex $i \in \mathcal{G}$) that we want to modify. We seek a way to regularize them by exploiting the graph structure, together with the outlier estimate obtained. Our regularization criterion is to achieve some level of compactness of bounded subsets: For a collection of subsets of the vertices with fixed size (corresponding to the volume of a subset) we want to find the subsets with the smallest size boundary. The proposed criterion rests on classical differential geometry of Riemannian manifolds, where in the most basic case, the most compact manifold that encloses a fixed area with minimum size boundary is a circle. Our criterion, quantified by the isoperimetric gap, generalizes this bias towards compactness to more general sets. In the continuous setting, for a Riemannian manifold the isoperimetric gap is zero when it is a sphere.

The pseudo-code for the suggested framework which combines the two steps - outlier estimation and regularization using the Isoperimetric Loss (IPL) is provided below in Algorithm 1.

We now define the isoperimetric inequality, that holds for general weighted graphs, and the associated isoperimetric dimension:

Definition 1 *[2]. We say that a graph $\mathcal{G}$ has an isoperimetric dimension δ with a constant c_δ, if, for every bounded subset X of $\mathcal{V}$, the number of edges between X and the complement of X, $\mathcal{V} \setminus \mathcal{X}$ satisfies*

$$\mu(\partial X) \geq c_\delta (\mu(X))^{1-\frac{1}{\delta}} \tag{8}$$

where ∂X denotes the boundary of X.

The isoperimetric inequality relates the volume of a bounded set to the area of its boundary, and is satisfied as an equality by a sphere.

Let $B_r(\xi)$ be the ball around $\xi \in \mathcal{V}$ of radius r, that is, the set of nodes within a distance $d_{\mathcal{G}}$ less than r. Given $\xi \in \mathcal{V}$ and a node x in the boundary of $X = B_r(\xi)$, we define

$$\mu_x^{(\xi)} = \sum_{\substack{x \sim y \\ d_G(y,\xi) < d_G(x,\xi)}} w_{xy} \tag{9}$$

as the flow from x towards ξ, that is the sum of weights of edges connecting x with points close to ξ. For a fixed r, we define

$$\mu_r^{(\xi)} = \sum_{\substack{x \\ d_G(x,\xi)=r}} \mu_x^{(\xi)} \tag{10}$$

which is the flow from all points at a distance r from ξ, towards ξ, i.e., the flow through the boundary of $B_r(\xi)$. We abuse the notation and use the subscript to indicate the radius of the ball, rather than a particular point on its boundary, and use context to disambiguate. With this notation, the isoperimetric inequality can be written in the form

$$\mu_r^{(\xi)} \geq c_\delta \left(\sum_{x,y \in B_r(\xi)} w_{x,y} \right)^{1-\frac{1}{\delta}} \tag{11}$$

The Isoperimetric Gap. The difference between the two sides in the isoperimetric inequality is called *isoperimetric gap*:

Definition 2. *The isoperimetric gap is defined as*

$$\beta(\xi; \delta, W) \doteq c_\delta \left(\sum_{x \ y \in B_r(\xi)} w_{x,y} \right)^{1-\frac{1}{\delta}} - \mu_r^{(\xi)} \tag{12}$$

which is negative and bounded above by $\mu_r^{(\xi)}$. The gap is zero when M is a sphere, and decreasing the gap is hypothesized to have regularization effects.

In general, a Riemannian manifold is equipped with a function defined on it and the graph, as its discrete approximation, produces isoperimetric inequalities such as (8). We utilize an additional function $f_{x,y}$ serving the role of the embedding distance.

Proposed Adaptive Regularization Approach. Minimizing the isoperimetric loss directly in the vertex domain entails solving the following optimization problem:

$$\min_{W \geq 0} \sum_{\xi \in V} \sum_{x,y \in B_r(\xi)} f_{x,y} w_{x,y} + \lambda \beta(\xi; \delta, W)$$
$$\text{s.t. } 0 \leq w_{x,y} \leq 1 \, \forall \, x, y \tag{13}$$

where the quantity to be minimized is called Isoperimetric Loss (IPL) and λ is a positive scalar tuning parameter.

Spectral Reduction of the IPL. In this work, we focus on a spectral reduction of the isoperimetric loss, which provides a faster alternative to solving (13) directly in the vertex domain. In addition, the direct method described above entails an approximation of the isoperimetric dimension. We can also approximate the IPL using a spectral method for the IPL loss which does not require an explicit estimate of the isoperimetric dimension. Specifically, we use the Spectral Graph Wavelet Transform [12] of the noisy graph signal $f(i)$, and set eigenvalues $\{\lambda_i\}$ of the unnormalized Laplacian $\mathbf{L} = \mathbf{D} - \mathbf{W}$ to zero beyond a nominal rank, and then recompute the adjacency matrix. Removing a number of bands in this way will correspond to shrinking the boundaries of each ball around each vertex i. More in detail, **Spectral Reduction** is performed as follows: Let $f(i)$ be a component of the given feature vector $\mathbf{x}_i$ and $\tilde{f}(i) = \sum f(j) \sum_{k=0}^{r} a_k \sum_{l=1}^{N} \lambda_l^k \phi_l(j)\phi_l(i) = \sum_{i=1}^{N} f(i) \sum_{k=0}^{r} a_k (\mathbf{L}^k)_{i,j}$ a filtered signal of a fixed dimension of $\mathbf{x}_i$. The coefficients a_k are constants of a polynomial function and for a specific choice correspond to spectral graph wavelet (SGW) coefficients. Note that the terms $(\mathbf{L}^r)_{i,j}$ can be interpreted as the localized spectral transform of the graph around the ball $B_r(\mathbf{x}_i)$, which are non-zero for all $\mathbf{x}_j$ in $B_r(\mathbf{x}_i)$ and vanish for all $\mathbf{x}_j \notin B_r(\mathbf{x}_i)$. With the SGW transform, we employ a redundant representation with $J = r$ polynomials $\kappa_{s(e)}(\lambda)$, $1 \leq e \leq J$ with corresponding scaling $s(e)$. Next choose the smallest $1 < r_0 < r$ such that the polynomial function $\kappa_{s(e)}(\lambda) = \sum_{k=1}^{r} a_k (\mathbf{L}^k)_{i,j}$ vanishes above $\lambda > d(i)$. Then, for all SGW coefficients $\tilde{f}(i)$ where the corresponding filter $\kappa_{s(e)}(\lambda) > 0$, $e \geq r_0$ for some $\lambda_l > d(i)$, we annihilate all terms $\tilde{f}(i)$, which has the effect of reducing the isoperimetric loss directly in the spectral domain. Take the

---

Algorithm 1. RIPL

Process 1: Outlier estimation.
Input: Set of points $\tilde{\mathbf{x}}$, voting scale σ
Step 1: Learn local geometric structure using (5)
Step 2: Using the output of step 1, estimate $\mathbf{f}$ - the likelihood of each point to be an inlier or outlier.
Output: $\hat{e}_{ik}(\mathbf{T}_i)$, $\alpha_{ik}(\mathbf{T}_i)$, estimated inliers $\mathbf{S}$, outliers $\mathbf{S}^c$, $\tilde{\mathbf{x}} = \mathbf{S} \cup \mathbf{S}^c$.
Process 2: Adaptive regularization:
While stopping criterion is not met:
Input: $\tilde{\mathbf{x}}$, k nearest neighbor parameter, σ_d (RBF Gaussian scale), $\mathbf{S}$(inliers), $\mathbf{S}^c$(outliers)
Step 1: Construct a weighted graph $\mathcal{G} = (\mathcal{V}, \mathcal{W})$ using (1)
Step 2: Using (7), partition $\mathcal{V}$ into $\mathbf{S}$ (inliers), $\mathbf{S}^c$ (outliers)
Step 3 Apply adaptive *inlier regularization* as follows:
If $\tilde{\mathbf{x}}_i \in \mathbf{S}$: apply isoperimetric loss regularization
Else if $\tilde{\mathbf{x}}_i \in \mathbf{S}^c$: no regularization is applied to $\tilde{\mathbf{x}}_i$ directly.
Output: Final partition of $\tilde{\mathbf{x}}$ into $\mathbf{S}$, $\mathbf{S}^c$, regularized inliers $\hat{\mathbf{x}}_i$

---

inverse transform to obtain the denoised signal and construct the new graph. Note that the spectral reduction method suggested is based on a local operator which does not require computing the eigensystem of the Laplacian, but rather only its largest eigenvalue. This makes the spectral reduction fast in comparison to other graph based method which are using the graph frequency information. The algorithm performs alternating steps in the aggregate minimizing a compound loss that we refer to as Robust Isoperimetric Loss optimization, or RIPL.

4   Experimental Results

4.1   Manifold Denoising

In this part, we first use our approach to learn the topology of an unknown surface in 3D, measured through noisy point samples, following which we can easily interpolate the inliers using customary manifold denoising tools. The synthetic dataset was sampled using 1,400 points for the Helix intersecting with a sphere, and 1,200 points for the half sphere on a plane. Quantitative results of our method in the case of a noisy half sphere on a plane, where ground truth is available, are shown in Table 1. In this case we compare our method to popular unsupervised denoising methods such as Laplacian-based regularization [13] and denoising using Spectral Graph Wavelets [7]. Our method can handle significant amount of noise whereas others, especially Laplacian-based regularization, fail.

The qualitative results we report show that our approach recovers the structural features such as singularities and discontinuities. Figures 1, 3, and 5 show representative examples from a large repertoire of synthetic data, as well as (pseudo-synthetic) real point clouds measured with range cameras or stereo, to which we have added a controlled amount of noise. 3D models for "Adirondack" and "Piano" from the Middlebury dataset [17] (Figs. 2 and 3) provide accurate

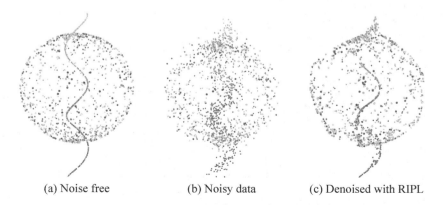

(a) Noise free   (b) Noisy data   (c) Denoised with RIPL

Fig. 1. Denoising results on noisy manifolds using the proposed graph construction.

ground truth while respecting the natural bias of real-world shapes. A further comparison with PCL outlier removal operator [16] is shown in Fig. 4: RIPL is able to better restore the fine structure of the scene. Quantitative experiments reflect the same behavior showing that the method is resilient to noise an outliers.

Complexity and Computation Time: The RIPL has a computational cost of $O(ND\log(N))$, which includes step 1 (outlier detection) and step 2 (IPL with spectral reduction). This includes fast computation of the k nearest neighbor graph using k-d tree, and the SGW transform which is $O(N)$ for each dimension of the manifold for sparse graphs, thus total complexity of $O(ND\log(N))$. The execution time of our unoptimized Matlab code implementation using Intel Core i7 7700 Quad-Core 3.6 GHz with 64B Memory on the "piano" scene in the Middlebury dataset using 340,000, 1.4 million, and 5.7 million points using $k = 10$ nearest neighbor graph takes ≈7.3, 32, and 140 s, respectively.

4.2   The RIPL Approach

We test RIPL on both synthetic and real datasets, in both low and high dimensions. For synthetic data, we use complex manifolds sampled non-uniformly and including complex structures such as singularities or self intersections, in addition to being contaminated with large amount of Gaussian noise. We use RIPL to identify outliers and smooth inlier noise. Table 1 shows the TPR (True Positive Rate) along with the FPR (False Positive Rate).

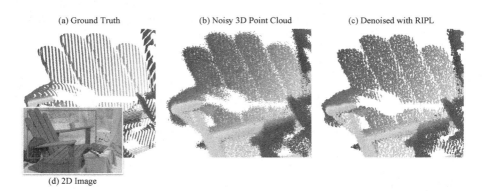

Fig. 2. 3D surface reconstruction of the "Adirondack" scene from the Middlebury dataset [17] (a) Ground truth. The red rectangle marks the detailed object whose 3D noisy point cloud is shown in (b); the denoised 3D surface reconstruction is presented in (c). RIPL removes inlier noise effectively. (Color figure online)

Table 1. A quantitative comparison measuring error of a half sphere on a plane using different amount of noise (standard deviation). In this example the two manifolds were sampled with total number of 1200 points and non-uniform sampling.

Method	Noise		
	0.03	0.06	0.08
Laplacian regular [13]	0.0105	0.014	0.017
Manifold den. w/ SGW [7]	0.005	0.007	0.01
RIPL (ours)	**0.003**	**0.005**	**0.007**

Table 2. Evaluation of outlier classification on a variety of complex manifolds: TPR corresponds correct detection of inliers noise, and FPR corresponds to percentage of inliers wrongly classified as outliers. The best performance is shown in boldface.

Data	TPR	FPR
Sphere and plane	91.50	86.22
Two circles	87.03	70.00
Fish bowl	85.33	80.01
Motion segm. (155 bench.)	**100.0**	**94.30**

4.3   Experiments with Real Data: Application to Motion Segmentation

We evaluate the robust topology estimation method on the Hopkins155 motion database,[1] where the goal is to segment a video sequence into multiple spatiotemporal regions corresponding to different rigid-body motions. In realistic scenarios, the data points can often be corrupted due to the limitation of the tracker. Sparse Subspace Clustering (SSC) [9] and its more recent variations can handle corrupted trajectories but not large amount of outliers. To simulate a more challenging scenario, all inliers were corrupted with noise of standard variation 0.1, while outliers were generated by corrupting each entry with noise

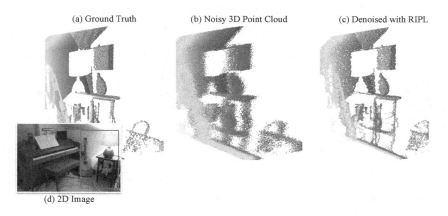

(a) Ground Truth   (b) Noisy 3D Point Cloud   (c) Denoised with RIPL

(d) 2D Image

Fig. 3. 3D surface reconstruction of the "piano" scene from the Middlebury dataset [17]. The red rectangle in (d) marks the detailed object whose 3D ground truth is shown in (a); the noisy 3D surface (b) and the denoised with RIPL (c). (Color figure online)

---

[1] www.vision.jhu.edu/data/hopkins155.

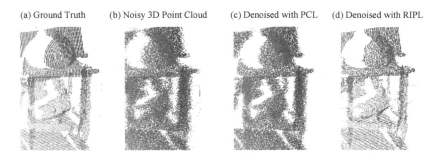

Fig. 4. Another detail of the "piano" scene from the Middlebury dataset: (a) ground-truth (b) noisy 3D point cloud (c) denoised with PCL [16] (d) denoised with RIPL.

of standard variation 0.3. In this case, the number of outliers was equal to the number of inliers. Results are shown in Table 2.

4.4 Experiments with Real Data: Application with 3D Laser Scan Data

We used the publicly available data table_scene_lms400.pcd from the Point Cloud Library (PCL [16]) which is the standard "de facto" for point cloud processing widely used in the industry. We ran the default statistical removal operator available off-the-shelf in PCL and compare it to our proposed RIPL. We motivate this experiment because laser scans produce point clouds with different point densities and their measurement errors can generate sparse outliers. As a result, these outliers points can impair the estimation of surface normals, mesh reconstruction, and curvature variations, which in turn could make difficult the registration, matching, and recognition of point clouds. To solve this problem and show that RIPL exhibits better performance than off-the-shelf methods, we compare RIPL with the PCL statistical removal operator. Figure 6 shows that RIPL better removes outliers in the `table_scene_lms400` point cloud used:

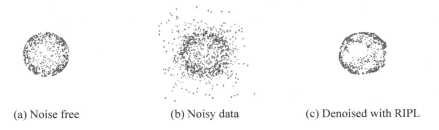

Fig. 5. (a) Noise free fish bowl (b) Noisy fish bowl input (c) the denoised output of RIPL. The outliers are removed and "noisy inlier" points are denoised. The blue color in the denoised figure shows the denoised output. Red color correspond are all outliers detected and removed. (Color figure online)

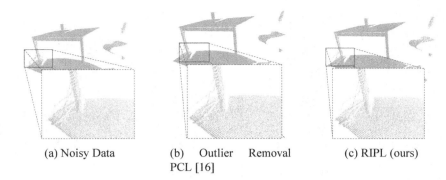

(a) Noisy Data   (b) Outlier Removal   (c) RIPL (ours)
                         PCL [16]

Fig. 6. Results of RIPL method on the Table point cloud compared with a standard outlier removal operator: (a) the noisy point cloud (b) the result by the outlier removal operator from the PCL [16] (c) our result; we can clearly see that RIPL better preserves local structure while removing outliers, while the method from PCL [16] tends to eliminate important structure.

zoomed views of Fig. 6 show that sparse outliers are removed for both RIPL and the PCL statistical operator yet RIPL keeps fine-grained structure that otherwise it is removed by off-the-shelf methods.

5 Discussion

RIPL is a method to denoise a point cloud and remove outliers that leverages the isoperimetric loss and an outlier identification stage, in an alternating fashion. The RIPL computational cost is of $O(ND\log(N))$, which includes fast computation of the k nearest neighbor graph and the IPL spectral reduction which is $O(N)$ per dimension for sparse graphs. Limitations includes processing very large graphs which consists more than dozen millions of points. Failure modes include cases when outlier noise is above 40% and inlier noise is significant. In such cases we observe that balancing between inlier and outliers denoising becomes increasingly challenging, and results in either severe over-smoothing or large errors in outlier identification. A possible direction for future research is to construct graph with negative weights to amplify dissimilarity between the estimated inliers and outliers, and incorporating additional measures of the graph topology in our scheme.

Acknowledgments. This work was supported in part by ONR grant #N00014-19-1-2229 and ARO grant #W911NF-17-1-0304.

References

1. Candès, E.J., Li, X., Ma, Y., Wright, J.: Robust principal component analysis? J. ACM (JACM) **58**(3), 11:1–11:37 (2011). Article No. 11
2. Chung, F., Grigor'yan, A., tung Yau, S.: Higher eigenvalues and isoperimetric inequalities on Riemannian manifolds and graphs (1999)
3. Deutsch, S., Bertozzi, A., Soatto, S.: Zero shot learning with the isoperimetric loss. Technical report, University of California, Los Angeles (UCLA) (2018)
4. Deutsch, S., Medioni, G.: Intersecting manifolds: detection, segmentation, and labeling. In: Proceedings of the Twenty-Fourth International Joint Conference on Artificial Intelligence (2015)
5. Deutsch, S., Medioni, G.: Unsupervised learning using the tensor voting graph. In: Scale Space and Variational Methods in Computer Vision (SSVM) (2015)
6. Deutsch, S., Medioni, G.G.: Learning the geometric structure of manifolds with singularities using the tensor voting graph. J. Math. Imaging Vis. **57**(3), 402–422 (2017)
7. Deutsch, S., Ortega, A., Medioni, G.: Manifold denoising based on spectral graph wavelets. In: International Conference on Acoustics, Speech and Signal Processing (ICASSP) (2016)
8. Deutsch, S., Ortega, A., Medioni, G.G.: Robust denoising of piece-wise smooth manifolds. In: 2018 IEEE International Conference on Acoustics, Speech and Signal Processing, ICASSP 2018, Calgary, AB, Canada, 15–20 April 2018, pp. 2786–2790 (2018)
9. Elhamifar, E., Vidal, R.: Sparse subspace clustering. In: CVPR, pp. 2790–2797 (2009)
10. Fischler, M.A., Bolles, R.C.: Random sample consensus: a paradigm for model fitting with applications to image analysis and automated cartography. Commun. ACM **24**, 381–395 (1981)
11. Golub, G.H., Van Loan, C.F.: Matrix Computations, 3rd edn. The Johns Hopkins University Press, Baltimore (1996)
12. Hammond, D.K., Vandergheynst, P., Gribonval, R.: Wavelets on graphs via spectral graph theory. Appl. Comput. Harmon. Anal. **30**, 129–150 (2011)
13. Hein, M., Maier, M.: Manifold denoising, pp. 561–568 (2007)
14. Liu, G., Lin, Z., Yu, Y.: Robust subspace segmentation by low-rank representation (2010)
15. Mordohai, P., Medioni, G.: Dimensionality estimation, manifold learning and function approximation using tensor voting. J. Mach. Learn. Res. **11**, 411–450 (2010)
16. Rusu, R.B., Cousins, S.: 3D is here: point cloud library (PCL). In: IEEE International Conference on Robotics and Automation (ICRA), Shanghai, China, 9–13 May 2011
17. Scharstein, D., et al.: High-resolution stereo datasets with subpixel-accurate ground truth. In: Jiang, X., Hornegger, J., Koch, R. (eds.) GCPR 2014. LNCS, vol. 8753, pp. 31–42. Springer, Cham (2014). https://doi.org/10.1007/978-3-319-11752-2_3
18. Suter, D., Chin, T., Wang, H.: Simultaneously fitting and segmenting multiple-structure data with outliers. IEEE Trans. Pattern Anal. Mach. Intell. **34**, 1177–1192 (2011)
19. Zhang, Z., Zha, H.: Principal manifolds and nonlinear dimension reduction via local tangent space alignment. SIAM J. Sci. Comput. **26**, 313–338 (2002)

Deep Eikonal Solvers

Moshe Lichtenstein$^{(\boxtimes)}$, Gautam Pai, and Ron Kimmel

Technion - Israel Institute of Technology, Haifa, Israel
{smosesli,paigautam,ron}@cs.technion.ac.il

Abstract. A deep learning approach to numerically approximate the solution to the Eikonal equation is introduced. The proposed method is built on the fast marching scheme which comprises of two components: a local numerical solver and an update scheme. We replace the formulaic local numerical solver with a trained neural network to provide highly accurate estimates of local distances for a variety of different geometries and sampling conditions. Our learning approach generalizes not only to flat Euclidean domains but also to curved surfaces enabled by the incorporation of certain invariant features in the neural network architecture. We show a considerable gain in performance, validated by smaller errors and higher orders of accuracy for the numerical solutions of the Eikonal equation computed on different surfaces. The proposed approach leverages the approximation power of neural networks to enhance the performance of numerical algorithms, thereby, connecting the somewhat disparate themes of numerical geometry and learning.

Keywords: The Eikonal equation · Deep learning for PDE · Geodesic distance

1 Introduction

Fast and accurate computation of distances is fundamental to innumerable problems in computer science. Specifically, in the sub-fields of computer vision, geometry processing and robotics, distance computation is imperative for applications like navigating robots [10,13,14], video object segmentation [28], image segmentation [3] and shape matching [30]. A distance function has a gradient with unit magnitude at every point in the domain, therefore, they are computed by estimating the viscosity solutions to a Hamilton-Jacobi type non-linear PDE called the Eikonal equation. Fast marching methods are the prominent numerical schemes to estimate distance functions on discretized Euclidean domains as well as triangulated curved surfaces. Fast marching methods are known to have a $\mathcal{O}(N \log N)$ computational complexity (for a discrete domain consist in N sample points). Different versions of the fast marching algorithm have been developed yielding different accuracies $\mathcal{O}(h)$ [22], $\mathcal{O}(h^2)$ [23] and $\mathcal{O}(h^3)$ [1] for 2D Cartesian grids and $\mathcal{O}(h)$ for triangulated surfaces [12], where h is the resolution of the discretization. A fast marching method comprises of a local numerical

© Springer Nature Switzerland AG 2019
J. Lellmann et al. (Eds.): SSVM 2019, LNCS 11603, pp. 38–50, 2019.
https://doi.org/10.1007/978-3-030-22368-7_4

solver and an update step. The local numerical solver approximates the gradient locally to estimate the distance of a point in the advancing unit speed wavefront. Similar to the strategy employed in the celebrated Dijkstra's algorithm [6], the update step involves selecting the least distant point in order to propagate the wavefront. The accuracy of fast marching scheme relies heavily on the local solver which is responsible for approximations to the gradient at the point of the advancing front [22, 26].

The success of deep learning has shown that neural networks can be utilized as powerful function approximators of complex attributes governing various visual and auditory phenomena. The availability of large amounts of data and computational power, coupled with parallel streaming architectures and improved optimization techniques, have led to computational frameworks that efficiently exploit their representational power. However, the utility of neural networks to provide accurate solutions to complex PDE's is still a very nascent topic in computational research. Notable efforts include methods like [24] which attempt to solve high-dimensional parabolic partial differential equations, whereas [18] demonstrates the learning of differential operators from data as convolution kernels.

The main contribution of this paper is to develop a deep-learning infrastructure that enables accurate distance computation. Using the identical computational schematic of the fast marching method, the basic premise of this paper is that one can *learn* the local numerical solver by training a neural network and use it in conjunction with the upwind update scheme. Experiments in Sects. 3.2 and 4.1 demonstrate lower errors and higher order of accuracy in the solutions. Importantly, we show that our method generalizes to different geometries and sampling conditions.

2   Background

2.1   The Eikonal Equation

Given a domain $\Omega \in \mathbb{R}^n$ and a curve or surface $\Gamma \subset \Omega$, we wish to find the distance, denoted by $u(x)$, for each $x \in \Omega$ from the given Γ. The distance function satisfies

$$\begin{aligned} |\nabla u(x)| &= 1, \quad x \in \Omega \setminus \Gamma \\ u(x) &= 0, \quad x \in \Gamma, \end{aligned} \tag{1}$$

where ∇ refers to the gradient operator. Equation (1) can be generalized to arbitrary boundary conditions $u(x) = g(x)$, $x \in \Gamma$, and also to weighted domains where the local metric at a point x is defined by a positive scalar function $F : \Omega \to \mathbb{R}^+$. As shown in [4, 8, 9], the viscosity solution for Eq. (1) is unique.

Next, we consider curved manifolds, specifically, surfaces embedded in $\mathbb{R}^3$. Let $\mathcal{M}$ be a Riemannian manifold with metric tensor G. The geodesic distance function $u : \mathcal{M} \to \mathbb{R}^+$ satisfies

$$|\nabla_G u(x)| = 1, \quad x \in \mathcal{M} \setminus \Gamma$$

$$u(x) = 0, \quad x \in \Gamma, \tag{2}$$

where ∇_G represents the gradient operator defined on the Riemannian manifold.

2.2 Numerical Approximation

Consider the task of numerically approximating the function $u(x)$ in Eq. (1). Let $u_{i,j} \equiv u(ih, jh)$, where h defines the distance between neighboring grid points along the x and y axis. The following numerical approximation is shown [19,21] to pick the unique viscosity solution of Eq. (1),

$$|\nabla u_{i,j}|^2 \approx \max(D_{i,j}^{-x} u, -D_{i,j}^{+x} u, 0)^2 + \max(D_{i,j}^{-y} u, -D_{i,j}^{+y} u, 0)^2. \tag{3}$$

Where $\{D_{i,j}^{+x}, D_{i,j}^{-x}\}, \{D_{i,j}^{+y}, D_{i,j}^{-y}\}$ are the forward and backward difference operators for point (i,j) along the x and y directions, respectively. Based on approximation (3), the solution of $u_{i,j}$, given the u values at its four neighbors $\{u_{i+1,j}, u_{i-1,j}, u_{i,j+1}, u_{i,j-1}\}$ requires a solution of the quadratic equation

$$\max(D_{i,j}^{-x} u, -D_{i,j}^{+x} u, 0)^2 + \max(D_{i,j}^{-y} u, -D_{i,j}^{+y} u, 0)^2 = 1. \tag{4}$$

Typically, the approximation in Eq. (3) is based on first order Taylor series expansion of $u_{i,j}$. However, more sophisticated approximations such as those using two points [23] or three points [1] away from (i,j) in each direction, or using diagonal neighbors [7] yield more accurate solutions. Unlike Euclidean domains which enjoy the benefit of regular sampling, developing local approximations similar to Eq. (3) for curved manifolds are evidently more challenging due to the lack of a universal regular sampling strategy.

2.3 Fast Eikonal Solvers

Fast Eikonal solvers estimate distance functions in linear or quasilinear time. They typically comprise of two computational parts: a local numerical solver and an ordering/update method. The local solver provides an approximation for the distance function at a desired point and the ordering method chooses the next point to approximate. Fast marching methods [22,26] is the most prominent subclass of fast Eikonal solvers that use approximations similar to Eq. (3) for computing the distance function at a point using the known distances of its neighbors. By employing Dijkstra's ordering strategy, fast marching schemes display an *upwind* nature, where the solution can be seen to grow outward from the source, equivalent to the propagation of a unit speed wavefront (See Fig. 1). The other prominent subclass of fast Eikonal solvers are the fast sweeping methods [11,29,32,33], iterative schemes that use alternating sweeping ordering. Some variants [17] of fast sweeping establish their local solver on finite element techniques, however most of them use the same upwind difference local solver as fast marching.

Kimmel and Sethian [12] extended the fast marching scheme to approximate geodesic distances on triangulated surfaces by treating the sampled manifold as

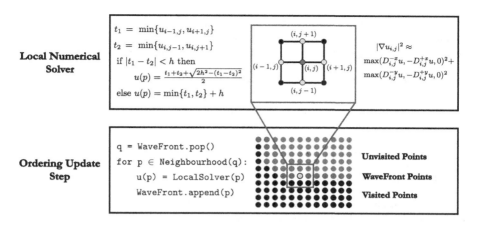

Fig. 1. Schematic for Fast Marching: The numerical solver approximates the gradient locally while the ordering scheme advances the unit speed wavefront. The suggested method employs a trained neural network as a Local Solver. (Color figure online)

Algorithm 1. Eikonal Estimation on Discretized Domain

1: **Initialize:**
2: $u(p) = 0$, Tag p as *visited*; $\forall p \in \mathfrak{s}$
3: $u(p) = \infty$, Tag p as *unvisited*; $\forall p \in \mathcal{S} \setminus \mathfrak{s}$
4: **while** there is a non-*visited* point **do**
    Denote the points adjacent to the newly *visited* points as $\mathcal{A}$
5:    **for all** $p \in \mathcal{A}$ **do**
6:        Estimate $u(p)$ based on *visited* points.
7:        Tag p as *wavefront*
8:    Tag the least distant *wavefront* point as *visited*.
9: **return** u

---

piecewise planar. In order to estimate $u(p)$ for some mesh point p, each triangle involving p is processed independently. Given a triangle comprising of the points p_1, p_2, p_3, the distance $u(p_1)$ is estimated based on $u(p_2), u(p_3)$ by approximating Eq. (1) on the triangle's plane. The minimum of the solutions computed on all the triangles involving p is chosen as the final distance assigned to p. This algorithm is analogous to the first-order fast marching method that operates on regular grids and therefore has an accuracy $\mathcal{O}(h)$.

3   Deep Eikonal Solver

We present a fast Eikonal solver estimating the geodesic distance for a discrete set of points $\mathcal{S}$ from a subset of source points $\mathfrak{s} \subset \mathcal{S}$. The local solver (step 6 in Algorithm 1) in our scheme is represented by a neural network. The process of choosing subsequent points for evaluation is similar to Dijkstra's algorithm [6], according to which each point $p \in \mathcal{S}$ is tagged into one of three disjoint sets,

1. **Visited:** where $u(p)$ is already computed (Black points in Fig. 1).
2. **WaveFront:** where $u(p)$ calculation is in process. (Red points in Fig. 1).
3. **Unvisited:** where $u(p)$ is yet to be determined (Green points in Fig. 1).

Since we use the same update method used in the fast marching scheme, the computational complexity of our method is $\mathcal{O}(N \log N)$. In Sect. 3.1 we describe the general philosophy behind training the network. In Sects. 3.2 and 4.1 we elaborate on the exact procedure and architecture of the network specific to the corresponding domain.

3.1   Training the Local Solver

Our basic premise is that the local solver can be *learned* using a neural network. The input to the local solver which is in action at a certain point $p \in$ WaveFront is the set of points in its neighborhood denoted by $\mathcal{N}(p) = \{p_1, p_2, \ldots, p_M\}$ and their corresponding distances $\{u(p_1), u(p_2), \ldots, u(p_M)\}$. Based on the local geometry of the Visited points in p's neighborhood, an estimate of the local distance is outputted by the local solver. See Figs. 1, 2 and 5 for a visual schematic.

Following a supervised training procedure, we train the local solver with examples containing the *ground truth* distances. By *ground truth* we refer to the most accurate available solution for distance (elaborated further in Sects. 3.2 and 4). For a given point p, denote this distance as $u_{gt}(p)$. The network inputs neighborhood information: $\{p_1, u_{gt}(p_1), \ldots, p_M, u_{gt}(p_M)\}$ and is trained to minimize the difference between its output and the corresponding ground truth distance at the point $p : u_{gt}(p)$.

To simulate the propagating unit speed wavefront in various representative scenarios, we employ the following strategy. We develop a variety of different sources and construct a dataset of local neighborhood patches at different locations relative to the sources. We allow a patch point to participate in the prediction of $u(p)$ according to the following rule.

$$q \in \text{Visited if } u_{gt}(q) < u_{gt}(p). \tag{5}$$

Therefore, given a point p and a patch of points constituting its neighborhood: $\{p_1, p_2, \ldots, p_M\}$, the network inputs $\{p_1, u_{gt}(p_1), \ldots, p_M, u_{gt}(p_M)\}$, where all points that are declared Visited according to rule 5 input their respective ground truth distance. For points $q \in \mathcal{N}(p)$ that do not satisfy condition 5, we simulate that the wavefront starting from the source arrives at their location later than it arrives to p. Hence, such points are not considered Visited and they input a constant value to the network.

$$\text{If } q \notin \text{Visited}; \ u_{gt}(q) = C \tag{6}$$

By employing the rules 5 and 6 stated above, we create a rich database of such local patches coupled with the desired outputs. The parameters of the network Θ are optimized to minimize the MSE loss

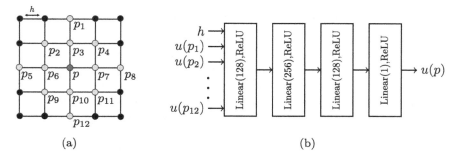

(a)   (b)

Fig. 2. Deep Eikonal Solver for Cartesian Grids: (a) For each point p, the network inputs the distances of its neighbors (colored yellow) along with the spacing h. (b) The network architecture with four fully connected layers that processes the local information from (a) and output the distance estimate $u(p)$. (Color figure online)

Fig. 3. Results on Euclidean grids: Iso-contours of the Eikonal solution on 50×50 grid. The magnified plots show that the Deep Eikonal solver demonstrates better fidelity to the ground truth compared to the 1st and 2nd order fast marching schemes.

$$L(\Theta) = \sum_i \Big(f_\Theta \big(p_{i1}, u_{gt}(p_{i1}), \dots, p_{iM}, u_{gt}(p_{iM}) \big) - u_{gt}(p_i) \Big)^2, \qquad (7)$$

over a dataset comprising of a variety of points p_i and their corresponding neighborhoods.

The intuition behind our approach is twofold. First, we expect that using a larger local support leads to a more accurate estimate of the solution to the differential equation. Secondly, by training the network to follow a *ground truth* distance, we avoid the need to develop a sophisticated formula for locally approximating the gradient with high accuracy. Rather, we expect the network to *learn* the necessary relationship between the local patch and the ground truth distance. As demonstrated in our results section, our learning-based approach generalizes to different scenarios including different shapes and different datasets.

3.2   Deep Eikonal Solver for Cartesian Grids

We begin by evaluating our scheme on a Euclidean domain sampled regularly from $\Omega = [0,1]^2$. As the network's input patch for point $p : \mathcal{N}(p)$, we choose all the points located within a radius of $2h$ from p (yellow points in Fig. 2). We use a multilayer perceptron architecture (henceforth abbreviated as mlp) with four fully connected layers having number of nodes: $mlp(13, 128, 256, 128, 1)$ respectively. After each linear stage we applied ReLU function as the non-linearity. Since the grid spacing is uniform in the Cartesian case, h encodes all necessary spatial information regarding neighborhood $\mathcal{N}(p)$. Accordingly, we input *only* the grid spacing h to the network instead of the explicit coordinates $\{p_1, p_2, \ldots, p_{12}\}$. We trained our network by generating 10,000 synthetic examples constructed as per the strategy enumerated in Sect. 3.1. We design a variety of different source configurations like points, circles, arbitrary curves etc, and construct different local patches using the ground truth distance from these sources: $\mathfrak{s}$. For the Cartesian case, we evaluate the ground truth distance as

$$u_{gt}(p) = \min\{\|p - s\|, s \in \mathfrak{s}\}. \qquad (8)$$

We pipeline each example by subtracting $\min\{u_{gt}(q), q \in \mathcal{N}(p)\}$ from each distance and by scaling the example to mean magnitude 0.5, thereby simplifying the diversity of inputs prior to its feeding into the network. Note, that the subtraction is equivalent to solving Eq. (1) with corresponding initial condition $g'(x) = g(x) - \min\{u_{gt}(q), q \in \mathcal{N}(p)\}$ and the scaling is equivalent to solving Eq. (1) on correspondingly scaled coordinates. Before proceeding with the update step, the network's output is re-scaled and the bias is added to achieve the distance estimate.

We test our method using the benchmark shown in [23] comprising of two point sources. The learned scheme shows a superior and more accurate approximation of the distance function compared to fast marching local solvers (See Fig. 3). We measure the order of accuracy r of the proposed scheme by evaluating the error ϵ as a function of the grid spacing h [15]. Let u_{gt} be the ground truth and u_h represent the solution of a numerical scheme on a grid spacing h.

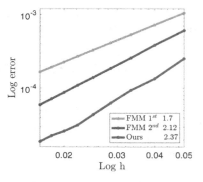

$$\epsilon(h) = |u_h - u_{gt}| \approx Ch^r + \mathcal{O}(h^{r+1}) \quad (9)$$

$$\log(\epsilon) \approx \log(C) + r\log(h) + \mathcal{O}(h), \quad (10)$$

Fig. 4. Order of Accuracy: The slope of each error plot is the order of accuracy of the corresponding scheme.

where C is a constant. We evaluated our scheme for a range of grid sizes h and plotted $\log \epsilon$ as a function of $\log h$. The slope of this line gives the order of accuracy r. From Fig. 4 we can see that our local solver has a higher order of accuracy as compared to the classical fast marching local solvers in addition to lower errors.

4   Deep Eikonal Solver for Triangulated Meshes

For triangulated meshes, we chose the second-ring neighbors of a point p to constitute the local patch $\mathcal{N}(p)$ (See Fig. 5). In Euclidean domains, the regular sampling allows us to encode the spatial information of a patch using only the grid spacing h. However, the unordered nature of the sampled points within a mesh patch demands an explicit description of the spatial position associated with each point in a given patch. For each point p_i, we construct a vector comprising of four values: The 3D coordinates along with the distance $u(p_i)$, namely,

$$V_i = \big(x(p_i) - x(p), y(p_i) - y(p), z(p_i) - z(p), u(p_i)\big). \tag{11}$$

The architecture we employ (see Fig. 5) has the following structure. It comprises of M input vectors, each encoding the information about a single point in the neighborhood point of some patch as shown in Fig. 5. Each vector V_i is processed independently in linear layers $mlp(4, 64, 128, 512, 1024)$ which produces a 1024-dimensional feature vector for each point. These 1024-dimensional feature vectors are max-pooled to generate a single vector of the same dimension and finally passed through a $mlp(512, 256, 1)$ which generates the desired output distance $u(p)$. Our architecture is motivated from previous deep learning frameworks for point clouds like [20]. This scheme of learning functions over non-Euclidean domains was analytically validated in [31].

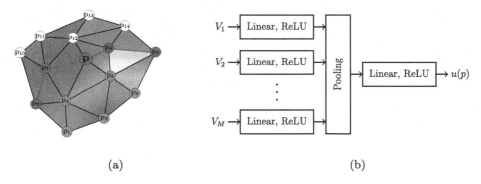

(a)   (b)

Fig. 5. Deep Eikonal Solver for triangulated meshes: (a) shows a local patch on a surface, the red line corresponds to the advancing WaveFront passing from Visited (green) to Unvisited (white). (b) shows the network architecture (described in Sect. 4.1) for processing the local information. (Color figure online)

4.1   Experimental Results

We create $100,000$ training examples from 8 TOSCA shapes [2] using the methodology described in Sect. 3.1. In order to train for rotation invariance, we augment the data by multiplying each example by a random rotation matrix. As a pre-processing stage, we practice the same pipeline described in Sect. 3.2.

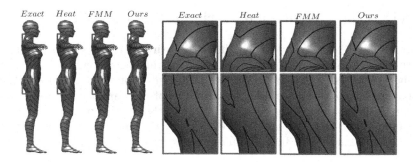

Fig. 6. Intra-Dataset Generalization: Iso-contours shown for (left to right) the polyhedral scheme, heat method, fast marching and our method. The network in our scheme was trained on patches extracted from 8 TOSCA shapes. The evaluation was conducted on a separate test set.

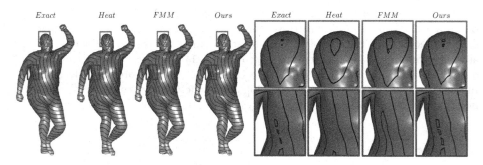

Fig. 7. Inter-Dataset Generalization: Iso-contours shown for (left to right) the polyhedral scheme, heat method, fast marching and our method. The evaluation was conducted on SHREC whereas the network was trained with TOSCA.

Table 1. Intra-Dataset Generalization: quantitative evaluation conducted on TOSCA. The errors were computed relative to the polyhedral scheme.

Model	L_1			L_∞		
	Heat	FMM	Ours	Heat	FMM	Ours
Cat	0.2237	0.0060	**0.0011**	0.7775	0.0493	**0.0352**
Centaur	0.1185	0.0131	**0.0042**	0.3504	0.1622	**0.1153**
Dog	0.0562	0.0121	**0.0022**	0.3393	0.2291	**0.1034**
Michael	0.0625	0.0082	**0.0015**	0.4196	0.2855	**0.1042**
Victoria	0.1129	0.0087	**0.0015**	0.6174	0.1885	**0.0764**
Wolf	0.0254	0.0164	**0.0054**	0.2721	0.1465	**0.0763**

We compare our Deep Eikonal solver to two different axiomatic approaches: our direct competitor, the fast marching method and recent geodesic approximation, the heat method [5]. The heat method uses the heat equation solutions to estimate geodesics on surfaces, motivated by Varadhan's formula [27], that connects the two. As shown in Figs. 6 and 7, the Deep Eikonal solver approximates the geodesic distance more accurately than the axiomatic approaches. Fast marching and heat method yield smooth but inaccurate approximations in comparison to the polyhedral scheme [25] which is known to be the most accurate scheme for computing distances on surfaces. Tables 1 and 2 show quantitative results on TOSCA [2] and SHREC [16] databases respectively. The error at point p was computed as $|\frac{u(p)-u_{gt}(p)}{u_{gt}(p)}|$ where the ground truth is taken as the polyhedral scheme solution. To evaluate our method's order of accuracy as described in Sect. 3.2, we use different resolutions of unit sphere, since the geodesic distance on a sphere can be computed analytically. Figure 8 demonstrates that the order of accuracy of the Deep Eikonal solver is comparable with the polyhedral scheme accuracy. Finally, we test the robustness of our method to noise in the sampled manifold. Figure 9 demonstrates that the isolines for our method remain robust to noise in comparison to the fast marching.

Table 2. Inter-Dataset Generalization: quantitative evaluation conducted on SHREC. The errors were computed relative to the polyhedral scheme.

Model	L_1			L_∞		
	Heat	FMM	Ours	Heat	FMM	Ours
Male 1	0.0288	0.0105	**0.0027**	0.2570	0.0981	**0.0681**
Male 2	0.0152	0.0122	**0.0026**	0.1553	0.1694	**0.0296**
Male 3	0.0274	0.0105	**0.0048**	0.1926	0.1479	**0.0762**
Female 1	0.0236	0.0089	**0.0034**	0.3184	0.1535	**0.0504**
Female 2	0.0229	0.0083	**0.0019**	0.1350	0.0970	**0.0383**
Female 3	0.0410	0.0087	**0.0026**	0.4205	0.2070	**0.0671**

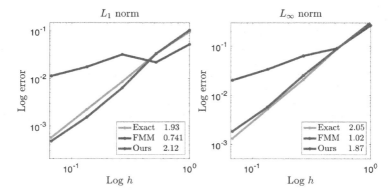

Fig. 8. Order of Accuracy: The plots shows the edge resolution impact on the error. Each scheme's accuracy associated with its corresponding slope. We construct a unit sphere mesh with various edge length using Loop's subdivision scheme.

FMM   Ours

Fig. 9. Robustness to noise: Equidistant isolines drawn on cat and its noisy version. The black isolines were computed on the noisy cat while the colorful isolines were computed on the original cat.

5   Conclusion

In this paper, we introduce a new methodology by which we train a neural network to numerically solve the Eikonal equation on regularly and non-regularly sampled domains. We develop a numerical scheme which uses a data-centric learning-based computation in conjunction with a well-established axiomatic update method. In comparison to the axiomatic counterparts, we demonstrate that our hybrid scheme results in a considerable gain in performance measured by lower errors and larger orders of accuracy tested on a variety of different settings. Our approach advocates combining the best of both worlds: approximation power of neural networks combined by meaningful axiomatic numerical computations in order to achieve performance as well as better understanding of computational algorithms.

References

1. Ahmed, S., Bak, S., McLaughlin, J., Renzi, D.: A third order accurate fast marching method for the Eikonal equation in two dimensions. SIAM J. Sci. Comput. **33**(5), 2402–2420 (2011)
2. Bronstein, A.M., Bronstein, M.M., Kimmel, R.: Numerical Geometry of Non-Rigid Shapes. Monographs in Computer Science. Springer, New York (2008). https://doi.org/10.1007/978-0-387-73301-2
3. Chen, D., Cohen, L.D.: Fast asymmetric fronts propagation for image segmentation. J. Math. Imaging Vis. **60**(6), 766–783 (2018)
4. Crandall, M.G., Lions, P.L.: Viscosity solutions of Hamilton-Jacobi equations. Trans. Am. Math. Soc. **277**(1), 1–42 (1983)
5. Crane, K., Weischedel, C., Wardetzky, M.: Geodesics in heat: a new approach to computing distance based on heat flow. ACM Trans. Graph. (TOG) **32**(5), 152 (2013)
6. Dijkstra, E.W.: A note on two problems in connexion with graphs. Numer. Math. **1**(1), 269–271 (1959)
7. Hassouna, M.S., Farag, A.A.: Multistencils fast marching methods: A highly accurate solution to the Eikonal equation on cartesian domains. IEEE Trans. Pattern Anal. Mach. Intell. **29**(9), 1563–1574 (2007)
8. Ishii, H.: Uniqueness of unbounded viscosity solution of Hamilton-Jacobi equations. Indiana Univ. Math. J. **33**(5), 721–748 (1984)

9. Ishii, H.: A simple, direct proof of uniqueness for solutions of the Hamilton-Jacobi equations of eikonal type. Proc. Am. Math. Soc., 247–251 (1987)
10. Kimmel, R., Kiryati, N., Bruckstein, A.M.: Multivalued distance maps for motion planning on surfaces with moving obstacles. IEEE Trans. Robot. Autom. **14**(3), 427–436 (1998)
11. Kimmel, R., Maurer, R.P.: Method of computing sub-pixel euclidean distance maps. US Patent 7,113,617, September 26 2006
12. Kimmel, R., Sethian, J.A.: Computing geodesic paths on manifolds. Proc. Nat. Acad. Sci. **95**(15), 8431–8435 (1998)
13. Kimmel, R., Sethian, J.A.: Optimal algorithm for shape from shading and path planning. J. Math. Imaging Vis. **14**(3), 237–244 (2001)
14. Lee, S.K., Fekete, S.P., McLurkin, J.: Structured triangulation in multi-robot systems: coverage, patrolling, voronoi partitions, and geodesic centers. Int. J. Robot. Res. **35**(10), 1234–1260 (2016)
15. LeVeque, R.J.: Finite difference methods for differential equations. Draft Version Use AMath **585**(6), 112 (1998)
16. Li, B., et al.: A comparison of 3D shape retrieval methods based on a large-scale benchmark supporting multimodal queries. Comput. Vis. Image Underst. **131**, 1–27 (2015)
17. Li, F., Shu, C.W., Zhang, Y.T., Zhao, H.: A second order discontinuous Galerkin fast sweeping method for Eikonal equations. J. Comput. Phys. **227**(17), 8191–8208 (2008)
18. Long, Z., Lu, Y., Ma, X., Dong, B.: PDE-net: Learning PDEs from data. arXiv preprint arXiv:1710.09668 (2017)
19. Osher, S., Sethian, J.A.: Fronts propagating with curvature-dependent speed: algorithms based on Hamilton-Jacobi formulations. J. Comput. Phys. **79**(1), 12–49 (1988)
20. Qi, C.R., Su, H., Mo, K., Guibas, L.J.: Pointnet: deep learning on point sets for 3D classification and segmentation. In: IEEE Proceedings of Computer Vision and Pattern Recognition (CVPR), vol. 1, no. 2, p. 4 (2017)
21. Rouy, E., Tourin, A.: A viscosity solutions approach to shape-from-shading. SIAM J. Numer. Anal. **29**(3), 867–884 (1992)
22. Sethian, J.A.: A fast marching level set method for monotonically advancing fronts. Proc. Nat. Acad. Sci. **93**(4), 1591–1595 (1996)
23. Sethian, J.A.: Level Set Methods and Fast Marching Methods: Evolving Interfaces in Computational Geometry, Fluid Mechanics, Computer Vision, and Materials Science, vol. 3. Cambridge University Press, Cambridge (1999)
24. Sirignano, J., Spiliopoulos, K.: DGM: a deep learning algorithm for solving partial differential equations. J. Comput. Phys. **375**, 1339–1364 (2018)
25. Surazhsky, V., Surazhsky, T., Kirsanov, D., Gortler, S.J., Hoppe, H.: Fast exact and approximate geodesics on meshes. ACM Trans. Graph. (TOG) **24**, 553–560 (2005)
26. Tsitsiklis, J.N.: Efficient algorithms for globally optimal trajectories. IEEE Trans. Autom. Control **40**(9), 1528–1538 (1995)
27. Varadhan, S.R.S.: On the behavior of the fundamental solution of the heat equation with variable coefficients. Commun. Pure Appl. Math. **20**(2), 431–455 (1967)
28. Wang, W., Shen, J., Porikli, F.: Saliency-aware geodesic video object segmentation. In: Proceedings of the IEEE Conference on Computer Vision and Pattern Recognition, pp. 3395–3402 (2015)

29. Weber, O., Devir, Y.S., Bronstein, A.M., Bronstein, M.M., Kimmel, R.: Parallel algorithms for approximation of distance maps on parametric surfaces. ACM Trans. Graph. (TOG) **27**(4), 104 (2008)
30. Younes, L.: Spaces and manifolds of shapes in computer vision: an overview. Image Vis. Comput. **30**(6–7), 389–397 (2012)
31. Zaheer, M., Kottur, S., Ravanbakhsh, S., Poczos, B., Salakhutdinov, R.R., Smola, A.J.: Deep sets. In: Advances in Neural Information Processing Systems, pp. 3391–3401 (2017)
32. Zhao, H.K., Osher, S., Merriman, B., Kang, M.: Implicit and nonparametric shape reconstruction from unorganized data using a variational level set method. Comput. Vis. Image Underst. **80**(3), 295–314 (2000)
33. Zhao, H.: A fast sweeping method for Eikonal equations. Math. Comput. **74**(250), 603–627 (2005)

A Splitting-Based Algorithm
for Multi-view Stereopsis
of Textureless Objects

Jean Mélou[1,3]([⊠]), Yvain Quéau[2], Fabien Castan[3], and Jean-Denis Durou[1]

[1] IRIT, UMR CNRS 5505, Université de Toulouse, Toulouse, France
`jean.melou@mikrosimage.com`
[2] GREYC, UMR CNRS 6072, Caen, France
[3] Mikros Image, Paris, France

Abstract. We put forward a simple, yet effective splitting strategy for multi-view stereopsis. It recasts the minimization of the classic photo-consistency + gradient regularization functional as a sequence of simple problems which can be solved efficiently. This framework is able to handle various photo-consistency measures and regularization terms, and can be used for instance to estimate either a minimal-surface or a shading-aware solution. The latter makes the proposed approach very effective for dealing with the well-known problem of textureless objects 3D-reconstruction.

Keywords: Multi-view stereo · 3D-reconstruction ·
Shape-from-shading

1 Introduction

Multi-view stereopsis consists in reconstructing dense 3D-geometry from multi-view images. A common approach to this problem is to estimate a mapping (depth) between pixels in a reference view and 3D-geometry, by maximizing the photo-consistency of the reference image with the others. To measure photo-consistency, the reference image is warped to the other views using the estimated depth map and the (known) relative poses, and compared against the target images. However, photo-consistency is not significant in textureless areas (see Fig. 1): the optimization problem needs to be regularized by constraining variations in the 3D-geometry. If $z : \Omega \subset \mathbb{R}^2 \to \mathbb{R}^+ \backslash \{0\}$ denotes the unknown depth map, with Ω the image domain, surface variations under perspective projection can be measured as a function of $\frac{\nabla z}{z} = \nabla \log z : \Omega \to \mathbb{R}^2$. Multi-view stereo can then be formulated in a classic and generic manner [13] as the minimization of the sum of a fidelity term f inversely proportional to photo-consistency, and of a regularization term g. We thus consider in this work the following variational problem:

$$\min_z f(z) + g(\nabla \log z) \tag{1}$$

(the choice of applying regularization in log-space is discussed in Sect. 2).

© Springer Nature Switzerland AG 2019
J. Lellmann et al. (Eds.): SSVM 2019, LNCS 11603, pp. 51–63, 2019.
https://doi.org/10.1007/978-3-030-22368-7_5

Possible choices for f and g are discussed in Sect. 2. Section 3 introduces our main contribution: a generic multi-view stereo algorithm, which recasts (1) as a sequence of nonlinear-yet-local and global-yet-linear problems, both of which can be solved efficiently. We present in Sect. 4 appropriate regularizers for the 3D-reconstruction of textureless objects, before empirically evaluating the potential of our algorithm in Sect. 5. Eventually, our conclusions are drawn and future research directions are suggested in Sect. 6.

Fig. 1. Given a reference view of a textureless object (first column), and a set of $t \geq 1$ target views (second column), multi-view stereopsis based solely on photo-consistency optimization fails to estimate a reasonable mapping (depth) between the reference view and 3D-geometry (third column). Much more satisfactory results are obtained when introducing shading-aware and/or minimal-surface regularizations (fourth column).

2   Preliminaries

In this work we focus on solving the discrete counterpart of (1). In the following, z is thus a vector in $\mathbb{R}^p$ containing the p unknown depth values, $\log z$ is to be understood in a element-wise manner and $\nabla \in \mathbb{R}^{2p \times p}$ is a first-order, forward finite differences matrix such that $\nabla z \in \mathbb{R}^{2p}$ approximates the depth gradient.

Fidelity Term. Let $\pi_z^{-1}(i)$ be the back-projection from the i-th pixel, $i \in \{1, \ldots, p\}$, in the reference view to its conjugate 3D-point, given a depth map z and the (known) intrinsics of the camera. Let $\left\{\pi^j\right\}_{j \in \{1, \ldots, t\}}$ be the projections from 3D-points to pixels in the $t \geq 1$ other cameras (hereafter "target cameras"), using the (known) camera poses and their (known) intrinsics. Let $v_i \in \mathbb{R}^m$ be a m-dimensional feature vector at pixel i in the reference view. Such a vector can be the brightness value in pixel i ($m = 1$), the RGB values in that pixel ($m = 3$), the concatenation of brightness values in a 3×3 neighborhood centered in i ($m = 9$), etc. For a given target camera $j \in \{1, \ldots, t\}$, photo-consistency measures the adequation between v_i and the feature vector $v^j_{\pi^j \circ \pi_z^{-1}(i)} \in \mathbb{R}^m$ at the matched pixel $\pi^j \circ \pi_z^{-1}(i)$ in the target view, in the sense of some loss function ρ. The fidelity term can then be constructed by averaging the photo-consistency contributions from all target cameras and summing over all pixels:

$$f(z) = \frac{1}{t} \sum_{i=1}^{p} \sum_{j=1}^{t} \rho \left(v_i, v_{\pi^j \circ \pi_z^{-1}(i)}^j \right). \tag{2}$$

One could consider as loss function ρ the normalized sum of squared deviations $\rho_{SSD}(x, y) = \frac{1}{m} \sum_{c=1}^{m} (x_c - y_c)^2$ or a robust variant of it, and then linearize (2) using first-order Taylor expansion as in [7,12]. However, linearization requires small depth increments. Robustness can also be reached by replacing SSD with the normalized sum of absolute deviations $\rho_{SAD}(x, y) = \frac{1}{m} \sum_{c=1}^{m} |x_c - y_c|$ or a loss function based on the zero-mean normalized cross-correlation such as $\rho_{ZNCC}(x, y) = \frac{1}{2} \left[1 - \frac{(x-\bar{x})^\top (y-\bar{y})}{\|x-\bar{x}\|\|y-\bar{y}\|} \right]$ (see [5, Chapter 2]). Photo-consistency measures are then further normalized within $(0, 1)$ using a nonlinear operator, e.g., the exponential transform $\rho(x, y) := 1 - \exp \left\{ -\frac{\rho(x,y)^2}{\sigma^2} \right\}$ with user-defined parameter σ. With all these choices, the fidelity term may become nonlinear, non-smooth and non-convex, and the optimization tedious. Therefore, minimization of f is usually carried out using bruteforce grid-search over the sampled depth space. This "winner-takes-all" strategy was first advocated in [8]. Despite its simplicity, it is remarkably efficient, and impressive depth map reconstructions of highly textured scenes have long been demonstrated [6].

Regularization. Nevertheless, in textureless areas, the fidelity term degenerates (in each pixel, there are multiple depth values for which it is globally minimized). Obviously, increasing the number of views will have no effect on this issue, and one should rather rely on regularization. For instance, one may introduce a total variation prior [16]. However, under perspective projection total variation does not enforce physically sound constraints on the geometry: smoothing should rather be carried out using the minimal surface prior [7]. Yet, this would turn the regularizer into a bilinear form involving both the depth z and its gradient ∇z, making the optimization challenging. Re-parameterization avoids this issue: introducing the change or variable $\tilde{z} = \sqrt{z}$, the total surface area can be rewritten as a function $g(\nabla \tilde{z})$ [7]. A logarithmic change of variable $\tilde{z} = \log z$ can also be considered for the same purpose, as well as to derive a physics-based regularization term based on shape-from-shading under natural illumination [14]. This might be particularly interesting for us, because shape-from-shading algorithms typically assume constant scene reflectance, which is essentially another wording for textureless scenes. Because we believe it is interesting to compare smoothness-based and physics-based priors for multi-view stereopsis of poorly textured objects within the same numerical framework, we opt for the logarithmic change of variable in the regularization term, which explains the form of the variational model (1). Shading-aware multi-view stereo has long been identified as a promising track [3], and theoretical guarantees on uniqueness exist [4]. Still, there is a lack of practical numerical solutions. Jin *et al.* presented in [9] a variational one, yet it assumes a single, infinitely distant light source, while we are rather interested in natural illumination. Other methods combining stereo and shading information under natural illumination have also recently been developed [10,11,17], but they only consider photometry as a way to refine an existing multi-view 3D-reconstruction, which remains the baseline of the process. We would rather like to follow an end-to-end joint approach, as for instance in the very recent work [12]. In comparison with [12], the approach presented in the next section avoids linearization of the fidelity term and is therefore slightly more generic, since any robust photo-consistency measure (including those based on non-differentiable or non-convex loss functions) can be considered.

3   A Generic Splitting Strategy for Multi-view Stereo

In this section, we show how to turn the discrete counterpart of the variational problem (1) into the simpler problem (6), and we introduce Algorithm 1 for solving the latter.

Proposed Variational Model. As discussed in the previous section, the fidelity term f in (1) is often chosen as a robust non-smooth cost function, hence the non-regularized problem is already challenging. The coupling induced by the gradient operator in the regularization term g makes things even worse. We separate those difficulties by splitting the optimization over f and g. Introducing an auxiliary variable $u = z \in \mathbb{R}^p$, (1) is equivalently rewritten as follows:

$$\min_{u,z} \quad f(u) + g(\nabla \log z)$$
$$\text{s.t.} \quad u = z \tag{3}$$

In (3), the u-subproblem is still non-smooth and possibly non-convex, but at least it is now small-scale (and hence, parallelizable) since f is separable (each term in the outer sum in (2) only involves the depth in a single pixel). Minimization of f can be carried out using bruteforce grid-search over a set of sampled depth values. Moreover, assuming g is smooth, its minimization can be achieved using gradient-based optimization. However, the hard constraint $u = z$ would prevent z from capturing thin surface variations. Thus, we relax the hard constraint $u = z$ in (3) into a quadratic penalization term:

$$\min_{u,z} \quad f(u) + g(\nabla \log z) + \beta \|\log u - \log z\|^2, \tag{4}$$

with $\beta > 0$ a tunable hyper-parameter. Let us remark that the penalization is applied in log-space: in this way, z appears in (4) only through its logarithm. We can thus equivalently optimize over $\tilde{z} = \log z$, and recover $z = \exp \tilde{z}$ at the end of the process. The new optimization problem becomes:

$$\min_{u,\tilde{z}} \quad f(u) + g(\nabla \tilde{z}) + \beta \|\log u - \tilde{z}\|^2. \tag{5}$$

As mentioned in the previous section, recent studies have advocated in favor of nonlinear regularization terms g, and thus the $\tilde{z}$-subproblem in (5) remains challenging. We simplify it through a second splitting: introducing an auxiliary variable $\theta = \nabla \tilde{z} \in \mathbb{R}^{2p}$, Problem (5) is turned into the following, equivalent one:

$$
\min_{u, \theta, \tilde{z}} \ f(u) + g(\theta) + \beta \left\| \log u - \tilde{z} \right\|^2 \\
\text{s.t.} \ \ \theta = \nabla \tilde{z}
\tag{6}
$$

Numerical Solving of (6). The linear constraint in (6) could be handled, e.g., by resorting to an augmented Lagrangian approach, but in this preliminary work we rather follow a simpler strategy consisting in approximating the solution of (6) by iteratively solving quadratically-penalized problems of the form

$$
\min_{u, \theta, \tilde{z}} f(u) + g(\theta) + \alpha^{(k)} \left\| \theta - \nabla \tilde{z} \right\|^2 + \beta \left\| \log u - \tilde{z} \right\|^2 ,
\tag{7}
$$

with values of $\alpha^{(k)} > 0$ increasing to infinity with iterations k. We want the hard constraint in (6) to be satisfied at convergence i.e., when $k \to +\infty$, in contrast with the one in (3) which we purposely replaced by a quadratic penalization with fixed parameter β. For each value $\alpha^{(k)}$, we approximately solve (7) by one sweep of alternating optimization. As discussed above, the u-subproblem can be solved by grid-search. We focus on smooth and separable regularizers g (cf. Sect. 4), so the θ-subproblem can be solved using parallelized gradient-based iterations. Eventually, the $\tilde{z}$-subproblem is a sparse linear least-squares problem which can be solved using conjugate gradient. We repeat this process until the relative residual between two estimates of $z = \exp \tilde{z}$ falls below a threshold set to 10^{-4}. This algorithm is sketched in Algorithm 1. Intuitively, it iteratively estimates a rough depth map by optimizing photo-consistency (Eq. (8)), then regularizes the depth variations (Eq. (9)), and integrates the refined gradient into the log-depth map (Eq. (10)). The values $\alpha^{(0)} = 1$ and $\beta = 0.1$ were empirically found to yield reasonable results and were used in all experiments. As initial depth map $z^{(0)}$, a fronto-parallel plane was always considered, with depth values taken as the mean of the ground-truth ones.

4   Regularizers for Textureless Multi-view Stereopsis

Given that f and g might be non-convex, it seems difficult to draw a theoretical convergence analysis of Algorithm 1. We thus leave this analysis for the future, and rather focus in this exploratory work on evaluating the efficiency of the algorithm on real-world multi-view stereo problems. In particular, we now turn our attention to the challenging problem of reconstructing poorly textured objects, and discuss, in view of this, the choice of a suitable regularizer. This requires clarifying the notion of "textureless" objects, hence let us first recall some photometric notions.

Lambertian Image Formation Model. The fidelity term in (2) is derived from the common assumption that the brightness of a surface patch is invariant to changes in the viewing angle. In other terms, the surface is assumed to be Lambertian,

---

input : Initial depth map $z^{(0)}$, $\alpha^{(0)} > 0$, $\beta > 0$

output: Refined depth map z

$\tilde{z}^{(0)} = \log z^{(0)}$, $k = 0$, $r^{(0)} = +\infty$;

while $r^{(k)} > 10^{-4}$ **do**

    // Photo-consistency optimization

    $u^{(k+1)} = \underset{u}{\operatorname{argmin}} \; f(u) + \beta \left\| \log u - \tilde{z}^{(k)} \right\|^2$;   (8)

    // Regularization of depth variations

    $\theta^{(k+1)} = \underset{\theta}{\operatorname{argmin}} \; g(\theta) + \alpha^{(k)} \left\| \theta - \nabla \tilde{z}^{(k)} \right\|^2$;   (9)

    // Integration

    $\tilde{z}^{(k+1)} = \underset{\tilde{z}}{\operatorname{argmin}} \; \alpha^{(k)} \left\| \nabla \tilde{z} - \theta^{(k+1)} \right\|^2 + \beta \left\| \tilde{z} - \log u^{(k+1)} \right\|^2$;   (10)

    // Auxiliary updates

    $\alpha^{(k+1)} = 1.5\,\alpha^{(k)}$; $z^{(k+1)} = \exp \tilde{z}^{(k+1)}$; $r^{(k)} = \dfrac{\left\| z^{(k+1)} - z^{(k)} \right\|}{\left\| z^{(k)} \right\|}$; $k = k+1$;

end

Algorithm 1. Generic splitting strategy for multi-view stereo.

---

and its reflectance is characterized by the albedo. Assuming a single point light source at infinity, the brightness I_i in the reference view at pixel i is then the product of albedo and shading:

$$I_i = a_i \max \left\{ 0, n_i^\top l \right\}, \tag{11}$$

with $a_i > 0$ the albedo at the 3D-point $\pi_z^{-1}(i)$ conjugate to pixel i, $n_i \in \mathbb{S}^2 \subset \mathbb{R}^3$ the unit-length surface normal at this 3D-point, and $l \in \mathbb{R}^3$ the lighting vector (in intensity and direction). The surface normal depends on the gradient of the log-depth map i.e., on θ, according to (see, for instance, [14]):

$$n_i := n(\theta_i) = \frac{1}{d(\theta_i)} \begin{bmatrix} \mathrm{f}\,\theta_i \\ -1 - [x,y]^\top \cdot \theta_i \end{bmatrix}, \tag{12}$$

where $\theta_i = \begin{bmatrix} \theta_i^1 \\ \theta_i^2 \end{bmatrix} \in \mathbb{R}^2$ denotes the depth gradient in pixel i (vector $\theta \in \mathbb{R}^{2n}$ introduced in (6) is thus the concatenation of all θ_i, $i \in \{1, \ldots, p\}$), $\mathrm{f} > 0$ is the focal length of the perspective camera, and $(x, y) \in \mathbb{R}^2$ are the centered coordinates of pixel i. The unit-length constraint on n_i is ensured thanks to the normalization by

$$d(\theta_i) = \sqrt{\mathrm{f}^2 \|\theta_i\|^2 + \left(1 + [x,y]^\top \cdot \theta_i\right)^2}. \tag{13}$$

Model (11) is valid for a single light source at infinity, which is rather unrealistic in practical scenarios. However, natural illumination can be represented as a collection of infinitely-distant light sources, and the brightness at pixel i is then obtained by integrating the right-hand side of Eq. (11) over the upper hemisphere. Approximating this integral using second-order spherical harmonics, one obtains (see [2] for details):

$$I_i = a_i \, \tilde{n}_i^\top \tilde{l}, \tag{14}$$

with $\tilde{l} \in \mathbb{R}^9$ a low-order lighting representation which can be calibrated beforehand using a reference object with known geometry and reflectance, and $\tilde{n}_i \in \mathbb{R}^9$ a "pseudo-normal" vector depending solely on the three components of $n_i = \left[n_i^1, n_i^2, n_i^3\right]^\top$ and thus, again, on θ_i:

$$\tilde{n}_i := \tilde{n}(\theta_i) = \begin{bmatrix} n_i \\ 1 \\ n_i^1 \, n_i^2 \\ n_i^1 \, n_i^3 \\ n_i^2 \, n_i^3 \\ \left(n_i^1\right)^2 - \left(n_i^2\right)^2 \\ 3\left(n_i^3\right)^2 - 1 \end{bmatrix} \overset{(12)}{=} \begin{bmatrix} \frac{f\,\theta_i}{d(\theta_i)} \\ \frac{-1 - [x,y]^\top \cdot \theta_i}{d(\theta_i)} \\ 1 \\ \frac{f^2 \theta_i^1 \theta_i^2}{d(\theta_i)^2} \\ \frac{f\theta_i^1\left(-1 - [x,y]^\top \cdot \theta_i\right)}{d(\theta_i)^2} \\ \frac{f\theta_i^2\left(-1 - [x,y]^\top \cdot \theta_i\right)}{d(\theta_i)^2} \\ \frac{f^2\left(\left(\theta_i^1\right)^2 - \left(\theta_i^2\right)^2\right)}{d(\theta_i)^2} \\ \frac{3\left(-1 - [x,y]^\top \cdot \theta_i\right)^2}{d(\theta_i)^2} - 1 \end{bmatrix}. \tag{15}$$

In highly-textured scenes, the albedo values a_i in (14) strongly differ from one pixel i to another. As a consequence, so do the brightness values I_i and the feature vectors v_i, which makes the optimization of the fidelity term $f(z)$ in (2) meaningful, even in the absence of regularization. However, in textureless scenes the albedo is uniform, say equal to one:

$$a_i = 1 \quad \forall i \in \{1, \dots, p\}, \tag{16}$$

and thus, according to (14), brightness variations are purely geometric i.e., due to variations in n_i. Such variations may be extremely subtle and thus unsuitable for use in a fidelity term such as (2), and regularization is required. Next we discuss two possible choices of regularizers.

Shading-Aware Regularization. Equation (14) can serve as a guide in multi-view stereo, in order to let the Lambertian image formation model disambiguate the matching problem in textureless areas. For instance, if we assume that the Lambertian model is satisfied up to a homoskedastic, zero-mean Gaussian noise, we can minimize the difference between both sides of Eq. (14) in the sense of the quadratic loss, in the spirit of the variational approach to shape-from-shading under natural illumination introduced in [14]. This yields the following regularization function (recall that $a_i = 1$):

$$g_{\text{Shading}}(\theta) = \lambda \sum_{i=1}^{p} \left(\tilde{n}(\theta_i)^\top \tilde{l} - I_i\right)^2, \tag{17}$$

with $\lambda > 0$ a tunable hyper-parameter. Using the regularization (17) in Algorithm 1 yields a solution to shading-aware multi-view stereo. Let us emphasize that $g_{\text{Shading}}(\theta)$ is smooth and separable (each term in the sum involves only $\theta_i = [\theta_i^1, \theta_i^2]^\top \in \mathbb{R}^2$), so (9) can be recast as a series of p two-dimensional nonlinear problems which can be solved in parallel using, e.g., BFGS iterations.

Minimal-Surface Regularization. The latter regularizer requires knowledge of the lighting vector $\tilde{l}$. In some situations, calibrating lighting might be tedious or impossible, and one may prefer not to use an explicit image formation model. In such cases, it is possible to simply limit the surface variations, for instance by penalizing the total surface area. Following [14], this can be achieved by penalizing the ℓ^1-norm of the map d defined in Eq. (13), which yields the following minimal-surface regularizer:

$$g_{\text{MS}}(\theta) = \mu \sum_{i=1}^{p} d(\theta_i). \tag{18}$$

Again, $g_{\text{MS}}(\theta)$ is smooth and separable, so (9) can be solved using parallelized BFGS iterations.

Combined Regularization. Obviously, the minimal-surface regularizer (18) will tend to favor smooth surfaces and may miss thin structures. Conversely, the shading-aware regularizer (17) will tend to explain all thin brightness variations in terms of surface variations, which may be a source of noise misinterpretation. Therefore, it might be interesting to combine both shading-aware and minimal-surface regularizations, and the experiments in the next section are carried out using the following regularizer:

$$g(\theta) = \lambda \underbrace{\sum_{i=1}^{p} \left(\tilde{n}(\theta_i)^\top \tilde{l} - I_i \right)^2}_{g_{\text{Shading}}(\theta)} + \mu \underbrace{\sum_{i=1}^{p} d(\theta_i)}_{g_{\text{MS}}(\theta)}, \tag{19}$$

which remains smooth and separable, and yields the shading-aware solution if $\lambda > 0$ and $\mu = 0$, and the minimal-surface one if $\lambda = 0$ and $\mu > 0$.

5   Experimental Results

In all our experiments, the feature vectors v_i are the concatenation of brightness values in a 3×3 neighborhood. Unless stated otherwise, the loss function ρ in (2) is the exponential-transformed SAD (with $\sigma = 0.2$). We first test our model on a synthetic dataset, using a renderer to generate images of size 540×540 of the well-known "Stanford's Bunny" with uniform albedo, knowing the lighting $\tilde{l}$ and the camera parameters. Gaussian noise with standard deviation equal to 1% of the maximum intensity is added, in order to get closer to real images.

Fig. 2. Top row: in shape-from-shading, thin details in the input synthetic image (left) are finely recovered (center, the estimated depth is rendered frontally), yet the overall shape is biased due to the concave/convex ambiguity (right, rendering from another viewing angle). Bottom row: non-regularized multi-view stereo ($t = 1$), where the target synthetic image (left) is generated by translating the perspective camera. The overall shape is reasonable, yet thin details are missing and artifacts appear because photo-consistency degenerates in textureless areas (same viewing angles as above).

To preface, we highlight in Fig. 2 the main issue of single-view shape-from-shading (i.e., $f(z) = 0$, $\lambda = 1$ and $\mu = 0$): though the reference image is well explained, the resulting depth map is obviously prone to the concave/convex ambiguity. Non-regularized multi-view stereo (i.e., $f(z) = (2)$, and $\lambda = \mu = 0$) is not satisfactory either: adding a second view (here, a simple translation of the perspective camera) and optimizing photo-consistency results in a noisy surface due to ambiguities in matching textureless patches.

As shown on the top row of Fig. 5, results improve with the introduction of regularization. When they are not set to zero, the hyper-parameters are set to $\lambda = 5.10^{-4}$ and $\mu = 5.10^{-5}$ (those values were determined empirically). As expected, minimal-surface regularization allows to estimate a noise-free depth map which is globally reasonable, yet fine-scale details are missing. Using shading-aware regularization, fine-scale details are recovered, but a single target view ($t = 1$) is not enough to remove all concave/convex ambiguities. On the other hand, a joint approach gives satisfactory results, since the advantages of both regularization terms are combined.

Let us also remark that, since we did not explicitly take into account visibility issues, the estimated depth is not valid around parts which are not visible in the target image: see, for instance, the right edge of the bunny on the top row of Fig. 5. To deal with visibility issues, we can simply increase the number t of target images ($t = 6$ in the example on the bottom row of Fig. 5) so that each part in the reference view is covered in a few target pictures, the remaining occlusions being treated as outliers in the robust fidelity term $f(z)$. This largely improves the results, as confirmed when evaluating the root mean squared error

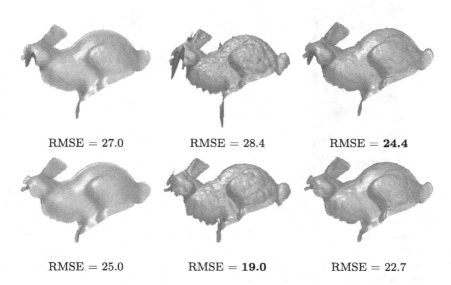

RMSE = 27.0   RMSE = 28.4   RMSE = **24.4**

RMSE = 25.0   RMSE = **19.0**   RMSE = 22.7

Fig. 3. RMSE for different regularization terms and for different numbers t of target images. If the combined approach gives better results with a low number of images, shading-aware regularization alone works better as soon as $t > 3$.

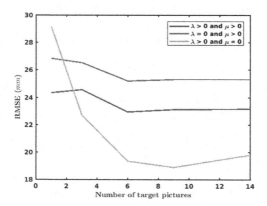

Fig. 4. Real-world multi-view stereo with $t = 6$ target views (two of them are shown in Fig. 1), using SAD (top row) or ZNCC (bottom row) loss functions. From left to right: minimal-surface regularization ($\lambda = 0$, $\mu = 1.10^{-5}$), shading-aware regularization ($\lambda = 5.10^{-3}$, $\mu = 0$), and combined regularization ($\lambda = 5.10^{-3}$, $\mu = 1.10^{-5}$).

(RMSE, expressed in millimeters, knowing that the ground truth values stand within a 800-millimeter interval) with respect to ground truth. Let us remark that shading-aware regularization alone seems sufficient and minimal-surface regularization tends to smooth out fine-scale details.

This is confirmed by Fig. 3: increasing the number t of target views removes all the concave/convex ambiguities of shape-from-shading, so minimal-surface regularization should be decreased, since it is not physics-based and tends to systematically flatten the surface.

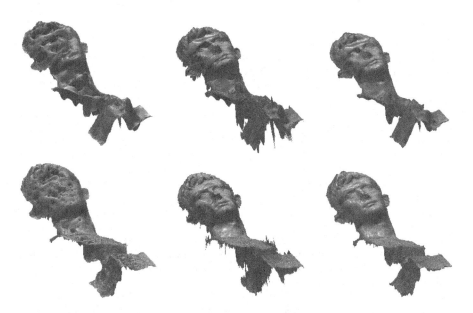

Fig. 5. Top row: Multi-view stereo with $t = 1$ target view (the two images are those of Fig. 2). Bottom row: Multi-view stereo with $t = 6$ target views. From left to right: minimal-surface regularization (λ set to zero), shading-aware regularization (μ set to zero), and combined regularization ($\lambda > 0$ and $\mu > 0$). The errors due to occlusions on the first row have largely been reduced on the second row.

Finally, we put this work in real context, using the Augustus dataset from [17]. We used an existing photogrammetric pipeline [1] to estimate the camera parameters, as well as a rough depth map from which we could estimate lighting and the position of the initial plane (let us emphasize that this rough depth map was not used any further, e.g., as initial estimate). To demonstrate the ability of our framework to handle various photo-consistency measures, we show results obtained with the exponential-transformed SAD or ZNCC loss functions. From the results in Fig. 4, the last reconstruction (bottom right), which uses exponential-transformed ZNCC and combined regularization, is the most satisfactory, at least from a qualitative point of view.

6   Conclusion and Perspectives

We have introduced a generic splitting algorithm for multi-view stereo. It handles a broad class of photo-consistency measures and regularization terms, and is a suitable approach to 3D-reconstruction of textureless objects with few parameters to tune. Now, the proposed numerical scheme could be extended to the use of higher order regularization terms [15], to a joint estimation of depth, reflectance and lighting as in [12], and the whole approach could be turned into a volumetric one in order to recover a full 3D-model, as for instance in [11].

References

1. AliceVision. https://github.com/alicevision/AliceVision
2. Basri, R., Jacobs, D.P.: Lambertian reflectances and linear subspaces. PAMI **25**(2), 218–233 (2003)
3. Blake, A., Zisserman, A., Knowles, G.: Surface descriptions from stereo and shading. IVC **3**(4), 183–191 (1985)
4. Chambolle, A.: A uniqueness result in the theory of stereo vision: coupling shape from shading and binocular information allows unambiguous depth reconstruction. Annales de l'IHP - Analyse non linéaire **11**(1), 1–16 (1994)
5. Furukawa, Y., Hernández, C.: Multi-view stereo: a tutorial. Found. Trends Comput. Graph. Vis. **9**(1–2), 1–148 (2015)
6. Goesele, M., Curless, B., Seitz, S.M.: Multi-view stereo revisited. In: Proceedings of CVPR (2006)
7. Graber, G., Balzer, J., Soatto, S., Pock, T.: Efficient minimal-surface regularization of perspective depth maps in variational stereo. In: Proceedings of CVPR (2015)
8. Hernández, C., Schmitt, F.: Silhouette and stereo fusion for 3D object modeling. CVIU **96**(3), 367–392 (2004)
9. Jin, H., Cremers, D., Wang, D., Yezzi, A., Prados, E., Soatto, S.: 3-D reconstruction of shaded objects from multiple images under unknown illumination. IJCV **76**(3), 245–256 (2008)
10. Langguth, F., Sunkavalli, K., Hadap, S., Goesele, M.: Shading-aware multi-view Stereo. In: Proceedings of ECCV (2016)
11. Maier, R., Kim, K., Cremers, D., Kautz, J., Nießner, M.: Intrinsic3D: high-quality 3D reconstruction by joint appearance and geometry optimization with spatially-varying lighting. In: Proceedings of ICCV (2017)
12. Maurer, D., Ju, Y.C., Breuß, M., Bruhn, A.: Combining shape from shading and stereo: a joint variational method for estimating depth, illumination and albedo. IJCV **126**(12), 1342–1366 (2018)
13. Newcombe, R.A., Lovegrove, S.J., Davison, A.J.: DTAM: dense tracking and mapping in real-time. In: Proceedings of ICCV, pp. 2320–2327 (2011)
14. Quéau, Y., Mélou, J., Castan, F., Cremers, D., Durou, J.D.: A variational approach to shape-from-shading under natural illumination. In: Proceedings of EMMCVPR (2017)

15. Schroers, C., Hafner, D., Weickert, J.: Multiview depth parameterisation with second order regularisation. In: Aujol, J.-F., Nikolova, M., Papadakis, N. (eds.) SSVM 2015. LNCS, vol. 9087, pp. 551–562. Springer, Cham (2015). https://doi.org/10.1007/978-3-319-18461-6_44
16. Wendel, A., Maurer, M., Graber, G., Pock, T., Bischof, H.: Dense reconstruction on-the-fly. In: Proceedings of CVPR (2012)
17. Zollhöfer, M., et al.: Shading-based refinement on volumetric signed distance functions. ACM Trans. Graph. **34**(4), 96:1–96:14 (2015)

Inpainting, Interpolation and Compression

Pseudodifferential Inpainting: The Missing Link Between PDE- and RBF-Based Interpolation

Matthias Augustin$^{(\boxtimes)}$, Joachim Weickert, and Sarah Andris

Mathematical Image Analysis Group, Faculty of Mathematics and Computer Science, Campus E1.7, Saarland University, 66041 Saarbrücken, Germany
{augustin,weickert,andris}@mia.uni-saarland.de

Abstract. Inpainting with partial differential equations (PDEs) has been used successfully to reconstruct missing parts of an image, even for sparse data. On the other hand, sparse data interpolation is a rich field of its own with different methods such as scattered data interpolation with radial basis functions (RBFs).

The goal of this paper is to establish connections between inpainting with linear shift- and rotation-invariant differential operators and interpolation with radial basis functions. The bridge between these two worlds is built by generalising inpainting methods to handle pseudodifferential operators and by considering their Green's functions. In this way, we find novel relations of various multiquadrics to pseudodifferential operators. Moreover, we show that many popular radial basis functions are related to processes from the diffusion and scale-space literature. We present a single numerical algorithm for all methods. It combines conjugate gradients with pseudodifferential operator evaluations in the Fourier domain. Our experiments show that the linear PDE- and the RBF-based communities have developed concepts of comparable quality.

Keywords: Inpainting · Sparse data · Partial differential equations · Pseudodifferential operators · Scattered data interpolation · Radial basis functions · Green's functions

1 Introduction

The problem of restoring damaged or lost parts of an image is known as inpainting. Solution strategies based on partial differential equations (PDEs) are very popular [2,22,26] since they can fill in missing (non-textured) data in a visually plausible way. This may even hold when the data become sparse. However, in this case one achieves best approximation quality if one optimises the inpainting data; see e.g. [1,6,13,21]. This idea is used successfully in inpainting-based lossy image compression [12]. Although nonlinear anisotropic diffusion methods perform best in this application [12,24], linear operators such as homogeneous

© Springer Nature Switzerland AG 2019
J. Lellmann et al. (Eds.): SSVM 2019, LNCS 11603, pp. 67–78, 2019.
https://doi.org/10.1007/978-3-030-22368-7_6

diffusion or biharmonic inpainting are often preferred: They are simpler and parameter-free, easier to analyse, and may give rise to faster algorithms [20,23].

Sparse inpainting can also be seen as a scattered data interpolation problem [11,29]. A popular approach in this field is the interpolation with radial basis functions (RBFs) [4]. Although these methods are most popular in geometric modelling and geosciences, they have also been used for image reconstruction; see e.g. [14,16,28]. However, to our knowledge, a systematic connection between linear PDE-based inpainting and scattered data interpolation cannot be found in the literature so far, and it is unclear how both paradigms perform in comparison.

Our Contribution. Our goal is to address these problems. We establish a general connection between inpainting with shift- and rotation-invariant PDEs and RBF-based interpolation. Since shift-invariant linear operators perform a pointwise multiplication in Fourier space, they are pseudodifferential operators. This motivates us to introduce the concept of pseudodifferential inpainting. By considering the Green's functions of rotationally invariant pseudodifferential operators, we derive the desired link to RBFs. We identify popular RBFs with pseudodifferential operators that are diffusion and scale-space processes, and we evaluate the performance of pseudodifferential inpainting of sparse data.

Related Work. Pseudodifferential operators are not a novelty for researchers working on scale-spaces and variational methods. Already in 1988, Yuille and Grzywacz [30] have expressed Gaussian convolution as regularisation with a pseudodifferential operator. Pseudodifferential operators are also a natural concept for the class of α-scale-spaces (see e.g. Duits et al. [8]), which comprises the Poisson scale-space of Felsberg and Sommer [10]. Other scale-spaces that involve pseudodifferential operators and are relevant for our paper are the Bessel scale-spaces of Burgeth et al. [5]. More recently, Schmidt and Weickert [25] have introduced general shift-invariant linear scale-spaces in terms of pseudodifferential operators and identified their corresponding morphological evolutions. However, to the best of our knowledge, none of these pseudodifferential operators have been used in inpainting so far.

To solve harmonic and biharmonic inpainting problems, Hoffmann et al. [14] have used linear combinations of Green's functions. Their considerations are our point of departure towards more general inpainting operators.

The connection between specific RBFs and partial differential operators, more precisely variational minimisation problems, was already used by Duchon [7] to establish thin-plate splines. Later on, several researchers derived suitable kernels for certain interpolation and approximation problems; see e.g. [9,16,19]. However, most publications are either specialised on one or two types of RBFs or consider the connection in a rather abstract setting. What is missing is a practical method that directly translates between an arbitrary (pseudo-)differential operator and a radial basis function. Our paper closes this gap.

Organisation of the Paper. In Sect. 2, we review the framework of harmonic inpainting and extend it to pseudodifferential inpainting. RBF interpolation is sketched in Sect. 3. We connect both worlds in Sect. 4 via the concept of Green's functions. Our numerical method, based on conjugate gradients and Fourier techniques, is described in Sect. 5, followed by a discussion of experimental results in Sect. 6. Section 7 gives a summary and an outlook.

2   From Harmonic to Pseudodifferential Inpainting

Let us consider a rectangular image domain $\Omega = [0, a] \times [0, b] \subset \mathbb{R}^2$ and a greyscale image $f \colon \Omega \to \mathbb{R}$, which is only known on a subdomain $K \subset \Omega$. A possible way to recover the missing data is so-called harmonic inpainting, which can be formulated as follows: Keep the data where it is known and solve the Laplace equation where no data is known, i.e.,

$$u = f \qquad \text{on } K, \tag{1}$$
$$-\Delta u = 0 \qquad \text{on } \Omega \setminus K, \tag{2}$$

with reflecting boundary conditions on $\partial\Omega$. This approach minimises the energy

$$E(u) = \int_K (u - f)^2 \, \mathrm{d}\boldsymbol{x} + \int_{\Omega \setminus K} |\boldsymbol{\nabla} u|^2 \, \mathrm{d}\boldsymbol{x}, \tag{3}$$

where $|\cdot|$ denotes the Euclidean norm and $\boldsymbol{\nabla}$ the nabla operator in $\mathbb{R}^2$.

We define the Fourier transform as

$$\widehat{u}(\boldsymbol{\zeta}) = \mathcal{F}[u](\boldsymbol{\zeta}) := \int_{\mathbb{R}^2} u(\boldsymbol{x}) \, \exp\!\left(-\mathrm{i} \, 2\pi \, \boldsymbol{\zeta}^T \boldsymbol{x}\right) \mathrm{d}\boldsymbol{x}. \tag{4}$$

The action of a linear, shift-, and rotation-invariant operator can be characterised by a factor $\widehat{p}(|\boldsymbol{\zeta}|)$ in the Fourier domain. This factor is called a *symbol*. Given a symbol, we can define a *pseudodifferential operator* $p(-\Delta)$ by reversing the Fourier transform, i.e.,

$$p(-\Delta)\, u(\boldsymbol{x}) := \int_{\mathbb{R}^2} \widehat{p}(|\boldsymbol{\zeta}|)\, \widehat{u}(\boldsymbol{\zeta}) \, \exp\!\left(\mathrm{i} \, 2\pi \, \boldsymbol{\zeta}^T \boldsymbol{x}\right) \mathrm{d}\boldsymbol{\zeta}. \tag{5}$$

For example, if we choose as operator the negative Laplacian $(-\Delta)$, its symbol is given by

$$\widehat{p}^{(-\Delta)}(\boldsymbol{\zeta}) = 4\pi^2 \, |\boldsymbol{\zeta}|^2 \,. \tag{6}$$

This motivates the notation $p(-\Delta)$ for the pseudodifferential operators which we consider here.

An inpainting problem with a pseudodifferential operator reads

$$u = f \qquad \text{on } K, \tag{7}$$
$$p(-\Delta)\, u = 0 \qquad \text{on } \Omega \setminus K, \tag{8}$$

with reflecting boundary conditions.

Our pseudodifferential inpainting framework comprises naturally integer powers of the negative Laplacian. This so-called multiharmonic inpainting includes e.g. biharmonic and triharmonic inpainting. However, also non-integer powers $\alpha > 0$ are allowed, leading to α-harmonic inpainting. The special case $\alpha = 0.5$ yields Poisson inpainting. Also other pseudodifferential operators from scale-space theory can be used, for instance the Bessel operators. These and more examples of pseudodifferential operators and their symbols are listed in Table 1.

To interpret pseudodifferential inpainting in terms of energy minimisation, we allow only symbols which are nonnegative. Furthermore, we assume that the value zero is attained at most for $\boldsymbol{\zeta} = \mathbf{0}$. Then the root of a pseudodifferential operator $p(-\Delta)$ is defined via

$$\sqrt{p(-\Delta)}\, u(\boldsymbol{x}) := \int_{\mathbb{R}^2} \sqrt{\widehat{p}(|\boldsymbol{\zeta}|)}\, \widehat{u}(\boldsymbol{\zeta})\, \exp\!\left(\mathrm{i}\, 2\pi\, \boldsymbol{\zeta}^T \boldsymbol{x}\right) \mathrm{d}\boldsymbol{\zeta}, \tag{9}$$

and pseudodifferential inpainting minimises the energy functional

$$E(u) = \int_K (u - f)^2\, \mathrm{d}\boldsymbol{x} + \int_{\Omega \backslash K} \left(\sqrt{p(-\Delta)}\, u(\boldsymbol{x})\right)^2 \mathrm{d}\boldsymbol{x} \tag{10}$$

with reflecting boundary conditions. Table 1 also lists the corresponding penalising function

$$\Psi(u) := \left(\sqrt{p(-\Delta)}\, u(\boldsymbol{x})\right)^2 \tag{11}$$

for each pseudodifferential operator.

Note that for a given pseudodifferential operator, the corresponding energy functional is not unique. For instance, for harmonic inpainting, Eq. (10) gives

$$E(u) = \int_K (u - f)^2\, \mathrm{d}\boldsymbol{x} + \int_{\Omega \backslash K} \left(\sqrt{-\Delta}\, u(\boldsymbol{x})\right)^2 \mathrm{d}\boldsymbol{x}, \tag{12}$$

which obviously differs from the energy functional (3). Here, the square root of the Laplacian $\sqrt{-\Delta}$ is the pseudodifferential operator defined by having $\sqrt{\widehat{p}^{(-\Delta)}}(\boldsymbol{\zeta})$ as its symbol.

3   Interpolation with Radial Basis Functions

In the sparse interpolation problem we have in mind, K is a set of finitely many distinct pixels $\boldsymbol{x}_0, ..., \boldsymbol{x}_N$ at which the image f is known. For RBF interpolation, the interpolating function u is obtained from the ansatz

$$u(\boldsymbol{x}) = \sum_{j=0}^N c_j\, g(|\boldsymbol{x} - \boldsymbol{x}_j|), \tag{13}$$

with a so-called *radial basis function* $g\colon \mathbb{R}_0^+ \to \mathbb{R}$. Popular choices include thin plate splines, polyharmonic splines, Matérn kernels, multiquadrics (MQs), inverse MQs, and inverse cubic MQs. Their formulas are displayed in Table 1.

The unknown coefficients $c_0,...,c_N \in \mathbb{R}$ are computed as solutions to the interpolation problem $u(\boldsymbol{x}_k) = f(\boldsymbol{x}_k)$ for all k. This yields the linear system

$$\sum_{j=0}^{N} c_j\, g(|\boldsymbol{x}_k - \boldsymbol{x}_j|) = f(\boldsymbol{x}_k) \qquad \text{for all } k = 0,\ldots,N. \tag{14}$$

A common condition to guarantee that this system has a unique solution is that its symmetric matrix $(g(|\boldsymbol{x}_k - \boldsymbol{x}_j|))_{j,k=0}^{N}$ is positive definite. Due to Bochner's theorem, this is equivalent to g having a positive (generalised) Fourier transform [29]. If the radial basis function has no compact support, the system matrix is dense. Then the numerical solution of the linear system is slow for large N.

Requiring a positive definite system matrix can be relaxed to positive semidefiniteness. For details and further information on RBF interpolation we refer to the monographs [4, 29].

4 Connecting both Worlds

Let us now establish a bridge between the pseudodifferential inpainting (8) and the RBF interpolation (13). Solving (8) requires inversion of a pseudodifferential operator, while (13) has a convolution-like structure. Thus, we employ the concept of *Green's functions*. The idea behind Green's functions is to define an inverse to a differential or pseudodifferential operator in the form of a convolution

$$(v \circledast g)(x,y) := \int_0^a \int_0^b v(s,t)\, g(x-s, y-t)\, \mathrm{d}t\, \mathrm{d}s. \tag{15}$$

A Green's function to the operator $p(-\Delta)$ is defined as a function g for which holds

$$p(-\Delta)\,(v \circledast g)(x) = v(x) \tag{16}$$

for all functions v which are orthogonal to the nullspace of $p(-\Delta)$ for all $\boldsymbol{x} \in \Omega$. Due to the convolution theorem, Fourier transform turns Eq. (16) into

$$\widehat{p}(|\boldsymbol{\zeta}|)\,(\widehat{g}(|\boldsymbol{\zeta}|)\,\widehat{v}(\boldsymbol{\zeta})) = \widehat{v}(\boldsymbol{\zeta}) \qquad \text{for all } \boldsymbol{\zeta} \in \mathbb{R}^2. \tag{17}$$

At this point, the fact that v is orthogonal to the nullspace of $p(-\Delta)$ comes into play, as this means that $\widehat{v}(\boldsymbol{\zeta})$ equals zero whenever $\widehat{p}(|\boldsymbol{\zeta}|)$ equals zero, such that Eq. (17) can still be satisfied in these cases. We can now obtain a Green's function g to $p(-\Delta)$ by defining its (generalised) Fourier transform to be

$$\widehat{g}(|\boldsymbol{\zeta}|) := \begin{cases} 0, & \text{if } \widehat{p}(|\boldsymbol{\zeta}|) = 0, \\ \frac{1}{\widehat{p}(|\boldsymbol{\zeta}|)}, & \text{else,} \end{cases} \tag{18}$$

and the condition that g satisfies the reflecting boundary conditions on $\partial\Omega$. This shows that Green's functions are pseudoinverses to pseudodifferential operators.

We can also read Eq. (18) the other way around: Given a radial basis function $g(|\boldsymbol{x}|)$, compute its (generalised) Fourier transform $\widehat{g}(|\boldsymbol{\zeta}|)$ and use Eq. (18) to find the corresponding pseudodifferential operator $p(-\Delta)$ such that the chosen radial basis function is a Green's function to the newly defined pseudodifferential operator. Thus, Eq. (18) establishes a simple and elegant one-to-one mapping between pseudodifferential operators and radial basis functions. Consequently, pseudodifferential inpainting on sparse data is equivalent to sparse interpolation with radial basis functions.

Table 1 lists a number of pseudodifferential operators and their corresponding RBFs. For the sake of recognisability, we display the version of the functions which does not obey any boundary conditions. This is equivalent to considering a free-space problem, i.e., $\Omega = \mathbb{R}^2$. Moreover, the radial basis functions may differ from the Green's functions by a constant factor, which does not matter in applications.

Our results prove that many RBF concepts are equivalent to inpainting with well-known scale-space operators, ranging from α-scale-spaces to Bessel scale-spaces. Moreover, they also establish additional interesting findings, such as the interpretations of various MQs in terms of pseudodifferential operators.

One of the most important columns in Table 1 is the smoothness of the RBF, since (13) implies that it immediately carries over to the smoothness of the interpolant. On one hand, this column confirms some known facts such as the logarithmic singularity in the Green's function for harmonic inpainting, and the C^1-smoothness of biharmonic inpainting. On the other hand, Table 1 displays also many smoothness results that are not well-known for the corresponding pseudodifferential operators. Note that all smoothness results hold only in 2D: For instance, the Green's function for harmonic inpainting is continuous with a kink in 1D, and it has a singularity of type $|\boldsymbol{x}|^{2-d}$ in $\mathbb{R}^d$ for $d \geq 3$.

5 One Numerical Algorithm for All Approaches

Interestingly, our unifying framework for pseudodifferential inpainting also carries over to the discrete setting, where a single algorithm handles all approaches.

We replace the continuous image domain Ω by a regular Cartesian grid with n_x and n_y pixels in x- and y-direction. The corresponding grid sizes are h_x and h_y. Our discrete image is reflected in x- and y-direction to implement reflecting boundary conditions. The data on the resulting domain are then extended periodically such that we can apply the discrete Fourier transform.

We discretise the negative Laplacian by the usual five-point stencil with symbol

$$\widehat{p}_{\ell,m}^{(5)} = \left(\frac{2}{h_x} \sin\left(\frac{\ell\pi}{n_x} \right) \right)^2 + \left(\frac{2}{h_y} \sin\left(\frac{m\pi}{n_y} \right) \right)^2. \tag{19}$$

This ensures that Fourier and finite difference techniques produce identical results. To discretise other pseudodifferential operators, we substitute $4\pi^2 |\boldsymbol{\zeta}|^2$

Table 1. Pseudodifferential operators, their symbols, radial basis functions along with their smoothness, and the penalising functions from (11). Negative signs are included where needed to obtain positive definite functions. Here, $t \in \mathbb{R}^+$, $\lfloor \cdot \rfloor$ is the floor function, K_ν is the modified Bessel function of the second kind of order $\nu \in \mathbb{R}$, ${}_2F_0(a,b;0;z)$ is a generalised hypergeometric function, and $\beta_n = \frac{{}_2F_0(1,3-n;0;1)}{(n-3)!}$.

Operator $p(-\Delta)$	Symbol $\hat{p}$	Radial basis function $g(|x|)$	Smoothness of $g(|x|)$	Penalising function $\Psi(u)$
$-\Delta$ harmonic [15]	$4\pi^2\|\zeta\|^2 =: \hat{p}(-\Delta)$	$-\ln(\|x\|)$ harmonic spline [19]	singular	$\left(\sqrt{-\Delta}\,u\right)^2$
$(-\Delta)^2$ biharmonic	$\left(\hat{p}(-\Delta)\right)^2$	$\|x\|^2\ln(\|x\|)$ thin plate spline [7]	C^1	$(\Delta u)^2$
$(-\Delta)^n,\ n \in \mathbb{N}\setminus\{1\}$ polyharmonic	$\left(\hat{p}(-\Delta)\right)^n$	$(-1)^n\|x\|^{2n-2}\ln(\|x\|)$ polyharmonic splines [29]	C^{2n-3}	$\left(\sqrt{(-\Delta)^n}\,u\right)^2$
$\sqrt{-\Delta}$ Poisson [3,10]	$\sqrt{\hat{p}(-\Delta)}$	$\|x\|^{-1}$	singular	$\left(\sqrt[4]{(-\Delta)}\,u\right)^2$
$(-\Delta)^\alpha,\ \alpha \in \mathbb{R}^+ \setminus \mathbb{N}$ α-harmonic [8]	$\left(\hat{p}(-\Delta)\right)^\alpha$	$(-1)^{\lceil\alpha\rceil}\|x\|^{2\alpha-2}$ α-harmonic splines [3,29]	singular for $\alpha \in (0,1)$, $C^{2\lceil\alpha\rceil-2}$ for $\alpha > 1$,	$\left(\sqrt{(-\Delta)^\alpha}\,u\right)^2$
$\left(1-\frac{t}{\alpha}\Delta\right)^\alpha,\ \alpha \in \mathbb{R}^+$ Bessel [5]	$\left(1+\frac{t}{\alpha}\hat{p}(-\Delta)\right)^\alpha$	$\left(\sqrt{\frac{\alpha}{t}}\,\|x\|\right)^{\alpha-1}K_{\alpha-1}\left(\sqrt{\frac{\alpha}{t}}\,\|x\|\right)$ Matérn kernels [27]	singular for $\alpha \le 1$, $C^{2\lceil\alpha\rceil-2}$ for $\alpha \in (1,\infty)\setminus\mathbb{N}$ $C^{2\alpha-3}$ for $\alpha \in \{2,3,...\}$	$\left(\sqrt{\left(1-\frac{t}{\alpha}\Delta\right)^\alpha}\,u\right)^2$
$\sum_{n=3}^\infty \beta_n t^n \left(\sqrt{-\Delta}\right)^n$	$\dfrac{\left(t\sqrt{\hat{p}(-\Delta)}\right)^3\exp\left(t\sqrt{\hat{p}(-\Delta)}\right)}{1+t\sqrt{\hat{p}(-\Delta)}}$	$-\sqrt{t^2+\|x\|^2}$ multiquadric (MQ) [29]	C^∞	$\left(\sqrt{\sum_{n=3}^\infty \beta_n t^n \left(\sqrt{-\Delta}\right)^n}\,u\right)^2$
$\sum_{n=0}^\infty \frac{t^n}{n!}\left(\sqrt{-\Delta}\right)^{n+1}$	$\sqrt{\hat{p}(-\Delta)}\exp\left(t\sqrt{\hat{p}(-\Delta)}\right)$	$t\big/\sqrt{t^2+\|x\|^2}$ inverse MQ [29]	C^∞	$\left(\sqrt{\sum_{n=0}^\infty \frac{t^n}{n!}\left(\sqrt{-\Delta}\right)^{n+1}}\,u\right)^2$
$\sum_{n=0}^\infty \frac{t^n}{n!}\left(\sqrt{-\Delta}\right)^n$	$\exp\left(t\sqrt{\hat{p}(-\Delta)}\right)$	$t\big/\left(t^2+\|x\|^2\right)^{\frac{3}{2}}$ inverse cubic MQ [29]	C^∞	$\left(\sqrt{\sum_{n=0}^\infty \frac{t^n}{n!}\left(\sqrt{-\Delta}\right)^n}\,u\right)^2$

in their symbol by $\widehat{p}_{\ell,m}^{(5)}$ and consider the results for all ℓ and m as eigenvalues of a matrix $\boldsymbol{A}$ whose eigenvectors are given by the discrete Fourier transform. Then the discrete inpainting problem can be written as

$$C\,(\boldsymbol{u} - \boldsymbol{f}) + (\boldsymbol{I} - \boldsymbol{C})\,\boldsymbol{A}\boldsymbol{u} = \boldsymbol{0}. \qquad (20)$$

Here the vectors $\boldsymbol{u}$ and $\boldsymbol{f}$ are discretisations of u and f, and $\boldsymbol{I}$ is the unit matrix. As $\boldsymbol{C}$ has 1 on the diagonal for mask points and 0 for non-mask points, the first term in Eq. (20) corresponds to the interpolation condition (7), whereas the second term corresponds to Eq. (8). Rewriting (20) yields the linear system

$$\big(\boldsymbol{C} + (\boldsymbol{I} - \boldsymbol{C})\,\boldsymbol{A}\big)\boldsymbol{u} = \boldsymbol{C}\boldsymbol{f}. \qquad (21)$$

Fig. 1. Test images *trui*, *peppers*, and *walter*.

By considering only non-mask points, Eq. (21) can be reduced to a system of linear equations with a symmetric positive definite matrix with arguments similar to the ones in [20]. In other words, substituting some rows of a positive (semi)definite matrix $\boldsymbol{A}$ by corresponding rows of a unit matrix yields a positive definite matrix. Thus, we can use a standard conjugate gradient (CG) solver. Its matrix–vector products with the circulant matrix $\boldsymbol{A} \in \mathbb{R}^{2n_x n_y \times 2n_x n_y}$ are computed in the Fourier domain with an effort that does not depend on the pseudodifferential operator. We stop the CG iterations when the Euclidean norm of the residual vector has dropped by a factor 10^{-20}.

6   Experiments

To evaluate the performance of the different pseudodifferential operators, we inpaint three greyscale images with size 256×256 and range $[0, 255]$: *trui*, *peppers*, and *walter* (Fig. 1). As known data, we use the grey values of each image at the locations given by a fixed random mask (Fig. 2 top left). Its density is 5%, i.e. we know the grey values for 5% of the pixels. The parameters α and t have been optimised by a simple grid search to produce the minimal mean squared error (MSE) w.r.t. the ground truth. Table 2 reports these errors for all three images, and Fig. 2 illustrates the inpainting results for *trui*.

We observe that Poisson inpainting performs much worse than all other approaches. It suffers from the strongly visible singularity of the RBF. The second worst is harmonic inpainting, whose logarithmic singularity is also visible.

Biharmonic, α-harmonic, MQ and inverse MQ inpainting produce very good results of comparable quality. For optimal parameters, their RBFs are at least in the smoothness class C^1 such that no singularities are visible. However, since these operators involve higher order powers of the Laplacian, they may violate a maximum-minimum principle. This becomes visible in over- and undershoots and ripple artifacts. These artifacts are slightly more pronounced for inverse cubic MQs, which is also confirmed by a somewhat worse MSE.

Bessel inpainting uses almost the same optimal α values as optimised α-harmonic inpainting, and its optimal t-values are large. In this setting both approaches are almost identical, since for large t, we have $\left(I - \frac{t}{\alpha}\Delta\right)^\alpha \approx (-\frac{t}{\alpha}\Delta)^\alpha$, and the latter is equivalent to α-harmonic inpainting.

The optimal α parameters for α-harmonic inpainting are close to 2. Thus, practitioners may prefer the parameter-free biharmonic inpainting as a method of choice. Since the corresponding thin plate splines can be seen as a rotationally invariant 2D extension of cubic spline interpolation, this also confirms earlier findings where cubic splines are reported as favourable interpolation methods [18]. Moreover, this indicates that on average, natural images can be approximated well by continuously differentiable functions.

The fact that biharmonic inpainting and the various MQs give results of similar quality is remarkable: It shows that the linear PDE community and the RBF community have reached a comparable level of maturity and sophistication, even without many interactions. Of course, for PDE-based inpainting, nonlinear methods may offer further improvements [12,24]. However, they cannot be treated adequately within a pseudodifferential framework, which is based on intrinsically linear concepts such as Fourier techniques.

Table 2. MSE results for the different inpainting approaches applied to the test images.

Operator	*trui*	*peppers*	*walter*
Harmonic	211.14	226.39	233.40
Biharmonic	136.65	152.91	121.58
Poisson	759.26	751.61	773.27
α-harmonic	135.89 ($\alpha = 1.86$)	**152.57** ($\alpha = 1.9$)	121.28 ($\alpha = 2.09$)
Bessel	135.79 ($\alpha = 1.89$, $t = 485$)	**152.57** ($\alpha = 1.9$, $t = 10^6$)	**121.23** ($\alpha = 2.13$, $t = 386$)
Multiquadric (MQ)	137.97 ($t = 5.13$)	166.83 ($t = 4.98$)	124.34 ($t = 5.9$)
Inverse MQ	**135.63** ($t = 3.89$)	161.99 ($t = 3.79$)	121.97 ($t = 4.63$)
Inverse cubic MQ	167.08 ($t = 7.05$)	198.23 ($t = 6.79$)	137.90 ($t = 7.11$)

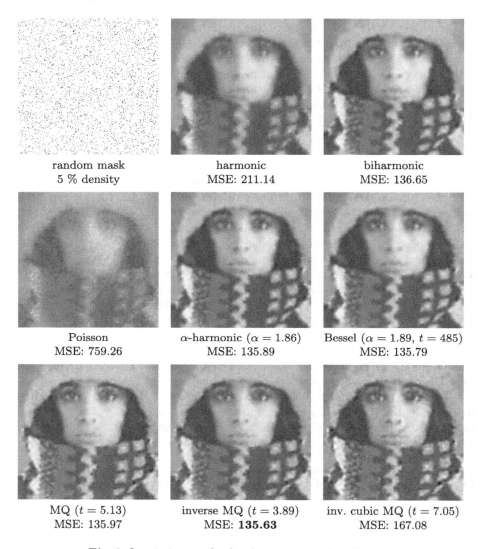

Fig. 2. Inpainting results for the image *trui* from Fig. 1.

7 Conclusions and Outlook

We have established pseudodifferential inpainting as a unifying concept that connects linear PDE-based inpainting and RBF interpolation. This framework is surprisingly simple and general: It can handle any linear shift- and rotation-invariant operator, not only analytically but also algorithmically. It enabled us to find a number of interesting, hitherto unknown insights and relations, ranging from smoothness results for all inpainting methods to connections between RBFs and scale-space operators. Last but not least, we have shown that the linear

PDE- and RBF-based communities have come up with approaches of similar maturity and quality.

Currently we are investigating additional RBFs such as truncated RBFs, and we are going to extend our evaluation to the setting of inpainting-based compression. The latter involves several new aspects, e.g. their performance for optimised data and sensitivity w.r.t. quantisation.

Acknowledgement. This project has received funding from the European Research Council (ERC) under the European Union's Horizon 2020 research and innovation programme (grant agreement no. 741215, ERC Advanced Grant INCOVID).

References

1. Belhachmi, Z., Bucur, D., Burgeth, B., Weickert, J.: How to choose interpolation data in images. SIAM J. Appl. Math. **70**(1), 333–352 (2009)
2. Bertalmío, M., Sapiro, G., Caselles, V., Ballester, C.: Image inpainting. In: Proceedings of SIGGRAPH 2000, New Orleans, LA, pp. 417–424, July 2000
3. Bucur, C.: Some observations on the Green function for the ball in the fractional Laplace framework. Commun. Pure Appl. Anal. **15**(2), 657–699 (2016)
4. Buhmann, M.D.: Radial Basis Functions: Theory and Implementations. Cambridge University Press, Cambridge (2003)
5. Burgeth, B., Didas, S., Weickert, J.: The Bessel scale-space. In: Olsen, O.F., Florack, L., Kuijper, A. (eds.) DSSCV 2005. LNCS, vol. 3753, pp. 84–95. Springer, Heidelberg (2005). https://doi.org/10.1007/11577812_8
6. Chen, Y., Ranftl, R., Pock, T.: A bi-level view of inpainting-based image compression. In: Kúkelová, Z., Heller, J. (eds.) Proceedings of 19th Computer Vision Winter Workshop, Křtiny, Czech Republic, pp. 19–26, February 2014
7. Duchon, J.: Interpolation des fonctions de deux variable suivant le principe de la flexion des plaques minces. Revue française d'automatique, informatique, recherche opérationelle. Analyse numérique **10**(12), 5–12 (1976)
8. Duits, R., Florack, L., de Graaf, J., ter Haar Romeny, B.: On the axioms of scale space theory. J. Math. Imaging Vis. **20**(3), 267–298 (2004)
9. Fasshauer, G.E., Ye, Q.: Reproducing kernels of generalized Sobolev spaces via a Green function approach with distributional operators. Numerische Mathematik **119**(3), 585–611 (2011)
10. Felsberg, M., Sommer, G.: The monogenic scale-space: a unifying approach to phase-based image processing in scale-space. J. Math. Imaging Vis. **21**(1), 5–26 (2004)
11. Franke, R., Nielson, G.M.: Scattered data interpolation and applications: A tutorial and survey. In: Hagen, H., Roller, D. (eds.) Geometric Modeling: Methods and Applications, pp. 131–160. Springer, Berlin (1991). https://doi.org/10.1007/978-3-642-76404-2_6
12. Galić, I., Weickert, J., Welk, M., Bruhn, A., Belyaev, A., Seidel, H.: Image compression with anisotropic diffusion. J. Math. Imaging Vis. **31**(2–3), 255–269 (2008)
13. Hoeltgen, L., Setzer, S., Weickert, J.: An optimal control approach to find sparse data for Laplace interpolation. In: Heyden, A., Kahl, F., Olsson, C., Oskarsson, M., Tai, X.-C. (eds.) EMMCVPR 2013. LNCS, vol. 8081, pp. 151–164. Springer, Heidelberg (2013). https://doi.org/10.1007/978-3-642-40395-8_12

14. Hoffmann, S., Plonka, G., Weickert, J.: Discrete Green's functions for harmonic and biharmonic inpainting with sparse atoms. In: Tai, X.-C., Bae, E., Chan, T.F., Lysaker, M. (eds.) EMMCVPR 2015. LNCS, vol. 8932, pp. 169–182. Springer, Cham (2015). https://doi.org/10.1007/978-3-319-14612-6_13

15. Iijima, T.: Basic theory on normalization of pattern (in case of typical one-dimensional pattern). Bull. Electrotechnical Lab. **26**, 368–388 (1962). (in Japanese)

16. Iske, A.: Multiresolution Methods in Scattered Data Modelling. Lecture Notes in Computational Science and Engineering, vol. 37. Springer, Heidelberg (2004). https://doi.org/10.1007/978-3-642-18754-4

17. Kozhekin, N., Savchenko, V., Senin, M., Hagiwara, I.: An approach to surface retouching and mesh smoothing. Visual Comput. **19**(8), 549–564 (2003)

18. Lehmann, T., Gönner, C., Spitzer, K.: Survey: Interpolation methods in medical image processing. IEEE Trans. Med. Imaging **18**(11), 1049–1075 (1999)

19. Madych, W.R., Nelson, S.A.: Polyharmonic cardinal splines. J. Approximation Theory **60**(2), 141–156 (1990)

20. Mainberger, M., Bruhn, A., Weickert, J., Forchhammer, S.: Edge-based image compression of cartoon-like images with homogeneous diffusion. Pattern Recogn. **44**(9), 1859–1873 (2011)

21. Mainberger, M., Hoffmann, S., Weickert, J., Tang, C.H., Johannsen, D., Neumann, F., Doerr, B.: Optimising spatial and tonal data for homogeneous diffusion inpainting. In: Bruckstein, A.M., ter Haar Romeny, B.M., Bronstein, A.M., Bronstein, M.M. (eds.) SSVM 2011. LNCS, vol. 6667, pp. 26–37. Springer, Heidelberg (2012). https://doi.org/10.1007/978-3-642-24785-9_3

22. Masnou, S., Morel, J.: Level lines based disocclusion. In: Proceedings of 1998 IEEE International Conference on Image Processing, . Chicago, IL, vol. 3, pp. 259–263, October 1998

23. Peter, P., Hoffmann, S., Nedwed, F., Hoeltgen, L., Weickert, J.: Evaluating the true potential of diffusion-based inpainting in a compression context. Signal Process. Image Commun. **46**, 40–53 (2016)

24. Schmaltz, C., Peter, P., Mainberger, M., Ebel, F., Weickert, J., Bruhn, A.: Understanding, optimising, and extending data compression with anisotropic diffusion. Int. J. Comput. Vis. **108**(3), 222–240 (2014)

25. Schmidt, M., Weickert, J.: Morphological counterparts of linear shift-invariant scale-spaces. J. Math. Imaging Vis. **56**(2), 352–366 (2016)

26. Schönlieb, C.B.: Partial Differential Equation Methods for Image Inpainting. Cambridge University Press, Cambridge (2015)

27. Stein, M.L.: Interpolation of Spatial Data - Some Theory for Kriging. Springer, New York (1999). https://doi.org/10.1007/978-1-4612-1494-6

28. Uhlir, K., Skala, V.: Radial basis function use for the restoration of damaged images. In: Wojciechowski, K., Smolka, B., Palus, H., Kozera, R., Skarbek, W., Noakes, L. (eds.) Computer Vision and Graphics, Computational Imaging and Vision, vol. 32, pp. 839–844. Springer, Dordrecht (2006). https://doi.org/10.1007/1-4020-4179-9_122

29. Wendland, H.: Scattered Data Approximation. Cambridge University Press, Cambridge (2005)

30. Yuille, A.L., Grzywacz, N.M.: A computational theory for the perception of coherent visual motion. Nature **333**, 71–74 (1988)

Towards PDE-Based Video Compression with Optimal Masks and Optic Flow

Laurent Hoeltgen, Michael Breuß$^{(\boxtimes)}$, and Georg Radow

Brandenburg University of Technology Cottbus-Senftenberg, 03046 Cottbus, Germany
{laurent.hoeltgen,breuss,radow}@b-tu.de

Abstract. Lossy image compression methods based on partial differential equations have received much attention in recent years. They may yield high quality results but rely on the computationally expensive task of finding optimal data.

For the possible extension to video compression, the data selection is a crucial issue. In this context one could either analyse the video sequence as a whole or perform a frame-by-frame optimisation strategy. Both approaches are prohibitive in terms of memory and run time.

In this work we propose to restrict the expensive computation of optimal data to a single frame and to approximate the optimal reconstruction data for the remaining frames by prolongating it by means of an optic flow field. We achieve a notable decrease in the computational complexity. As a proof-of-concept, we evaluate the proposed approach for multiple sequences with different characteristics. We show that the method preserves a reasonable quality in the reconstruction, and is very robust against errors in the flow field.

Keywords: Partial differential equations · Inpainting ·
Laplace interpolation · Optic flow · Video reconstruction

1 Introduction

Transform-based image and video compression algorithms are still the preferred choice in many applications [29]. However, there has been a surge in research on alternative approaches in recent years [2,12,17,27]. Especially partial differential equation (PDE)-based methods have proven to be a viable alternative in the context of image compression. To be on a competitive level with state-of-the-art codecs, these methods require sophisticated data optimisation schemes and fast numerical algorithms. The most important task is the choice of a small subset of pixels, often called mask, from which the original image can be accurately reconstructed by solving a PDE.

Especially this data selection problem has proven to be delicate. See [7,9, 13,14,34] for some strategies considered in the past. Most approaches are either very fast but yield suboptimal results or they are relatively slow and return very appropriate data. A thorough optimisation of a whole image sequence is

© Springer Nature Switzerland AG 2019
J. Lellmann et al. (Eds.): SSVM 2019, LNCS 11603, pp. 79–91, 2019.
https://doi.org/10.1007/978-3-030-22368-7_7

therefore computationally rather demanding and most approaches have resorted to a frame-by-frame consideration. Yet, even such a frame-wise tuning can be expensive, especially for longer videos.

In this work we discuss a simple and fast approach to skip the costly data selection in a certain number of frames. Instead we perform a significantly cheaper data transport along the temporal axis of the sequence. In order to evaluate this idea, we focus on the interplay between reconstruction quality and the accuracy of the transporting vector field. The actual data compression will be the subject of future research.

To give some more details of our approach, we consider an image sequence and compute a highly optimised pixel mask used for a PDE-based reconstruction within the first, single frame. Next, we seek the displacement field between the individual subsequent frames by means of a simple optic flow method. We shift the carefully selected pixels from the first frame according to this flow field and the shifted data is then used for the reconstruction process, in this case PDE-based inpainting. The effects of erroneous or suboptimal shifts of mask pixels on the resulting video reconstruction quality can then be evaluated.

The framework for video compression recently presented in [2] has some technical similarities to our approach. The conceptual difference is that in their work a reconstructed image is shifted via optic flow fields from the first to following frames. In contrast, we use optic flow fields only for the propagation of mask pixel and deal with an inpainting problem in each frame.

Our paper will be structured as follows. We will briefly describe the considered models and methods. Next we describe how they are concatenated in our strategy. Finally, all components are carefully evaluated, where we focus here on quality in terms of reconstruction error. Let us note again that we will not consider the impact on the file compression efficiency, as a detailed analysis of the complete, resulting data compression pipeline would be beyond the scope of this work.

2   Discussion of Considered Models and Methods

The recovery of images, as in a video sequence, by means of interpolation is commonly called inpainting. Since the main issue in our approach is concerned with the selection of data for a corresponding PDE-based inpainting task, it will be useful to elaborate on the problem in some detail. After discussing possible extensions from image to video inpainting, we consider optical flow.

2.1   Image Inpainting with PDEs

The inpainting problem goes back to the works of Masnou and Morel as well as Bertalmo and colleagues [4,23], although similar problems have been considered in other fields already before. There exist many inpainting techniques, often based on interpolation algorithms, but PDE-based approaches are among the most successful ones, see e.g. [15,16]. For the latter, strategies based on the

Laplacian are often advocated [6, 21, 26, 28]. Mathematically, the simplest model is given by the elliptic mixed boundary value problem

$$- \Delta u = 0 \text{ in } \Omega \setminus \Omega_K, \qquad u = f \text{ in } \partial\Omega_K, \qquad \partial_n u = 0 \text{ in } \partial\Omega \setminus \partial\Omega_K, \qquad (1)$$

Here, f represents known image data in a region $\Omega_K \subset \Omega$ (resp. on the boundary $\partial\Omega_K$) of the whole image domain Ω. Further, $\partial_n u$ denotes the derivative in outer normal direction. In an image compression context the image f is known on its whole domain Ω and one would like to identify the smallest set Ω_K that yields a good reconstruction when solving (1).

While solving (1) numerically is a rather straightforward task, finding an optimal subset Ω_K is much more challenging. Mainberger et al. [22] consider a combinatorial strategy while Belhachmi and colleagues [3] approach the topic from the analytic side. Recently [18], the "hard" boundary conditions in (1) have been replaced by softer weighting schemes. If we denote the weighting function by $c : \Omega \to \mathbb{R}$, then (1) becomes:

$$\begin{cases} c(x)(u(x) - f(x)) + (1 - c(x))(-\Delta)u(x) = 0, & \text{in } \Omega, \\ \partial_n u(x) = 0, & \text{in } \partial\Omega \setminus \partial\Omega_K. \end{cases} \qquad (2)$$

In the case where c is the indicator function of Ω_K, (2) coincides with the PDE in (1). Whenever $c(x) = 1$, we require $u(x) - f(x) = 0$ and $c(x) = 0$ implies $-\Delta u(x) = 0$.

Optimising a weighting function c which maps to $\mathbb{R}$ is notably simpler than solving a combinatorial optimisation problem when the mask c maps to $\{0, 1\}$. As the optimal set Ω_K is given by the support of the function c the benefit of the formulation (2) is that one may adopt ideas from sparse signal processing to find such a good mask. To this end, Hoeltgen et al. [18] following optimal control formulation:

$$\arg\min_{u,c} \left\{ \int_\Omega \frac{1}{2}(u(x) - f(x))^2 + \lambda|c(x)| + \frac{\varepsilon}{2}c(x)^2 \, dx \right\},$$
$$\begin{cases} c(x)(u(x) - f(x)) + (1 - c(x))(-\Delta)u(x) = 0, & \text{in } \Omega, \\ \partial_n u(x) = 0, & \text{in } \partial\Omega \setminus \partial\Omega_K. \end{cases} \qquad (3)$$

Equation (3) can be solved by an iterative linearisation of the PDE in terms of (u, c), followed by a primal-dual optimisation strategy such as [10] for the occurring convex problem with linear constraints. As reported in [18], a few hundred linearisations need to be performed to obtain a good solution. This also implies that an equal amount of convex optimisation problems need to be solved. Even if highly efficient solvers are used for the latter convex optimisation, the run time will still be considerable. An alternative approach for solving (3) was also presented in [24].

Besides optimising Ω_K (resp. c), it is also possible to optimise the Dirichlet boundary data in such a way that the global error is minimal. If $M(c)$ denotes the linear solution operator with mask c that yields the solution of (2), then we can write this tonal optimisation as

$$\arg\min_{g}\{\|M(c)g - f\|_2^2\} \ . \tag{4}$$

This idea has originally been presented in [22]. In [19] it is shown that there exists a dependence between non-binary optimal c (i.e. mapping to $\mathbb{R}$ instead of $\{0,1\}$) and optimal tonal values g. Efficient algorithms for solving (4) can be found in [19,22]. These algorithms are faster than solving (3), yet their run times still range from a few seconds to a minute.

2.2   Extension from Images to Videos

The mentioned strategies have so far been applied to grey-value or colour images almost exclusively. Yet extensions to video sequences would be rather straightforward. The simplest strategy would be to consider a frame-by-frame strategy. In (3) one could also extend the Laplacian into the temporal direction to compute an optimal mask in space-time. This would reduce the temporal redundancy (assuming that the content of subsequent frames does not change much) in the mask c compared to a frame-wise approach. Unfortunately, the latter strategy is prohibitively expensive. A one second long video sequence in 4K resolution (3860×2160 pixels) with a framerate of 60 Hz would require analysing approximately 500 million pixels. A frame-by-frame optimisation would be more memory efficient, since the whole sequence does not need to be loaded at once, but it would still require solving 60 expensive optimisation problems.

There exists an alternative approach which is commonly used in modern video compression codecs such as MPEG, see [30] for a general overview on the concepts and ideas. Instead of computing mask points for each frame, we compute a displacement field and shift mask points from one frame to the next.

2.3   Optical Flow

For the sake of simplicity we opt for the method of Horn and Schunck [20]. Given an image sequence $f(x,y,t)$, where x and y are the spatial dimensions and t the temporal dimension, this method computes a displacement field $(u(x,y), v(x,y))$ that maps the frame at time t onto the frame at time $t+1$ by minimising the energy functional

$$\int_{\Omega} (f_x u + f_y v + f_t)^2 + \alpha(|\nabla u|^2 + |\nabla v|^2)\, \mathrm{d}x\,\mathrm{d}y \tag{5}$$

where f_x, f_y, and f_t denote the partial derivatives of f with respect to x, y, and t and where $\Omega \subset \mathbb{R}^2$ denotes the image domain. The model of Horn and Schunck is very popular and highly efficient numerical schemes exist that are capable of solving (5) in real-time (30 frames per second), see [8]. Obviously, replacing already a single computation of c with the computation of a displacement field

(u, v) will save a significant amount of time. If the movements in the image sequence are small and smooth enough, it is very likely, that several masks c can be replaced by a flow field, thus saving even more run time.

3   Combining Optimal Masks with Flow Data

Given an image sequence f, we compute a sparse inpainting mask for the first frame with the method from [18]. According to the results in [19], we threshold the mask c and set all non-zero values to 1. Next, we compute the displacement field between all subsequent frames in the sequence by solving (5) for each pair of consecutive frames. The obtained flow fields (u, v) are rounded point-wise to their nearest integers to assert that they point exactly onto a grid point. Then, the mask points from the first frame are simply moved according to the displacement field. If the displacement points outside of the image or if it points onto a position where a mask point is already located, then we drop the current mask point. Since we are considering sparse sets of mask points, the probability of these events is rather low such that hardly any data gets lost over the course of action. Once the mask has been set for each frame, we perform a tonal optimisation of the data as discussed in [19]. The reconstruction can then simply be done by solving (2) for each frame. The complete procedure is also detailed in Algorithm 1.

Algorithm 1. Data Selection in Image Sequence

Data: Image sequence f
Result: Optimised data for the reconstruction

1 Optimise mask c for first frame of the sequence f by solving (3)
2 Binarise c by setting all non-zero entries to 1

3 Find optical flow between all frames of the sequence by minimising (5)
4 Round flow field entries to nearest integer
5 Transport c according to the flow field

6 Perform tonal optimisation by solving (4) on the mask locations in each frame

---

Instead of rounding the flow field vectors, one could also follow the idea to perform a forward warping [25] and spread a single mask point on all neighbouring mask points. With this strategy, flow fields that point to the same location would simply add up the mask values. Even though this appears as a mathematically clean approach, our experiments showed that the smearing of the mask values caused strong blurring effects in the reconstructions and lead to overall worse results.

The data that needs to be stored for the reconstruction consists of the mask point positions in the first frame, the flow fields that move the mask points along the image sequence (resp. the mask positions in the subsequent frames), and the corresponding tonal optimised pixel values. We emphasise that it is not necessary to store the whole displacement field but only at the locations of a mask point in each frame. Thus, the memory requirements for the storage remain the same as when optimising the mask in each frame. Yet, we are considerably faster. We also remark that the considered strategy is rather generic. One may exchange the mask selection algorithm and the optic flow computation with any other method that yields similar data.

4   Experimental Evaluation

To evaluate the proposed approach, we give further details on our experimental setup, including a rough comparison of runtimes for the different stages of Algorithm 1.

We discuss the influence of the quality of the flow fields at hand of an example. By comparing our approach to compression with fixed mask points, we derive some clues for typical use case scenarios. Then we proceed by evaluating the proposed method for a number of image sequences.

4.1   Methods Considered

As already mentioned, we compute the inpainting masks with the algorithm from [18] and use the LSQR-based algorithm from [19] for tonal optimisation. In terms of quality these methods are among the best performing ones for Laplace reconstruction. However, alternative solvers such as presented in [11,22] may be used as well.

For a reasonable comparison of simple optical flow methods we have resorted to the builtin Matlab implementation of the Horn and Schunck method [32] and a more sophisticated implementation available from [31]. The latter implementation additionally includes a coarse-to-fine warping strategy. Evaluations on the Yosemite sequence have shown that the latter is usually twice as accurate (see Fig. 1) as the builtin Matlab function, but it also exhibits slightly larger run times. However, the computation of an accurate displacement field is still significantly faster than a thorough optimisation of the mask point locations.

All methods have been implemented in Matlab. On a desktop computer with an Intel Xeon E5 CPU with 6 cores clocked at 1.60 GHz and 16 GB of memory the average run time of the Matlab optic flow implementation (10000 iterations at most) on the $512 \times 512 \times 10$ "Toy Vehicle" sequence from [1] was 41 s for each flow field between two frames. The implementation from [31] (8 coarse-to-fine levels with 10 warping steps at most) took 50 s. The tonal optimisation (360 iterations at most) took on average 32 s per frame. The optimal control based mask optimisation (1500 linearisation and 3000 primal dual iterations at most) required on average 6–30 s per linearisation and usually all 1500 linearisations are carried out. A complete optimisation takes therefore about 8 hours per frame. The large variations in the run times of the single linearisations stem from the fact that the sparser the mask becomes the more ill-posed the optimisation problem becomes and the more iterations are needed to achieve the desired accuracy. All in all, the mask optimisation is at least 600 times slower than the optic flow computation or the tonal optimisation.

4.2 Evaluation

We evaluate the proposed Algorithm 1 on several image sequences. First we consider the Yosemite sequence with clouds, available from [5]. Since the ground truth of the displacement field is completely known we can also analyse the impact of the quality of the flow on the reconstruction. Further, we evaluate the image sequences from the USC-SIPI Image Database [1]. The database contains four sequences of different length with varying image characteristics. For the latter sequences, no ground truth displacement field is known. As a such we can only report the reconstruction error in terms of squared error (MSE) and structural similarity index (SSIM) [33].

4.3 Influence of the Optical Flow

In Table 1 we present the evaluation of our approach on the Yosemite sequence for different choices of parameters of the mask optimisation algorithm and the corresponding reconstruction. In all these experiments we set μ to 1.25 (see [18] for a definition of this parameter) and ε to 10^{-9} in the mask optimisation algorithm. The regularisation weight in (5) was always optimised by means of a line search strategy.

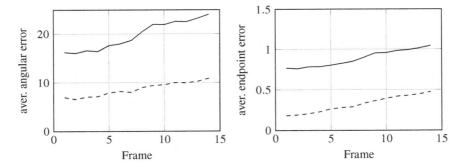

Fig. 1. Angular errors (in degree) and endpoint errors (in pixel width) in the optic flow field of the Yosemite sequence in-between frame i and $i + 1$ for the considered methods. The solid line corresponds to the implementation [32] and the dashed line to the implementation [31]. The regularisation weight was optimised for each pair of frames to minimise the angular error. The method from [31] is roughly twice as accurate as [32] but exhibits slightly higher run times.

The first column of the table lists the parameter λ which is responsible for the mask density and the second column contains the corresponding mask density in the first frame. The last five columns list the average reconstruction error over all 15 frames when (i) using an optimised mask obtained from the optimal control framework explained in [18] in all the frames, (ii) the optimised mask from the first frame shifted in accordance with the ground truth displacement field, (iii) the mask from the first frame shifted in accordance with the computed displacement fields for both considered implementations of the Horn and Schunck model, (iv) the mask from the first frame used for all subsequent frames (i.e. using a zero flow field), and (v) the mask from the first frame shifted by a random flow field within the same numerical range between each pair of frames as the ground truth.

All reconstructions in the upper half of the table have been done according to Algorithm 1. The lower half exhibits the same experiment but without the tonal optimisation in step 6 of Algorithm 1. Instead the original image data at the mask locations were used.

As expected, a higher mask density yields a smaller error in the reconstruction in all cases. Interestingly, we observe that computed flow fields are accurate enough to outperform in many cases the ground truth flow (rounded to the nearest grid point). The solution of the Horn and Schunck model in (5) involves the Laplacian and is a smooth flow field. We conjecture that, compared to the ground truth flow, this solution is more compatible with our choice for the inpainting procedure, which is also based on the Laplacian. The investigation of this possible synergy will require a more dedicated analysis in the future. When considering the plots in Fig. 2, one sees that there is a clear benefit to using computed flow fields in the first 7 or 8 frames of the sequence, when comparing to a flow field that is zero everywhere. Afterwards the iterative shifting of the masks has accumulated too many errors to outperform a zero flow. This suggests that the usage

Table 1. Evaluation of the Yosemite sequence. The density specifies the percentage of non-zero mask pixels in the first frame. The errors in the Horn & Schunck column correspond (in order) to the implementation from [31], and the builtin Matlab function [32]. The second-to-last column sets the flow field in every pixel to 0. The last column shows the error when using a random flow field in the same numerical range as the ground truth. The bottom part of the table represents the same experiments but without tonal optimisation in the reconstruction. The results show that on average, zero flow is almost as good as the methods where a computed flow field is used. These methods outperform zero flow especially in the first frames, cf. Fig. 2.

λ	Density	Average MSE with tonal optimisation				
		Optimal c	Exact shift	Horn & Schunck	Zero	Random
0.0030	15.75%	25.60	130.15	124.25/129.83	131.41	237.16
0.0063	9.26%	51.79	178.71	173.86/179.27	184.94	284.48
0.0125	5.51%	88.74	227.84	221.38/228.52	239.24	334.29
0.0250	3.11%	139.50	284.16	277.16/285.57	299.28	394.80
0.0030	15.75%	37.09	227.76	206.29/212.99	218.74	364.37
0.0063	9.26%	80.63	318.07	300.53/301.65	312.18	442.92
0.0125	5.51%	147.42	404.44	384.12/382.45	397.32	533.59
0.0250	3.11%	255.06	517.06	496.97/491.22	508.25	629.78

of a flow field is mostly beneficial for a short time prediction of the mask. Let us also note that the impact of the quality of the computed optical flow is visible over a shorter period within the first 5 frames.

Table 1 also shows that tonal optimisation has the expected beneficial influence. The tonal optimisation causes a global decrease in the error by as much as a factor 2, however it cannot compensate errors in the flow field.

Table 2. Evaluations of the MSE and SSIM on Image Sequences from the USC-SIPI Image Database [1]. An optimal mask is computed on the first frame and shifted according to the computed optic flow. The error for the "Toy Vehicle" sequence is not monotonically increasing due to strong occlusions in certain frames.

Sequence	Density	MSE				
		First	Last	Min.	Max	Mean
Yosemite	5.50%	88.38	307.85	88.38	307.85	225.52
	15.75%	25.33	187.10	25.33	187.10	129.83
Walter	5.18%	7.33	50.70	7.33	50.70	30.74
	27.06%	1.88	16.63	1.88	16.63	9.04
Toy Vehicle	3.00%	4.11	23.54	4.11	40.42	28.63
	20.00%	1.25	9.49	1.25	20.94	12.74
Plant (close)	3.23%	111.9	545.9	111.9	610.4	432.8
	5.36%	73.74	488.55	73.74	541.28	373.48
	19.05%	18.50	302.89	18.50	328.65	220.82
Plant (far)	3.65%	137.9	391.8	137.9	392.5	337.5
	6.64%	87.32	328.96	87.32	342.34	285.37
	23.18%	19.70	217.91	19.70	222.47	171.51
Sequence	Density	SSIM				
		First	Last	Min.	Max	Mean
Yosemite	5.50%	0.8196	0.5804	0.5804	0.8196	0.6509
	15.75%	0.9372	0.7721	0.7221	0.9372	0.7840
Walter	5.18%	0.9577	0.8867	0.8867	0.9577	0.9134
	27.06%	0.9810	0.9522	0.9522	0.9810	0.9640
Toy Vehicle	3.00%	0.9692	0.9343	0.9267	0.9692	0.9426
	20.00%	0.9856	0.9679	0.9584	0.9856	0.9695
Plant (close)	3.23%	0.7177	0.4250	0.4143	0.7177	0.4850
	5.36%	0.7843	0.4682	0.4599	0.7843	0.5370
	19.05%	0.9230	0.6356	0.6300	0.9230	0.7033
Plant (far)	3.65%	0.6739	0.4069	0.4069	0.6739	0.4533
	6.64%	0.7667	0.4672	0.4672	0.7667	0.5184
	23.18%	0.9300	0.6410	0.6410	0.9300	0.7033

4.4   Evaluation of the Reconstruction Error

Overall, the error evolution, as observed in the Yosemite sequence, is rather steady and predictable, even though such a behaviour can only be expected in well behaved sequences. The "Toy Vehicle" sequence from [1] exhibits strong occlusions and non-monotonic behaviour of the error, see Table 2. Nevertheless, the behaviour of the error evolution could be used to automatically detect frames after which a full mask optimisation becomes again necessary.

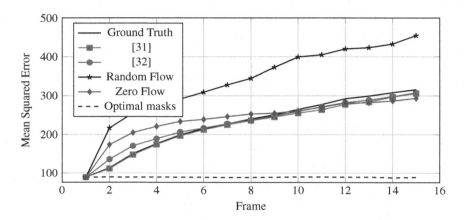

Fig. 2. Reconstruction error for the Yosemite sequence in each frame using a mask with density 5.5% shifted by different flow fields. The average angular error over all frames of the method from [31] is 8.59 and 5.27 if measured at mask points only. For the method from [32], the corresponding errors are 19.72 and 15.74. The error in the reconstruction is hardly influenced by the quality of the optic flow. The dashed line indicates the error in the reconstruction from an optimal mask.

Figure 3 presents an optimal mask for the last frame of the Yosemite sequence as well as the shifted mask. The corresponding reconstructions are also depicted. Fine details are lost with the reconstruction from the shifted mask. However, the overall structure of the scene remains preserved. We remark that the bright spots are due to our choice of the inpainting operator, see also [14].

Finally, Tab. 2 contains further evaluations of the MSE as well as the SSIM for the image sequences from [1]. Both measures show a similar behaviour. Denser masks have higher a SSIM (resp. lower MSE), and the SSIM decreases (resp. MSE increases) with the number of considered frames. The error evolution is usually monotone. However, if occlusions occur, then important mask pixels may be badly positioned or even completely absent. In that case notable fluctuations in the error will occur. This is especially visible in the "Toy Vehicle" sequence where the maximal error is not the error in the last frame.

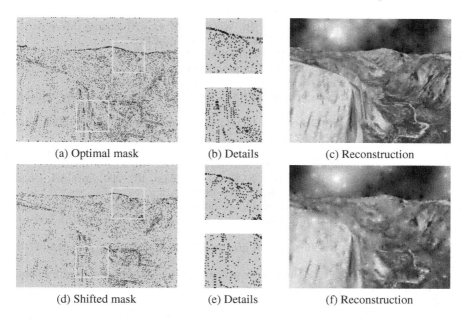

(a) Optimal mask   (b) Details   (c) Reconstruction

(d) Shifted mask   (e) Details   (f) Reconstruction

Fig. 3. *(a) and (d):* Inpainting masks (5.5% density) with *(b) and (e):* magnified details and *(c) and (f):* corresponding reconstructions for frame 15 of the Yosemite sequence. Black pixels indicate mask pixels, grey regions are to be inpainted. *Top:* optimal mask, *Bottom:* shifted mask.

5 Summary and Conclusion

Our work shows that it is possible to replace the expensive frame-wise computation of optimal inpainting data with the simple computation of a displacement field. Since run times to compute the latter are almost negligible when compared to the former, we gain a significant increase in performance. Our experiments demonstrate that simple and fast optic flow methods are sufficient for the task at hand, yet one may spend higher attention to movement of object boundaries.

In addition, the loss in accuracy along the temporal axis can easily be predicted. We may decide automatically when it becomes necessary to recompute an optimal mask while traversing the individual frames. We conjecture that the presented insights are certainly helpful in the future development of PDE-based video compression techniques.

References

1. The USC-SIPI image database (2014). http://sipi.usc.edu/database/
2. Andris, S., Peter, P., Weickert, J.: A proof-of-concept framework for PDE-based video compression. In: Proceedings of 32nd Picture Coding Symposium, IEEE (2016)

3. Belhachmi, Z., Bucur, D., Burgeth, B., Weickert, J.: How to choose interpolation data in images. SIAM J. Appl. Math. **70**(1), 333–352 (2009)
4. Bertalmío, M., Sapiro, G., Caselles, V., Ballester, C.: Image inpainting. In: Proceedings of 27th Annual Conference on Computer Graphics and Interactive Techniques, pp. 417–424. ACM Press/Addison-Wesley Publishing Company (2000)
5. Black, M.J.: Image sequences (2018). http://cs.brown.edu/people/mjblack/images.html
6. Bloor, M., Wilson, M.: Generating blend surfaces using partial differential equations. Comput. Aided Des. **21**(3), 165–171 (1989)
7. Brinkmann, E.-M., Burger, M., Grah, J.: Regularization with sparse vector fields: from image compression to TV-type reconstruction. In: Aujol, J.-F., Nikolova, M., Papadakis, N. (eds.) SSVM 2015. LNCS, vol. 9087, pp. 191–202. Springer, Cham (2015). https://doi.org/10.1007/978-3-319-18461-6_16
8. Bruhn, A., Weickert, J., Feddern, C., Kohlberger, T., Schnörr, C.: Variational optical flow computation in real time. IEEE Trans. Image Process. **14**(5), 608–615 (2003)
9. Carlsson, S.: Sketch based coding of grey level images. Signal Process. **15**, 57–83 (1988)
10. Chambolle, A., Pock, T.: A first order primal-dual algorithm for convex problems with applications to imaging. J. Math. Imaging Vis. **40**(1), 120–145 (2011)
11. Chen, Y., Ranftl, R., Pock, T.: A bi-level view of inpainting-based image compression. In: Kúkelová, Z., Heller, J. (eds.) Computer Vision Winter Workshop (2014)
12. Demaret, L., Iske, A., Khachabi, W.: Contextual image compression from adaptive sparse data representations. In: Gribonval, R. (ed.) Proceedings of SPARS 2009, Signal Processing with Adaptive Sparse Structured Representations Workshop (2009)
13. Facciolo, G., Arias, P., Caselles, V., Sapiro, G.: Exemplar-based interpolation of sparsely sampled images. In: Cremers, D., Boykov, Y., Blake, A., Schmidt, F.R. (eds.) EMMCVPR 2009. LNCS, vol. 5681, pp. 331–344. Springer, Heidelberg (2009). https://doi.org/10.1007/978-3-642-03641-5_25
14. Galić, I., Weickert, J., Welk, M., Bruhn, A., Belyaev, A., Seidel, H.-P.: Towards PDE-based image compression. In: Paragios, N., Faugeras, O., Chan, T., Schnörr, C. (eds.) VLSM 2005. LNCS, vol. 3752, pp. 37–48. Springer, Heidelberg (2005). https://doi.org/10.1007/11567646_4
15. Guillemot, C., Meur, O.L.: Image inpainting: overview and recent advances. IEEE Signal Process. Mag. **31**(1), 127–144 (2014)
16. Hoeltgen, L., et al.: Optimising spatial and tonal data for PDE-based inpainting. In: Bergounioux, M., Peyré, G., Schnörr, C., Caillau, J.B., Haberkorn, T. (eds.) Variational Methods, pp. 35–83. No. 18 in Radon Series on Computational and Applied Mathematics, De Gruyter (2016)
17. Hoeltgen, L., Peter, P., Breuß, M.: Clustering-based quantisation for PDE-based image compression. Signal Image Video Process. **12**(3), 411–419 (2018)
18. Hoeltgen, L., Setzer, S., Weickert, J.: An optimal control approach to find sparse data for laplace interpolation. In: Heyden, A., Kahl, F., Olsson, C., Oskarsson, M., Tai, X.-C. (eds.) EMMCVPR 2013. LNCS, vol. 8081, pp. 151–164. Springer, Heidelberg (2013). https://doi.org/10.1007/978-3-642-40395-8_12
19. Hoeltgen, L., Weickert, J.: Why does non-binary mask optimisation work for diffusion-based image compression? In: Tai, X.-C., Bae, E., Chan, T.F., Lysaker, M. (eds.) EMMCVPR 2015. LNCS, vol. 8932, pp. 85–98. Springer, Cham (2015). https://doi.org/10.1007/978-3-319-14612-6_7

20. Horn, B.K., Schunck, B.G.: Determining optical flow. Artif. Intell. **17**(1–3), 185–203 (1981)
21. Mainberger, M., Bruhn, A., Weickert, J., Forchhammer, S.: Edge-based compression of cartoon-like images with homogeneous diffusion. Pattern Recogn. **44**(9), 1859–1873 (2011)
22. Mainberger, M., Hoffmann, S., Weickert, J., Tang, C.H., Johannsen, D., Neumann, F., Doerr, B.: Optimising spatial and tonal data for homogeneous diffusion inpainting. In: Bruckstein, A.M., ter Haar Romeny, B.M., Bronstein, A.M., Bronstein, M.M. (eds.) SSVM 2011. LNCS, vol. 6667, pp. 26–37. Springer, Heidelberg (2012). https://doi.org/10.1007/978-3-642-24785-9_3
23. Masnou, S., Morel, J.M.: Level lines based disocclusion. In: Proceedings of 1998 IEEE International Conference on Image Processing, vol. 3, pp. 259–263. IEEE (1998)
24. Ochs, P., Chen, Y., Brox, T., Pock, T.: iPiano: inertial proximal algorithm for nonconvex optimization. SIAM J. Imaging Sci. **7**(2), 1388–1419 (2014)
25. Papenberg, N., Bruhn, A., Brox, T., Didas, S., Weickert, J.: Highly accurate optic flow computation with theoretically justified warping. Int. J. Comput. Vis. **67**(2), 141–158 (2006)
26. Peter, P., Hoffmann, S., Nedwed, F., Hoeltgen, L., Weickert, J.: From optimised inpainting with linear PDEs towards competitive image compression codecs. In: Bräunl, T., McCane, B., Rivera, M., Yu, X. (eds.) PSIVT 2015. LNCS, vol. 9431, pp. 63–74. Springer, Cham (2016). https://doi.org/10.1007/978-3-319-29451-3_6
27. Schmaltz, C., Peter, P., Mainberger, M., Ebel, F., Weickert, J., Bruhn, A.: Understanding, optimising, and extending data compression with anisotropic diffusion. Int. J. Comput. Vis. **108**(3), 222–240 (2014)
28. Shen, J., Chan, T.F.: Mathematical models for local nontexture inpaintings. SIAM J. Appl. Math. **62**(3), 1019–1043 (2002)
29. Strutz, T.: Bilddatenkompression. Vieweg (2002)
30. Sullivan, G.J., Wiegand, T.: Video compression - from concepts to the H. 264/AVC standard. Proc. IEEE. **93**, 18–31 (2005)
31. Sun, D.: (2018). http://research.nvidia.com/person/deqing-sun
32. The Mathworks Inc.: Compute optical flow using Horn-Schunck method (2018). https://de.mathworks.com/help/vision/ug/compute-optical-flow-using-horn-schunck-method.html
33. Wang, Z., Bovik, A.C., Sheikh, H.R., Simoncelli, E.P.: Image quality assessment: from error visibility to structural similarity. IEEE Trans. Image Process. **13**(4), 600–612 (2004)
34. Weinzaepfel, P., Jégou, H., Pérez, P.: Reconstructing an image from its local descriptors. In: Proceedings of 2011 IEEE Computer Society Conference on Computer Vision and Pattern Recognition, pp. 337–344. IEEE Computer Society Press (2011)

Compressing Audio Signals with Inpainting-Based Sparsification

Pascal Peter[✉], Jan Contelly, and Joachim Weickert

Mathematical Image Analysis Group, Faculty of Mathematics and Computer Science,
Saarland University, Campus E1.7, 66041 Saarbrücken, Germany
{peter,contelly,weickert}@mia.uni-saarland.de

Abstract. Inpainting techniques are becoming increasingly important for lossy image compression. In this paper, we investigate if successful ideas from inpainting-based codecs for images can be transferred to lossy audio compression. To this end, we propose a framework that creates a sparse representation of the audio signal directly in the sample-domain. We select samples with a greedy sparsification approach and store this optimised data with entropy coding. Decoding restores the missing samples with well-known 1-D interpolation techniques. Our evaluation on music pieces in a stereo format suggests that the lossy compression of our proof-of-concept framework is quantitatively competitive to transform-based audio codecs such as mp3, AAC, and Vorbis.

1 Introduction

Inpainting [23] originates from image restoration, where missing or corrupted image parts need to be filled in. This concept can also be applied for compression: Inpainting-based codecs [13] represent an image directly by a sparse set of known pixels, a so-called inpainting mask. This mask is selected and stored during encoding, and decoding involves a reconstruction of the missing image parts with a suitable interpolation algorithm. Such codecs [26,27] can reach competitive quality to JPEG [25] and JPEG2000 [31], which create sparsity indirectly via cosine or wavelet transforms.

In lossy audio compression, all state-of-the-art codecs use a time-frequency representation of the signal and are thereby also transform-based. This applies to mp3 (MPEG layer-III) [17], advanced audio coding (AAC) [18], and the open source alternative Vorbis [34]. They resemble the classic image codecs, whereas inpainting-based compression has so far not been explored for audio data. Therefore, we propose to select and store samples directly for sparse audio representations that act as known data for inpainting.

The transition from images to audio creates some unique challenges, since visual and audio data differ in many regards. As has been shown for 3-D medical images [27], the effectiveness of inpainting-based codecs increases with the dimensionality of the input data. Audio signals only have a single time dimension, but feature a high dynamic range compared to the 8-bit standard in images.

© Springer Nature Switzerland AG 2019
J. Lellmann et al. (Eds.): SSVM 2019, LNCS 11603, pp. 92–103, 2019.
https://doi.org/10.1007/978-3-030-22368-7_8

Moreover, more high-frequent changes can be expected in audio files. So far it is unknown how these differences affect the performance of interpolation and data selection strategies. In the following, we want to investigate the potential of inpainting-based audio compression.

Our Contribution. We propose a framework for lossy audio compression that is designed to transfer successful ideas from inpainting-based image compression to the audio setting. Based on this framework, we implement two proof-of-concept codecs that rely on different 1-D inpainting techniques: linear and cubic Hermite spline interpolation. Moreover, we integrate two core concepts from inpainting-based compression: sparsification [22] for the selection of known data locations, and tonal optimisation [22,27] of the corresponding values. Our input data, music pieces in a stereo format, contain significantly more data than standard test images. Therefore, we need to adapt the optimisation techniques to the audio setting. Localised inpainting allows us to decrease computation time significantly without affecting quality. Moreover, we propose a greedy sparsification approach with global error computation instead of the stochastic, local methods common in image compression. A combination of quantisation, run-length encoding (RLE) and context-mixing for storage of the known audio data complements these optimisation strategies. We compare our new codecs to mp3, AAC, and Vorbis w.r.t. the signal-to-noise ratio.

Related Work. The reconstructing of missing image areas was first investigated by Masnou and Morel [23] who referred to this as a disocclusion problem. Later, Bertalmío et al. [4] coined the term inpainting for this application of interpolation in image restoration. Many successful contemporary inpainting operators rely on partial differential equations (PDEs), for instance homogeneous diffusion [16] or edge-enhancing anisotropic diffusion (EED) [33] inpainting. An overview of methods can be found in the monograph of Schönlieb [28]. These methods achieve a filling-in effect based on physical propagation models. Another popular approach to inpainting are exemplar-based strategies that restore missing values by nonlocal copy-paste according to neighbourhood similarities [12]. For audio, the term inpainting is rarely used. However, interpolation is an important tool for signal restoration (e.g. [20]) or synthesis (e.g. [24]). Adler et al. [2] were the first to apply the core ideas of inpainting to audio signals: They presented a framework for filling in missing audio data from a sparse representation in the time domain. Their dictionary approach relies on discrete cosine and Gabor bases. There is also a vast number of publications that deal with specific scenarios such as removal of artefacts from damaged records [20] or noise in voice recognition applications [14] that rely on interpolation and signal reconstruction in a broader sense. Since a complete review is beyond the scope of this paper, we refer to the overview by Adler et al. [2].

It should be noted that while interpolation is not widely used for audio compression, linear prediction has been applied successfully, for instance by Schuller et al. [29]. However, the core technique behind common codecs are transforms. For instance, MPEG layer-III [17] (mp3) uses a modified discrete cosine transform (MDCT) on a segmented audio signal. The sophisticated non-uniform quantisation strategies and subsequent entropy coding are augmented with psychoacoustic analysis of the signal's Fourier transform. Advanced audio coding (AAC) [18], the successor of mp3, and the open source codec Vorbis [34] rely on the same basic principles. They also combine the MDCT with psychoacoustic modelling, but can achieve a better quality due to an increased flexibility of the encoder. A more detailed discussion of these codecs and a broader overview of the field can be found in the monograph by Spanias et al. [30].

A major competitor to transform-based audio compression arose from the so-called sinusoidal model [20,24]. It represents an audio signal as the weighted sum of wave functions that have time-adaptive parameters such as amplitude and frequency. Sinusoidal synthesis approaches have also been applied for compression. These use interpolation, but only in the domain of synthesis parameters. Such parametric audio compression [11] also forms the foundation of the MPEG4 HILN (harmonic and individual lines plus noise) standard [19]. HILN belongs to the class of object-based audio codecs: It is able to model audio files as a composition of semantic parts (e.g. chords). Vincent and Plumbley [32] transferred these ideas to a Bayesian framework for decomposing signals into objects.

Since we use ideas from image compression to build our framework, some inpainting-based codecs are closely related to our audio approach. The basic structure of the codec is inspired by the so-called exact mask approach by Peter et al. [26]. It is one of the few codecs that allows to choose and store known data without any positional restrictions. However, we do not use optimal control to find points as in [26]. Instead, we rely on sparsification techniques that resemble the probabilistic approach of Mainberger et al. [22]. Our greedy global sparsification differs from probabilistic sparsification in a few key points: In accordance to the findings of Adam et al. [1], we use global error computation instead of localised errors. Moreover, our approach does not rely on randomisation for spatial optimisation anymore, but is a deterministic, greedy process that always yields the same known data positions. Our choice of PAQ [21] for the storage of the sample data is motivated by several works that have evaluated different entropy coding techniques for inpainting-based compression [26,27].

Organisation of the Paper. We introduce our framework for inpainting-based audio compression in Sect. 2. Details on inpainting and data optimisation for two codecs follow in Sect. 3. These codecs are evaluated in Sect. 4, and we conclude our paper with a summary and outlook in Sect. 5.

2   A Framework for Inpainting-Based Audio Compression

Our proof-of-concept framework follows the common structure of current inpainting-based compression methods [26,27]. An input audio file $f: \Omega \subset \mathbb{N} \to \mathbb{Z}^c$ maps time coordinates $\Omega = \{1, ..., n\}$ to samples of the waveforms from $c \geq 1$ audio channels. Encoding aims to select and store a subset $K \subset \Omega$ of known data. During decoding, inpainting uses these data for a lossy restoration of the missing samples on $\Omega \setminus K$.

Input: Original audio signal $\boldsymbol{f} \in \mathbb{Z}^n$, desired final mask density d.
Initialisation: Mask $\boldsymbol{m} = (1, 1, ..., 1)^\top \in \{0, 1\}^n$ is full. Heap structure h is empty.
   Update set $A = \{1, ..., n\}$ contains all samples. Reconstruction error cache $\boldsymbol{e} = \boldsymbol{0}$.
Compute:
   do
      for all k in A **do**
         Remove the sample f_k temporarily from the mask (i.e. $m_k = 0$).
         Inpaint reconstruction $r(\boldsymbol{m}, \boldsymbol{f})$ with error e_k.
         Set $m_k \leftarrow 1$, add sample f_k to heap with reconstruction error $h_{e,k} \leftarrow e_k$.
      end for
      do
         Get the sample f_j from top of heap h.
         if $m_j = 1$ and $h_{e,j} = e_j$
         then remove f_j permanently from mask ($m_j \leftarrow 0$).
      while no sample has been permanently removed.
      Define A as set of samples whose error value is affected by removal of f_j.
   while pixel density of $\boldsymbol{m}$ larger than d.
Output: Mask $\boldsymbol{m}$ of density d.

Algorithm 1. Greedy Global Mask Sparsification.

In the following, we describe the individual steps of the encoding pipeline: This includes sample selection and optimisation, as well as efficient storage with prediction and entropy coding. Our optimisation are very flexible w.r.t. the inpainting method: We only assume that a deterministic inpainting algorithm computes a reconstruction $r(K, \boldsymbol{g}) \colon \Omega \to \mathbb{Z}$ from samples $\boldsymbol{g} \colon K \to \mathbb{Z}$ on the set K. In Sect. 3 we discuss the actual inpainting methods for our experiments in Sect. 4. Our codec is designed to be easily extendable with other inpainting techniques, such as the dictionary-based approach of Adler et al. [2].

Step 1: Sample Quantisation. First, we apply a coarse quantisation that reduces the number of sample values to $q \geq 2$. In order to adapt to the coding pipeline from image processing, this involves a global shift to a non-negative sample range $\{0, \ldots, p-1\}$ which is reversed again during decoding. A uniform quantisation partitions this sample range into q subintervals of length p/q, mapping to quantised values $\{0, \ldots, q-1\}$. For inpainting, we assign the quantisation index k to the corresponding quantised value ℓ from the original range:

$$\ell = \left\lfloor \frac{kp}{q} + \frac{1}{2} \right\rfloor. \tag{1}$$

All following optimisation steps use quantised values for inpainting and coding.

Step 2: Greedy Global Sparsification. A popular method for selecting the spatial location of known pixels in inpainting-based compression is probabilistic sparsification [22]: It starts with a full pixel mask, removes a set of randomly selected candidate points. and performs inpainting. A subset of candidates with the lowest local error are then permanently removed, since they are considered easy to reconstruct. We iterate these steps until the desired number of mask points, the target density, is reached. This method is easy to implement and supports all inpainting techniques. However, a recent analysis [1] revealed that the local error computation in this approach yields a suboptimal point selection. Therefore, we use a different sparsification strategy that relies on global error computation as proposed by Adam et al. [1]. Moreover, we remove the random component for candidate selection and obtain a *greedy global sparsification* that is described in Algorithm 1.

For each audio sample in the mask we compute the increase in the reconstruction error that would result from its removal. With this global reconstruction error, we sort the samples in a heap. Then we iteratively remove in every step the sample on top of the heap (i.e. with the lowest effect on the error) permanently from the mask. Afterwards, all mask samples that are affected by this removal are updated and reinserted into the heap. Which mask samples need to be updated depends on the inpainting approach (see Sect. 3). Note that, in order to avoid a costly purging of the heap in each iteration, the unmodified heap elements remain. If the sample at the top has been already removed or its error is not up-to-date, the algorithm moves on to the next one. For image compression, the global impact of individual changes in the mask cannot be considered due to runtime issues. In Sect. 3 we explain how we can reduce the computational load with an update strategy for the audio setting.

Step 3: Sample Optimisation. It is well-known in inpainting-based image compression that optimising not only the location of known data, but also the function value in the stored pixels can yield large improvements [7,15,22,27]. Since we aim for a flexible framework, we use the technique from [27], as it does not require a specific inpainting technique. It performs a random walk over all mask samples: If a change to the next higher or lower quantisation level improves the reconstruction, it is kept, otherwise it is reverted. As for sparsification, we address runtime questions in Sect. 3.

Step 4: Location Encoding. Current state-of-the-art codecs [26] employ block coding in 2-D to store *exact masks* with unrestricted placement of known points. A natural substitute for this in 1-D is run-length encoding (RLE) [5]. We represent the mask as a sequences of ones (known samples) and zeroes (unknown samples). In sparse masks, we expect isolated ones with long runs of zeroes inbetween. Therefore, we only encode runs of zeroes together with a terminating one. This allows us to store the mask as a sequence of 8bit symbols. Runs up to length 254 require only one symbol while longer runs are split accordingly (e.g. 300 is represented by 255, 45).

Step 5: Prediction and Entropy Encoding. Due to recurring patterns in audio files (in particular for music recordings), prediction can be used to achieve higher compression ratios. To this end, many publications on inpainting-based compression (e.g. [26,27]) apply the context-mixing algorithm PAQ [21]. It predicts the next bit in a stream containing different data types according to

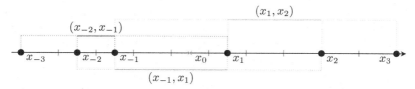

(a) local sparsification update for cubic Hermite spline

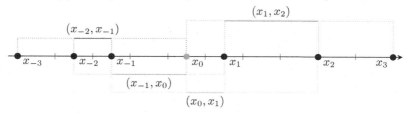

(b) local sample update for cubic Hermite spline

Fig. 1. Local Update Intervals for the cubic Hermite spline. The coloured solid lines mark the update intervals and the associated dotted lines show which known samples are involved in the corresponding reconstruction.

numerous predefined and learned contexts. The weighting of these contexts adapts to the local file content with a gradient descent on the coding cost. We use PAQ for an additional joint encoding of the output data from Steps 3–4.

3 Localised Sample Optimisation with 1-D Inpainting

So far, we have not specified concrete inpainting operators for our general framework. In the following, we transfer popular inpainting techniques from image compression to the audio setting. For these inpainting operators, we develop new techniques for the acceleration of the corresponding data optimisation.

Inpainting Techniques. For our first proof-of-concept implementation of the framework, we explore the potential of successful inpainting approaches from image compression. So far, three operators have shown convincing performance [7,13,27]: homogeneous diffusion [16], biharmonic [10], and edge-enhancing anisotropic diffusion (EED) inpainting [33]. EED has been particularly successful, since it allows to reconstruct image edges due to a direction dependent propagation. However, due to the 1-D nature of audio data, EED is not an option.

Homogeneous diffusion inpainting keeps all of the known data points on $K \subset \Omega$ unchanged, while the unknown data on $\Omega \setminus K$ must fulfil the Laplace equation $\Delta u = 0$ with $\Delta u = \partial_{xx} u + \partial_{yy} u$. In 1-D, this implies a vanishing second order derivative, which leads to a straightforward linear interpolation between the known data points. This comes down to a minimisation of the energy

$$E_L(u) = \int_{\Omega \setminus K} (u'(x))^2 \, dx. \tag{2}$$

In the following sections we benefit from the *compact support* of the corresponding interpolation function: For the reconstruction $u(x)$ at a location $x \in \Omega \setminus K$ in the inpainting domain, we need a small amount of neighbouring known values (x_k, u_k). In the following, the indices $k = \pm 1, \pm 2, \ldots$ denote the respective closest known samples in positive/negative x-direction. For linear interpolation, we only require the two known samples (x_{-1}, u_{-1}) and (x_1, u_1) to obtain the reconstruction

$$u(x) = \frac{x - x_{-1}}{x_1 - x_{-1}} u_1 + \left(1 - \frac{x - x_{-1}}{x_1 - x_{-1}}\right) u_{-1}. \tag{3}$$

Biharmonic inpainting is a higher-order approach that imposes the constraint $-\Delta^2 u = 0$ to the inpainted data, thereby providing a smoother reconstruction compared to the homogeneous case. Cubic splines are a natural 1-D counterpart to this approach. They have been originally motivated by a physical elasticity model for draftman's splines [9] and minimise the energy

$$E_{CS}(u) = \int_{\Omega \setminus K} (u''(x))^2 \, dx. \tag{4}$$

However, since we aim to reach a similar locality as for the linear interpolation, we consider a specific variant of cubic splines, the cubic Hermite spline interpolation [6] (Catmull-Rom spline). It yields an interpolant with C^1-smoothness using a finite support. Since it does not require equidistant sampling, it is therefore compatible with sparsification. With $\alpha := \frac{x - x_{-1}}{x_1 - x_{-1}}$, the interpolant of cubic Hermite spline interpolation is

$$u(x) = (2\alpha^3 - 3\alpha^2 + 1)u_{-1} + (\alpha^3 - 2\alpha^2 + \alpha)\frac{u_1 - u_{-2}}{x_1 - x_{-2}}$$
$$+ (-2\alpha^3 + 3\alpha^2)u_1 + (\alpha^3 - \alpha^2)\frac{u_2 - u_{-1}}{x_2 - x_{-1}}. \tag{5}$$

For the interpolation techniques above, we round to the next 16bit integer. Note that this rounding is explicitly not restricted to the quantisation levels according to Step 3 of our compression pipeline from Sect. 2. From a very small set of quantised values, the inpainting can potentially recover a much broader sample range, if the known data is chosen appropriately. In the following, we discuss how the locality of our inpainting methods can accelerate the data optimisation.

Local Interpolation Updates. Both the greedy sparsification from Algorithm 1 and the sample optimisation from Step 3 require a global reconstruction error. Recomputing the whole reconstruction after a change of a single mask point is a significant drawback of these approaches in 2-D. However, in our 1-D audio signal setting using interpolations methods with finite support, the influence of each mask sample is limited. In the following, we always assume a sample x_0 is changed by an optimisation algorithm, and $x_{-1}, x_{-2}, \ldots$ denote it left mask neighbours while $x_1, x_2, \ldots$ are its right mask neighbours.

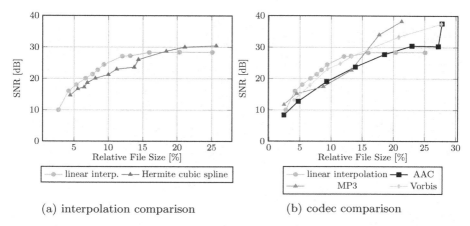

(a) interpolation comparison   (b) codec comparison

Fig. 2. SNR Comparisons. Fig. 2(a) reveals that linear interpolation mostly outperforms the cubic Hermite spline. In Fig. 2(b), our inpainting-based codec with linear
interpolation compares favourably with established methods like AAC, mp3, and Vorbis for low to medium compression ratios.

In a sparsification step, we remove the known sample value y_0 with time coordinate x_0 from the mask. For linear interpolation, this removal affects exactly
the reconstruction of the samples $x \in (x_{-1}, x_1)$, which are now reconstructed
with the known data x_{-1} and x_1. For the cubic Hermite spline, the situation
is similar, but due to the larger support, the interval (x_{-2}, x_2) is affected now.
Moreover, it has to be split into three subintervals that are inpainted with different combinations of the known data $x_{-3}, ..., x_3$ (see Fig. 1(a)).

The update strategy for sample optimisation follows the same principle, but
more subintervals need to be considered, since we now change the value y_0 at
location x_0 instead of removing the sample completely. Thus, for linear interpolation, the optimisation algorithm needs to recompute the intervals (x_{-1}, x_0) and
(x_0, x_1) with the new known sample (x_0, y_0). Since the cubic Hermite spline relies
on four samples, the sample change affects four intervals: (x_{-2}, x_{-1}), (x_{-1}, x_0),
(x_0, x_1), and (x_1, x_2). Figure 1(b) illustrates the associated samples. Note that
we also need the samples x_{-3} and x_3 to compute these reconstructions.

4   Experiments

Experimental Setup. We present detailed results for a royalty-free sound file
of the song *Exploring the Forest* [8] (linear 16bit pulse coded modulation (PCM)
with 44.100 kHz sampling rate and two channels). Results for additional music
pieces from a variety of genres as well as playable soundfiles are available online
as supplementary material[1]. As a quantitative measure, we use the signal to

---

[1] https://www.mia.uni-saarland.de/Publications/peter-ssvm19-supplement.zip.

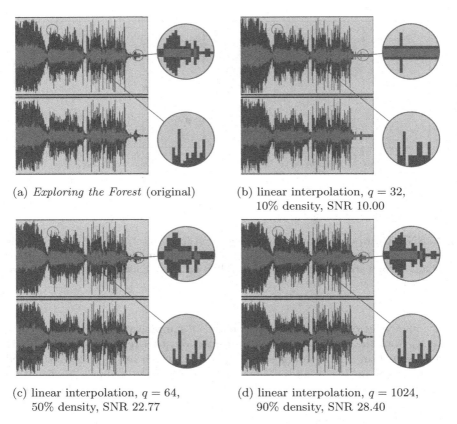

(a) *Exploring the Forest* (original)

(b) linear interpolation, $q = 32$, 10% density, SNR 10.00

(c) linear interpolation, $q = 64$, 50% density, SNR 22.77

(d) linear interpolation, $q = 1024$, 90% density, SNR 28.40

Fig. 3. Visual Comparison of Waveforms. Waveforms of the original file *Exploring the Forest* and the reconstructions corresponding to low, medium, and high compression ratios from Fig. 2.

noise ratio (SNR) that is defined by

$$\text{SNR}(f, g) = 10 \log_{10} \left(\frac{\sum_{i=1}^{n} f_i^2}{\sum_{i=1}^{n} (f_i - g_i)^2} \right). \tag{6}$$

Comparison of Inpainting Methods. In a first experiment, we compare the performance of the two inpainting methods from Sect. 3. A quantitative SNR comparison in Fig. 2(a) reveals that linear interpolation yields almost the same SNR and outperforms the Hermite cubic spline for small compressed file sizes. The increased smoothness of the cubic spline comes at the cost of over- and undershoots close to the known samples. These can only be compensated adequately if most of the samples are known. Moreover, for linear interpolation, the best density and quantisation parameters increase proportionally to the file size. Hermite cubic spline interpolation is more sensitive in this regard. Therefore, we choose linear interpolation for our comparison to established codecs.

Comparison to Established Codecs. Our second series of experiments evaluates the compression performance of our best inpainting codec (with linear interpolation) to the established codecs mp3, Vorbis, and AAC. Our evaluation in Fig. 2(b) yields a surprising result: For medium to high compression ratios, our codec surpasses all three transform-based approaches w.r.t. the SNR. This demonstrates that concepts from inpainting-based compression can be viable in an audio setting, even with simple inpainting methods. For small compression ratios, our codec falls slightly below the SNR of AAC. Inpainting-based methods show similar behaviour for near-lossless coding of images (see e.g. [27]). This is natural, since the impact of inpainting diminishes for dense masks.

Figure 3 provides a visualisation of the inpainting results for the lowest and highest compression ratios from Fig. 2(a) with linear interpolation. On first glance, even with a low density of 10 % and a very coarse quantisation ($q = 32$) the reconstructed waveform in Fig. 3(b) looks similar to the original in Fig. 3(a). However, some of the peaks are flattened (especially apparent at the end of the signal). On a temporal average, this is still close to the original signal in terms of SNR, but there are some audible artefacts like background noise. A higher density and finer quantisation leads to increasingly improved results in Fig. 3(c) and (d). Simple linear interpolation can reproduce the original waveform from carefully optimised known samples with surprising accuracy.

5   Conclusions and Outlook

Our modular framework for audio compression demonstrates the potential of inpainting with data optimisation for the sparse representation of sample data. Even with fairly simple ingredients, our proof-of-concept codecs are able to compete with established audio codecs w.r.t. quantitative analysis. In particular, this discovery is relevant for recent approaches in inpainting-based video compression [3]: Inpainting-based audio codecs would augment them in a natural way by offering a consistent way of encoding the corresponding audio tracks.

In our future work, we plan to investigate more sophisticated inpainting techniques that have been designed specifically for the audio setting (e.g. [2]), and address practical issues such as random access. Moreover, we will incorporate the psychoacoustic modelling used by transform-based codecs: Prefiltering the signal to eliminate frequencies that are unimportant for human perception might further improve the performance of inpainting-based audio compression.

Acknowledgements. We thank Jan Østergaard (Aalborg University) for the valuable discussions that allowed us to improve our work. This project has received funding from the European Research Council (ERC) under the European Union's Horizon 2020 research and innovation programme (grant agreement no. 741215, ERC Advanced Grant INCOVID).

References

1. Adam, R.D., Peter, P., Weickert, J.: Denoising by inpainting. In: Lauze, F., Dong, Y., Dahl, A.B. (eds.) SSVM 2017. LNCS, vol. 10302, pp. 121–132. Springer, Cham (2017). https://doi.org/10.1007/978-3-319-58771-4_10
2. Adler, A., Emiya, V., Jafari, M.G., Elad, M., Gribonval, R., Plumbley, M.D.: Audio inpainting. IEEE Trans. Audio Speech Lang. Process. **20**(3), 922–932 (2012)
3. Andris, S., Peter, P., Weickert, J.: A proof-of-concept framework for PDE-based video compression. In: Proceedings 32nd Picture Coding Symposium (PCS 2016), Nuremberg, Germany, pp. 1–5, December 2016
4. Bertalmío, M., Sapiro, G., Caselles, V., Ballester, C.: Image inpainting. In: Proceedings SIGGRAPH 2000, New Orleans, LI, pp. 417–424, July 2000
5. Capon, J.: A probabilistic model for run-length coding of pictures. IRE Trans. Inf. Theor. **5**(4), 157–163 (1959)
6. Catmull, E., Rom, R.: A class of local interpolating splines. In: Barnhill, R.E., Riesenfeld, R.F. (eds.) Computer Aided Geometric Design, pp. 317–326. Academic Press, New York (1974)
7. Chen, Y., Ranftl, R., Pock, T.: A bi-level view of inpainting-based image compression. In: Proceedings 19th Computer Vision Winter Workshop, Křtiny, Czech Republic, pp. 19–26, February 2014
8. Crowley, P.: Exploring the Forest (2013). Audio file, available under http://petercrowleyfantasydream.jimdo.com
9. de Boor, C.: A Practical Guide to Splines, Applied Mathematical Sciences, vol. 27. Springer, New York (1978)
10. Duchon, J.: Interpolation des fonctions de deux variables suivant le principe de la flexion des plaques minces. RAIRO Analyse Numérique **10**(3), 5–12 (1976)
11. Edler, B., Purnhagen, H.: Parametric audio coding. In: Proceedings International Conference on Communication Technology Proceedings (WCC-ICCT 2000), vol. 1, Beijing, China, pp. 614–617, August 2000
12. Efros, A.A., Leung, T.K.: Texture synthesis by non-parametric sampling. In: Proceedings Seventh IEEE International Conference on Computer Vision, vol. 2, Corfu, Greece, pp. 1033–1038, September 1999
13. Galić, I., Weickert, J., Welk, M., Bruhn, A., Belyaev, A., Seidel, H.P.: Image compression with anisotropic diffusion. J. Math. Imaging Vis. **31**(2–3), 255–269 (2008)
14. Gemmeke, J.F., Van Hamme, H., Cranen, B., Boves, L.: Compressive sensing for missing data imputation in noise robust speech recognition. IEEE J. Sel. Top. Sign. Process. **4**(2), 272–287 (2010)
15. Hoeltgen, L., et al.: Optimising spatial and tonal data for PDE-based inpainting. In: Bergounioux, M., Peyré, G., Schnörr, C., Caillau, J.P., Haberkorn, T. (eds.) Variational Methods in Image Analysis, De Gruyter, Berlin, pp. 35–83 (2017)
16. Iijima, T.: Basic theory on normalization of pattern (in case of typical one-dimensional pattern). Bull. Electrotech. Lab. **26**, 368–388 (1962). (in Japanese)
17. ISO/IEC: Information technology - coding of moving pictures and associated audio - part 3: Audio, Standard, ISO/IEC 11172–3 (1992)
18. ISO/IEC: Information technology - generic coding of moving pictures and associated audio - part 7: Avanced audio coding, Standard, ISO/IEC 13818–7 (1992)
19. ISO/IEC: Information technology - coding of audio-visual objects - part 3: Audio, standard, ISO/IEC 14496–3 (2001)
20. Maher, R.: A method for extrapolation of missing digital audio data. J. Audio Eng. Soc. **42**(5), 350–357 (1994)

21. Mahoney, M.: Adaptive weighing of context models for lossless data compression. Technical report, CS-2005-16, Florida Institute of Technology, Melbourne, FL, December 2005
22. Mainberger, M., et al.: Optimising spatial and tonal data for homogeneous diffusion inpainting. In: Bruckstein, A.M., ter Haar Romeny, B.M., Bronstein, A.M., Bronstein, M.M. (eds.) SSVM 2011. LNCS, vol. 6667, pp. 26–37. Springer, Heidelberg (2012). https://doi.org/10.1007/978-3-642-24785-9_3
23. Masnou, S., Morel, J.M.: Level lines based disocclusion. In: Proceedings 1998 IEEE International Conference on Image Processing, vol. 3, Chicago, IL, pp. 259–263, October 1998
24. McAulay, R., Quatieri, T.: Computationally efficient sine-wave synthesis and its application to sinusoidal transform coding. In: Proceedings International Conference on Acoustics, Speech, and Signal Processing (ICASSP-88), New York, NY, pp. 370–373, April 1988
25. Pennebaker, W.B., Mitchell, J.L.: JPEG: Still Image Data Compression Standard. Springer, New York (1992)
26. Peter, P., Hoffmann, S., Nedwed, F., Hoeltgen, L., Weickert, J.: Evaluating the true potential of diffusion-based inpainting in a compression context. Sign. Process. Image Commun. **46**, 40–53 (2016)
27. Schmaltz, C., Peter, P., Mainberger, M., Ebel, F., Weickert, J., Bruhn, A.: Understanding, optimising, and extending data compression with anisotropic diffusion. Int. J. Comput. Vis. **108**(3), 222–240 (2014)
28. Schönlieb, C.B.: Partial Differential Equation Methods for Image Inpainting, Cambridge Monographs on Applied and Computational Mathematics, vol. 29. Cambridge University Press, Cambridge (2015)
29. Schuller, G.D., Yu, B., Huang, D., Edler, B.: Perceptual audio coding using adaptive pre-and post-filters and lossless compression. IEEE Trans. Speech Audio Process. **10**(6), 379–390 (2002)
30. Spanias, A., Painter, T., Atti, V.: Audio Signal Processing and Coding. Wiley, Hoboken (2006)
31. Taubman, D.S., Marcellin, M.W. (eds.): JPEG 2000: Image Compression Fundamentals, Standards and Practice. Kluwer, Boston (2002)
32. Vincent, E., Plumbley, M.D.: Low bit-rate object coding of musical audio using Bayesian harmonic models. IEEE Trans. Audio Speech Lang. Process. **15**(4), 1273–1282 (2007)
33. Weickert, J.: Theoretical foundations of anisotropic diffusion in image processing. In: Kropatsch, W., Klette, R., Solina, F., Albrecht, R. (eds.) Theoretical Foundations of Computer Vision, Computing Supplement, vol. 11, pp. 221–236. Springer, Vienna (1996). https://doi.org/10.1007/978-3-7091-6586-7_13
34. Xiph.Org Foundation: Vorbis I specification (2015). https://xiph.org/vorbis/doc/Vorbis_I_spec.html

Alternate Structural-Textural Video Inpainting for Spot Defects Correction in Movies

Arthur Renaudeau[1(✉)], François Lauze[2], Fabien Pierre[3], Jean-François Aujol[4], and Jean-Denis Durou[1]

[1] IRIT, UMR CNRS 5505, Université de Toulouse, Toulouse, France
arthur.renaudeau@irit.fr
[2] DIKU, University of Copenhagen, Copenhagen, Denmark
[3] LORIA, UMR CNRS 7503, Université de Lorraine,
Inria projet Magrit, Nancy, France
[4] Univ. Bordeaux, Bordeaux INP, CNRS, IMB, UMR 5251, 33400 Talence, France

Abstract. We propose a new video inpainting model for movies restoration application. Our model combines structural reconstruction with a diffusion-based method and textural reconstruction with a patch-based method. Both proposed energies (one for each method) are alternatively minimized in order to preserve the overall structure while adding textural refinement. While the structural reconstruction is obtained jointly with optical flow computation with several proximal approaches, the textural reconstruction is processed by a variational non-local approach. Preliminary results on different Middlebury frames show quality improvement in the reconstruction.

1 Introduction

Video inpainting is a key issue for the movie industry, as it could help to automate the restoration of films that have suffered significant degradation (see Fig. 1), or the use of certain special effects that require the removal of elements for action scenes. Video inpainting, as every video processing, is increasingly being studied thanks to the power of processors and GPUs to perform large-scale calculations. Until now, video inpainting techniques have used separately diffusion-based methods with motion estimation, or patch-based methods with 3D patches to take into account temporal redundancy (similarity between consecutive frames), but without any explicit motion estimation this time. This motion can give a lot of information to recover data so it is a really good help for inpainting. However, in order to estimate motion, full data is needed and this is why this estimation must be processed at the same time as inpainting, which represents the main challenge.

In this paper, we aim to restore spot defects on previously digitized films. Each of these defects appears only in one frame and not in those located just before or just after. To eliminate them, our approach consists in combining

© Springer Nature Switzerland AG 2019
J. Lellmann et al. (Eds.): SSVM 2019, LNCS 11603, pp. 104–116, 2019.
https://doi.org/10.1007/978-3-030-22368-7_9

a diffusion-based video inpainting model which jointly computes optical flow, with a patch-based model with 2D patches and shift maps to the temporally neighbouring images. With this approach, our model only needs the two adjacent frames to reconstruct the damaged area. While each model taken separately has drawbacks in terms of reconstruction quality, combining them both gives better results.

Fig. 1. Example of digitized frames from an old movie of the Cinémathèque de Toulouse with a defect in the central frame.

After reviewing related approaches in Sect. 2, we present our combined model in Sect. 3. Our numerical strategy for solving this variational problem, which is presented in Sect. 4, is based on alternating optimization of the diffusion-based and patch-based models. Preliminary experiments on Middlebury frames with added defects are conducted in Sect. 5, which confirm the interest of using both models together.

2   Related Work

Inpainting is the name given to the technique of filling damaged or missing areas in an image. The term "inpainting" is only used from 2000 in [4], by analogy with the restoration process used in the field of art, after that of "disocclusion" in [21] in 1998. The first inpainting applications came from diffusion models for denoising, which date back to the early 1990s. This field of research has been very active in recent years, stimulated by many applications: removal of scratches or of text superimposed on an image, restoration of an altered image following a transmission, elimination of objects in an editing context for diminished reality.

Filling the area to be restored is an ill-posed inverse problem because there is not a single well-defined solution. It is therefore necessary to introduce a priori knowledge into the model. All existing methods are guided by the assumption that pixels located in known and missing parts of the image share the same statistical properties or geometric structures. This hypothesis is reflected in different local or global a priori assumptions, in order to obtain a restored image that is visually plausible. In diffusion-based inpainting, one wants to propagate the information contained in the pixels from the edge of the damaged area to the inside of this area. Total variation for inpainting was introduced in [11] to block the diffusion at the edges of the objects and recover piecewise constant data. The extension of diffusion-based inpainting to video started with [12,17,18] where motion is simultaneously estimated to fill the damaged area. From the well-known optical flow model of [16] with L^2 smooth regularization, [2] switched to L^1 norms to preserve discontinuities of the different motions, which was later solved using proximal algorithms in [23]. Very recently, [7] chose a complete TV-L^1 model to solve motion estimation and image reconstruction, using also proximal algorithms.

However, these diffusion-based models are limited because they cannot handle textures. This is why models based on full or partial patch copying have been developed (see more details in [8]) to keep details at high frequencies, starting with texture synthesis in [15] and then local patch-based inpainting (patches are only looked for in a neighbourhood of the defect area) in [13] with priority for filling based on the magnitude of the spatial gradient at the edges of the area to be filled. Finally, recent methods consider a mixture of patches following a spatial non-local search as in [1]. This patch-based approach for inpainting was also extended to videos in [19,22] with 3D patches to include some temporal similarity in patch comparisons. If these models yield better results in the recovery of texture, they are however highly dependent on the initial filling of the area, so as not to remain blocked in a local minimum for the solution, which usually fails at reconstructing regular structures. Moreover, considering the patch size is also an important criterion in order to recover textures with different statistical properties.

In the image inpainting context, the idea of mixing both approaches has already been developed in [14] where diffusion and texture filling are sequentially processed, in [6] where the image is decomposed into cartoon and texture before being filled separately, and in [9] where texture is filled guided by the level lines. In the video denoising context, [5] uses patches combined by the computation of a structural optical flow of [23]. Our aim in this paper is also to get the best of both worlds by combining diffusion to recover structure with patches for texture, with also diffusion-based and patch-based approaches for motion estimation.

3 Statement of the Problem

Let us define a sequence of 3 successive color frames $\{u_b, u, u_f\}$ as functions $\Omega \subset \mathbb{R}^2 \to \mathbb{R}^3$ with bounded variation, where u_b is the backward frame, u is the current frame containing the defect to be inpainted, and u_f is the forward frame. This defect area is defined by $O \subset \Omega$. The problem to be solved is as follows:

$$\left\{ v_b^*, \Gamma_b^*, u^*, v_f^*, \Gamma_f^* \right\} = \underset{v_b, \Gamma_b, u, v_f, \Gamma_f}{\operatorname{argmin}} \left\{ E_S(v_b, u, v_f) + E_T(\Gamma_b, u, \Gamma_f) \right\} \quad (1)$$

where E_S represents the energy for structural reconstruction, minimized channel by channel (or using the luminance channel for the motion estimation), and E_T the energy for textural reconstruction, minimized using color frames directly. The different variables are explained in the following subsections.

3.1 Structural Reconstruction Energy

The model chosen for E_S is based on the works of [7,18], adding a symmetry between the optical flows (reminding u is here only one channel):

$$E_S(v_b, u, v_f) = \mu \int_\Omega |\nabla u(x)| \; \mathrm{d}x + \lambda \int_\Omega |Jv_b(x)| \; \mathrm{d}x + \lambda \int_\Omega |Jv_f(x)| \; \mathrm{d}x$$
$$+ \int_\Omega |u_b(x + v_b(x)) - u(x)| \; \mathrm{d}x + \int_\Omega |u_f(x + v_f(x)) - u(x)| \; \mathrm{d}x \qquad (2)$$

under the constraint $u = u^0$ over $O^c = \Omega \backslash O$ to preserve the healthy part of the frame. The terms containing the motion field $v_b : \Omega \to \mathbb{R}^2$ (respectively v_f) represent the L^1 regularized optical flow constraint, proposed by [23], between the current frame u and the backward frame u_b (respectively the forward frame u_f), but using Jacobian matrices, as a rewriting of the formula in [10]. Every integral contains a discrete norm $|.|$ defined as $|M| = \sqrt{\sum_{i,j} m_{i,j}^2}$, which means either an absolute value, a vector norm or a Frobenius norm depending on the case. The parameters λ and μ are used to define the trade-off between data fitting and regularization.

3.2 Textural Reconstruction Energy

The second energy E_T is an extension to video of the work of [1], using directly the color frames (not channel by channel as for the structural energy). Here the search for optimal patches is no longer carried out in a spatial neighbourhood around the defect, but in a temporal neighbourhood (in the previous frame u_b and the next frame u_f). While in [19,22] the 3D patch search is not limited in time distance, here we focus only on 2D patches in the backward and forward frames. In the current frame u, the central pixel x of the patch $p_u(x)$ concerned by the search of the optimal patch in the neighbour frames u_b and u_f is in the area $\tilde{O}$ which is O expanded by half a patch width, in order to propagate patches containing sufficient healthy data:

$$E_T(\Gamma_b, u, \Gamma_f) = \int_{\tilde{O}} \omega(x) \, \varepsilon \Big[p_{u_b}(\Gamma_b(x)) - p_u(x) \Big] \; \mathrm{d}x$$
$$+ \int_{\tilde{O}} (1 - \omega(x)) \, \varepsilon \Big[p_{u_f}(\Gamma_f(x)) - p_u(x) \Big] \; \mathrm{d}x \qquad (3)$$

where Γ_b and Γ_f are shift maps, respectively, from u to u_b and from u to u_f, $\omega(x) \in [0,1]$ is a weight between the two possible reconstructions of u from forward or backward frames (see Sect. 4.3), and ε represents the chosen distance between patches. Here in (4), ε is a convolution between the squared difference of the patches $(p_{u_b}(\Gamma_b(x)) - p_u(x))^2$ and a Gaussian kernel g_a of standard deviation a for a non-local means reconstruction:

$$\varepsilon \Big[p_{u_b}(\Gamma_b(x)) - p_u(x) \Big] = \int_{\Omega_p} g_a(x_p) \, [u_b(\Gamma_b(x) - x_p) - u(x - x_p)]^2 \; \mathrm{d}x_p \qquad (4)$$

where $x_p \in \Omega_p$ denotes the coordinates of a pixel inside the patch $p_u(x)$ relative to its center x. Minimizing E_S and E_T at the same time is a very complex problem with no proof of existence and uniqueness of a solution for (1), but one only wishes to obtain an approximate numerical solution by minimizing E_S and E_T alternatively, using the result u of the minimization of one to initialize the other one.

4   Optimization

Applying inpainting to large defects or estimating motions requires a coarse-to-fine framework. In a video context, it is even more important to follow this strategy in order to initialize every variable correctly. The idea here is to downsample enough the frames to consider that motions are small enough between them. At such a resolution (the level $L \to L_{\max}$), considering Gaussian filtering to eliminate high frequencies in the downsampling step, the structural reconstruction works well whereas the textural one is not efficient. On the other hand, at higher resolution ($L \to 0$), we want to put more emphasis on texture. This is why our algorithm can choose a maximum resolution level $L_{\max}^{\text{texture}}$ to start texture reconstruction and a minimum resolution level $L_{\min}^{\text{structure}}$ to stop structural reconstruction, with $L_{\max}^{\text{texture}} \geq L_{\min}^{\text{structure}} - 1$ to ensure that at least one of the reconstructions is applied at each resolution. Consequently, at a given resolution level $L > 0$, our algorithm applies only one reconstruction or both in a row:

Algorithm 1. Reconstruction at resolution levels L and $L - 1$

	6: **if** $L \leq L_{\max}^{\text{texture}}$ **then**
1: **if** $L \geq L_{\min}^{\text{structure}}$ **then**	7: $\Gamma_b^*, \Gamma_f^* \leftarrow \underset{\Gamma_b, \Gamma_f}{\operatorname{argmin}} \left\{ E_T \left(\Gamma_b, u, \Gamma_f \right) \right\}$
2: $v_b^*, u^*, v_f^* \leftarrow \underset{v_b, u, v_f}{\operatorname{argmin}} \left\{ E_S \left(v_b, u, v_f \right) \right\}$	8: $\Gamma_b, u, \Gamma_f \leftarrow \text{upsampling}(\Gamma_b^*, u, \Gamma_f^*)$
3: $v_b, v_f \leftarrow \text{upsampling}(v_b^*, v_f^*)$	9: $u^* \leftarrow \underset{u}{\operatorname{argmin}} \left\{ E_T \left(\Gamma_b, u, \Gamma_f \right) \right\}$
4: $u \leftarrow u^*$	10: $u \leftarrow u^*$
5: **end if**	11: **else**
	12: $u \leftarrow \text{upsampling}(u)$
	13: **end if**

Notice that shift maps upsampling is carried out with the nearest neighbour interpolation, while the bicubic one is used for the other upsamplings. The different minimizations of u, v_b, v_f, Γ_b and Γ_f in Algorithm 1 are explained below.

4.1   Motion Estimation

In order to minimize E_S with respect to the motion vector v_b, we proceed as in [23] by linearizing inside the two absolute differences in (2). However, this linearization is only possible in the case of small displacements. This is why a constant motion vector v_b^0 is introduced, close to v_b, around which the latter is estimated, to get:

$$E_S(v_b, u, v_f) = \mu \int_\Omega |\nabla u(x)| \, \mathrm{d}x + \lambda \int_\Omega |Jv_b(x)| \, \mathrm{d}x + \lambda \int_\Omega |Jv_f(x)| \, \mathrm{d}x$$

$$+ \int_\Omega |\nabla u_b(x + v_b^0(x)) \cdot [v_b - v_b^0](x) + u_b(x + v_b^0(x)) - u(x)| \, \mathrm{d}x$$

$$+ \int_\Omega |\nabla u_f(x + v_f^0(x)) \cdot [v_f - v_f^0](x) + u_f(x + v_f^0(x)) - u(x)| \, \mathrm{d}x \tag{5}$$

where $\nabla u_b(x + v_b^0(x)) \cdot [v_b - v_b^0](x) + u_b(x + v_b^0(x)) - u(x)$ will appear as $\rho(u, v_b, u_b)$ afterwards (same goes for v_f). Minimizing with respect to the motion vector v_b leads to the form:

$$v_b^* = \underset{v_b}{\mathrm{argmin}} \, \underset{y}{\max} \int_\Omega |\rho(u, v_b, u_b)| \, \mathrm{d}x + \langle Jv_b \,|\, y \rangle - \iota_{B^\infty} \left(\frac{y}{\lambda} \right) \tag{6}$$

introducing the dual variable of v_b, $y : \Omega \to \mathbb{R}^{2 \times 2}$. This convex problem can be solved by the primal-dual algorithm of [10], noticing that $Jv_b = [\nabla v_{b,1}, \nabla v_{b,2}]^\top$ and so we get the adjoint operator $J^* y = -[\mathrm{div}([y_{1,1}, y_{1,2}]^\top), \mathrm{div}([y_{2,1}, y_{2,2}]^\top)]^\top$. Whereas the proximal operator associated to y is a projection onto the L^∞-norm ball, the proximal operator associated to v_b is a soft thresholding (see [23] for details):

$$\begin{cases} y^{(n+1)} \leftarrow \mathrm{prox}_{\lambda \sigma \iota_{B^\infty}} \left(y^{(n)} + \sigma J \bar{v}_b^{(n)} \right) \\ v_b^{(n+1)} \leftarrow \mathrm{prox}_{\tau \rho(u, -, u_b)} \left(v_b^{(n)} - \tau J^* y^{(n+1)} \right) \\ \bar{v}_b^{(n+1)} \leftarrow v_b^{(n+1)} + \theta \left(v_b^{(n+1)} - v_b^{(n)} \right) \end{cases} \tag{7}$$

where $\sigma, \tau > 0$ are time steps and $\theta \in [0, 1]$. The minimization of E_S with respect to v_f is carried out in a similar way.

4.2   Structural Reconstruction

After motion has been estimated, the inpainting process is obtained by minimizing:

$$u^* = \underset{u}{\mathrm{argmin}} \quad \int_\Omega |u_b(x + v_b(x)) - u(x)| \, \mathrm{d}x$$

$$+ \int_\Omega |u_f(x + v_f(x)) - u(x)| \, \mathrm{d}x + \mu \int_\Omega |\nabla u(x)| \, \mathrm{d}x \tag{8}$$

In order to rewrite the convex problem (8) with dual variables, the time variable must be clarified with respect to u, v_b and v_f. Indeed, taking 1 as the time step between two frames, and t as the current time for u, then u_b and u_f take the form:

$$u_b(x + v_b(x)) = u(x + v_b(x, t), t - 1) = u(\varphi_b(x, t))$$

$$u_f(x + v_f(x)) = u(x + v_f(x, t), t + 1) = u(\varphi_f(x, t)) \tag{9}$$

with φ_b and φ_f two transformations, which are similar to shift maps, with the hypothesis that $\varphi_b \circ \varphi_f = \varphi_f \circ \varphi_b = I_d$ almost everywhere. With these new notations, (8) can be rewritten as:

$$u^* = \underset{u}{\mathrm{argmin}} \max_z \left\langle \begin{array}{c} u \circ \varphi_b - u \\ u \circ \varphi_f - u \\ \nabla u \end{array} \middle| z \right\rangle - \iota_{B^\infty}(z_1) - \iota_{B^\infty}(z_2) - \iota_{B^\infty}\left(\frac{1}{\mu}\begin{bmatrix} z_3 \\ z_4 \end{bmatrix}\right) \quad (10)$$

introducing the dual variable of u, $z : \Omega \to \mathbb{R}^4$. Noting K the operator with respect to u in the inner product, it leads to the adjoint operator K^* as in [18]:

$$K^* z = \det\left(J\varphi_f\right) z_1 \circ \varphi_f - z_1 + \det\left(J\varphi_b\right) z_2 \circ \varphi_b - z_2 - \mathrm{div}([z_3, z_4]^\top) \quad (11)$$

Minimizing (10) can also be carried out, using the primal-dual algorithm of [10]:

$$\begin{cases} z_1^{(n+1)} \leftarrow \mathrm{prox}_{\sigma\iota_{B^\infty}}\left(z_1^{(n)} + \sigma\left(u \circ \varphi_b - \bar{u}^{(n)}\right)\right) \\ z_2^{(n+1)} \leftarrow \mathrm{prox}_{\sigma\iota_{B^\infty}}\left(z_2^{(n)} + \sigma\left(u \circ \varphi_f - \bar{u}^{(n)}\right)\right) \\ \begin{bmatrix} z_3^{(n+1)} \\ z_4^{(n+1)} \end{bmatrix} \leftarrow \mathrm{prox}_{\mu\,\sigma'\iota_{B^\infty}}\left(\begin{bmatrix} z_3^{(n)} \\ z_4^{(n)} \end{bmatrix} + \sigma'\nabla\bar{u}^{(n)}\right) \\ u^{(n+1)} \leftarrow u^{(n)} - \tau K^* z^{(n+1)} \\ \bar{u}^{(n+1)} \leftarrow u^{(n+1)} + \theta\left(u^{(n+1)} - u^{(n)}\right) \end{cases} \quad (12)$$

where $\sigma, \sigma', \tau > 0$ are time steps and $\theta \in [0, 1]$.

To minimize an energy similar to (2), [7] repeats alternating minimizations between inpainting u and estimating a unique motion vector field v until convergence. In our case, instead of doing such a thing, it was decided to minimize the three variables (v_b, u, v_f) together, by applying successive proximal steps on each variable. The three descents in v_b, v_f and u give thus better results in the reconstruction than the first option for an equivalent computation time.

4.3   Shift Maps Estimation and Textural Reconstruction

The textural reconstruction is based on the work of [1], where the shift maps Γ_b and Γ_f are estimated using the PatchMatch algorithm of [3], with a L^2-distance between patches. Consequently, for Γ_b (respectively Γ_f), $\forall x \in \tilde{O}$:

$$\Gamma_b(x) = \underset{x_b \in \Omega}{\mathrm{argmin}} \int_{\Omega_p} g_a(x_p)\, [u_b(x_b - x_p) - u(x - x_p)]^2 \, \mathrm{d}x_p \quad (13)$$

For each temporal neighbour frame u_b and u_f, each associated part of (3) can be rewritten, without considering the weight ω for now, as the extreme case of choosing only the nearest neighbour patch for every pixel in the defect area (see [1] for details), which introduces a Dirac δ:

$$E_T^b(u, \Gamma_b) = \int_{\tilde{O}} \int_{\Omega} \delta(\Gamma_b(x) - x_b) \, \varepsilon[p_{u_b}(x_b) - p_u(x)] \, \mathrm{d}x_b \, \mathrm{d}x \qquad (14)$$

With the changes of variables $x := x - x_p$ and $x_b := x_b - x_p$ which operate two translations, we obtain from (4) and (14):

$$E_T^b(u, \Gamma_b) = \int_O \int_{\Omega} m(x, x_b) \, [u_b(x_b) - u(x)]^2 \, \mathrm{d}x_b \, \mathrm{d}x \qquad (15)$$

with:

$$m(x, x_b) = \int_{\Omega_p} g_a(x_p) \, \delta(\Gamma_b(x + x_p) - (x_b + x_p)) \, \mathrm{d}x_p \qquad (16)$$

whose integral over Ω is equal to 1 since g_a is assumed normalized. By expanding the squared difference in (15), u is also the minimizer of the following energy, which is equal to $E_T^b(u, \Gamma_b)$ up to a constant:

$$\tilde{E}_T^b(u, \Gamma_b) = \int_O \left[u(x) - \int_{\Omega} m(x, x_b) \, u_b(x_b) \, \mathrm{d}x_b \right]^2 \mathrm{d}x \qquad (17)$$

which directly leads to the non-local means solution defined $\forall x \in O$:

$$u(x) = \int_{\Omega} m(x, x_b) \, u_b(x_b) \, \mathrm{d}x_b = \int_{\Omega_p} g_a(x_p) \, u_b(\Gamma_b(x + x_p) - x_p)) \, \mathrm{d}x_p \qquad (18)$$

This is the result for only one of the two neighbour frames. To take into account both frames, some weighted means are introduced with the weights $w(x)$ and $(1 - w(x))$. Choosing half of both results ($w(x) = 0.5$, named T_H afterwards) will cause some blur. However, taking only the best ($w(x) \in \{0, 1\}$, named T_B) will cause spatial artifacts. In this case, the choice of the weight 0 or 1 is given after both shift maps Γ_b and Γ_f having been estimated. Then, for each pixel to recover, the weight $w(x)$ is equal to 1 if the final distance $\varepsilon \left[p_{u_b}(\Gamma_b(x)) - p_u(x) \right]$ from the backward frame is smaller than the one from the forward frame, i.e. $\varepsilon \left[p_{u_f}(\Gamma_f(x)) - p_u(x) \right]$, and is equals to 0 otherwise. An intermediary solution can be to take a weighted mean (named T_W^α), using the following ratio containing both previous distances ε:

$$w(x) = \frac{\varepsilon \left[p_{u_f}(\Gamma_f(x)) - p_u(x) \right]^\alpha}{\varepsilon \left[p_{u_f}(\Gamma_f(x)) - p_u(x) \right]^\alpha + \varepsilon \left[p_{u_b}(\Gamma_b(x)) - p_u(x) \right]^\alpha} \qquad (19)$$

where the power $\alpha \in [0, +\infty)$. One can notice that this formula can generalize the two first choices as follows:

$$\lim_{\alpha \to 0} T_W^\alpha = T_H \quad \text{and} \quad \lim_{\alpha \to +\infty} T_W^\alpha = T_B. \qquad (20)$$

In the next section, the use of T_W without specifying α means that α is equal to 1, which is the case of a classical ratio between the distance from the forward frame and the sum of both distances.

5   Experiments

Our algorithm is implemented on Matlab and C, using the maximum level $L_{\max} = 10$ in multiresolution pyramid with a factor of $\sqrt{2}$. In order to test the algorithm, three consecutive Middlebury frames were used where a defect was artificially put in the central frame (see Fig. 2). The videos of the complete Middlebury sequences of 8 frames with the defect, and the reconstructions are available for downloading at the following link: *SSVM VideoInpainting*.

Fig. 2. Frames 8 to 10 from "RubberWhale" (top), "Evergreen" (middle) and "Dumptruck" (bottom) sequences with a defect highlighted in red in the central frame. (Color figure online)

5.1   Qualitative Results

Concerning the "RubberWhale" frames, results in Fig. 3 show that the textural methods (Fig. 3b to d) are inadequate to reconstruct the puppet and the backward properly, whatever the chosen patch. On the other hand, the structural reconstruction (Fig. 3e) is already good. However, our algorithm still gives some little improvement (Fig. 3f to h), in particular at the top of the head of the puppet where there is false color with the structure, due to the channel separation to inpaint, and at the corner of the wooden bar at the back which is less rounded.

In order to show more improvements, the two next sequences represent realistic scenes with more texture and larger motions. In the case of "Evergreen" frames for instance, results in Fig. 4 show that textural methods (Fig. 4b to d) cannot manage to reconstruct the tree branches properly. The best patch approach (Fig. 4b) leads

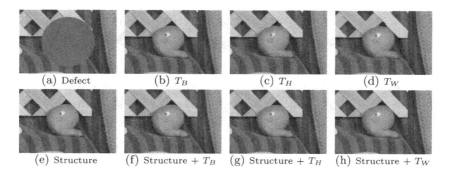

Fig. 3. "RubberWhale" reconstructions (zooms on the defect area).

to artifacts whereas the mean approaches (Fig. 4c and d) operates an averaging, as expected, which results in a lack of texture. The structural reconstruction (Fig. 4e) is better but still a little blurry because of the diffusivity and some branches seems a little bit stretched out. On the other hand, the combination of both reconstructions (Fig. 4f to h) succeeds in adding texture to structure.

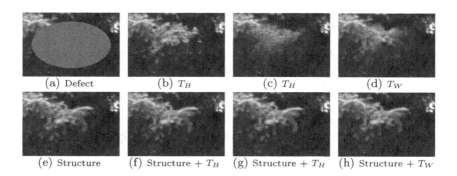

Fig. 4. "Evergreen" reconstructions (zooms on the defect area).

The results for "Dumptruck" frames in Fig. 5 show that the structural reconstruction (Fig. 5e) leads to a deformation of the truck, which seems to oscillate vertically in the video. Moreover, there is also some ghosting effect behind the car that goes to the right. Textural reconstructions using means (Fig. 5c and d) lead, as expected, to blurry reconstructions of the truck. Even if using the best patch from both adjacent frames (Fig. 5b), whose result seems to be really good on the static reconstructed frame, the video shows that the algorithm chooses the closest frame in terms of patch distance and stays locked, this is why a lack of motion appears in the defect area in the video. Using our algorithm (Fig. 5f to h), we obtain better results with the textural refinement, which permits to keep a good optical flow. Only the ghosting effect remains, due to the large displacement of the car.

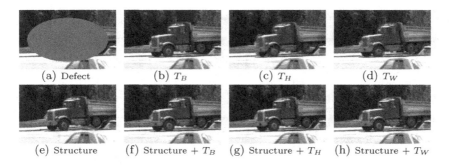

(a) Defect (b) T_B (c) T_H (d) T_W

(e) Structure (f) Structure + T_B (g) Structure + T_H (h) Structure + T_W

Fig. 5. "Dumptruck" reconstructions (zooms on the defect area).

5.2 Quantitative Results

Regarding the PNSR and SSIM quality metrics for the "RubberWhale" frame reconstructions in Table 1 (up), it leads to the same interpretation: the structural reconstruction performs as well as our algorithm. For the "Evergreen" frame reconstructions, the first visual interpretations are also validated by the quality metrics in Table 1 (middle) where structural reconstruction performs better than textural reconstruction, and our algorithm gives the best results with a certain gap. Moreover, choosing any of the three textural reconstructions

Table 1. PSNR - SSIM for the different reconstructions of the "RubberWhale" frame (top), the "Evergreen" frame (middle) and the "Dumptruck" frame (bottom).

| *RubberWhale* | - | Textural reconstructions | | |
		T_B	T_H	T_W
-	16.34 - 0.915	38.46 - 0.994	41.12 - 0.996	39.92 - 0.995
Structural reconstruction	**48.49 - 0.999**	46.99 - 0.999	**48.81 - 0.999**	**48.59 - 0.999**

| *Evergreen* | - | Textural reconstructions | | |
		T_B	T_H	T_W
-	18.66 - 0.943	30.94 - 0.978	33.37 - 0.980	33.56 - 0.982
Structural reconstruction	37.24 - 0.991	**40.37 - 0.994**	**40.47 - 0.994**	**40.50 - 0.994**

| *Dumptruck* | - | Textural reconstructions | | |
		T_B	T_H	T_W
-	17.53 - 0.937	33.16 - 0.987	34.35 - 0.989	34.31 - 0.988
Structural reconstruction	27.76 - 0.975	**38.11 - 0.994**	**38.87 - 0.994**	**38.58 - 0.994**

with the structural one has no longer a real impact on the result. Concerning the "Dumptruck" frame reconstructions, by looking at the quality metrics in Table 1 (bottom), our algorithm leads to a significant improvement compared to the separate structural and textural reconstructions.

6   Conclusion and Perspectives

In this paper, we have shown that combining structural reconstruction based on diffusion approaches and textural reconstruction leads to better results in terms of visual and metrics quality. To go further, it would be interesting to refine the optical flow model using total generalized variation as in [20]. Other non-local textural reconstructions could also be processed, as with the median filter and also using patch gradients or other new patch regularizations.

References

1. Arias, P., Facciolo, G., Caselles, V., Sapiro, G.: A variational framework for exemplar-based image inpainting. IJCV **93**, 319–347 (2011)
2. Aubert, G., Deriche, R., Kornprobst, P.: Computing optical flow via variational techniques. SIAM J. Appl. Math. **60**, 156–182 (1999)
3. Barnes, C., Shechtman, E., Finkelstein, A., Goldman, D.B.: PatchMatch: a randomized correspondence algorithm for structural image editing. ACM TOG **28** (2009)
4. Bertalmio, M., Sapiro, G., Caselles, V., Ballester, C.: Image inpainting. In: Proceedings SIGGRAPH (2000)
5. Buades, A., Lisani, J.L.: Video Denoising with optical flow estimation. IPOL **8**, 142–166 (2018)
6. Bugeau, A., Bertalmio, M.: Combining texture synthesis and diffusion for image inpainting. In: Proceedings VISAPP (2009)
7. Burger, M., Dirks, H., Schönlieb, C.B.: A variational model for joint motion estimation and image reconstruction. SIAM J. Imag. Sc. **11**, 94–128 (2018)
8. Buyssens, P., Daisy, M., Tschumperlé, D., Lézoray, O.: Exemplar-based inpainting: technical review and new heuristics for better geometric reconstructions. IEEE TIP **24**, 1809–1824 (2015)
9. Cao, F., Gousseau, Y., Masnou, S., Pérez, P.: Geometrically guided exemplar-based inpainting. SIAM J. Appl. Math. **4**, 1143–1179 (2011)
10. Chambolle, A., Pock, T.: A first-order primal-dual algorithm for convex problems with applications to imaging. JMIV **40**, 120–145 (2011)
11. Chan, T.F., Shen, J.: Local inpainting models and TV inpainting. SIAM J. Appl. Math. **62** (2001)
12. Cocquerez, J.P., Chanas, L., Blanc-Talon, J.: Simultaneous inpainting and motion estimation of highly degraded video-sequences. In: Bigun, J., Gustavsson, T. (eds.) SCIA 2003. LNCS, vol. 2749, pp. 685–692. Springer, Heidelberg (2003). https://doi.org/10.1007/3-540-45103-X_91
13. Criminisi, A., Pérez, P., Toyama, K.: Region filling and object removal by exemplar-based image inpainting. IEEE TIP **13**, 1200–1212 (2004)

14. Do, V., Lebrun, G., Malapert, L., Smet, C., Tschumperlé, D.: Inpainting d'images couleurs par lissage anisotrope et synthèse de textures. In: Proceedings RFIA (2006)
15. Efros, A.A., Leung, T.K.: Texture synthesis by non-parametric sampling. In: Proceedings ICCV (1999)
16. Horn, B.K., Schunck, B.G.: Determining optical flow. AI **17**, 185–203 (1981)
17. Lauze, F., Nielsen, M.: A variational algorithm for motion compensated inpainting. In: Proceedings BMVC (2004)
18. Lauze, F., Nielsen, M.: On variational methods for motion compensated inpainting. arXiv preprint (2009)
19. Le, T., Almansa, A., Gousseau, Y., Masnou, S.: Motion-consistent video inpainting. In: Proceedings ICIP (2017)
20. March, R., Riey, G.: Analysis of a variational model for motion compensated inpainting. Inverse Prob. Imaging **11**, 997–1025 (2017)
21. Masnou, S., Morel, J.M.: Level lines based disocclusion. In: Proceedings ICIP (1998)
22. Newson, A., Almansa, A., Fradet, M., Gousseau, Y., Pérez, P.: Video inpainting of complex scenes. SIAM J. Imag. Sci. **7** (2014)
23. Zach, C., Pock, T., Bischof, H.: A duality based approach for realtime TV-L^1 optical flow. In: Hamprecht, F.A., Schnörr, C., Jähne, B. (eds.) DAGM 2007. LNCS, vol. 4713, pp. 214–223. Springer, Heidelberg (2007). https://doi.org/10.1007/978-3-540-74936-3_22

Inverse Problems in Imaging

Iterative Sampled Methods for Massive and Separable Nonlinear Inverse Problems

Julianne Chung[✉], Matthias Chung, and J. Tanner Slagel

Department of Mathematics, Virginia Tech, Blacksburg, VA 24060, USA
{jmchung,mcchung,slagelj}@vt.edu

Abstract. In this paper, we consider iterative methods based on sampling for computing solutions to separable nonlinear inverse problems where the entire dataset cannot be accessed or is not available all-at-once. In such scenarios (e.g., when massive amounts of data exceed memory capabilities or when data is being streamed), solving inverse problems, especially nonlinear ones, can be very challenging. We focus on separable nonlinear problems, where the objective function is nonlinear in one (typically small) set of parameters and linear in another (larger) set of parameters. For the linear problem, we describe a limited-memory sampled Tikhonov method, and for the nonlinear problem, we describe an approach to integrate the limited-memory sampled Tikhonov method within a nonlinear optimization framework. The proposed method is computationally efficient in that it only uses available data at any iteration to update both sets of parameters. Numerical experiments applied to massive super-resolution image reconstruction problems show the power of these methods.

Keywords: Tikhonov regularization · Sampled methods ·
Variable projection · Kaczmarz methods · Super-resolution ·
Medical imaging and other applications

1 Introduction

Advanced tools for image reconstruction are essential in many scientific applications ranging from biomedical to geophysical imaging [11]. A major challenge in many of the newer imaging systems is that the entire dataset cannot be accessed or is not available *all-at-once*. For example, faster scan speeds on recently-developed micro-tomography instruments have resulted in very large datasets [14]. Using standard image reconstruction techniques to analyze the massive amounts of data is computationally intractable. Another example arises in streaming-data problems or automated pipelines, where immediate feedback (e.g., a partial reconstruction) may be needed to inform the data acquisition process [21]. These scenarios are becoming common in many applications, thereby

© Springer Nature Switzerland AG 2019
J. Lellmann et al. (Eds.): SSVM 2019, LNCS 11603, pp. 119–130, 2019.
https://doi.org/10.1007/978-3-030-22368-7_10

motivating the need for further developments on sampled iterative methods for image reconstruction.

We consider image reconstruction problems where the underlying model is separable and nonlinear. For the case where observations are available all-at-once, the data acquisition process can be modeled as,

$$\mathbf{b} = \mathbf{A}(\mathbf{y}_{\text{true}})\mathbf{x}_{\text{true}} + \boldsymbol{\epsilon} \,, \tag{1}$$

where $\mathbf{x}_{\text{true}} \in \mathbb{R}^n$ contains the desired image, $\mathbf{y}_{\text{true}} \in \mathbb{R}^p$ contains the desired forward model parameters, $\mathbf{A}(\cdot) : \mathbb{R}^p \to \mathbb{R}^{m \times n}$ is a nonlinear operator describing the forward model, $\boldsymbol{\epsilon} \in \mathbb{R}^m$ contains noise or measurement errors (typically treated as a realization from a Gaussian distribution with zero mean), and $\mathbf{b} \in \mathbb{R}^m$ contains the observations. It is often assumed that $\mathbf{A}(\mathbf{y}_{\text{true}})$ is known, in which case we have a linear model. However, in many realistic scenarios, parameters $\mathbf{y}_{\text{true}}$ must be estimated from the data. Here we assume that the parameterization of the model $\mathbf{A}(\cdot)$ is known and that the number of parameters in $\mathbf{y}_{\text{true}}$ is significantly smaller than the number of unknowns in $\mathbf{x}_{\text{true}}$, i.e., $p \ll n$. An example of a separable nonlinear inverse problem of this form arises in super-resolution image reconstruction, see Sect. 4.

Since image reconstruction problems are usually ill-posed, small errors in the data can result in very large errors in the solution. Thus regularization is need to compute a reasonable solution, and here we consider the widely-used Tikhonov regularization method. That is, we are interested in Tikhonov-regularized optimization problems of the form,

$$\min_{\mathbf{x},\mathbf{y}} f(\mathbf{x}, \mathbf{y}) = \|\mathbf{A}(\mathbf{y})\mathbf{x} - \mathbf{b}\|_2^2 + \lambda \|\mathbf{x}\|_2^2 \,, \tag{2}$$

where $\lambda > 0$ is a regularization parameter that balances the data-fit and the regularization term. Note that (2) is separable[1] since the objective function f is a linear function in terms of $\mathbf{x}$ and a nonlinear function in terms of $\mathbf{y}$. Previously developed numerical optimization methods for (2) have been investigated and range from fully decoupled approaches (e.g., alternating optimization) to fully coupled (e.g, nonlinear) approaches [6]. A popular alternative is the *variable projection* method [9, 19], where the linear parameters are mathematically eliminated and a nonlinear optimization scheme is used to solve the reduced optimization problem. These methods have been investigated for various image processing applications, see e.g., [2, 7, 12]. However, all of these methods require all-at-once access to the data to perform full matrix-vector multiplications with $\mathbf{A}(\mathbf{y})$, and hence they cannot be used for massive or streaming problems.

In this paper, we develop an iterative sampled method to estimate a solution for (2) in the case of massive or streaming data. The method follows a variable projection approach by first mathematically eliminating the linear variables. However, to address massive or streaming data, we use recently-developed sampled Tikhonov methods to approximate the regularized linear problem and use

---

[1] This is sometimes referred to as partially separable [18].

a sampled Gauss-Newton method to approximate the nonlinear variables. Sampled Tikhonov methods are simple and have favorable convergence properties [22]. Also, limited-memory variants can reduce computational costs.

An outline of the paper is as follows. In Sect. 2 we provide an overview of sampled Tikhonov methods and provide a numerical exploration of the convergence properties of the limited-memory variants. Then in Sect. 3 we describe iterative sampled methods for separable nonlinear inverse problems, where the sampled Tikhonov methods from Sect. 2 are integrated within a nonlinear optimization framework for updating estimates of $\mathbf{x}_{\text{true}}$ and $\mathbf{y}_{\text{true}}$. Numerical results from super-resolution imaging are presented in Sect. 4, and conclusions and future work are presented in Sect. 5.

2   Sampled Tikhonov Methods for Linear Inverse Problems

Suppose $\mathbf{y}$ is fixed and consider computing the Tikhonov solution,

$$\mathbf{x}_{\text{tik}} = \arg\min_{\mathbf{x}} \|\mathbf{A}\mathbf{x} - \mathbf{b}\|_2^2 + \lambda \|\mathbf{x}\|_2^2, \tag{3}$$

for the case where all of $\mathbf{A}$ and $\mathbf{b}$ are not available at once. For linear problems, we can use *sampled limited-memory Tikhonov* (slimTik) methods, in which we iteratively solve a sequence of sampled least-squares problems [22]. These sampled Tikhonov methods can be interpreted as extensions of block Kaczmarz type methods and are related to recursive least squares methods for ill-posed problems, randomized least-squares solvers, and stochastic optimization methods for convex programs, see e.g., [1,3,8,10,16,17,24].

Let's assume that matrix $\mathbf{A}$ and vector $\mathbf{b}$ can be partitioned into M blocks,

$$\mathbf{A} = \begin{bmatrix} \mathbf{A}^{(1)} \\ \vdots \\ \mathbf{A}^{(M)} \end{bmatrix} \quad \text{and} \quad \mathbf{b} = \begin{bmatrix} \mathbf{b}^{(1)} \\ \vdots \\ \mathbf{b}^{(M)} \end{bmatrix}. \tag{4}$$

For simplicity we assume that all blocks have the same dimension, i.e., $\mathbf{A}^{(i)} \in \mathbb{R}^{\ell \times n}$ and $\mathbf{b}^{(i)} \in \mathbb{R}^{\ell}$, $i = 1, \dots, M$, with $\ell = m/M$. Then for an arbitrary initial guess $\mathbf{x}_0 \in \mathbb{R}^n$ and $\lambda > 0$, the k-th slimTik iterate can be written as

$$\mathbf{x}_k = \mathbf{x}_{k-1} - \mathbf{s}_k$$

with

$$\mathbf{s}_k = \left(\tfrac{k\lambda}{M}\mathbf{I}_n + \mathbf{M}_k^\top \mathbf{M}_k + \mathbf{A}_k^\top \mathbf{A}_k\right)^{-1} \left(\mathbf{A}_k^\top (\mathbf{A}_k \mathbf{x}_{k-1} - \mathbf{b}_k) + \tfrac{\lambda}{M}\mathbf{x}_{k-1}\right),$$

where $\mathbf{A}_k = \mathbf{A}^{(((k-1) \bmod M)+1)}$, $\mathbf{b}_k = \mathbf{b}^{(((k-1) \bmod M)+1)}$, and matrix $\mathbf{M}_k = \left[\mathbf{A}_{k-r}^\top, \dots, \mathbf{A}_{k-1}^\top\right]^\top$ collects the previously-accessed matrix blocks up to *memory level* $r \in \mathbb{N}$. Note that mod is the modulo operation, and hence we are sampling with *cyclic control*. Thus, after all blocks in ascending order have been visited, the algorithm continues with the first block; other sampling strategies are possible, but these investigations are beyond the scope of this paper, see [4]. The memory level is assumed to be constant, and for the first r iterates, the blocks of $\mathbf{A}$ and $\mathbf{b}$ with negative indices are set to zero. Notice that the update $\mathbf{s}_k$ can be computed efficiently by solving the regularized least-squares problem,

$$
\mathbf{s}_k = \arg\min_{\mathbf{s}} \left\| \begin{bmatrix} \mathbf{M}_k \\ \mathbf{A}_k \\ \sqrt{\frac{k\lambda}{M}}\mathbf{I}_n \end{bmatrix} \mathbf{s} - \begin{bmatrix} \mathbf{0} \\ \mathbf{A}_k\mathbf{x}_{k-1} - \mathbf{b}_k \\ \sqrt{\frac{\lambda}{kM}}\mathbf{x}_{k-1} \end{bmatrix} \right\|_2^2,
$$

using iterative methods such as LSQR [20]. The `slimTik` method with memory r is an approximation of the full memory method where $r = M - 1$, for which it can be shown that $\mathbf{x}_M = \mathbf{x}_{\text{tik}}$, for details see [22]. Hence, the full memory `slimTik` method converges after M iterations to the Tikhonov solution (3), with the corresponding regularization parameter λ.

We illustrate convergence for an example from the *Regularization Tools* toolbox [11]. We use the `gravity` example which provides a matrix $\mathbf{A} \in \mathbb{R}^{1000 \times 1000}$ and a vector $\mathbf{x}_{\text{true}}$. We partition $\mathbf{A}$ into $M = 100$ blocks with $\ell = 10$ and let $\lambda = 0.0196$. We simulate observed data by adding Gaussian white noise with zero mean such that the noise level is 0.01, i.e., $\mathbf{b} = \mathbf{A}\mathbf{x}_{\text{true}} + \boldsymbol{\epsilon}$ where $\frac{\|\boldsymbol{\epsilon}\|_2}{\|\mathbf{A}\mathbf{x}_{\text{true}}\|_2} = 0.01$. First, we run `slimTik` for one epoch ($k = M$) with memory levels $r = 0, \ldots, M - 1$, and we report the relative error between the reconstructions $\mathbf{x}_M$ and the Tikhonov solution $\mathbf{x}_{\text{tik}}$ in the left panel of Fig. 1. Note that for full memory (i.e., $r = M-1$), the relative error is within machine precision. Also, for lower memory levels, the reconstructions $\mathbf{x}_M$ are close to the Tikhonov solution. The right panel of Fig. 1 illustrates the asymptotic convergence of `slimTik` for memory levels $r = 0, 2, 4, 6$, and 8, where we also compare to a standard sampled gradient (`sg`) method without regularization. Errors are plotted after each full epoch. Empirically, we observe that the iterates $\mathbf{x}_k$ converge to $\mathbf{x}_{\text{tik}}$ as $k \to \infty$, using cyclic control. Asymptotic convergence of these methods using cyclic, random cyclic, or fully random control has not yet been studied and is current research. Asymptotic convergence for consistent systems using cyclic or random control has been shown in [4,13,16,23].

3   Iterative Sampled Methods for Separable Nonlinear Inverse Problems

Next for separable nonlinear inverse problems of the form (2), we describe an iterative sampled approach that integrates `slimTik` within a nonlinear optimization framework so that both sets of parameters can be updated as data become available. Similar to the mathematical description in Sect. 2, we assume that $\mathbf{A}$ and $\mathbf{b}$ can be split into blocks as in (4), but where $\mathbf{A}^{(i)}(\cdot) : \mathbb{R}^p \to \mathbb{R}^{\ell \times n}$, $i = 1, \ldots, M$, with $\ell = m/M$. For an initial guess of the linear parameters

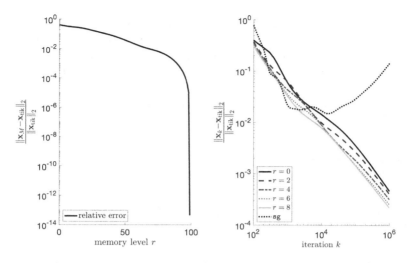

Fig. 1. Convergence of the `slimTik` method. The plot in the left panel contains the relative errors between the iterates after one epoch and the Tikhonov solution, for different memory levels. The plot in the right panel illustrates asymptotic convergence of the `slimTik` method for memory levels $r = 0, 2, 4, 6$, and 8. For comparison we include relative errors for a sample gradient method.

$\mathbf{x}_0 \in \mathbb{R}^n$, nonlinear parameters $\mathbf{y}_0 \in \mathbb{R}^p$, and $\lambda > 0$, the k-th iterate of the separable nonlinear `slimTik` (`sn-slimTik`) method can be written as

$$\mathbf{x}_k = \mathbf{x}_{k-1} - \mathbf{s}_k$$
$$\mathbf{y}_k = \mathbf{y}_{k-1} - \alpha_k \left(\mathbf{J}_k^\top \mathbf{J}_k \right)^\dagger \mathbf{J}_k^\top \mathbf{r}_k \left(\mathbf{y}_{k-1} \right) \tag{5}$$

with

$$\mathbf{s}_k = \arg\min_{\mathbf{s}} \left\| \begin{bmatrix} \mathbf{M}_k \left(\mathbf{y}_{k-1} \right) \\ \mathbf{A}_k \left(\mathbf{y}_{k-1} \right) \\ \sqrt{\frac{k\lambda}{M}} \mathbf{I}_n \end{bmatrix} \mathbf{s} - \begin{bmatrix} \mathbf{0} \\ \mathbf{A}_k \left(\mathbf{y}_{k-1} \right) \mathbf{x}_{k-1} - \mathbf{b}_k \\ \sqrt{\frac{\lambda}{kM}} \mathbf{x}_{k-1} \end{bmatrix} \right\|_2^2,$$

where $\mathbf{A}_k \left(\cdot \right) = \mathbf{A}^{(((k-1) \bmod M)+1)} \left(\cdot \right)$, $\mathbf{b}_k = \mathbf{b}^{(((k-1) \bmod M)+1)}$, and $\mathbf{M}_k \left(\cdot \right) = \left[\mathbf{A}_{k-r} \left(\cdot \right)^\top, \ldots, \mathbf{A}_{k-1} \left(\cdot \right)^\top \right]^\top$ for chosen memory level $r \in \mathbb{N}$. The blocks of $\mathbf{A}$ and $\mathbf{b}$ with negative indices are set to the zero function and zero vector, respectively. Here $\mathbf{r}_k(\cdot) : \mathbb{R}^p \to \mathbb{R}^{\ell(r+1)}$ is the sample residual function defined as

$$\mathbf{r}_k \left(\mathbf{y} \right) = \begin{bmatrix} \mathbf{A}_{k-r} \left(\mathbf{y} \right) \\ \vdots \\ \mathbf{A}_{k-1} \left(\mathbf{y} \right) \\ \mathbf{A}_k \left(\mathbf{y} \right) \end{bmatrix} \mathbf{x}_k - \begin{bmatrix} \mathbf{b}_{k-r} \\ \vdots \\ \mathbf{b}_{k-1} \\ \mathbf{b}_k \end{bmatrix},$$

$\mathbf{J}_k$ is the Jacobian of $\mathbf{r}_k$ evaluated at $\mathbf{y}_{k-1}$, and α_k is the step size determined by a line search method [18]. The Jacobian can be approximated with finite differences or found analytically. Note that † represents the pseudo-inverse in (5) and is required since $\mathbf{J}_k$ might not have full column rank. Also, as with any nonlinear, nonconvex optimization method, the initial guess must be within the basin of attraction of the desired minimizer. A summary of the sn-slimTik algorithm is provided below.

Algorithm 1. sn-slimTik

1: Inputs: $\mathbf{x}_0$, $\mathbf{y}_0$, r, λ, M

2: **for** $k = 1, 2, \ldots$ **do**

3:    Get $\mathbf{A}_k\left(\mathbf{y}_{k-1}\right)$, $\mathbf{b}_k$, and $\mathbf{M}_k\left(\mathbf{y}_{k-1}\right)$

4:    $\mathbf{s}_k = \arg\min\limits_{\mathbf{s}} \left\| \begin{bmatrix} \mathbf{M}_k\left(\mathbf{y}_{k-1}\right) \\ \mathbf{A}_k\left(\mathbf{y}_{k-1}\right) \\ \sqrt{\frac{k\lambda}{M}}\mathbf{I}_n \end{bmatrix} \mathbf{s} - \begin{bmatrix} \mathbf{0} \\ \mathbf{A}_k\left(\mathbf{y}_{k-1}\right)\mathbf{x}_{k-1} - \mathbf{b}_k \\ \sqrt{\frac{\lambda}{kM}}\mathbf{x}_{k-1} \end{bmatrix} \right\|_2^2$

5:    $\mathbf{x}_k = \mathbf{x}_{k-1} - \mathbf{s}_k$

6:    $\mathbf{y}_k = \mathbf{y}_{k-1} - \alpha_k \left(\mathbf{J}_k^\top \mathbf{J}_k\right)^\dagger \mathbf{J}_k^\top \mathbf{r}_k\left(\mathbf{y}_{k-1}\right)$

7: **end for**

---

4   Numerical Results

In this section, we provide numerical results for super-resolution image reconstruction, which can be represented as a separable nonlinear inverse problem [5]. Suppose we have M low-resolution images. The underlying model for super-resolution imaging can be represented as (1), where $\mathbf{x}_{\text{true}}$ contains the high-resolution (HR) image, and $\mathbf{b}$ and $\mathbf{A}(\mathbf{y}_{\text{true}})$ can be partitioned as in (4), where $\mathbf{b}^{(i)}$ contains the i-th low-resolution (LR) image and $\mathbf{A}^{(i)}(\cdot) : \mathbb{R}^p \to \mathbb{R}^{\ell \times n}$. More specifically, if we assume that the deformation for each LR image is affine (e.g., can be described with at most 6 parameters) and independent of the parameters for the other images, then we can partition $\mathbf{y}$ as

$$\mathbf{y} = \begin{bmatrix} \mathbf{y}^{(1)} \\ \vdots \\ \mathbf{y}^{(M)} \end{bmatrix}$$

and have $\mathbf{A}^{(i)}(\mathbf{y}) = \mathbf{RS}(\mathbf{y}^{(i)})$ where $\mathbf{R}$ is a restriction matrix that takes a HR image to a LR one and $\mathbf{S}(\mathbf{y}^{(i)})$ represents an affine transformation defined by parameters in $\mathbf{y}^{(i)}$. Then the goal is to solve (2) to estimate the HR image as well as update the transformation parameters.

We will investigate iterative sampled methods for super-resolution problems with massive or streaming data, but first we investigate a smaller problem where all of the data can be accessed at once. In Experiment 1, we compare our proposed sn-slimTik method with different memory levels to the results from the variable projection method. We show that with relatively modest memory levels, our approaches can achieve reconstructions with similar quality to full-memory reconstructions in comparable time. Then in Experiment 2, we consider a very large streaming super-resolution problem, where both the resolution of the images as well as the number of LR images present a computational bottleneck.

In both experiments, we initialize $\mathbf{x}_0 = \mathbf{0}$, and $\mathbf{y}_0$ is obtained by adding Gaussian white noise with zero mean to $\mathbf{y}_{\text{true}}$ where the variance is $2.45 \cdot 10^{-3}$ in Experiment 1 and $4.48 \cdot 10^{-4}$ in Experiment 2. We set the regularization parameter in advance, but mention that methods for updating the regularization parameter can be found in [22].

Algorithm 2. `variable projection`

1: Inputs: $\mathbf{y}_0, \lambda$
2: **for** $k = 1, 2, \ldots$ **do**
3: $\quad \mathbf{x}_k = \underset{\mathbf{x}}{\arg\min} \left\| \begin{bmatrix} \mathbf{A}(\mathbf{y}_{k-1}) \\ \sqrt{\lambda}\mathbf{I}_n \end{bmatrix} \mathbf{x} - \begin{bmatrix} \mathbf{b} \\ \mathbf{0} \end{bmatrix} \right\|_2^2$
4: $\quad \tilde{\mathbf{r}}_k(\mathbf{y}_{k-1}) = \mathbf{A}(\mathbf{y}_{k-1})\mathbf{x}_k - \mathbf{b}$
5: $\quad \mathbf{y}_k = \mathbf{y}_{k-1} - \alpha_k \left(\tilde{\mathbf{J}}_k^\top \tilde{\mathbf{J}}_k \right)^\dagger \tilde{\mathbf{J}}_k^\top \tilde{\mathbf{r}}_k(\mathbf{y}_{k-1})$
6: **end for**

4.1 Experiment 1: Comparing `sn-slimTik` to Variable Projection

Both `sn-slimTik` and variable projection are iterative methods that update $\mathbf{x}$ and $\mathbf{y}$. However, the variable projection method requires access to all data at once and thus may be infeasible for massive or streaming problems. The goal of this experiment is to show that we can achieve similar reconstructions as existing methods, but without the need to access all data and matrices at once.

For completeness, we provide in Algorithm 2 the basic variable projection algorithm [9,19], which is a Gauss-Newton algorithm applied to the problem,

$$\min_{\mathbf{y}} f(\mathbf{x}(\mathbf{y}), \mathbf{y}).$$

Here $\tilde{\mathbf{J}}_k$ is the Jacobian of $\mathbf{A}(\mathbf{y})\mathbf{x}_k - \mathbf{b}$ with respect to $\mathbf{y}$ at $\mathbf{y}_{k-1}$, and α_k is a line search parameter. Analytical methods can be used to obtain the Jacobian, see [5]. Notice that each iteration of the variable projection algorithm requires access to the entire data set $\mathbf{b}$ as well as matrix $\mathbf{A}(\mathbf{y})$ in order to solve the linear least squares problem in step 3. For our experiments, we use the LSQR method to solve the linear Tikhonov problem, where each *iteration* of LSQR requires a matrix-vector multiplication with $\mathbf{A}(\mathbf{y}_{k-1})$ and $\mathbf{A}(\mathbf{y}_{k-1})^\top$. Each multiplication requires access to all of the data, and thus, in terms of data access, is equivalent to one epoch of `slimTik`.

For this experiment, the goal is to recover a HR image that contains 512^2 pixels from a set of $M = 100$ LR images, each containing 128^2 pixels, i.e., $\mathbf{A}(\mathbf{y}) \in \mathbb{R}^{100 \cdot 128^2 \times 512^2}$. The HR image is of an astronaut and was obtained from NASA's website [15]. The HR image and three of the simulated LR images are provided in Fig. 2. The noise level for each LR image was set to 0.01, and the regularization parameter was set to $\lambda = 8 \cdot 10^{-2}$.

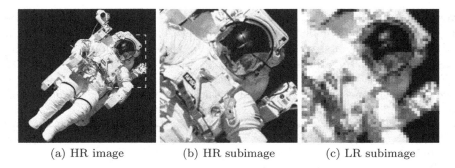

(a) HR image   (b) HR subimage   (c) LR subimage

Fig. 2. Super-resolution imaging example. The high-resolution (HR) image and a subimage corresponding to the yellow box are provided in (a) and (b) respectively. The subimage of one of the low-resolution (LR) images is provided in (c).

In Fig. 3, we provide relative error norms for the reconstructions and relative error norms for the affine parameters,

$$\frac{\|\mathbf{x}_k - \mathbf{x}_{\text{true}}\|_2}{\|\mathbf{x}_{\text{true}}\|_2} \quad \text{and} \quad \frac{\|\mathbf{y}_k - \mathbf{y}_{\text{true}}\|_2}{\|\mathbf{y}_{\text{true}}\|_2}, \tag{6}$$

respectively. We compare the sn-slimTik method with memory levels $r = 0, 1$, and 5 for 5 epochs (100 iterations correspond to one epoch), and provide results for 4 iterations of the variable projection method for comparison.

Following the discussion above, it is difficult to provide a fair comparison since each variable projection iteration requires a linear solve and here we use 20 LSQR iterations for each outer iteration. Performing one LSQR iteration requires the same memory access as 100 iterations of sn-slimTik with any memory level. Thus, in Fig. 3 we plot the relative reconstruction error norms for variable projection only after every 100 iterations of sn-slimTik. We see that for both parameters sets, sn-slimTik produces relative reconstruction errors that are comparable to the variable projection method. For this experiment variable projection took 644 s, sn-slimTik took 366 s with memory 0, 800 s with memory 1, and 2,570 s with memory 5.

Sub-images of sn-slimTik reconstructions at iterations $k = 100$ and 200 with memory parameters 0, 1, and 5 are provided in Fig. 4. We note that for $k = 1$ all three reconstructions are identical since all of them only have access to the first LR image. Reconstructions after 100 iterations are also similar, but after 200 iterations, we see that sn-slimTik with memory level 5 produces a better reconstruction. These results show that including memory in the slimTik algorithm may be beneficial, and results are comparable to those of variable projection.

4.2 Experiment 2: sn-slimTik for a Massive Problem

Next we consider a very large *streaming* super-resolution problem, where the goal is to reconstruct a HR image of 1024^2 pixels from 300 LR images of 64^2

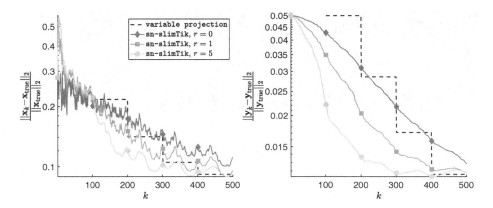

Fig. 3. Relative reconstruction error norms for the image $\mathbf{x}_k$ (left) and the nonlinear parameters $\mathbf{y}_k$ (right) for variable projection and `sn-slimTik` for various memory levels. Note that variable projection errors are only provided after every 100 iterations of `sn-slimTik`.

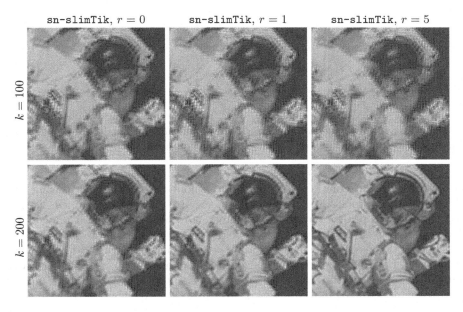

Fig. 4. Sub-images of `sn-slimTik` reconstructions for memory levels $r = 0, 1$, and 5 for iterates within the first two epochs of data access.

pixels that are being observed in time. The HR image comes from NASA [15] and is depicted, along with three of the LR images, in Fig. 5. For this example, once all data has been accessed, $\mathbf{A} \in \mathbb{R}^{300 \cdot 64^2 \times 1{,}024^2}$ is too large to store in memory. Furthermore, in many streaming scenarios, we would like to be able to compute

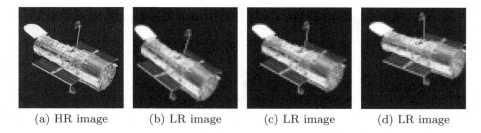

(a) HR image   (b) LR image   (c) LR image   (d) LR image

Fig. 5. Streaming super-resolution imaging example. The high-resolution (1,024 × 1,024) image is provided in (a), along with three of the low-resolution (64 × 64) images in (b)–(d).

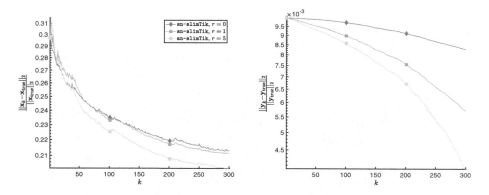

Fig. 6. Relative reconstruction errors for both the linear (left) and nonlinear (right) parameters for the streaming data super-resolution problem.

partial image reconstructions and update the nonlinear parameters during the data acquisition process, e.g., while LR images are still being streamed. Notice that the variable projection method requires us to wait until all LR images are observed, and even then it may be too costly to access all of $\mathbf{A}$ at once.

Thus, in this experiment, we consider the `sn-slimTik` method with memory levels $r = 0, 1$, and 5. We run 300 iterations (e.g., accessing one epoch of the data) and set the noise level for each LR image to be 0.01 and $\lambda = 5 \cdot 10^{-3}$. In Fig. 6 we provide the relative reconstruction errors for $\mathbf{x}_k$ and $\mathbf{y}_k$. We observe that a higher memory level corresponds to improved estimates of the nonlinear parameters and the reconstructions. In Fig. 7, we provide sub-images of absolute errors images of the reconstructions, computed as $|\mathbf{x}_{300} - \mathbf{x}_{\mathrm{true}}|$, in inverted colormap so that white corresponds to small absolute errors. These images show that `sn-slimTik` methods produce better reconstructions with increased memory level, but an increased memory level comes with an increase in computation time. For this example, the CPU times for `sn-slimTik` are 1,035, 1,954, and 5,858 s for memory levels of 0, 1, and 5, respectively.

error sub-image $r = 0$ error sub-image $r = 1$ error sub-image $r = 5$

Fig. 7. Sub-image of absolute error images for `sn-slimTik` reconstructions with different memory levels.

5 Conclusions

In this work we introduced the `sn-slimTik` method, which is a sample based iterative algorithm to approximate the solution of a separable nonlinear inverse problem, for the case where the data cannot be accessed all-at-once. The method combines limited-memory sampled Tikhonov methods, which were developed for linear inverse problems, within a nonlinear optimization framework. Numerical results on massive super-resolution problems show that results are comparable to those from variable projection, when all data can be accessed at once. When this is not the case (e.g., streaming or massive data), the `sn-slimTik` method can effectively and efficiently update both sets of parameters.

A future area of research is to develop a theoretical analysis of the convergence properties of `sn-slimTik` and `slimTik` methods, including asymptotic convergence and a mean squared error analysis. Furthermore, future investigations should incorporate importance sampling, where the sampling strategy can be adapted as data become available.

Acknowledgements. We gratefully acknowledge support by the National Science Foundation under grants NSF DMS 1723005 and NSF DMS 1654175.

References

1. Andersen, M.S., Hansen, P.C.: Generalized row-action methods for tomographic imaging. Numer. Algorithms **67**(1), 121–144 (2014)
2. Berisha, S., Nagy, J.G., Plemmons, R.J.: Estimation of atmospheric PSF parameters for hyperspectral imaging. Numer. Linear Algebra Appl. (2015)
3. Björck, A.: Numerical Methods for Least Squares Problems. SIAM (1996)
4. Chung, J., Chung, M., Slagel, J.T., Tenorio, L.: Stochastic Newton and quasi-Newton methods for large linear least-squares problems. arXiv preprint arXiv:1702.07367 (2017)
5. Chung, J., Haber, E., Nagy, J.G.: Numerical methods for coupled super-resolution. Inverse Prob. **22**, 1261–1272 (2006)
6. Chung, J., Nagy, J.G.: An efficient iterative approach for large-scale separable nonlinear inverse problems. SIAM J. Sci. Comput. **31**(6), 4654–4674 (2010)

7. Cornelio, A., Piccolomini, E.L., Nagy, J.G.: Constrained variable projection method for blind deconvolution. J. Phys. Conf. Ser. **386**, 012005 (2012)
8. Escalante, R., Raydan, M.: Alternating Projection Methods, vol. 8. SIAM (2011)
9. Golub, G., Pereyra, V.: Separable nonlinear least squares: the variable projection method and its applications. Inverse Prob. **19**, R1–R26 (2003)
10. Gower, R.M., Richtárik, P.: Randomized iterative methods for linear systems. SIAM J. Matrix Anal. Appl. **36**(4), 1660–1690 (2015)
11. Hansen, P.C.: Discrete Inverse Problems: Insight and Algorithms. SIAM (2010)
12. Herring, J., Nagy, J., Ruthotto, L.: LAP: a linearize and project method for solving inverse problems with coupled variables. Sampling Theor. Sign. Image Process. **17**(2), 127–151 (2018)
13. Kaczmarz, S.: Angenäherte Auflösung linearer Gleichungssysteme. Bulletin International de l'Académie Polonaise des Sciences et des Lettres. Classe des Sciences Mathématiques et Naturelles. Série A, Sciences Mathématiques, pp. 355–357 (1937)
14. Marchesini, S., et al.: SHARP: a distributed GPU-based ptychographic solver. J. Appl. Crystallogr. **49**(4), 1245–1252 (2016)
15. NASA: Images from NASA webpage. https://www.nasa.gov. Accessed 10 Jan 2019
16. Needell, D., Tropp, J.A.: Paved with good intentions: analysis of a randomized block Kaczmarz method. Linear Algebra Appl. **441**, 199–221 (2014)
17. Needell, D., Zhao, R., Zouzias, A.: Randomized block Kaczmarz method with projection for solving least squares. Linear Algebra Appl. **484**, 322–343 (2015)
18. Nocedal, J., Wright, S.J.: Numerical Optimization, 2nd edn. Springer, New York (2006). https://doi.org/10.1007/978-0-387-40065-5
19. O'Leary, D.P., Rust, B.W.: Variable projection for nonlinear least squares problems. Comput. Optim. Appl. **54**(3), 579–593 (2013)
20. Paige, C.C., Saunders, M.A.: LSQR: an algorithm for sparse linear equations and sparse least squares. ACM Trans. Math. Softw. **8**(1), 43–71 (1982)
21. Parkinson, D.Y., et al.: Machine learning for micro-tomography. In: Developments in X-Ray Tomography XI, vol. 10391, p. 103910J. International Society for Optics and Photonics (2017)
22. Slagel, J.T., Chung, J., Chung, M., Kozak, D., Tenorio, L.: Sampled Tikhonov regularization for large linear inverse problems. In: Inverse Problems (2019, to appear)
23. Strohmer, T., Vershynin, R.: A randomized Kaczmarz algorithm with exponential convergence. J. Fourier Anal. Appl. **15**(2), 262–278 (2009)
24. Zouzias, A., Freris, N.M.: Randomized extended Kaczmarz for solving least squares. SIAM J. Matrix Anal. Appl. **34**(2), 773–793 (2013)

Refitting Solutions Promoted by ℓ_{12} Sparse Analysis Regularizations with Block Penalties

Charles-Alban Deledalle[1,2] , Nicolas Papadakis[2(✉)] , Joseph Salmon[3], and Samuel Vaiter[4]

[1] Department of Electrical and Computer Engineering, University of California, San Diego, La Jolla, USA
[2] CNRS, Univ. Bordeaux, IMB, UMR 5251, 33400 Talence, France
nicolas.papadakis@math.u-bordeaux.fr
[3] IMAG, Univ. Montpellier, CNRS, Montpellier, France
[4] IMB, CNRS, Université de Bourgogne, 21078 Dijon, France

Abstract. In inverse problems, the use of an ℓ_{12} analysis regularizer induces a bias in the estimated solution. We propose a general refitting framework for removing this artifact while keeping information of interest contained in the biased solution. This is done through the use of refitting block penalties that only act on the co-support of the estimation. Based on an analysis of related works in the literature, we propose a new penalty that is well suited for refitting purposes. We also present an efficient algorithmic method to obtain the refitted solution along with the original (biased) solution for any convex refitting block penalty. Experiments illustrate the good behavior of the proposed block penalty for refitting.

Keywords: Total variation · Bias correction · Refitting

1 Introduction

We consider linear inverse problems of the form $y = \Phi x + w$, where $y \in \mathbb{R}^p$ is an observed degraded image, $x \in \mathbb{R}^n$ the unknown clean image, $\Phi : \mathbb{R}^n \to \mathbb{R}^p$ a linear operator and $w \in \mathbb{R}^p$ a noise component, typically a zero-mean white Gaussian random vector with standard deviation $\sigma > 0$. To reduce the effect of noise and the potential ill-conditioning of Φ, we consider a regularized least square problem with an ℓ_{12} structured sparse analysis term of the form

$$\hat{x} \in \underset{x}{\operatorname{argmin}} \ \tfrac{1}{2}\|\Phi x - y\|^2 + \lambda\|\Gamma x\|_{1,2} \ . \tag{1}$$

where $\lambda > 0$ is a regularization parameter, $\Gamma : \mathbb{R}^n \to \mathbb{R}^{m \times b}$ is a linear analysis operator mapping an image over m blocks of size b and $\|z\|_{1,2} = \sum_{i=1}^{m} \|z_i\| = \sum_{i=1}^{m} (\sum_{j=1}^{b} z_{i,j}^2)^{1/2}$, with $z_i = \{z_{i,j}\}_{j=1}^{b} \in \mathbb{R}^b$. This model is known to recover co-sparse solutions, *i.e.*, such that $(\Gamma x)_i = 0_b$ for most blocks $1 \leqslant i \leqslant m$.

© Springer Nature Switzerland AG 2019
J. Lellmann et al. (Eds.): SSVM 2019, LNCS 11603, pp. 131–143, 2019.
https://doi.org/10.1007/978-3-030-22368-7_11

A typical example is the one of isotropic total-variation (TViso) with $\Gamma = \nabla$ being the operator which extracts $m = n$ image gradient vectors of size $b = 2$ (for volumes $b = 3$, and so on). The anisotropic total-variation is another example corresponding to Γ the operator which concatenates the vertical and horizontal components of the gradients into a vector of size $m = 2n$, hence $b = 1$.

1.1 Refitting

The co-support of an image x (or support of Γx) is the set of its non-zero blocks:

$$\operatorname{supp}(\Gamma x) = \{1 \leqslant i \leqslant m \ : \ (\Gamma x)_i \neq 0_b\} \ . \tag{2}$$

While in some cases, the estimate $\hat{x}$ obtained by structured sparse analysis regularization (1) recovers correctly the co-support of the underlying signal x, it nevertheless suffers from a systematical bias in the estimated amplitudes $\hat{x}_i$. With TViso, this bias is reflected by a loss of contrast (see Fig. 1(b)). A standard strategy to reduce this effect, called refitting [1,8,9,11], consists in approximating y through Φ by an image sharing the same co-support as $\hat{x}$:

$$\tilde{x}^{\text{supp}} \in \underset{x; \ \operatorname{supp}(\Gamma x) \subseteq \hat{\mathcal{I}}}{\operatorname{argmin}} \ \tfrac{1}{2}\|\Phi x - y\|^2 \ , \tag{3}$$

where $\hat{\mathcal{I}} = \operatorname{supp}(\Gamma \hat{x})$. While this strategy works well for blocks of size $b = 1$, it suffers from an excessive increase of variance whenever $b \geqslant 2$, $e.g.$, for TViso. This is due to the fact that solutions do not only present sharp edges, but may involve gradual transitions. To cope with this issue, additional features of $\hat{x}$ than its co-support must be preserved by a refitting procedure. For the LASSO ($\Gamma = \text{Id}$, $m = n$ and $b = 1$), a pointwise preservation of the sign of $\hat{x}_i$ onto the support improves the numerical performances of the refitting [5].

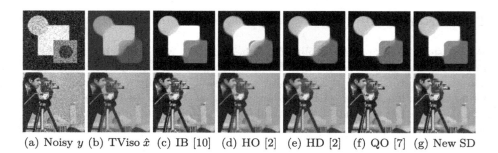

(a) Noisy y (b) TViso $\hat{x}$ (c) IB [10] (d) HO [2] (e) HD [2] (f) QO [7] (g) New SD

Fig. 1. Comparison of standard refitting approaches with the proposed SD model.

1.2   Outline and Contributions

In this paper, we introduce a new framework for refitting solutions promoted by ℓ_{12} structured sparse analysis. In Sect. 2, we present related works, illustrated in Fig. 1, that include Bregman iterations [10] or debiasing approaches [2,7]. In Sect. 3, we describe our general variational refitting method for block penalties and show how the works presented in Sect. 2 can be described with such a framework. We discuss suitable properties a refitting block penalty should satisfy and introduce the Soft-penalized Direction model (SD), a new flexible refitting block penalty inheriting the advantages of the Bregman-based approaches. In Sect. 4, we propose a stable and one-step algorithm to compute our refitting strategy for any convex refitting block penalty, including the models in [2,7]. Experiments in Sect. 5 illustrate the practical benefits for the SD refitting and the potential of our framework for image processing.

2   Related Re-fitting Works

We first present some properties of Bregman divergences used all along the paper.

2.1   Properties of Bregman Divergence of ℓ_{12} Structured Regularizers

In the literature [2,10], Bregman divergences have proven to be well suited to measure the discrepancy between the biased solution $\hat{x}$ and its refitting $\tilde{x}$. We recall that for a convex function ψ, the associated (generalized) Bregman divergence between x and $\hat{x}$ is, for any subgradient $\hat{p} \in \partial \psi(\hat{x})$: $D_\psi(x,\hat{x}) = \psi(x) - \psi(\hat{x}) - \langle \hat{p}, x - \hat{x} \rangle \geqslant 0$. If ψ is an absolutely 1-homogeneous function $(\psi(\alpha x) = |\alpha|\psi(x), \forall \alpha \in \mathbb{R})$ then $p \in \partial \psi(x) \Rightarrow \psi(x) = \langle p, x \rangle$ and the Bregman divergence simplifies into $D_\psi(x,\hat{x}) = \psi(x) - \langle \hat{p}, x \rangle \geqslant 0$. For regularizers of the form $\psi(x) = \|\Gamma x\|_{1,2}$, a subgradient $\hat{p} \in \partial \|\Gamma \cdot \|_{1,2}(\hat{x})$ satisfies $\hat{p} = \Gamma^\top \hat{z}$ with [3]:

$$\begin{cases} \hat{z}_i = \frac{(\Gamma\hat{x})_i}{\|(\Gamma\hat{x})_i\|} & \text{if } i \in \hat{\mathcal{I}} = \operatorname{supp}(\Gamma\hat{x}) \ , \\ \|\hat{z}_i\| \leqslant 1 & \text{otherwise} \ , \end{cases} \tag{4}$$

where we have $\Gamma^\top \hat{z} \in \partial \|\Gamma \cdot \|_{1,2}(\hat{x}) \Leftrightarrow \hat{z}_i \in \partial \| \cdot \|((\Gamma\hat{x})_i), \forall i \in [m]$. Now denoting

$$D_i(\Gamma x) = \|(\Gamma x)_i\| - \langle \hat{z}_i, (\Gamma x)_i \rangle = D_{\|\cdot\|}((\Gamma x)_i, (\Gamma\hat{x})_i), \tag{5}$$

there exists an interesting relation between the global Bregman divergence on x's and the local ones on $(\Gamma x)_i$'s:

$$D_{\|\Gamma\cdot\|_{1,2}}(x,\hat{x}) = \sum_{i=1}^{m} (\|(\Gamma x)_i\| - \langle \hat{z}_i, (\Gamma x)_i \rangle) = \sum_{i=1}^{m} D_i(\Gamma x) \ . \tag{6}$$

Combining relations (4) and (5), we see that on the co-support $i \in \hat{\mathcal{I}}$, such divergence measures the fit of directions between $(\Gamma x)_i$ and $(\Gamma\hat{x})_i$:

$$\forall i \in \hat{\mathcal{I}}, \ D_i(\Gamma x) = 0 \text{ iff } \exists \alpha_i \geqslant 0 \text{ such that } (\Gamma x)_i = \alpha_i(\Gamma\hat{x})_i \ . \tag{7}$$

This divergence also partially captures the co-support as we have from (5)

$$\forall i \in \hat{\mathcal{I}}^c, \; D_i(\Gamma x) = 0 \text{ with } \hat{z}_i \in \partial \| \cdot \| ((\Gamma \hat{x})_i) \text{ s.t. } \|\hat{z}_i\| < 1 \text{ iff } (\Gamma x)_i = 0_b \; . \quad (8)$$

2.2   Bregman-Based Refitting

We now review some refitting methods based on Bregman divergences.

Flexible Iterative Bregman Regularization. The Bregman process [10] reduces the bias of solutions of (1) by successively solving problems of the form

$$\tilde{x}_{l+1} \in \operatorname*{argmin}_{x} \tfrac{1}{2} \| \Phi x - y \|^2 + \lambda D_{\| \Gamma \cdot \|_{1,2}}(x, \tilde{x}_l). \quad (9)$$

We here consider a fixed λ, but different refitting strategies can be considered with decreasing parameters λ_l as in [12,13].

Setting $\tilde{x}_0 = 0_n$, we have $0_n \in \partial \| \Gamma \cdot \|_{1,2}(0_n)$ so that $D_{\| \Gamma \cdot \|_{1,2}}(x, 0_n) = \| \Gamma x \|_{1,2}$ and we recover the biased solution of (1) as $\tilde{x}_1 = \hat{x}$. We denote by $\tilde{x}^{\mathrm{IB}} = \tilde{x}_2$ the refitting obtained after 2 steps of the Iterative Bregman (IB) procedure (9):

$$\tilde{x}^{\mathrm{IB}} = \tilde{x}_2 \in \operatorname*{argmin}_{x} \tfrac{1}{2} \| \Phi x - y \|^2 + \lambda D_{\| \Gamma \cdot \|_{1,2}}(x, \hat{x}) \; . \quad (10)$$

As underlined in relation (7), by minimizing $D_{\| \Gamma \cdot \|_{1,2}}(x, \hat{x})$, one aims at preserving the direction of $\Gamma \hat{x}$ on the support $\hat{\mathcal{I}}$, without ensuring $\operatorname{supp}(\Gamma \tilde{x}^{\mathrm{IB}}) \subseteq \hat{\mathcal{I}}$. This issue can be observed in the background of *Cameraman* in Fig. 1(c), where noise reinjection is visible. For the iterative framework, the co-support of the previous solution may indeed not be preserved ($\|(\Gamma \tilde{x}_l)_i\| = 0 \not\Rightarrow \|(\Gamma \tilde{x}_{l+1})_i\| = 0$) and can hence grow. The support of Γx_0 for $\tilde{x}_0 = 0_n$ is for instance totally empty whereas the one of $\hat{x} = \tilde{x}_1$ may not (and should not) be empty. For $l \to \infty$, the process actually converges to some x such that $\Phi x = y$.

Hard-Constrained Refitting Without Explicit Support Identification. In order to respect the support of the biased solution $\hat{x}$ and to keep track of the direction $\Gamma \hat{x}$ during the refitting, the authors of [2] proposed the following model:

$$\tilde{x}^{\mathrm{HD}} \in \operatorname*{argmin}_{x; \hat{p} \in \partial \| \Gamma \cdot \|_{1,2}(x)} \tfrac{1}{2} \| \Phi x - y \|^2 \; , \quad (11)$$

for $\hat{p} \in \partial \| \Gamma \cdot \|_{1,2}(\hat{x})$. This model enforces the Bregman divergence to be 0, since $\hat{p} \in \partial \| \Gamma \cdot \|_{1,2}(x) \Rightarrow \| \Gamma x \|_{1,2} = \langle \hat{p}, x \rangle \Rightarrow D_{\| \Gamma \cdot \|_{1,2}}(x, \hat{x}) = 0$.

We see from (7) that for $i \in \hat{\mathcal{I}}$, the direction of $(\Gamma \hat{x})_i$ is preserved in the refitted solution. Following relations (4) and (8), the co-support is also preserved for any $i \in \hat{\mathcal{I}}^c$ such that $\|\hat{z}_i\| < 1$. Note though that extra elements in the co-support $\Gamma \tilde{x}^{\mathrm{HD}}$ may be added at coordinates $i \in \hat{\mathcal{I}}^c$ such that $\|\hat{z}_i\| = 1$. We denote this model as HD, for Hard-constrained Direction. To get ride of the direction dependency, a Hard-constrained Orientation (HO) model is also proposed in [2]:

$$\tilde{x}^{HO} \in \underset{x; \pm\hat{p}\in\partial\|\Gamma\cdot\|_{1,2}(x)}{\operatorname{argmin}} \frac{1}{2}\|\Phi x - y\|^2 \ . \tag{12}$$

The orientation model may nevertheless involve contrast inversions between biased and refitted solutions, as shown in Fig. 1(d) with the banana dark shape in the white region. In practice, relaxations are used in [2] by solving, for a large value $\gamma > 0$

$$\tilde{x}^{HD}_{\gamma} \in \underset{x}{\operatorname{argmin}} \frac{1}{2}\|\Phi x - y\|^2 + \gamma D_{\|\Gamma\cdot\|_{1,2}}(x, \hat{x}) \ . \tag{13}$$

The main advantage of this refitting strategy is that no support identification is required since everything is implicitly encoded in the subgradient $\hat{p}$. This makes the process stable even if the estimation of $\hat{x}$ is not highly accurate. The support of $\Gamma\hat{x}$ is nevertheless only approximately preserved, since the constraint $D_{\|\Gamma\cdot\|_{1,2}}(x, \hat{x}) = 0$ can never be ensured numerically with a finite value of γ. Finally, as shown in Fig. 1(d–e), such constrained approaches lack of flexibility since the orientation of $\Gamma\tilde{x}$ cannot deviate from the one of $\Gamma\hat{x}$ (for complex signals, such as *Cameraman*, amplitudes remain significantly biased and less details are recovered).

2.3   Flexible Quadratic Refitting Without Support Identification

We now describe an alternative way for performing variational refitting. When specialized to $\ell_{1,2}$ sparse analysis regularization, CLEAR, a general refitting framework [7], consists in computing

$$\tilde{x}^{QO} \in \underset{x;\ \operatorname{supp}(\Gamma x)\subseteq\hat{\mathcal{I}}}{\operatorname{argmin}} \frac{1}{2}\|\Phi x - y\|^2 + \sum_{i\in\hat{\mathcal{I}}} \frac{\lambda}{2\|(\Gamma\hat{x})_i\|} \left\| (\Gamma x)_i - \left\langle (\Gamma x)_i, \frac{(\Gamma\hat{x})_i}{\|(\Gamma\hat{x})_i\|} \right\rangle \frac{(\Gamma\hat{x})_i}{\|(\Gamma\hat{x})_i\|} \right\|^2 . \tag{14}$$

This model promotes refitted solutions preserving to some extent the orientation $\Gamma\hat{x}$ of the biased solution. It also shrinks the amplitude of Γx all the more that the amplitude of $\Gamma\hat{x}$ are small. As this model penalizes changes of orientation, we refer to it as QO for *Quadratic-penalized Orientation*. This penalty does not promote any kind of direction preservation, and as for the HO model, contrast inversions may be observed between biased and refitted solutions (see Fig. 1(f)). The quadratic term also over-penalizes large changes of orientation.

3   Refitting with Block Penalties

As mentioned in the previous section, the methods of [2,7] have proposed variational refitting formulations that not only aim at preserving the co-support $\hat{\mathcal{I}}$

but also the orientation of $(\Gamma \hat{x})_{i \in \hat{\mathcal{I}}}$. In this paper, we propose to express these (two-steps) refitting procedures in the following general framework

$$\tilde{x}^{\varphi} \in \underset{x;\ \mathrm{supp}(\Gamma x) \subseteq \hat{\mathcal{I}}}{\mathrm{argmin}} \ \tfrac{1}{2} \|\Phi x - y\|^2 + \sum_{i \in \hat{\mathcal{I}}} \varphi((\Gamma x)_i, (\Gamma \hat{x})_i) \ , \qquad (15)$$

where $\varphi : \mathbb{R}^b \times \mathbb{R}^b \to \mathbb{R}$ is a refitting block penalty ($b \geqslant 1$ is the size of the blocks) promoting Γx to share information with $\Gamma \hat{x}$ in some sense to be specified. To refer to some features of the vector $\Gamma \hat{x}$, let us first define properly the notions of relative orientation, direction and projection between two vectors.

Definition 1. *Let z and $\hat{z}$ being two vectors in $\mathbb{R}^b$, we define*

$$\cos(z, \hat{z}) = \left\langle \tfrac{z}{\|z\|}, \tfrac{\hat{z}}{\|\hat{z}\|} \right\rangle = \tfrac{1}{\|z\|\|\hat{z}\|} \sum_{j=1}^{b} z_j \hat{z}_j \ , \qquad (16)$$

and $\quad P_{\hat{z}}(z) = \left\langle z, \tfrac{\hat{z}}{\|\hat{z}\|} \right\rangle \tfrac{\hat{z}}{\|\hat{z}\|} = \tfrac{\|z\|}{\|\hat{z}\|} \cos(z, \hat{z}) \hat{z} \ , \qquad (17)$

where $P_{\hat{z}}(z)$ is the orthogonal projection of z onto $\mathrm{Span}(\hat{z})$ (i.e., the orientation axis of $\hat{z}$). We say that z and $\hat{z}$ share the same orientation (resp. direction), if $|\cos(z, \hat{z})| = 1$ (resp. $\cos(z, \hat{z}) = 1$).

Thanks to Definition 1, we can now reformulate the previous refitting models in terms of block penalties. The Hard-constrained refitting models preserving Direction (11) and Orientation (12) of [2] as well as the flexible Quadratic model of CLEAR [7] correspond to the following block penalties:

$$\varphi_{\mathrm{HD}}(z, \hat{z}) = \iota_{\{z \in \mathbb{R}^b : \cos(z, \hat{z}) = 1\}} \qquad (18)$$

$$\varphi_{\mathrm{HO}}(z, \hat{z}) = \iota_{\{z \in \mathbb{R}^b : |\cos(z, \hat{z})| = 1\}} \qquad (19)$$

$$\text{and} \quad \varphi_{\mathrm{QO}}(z, \hat{z}) = \tfrac{\lambda}{2\|\hat{z}\|} \|z - P_{\hat{z}}(z)\|^2 = \tfrac{\lambda}{2} \tfrac{\|z\|^2}{\|\hat{z}\|} (1 - \cos^2(z, \hat{z})) \ . \qquad (20)$$

where $\iota_{\mathcal{C}}$ is the $0/+\infty$ indicator function of a set $\mathcal{C}$. These block penalties are either insensitive to directions (HO and QO) or intolerant to small changes of orientations (HD and HO), hence not satisfying (*cf.* drawbacks visible in Fig. 1).

When $b = 1$, the orientation-based penalties (QO and HO) have absolutely no effect while the direction-based penalty HD preserves the sign of $(\Gamma \hat{x})_i$. In this paper, we argue that the direction of $(\Gamma \hat{x})_i$, for any $b \geqslant 1$, carries important information that is worth preserving when refitting, at least to some extent.

3.1   Desired Properties of Refitting Block Penalties

To compute global optimum of the refitting model (15), we only consider convex refitting block penalties $z \mapsto \varphi(z, \hat{z})$. Hence, we now introduce properties a block penalty φ should satisfy for refitting purposes:

(P1) φ is convex, non negative and $\varphi(z, \hat{z}) = 0$, if $\cos(z, \hat{z}) = 1$ or $\|z\| = 0$,

(P2) $\varphi(z', \hat{z}) \geqslant \varphi(z, \hat{z})$ if $\|z'\| = \|z\|$ and $\cos(z, \hat{z}) \geqslant \cos(z', \hat{z})$,

(P3) $z \mapsto \varphi(z, \hat{z})$ is continuous,

Property (P1) stipulates that no configuration can be more favorable than z and $\hat{z}$ having the same direction. Hence, the direction of the refitted solution should be encouraged to follow the one of the biased solution. Property (P2) imposes that for a fixed amplitude, the penalty should be increasing w.r.t. the angle formed with $\hat{z}$. Property (P3) enforces refitting that can continuously adapt to the data and be robust to small perturbations.

Table 1. Properties satisfied by the considered block penalties φ.

Properties	HO	HD	QO	SD
1	√	√	√	√
2		√		√
3			√	√

3.2 A New Flexible Refitting Block Penalty

We now introduce our refitting block penalty designed to preserve the desired features of $\hat{z} = \Gamma\hat{x}$ in a simple way. The *Soft-penalized Direction* penalty reads

$$\varphi_{\mathrm{SD}}(z, \hat{z}) = \lambda\|z\|(1 - \cos(z, \hat{z})) \ . \tag{21}$$

The properties of the different studied block penalties are presented in Table 1. The proposed SD model is the only one satisfying all the desired properties. As illustrated in Fig. 2, it is a continuous penalization that increases continuously with respect to the absolute angle between z and $\hat{z}$.

3.3 SD Block Penalty: The Best of Both Bregman Worlds

Denoting $D_i(\Gamma x) = D_{\|\cdot\|}((\Gamma x)_i, (\Gamma\hat{x})_i)$ as introduced in Sect. 2.1, the refitting models given in (10) and (11) can be expressed as

$$\tilde{x}^{\mathrm{IB}} \in \operatorname*{argmin}_x \tfrac{1}{2}\|\Phi x - y\|^2 + \lambda\sum_{i=1}^{m} D_i(\Gamma x) \ , \tag{22}$$

$$\tilde{x}^{\mathrm{HD}} \in \operatorname*{argmin}_x \tfrac{1}{2}\|\Phi x - y\|^2 \qquad\qquad \text{s.t. } D_i(\Gamma x) = 0, \ \forall i \in [m] \ . \tag{23}$$

Observing that the new SD block penalty (21) may be rewritten as $\varphi_{\mathrm{SD}}(z, \hat{z}) = \lambda\left(\|z\| - \langle z, \hat{z}/\|\hat{z}\|\rangle\right) = \lambda D_{\|\cdot\|}(z, \hat{z})$, the SD refitting model is

$$\tilde{x}^{\mathrm{SD}} \in \operatorname*{argmin}_x \tfrac{1}{2}\|\Phi x - y\|^2 + \lambda\sum_{i\in\hat{\mathcal{I}}} D_i(\Gamma x), \quad \text{s.t. } D_i(\Gamma x) = 0, \ \forall i \in \hat{\mathcal{I}}^c \ . \tag{24}$$

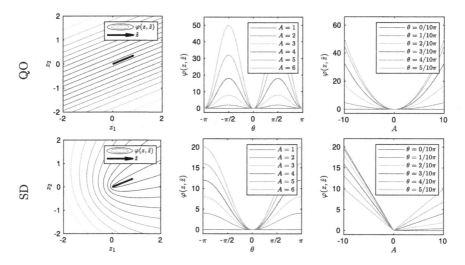

Fig. 2. Illustration of continuous block penalties QO and SD: (left) 2D level lines of φ for $z = (z_1, z_2) = A \left(\begin{smallmatrix} \cos\theta & -\sin\theta \\ \sin\theta & \cos\theta \end{smallmatrix} \right) \hat{z}$, (middle) evolution regarding θ and (right) A.

With such reformulations, connections between refitting models (22), (23) and (24) can be clarified. The solution $\tilde{x}^{IB}$ [10] is too relaxed, as it only penalizes the directions $(\varGamma x)_i$ using $(\varGamma \hat{x})_i$, without aiming at preserving the co-support of $\hat{x}$. The solution $\tilde{x}^{HD}$ [2] is too constrained: the direction within the co-support is required to be preserved exactly. Our proposed refitting $\tilde{x}^{SD}$ lies in-between: it preserves the co-support, while authorizing some directional flexibility, as illustrated by the sharper square edges in Fig. 1(g).

With respect to the Hard-constrained approach [2], an important difference is that we consider local inclusions of subgradients of the function $\lambda \| \cdot \|_{1,2}$ at point $(\varGamma x)_i$ instead of the global inclusion of subgradients of the function $\lambda \| \varGamma \cdot \|_{1,2}$ at point x as in HD (11) and HO (12). Such a change of paradigm allows to adapt the refitting locally by preserving the co-support while including the flexibility of the original Bregman approach [10].

4   Refitting in Practice

We first describe how computing $\hat{x}$ and its refitting $\tilde{x}$ in two successive steps and then propose a new numerical scheme for the joint computation of $\hat{x}$ and $\tilde{x}$.

4.1   Biased Problem and Posterior Refitting

To obtain $\hat{x}$ solution of (1), we consider the primal dual formulation that reads

$$\min_x \max_z \tfrac{1}{2}\|\varPhi x - y\|^2 + \langle \varGamma x, z \rangle - \iota_{B_2^\lambda}(z) \ , \tag{25}$$

where $\iota_{B_2^\lambda}$ is the indicator function of the ℓ_2 ball of radius λ that is 0 if $\|z_i\| \leqslant \lambda$ for all $i \in [m]$ and $+\infty$ otherwise. This problem can be solved with the iterative primal-dual algorithm [4] presented in the left part of Algorithm (29). For positive parameters satisfying $\tau\sigma < \|\Gamma\|$ and $\theta \in [0,1]$, the iterates $(\hat{z}^k, \hat{x}^k)$ converge to a saddle point $(\hat{z}, \hat{x})$ satisfying $\hat{z}_i \in \partial\|\cdot\|((\Gamma\hat{x})_i)$, $\forall i$.

Assume that the co-support $\hat{\mathcal{I}} = \text{supp}(\Gamma\hat{x})$ of the biased solution has been identified, a posterior refitting can be obtained by solving (15) for any refitting block penalty φ. To that end, we write the characteristic function of co-support preservation as $\sum_{i \in \hat{\mathcal{I}}^c} \iota_{\{0\}}(z_i)$, where $\iota_{\{0\}}(z) = 0$ if $z = 0$ and $+\infty$ otherwise. By introducing the convex function

$$\omega_\varphi(z, \hat{z}, \hat{\mathcal{I}}) = \sum_{i \in \hat{\mathcal{I}}^c} \iota_{\{0\}}(z_i) + \sum_{i \in \hat{\mathcal{I}}} \varphi(z_i, \hat{z}_i) \ , \tag{26}$$

the general refitting problem (15) can be expressed as

$$\tilde{x}^\varphi \in \underset{x}{\text{argmin}} \ \tfrac{1}{2}\|\Phi x - y\|^2 + \omega_\varphi(\Gamma x, \Gamma\hat{x}, \hat{\mathcal{I}}) \ . \tag{27}$$

Subsequently, we can consider its primal dual formulation

$$\min_x \max_z \ \tfrac{1}{2}\|\Phi x - y\|^2 + \langle \Gamma x, z \rangle - \omega_\varphi^*(z, \Gamma\hat{x}, \hat{\mathcal{I}}) \ , \tag{28}$$

where $\omega_\varphi^*(z, \Gamma\hat{x}, \hat{\mathcal{I}}) = \sup_x \langle z, x \rangle - \omega_\varphi(x, \Gamma\hat{x}, \hat{\mathcal{I}})$ is the convex conjugate, with respect to the first argument, of $\omega_\varphi(\cdot, \Gamma\hat{x}, \hat{\mathcal{I}})$. Such problem can again be solved with the primal-dual algorithm [4]. The crucial point is to have an accurate estimation of the vector $\Gamma\hat{x}$ and its support $\hat{\mathcal{I}}$. Yet, it is well known that estimating $\text{supp}(\Gamma\hat{x})$ from an estimation $\hat{x}^k$ is not stable numerically: the support $\text{supp}(\Gamma\hat{x}^k)$ can be far from $\text{supp}(\Gamma\hat{x})$ even though $\hat{x}^k$ is arbitrarily close to $\hat{x}$.

4.2   Joint-Refitting Algorithm

We now introduce a general algorithm aiming to jointly solve the original problem (1) and the refitting one (15) for any refitting block penalty φ. This framework has been developed for stable projection onto the support in [6] and later extended to refitting with the Quadratic Orientation penalty in [7]. The strategy consists in solving in parallel the two problems (25) and (28).

Two iterative primal-dual algorithms are used for the biased variables $(\hat{z}^k, \hat{x}^k)$ and the refitted ones $(\tilde{z}^k, \tilde{x}^k)$. Let us now present the whole algorithm:

$$\hat{z}_i^{k+1} = \frac{\hat{z}_i^k + \sigma(\Gamma\hat{v}^k)_i}{\max(\lambda, \|\hat{z}_i^k + \sigma(\Gamma\hat{v}^k)_i\|)} \qquad \left| \begin{aligned} \hat{\mathcal{I}}^k &= \{i \in [m] \ : \ \|\hat{z}_i^k + \sigma(\Gamma\hat{v}^k)_i\| > \lambda\} \\ \tilde{z}^{k+1} &= \text{prox}_{\sigma\omega_\varphi^*(\cdot, \Psi(\hat{z}^k, \hat{v}^k), \hat{\mathcal{I}}^k)}(\tilde{z}^k + \sigma\Gamma\tilde{v}^k) \end{aligned} \right.$$

$$\hat{x}^{k+1} = \Phi_\tau^- \left(\hat{x}^k + \tau(\Phi^\top y - \Gamma^\top \hat{z}^{k+1}) \right) \qquad \tilde{x}^{k+1} = \Phi_\tau^- \left(\tilde{x}^k + \tau(\Phi^\top y - \Gamma^\top \tilde{z}^{k+1}) \right),$$

$$\hat{v}^{k+1} = \hat{x}^{k+1} + \theta(\hat{x}^{k+1} - \hat{x}^k) \qquad\qquad \tilde{v}^{k+1} = \tilde{x}^{k+1} + \theta(\tilde{x}^{k+1} - \tilde{x}^k), \tag{29}$$

with the operator $\Phi_\tau^- = (\text{Id} + \tau\Phi^\top\Phi)^{-1}$ and the auxiliary variables $\hat{v}^k$ and $\tilde{v}^k$ that accelerate the algorithm. Following [4], for any positive scalars τ and σ

Table 2. Convex conjugates and proximal operators of the studied block penalties φ.

φ	$\varphi^*(z, \hat{z})$		$\text{prox}_{\sigma\varphi^*}(z_0, \hat{z})$	
HO	$\begin{cases} 0, & \text{if} \quad \cos(z, \hat{z}) = 0 \\ +\infty, & \text{otherwise} \end{cases}$		$z_0 - P_{\hat{z}}(z_0)$	
HD	$\begin{cases} 0, & \text{if} \quad \cos(z, \hat{z}) \leqslant 0 \\ +\infty, & \text{otherwise} \end{cases}$		$\begin{cases} z_0 - P_{\hat{z}}(z_0), & \text{if} \quad \langle z_0, \hat{z} \rangle \geqslant 0 \\ z_0, & \text{otherwise} \end{cases}$	
QO	$\begin{cases} \frac{\|\hat{z}\|}{2\lambda}\|z\|^2, & \text{if} \quad \cos(z, \hat{z}) = 0 \\ +\infty, & \text{otherwise} \end{cases}$		$\frac{\lambda}{\lambda + \sigma\|\hat{z}\|}\left(z_0 - P_{\hat{z}}(z_0)\right)$	
SD	$\begin{cases} 0, & \text{if} \quad \|z + \lambda\frac{\hat{z}}{\|\hat{z}\|}\| \leqslant \lambda^a \\ +\infty, & \text{otherwise} \end{cases}$		$\lambda\left(\frac{z_0 + \lambda\frac{\hat{z}}{\|\hat{z}\|}}{\max(\lambda, \|z_0 + \lambda\frac{\hat{z}}{\|\hat{z}\|}\|)} - \frac{\hat{z}}{\|\hat{z}\|}\right)$	

^{a}: Note that the condition implies that $\cos(z, \hat{z}) \leqslant 0$.

satisfying $\tau\sigma\|\Gamma^\top\Gamma\| < 1$ and $\theta \in [0, 1]$, the estimates $(\hat{z}^k, \hat{x}^k, \hat{v}^k)$ of the biased solution converge to $(\hat{z}, \hat{x}, \hat{x})$, where $(\hat{z}, \hat{x})$ is a saddle point of (25).

In the right part of Algorithm (29), we rely on the proximal operator that is, for a convex function ψ and at point z_0, $\text{prox}_{\sigma\psi}(z) = \text{argmin}_z \frac{1}{2\sigma}\|z - z_0\|^2 + \psi(z)$. From the block structure of the function ω_φ defined in (26), the computation of its proximal operator may be realized pointwise. Since $\iota_{\{0\}}(z)^* = 0$, we have

$$\text{prox}_{\sigma\omega_\varphi^*}(z^0, \Gamma\hat{x}, \hat{\mathcal{I}})_i = \begin{cases} \text{prox}_{\sigma\varphi^*}(z_i^0, (\Gamma\hat{x})_i), & \text{if } i \in \hat{\mathcal{I}}, \\ z_i^0, & \text{otherwise}. \end{cases} \tag{30}$$

Table 2 gives the expressions of the dual functions φ^* with respect to their first variable and their related proximal operators $\text{prox}_{\sigma\varphi^*}$ for the refitting block penalties considered in this paper.

The idea behind this joint-refitting algorithm is to perform online co-support detection using the dual variable $\hat{z}^k$ of the biased variable $\hat{x}^k$. From relations in (4), we expect at convergence $\hat{z}^k$ to saturate on the support of $\Gamma\hat{x}$ and to satisfy the optimality condition $\hat{z}_i^k = \lambda\frac{(\Gamma\hat{x})_i}{\|(\Gamma\hat{x})_i\|}$. In practice, the norm of the dual variable $\hat{z}_i^k$ saturates to λ relatively fast onto $\hat{\mathcal{I}}$. As a consequence, it is far more stable to detect the support of $\Gamma\hat{x}$ with the dual variable $\hat{z}^k$ than with the vector $\Gamma\hat{x}^k$ itself. In the first step of Algorithm (29), the condition $\|\hat{z}_i^k + \sigma(\Gamma\hat{v}^k)_i\| > \lambda$ is thus used to detect elements of the support $i \in \hat{\mathcal{I}}^k$ of $\Gamma\hat{x}^k$ along iterations[1]. The function $\Psi(\hat{z}^k, \hat{v}^k)$ aims at approximating $\Gamma\hat{x}$ with the current values of the available variables $(\hat{z}^k, \hat{v}^k)$ of the biased problem that is solved simultaneously. Following [7], the function Ψ can be chosen as

$$\Psi(\hat{z}^k, \hat{v}^k)_i = \frac{\|\hat{\nu}_i^k\| - \lambda}{\sigma\|\hat{\nu}_i^k\|}\hat{\nu}_i^k \quad \text{where} \quad \hat{\nu}_i^k = (\hat{z}^k + \sigma\Gamma\hat{v}^k)_i. \tag{31}$$

that satisfies $(\Psi(\hat{z}, \hat{v}))_i = (\Gamma\hat{x})_i$ at convergence, while appearing to give very stable online estimations of the direction of $\Gamma\hat{x}$ through $\hat{z}^k$.

---

[1] As in [2], extended support $\|\hat{z}_i\| = \lambda$ can be tackled by testing $\|(\hat{z}^k + \sigma\Gamma\hat{v}^k)_i\| \geqslant \lambda$.

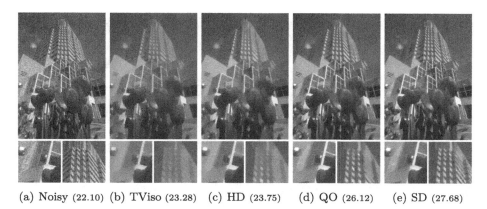

(a) Noisy (22.10) (b) TViso (23.28)   (c) HD (23.75)   (d) QO (26.12)   (e) SD (27.68)

Fig. 3. (a) An 8bit color image corrupted by Gaussian noise with standard deviation $\sigma = 20$. (b) Solution of TViso. Debiased solution with (c) HD, (d) QO and (e) SD. The Peak Signal to Noise Ratio (PSNR) is indicated in brackets below each image.

This joint-estimation considers at every iteration k different refitting functions $\omega_\varphi^*(., \Psi(\hat{z}^k, \hat{v}^k), \hat{\mathcal{I}}^k)$ in (27). For fixed values of $\Psi(\hat{z}^k, \hat{v}^k)$ and $\hat{\mathcal{I}}^k$, the refitted variables $(\tilde{z}^k, \tilde{x}^k)$ in the right part of the Algorithm (29) converges since it exactly corresponds to the primal-dual algorithm [4] applied to the problem (28). However, unless $b = 1$ (see [6]), we do not have guarantee of convergence of the presented scheme with a varying ω_φ^*. As in [7], we nevertheless observe convergence and a very stable behavior for this algorithm.

In addition to its better numerical stability, the running time of joint-refitting is more interesting than the posterior approach. In Algorithm (29), the refitting variables at iteration k require the biased variables at the same iteration and the whole process can be realized in parallel without significantly affecting the running time of the original biased process. On the other hand, posterior refitting is necessarily sequential and the running time is doubled in general.

5   Results

We considered TViso regularization of degraded color images. We defined blocks obtained by applying $\Gamma = [\nabla_x^R, \nabla_y^R, \nabla_x^G, \nabla_y^G, \nabla_x^B, \nabla_y^B]$ where $m = n$, $b = 6$, and ∇_d^C denotes the gradient in the direction $d \in \{x, y\}$ for the color channel $C \in \{R, G, B\}$. We first focused on a denoising problem $y = x + w$ where x is an 8bit color image and w is an additive white Gaussian noise with standard deviation $\sigma = 20$. We next focused on a deblurring problem $y = \Phi x + w$ where x is an 8bit color image, Φ is a convolution simulating a directional blur, and w is an additive white Gaussian noise with standard deviation $\sigma = 2$. We chose $\lambda = 4.3\sigma$. We applied the iterative primal-dual algorithm with our joint-refitting (Algorithm (29)) for $1,000$ iterations, with $\tau = 1/4$, $\sigma = 1/6$ and $\theta = 1$.

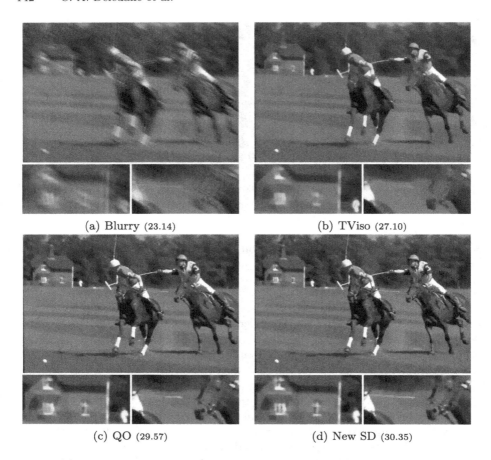

(a) Blurry (23.14)   (b) TViso (27.10)

(c) QO (29.57)   (d) New SD (30.35)

Fig. 4. (a) An 8bit color image corrupted by a directional blur and Gaussian noise with standard deviation $\sigma = 2$. (b) Solution of TViso. Debiased solution with (d) QO and (e) SD. The PSNR is indicated in brackets below each image.

Results are provided on Figs. 3 and 4. Comparisons of refitting with our proposed SD block penalty, HD (only for denoising) and QO are provided. Using our proposed SD block penalty offers the best refitting performances in terms of both visual and quantitative measures. The loss of contrast of TViso is well-corrected, amplitudes are enhanced while smoothness and sharpness of TViso is preserved. Meanwhile, the approach does not create artifacts, invert contrasts, or reintroduce information that were not recovered by TViso.

6   Conclusion

In this work, we have reformulated the refitting problem of solutions promoted by ℓ_{12} structured sparse analysis in terms of block penalties. We have introduced a new block penalty that interpolates between Bregman iterations [10] and direction preservation [2].This framework easily allows the inclusion of additional desirable properties of refitted solutions as well as new penalties that may increase refitting performances. In order to take advantage of our efficient joint-refitting algorithm, it is important to consider *simple* block penalty functions, which proximal operator can be computed explicitly or at least easily. Refitting in the case of other regularizers and loss functions will be investigated in the future.

Acknowledgement. This project has been carried out with support from the French State, managed by the French National Research Agency (ANR-16-CE33-0010-01). This work was supported by the European Union's Horizon 2020 research and innovation programme under the Marie Skłodowska-Curie grant agreement No 777826.

References

1. Belloni, A., Chernozhukov, V.: Least squares after model selection in high-dimensional sparse models. Bernoulli **19**(2), 521–547 (2013)
2. Brinkmann, E.-M., Burger, M., Rasch, J., Sutour, C.: Bias-reduction in variational regularization. J. Math. Imaging Vis. (2017)
3. Burger, M., Gilboa, G., Moeller, M., Eckardt, L., Cremers, D.: Spectral decompositions using one-homogeneous functionals. SIAM J. Imaging Sci. **9**(3), 1374–1408 (2016)
4. Chambolle, A., Pock, T.: A first-order primal-dual algorithm for convex problems with applications to imaging. J. Math. Imaging Vis. **40**, 120–145 (2011)
5. Chzhen, E., Hebiri, M., Salmon, J.: On lasso refitting strategies. Bernouilli (2019)
6. Deledalle, C.-A., Papadakis, N., Salmon, J.: On debiasing restoration algorithms: applications to total-variation and nonlocal-means. In: Aujol, J.-F., Nikolova, M., Papadakis, N. (eds.) SSVM 2015. LNCS, vol. 9087, pp. 129–141. Springer, Cham (2015). https://doi.org/10.1007/978-3-319-18461-6_11
7. Deledalle, C.-A., Papadakis, N., Salmon, J., Vaiter, S.: CLEAR: covariant least-square re-fitting with applications to image restoration. SIAM J. Imaging Sci. **10**(1), 243–284 (2017)
8. Efron, B., Hastie, T., Johnstone, I., Tibshirani, R.: Least angle regression. Ann. Stat. **32**(2), 407–499 (2004)
9. Lederer, J.: Trust, but verify: benefits and pitfalls of least-squares refitting in high dimensions. arXiv preprint arXiv:1306.0113 (2013)
10. Osher, S., Burger, M., Goldfarb, D., Xu, J., Yin, W.: An iterative regularization method for total variation-based image restoration. Multiscale Model. Simul. **4**(2), 460–489 (2005)
11. Rigollet, P., Tsybakov, A.B.: Exponential screening and optimal rates of sparse estimation. Ann. Stat. **39**(2), 731–771 (2011)
12. Scherzer, O., Groetsch, C.: Inverse scale space theory for inverse problems. In: Kerckhove, M. (ed.) Scale-Space 2001. LNCS, vol. 2106, pp. 317–325. Springer, Heidelberg (2001). https://doi.org/10.1007/3-540-47778-0_29
13. Tadmor, E., Nezzar, S., Vese, L.: A multiscale image representation using hierarchical (BV, L2) decompositions. Multiscale Model. Simul. **2**(4), 554–579 (2004)

An Iteration Method for X-Ray CT Reconstruction from Variable-Truncation Projection Data

Limei Huo[1], Shousheng Luo[1,2]([✉]), Yiqiu Dong[3], Xue-Cheng Tai[4],
and Yang Wang[5]

[1] School of Mathematics and Statistics, Data Analysis Technology Lab,
Henan University, Kaifeng 475001, Henan, China
limei.huo@qq.com, sluo@henu.edu.cn
[2] Beijing Computational Science Research Center, Beijing 100193, China
sluo1983@csrc.ac.cn
[3] Department of Applied Mathematics and Computer Science,
Technical University of Denmark, 2800 Kongens Lyngby, Denmark
yido@dtu.dk
[4] Department of Mathematics, Hong Kong Baptist University,
Kowloon Tong, Hong Kong
xuechengtai@hkbu.edu.hk
[5] Department of Mathematics, Hong Kong University of Science and Technology,
Clear Water Bay, Kowloon, Hong Kong
yangwang@ust.hk

Abstract. In this paper, we investigate the in-situ X-ray CT reconstruction from occluded projection data. For each X-ray beam, we propose a method to determine whether it passes through a measured object by comparing the observed data before and after the measured object is placed. Therefore, we can obtain a prior knowledge of the object, that is some points belonging to the background, from the X-ray beam paths that do not pass through the object. We incorporate this prior knowledge into the sparse representation method for in-situ X-ray CT reconstruction from occluded projection data. In addition, the regularization parameter can be determined easily using the artifact severity estimation on the identified background points. Numerical experiments on simulated data with different noise levels are conducted to verify the effectiveness of the proposed method.

Keywords: In-situ X-ray CT · Occluded projection data ·
Sparse representation · Noise estimation

S. Luo was supported by the National Natural Science Foundation of China via Grant A011703. Programs for Science and Technology Development of He'nan Province (192102310181). Y. Dong was supported by the National Natural Science Foundation of China via Grant 11701388. X. C. Tai was supported by the startup grant at Hong Kong Baptist University, grants RG(R)-RC/17-18/02-MATH and FRG2/17-18/033. Y. Wang was supported in part by the Hong Kong Research Grant Council grants 16306415 and 16308518.

© Springer Nature Switzerland AG 2019
J. Lellmann et al. (Eds.): SSVM 2019, LNCS 11603, pp. 144–155, 2019.
https://doi.org/10.1007/978-3-030-22368-7_12

1   Introduction

The X-ray computed tomography technology has experienced a rapid development in the past four decades due to its ability to reconstruct the inner structure
of object noninvasively [13, 16]. Recently, in-situ X-ray CT (ISXCT) has become
an useful technique to investigate the processes occurring in the material sample under various environment conditions, such as high pressure and tempera-
ture [7, 18]. The ISXCT imaging system is often complemented by an additional
equipment to control the in-situ environment, furnaces for example. Such device
is usually equipped outside the X-ray beam from source to detector to avoid any
effect on the acquired data. However, this is unavoidable for some cases, and the
X-ray beam at some directions will be either fully or partially occluded by the
device [18]. Because the missing data vary at different projection directions, it
is called variable truncation (VT) data [6], which can be viewed as the combination of limited angle data and truncation data. Evidently, the reconstruction
images from the VT projection data often suffer from strong artifacts [5, 12, 20].
Figure 1 is an example, and see [6] for more details about it.

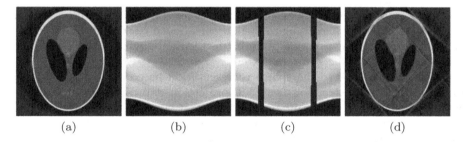

(a)   (b)   (c)   (d)

Fig. 1. Example: Original phantom (a), its Radon transform (b) and the reconstruction
image (d) from the VT data (c).

Various methods for the X-ray CT reconstruction from the incomplete projection data have been proposed [3, 4, 12, 20]. These methods can be divided into
two categories, analytic methods and algebraic methods. Analytic methods complete the missing data using different estimate method, interpolation for example, and then reconstruct the attenuation coefficient of the measured object by
filtered back-projection (FBP) algorithm from the filled projection data [6, 20].
The reconstruction images by these methods usually suffer from artifacts because
it is impossible to estimate the missing data accurately and precisely.

Differently, algebraic methods formulate the X-ray CT reconstruction problem as a large scale linear system by completely omitting the missing data
[1, 4, 12]. In general, such linear system is ill-conditioned. In order to obtain a stable solution, regularization techniques such as total variation minimization and
sparse representation [10, 14, 15, 19, 23] are commonly applied. These methods

are usually formulated as an optimization problem consisting of regularization term and fidelity term balanced by an user-specified parameter.

The regularization term, which represents the prior information of the objects, plays an important role in the performance of the algebraic methods. It is obvious that more prior knowledge we have, better reconstruction qualities we may obtain. In the data acquisition procedure, the original X-ray energies are recorded before the object is placed, and the attenuated X-ray energies are recorded after placing the object. Traditionally, these two observed data sets are only used to calculate the integration (Radon transform) of the object attenuation map along all X-ray beam paths. In fact, we can learn more from these acquired data sets. By comparing the two data sets, we can identify some X-ray paths not passing through the object, i.e. all the points on these paths belong to the background, so the attenuation coefficients should be zero. We can use this information properly to improve the X-ray CT reconstruction results, especially for the incomplete data case.

The parameter which controls the balance between the regularization term and the fidelity term is critical for the performance of the algebraic methods. However, it is a challenging work to select the parameter properly in practice. A well-known fact is that this parameter is closely related to the artifact severity. Therefore, if we can properly estimate the artifact severity, we could determine this parameter easily. According to the aforementioned background prior knowledge, we propose a way to estimate the artifact variance, which is a measurement of the severity of artifacts in some extent.

In this paper, we use the sparse representation based on dictionary learning for ISXCT reconstruction from VT projection data. The method used in this work consists of two main steps. In the first step, an adaptive dictionary is learned from a given initial or the estimated attenuation coefficient map at previous iteration. Accordingly, processed patches are obtained by sparse representation under the learned dictionary. In the second step, the attenuation map is reconstructed by solving an optimization problem containing two quadratic terms. The first one is a weighted least square term for data fitting, and the second one measures the deviation of each patch from the corresponding processed patch. In each step, the parameter is determined by the estimated artifact variance.

The rest of this paper is organized as follows. In Sect. 2, we describe the details for background prior knowledge determination. The reconstruction method is presented in Sect. 3. In Sect. 4, some experiments on simulated data are illustrated to show the effectiveness of the proposed method. We conclude this paper in Sect. 5.

2   Prior Knowledge of Background

In this section, we review the mathematical principles of X-ray CT briefly, and present the method to identify the imaging object background, which will be incorporated into our X-ray CT reconstruction method introduced in the next section.

Here we only consider 2-dimensional case. The path of an X-ray beam in the plane can be denoted by $L(\theta(\alpha), s) = \{s\theta(\alpha) + t\theta^{\perp}(\alpha)|t \in \mathbb{R}\}$ where $\theta(\alpha) = (\cos \alpha, \sin \alpha)^T$ and $\theta^{\perp}(\alpha) = (-\sin \alpha, \cos \alpha)^T \in \mathbb{S}^1$ with $\alpha \in [0, \pi)$ are perpendicular to each other, and $s \in \mathbb{R}$ is the deviation from the origin point with respect to θ. Obviously, all X-ray beams can be indexed by $(\theta(\alpha), s)$ with $\alpha \in [0, \pi)$ and $s \in \mathbb{R}$. In the following, we will omit the variable α and let $\mathbb{S} = \{(\alpha, s)|\alpha \in [0, \pi), s \in \mathbb{R}\}$ for simplicity. Suppose $f(x)$ is the attenuation coefficient at $x = (x_1, x_2)^T \in D \subset \mathbb{R}^2$, where D is the object domain, and $E^0(\theta, s)$ denotes the recorded energy before the measured object placed. By Lambert-Beer's law, the detected energy along $L(\theta, s)$ after placing the measured object is

$$E^1(\theta, s) \sim \text{Poisson}[E^0(\theta, s) \exp(-Rf(\theta, s))], \tag{1}$$

where "Poisson" means the Poisson process from photon counting system, and Rf denotes the Radon transform of f, i.e.

$$Rf(\theta, s) = \int_{-\infty}^{+\infty} f(s\theta + t\theta^{\perp})dt. \tag{2}$$

Thus we can get the Radon transform of f from $E^1(\theta, s)$ and $E^0(\theta, s)$ by

$$Rf(\theta, s) = g(\theta, s) + n(\theta, s), \tag{3}$$

where $g(\theta, s) = \ln(E^0(\theta, s)/E^1(\theta, s))$ and n is the noise.

If the imaging system can record $E^1(\theta, s)$ for $(\alpha, s) \in \mathbb{S}$, we can calculate $Rf(\theta, s)$ for all $(\alpha, s) \in \mathbb{S}$, which is called the complete projection data. Otherwise, the projection data is incomplete. The VT problem we concern in this work is the case that $Rf(\theta, s)$ (i.e. $E^1(\theta, s)$) are completely or partly missing at some angles. If Rf is complete, we can reconstruct f using the FBP algorithm efficiently. However, when Rf (VT data for instance) is incomplete, the reconstruction using FBP suffers from artifacts. The existed methods for incomplete problems try to either fill the missing data in Rf or reconstruct f directly from the incomplete data Rf.

In fact, besides Rf we can obtain more information on f from the observed data $E^0(\theta, s)$ and $E^1(\theta, s)$. If X-ray beam indexed by (θ, s) goes through the object, $E^1(\theta, s)$ will be much lower than $E^0(\theta, s)$ because of the object attenuation. In other words, if $E^1(\theta, s)$ is not much less than $E^0(\theta, s)$ quantitatively, we deem that the corresponding X-ray path $L(\theta, s)$ does not pass through the object. We define $T = \{(\theta, s)|E^1(\theta, s) > (1-\rho)E^0(\theta, s)\}$ as the set of X-ray paths that do not pass though the object domain, where $\rho > 0$ is a small number to suppress the noise effect. Let $S = \{x|x \in L(\theta, s), (\theta, s) \in T\}$ mark the estimate part background. It is obvious that $f(x) = 0$ for all $x \in S$ ideally.

In order to implement the determination of background prior and introduce the proposed model, we describe the data acquisition discretely. In practice, only finite number of data $E^v(\theta_i, s_j), v = 0, 1$, for (θ_i, s_j) with $i = 1, 2, \cdots, q, j = 1, 2, \cdots, p$, are recorded, so the data $g \in \mathbb{R}^M$ with $M = pq$. Discretizing the reconstructed domain Ω enclosing the image domain D, we can approximate the reconstruction image by an array in $\mathbb{R}^{m \times n}$ or $\mathbb{R}^N$ with $N = mn$ by stacking the columns. We define the attenuation coefficients $f \in \mathbb{R}^N$ with the same notation as in the continuous case for the simplicity reason. The Radon transform $Rf(\theta_i, s_j)$ along each line $L(\theta_i, s_j), i = 1, 2, \cdots, q, j = 1, 2, \cdots, p$, is computed by

$$Rf(\theta_i, s_j) = \sum_{l=1}^{N} a_l(\theta_i, s_j) f_l, \tag{4}$$

where $a_l(\theta_i, s_j), l = 1, 2, \cdots, N$, is the intersection length of line $L(\theta_i, s_j)$ with the l-th component of f. Similar as in (3), the X-ray CT reconstruction can be algebraically formulated as an inverse problem

$$Af = g + n, \tag{5}$$

for solving f from the given data g, where $A = (a_{kl}) \in \mathbb{R}^{M \times N}$ with a_{kl} denotes the intersection length of the k-th X-ray beam with the l-th component of f, $k = 1, 2, \cdots M, l = 1, 2, \cdots, N$. For the incomplete data, it means that some components of g are missing, and the corresponding equations in (5) are removed. We also denote the imaging system for incomplete data by (5) for simplicity.

In this discrete case, let $T = \{(\theta_i, s_j) | E^1(\theta_i, s_j) \geq (1 - \rho) E^0(\theta_i, s_j)\}$ be the index set of the X-ray paths that do not go through the object domain D. Then the point set belonging to the background is determined by

$$S = \{l | \sum_{k \in T} a_{kl} > 0\}, \tag{6}$$

and we have $f_l = 0$ for $l \in S$. This background information of f is very helpful to the X-ray CT reconstruction from incomplete data.

3   CT Reconstruction Based on Sparse Representation

Due to the effect of noise n and the ill-posedness of the inverse problem defined in (5) for VT data, prior knowledge of f is usually incorporated by using regularization technique in order to obtain a desirable reconstruction. Sparse representation under dictionary learning is an adaptive and very successful method, and it has been widely used in image restoration and image reconstruction [2,17,21]. In this section, we introduce a new iteration method (see Algorithm 1) based on dictionary learning for ISXCT reconstruction from VT projection data.

The proposed algorithm consists of two main steps, and each step corresponds to solve an optimization problem. One is for learning the dictionary (see (7) in Algorithm 1), and the other is for reconstructing the attenuation coefficients (see (9) in Algorithm 1). Since we are able to estimate the artifact variance from the background prior knowledge, the proposed method successfully eases the dilemma of parameter selection.

In Algorithm 1, Q_s is the operator to extract the s-th patch with size $b \times b$ from f and stack the columns as a vector in $\mathbb{R}^d$ with $d = b^2$. $D^t \in \mathbb{R}^{d \times h}$ is

---

Algorithm 1. Iteration method based on dictionary learning

1: **Inputs:** Initial estimation f^0, total iteration number Num^1, tolerance $\epsilon_1 > 0$, $\text{res}_0^1 > \epsilon_1$, and parameters $r_i > 0, i = 1, 2$. Set $t = 0$.

2: **While** $t < \text{Num}^1$ and $\text{res}_t^1 > \epsilon_1$

3: Estimate the artifact variance σ_t of f^t.

4: Learn dictionary D^t for sparse representation

$$\{D^t, \boldsymbol{\beta}^t\} = \arg\min_{D, \boldsymbol{\beta}} \sum_{s=1}^{P} \left[\|Q_s f^t - D\beta_s\|_2^2 + r_1^t \|\beta_s\|_0 \right]. \tag{7}$$

5: Solve sparse representation coefficient $\alpha_s^t, s = 1, 2, \cdots, P$ for each patch under dictionary D^t

$$\alpha_s^t = \arg\min_{\alpha_s} \left[\|Q_s f^t - D^t \alpha_s\|_2^2 + r_1^t \|\alpha_s\|_0 \right]. \tag{8}$$

6: Reconstruct attenuation map by sparse representation with dictionary D^t, $\boldsymbol{\alpha}^t$

$$f^{t+1} = \arg\min_{f \geq 0} \{ \frac{1}{2} \|Af - g\|_W^2 + r_2^t \sum_{s=1}^{P} \|Q_s f - D^t \alpha_s^t\|_2^2 \}. \tag{9}$$

7: $\text{res}_{t+1}^1 = \sqrt{\frac{1}{N} \sum_{l=1}^{N} (f_l^{t+1} - f_l^t)^2}$.

8: $t = t + 1$.

9: **End(While)**

10: Set $f_l^t = 0$ for $l \in S$, and output f^t.

---

the learnt dictionary for the sparse representation of the patches, where h is the number of elements in the dictionary. We define $\boldsymbol{\beta}^t = [\beta_1^t, \cdots, \beta_P^t]$ (same meaning for $\boldsymbol{\alpha}^t$), where β_s^t (also α_s^t) is the representation coefficient of $Q_s f^t$ under the learned dictionary D^t. Here the zero norm of α_s^t (β_s^t) is to facilitate the sparsity of it. $r_i > 0, i = 1, 2$ are two regularization parameters, which will discuss in the following.

For the optimization problems (7) and (8), we apply the K-SVD algorithm in [15] to solve them. The problem (9) is solved by primal-dual algorithm [9], although it is possible to adopt more efficient algorithms, such as BFGS or conjugate gradient algorithm, because both terms in (9) are quadratic.

3.1 Determinations of r_1^t and r_2^t

The parameters $r_i, i = 1, 2$ play an important role in qualities of the reconstruction results. Usually, the artifacts in the reconstruction map, also in the patches, lack self-similarity, and would be removed with the parameters $r_i, i = 1, 2$ increasing. Therefore, the parameters $r_i, i = 1, 2$ should proportionally increase with the artifacts severity to eliminate them. However, it is a challenging task to get an accurate estimate of the artifacts.

To some extent, the artifacts variance is an index for the severity of artifact because the variance measures the deviation of the reconstruction from the true attenuation map. According to the discussion in Sect. 2, the artifacts variance (severity) of f^t on the identified background region S is computed by

$$\sigma_t = \sqrt{\frac{1}{|S|} \sum_{l \in S} |f_l^t|^2}, \tag{10}$$

because the values of f_l ($l \in S$) are zeros. We can deem σ_t as the artifact variance of f^t under the assumption that the artifact in the reconstructed map is uniform.

We can design different schemes to determine the parameters $r_i, i = 1, 2$ in (7) and (9) using the estimated artifact variance. In this work, these two parameters are computed by simple linear schemes as follows:

$$r_1^t = \lambda_1 \sigma_t,$$
$$r_2^t = \lambda_2 \sigma_t,$$

where $\lambda_i > 0, i = 1, 2$ are two parameters. In our opinion, it is valuable to design more robust and adaptive schemes for $r_i, i = 1, 2$ determination. And we will study more robust methods for the parameters determination in the future.

3.2   Weight Matrix W

The matrix $W = \mathrm{diag}(w_1, w_2, \cdots, w_M)$ in (9) is usually obtained by the Taylor's expansion of the likelihood function of Poisson distribution in (1) [8,23]

$$w_i = (E_i^1)^2 / (E_i^1 + \delta^2), \text{ for all } i = 1, 2 \cdots, M, \tag{11}$$

where δ^2 is the noise variance of n in (5). In general, $w_i \approx E_i^1, i = 1, 2, \cdots, M$ are very large because δ is relatively small (about 10^2) comparing with E_i^1 (about 10^7) in practice. However, the values of g, A and f are relatively small. It leads to that the reconstruction is very sensitive to the parameter selection. Thus, we introduce a new weight computing scheme as follows

$$w_i = \beta_0 + \beta_1 \arctan(E_i^1 / \max_j \{E_j^1\}), \tag{12}$$

where $\beta_0, \beta_1 \geq 0$ are user-specified parameters. Note that the weights calculated by (12) are independent of the amount of the recorded energy E^1.

4   Numerical Experiments

In this section, we will present the results on simulated data of Shepp-Logan phantom polluted by different noise variances. We compare the results with the ones by the FBP method from the zero filling projection data and TV regularization method. The structure similarity (SSIM) indices [22] of the reconstructions are computed to illustrate the effectiveness of the proposed method quantitatively. All the parameters in the model and algorithm are kept the same choices $\beta_0 = 0$, $\beta_1 = 1$, $\lambda_1 = 1.15$, $\lambda_2 = 0.06$, Num[1] $= 6, \epsilon_1 = 1 \times 10^{-4}$ and $\rho = 0.01$ for all experiments.

Besides the reconstruction by (9), we can get a denoised version of f^t using the learnt dictionary and coefficients α_s^t for $Q_s f^t, s = 1, 2, \cdots, P$ (8) [11,15]. In the following, we name the reconstructions by (9) and the dictionary denoising method (8) as DictRect and DictDen, respectively. The initial value of f^0 in the iteration procedure of Algorithm 1 is obtained by TV method.

The noiseless simulated data $E^1(\theta_i, s_j)$ are generated by (1) for 180 angles at $1°$ increments from $1°$ to $180°$ with 256 equal-distance parallel X-ray beams, and the initial X-ray energy $E_i^0 = 5 \times 10^5, i = 1, 2 \cdots, M$, and Gaussian noise with zero mean and different variances are added. Then the Radon transform g of f is calculated, and occluded data are removed (see Fig. 1).

In order to show the effectiveness of background point determination method and robustness to noise and ρ, we count the number N_1 of points that are determined in the background by the proposed method but should belong to the object domain, and the number N_2 of the determined points in background. The results for different values of ρ and Gaussian noise variances are presented in Table 1. One can see that N_1 is robust to noise pollution and choice of ρ, and N_2 will increase with the increasing of ρ since the larger ρ is the larger the set T. Although we cannot identify all the background points, (10) can yields a good approximation of the noise level since the number N_2 is large enough.

Table 1. Robustness of the proposed method for background points determination.

ρ		0.001	0.01	0.02	0.03	0.05
0%	N_1	0	0	0	0	36
	N_2	33128	33160	33160	33160	33196
5%	N_1	0	0	0	2	46
	N_2	33098	33160	33160	33162	33206
11%	N_1	0	0	1	3	37
	N_2	33091	33138	33158	33163	33197

Table 2. SSIM values of reconstruction results in Fig. 2.

Noise	FBP	TV	DictRec	DictDen
0%	0.4838	0.7712	0.9446	0.9528
5%	0.4428	0.7389	0.9379	0.9532
11%	0.2053	0.3616	0.7661	0.8164

Figure 2 displays the attenuation maps reconstructed from the VT projection data set with 0%, 5% and 11% Gaussian noise as images. The SSIM values of

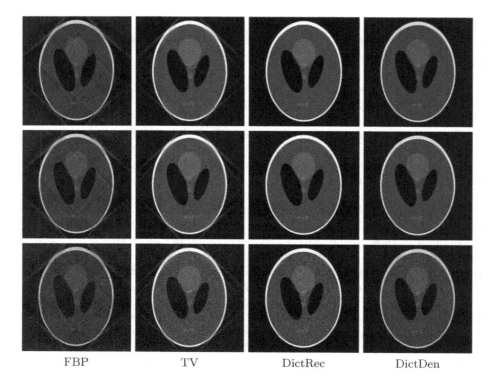

FBP   TV   DictRec   DictDen

Fig. 2. Reconstruction results with different noise levels. From top to bottom, the Gaussian noise level of each row is 0%, 5%, and 11%, respectively.

the attenuation maps in Fig. 2 are tabulated in Table 2. In order to compare the maps reconstructed by different methods clearly and visually, the central vertical line of the maps in Fig. 2 are profiled in Fig. 3.

By comparing the results in Fig. 2, we can easily obtain that the proposed method is much better than FBP and TV regularization method. The DictRec, DictDen and TV method could efficiently suppress the striated artifacts that present in the FBP results, and the DictRec and DictDen eliminate the fake edges that the TV reconstructions suffer from visually. The SSIM values presented in Table 2 also verify the observations above quantitatively. The superiority of the proposed method is more clear in the profiles comparison, see Fig. 3. In addition, the DictDen is superior to DictRec since it can be viewed as DictRec followed by an additional denoising step which can remove some noise and improve the reconstruction quality. In addition, the qualities of DictRec and DictDen reconstructions for heavy pollution projection with 11% Gaussian noise are not desired yet by comparing the profiles in Fig. 3. Therefore, methods need to be studied and developed to properly employ the estimated background knowledge and artifact variance.

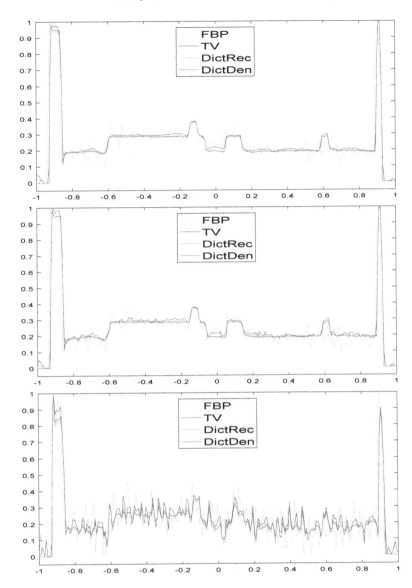

Fig. 3. Profiles of the vertical central lines of the maps in Fig. 2. From top to bottom, the Gaussian noise levels are 0%, 5% and 11%, respectively.

5   Conclusions and Future Work

In this paper, we explored a method to identify some points belonging to the object background in the reconstruction domain. This information is employed to estimate the artifact severity in the reconstruction result, which is incorporated into the sparse representation method for X-ray CT reconstruction. Experiments show the effectiveness of the proposed methods. In the future, we will develop methods employing the background prior knowledge effectively and appropriately to discover more information about the attenuation map and to improve the quality of reconstructions.

References

1. Andersen, A.H., Kak, A.C.: Simultaneous algebraic reconstruction technique (SART): a superior implementation of the ART algorithm. Ultrason. Imaging **6**(1), 81–94 (1984)
2. Bao, C., Cai, J.F., Ji, H.: Fast sparsity-based orthogonal dictionary learning for image restoration. In: IEEE International Conference on Computer Vision (2013)
3. Batenburg, K.J., Sijbers, J.: DART: a practical reconstruction algorithm for discrete tomography. IEEE Trans. Image Process. **20**(9), 2542–2553 (2011)
4. Boas, F.E., Fleischmann, D.: Evaluation of two iterative techniques for reducing metal artifacts in computed tomography. Int. J. Med. Radiol. **259**(3), 894–902 (2011)
5. Boas, F.E., Fleischmann, D.: CT artifacts: causes and reduction techniques. Imaging Med. **4**(2), 229–240 (2012)
6. Borg, L., Jørgensen, J.S., Frikel, J., Sporring, J.: Reduction of variable-truncation artifacts from beam occlusion during in situ x-ray tomography. Measur. Sci. Technol. **28** (2017). 12004(19)
7. Buffiere, J.Y., Maire, E., Adrien, J., Masse, J.P., Boller, E.: In situ experiments with X-ray tomography: an attractive tool for experimental mechanics. Exp. Mech. **50**(3), 289–305 (2010)
8. Cai, A., Li, L., Zheng, Z., Wang, L., Yan, B.: Block-matching sparsity regularization-based image reconstruction for low-dose computed tomography. Med. Phys. **45**(6), 2439–2452 (2018)
9. Chambolle, A., Pock, T.: A first-order primal-dual algorithm for convex problems with applications to imaging. J. Math. Imaging Vis. **40**(1), 120–145 (2011)
10. Debatin, M., Hesser, J.: Accurate low-dose iterative CT reconstruction from few projections by generalized anisotropic total variation minimization for industrial CT. J. X-ray Sci. Technol. **23**(6), 701–726 (2015)
11. Elad, M., Aharon, M.: Image denoising via sparse and redundant representations over learned dictionaries. IEEE Trans. Image Process. **15**(12), 3736–3745 (2006)
12. Gjesteby, L., et al.: Metal artifact reduction in CT: where are we after four decades? IEEE Access **4**(99), 5826–5849 (2016)
13. Herman, G.T.: Fundamentals of Computerized Tomography: Image Reconstruction From Projections. Springer, London (2009). https://doi.org/10.1007/978-1-84628-723-7
14. Hu, H., Wohlberg, B., Chartrand, R.: Task-driven dictionary learning for inpainting. In: IEEE International Conference on Acoustics, Speech and Signal Processing, pp. 3543–3547 (2014)
15. Michal, A., Michael, E., Alfred, B.: K-SVD: an algorithm for designing over complete dictionaries for sparse representation. IEEE Trans. Sign. Process. **54**, 4311–4322 (2006)
16. Natterer, F., Wubbeling, F., Wang, G.: Mathematical Methods in Image Reconstruction. SIAM, Philadelphia (2001)

17. Soltani, S., Kilmer, M.E., Hansen, P.C.: A tensor-based dictionary learning approach to tomographic image reconstruction. BIT Numer. Math. **56**(4), 1425–1454 (2016)
18. Terzi, S., et al.: In situ x-ray tomography observation of inhomogeneous deformation in semi-solid aluminum alloys. Scripta Mater. **61**(5), 449–452 (2009)
19. Tian, Z., Xun, J., Yuan, K., Jiang, S.: TU-B-201B-03: CT reconstruction from undersampled projection data via edge-preserving total variation regularization. Med. Phys. **37**(6Part6), 3378 (2010)
20. Veldkamp, W.J.H., Joemai, R.M.S., van der Molen, G., Jacob, G.: Development and validation of segmentation and interpolation techniques in sinograms for metal artifact suppression in CT. Med. Phys. **37**(2), 620–628 (2010)
21. Wang, B., Li, L.: Recent development of dual-dictionary learning approach in medical image analysis and reconstruction. Comput. Math. Methods Med. **2015**(2), 1–9 (2015)
22. Wang, Z., Bovik, A.C., Sheikh, H.R., Simoncelli, E.P.: Image quality assessment: from error visibility to structural similarity. IEEE Trans. Image Process. **13**(4), 600–612 (2004)
23. Xu, Q., Yu, H., Mou, X., Zhang, L., Hsieh, J., Wang, G.: Low-dose x-ray CT reconstruction via dictionary learning. IEEE Trans. Med. Imaging **31**(9), 1682–1697 (2012)

A New Iterative Method for CT Reconstruction with Uncertain View Angles

Nicolai André Brogaard Riis[✉] and Yiqiu Dong

Department of Applied Mathematics and Computer Science, Technical University of Denmark, Richard Petersens Plads, 2800 Kongens Lyngby, Denmark
nabr@dtu.dk

Abstract. In this paper, we propose a new iterative algorithm for Computed Tomography (CT) reconstruction when the problem has uncertainty in the view angles. The algorithm models this uncertainty by an additive model-discrepancy term leading to an estimate of the uncertainty in the likelihood function. This means we can combine state-of-the-art regularization priors such as total variation with this likelihood. To achieve a good reconstruction the algorithm alternates between updating the CT image and the uncertainty estimate in the likelihood. In simulated 2D numerical experiments, we show that our method is able to improve the relative reconstruction error and visual quality of the CT image for the uncertain-angle CT problem.

Keywords: Computed Tomography · Uncertain view angles · Model error · Variational methods · Total variation · Model discrepancy

1 Introduction

In this paper, we consider Computed Tomography (CT) where the geometry of the physical set-up is only known approximately. The goal is to achieve reconstructions that are stable in the presence of uncertainties in the geometric parameters. We restrict our attention to uncertainty in the view angles. We assume that the actual view angles are realizations of some known probability distribution $\pi_{\text{angles}}(\cdot)$ and that the measured sinogram is corrupted by additive Gaussian noise with known mean and covariance.

With the above assumptions, we formulate the CT reconstruction problem under uncertain view angles as estimating the unknown attenuation coefficient image $\mathbf{x} \in \mathbb{R}^n$ from a measured (noisy) sinogram $\mathbf{b} \in \mathbb{R}^m$ following the model

$$\mathbf{b} = \mathbf{A}(\boldsymbol{\theta})\,\mathbf{x} + \mathbf{e}, \quad \boldsymbol{\theta} \sim \pi_{\text{angles}}(\boldsymbol{\theta}), \, \mathbf{e} \sim \mathcal{N}(\boldsymbol{\mu}_{\mathbf{e}}, \mathbf{C}_{\mathbf{e}}), \tag{1}$$

The work was supported by the National Natural Science Foundation of China via Grant 11701388.

J. Lellmann et al. (Eds.): SSVM 2019, LNCS 11603, pp. 156–167, 2019.
https://doi.org/10.1007/978-3-030-22368-7_13

where $\mathbf{e} \in \mathbb{R}^m$ is the additive Gaussian noise with mean $\boldsymbol{\mu_e}$ and symmetric positive definite covariance $\mathbf{C_e}$. The parameterized matrix $\mathbf{A}(\boldsymbol{\theta}) \in \mathbb{R}^{m \times n}$ is the discrete approximation of the Radon transform with view angles $\boldsymbol{\theta} \in \mathbb{R}^q$. The measurement at detector element l from the view angle θ_i with $i = 1, \dots, q$, i.e., $(\mathbf{A}(\theta_i)\mathbf{x})_l$, is a discretization of $(\mathcal{R}f)(\theta_i, s_l) = \int_{\mathbb{R}} f(s_l V(\theta_i) + t V^\perp(\theta_i)) \, dt$, where s_l gives the position of the lth pixel on the detector with $l = 1, \cdots, p$, and $m = qp$. Moreover, the function f is the continuous representation of $\mathbf{x}$, $V(\theta) = (\cos \theta, \sin \theta)$ and $V^\perp(\theta) = V(\theta + \pi/2)$. For more details on the mathematical model of CT see e.g. [1]. We emphasize that the goal in this work is to estimate the CT image $\mathbf{x}$ from a measured sinogram $\mathbf{b}$ according to the model (1) with uncertain view angles $\boldsymbol{\theta}$ and noise $\mathbf{e}$. Here, $\boldsymbol{\theta}$ and $\mathbf{e}$ are considered as nuisance or uninteresting parameters, and they are only taken into account when reconstructing $\mathbf{x}$ without being explicitly estimated.

1.1   Previous Work

Many variational methods have been proposed for CT reconstruction, see e.g., [2–4]. In general, variational models in these methods consist of a data-fitting term and a regularization term, and these two terms are balanced by a regularization parameter. In order to deal with the ill-posedness in CT reconstruction problems, the choice of regularization is very important. Different regularization techniques have been applied, for example, total variation (TV) regularization [5] and framelet representations [2]. But these methods do not take parametric uncertainty such as uncertainty in view angles into account. Therefore, good performance of the methods is not guaranteed if the view angles are uncertain.

The CT reconstruction problem with uncertain view angles in (1) is generally solved by estimating the view angles from some measurements. Geometrical calibration of models in CT has been studied, see e.g. [6] for a review. Typically such methods are based on reference objects or reference instruments for the calibration. Recently, in the case of uncertain or unknown view angles a few reconstruction methods that only use the measured sinogram without reference objects or instruments have been proposed, see [7–9]. These methods aim to estimate the view angles $\boldsymbol{\theta}$ in addition to the CT image $\mathbf{x}$, and can be categorized into two groups: (1) Estimating view angles directly from projection data and then estimating the CT image and (2) simultaneously estimating view angles and CT image.

In [9] it has been shown that if the scanned objects are asymmetrical then view angles can be uniquely determined by sinogram measurements. According to this result, we can estimate angles directly from complete measurements. However, if the object is partly symmetrical or the measurements are not sufficient, we cannot obtain an accurate angle estimation, see [8]. Then, due to error propagation, an inaccurate angle estimation would lead to an unsatisfactory reconstruction.

The simultaneous methods such as Bayesian sampling-based methods [7] can effectively avoid error propagation, but they are limited by computational complexity and generally require many evaluations of the forward model (1), which makes them unfeasible for large-scale problems.

There are also a few methods for characterizing and reducing model errors in general inverse problems, see [10–15]. Most of these methods are based on the statistical description of the model error in a Bayesian setting. This leads to a natural way of incorporating uncertainties and modelling errors in the model. However, full Bayesian methods also suffer from computational complexity issues except in cases when the object is assumed to follow a Gaussian distribution, in which case closed-form solutions exist.

1.2   Our Contribution

In this paper, we propose a new iterative algorithm for CT reconstruction with uncertain view angles. The main step in the algorithm is based on a variational method, which combines the state-of-the-art regularization such as TV with a modified data-fitting term, that includes the uncertainty in the view angles via a model-discrepancy term. Since the model-discrepancy term depends on the estimated reconstruction, we update it and the reconstruction alternately. The simulated numerical results show that the new algorithm is able to reduce the relative error and improve the visual quality of the reconstructions.

2   Our Method

The CT reconstruction with uncertain view angles is formulated in (1) with an assumption of the probability distribution on the view angles. By including the known expected view angles, $\hat{\boldsymbol{\theta}}$, we can reformulate the problem as follows.

$$\mathbf{b} = \mathbf{A}(\hat{\boldsymbol{\theta}})\,\mathbf{x} + \boldsymbol{\eta} + \mathbf{e}, \quad \mathbf{e} \sim \mathcal{N}(\boldsymbol{\mu_e}, \mathbf{C_e}), \tag{2}$$

where the new random variable $\boldsymbol{\eta} = \boldsymbol{\eta}(\boldsymbol{\theta}, \mathbf{x}) = \mathbf{A}(\boldsymbol{\theta})\mathbf{x} - \mathbf{A}(\hat{\boldsymbol{\theta}})\mathbf{x}$ with $\boldsymbol{\theta} \sim \pi_{\mathrm{angles}}(\boldsymbol{\theta})$ models the uncertainties associated with the view angles. Note that (2) is consistent with (1). By this reformulation, we basically shift the uncertainties in the view angles to the model-discrepancy term $\boldsymbol{\eta}$, which will be used to derive our variational model.

Defining modelling errors as an additive model-discrepancy term was first applied in [16] in the field of model calibration of physical and computer models. The distribution of $\boldsymbol{\eta}$ was assumed as a Gaussian Process and determined as a model correction term in addition to $\mathbf{x}$. In [11,17], this idea was applied in Bayesian inverse problems and named as the Approximation Error Approach (AEA). The main differences are that in the AEA $\boldsymbol{\eta}$ is used to represent the difference between two grid systems instead of a model discrepancy and it is marginalized out in the likelihood function. The outputs of the AEA are the distributions of $\mathbf{x}$ and $\boldsymbol{\eta}$. To further improve the results, in [14] an iterative scheme was introduced where the distributions of $\mathbf{x}$ and $\boldsymbol{\eta}$ are updated alternately.

Inspired by the ideas of the AEA, we derive the likelihood according to the model (2) by marginalizing out both $\boldsymbol{\eta}$ and $\mathbf{e}$. Define $\boldsymbol{\nu} = \boldsymbol{\eta} + \mathbf{e}$, and the likelihood is given by

$$\pi(\mathbf{b}\,|\,\mathbf{x}) = \int_{\mathbb{R}^m} \pi(\mathbf{b}, \boldsymbol{\nu}\,|\,\mathbf{x})\mathrm{d}\boldsymbol{\nu} = \int_{\mathbb{R}^m} \pi(\mathbf{b}\,|\,\mathbf{x}, \boldsymbol{\nu})\pi(\boldsymbol{\nu}\,|\,\mathbf{x})\mathrm{d}\boldsymbol{\nu} = \pi_{\boldsymbol{\nu}\,|\,\mathbf{x}}(\mathbf{b} - \mathbf{A}(\hat{\boldsymbol{\theta}})\,\mathbf{x}\,|\,\mathbf{x}). \quad (3)$$

The conditional distribution of $\boldsymbol{\nu}\,|\,\mathbf{x}$ may be rather complicated, but we can approximate it by a simpler distribution such as a Gaussian. Gaussian approximations has been shown experimentally to be reasonable in many applications [11,13,14,17,18]. Here, we assume that $\boldsymbol{\eta}\,|\,\mathbf{x}$ follows a Gaussian distribution $\mathcal{N}(\boldsymbol{\mu}_{\boldsymbol{\eta}\,|\,\mathbf{x}}, \mathbf{C}_{\boldsymbol{\eta}\,|\,\mathbf{x}})$ with mean $\boldsymbol{\mu}_{\boldsymbol{\eta}\,|\,\mathbf{x}}$ and covariance $\mathbf{C}_{\boldsymbol{\eta}\,|\,\mathbf{x}}$ and $\mathbf{e}$ is independent of $\mathbf{x}$. Then we obtain the negative log-likelihood function

$$-\log \pi(\mathbf{b}\,|\,\mathbf{x}) \propto \frac{1}{2}\|\mathbf{b} - \mathbf{A}(\hat{\boldsymbol{\theta}})\,\mathbf{x} - \boldsymbol{\mu}_{\boldsymbol{\nu}\,|\,\mathbf{x}}\|^2_{\mathbf{C}^{-1}_{\boldsymbol{\nu}\,|\,\mathbf{x}}} = \frac{1}{2}\|\mathbf{L}_{\boldsymbol{\nu}\,|\,\mathbf{x}}(\mathbf{b} - \mathbf{A}(\hat{\boldsymbol{\theta}})\,\mathbf{x} - \boldsymbol{\mu}_{\boldsymbol{\nu}\,|\,\mathbf{x}})\|^2_2, \quad (4)$$

where $\boldsymbol{\mu}_{\boldsymbol{\nu}\,|\,\mathbf{x}} = \boldsymbol{\mu}_{\mathbf{e}} + \boldsymbol{\mu}_{\boldsymbol{\eta}\,|\,\mathbf{x}}$, $\mathbf{C}_{\boldsymbol{\nu}\,|\,\mathbf{x}} = \mathbf{C}_{\mathbf{e}} + \mathbf{C}_{\boldsymbol{\eta}\,|\,\mathbf{x}}$ is the combined covariance of the measurement noise and model discrepancy, and $\mathbf{L}^T_{\boldsymbol{\nu}\,|\,\mathbf{x}}\mathbf{L}_{\boldsymbol{\nu}\,|\,\mathbf{x}} = \mathbf{C}^{-1}_{\boldsymbol{\nu}\,|\,\mathbf{x}}$ is the Cholesky factorization of the inverse covariance. Applying regularization techniques, we can formulate a variational model for (2) that gives a stable solution with respect to the uncertain view angles and measurement noise using the likelihood (4). TV regularization has shown good performance in CT [2], and thus we use it as regularization term and obtain the following variational model

$$\min_{\mathbf{x} \geq 0} \frac{1}{2}\|\mathbf{L}_{\boldsymbol{\nu}\,|\,\mathbf{x}}(\mathbf{b} - \mathbf{A}(\hat{\boldsymbol{\theta}})\,\mathbf{x} - \boldsymbol{\mu}_{\boldsymbol{\nu}\,|\,\mathbf{x}})\|^2_2 + \lambda \mathrm{TV}(\mathbf{x}), \quad (5)$$

with minimizer $\mathbf{x}_{\mathrm{STV}}$ and regularization parameter $\lambda > 0$. We use $\mathrm{TV}(\mathbf{x}) = \sum_i \|[\nabla\mathbf{x}]_i\|_2$, where $\|[\nabla\mathbf{x}]_i\|_2 = \sqrt{(\nabla_h\mathbf{x})^2_i + (\nabla_v\mathbf{x})^2_i}$, with $(\nabla_h\mathbf{x})_i$ and $(\nabla_v\mathbf{x})_i$ denoting the derivatives of $\mathbf{x}_i$ along horizontal and vertical directions with symmetric boundary condition, respectively. A non-negativity constraint is added because the attenuation coefficients $\mathbf{x}$ cannot be negative.

2.1   An Iterative Algorithm

The variational model defined in (5) still leaves the question of how the mean and covariance of $\boldsymbol{\eta}\,|\,\mathbf{x}$ are determined. Given a reconstruction $\mathbf{x}$, one can generate N_{samp} samples of $\boldsymbol{\eta}\,|\,\mathbf{x}$ by drawing samples $\boldsymbol{\theta}^s$ following the distribution $\pi_{\mathrm{angles}}(\boldsymbol{\theta})$, and then evaluate the model discrepancy by

$$\boldsymbol{\eta}^s = \mathbf{A}(\boldsymbol{\theta}^s)\,\mathbf{x} - \mathbf{A}(\hat{\boldsymbol{\theta}})\,\mathbf{x}, \quad \boldsymbol{\theta}^s \sim \pi_{\mathrm{angles}}(\boldsymbol{\theta}), \ s = 1, \ldots, N_{\mathrm{samp}}. \quad (6)$$

The sample mean and covariance given $\mathbf{x}$ can then be calculated by

$$\boldsymbol{\mu}^{\mathrm{samp}}_{\boldsymbol{\eta}\,|\,\mathbf{x}} = \frac{1}{N_{\mathrm{samp}}} \sum_{s=1}^{N_{\mathrm{samp}}} \boldsymbol{\eta}^s, \quad (7)$$

and

$$\mathbf{C}_{\eta \mid \mathbf{x}}^{\text{samp}} = \frac{1}{N_{\text{samp}} - 1} \sum_{s=1}^{N_{\text{samp}}} (\boldsymbol{\eta}^s - \boldsymbol{\mu}_{\eta \mid \mathbf{x}})(\boldsymbol{\eta}^s - \boldsymbol{\mu}_{\eta \mid \mathbf{x}})^T. \tag{8}$$

If we have a good estimate of $\mathbf{x}$, we can obtain good samples of the model discrepancy, and then we can use the sample mean and covariance in the model (5) to further improve the reconstruction result. This leads to an iterative scheme for alternately updating the estimate of $\mathbf{x}$ and the estimates of mean and covariance of $\boldsymbol{\eta} \mid \mathbf{x}$. The iterative scheme is shown in Algorithm 1.

Algorithm 1. Iterative update of reconstruction and likelihood

Inputs: $\mathbf{b}$, λ, $\hat{\boldsymbol{\theta}}$, and $\pi_{\text{angles}}(\boldsymbol{\theta})$. Initial choice of $\mathbf{L}_{\nu \mid \mathbf{x}}^{[0]}$ and $\boldsymbol{\mu}_{\nu \mid \mathbf{x}}^{[0]}$.

Output: $\mathbf{x}_{\text{STV}}^{[K]}$

1: **for** $k = 1, 2, \ldots, K$
2:     $\mathbf{x}_{\text{STV}}^{[k+1]} = \arg\min_{\mathbf{x} \geq 0} \frac{1}{2} \|\mathbf{L}_{\nu \mid \mathbf{x}}^{[k]}(\mathbf{b} - \mathbf{A}(\hat{\boldsymbol{\theta}})\mathbf{x} - \boldsymbol{\mu}_{\nu \mid \mathbf{x}}^{[k]})\|_2^2 + \lambda \text{TV}(\mathbf{x})$
3:     **for** $s = 1, 2, \ldots, N_{\text{samp}}$
4:        $\boldsymbol{\eta}^s = \mathbf{A}(\boldsymbol{\theta}^s)\mathbf{x}_{\text{STV}}^{[k+1]} - \mathbf{A}(\hat{\boldsymbol{\theta}})\mathbf{x}_{\text{STV}}^{[k+1]}$ for $\boldsymbol{\theta}^s \sim \pi_{\text{angles}}(\boldsymbol{\theta})$
5:     **end**
6:     Estimate $\boldsymbol{\mu}_{\eta \mid \mathbf{x}}$ and $\mathbf{C}_{\eta \mid \mathbf{x}}$ according to (7) and (8)
7:     $\boldsymbol{\mu}_{\nu \mid \mathbf{x}}^{[k+1]} = \boldsymbol{\mu}_{\mathbf{e}} + \boldsymbol{\mu}_{\eta \mid \mathbf{x}}$
8:     $\mathbf{L}_{\nu \mid \mathbf{x}}^{[k+1]} = \text{chol}\left((\mathbf{C}_{\mathbf{e}} + \mathbf{C}_{\eta \mid \mathbf{x}})^{-1}\right)$
9: **end**

Here, $\text{chol}(\mathbf{C}^{-1})$ gives the Cholesky factorization of the inverse covariance $\mathbf{C}^{-1}$, i.e., $\mathbf{L}^T \mathbf{L} = \mathbf{C}^{-1}$. In the initialization, we use the measurement noise mean and covariance to initialize $\mathbf{L}_{\nu \mid \mathbf{x}}^{[0]} = \mathbf{L}_{\mathbf{e}}$ and $\boldsymbol{\mu}_{\nu \mid \mathbf{x}}^{[0]} = \boldsymbol{\mu}_{\mathbf{e}}$, where $\mathbf{L}_{\mathbf{e}}^T \mathbf{L}_{\mathbf{e}} = \mathbf{C}_{\mathbf{e}}^{-1}$.

Compared with the AEA proposed in [14], our method has two main differences. First, our method deals with the uncertainties in the model parameters and the accurate model is unknown, while in the AEA the accurate model is known and the discrepancy is between two different grid systems. Secondly, in our method, we apply a variational method incorporated with regularization techniques to obtain a reconstruction result, which can be solved by many advanced optimization methods, while in the AEA the distribution of the reconstruction is obtained by applying Bayesian inversion methods, which leads to much higher computational complexity.

2.2 Approximation of $\mathbf{L}_{\nu \mid \mathbf{x}}$

Because of the computational complexity in calculating the inverse and Cholesky factorization of the covariance matrix $\mathbf{C}_{\mathbf{e}} + \mathbf{C}_{\eta \mid \mathbf{x}}$, our method can be of limited use for solving large-scale CT problems. To overcome this limitation, we

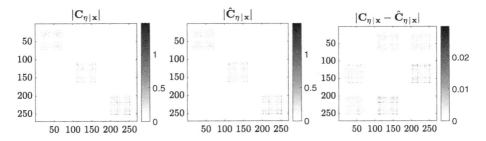

Fig. 1. An example of the absolute value of the full covariance $\mathbf{C}_{\eta\,|\,\mathbf{x}}$, its approximation $\hat{\mathbf{C}}_{\eta\,|\,\mathbf{x}}$ according to (9), and their absolute difference for the uncertain angle CT problem (2). We conclude that the approximation is reasonable for this problem.

approximate the covariance matrix $\mathbf{C}_{\eta\,|\,\mathbf{x}} \in \mathbb{R}^{m\times m}$ by a block diagonal matrix $\hat{\mathbf{C}}_{\eta\,|\,\mathbf{x}}$ given by

$$
\hat{\mathbf{C}}_{\eta\,|\,\mathbf{x}} =
\begin{bmatrix}
\mathbf{C}_{\eta\,|\,\mathbf{x},11} & & & \\
& \mathbf{C}_{\eta\,|\,\mathbf{x},22} & & \\
& & \ddots & \\
& & & \mathbf{C}_{\eta\,|\,\mathbf{x},qq}
\end{bmatrix},
\tag{9}
$$

where $\mathbf{C}_{\eta\,|\,\mathbf{x},ii} \in \mathbb{R}^{p\times p}$ are the block diagonal parts of $\mathbf{C}_{\eta\,|\,\mathbf{x}}$, $q = m/p$ is the number of view angles and p is the number of detector pixels. Then, if the Gaussian measurement noise $\mathbf{e}$ is i.i.d., i.e., $\mathbf{C}_{\mathbf{e}} = \sigma^2 \mathbf{I}_m$, we can compute the Cholesky factorization of the approximate inverse covariance $(\mathbf{C}_{\mathbf{e}} + \hat{\mathbf{C}}_{\eta\,|\,\mathbf{x}})^{-1}$ block-wise as follows

$$
\hat{\mathbf{L}}_{\nu\,|\,\mathbf{x}} =
\begin{bmatrix}
\mathrm{chol}((\mathbf{C}_{\eta\,|\,\mathbf{x},11} + \sigma^2 \mathbf{I}_p)^{-1}) & & \\
& \ddots & \\
& & \mathrm{chol}((\mathbf{C}_{\eta\,|\,\mathbf{x},qq} + \sigma^2 \mathbf{I}_p)^{-1})
\end{bmatrix}.
\tag{10}
$$

If the full covariance was used multiplication of an vector with $\mathbf{L}_{\nu\,|\,\mathbf{x}}$ would require $\mathcal{O}(m^2) = \mathcal{O}(p^2q^2)$ operations, whereas multiplication with $\hat{\mathbf{L}}_{\nu\,|\,\mathbf{x}}$ would only be $\mathcal{O}(p^2q)$ operations. Additionally, the matrix inversion and Cholesky factorization is reduced from $\mathcal{O}(m^3) = \mathcal{O}(p^3q^3)$ to $\mathcal{O}(p^3q)$ operations.

In Fig. 1, we show the absolute values of the full covariance $\mathbf{C}_{\eta\,|\,\mathbf{x}}$, its approximation $\hat{\mathbf{C}}_{\eta\,|\,\mathbf{x}}$ according to (9), and their absolute difference. The values in the off-diagonal parts are much smaller than those in the block diagonal parts of $\mathbf{C}_{\eta\,|\,\mathbf{x}}$. Hence, the approximation is reasonable for this problem. In the following experiments line 8 in Algorithm 1 is therefore approximated by (10).

3   Numerical Experiments

In this section, we present simulated 2D numerical results to show the performance of our method. The experiments are carried out in MATLAB and we use

Table 1. The physical parameters in the simulated CT experiments.

Parameter	Value
Beam type	Fan-beam
Reconstruction domain size	50 cm × 50 cm
Source to center distance	50 cm
Source to detector distance	100 cm
Detector length	130 cm
Small example	
Image pixels	$n = 45 \times 45$
Detector pixels	$p = 90$
Number of view angles	$q = 90$
View angle standard deviation	$\delta = 1.2°$
Larger example	
Image pixels	$n = 135 \times 135$
Detector pixels	$p = 270$
Number of projection angles	$q = 270$
View angle standard deviation	$\delta = 0.4°$

the ASTRA toolbox [19] and "Spot Operators" [20] for matrix-free forward- and back-projections, i.e., for multiplication with $\mathbf{A}(\boldsymbol{\theta})$ and $\mathbf{A}(\boldsymbol{\theta})^T$. In the simulated CT problems arising from (1), the physical parameters are shown in Table 1. In both examples, the distribution of the view angles is assumed to be i.i.d. Gaussian with equidistant view angles from 0° to 360°, denoted by $\boldsymbol{\theta}^{\mathrm{equid}}$, as mean and δ^2 as variance. These examples illustrate the physical case where the measurements are acquired at equidistant view angles, but each measurement is associated with some independent uncertainty. Note that the "small example" has 90 view angles and standard deviation $\delta = 1.2°$, whereas the "larger example" has 270 view angles with $\delta = 0.4°$. This is to ensure that the view angles are unlikely to switch positions relative to each other from the added uncertainty. In our numerical tests, we generate the measurements according to

$$\mathbf{b} = \mathbf{A}(\boldsymbol{\theta}^{\mathrm{machine}})\bar{\mathbf{x}} + \mathbf{e}, \qquad (11)$$

where $\bar{\mathbf{x}}$ is either the Shepp-Logan or Grains phantom generated from AIR Tools II [21], and $\boldsymbol{\theta}^{\mathrm{machine}}$ denotes the actual view angles, which is a realization of $\mathcal{N}(\boldsymbol{\theta}^{\mathrm{equid}}, \delta^2 \mathbf{I})$. Here $\mathbf{e} \sim \mathcal{N}(\mathbf{0}, \sigma^2 \mathbf{I})$, where $\sigma = 0.005 \|\mathbf{A}(\boldsymbol{\theta}^{\mathrm{machine}})\bar{\mathbf{x}}\|_2 / \sqrt{m}$. We solve the TV minimization problem using the Chambolle-Pock algorithm in [22] and stopping when the relative change in the objective function is below 10^{-6}. In our method, we set the maximum iteration number $K = 10$ and the number of samples $N_{\mathrm{samp}} = 5000$.

We compare our results with the non-negative TV reconstruction that does not take the uncertainty in the view angles into account, i.e.,

$$\mathbf{x}_{\text{TV}} = \arg\min_{\mathbf{x}\geq 0} \frac{1}{2\sigma^2}\|(\mathbf{b} - \mathbf{A}(\hat{\boldsymbol{\theta}})\,\mathbf{x})\|_2^2 + \lambda\text{TV}(\mathbf{x}), \tag{12}$$

where $\hat{\boldsymbol{\theta}} = \boldsymbol{\theta}^{\text{equid}}$. In addition, we also show the results from the non-negative TV reconstruction with the actual view angles $\boldsymbol{\theta}^{\text{machine}}$, which would be the best-case scenario:

$$\mathbf{x}_{\text{TV-opt}} = \arg\min_{\mathbf{x}\geq 0} \frac{1}{2\sigma^2}\|(\mathbf{b} - \mathbf{A}(\boldsymbol{\theta}^{\text{machine}})\,\mathbf{x})\|_2^2 + \lambda\text{TV}(\mathbf{x}). \tag{13}$$

3.1   The Small Example

In Fig. 2 we show the expected view angles $\hat{\boldsymbol{\theta}} = \boldsymbol{\theta}^{\text{equid}}$ and 3 realizations of $\pi_{\text{angles}}(\boldsymbol{\theta}) = \mathcal{N}(\boldsymbol{\theta}^{\text{equid}}, \delta^2 \mathbf{I})$, i.e. 3 examples of $\boldsymbol{\theta}^{\text{machine}}$. The realizations are used to generate noisy sinograms according to (11). We compare the reconstructions $\mathbf{x}_{\text{STV}}$ from our method (Algorithm 1) with the ones obtained by solving (12) and (13). In the left column of Fig. 3 we plot the relative error $\frac{\|\mathbf{x}-\bar{\mathbf{x}}\|_2}{\|\bar{\mathbf{x}}\|_2}$ of the three methods with the regularization parameter λ varying from 10^{-6} to 10^{-2}. We can see that except for large λ, where the influence of the data-fitting term becomes weak, the reconstructions from our method has lower relative errors compared to $\mathbf{x}_{\text{TV}}$ from (12). With the optimal λ choice, which gives the smallest relative error, the improvement by our method is significant. It shows the importance of taking the uncertainty in the view angles into account. In the right column of Fig. 3 we numerically show the convergence of the relative errors in our method.

In order to visually compare the reconstructions from these three methods, in Fig. 4 we show the reconstruction results for the same λ values, which corresponds to the optimal choice in (13). It is clear that our method can effectively reduce the artifacts due to the uncertain view angles.

For this small example, in our method we can also compute the full covariance $\mathbf{C}_{\eta\,|\,\mathbf{x}} \in \mathbb{R}^{m\times m}$ instead of using the approximation $\hat{\mathbf{C}}_{\eta\,|\,\mathbf{x}}$ introduced in Sect. 2.2. Since the relative errors by using the full covariance are almost identical to using the approximation, we do not show them here.

Fig. 2. For the small experiment in Table 1. From left to right: The expected view angles $\hat{\boldsymbol{\theta}} = \boldsymbol{\theta}^{\text{equid}}$ and three realizations of $\pi_{\text{angles}}(\boldsymbol{\theta})$.

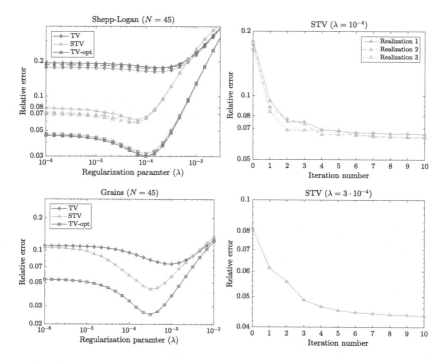

Fig. 3. For $N = 45$ in Table 1. Top: Shepp-Logan. Bottom: Grains. Left: Relative error vs. regularization parameter. Right: Relative error vs. iteration number in Algorithm 1.

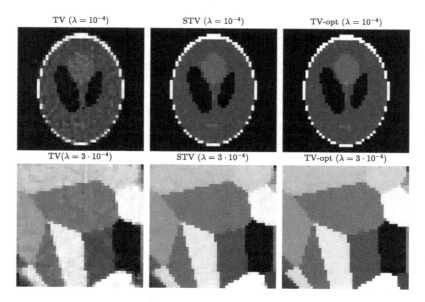

Fig. 4. Reconstructions for $N = 45$ in Table 1. Top: Shepp-Logan. Bottom: Grains.

3.2   The Larger Example

We also compute a larger example according to the parameters in Table 1. In Fig. 5 we show a zoomed part of the expected view angles $\hat{\boldsymbol{\theta}}$ as well as a realization of $\pi_{\mathrm{angles}}(\boldsymbol{\theta})$ that is used to generate the data. In Fig. 6 we show the plots of the relative errors with different choices of λ and along the iterations in our method. In this case the difference between $\mathbf{x}_{\mathrm{TV}}$ from solving (12) and $\mathbf{x}_{\mathrm{TV-opt}}$ from solving (13) is not as big as in the small example, and the main reason is that the variance δ^2 in the view angles is much smaller. However, we can still clearly see that our method improves the reconstruction quality in terms of relative error. To compare the reconstruction visually, in Fig. 7 we give the reconstruction results from three methods.

Fig. 5. For the larger experiment in Table 1. Left: the expected view angles $\hat{\boldsymbol{\theta}}$. Right: a realization of $\pi_{\mathrm{angles}}(\boldsymbol{\theta})$. Here we zoomed in on a part of the view angles.

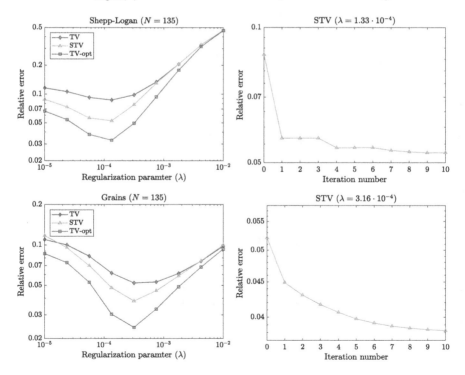

Fig. 6. For the experiment with $N = 135$ in Table 1. Top: Shepp-Logan. Bottom: Grains. Left: Relative error vs. regularization parameter. Right: Relative error vs. iteration number in Algorithm 1.

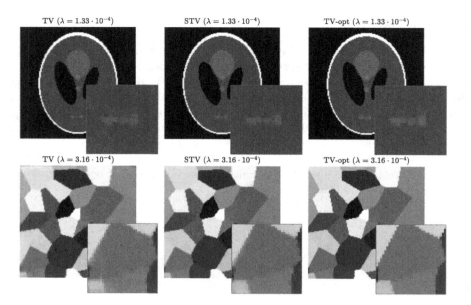

Fig. 7. Reconstructions for $N = 135$ in1 Table 1. Top: Shepp-Logan. Bottom: Grains.

Based on our tests, if we increase the image size n and keep the same number of measurements, the quality of the reconstruction by our method gets closer to the one from (12). The reason is that the reconstruction problem becomes more ill-posed and therefore more difficult to deal with. In this case, we would need a better initial guess on $\hat{\mathbf{L}}_{\nu \mid \mathbf{x}}^{[0]}$ and $\boldsymbol{\mu}_{\nu \mid \mathbf{x}}^{[0]}$ in order to obtain a good estimate of $\bar{\mathbf{x}}$. Another idea would be to update the estimate of $\hat{\boldsymbol{\theta}}$ and $\pi_{\mathrm{angles}}(\cdot)$ in each iteration. We leave this to the future study.

4   Conclusion

We proposed a new iterative algorithm for the uncertain angle CT problem. The method models the uncertainty of the view angles in the likelihood function. We showed numerically that combining this likelihood with a strong prior such as total variation can significantly improve the relative error and visual quality of reconstructions. Furthermore, we showed a method for approximating the likelihood by a block-diagonal approximation of the covariance, which leads to an algorithm that can run on large-scale CT problems.

References

1. Natterer, F.: The Mathematics of Computerized Tomography. Wiley, Chicago (1986)
2. Benning, M., Burger, M.: Modern regularization methods for inverse problems. Acta Numerica **27**, 1–111 (2018)

3. Riis, N.A.B., Frøsig, J., Dong, Y., Hansen, P.C.: Limited-data X-ray CT for underwater pipeline inspection. Inverse Prob. **34**(3), 034002 (2018)
4. Vandeghinste, B., et al.: Iterative CT reconstruction using shearlet-based regularization. IEEE Tran. Nucl. Sci. **60**(5), 3305–3317 (2013)
5. Rudin, L.I., Osher, S., Fatemi, E.: Nonlinear total variation based noise removal algorithms. Physica D **60**(1–4), 259–268 (1992)
6. Ferrucci, M., Leach, R.K., Giusca, C., Carmignato, S., Dewulf, W.: Towards geometrical calibration of X-ray computed tomography systems-a review. Measur. Sci. Technol. **26**(9), 092003 (2015)
7. Mallick, S.P., Agarwal, S., Kriegman, D.J., Belongie, S.J., Carragher, B., Potter, C.S.: Structure and view estimation for tomographic reconstruction: a bayesian approach. In: Proceedings of the IEEE Computer Society Conference on Computer Vision and Pattern Recognition, vol. 2(1), pp. 2253–2260 (2006). 1641029
8. Fang, Y., Murugappan, S., Ramani, K.: Estimating view parameters from random projections for tomography using spherical MDS. BMC Med. Imaging **10**(1), 12 (2010)
9. Basu, S., Bresler, Y.: Uniqueness of tomography with unknown view angles. IEEE Trans. Image Process. **9**(6), 1094–1106 (2000)
10. Korolev, Y., Lellmann, J.: Image reconstruction with imperfect forward models and applications in deblurring. SIAM J. Imaging Sci. **11**(1), 197–218 (2018)
11. Kaipio, J., Somersalo, E.: Statistical and Computational Inverse Problems. Springer, New York (2005). https://doi.org/10.1007/b138659
12. Madsen, R.B., Hansen, T.M.: Estimation and accounting for the modeling error in probabilistic linearized amplitude variation with offset inversion. Geophysics **83**(2), N15–N30 (2018)
13. Hansen, T.M., Cordua, K.S., Holm Jacobsen, B., Mosegaard, K.: Accounting for imperfect forward modeling in geophysical inverse problems exemplified for crosshole tomography. Geophysics **79**(3), H1–H21 (2014)
14. Calvetti, D., Dunlop, M., Somersalo, E., Stuart, A.: Iterative updating of model error for bayesian inversion. Inverse Prob. **34**(2), 025008 (2018)
15. Kolehmainen, V., Tarvainen, T., Arridge, S.R., Kaipio, J.P.: Marginalization of uninteresting distributed parameters in inverse problems - application to diffuse optical tomography. Int. J. Uncertainty Quantification **1**(1), 1–17 (2011)
16. Kennedy, M.C., O'Hagan, A.: Bayesian calibration of computer models. J. Roy. Stat. Soc. Ser. B: Stat. Methodol. **63**(3), 425–450 (2001)
17. Kaipio, J., Somersalo, E.: Statistical inverse problems: discretization, model reduction and inverse crimes. J. Comput. Appl. Math. **198**(2), 493–504 (2007)
18. Nissinen, A., Heikkinen, L.M., Kaipio, J.P.: The bayesian approximation error approach for electrical impedance tomography - experimental results. Measur. Sci. Technol. **19**(1), 015501 (2008)
19. van Aarle, W., et al.: Fast and flexible X-ray tomography using the ASTRA toolbox. Opt. Express **24**(22), 25129–25147 (2016). www.astra-toolbox.com
20. van den Berg, E., Friedlander, M.P.: Spot – A Linear-Operator Toolbox: MATLAB software. www.cs.ubc.ca/labs/scl/spot/
21. Hansen, P.C., Jørgensen, J.S.: AIR tools II: algebraic iterative reconstruction methods, improved implementation. Numer. Algorithms **79**(1), 107–137 (2017)
22. Chambolle, A., Pock, T.: A first-order primal-dual algorithm for convex problems with applications to imaging. J. Math. Imaging Vis. **40**(1), 120–145 (2011)

Optimization Methods in Imaging

Time Discrete Geodesics in Deep Feature Spaces for Image Morphing

Alexander Effland[1]([✉]) [iD], Erich Kobler[1] [iD], Thomas Pock[1] [iD],
and Martin Rumpf[2]

[1] Institute of Computer Graphics and Vision, Graz University of Technology,
Graz, Austria
{alexander.effland,erich.kobler,pock}@icg.tugraz.at
[2] Institute for Numerical Simulation, University of Bonn, Bonn, Germany
martin.rumpf@ins.uni-bonn.de

Abstract. Image morphing in computer vision amounts to computing a visually appealing transition of two images. A prominent model for image morphing originally proposed by Trouvé, Younes and coworkers is image metamorphosis. Here, the space of images is endowed with a Riemannian metric that separately quantifies the contributions due to transport and image intensity variations along a transport path. Geodesic curves in this Riemannian space of images give rise to morphing transitions. The classical metamorphosis model considers images as square-integrable functions on some image domain and thus is non-sensitive to image features such as sharp interfaces or fine texture patterns. To resolve this drawback, we treat images not as intensity maps, but rather as maps into some feature space that encode local structure information. In the simplest case, color intensities are such feature vectors. To appropriately treat local structures and semantic information, deep convolutional neural network features are investigated. The resulting model is formulated directly in terms of a variational time discretization developed for the classical metamorphosis model by Berkels, Effland and Rumpf. The key ingredient is a mismatch energy that locally approximates the squared Riemannian distance and consists of a regularization energy of the time discrete flow and a dissimilarity energy that measures the feature vector modulation along discrete transport paths. The spatial discretization is based on a finite difference and a stable spline interpolation. A variety of numerical examples demonstrates the robustness and versatility of the proposed method for real images using a variant of the iPALM algorithm for the minimization of the fully discrete energy functional.

Keywords: Image metamorphosis · Image morphing ·
Image registration · Feature space · Convolutional neural network ·
iPALM algorithm

A. Effland, E. Kobler and T. Pock acknowledge support from the European Research Council under the Horizon 2020 program, ERC starting grant HOMOVIS, No. 640156. M. Rumpf acknowledges support of the Hausdorff Center for Mathematics and the Collaborative Research Center 1060 funded by the German Research Foundation.

J. Lellmann et al. (Eds.): SSVM 2019, LNCS 11603, pp. 171–182, 2019.
https://doi.org/10.1007/978-3-030-22368-7_14

1   Introduction

Computing smooth transitions joining two images also known as image morphing is a numerically challenging task. An established framework for image morphing is the metamorphosis model proposed by Trouvé, Younes and coworkers [17, 18] as a generalization of the flow of diffeomorphism [10]. Here, the temporal evolution of each image pixel along the morphing sequence is determined by a Riemannian metric in image space. In detail, this metric associates a cost both to the transport of image intensities via a viscous flow and the image intensity variation along the motion path. A morphing sequence for two given input images is defined as the corresponding geodesic curve in the manifold of images, which is given as a minimizer of the path energy in the manifold of images among all regular curves joining the input images.

Starting from the general framework for variational time discretization in geodesic calculus [14], a time discrete metamorphosis model was proposed in [4], which Γ-converges to the original metamorphosis model. However, the classical metamorphosis model, its time discrete counterpart and the spatial discretization based on finite elements in [4] exhibit several drawbacks:

- Comparing images in their original gray- or color space is not invariant to natural radiometric transformations caused by lighting changes, material changes, shadows etc. and hence leads to a blending along the geodesic path instead of geometric transformations.
- Texture patterns, which are important for a natural appearance of images, are often destroyed along the geodesic path due to the color-based matching.
- Sharp interfaces (object boundaries), which usually coincide with depth discontinuities of a scene, are in general not preserved along a geodesic path because of a too strong smoothness assumption of the deformation field.

To address the aforementioned issues, we advocate in this paper a multiscale feature space model proposed in [16], which is closely related to the ImageNet classification approaches shown in [8,15]. This model takes into account a deep convolutional neural network feature space with 19 weight layers using small 3×3-convolution filters and allows to compute a local feature classification for each of the two input images. Here, each of these feature maps is considered as a pointwise map into some higher-dimensional feature space consisting of vectors in $\mathbb{R}^C$, where C ranges from 64 to 512 depending on the considered scale. Now, the metamorphosis model is applied in this feature space and the resulting deformation fields are used to construct the actual image morphing sequence. In addition, we consider an anisotropic regularization for the time discrete deformation sequence, where the anisotropy reflects the edge structure of the underlying images. The resulting time discrete model is finally discretized using finite differences and a third order B-spline interpolation of the deformations for the required warping of feature vectors. This deep feature based model allows not only for an improved detail preservation along motion paths, but also enables a robust computation of time discrete geodesic paths that is significantly less sensitive to color intensity modulations due to the exploitation of semantic information.

This paper is organized as follows: in Sect. 2, some fundamental properties of the original metamorphosis model as well as the time discretization are presented. The extension of the time discrete metamorphosis model to feature spaces is discussed in Sect. 3, and the fully discrete metamorphosis model in feature spaces is described in Sect. 4. In Sect. 5, the iPALM algorithm is adapted to compute geodesic curves. Finally, in Sect. 6 several examples demonstrate the applicability of the proposed methods to real image data.

2   Review of the Metamorphosis Model

Let us briefly review the flow of diffeomorphism approach and its generalization, the metamorphosis model [3,7,9]. The starting point of the theory is the diffeomorphic flow paradigm by Arnold [1], where the temporal evolution of image intensities is governed by a flow defined via a family of diffeomorphisms $(\psi_t)_{t \in [0,1]} : \overline{\Omega} \to \overline{\Omega}$. Here, one assumes at first that the intensity is constant along trajectories on an image domain $\Omega \subset \mathbb{R}^2$ with Lipschitz boundary. The underlying *Riemannian metric g* and the associated *(continuous) path energy* $\mathcal{E}_\psi$ solely depend on the *Eulerian velocity field* $v(t,x) = \dot{\psi}_t \circ \psi_t^{-1}(x)$ and are defined as

$$g_{\psi_t}(\dot{\psi}_t, \dot{\psi}_t) = \int_\Omega L[v,v] \, dx \,, \quad \mathcal{E}_\psi[(\psi_t)_{t \in [0,1]}] = \int_0^1 g_{\psi_t}(\dot{\psi}_t, \dot{\psi}_t) \, dt \,,$$

respectively. Note that g_{ψ_t} represents a Riemannian metric on the space of diffeomorphic transformations $u(t) = u_0 \circ \psi_t^{-1}$ of a fixed input image u_0. The brightness constancy assumption along motion paths readily implies that the material derivative $z = \dot{u} + v \cdot \nabla u$ vanishes.

In the metamorphosis model, this assumption is dropped. Instead, the squared material derivative of the image path, which quantifies the magnitude of the image intensity variation, is added as a penalization term in the metric leading to the *metamorphosis path energy*

$$\mathcal{E}[u] := \int_0^1 \left(\inf_{(v,z)} \int_\Omega L[v,v] + \frac{1}{\delta} z^2 \, dx \right) dt \tag{1}$$

for an image curve $u \in L^2([0,1], L^2(\Omega))$ and a fixed constant $\delta > 0$. If the quadratic form is coercive on $H^m(\Omega, \mathbb{R}^2) \cap H_0^1(\Omega, \mathbb{R}^2)$, which for instance holds for $L[v,v] = |Dv|^2 + \gamma |D^m v|^2$, then one naturally considers tupels of velocity fields/material derivatives $(v,z) \in H_0^m(\Omega, \mathbb{R}^2) \times L^2(\Omega, \mathbb{R})$ such that $\dot{u} + v \cdot \nabla u = z$. For a rigorous formulation, one considers the last equation in a (weak) distributional sense, which allows for material derivatives $z \in L^2([0,1], L^2(\Omega))$ [17].

Then, image morphing of two images $u_A, u_B \in L^2(\Omega)$ amounts to computing shortest geodesic paths of images $u \in L^2([0,1], L^2(\Omega))$ as minimizers of the above path energy (1) in the class of regular curves with $u(0) = u_A$ and $u(1) = u_B$.

Under quite general assumptions, the existence of geodesic paths for arbitrary input images was shown in [17]. A robust and numerically effective variational time discretization of the metamorphosis model was proposed in [4]. The model makes use of the general discrete geodesic calculus analyzed in [14] and introduces an approximation $\mathbf{W}[u, \tilde{u}]$ of the squared Riemannian distance between two images $u, \tilde{u} \in L^2(\Omega)$, i.e.

$$\mathbf{W}[u, \tilde{u}] = \min_{\phi} \int_{\Omega} |D\phi - \mathbb{1}|^2 + \gamma |D^m \phi|^2 + \tfrac{1}{\delta}(\tilde{u} \circ \phi - u)^2 \, dx, \qquad (2)$$

where ϕ is contained in the set of admissible deformations in $H^m(\Omega, \Omega)$ with $\det(D\phi) > 0$ a.e. and $\phi = \mathrm{Id}$ on $\partial\Omega$ with $m > 2$. Then, the discrete path energy on a discrete image path $(u_0, \ldots, u_K) \in (L^2(\Omega))^{K+1}$ for $K \geq 2$ is defined as $\mathbf{E}_K[(u_0, \ldots, u_K)] = K \sum_{k=1}^{K} \mathbf{W}[u_{k-1}, u_k]$, and discrete geodesic curves joining u_A and u_B are minimizers of $\mathbf{E}_K$ subject to the constraints $u_0 = u_A$ and $u_K = u_B$. The existence of discrete geodesic curves for arbitrary input images and the Mosco–convergence of the time discrete to the time continuous metamorphosis model are proven in the case of gray-valued images in [4]. The case of manifold-valued images is investigated in [11].

With a slight misuse of notation, the deformations over which one minimizes in (2) can be incorporated in the energy writing $\mathbf{W}[u, \tilde{u}, \phi]$, which then splits into a regularization energy $\mathbf{R}[\phi]$ for the deformation ϕ composed by the first two terms in (2) and a matching energy $\mathbf{D}[u, \tilde{u}, \phi]$ consisting of the last term in (2). Correspondingly, the path energy can be rephrased as $\mathbf{E}_K[(u_k)_{k=0}^{K}, (\phi_k)_{k=1}^{K}] = K \sum_{k=1}^{K} \mathbf{W}[u_{k-1}, u_k, \phi_k]$, where we minimize over image intensity maps $(u_k)_{k=1}^{K-1}$ and deformations $(\phi_k)_{k=1}^{K}$.

3   Time Discrete Metamorphosis Model in Feature Space

Now, we generalize the metamorphosis model to a space of feature vectors instead of the space of image intensities. The core idea is to replace the image intensity mismatch functional $\mathbf{D}[u, \tilde{u}, \phi]$ of the previous section by the mismatch functional

$$\mathbf{D}[f, \tilde{f}, \phi] = \frac{1}{\delta} \int_{\Omega} \|\tilde{f} \circ \phi - f\|^2 \, dx \qquad (3)$$

for two feature maps $f, \tilde{f} \in L^2(\Omega, \mathbb{R}^C)$ with $C \geq 1$ feature channels. The *time discrete path energy in feature space*

$$\mathbf{E}_K[(f_k)_{k=0}^{K}, (\phi_k)_{k=1}^{K}, (a_k)_{k=1}^{K}] = K \sum_{k=0}^{K-1} \mathbf{R}[\phi_{k+1}, a_{k+1}] + \mathbf{D}[f_k, f_{k+1}, \phi_{k+1}].$$

for features $(f_k)_{k=0}^{K} \in L^2(\Omega, \mathbb{R}^C)^{K+1}$, deformations $(\phi_k)_{k=1}^{K} \in (H^m(\Omega, \Omega))^K$ (subject to $\det(D\phi_k) > 0$ and $\phi = \mathrm{Id}$ on $\partial\Omega$) and positive weights $(a_k)_{k=1}^{K} \in (L^\infty(\Omega, (0, 1]))^K$ is composed of a regularization functional $\mathbf{R}$ and the mismatch functional $\mathbf{D}$ (3). Here, $f_0 = \mathbf{F}[u_0]$ and $f_K = \mathbf{F}[u_K]$, where $\mathbf{F} : L^2(\Omega, \mathbb{R}^3) \rightarrow$

$L^2(\Omega, \mathbb{R}^C)$ refers to a continuous feature extraction operator. In general, the high-dimensional features encode local image patterns describing the local structure of the image as a superposition on different levels of a multiscale image approximation. As a trivial example, the RGB color model with $C = 3$ and $\mathbf{F} = \mathrm{Id}$ is embedded in this new approach, where image intensities and feature channels coincide. As announced, our actual focus is on features on discrete pixel images extracted from two images to be matched via the VGG network with 19 convolutional layers introduced in [16], which is discussed in more detail in a separate paragraph below.

Finally, given the input images $u_0 = u_A$, $u_K = u_B$ and an already computed set of optimal deformations $(\phi_k)_{k=1}^K$ in the feature space model, one can compute a sequence of image intensities $(u_k)_{k=0}^K$ via a minimization of the original mismatch energy

$$\sum_{k=0}^{K-1} \mathbf{D}[u_k, u_{k+1}, \phi_{k+1}],$$

of the time discrete metamorphosis model subject to the constraints $u_0 = u_A$ and $u_K = u_B$. The solutions $u_k = \mathbf{M}_k[u_A, u_B, (\phi_k)_{k=1}^K]$ for $k = 1, \ldots, K-1$ of this block diagonal, quadratic minimization problem can be easily computed. Let us remark that it is in general not possible to compute a time discrete image intensity morphing from the originally computed features maps since $\mathbf{F}$ is not invertible. Thus, for the actual intensity morphing we solely use the optimal deformations.

To properly treat image structures such as sharp edges or corners, all integrands appearing in the regularization energy are rescaled by positive weight functions $(a_k)_{k=1}^K$ reflecting anisotropy. We consider the particular anisotropy $a_k = \mathbf{A}_{\alpha,\sigma,\rho}[\mathbf{M}_k[u_A, u_B, (\phi_k)_{k=0}^K]]$. Following [12], the anisotropy operator is given by $\mathbf{A}_{\alpha,\sigma,\rho}[u] := \exp(-\frac{\|\mathcal{G}_\rho[D\mathcal{G}_\sigma[u]]\|_2^2}{\alpha})$, where $\mathcal{G}_\sigma$ is a Gaussian operator with standard deviation $\sigma > 0$. Throughout this paper, we use

$$\mathbf{R}[\phi, a] = \frac{\mu}{2} \left\| \sqrt{a}(D\phi - \mathbb{1}) \right\|_{L^2(\Omega)}^2 + \lambda \left\| a \left(e^{(\log \det D\phi)^2} - 1 \right) \right\|_{L^1(\Omega)} \qquad (4)$$

for fixed $\mu, \lambda > 0$. The first term penalizes large deviations of the deformation from the identity mapping. Moreover, the second term enforces the positivity of the determinant of the Jacobian matrix of a deformation and favors a balance of shrinkage and growth as advocated in [5,6]. Here, the positivity constraint of the determinant of the Jacobian of the deformations prohibits interpenetration of matter [2].

Let us remark that due to the incorporation of this anisotropy already the simple RGB model performs better than the original time discrete metamorphosis model proposed in [4] in regions with strong spatially varying deformations. For the latter, the time discrete energy is known to converge in the Mosco sense to the time continuous metamorphosis energy. For the general case described here, a rigorous justification of a corresponding convergence result is still open.

4   Fully Discrete Metamorphosis Model in Feature Space

In this section, the fully discrete metamorphosis model in feature space is developed. To this end, we define the computational domain

$$\Omega_{MN} = \left\{ \tfrac{0}{M-1}, \tfrac{1}{M-1}, \ldots, \tfrac{M-1}{M-1} \right\} \times \left\{ \tfrac{0}{N-1}, \tfrac{1}{N-1}, \ldots, \tfrac{N-1}{N-1} \right\} \quad \text{for } M, N \geq 3$$

with discrete boundary $\partial\Omega_{MN} = \Omega_{MN} \backslash \{ \tfrac{1}{M-1}, \ldots, \tfrac{M-2}{M-1} \} \times \{ \tfrac{1}{N-1}, \ldots, \tfrac{N-2}{N-1} \}$ and $\|f\|_{L^p(\Omega_{MN})}^p = \tfrac{1}{MN} \sum_{(i,j)\in\Omega_{MN}} \|f(i,j)\|_2^p$. The set of admissible deformations is

$$\mathcal{D} = \left\{ \phi : \Omega_{MN} \to [0,1]^2 : \phi = \mathrm{Id} \text{ on } \partial\Omega_{MN}, \ \det(D\phi) > 0 \right\} .$$

Furthermore, the discrete Jacobian operator of ϕ at $(i,j) \in \Omega_{MN}$ is defined as the forward finite difference operator with Neumann boundary conditions. The discrete image space and the discrete feature space are given by $\mathcal{I} = \{u : \Omega_{MN} \to \mathbb{R}^3\}$ and $\mathcal{F} = \{f : \Omega_{MN} \to \mathbb{R}^C\}$, respectively.

Next, we introduce the spatial warping operator $\mathbf{T}$ that approximates the pullback of a feature channel $\tilde{f}^c \circ \phi$ at a point $(x,y) \in \Omega_{MN}$ by

$$\mathbf{T}[f^c, \phi](x,y) = \sum_{(i,j)\in\Omega_{MN}} s(\phi_1(x,y) - i) s(\phi_2(x,y) - j) f^c(i,j),$$

where s is the third order B-spline interpolation kernel. Then, the fully discrete mismatch term reads as

$$\mathbf{D}_{MN}[f, \tilde{f}, \phi] = \frac{1}{2C} \sum_{c=1}^{C} \left\| \mathbf{T}[\tilde{f}^c, \phi] - f^c \right\|_{L^2(\Omega_{MN})}^2 .$$

A fully discrete regularization functional $\mathbf{R}_{MN}$ is adapted from (4) in a straightforward manner. Finally, a discrete geodesic path in feature space on a specific multiscale level of a feature hierarchy is the first component $(f_k)_{k=0}^K$ of a minimizer of the resulting fully discrete path energy $\mathbf{E}_{K,MN}$ on discrete feature vector fields, discrete deformations and discrete anisotropy weights, subject to $f_0 = \mathbf{F}[u_0]$, $f_K = \mathbf{F}[u_K]$ and the discrete counterpart of the above constraint on the anisotropy.

Simple RGB Model. For the trivial feature operator $\mathbf{F} = \mathrm{Id}$ with $C = 3$ the features coincide with the RGB images $f_0 = u_A$ and $f_K = u_B$. To improve the numerical computation of minimizing deformations, a multilevel scheme is incorporated. In detail, on the coarse computational domain of size $M_{\mathrm{init}} \times N_{\mathrm{init}}$ with $M_{\mathrm{init}} = 2^{-(L-1)}M$ and $N_{\mathrm{init}} = 2^{-(L-1)}N$ for a given $L > 0$, a time discrete geodesic sequence is computed for suitably restricted input images. In subsequent prolongation steps, the width and the height of the computational domain are successively doubled and the initial deformations and images are obtained via a bilinear interpolation of the preceding coarse scale solutions.

Deep Feature Space. In the second model, we extract image features using convolutional neural networks to incorporate semantic information in the image morphing. To this end, we use the prominent VGG network with 19 convolutional layers as presented in [16], which is designed for localization and classification of objects in natural images. The building blocks of this network are convolutional layers with subsequent ReLU nonlinear activation functions and max pooling layers. In particular, the max pooling layers canonically yield a multiscale semantic decomposition of images. Consequently, we use this multiscale feature approach of the VGG network rather than the multiscale image pyramid via downsampling. For a given scale $l \in \{0, \ldots, 4\}$, the features of the fixed input images u_A, u_B are computed by $f_0 = \mathbf{F}^l[u_A]$, $f_K = \mathbf{F}^l[u_B]$, where the operator $\mathbf{F}^l[u]$ extracts the features of a layer as specified in the following table for an image $u \in \mathcal{I}$, the multiscale index l, the number of feature channels C and the spatial resolution $M \times N$:

l	Layer	C	$M \times N$
0	$\mathrm{conv}_{1,2}$	64	512×512
1	$\mathrm{conv}_{2,2}$	128	256×256
2	$\mathrm{conv}_{3,4}$	256	128×128
3	$\mathrm{conv}_{4,4}$	512	64×64
4	$\mathrm{conv}_{5,4}$	512	32×32

Then, we compute a time discrete geodesic path in feature space by minimizing the discrete path energy $\mathbf{E}_{K,MN}$ for each multilevel separately. For the deformations, we again use a prolongation based on bilinear interpolation. However, the features $(f_k)_{k=1}^{K-1}$ of the subsequent multilevel cannot be computed by bilinear prolongation since features on each multilevel are not necessarily related. To retrieve the anisotropic regularization weights a_k on each scale, we compute the optimal time discrete image path for given deformations $(\phi_k)_{k=1}^{K}$ and fixed input images $u_0, u_K \in \mathcal{I}$ on each scale l separately.

5   Numerical Optimization Using the iPALM Algorithm

In this section, we discuss the computation of time discrete geodesic sequences both for the RGB and the deep feature model. To this end, a variant of the inertial proximal alternating linearized minimization algorithm (iPALM, [13]) is used. Several numerical experiments indicate that a direct gradient based minimization of the data mismatch term $\mathbf{D}_{MN}$ with respect to the deformations is challenging due to the sensitivity of the warping operator to small perturbations of the deformations. Thus, to enhance the stability of the algorithm the warping operator is linearized w.r.t. the deformation at $\tilde{\phi} \in \mathcal{D}$, which is chosen as the deformation of the previous iteration step in the algorithm. To further improve the stability of the algorithm, the linearization is based on the gradient $\Lambda_c(f, \tilde{f}, \tilde{\phi}) = \frac{1}{2}(D\mathbf{T}[\tilde{f}^c, \tilde{\phi}] + Df^c)$, which yields the modified mismatch energy

$$\widetilde{\mathbf{D}}_{MN}[f, \tilde{f}, \phi, \tilde{\phi}] = \frac{1}{2C} \sum_{c=1}^{C} \left\| \mathbf{T}[\tilde{f}^c, \tilde{\phi}] + \left\langle \Lambda_c(f, \tilde{f}, \tilde{\phi}), \phi - \tilde{\phi} \right\rangle - f^c \right\|_{L^2(\Omega_{MN})}^2 .$$

Recall that the proximal mapping of a function $f : \Omega_{MN} \to (-\infty, \infty]$ for $\tau > 0$ is given as $\mathrm{prox}_\tau^f(x) := \mathrm{argmin}_{y:\Omega_{MN}\to(-\infty,\infty]}(\frac{\tau}{2}\|x - y\|_{L^2(\Omega_{MN})}^2 + f(y))$. Then, the proximal operator of $\widetilde{\mathbf{D}}_{MN}$ with respect to the deformation ϕ is given by

$$\mathrm{prox}_\tau^{\widetilde{\mathbf{D}}}[\phi] = \left(\mathbb{1} + \tfrac{1}{\tau C}\Lambda_c(f, \tilde{f}, \tilde{\phi})\Lambda_c(f, \tilde{f}, \tilde{\phi})^\top\right)^{-1} \left(\phi - \tfrac{1}{\tau C}\left(\Lambda_c(f, \tilde{f}, \tilde{\phi})\mathbf{T}[\tilde{f}^c, \tilde{\phi}]\right.\right.$$
$$\left.\left. - \Lambda_c(f, \tilde{f}, \tilde{\phi})\Lambda_c(f, \tilde{f}, \tilde{\phi})^\top\tilde{\phi} - \Lambda_c(f, \tilde{f}, \tilde{\phi})f\right)\right),$$

---

for $j = 1$ **to** J **do**

  for $k = 1$ **to** K **do**

    /* **update anisotropy**   */

    $a_k^{[j+1]} = \mathbf{A}_{\alpha,\sigma,\rho}[u_k^{[j]}];$

    /* **update deformation**   */

    $\phi_k^{[j+1]} = \mathrm{prox}_{L[\phi_k^{[j]}]}^{\widetilde{\mathbf{D}}_{MN}} \left[\tilde{\phi}_k^{[j]} - \frac{1}{L[\phi_k^{[j]}]}\nabla_{\phi_k}\mathbf{R}_{MN}[\tilde{\phi}_k^{[j]}, a_k^{[j+1]}]\right];$

    if $k < K$ **then**

      /* **update features**   */

      $f_k^{[j+1]} = \tilde{f}_k^{[j]} - \frac{1}{L[f_k^{[j]}]}\nabla_{f_k}\mathbf{E}_{K,MN}[\hat{f}_k^{[j]}, (\phi_1^{[j+1]}, \ldots, \phi_k^{[j+1]}, \phi_{k+1}^{[j]}, \ldots, \phi_K^{[j]}),$

                                         $(a_1^{[j+1]}, \ldots, a_k^{[j+1]}, a_{k+1}^{[j]}, \ldots, a_K^{[j]})];$

      /* **update image (only for deep feature model)**   */

      $u_k^{[j+1]} = \tilde{u}_k^{[j]} - \frac{1}{L[u_k^{[j]}]}\nabla_{u_k}\mathbf{E}_{K,MN}[\hat{u}_k^{[j]}, (\phi_1^{[j+1]}, \ldots, \phi_k^{[j+1]}, \phi_{k+1}^{[j]}, \ldots, \phi_K^{[j]})];$

    end

  end

end

Algorithm 1. Algorithm for minimizing $\mathbf{E}_{K,MN}$ on one multilevel.

---

where the function values on $\partial\Omega_{MN}$ remain unchanged. Algorithm 1 summarizes the iteration steps for the minimization of the path energy $\mathbf{E}_{K,MN}$, where for a specific optimization variable f the extrapolation with $\beta > 0$ of the kth path element in the jth iteration step reads as follows:

$$\tilde{f}_k^{[j]} = f_k^{[j]} + \beta(f_k^{[j]} - f_k^{[j-1]}), \quad \hat{f}_k^{[j]} = (f_1^{[j+1]}, \ldots, f_{k-1}^{[j+1]}, \tilde{f}_k^{[j]}, f_{k+1}^{[j]}, \ldots, f_{K-1}^{[j]}).$$

Furthermore, we denote by $L[f]$ the Lipschitz constant of the function f, which is determined by backtracking.

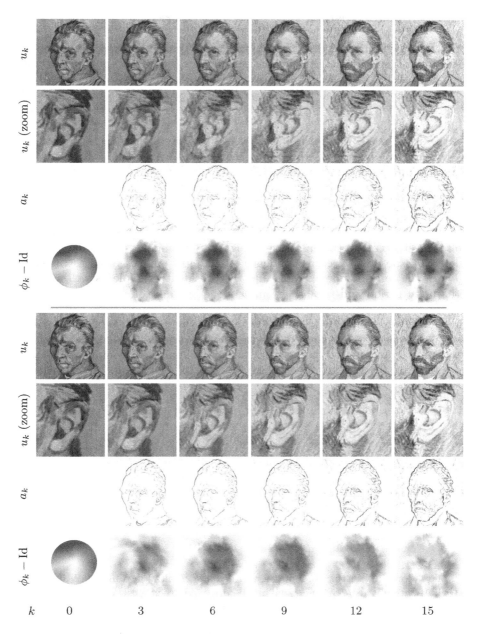

Fig. 1. Time discrete geodesic sequences of self-portraits by van Gogh for the RGB feature (first row) and deep feature model (fifth row) along with a zoom of the ear region with magnification factor 4 (second/sixth row) and the associated sequences of anisotropy weights (third/seventh row) and color-coded displacement fields $\phi_k - \mathrm{Id}$ (fourth/eighth row).

6 Numerical Results

In this section, numerical results for both the RGB and the deep feature model are shown. Figure 1 depicts the geodesic sequences for two self-portraits by van Gogh[1] for $k \in \{0, 3, 6, 9, 12, 15\}$ obtained with the simple RGB model (first row) and the deep feature model (fifth row) along with a zoom (magnification factor 4) of the ear region (second/sixth row) as well as the corresponding sequences of anisotropy weights (third/seventh row) and color-coded displacements fields (fourth/eighth row), where the hue refers to the direction and the intensity is proportional to its norm as indicated by the leftmost color wheel. All parameters used in the computation are specified in Table 1 and were chosen experimentally. Figure 2 shows the corresponding results for two paintings of US presidents[2] and two animals[3], respectively.

In all numerical results, it becomes apparent that the displacement fields evolve over time and exhibit large gradients in the proximity of image interfaces due to the structure of the anisotropy weights. Both models fail to match image regions with no obvious correspondence of the input images (such as the cloth region of the self-portrait example and the president example as well as parts of the body region and the background in the animal example), which results in image blending. However, the model incorporating deep features significantly outperforms the simple RGB model in regions where image semantics matter (such as the cheek and the ear in the van Gogh example, the hair region of the president example as well as the teeth for the animals). To compute a visually appealing time discrete geodesic sequence, a fourth color channel representing a

Table 1. Collection of all parameters for both models in all examples considered.

		$M \times N$	K	L	μ	λ	α	σ	ρ	β	J
van Gogh	(RGB)	496×496	15	5	0.05	0.05	1000	0.5	2	$\frac{1}{\sqrt{2}}$	250
	(deep)				0.002	0.002					
Presidents	(RGB)	512×512	15	5	0.1	0.1	1000	0.5	2	$\frac{1}{\sqrt{2}}$	250
	(deep)				0.002	0.002					
Animals	(RGB)	512×512	15	5	0.05	0.05	1000	0.5	2	$\frac{1}{\sqrt{2}}$	250
	(deep)				0.002	0.002					

---

[1] Public domain, https://commons.wikimedia.org/wiki/File:Vincent_Willem_van_Gogh_102.jpg; https://commons.wikimedia.org/wiki/File:SelbstPortrait_VG2.jpg.

[2] First painting by Gilbert Stuart (public domain), https://commons.wikimedia.org/wiki/File:Gilbert_Stuart_Williamstown_Portrait_of_George_Washington.jpg; second painting by Rembrandt Peale (public domain), https://commons.wikimedia.org/wiki/File:Thomas_Jefferson_by_Rembrandt_Peale,_1800.jpg.

[3] First photo detail by Domenico Salvagnin (CC BY 2.0), https://commons.wikimedia.org/wiki/File:Yawn!!!_(331702223).jpg; second photo detail by Eric Kilby (CC BY-SA 2.0), https://commons.wikimedia.org/wiki/File:Panthera_tigris_-Franklin_Park_Zoo,_Massachusetts,_USA-8a_(2).jpg.

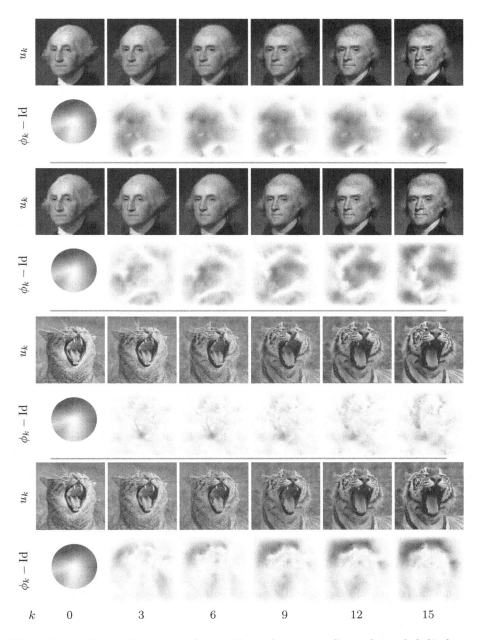

Fig. 2. Pairs of time discrete geodesic paths and corresponding color-coded displacement fields for paintings of US presidents with RGB features (first/second row) and deep features (third/fourth row) as well as for animals with RGB features (fifth/sixth row) and deep features (seventh/eighth row).

manual segmentation of image regions and a color adaptation of the van Gogh self-portraits are required in [4]. This is obsolete in the proposed deep feature based model due to the incorporation of semantic information.

References

1. Arnold, V.: Sur la géométrie différentielle des groupes de Lie de dimension infinie et ses applications à l'hydrodynamique des fluides parfaits. Ann. Inst. Fourier (Grenoble) **16**(fasc. 1), 319–361 (1966)
2. Ball, J.M.: Global invertibility of Sobolev functions and the interpenetration of matter. Proc. R. Soc. Edinb. **88A**, 315–328 (1981)
3. Beg, M.F., Miller, M.I., Trouvé, A., Younes, L.: Computing large deformation metric mappings via geodesic flows of diffeomorphisms. Int. J. Comput. Vis. **61**(2), 139–157 (2005)
4. Berkels, B., Effland, A., Rumpf, M.: Time discrete geodesic paths in the space of images. SIAM J. Imaging Sci. **8**(3), 1457–1488 (2015)
5. Burger, M., Modersitzki, J., Ruthotto, L.: A hyperelastic regularization energy for image registration. SIAM J. Sci. Comput. **35**(1), B132–B148 (2013)
6. Droske, M., Rumpf, M.: A variational approach to nonrigid morphological image registration. SIAM J. Appl. Math. **64**(2), 668–687 (2003)
7. Dupuis, P., Grenander, U., Miller, M.I.: Variational problems on flows of diffeomorphisms for image matching. Q. Appl. Math. **56**(3), 587–600 (1998)
8. Krizhevsky, A., Sutskever, I., Hinton, G.E.: Imagenet classification with deep convolutional neural networks. In: IEEE Conference on Neural Information Processing Systems NIPS, pp. 1097–1105 (2012)
9. Miller, M.I., Trouvé, A., Younes, L.: On the metrics and Euler-Lagrange equations of computational anatomy. Ann. Rev. Biomed. Eng. **4**(1), 375–405 (2002)
10. Miller, M.I., Younes, L.: Group actions, homeomorphisms, and matching: a general framework. Int. J. Comput. Vis. **41**(1–2), 61–84 (2001)
11. Neumayer, S., Persch, J., Steidl, G.: Morphing of manifold-valued images inspired by discrete geodesics in image spaces. SIAM J. Imaging Sci. **11**(3), 1898–1930 (2018)
12. Perona, P., Malik, J.: Scale-space and edge detection using anisotropic diffusion. IEEE Trans. Pattern Anal. Mach. Intell. **12**(7), 629–639 (1990)
13. Pock, T., Sabach, S.: Inertial proximal alternating linearized minimization (iPALM) for nonconvex and nonsmooth problems. SIAM J. Imaging Sci. **9**(4), 1756–1787 (2016)
14. Rumpf, M., Wirth, B.: Variational time discretization of geodesic calculus. IMA J. Numer. Anal. **35**(3), 1011–1046 (2015)
15. Sermanet, P., Eigen, D., Zhang, X., Mathieu, M., Fergus, R., Lecun, Y.: Overfeat: integrated recognition, localization and detection using convolutional networks. In: ICLR (2014)
16. Simonyan, K., Zisserman, A.: Very deep convolutional networks for large-scale image recognition. In: ICLR (2015)
17. Trouvé, A., Younes, L.: Local geometry of deformable templates. SIAM J. Math. Anal. **37**(1), 17–59 (2005)
18. Trouvé, A., Younes, L.: Metamorphoses through lie group action. Found. Comput. Math. **5**(2), 173–198 (2005)

Minimal Lipschitz Extensions for Vector-Valued Functions on Finite Graphs

Johannes Hertrich[1], Miroslav Bačák[3], Sebastian Neumayer[1], and Gabriele Steidl[1,2(✉)]

[1] TU Kaiserslautern, Kaiserslautern, Germany
steidl@mathematik.uni-kl.de
[2] Fraunhofer ITWM, Kaiserslautern, Germany
[3] MPI, Leipzig, Germany

Abstract. This paper deals with extensions of vector-valued functions on finite graphs fulfilling distinguished minimality properties. We show that so-called lex and L-lex minimal extensions are actually the same and call them minimal Lipschitz extensions. We prove that the minimizers of functionals involving grouped ℓ_p-norms converge to these extensions as $p \to \infty$. Further, we examine the relation between minimal Lipschitz extensions and iterated weighted midrange filters and address their connection to ∞-Laplacians for scalar-valued functions. A convergence proof for an iterative algorithm proposed in [9] for finding the zero of the ∞-Laplacian is given.

Keywords: Lipschitz extensions · ∞-Laplacian · p-Laplacian · Midrange filter · ∞-Harmonic functions

1 Introduction

The topic of minimal Lipschitz extensions on graphs appears in various fields of mathematics and computer science from different points of view and with different notation. Results can be found in approximation theory [7], discrete mathematics/graph theory [18], data and image processing [2,4,6,10,21,24], and mathematical morphology [22] to mention only a few. While many results are available for scalar-valued functions, only few is proved for the vector-valued case. In this paper, we want to clarify some aspects and to highlight relations to some known concepts in image processing as midrange filters and ∞-Laplacians.

Let $G := (V, E, w)$ with $M := |E|$ be an undirected, connected, weighted graph with weight function $w \colon E \to [0, 1]$ and $\emptyset \neq U \subset V$, where $(u, v) \notin E$ if $u, v \in U$. We write $u \sim v$ if and only if $(u, v) \in E$ and suppose that $w(u, v) = 0$ if and only if $(u, v) \notin E$. Since the graph is not directed, the weights are symmetric, i.e. $w(u, v) = w(v, u)$. The set of functions $f \colon V \to \mathbb{R}^m$ is denoted by $\mathcal{H}(V)$. For

Funding by the DFG within the Research Training Group 1932, and within project STE 571/13-1 is gratefully acknowledged.

J. Lellmann et al. (Eds.): SSVM 2019, LNCS 11603, pp. 183–195, 2019.
https://doi.org/10.1007/978-3-030-22368-7_15

a given function $g \colon U \to \mathbb{R}^m$, let $\mathcal{H}_g(V)$ denote those functions $f \in \mathcal{H}(V)$ with $f \restriction_U = g$, which are called *extensions* of g.

The *2-p-norm*, $p \in [1, \infty)$ and the *2-∞-norm* are defined for $x = (x_1, \ldots, x_n) \in \mathbb{R}^{mn}$ as $\|x\|_{2,p} := \left(\sum_{i=1}^n |x_i|^p \right)^{1/p}$, $\|x\|_{2,\infty} := \max_{i=1,\ldots,n} |x_i|$, where $|x_i|$ denotes the Euclidean norm of $x_i \in \mathbb{R}^m$. In [11], the discrete gradient operator on graphs $\nabla_w \colon \mathcal{H}(V) \to \mathbb{R}^{2mM}$ was introduced by $\nabla_w f(u) := (\partial_v f(u))_{v \sim u}$, where $\partial_v f(u) := w(u,v)(f(v) - f(u)) \in \mathbb{R}^m$. Let $g \colon U \to \mathbb{R}^m$ be given. For $f \in \mathcal{H}_g(V)$, we are interested in the *anisotropic energies of the p-Laplacians*

$$E_p f := \|\nabla_w f\|_{2,p}^p = \sum_{u \in V} \left(\sum_{v \sim u} w(u,v)^p |f(u) - f(v)|^p \right), \tag{1}$$

$$E_\infty f := \|\nabla_w f\|_{2,\infty} = \max_{u \in V} \left(\max_{v \sim u} w(u,v) |f(u) - f(v)| \right).$$

The functionals E_p, $p \in (1, \infty)$ are strictly convex and hence they have a unique global minimum f_p. Besides E_p, $p \in [1, \infty)$ the functional

$$E_{\infty,p} f := \left\| \left(\|\nabla_w f(u)\|_{2,\infty} \right)_{u \in V \setminus U} \right\|_p^p = \sum_{u \in V \setminus U} \left(\max_{v \sim u} w(u,v)^p |f(u) - f(v)|^p \right)$$

was considered in the literature [23]. This functional is not strictly convex for $p \in (1, \infty)$, but has nevertheless a unique minimizer $f_{\infty,p} \in \mathcal{H}_g(V)$ of $E_{\infty,p}$, see [12]. In contrast to E_p or $E_{\infty,p}$, $p \in (1, \infty)$, the functional E_∞ has in general many minimizers. Using Γ-convergence arguments, it is not hard to show that every cluster point of the sequence of minimizers of E_p, resp. $E_{\infty,p}$ is a minimizer of E_∞. In this paper, we want to accent minimizers of E_∞ with distinguished properties.

Contribution. In Sect. 2, we prove that so-called lex and L-lex minimal extensions coincide. We show that the minimizers of E_p, resp. $E_{\infty,p}$ converge to these extensions as $p \to \infty$. Section 3 deals with the relation between iterated midrange filters and minimal Lipschitz extensions. In Sect. 4, we consider ∞-Laplacians for scalar-valued functions and provide a convergence proof for an iterative algorithm of Elmoataz et al. [9] to compute minimal Lipschitz extensions.

2   Minimal Lipschitz Extensions and Approximations

For $u \in V$ and $v \sim u$, let $lf(u,v) := w(u,v)|f(u) - f(v)| = |\partial_v f(u)|$ and $\text{lex}(f) \in \mathbb{R}^M$ be the vector with entries $(lf(u,v))_{v \sim u}$ in nonincreasing order. For $u \in V \setminus U$, define $Lf(u) := \max_{v \sim u} lf(u,v) = \|\nabla_w f(u)\|_{2,\infty}$ and the vector $L\text{-lex}(f) \in \mathbb{R}^{|V \setminus U|}$ with entries $(Lf(u))_{u \in V \setminus U}$ in nonincreasing order. A function $f \in \mathcal{H}_g(V)$ is called lex *minimal extension* (of g) if $\text{lex}(f) \leq \text{lex}(h)$ for every $h \in \mathcal{H}_g(V)$, and L-lex *minimal extension* (of g) if $L\text{-lex}(f) \leq L\text{-lex}(h)$ for every $h \in \mathcal{H}_g(V)$, where $\leq$ is the lexicographical ordering. The first notation can be found, e.g., in [14], the L-lex *minimal extension* was called *tight extension* in [23]. By definition we have $\text{lex}_1(f) = L\text{-lex}_1(f)$, where the subscript denotes the coordinate index. The existence of L-lex minimal extensions in the non-weighted case was shown in [23, Theorem 1.2]. The existence in the weighted setting as well as for lex minimal extensions can be proved similarly; see [12]. To show that lex and L-lex minimal extensions coincide we need the following lemma.

Lemma 1. *Let $f \in \mathcal{H}_g(V)$ be a lex minimal extension and $\tilde{f} \in \mathcal{H}_g(V)$ with $f \neq \tilde{f}$ and $\mathrm{lex}_1(f) = \mathrm{lex}_1(\tilde{f})$. Let $k \in \{1, \ldots, M\}$ be the largest index such that $(\mathrm{lex}_i(f))_{i=1}^k = (\mathrm{lex}_i(\tilde{f}))_{i=1}^k$ and $K := \mathrm{lex}_k(f)$. Then f and $\tilde{f}$ coincide on the set*

$$W := \{u \in V \setminus U : \exists v \sim u \quad \text{such that} \quad lf(u,v) \geq K\}.$$

Proof. 1. First, we prove that $f(u) - f(v) = \tilde{f}(u) - \tilde{f}(v)$ for all $u \sim v$ with $lf(u,v) \geq K$. Consider $u \sim v$ with $lf(u,v) \geq K$. We suppose that if $lf(u,v) = lf(\tilde{u}, \tilde{v}) = l\tilde{f}(u,v) = l\tilde{f}(\tilde{u}, \tilde{v})$ and those values appear at positions i and j in $\mathrm{lex}(f)$ and $\mathrm{lex}(\tilde{f})$, then the corresponding values $lf(u,v)$ and $l\tilde{f}(u,v)$ have the same position. Let C be the largest value in $(\mathrm{lex}_i(f))_{i=1}^k$ with $f(u) - f(v) \neq \tilde{f}(u) - \tilde{f}(v)$. For $u \sim v$ with $lf(u,v) > C$, we have $f(u) - f(v) = \tilde{f}(u) - \tilde{f}(v)$ and for $lf(u,v) = C$ the relation $|f(u) - f(v)| \geq |\tilde{f}(u) - \tilde{f}(v)|$, where at least one $u \sim v$ with $f(u) - f(v) \neq \tilde{f}(u) - \tilde{f}(v)$ exists. For $u \sim v$ with $lf(u,v) < C$ we have $l\tilde{f}(u,v) \leq C$. For $h := \frac{1}{2}(f + \tilde{f}) \in \mathcal{H}_g(V)$ and all $u \sim v$, we obtain that

$$|h(u) - h(v)| = \tfrac{1}{2}|f(u) - f(v) + \tilde{f}(u) - \tilde{f}(v)| \leq \tfrac{1}{2}(|f(u) - f(v)| + |\tilde{f}(u) - \tilde{f}(v)|)$$

and consequently $lh(u,v) \leq lf(u,v)$ whenever $lf(u,v) \geq C$. Further, for two vectors $a, b \in \mathbb{R}^m$ with $|a| = |b|$ we have $|a + b| = |a| + |b|$ if and only if $a = b$. For $u \sim v$ with $f(u) - f(v) \neq \tilde{f}(u) - \tilde{f}(v)$ this implies $lh(u,v) < lf(u,v)$, i.e. strict inequality holds at least once where $lf(u,v) = C$. Thus $\mathrm{lex}(h) < \mathrm{lex}(f)$ which contradicts the lex minimality of f. Hence $f(u) - f(v) = \tilde{f}(u) - \tilde{f}(v)$ for all $u \sim v$ with $lf(u,v) \geq \mathrm{lex}_k(f)$.

2. Next, we show that there exits $u \sim v$ with $v \in U$ such that $lf(u,v) = \mathrm{lex}_1(f)$. Assume in contrary that this is not the case, i.e., there exists $\delta > 0$ such that $lf(u,v) \leq \mathrm{lex}_1(f) - \delta$ for all $u \sim v$ with $v \in U$. For $\varepsilon > 0$, consider the function $\hat{f} \in \mathcal{H}_g(V)$ with $\hat{f}(u) := (1 - \varepsilon)f(u)$, $u \in V \setminus U$. Then we have for all $u_1, u_2 \in V \setminus U$ with $u_1 \sim u_2$ that $l\hat{f}(u_1, u_2) = (1 - \varepsilon)lf(u_1, u_2) < lf(u_1, u_2)$ and for all $u \in V \setminus U$ and $v \in U$ that

$$l\hat{f}(u,v) = w(u,v)|(1 - \varepsilon)f(u) - f(v)| \leq lf(u,v) + \varepsilon w(u,v)|f(u)| < \mathrm{lex}_1 f$$

for $\varepsilon < \delta / \max_{u \sim v} (w(u,v)|f(u)|)$. Thus $\mathrm{lex}\hat{f} < \mathrm{lex}f$ which is a contradiction.

3. For $u \sim v$ with $v \in U$ such that $lf(u,v) = \mathrm{lex}_1(f)$, Part 1 of the proof implies that $f(u) = \tilde{f}(u)$. Choose one such u and set $U_1 := U \cup \{u\}$, extend g to g_1 on U_1 by $g_1(u) := f(u)$. Cut all edges $u \sim v$ with $v \in U$ and remove the corresponding entries in $\mathrm{lex}(f)$ and $\mathrm{lex}(\tilde{f})$. Note that only entries with the same value are removed, including the first one. Consider f and $\tilde{f}$ as extensions of g_1, where f is still the lex minimal extension of g_1.

The whole procedure is repeated with respect to the new first component $\mathrm{lex}_1(f)$ and so on until all edges with $lf(u,v) \geq K$ are removed. This yields the assertion. $\qquad\square$

Now we can prove equivalence of lex and L-lex minimal extensions.

Theorem 1. *There exists a unique lex/L-lex minimal extension $f \in \mathcal{H}_g(V)$ and both extensions coincide.*

We call the lex/L-lex minimal extension *minimal Lipschitz extension.*

Proof. 1. The uniqueness of the L-lex minimal extension was shown in [23]. To show the uniqueness of the lex minimal extension assume in contrary that there exist two different lex minimal extensions f and $\tilde{f}$. Then $\text{lex}(f) = \text{lex}(\tilde{f})$ and Lemma 1 with $k = M$ yields the contradiction $f = \tilde{f}$.

2. Let $f \in \mathcal{H}_g(V)$ be the lex minimal extension and $\tilde{f} \in \mathcal{H}_g(V)$ the L-lex minimal extension. Assume that $f \neq \tilde{f}$. Then $\text{lex}(f) < \text{lex}(\tilde{f})$ and $L\text{-lex}(f) > L\text{-lex}(\tilde{f})$. If $L\text{-lex}_1(f) = \text{lex}_1(f) < \text{lex}_1(\tilde{f}) = L\text{-lex}_1(\tilde{f})$ we have a contradiction. Thus, $\text{lex}_1(f) = \text{lex}_1(\tilde{f})$ and we can apply Lemma 1 which implies that f and $\tilde{f}$ coincide on the set W. Now we can consider f and $\tilde{f}$ as extensions of g extended to $U \cup W$, where the edges between vertices in $U \cup W$ and the corresponding entries in $\text{lex}(f)$ and $\text{lex}(\tilde{f})$ are removed. But for this new constellation we have $\text{lex}_1(f) < \text{lex}_1(\tilde{f})$ which again contradicts the L-lex minimality of $\tilde{f}$. $\square$

So far, there do not exist methods for computing minimal Lipschitz extensions of vector-valued functions. For scalar-valued functions, a combinatorial algorithm running in expected time $\mathcal{O}(N(M + N \log N))$ was given in [14] along with a faster variant in practice. For the special case of non-weighted graphs a polynomial-time algorithm was already presented in [15,18]. Later we will deal with iterative algorithms.

However, minimal Lipschitz extensions of vector-valued functions can be approximated. In this direction, the following theorem is just a generalization of [23, Theorem 1.3] for our weighted gradients; see [12].

Theorem 2. *For any $m \geq 1$, the whole sequence of minimizers $\{f_{\infty,p}\}_p$ of $E_{\infty,p}$ subject to $f \upharpoonright_U = g$ converges to the minimal Lipschitz extension of g.*

Next, we consider the sequence $\{f_p\}_p$ of minimizers of E_p. For real-valued functions, i.e., $m = 1$, its convergence to the lex minimal extension of g can be shown by applying the following classical result from approximation theory concerning the convergence of Pólya's algorithm [19]. In [7] it was proved that for an affine subspace $\mathcal{K} \subset \mathbb{R}^M$, the sequence of L_p approximations $\{x_p\}_p$ given by

$$x_p := \underset{x \in \mathcal{K}}{\arg \min} \|x - z\|_p, \quad p \in (1, \infty),$$

converges for $p \to \infty$ to those minimizer $\hat{x}_\infty \in \mathcal{K}$ of $\|x - z\|_\infty$ with the following property: For every minimizer $x_\infty \in \mathcal{K}$ of $\|x - z\|_\infty$ consider the vector $\sigma(x_\infty)$ whose coordinates $|x_{\infty,i} - z_i|$ are arranged in nonincreasing order. Then $\hat{x}_\infty$ is the minimizer with smallest lexicographical ordering with respect to σ. Indeed this vector is uniquely determined and called *strict uniform approximation* of

z. Concerning the convergence rate it was shown in [20], see also [8], that there exist constants $0 < C_1, C_2 < \infty$ and $a \in [0,1]$ depending on $\mathcal{K}$, such that

$$C_1 \frac{1}{p} a^p \leq \|x_p - \hat{x}_\infty\|_\infty \leq C_2 \frac{1}{p} a^p.$$

Concerning our setting, we rewrite $E_p f = 2 \, \|Af \restriction_{V \setminus U} + b\|_p^p$, where $A \in \mathbb{R}^{M, |V \setminus U|}$ is the matrix representing the linear operator ∇_w on functions restricted to $V \setminus U$ and $b \in \mathbb{R}^M$ accounts for the fixed values g on U. Then, considering $\mathcal{K} := \{Ay + b \colon y \in \mathbb{R}^{|V \setminus U|}\}$ and $z = 0$, we obtain the desired result. For $m \geq 2$, the convergence of the minimizers of E_p, $p \in [1, \infty)$ cannot be deduced in this way, so we prove the next theorem.

Theorem 3. *For any $m \geq 1$, the whole sequence of minimizers $\{f_p\}_p$ of E_p subject to $f \restriction_U = g$ converges to the* lex *minimal extension of g.*

Proof. There exists a convergent subsequence $\{f_{p_j}\}_j$ of $\{f_p\}_p$ with limit $\tilde{f}$. Let $f \in \mathcal{H}_g(V)$ be the lex minimal extension and assume that $f \neq \tilde{f}$. Choose $k \in \{1, \ldots, M\}$ as the largest index such that $(\mathrm{lex}_i(f))_{i=1}^k = (\mathrm{lex}_i(\tilde{f}))_{i=1}^k$, which exists since $\tilde{f}$ is a minimizer of E_∞. By Lemma 1 the functions f and $\tilde{f}$ coincide on the set $W := \{u \in V \setminus U \colon \exists v \sim u \ \text{ such that } \ lf(u,v) \geq \mathrm{lex}_k(f)\}$. Define

$$h_p(u) := \begin{cases} f_p(u) & \text{if } u \in W \cup U, \\ f(u) & \text{otherwise.} \end{cases}$$

Since $f(u) = \tilde{f}(u)$ for $u \in W$, we have that $h_{p_j} \to f$ as $j \to \infty$. For $u \in W \cup U$ and $v \in W$ with $u \sim v$ it holds

$$w(u,v)|f_{p_j}(u) - f_{p_j}(v)| = w(u,v)|h_{p_j}(u) - h_{p_j}(v)|$$

and for $u \in V$ and $v \in V \setminus (W \cup U)$ with $u \sim v$,

$$w(u,v)|h_{p_j}(u) - h_{p_j}(v)| \to w(u,v)|f(u) - f(v)| \leq \mathrm{lex}_{k+1}(f).$$

Setting $\delta := \mathrm{lex}_{k+1}(\tilde{f}) - \mathrm{lex}_{k+1}(f) > 0$, we obtain for sufficiently large j and $u \in V$, $v \in V \setminus (W \cup U)$ with $u \sim v$ that

$$w(u,v)|h_{p_j}(u) - h_{p_j}(v)| \leq \mathrm{lex}_{k+1}(f) + \frac{\delta}{4}.$$

On the other hand, there exist $u \in V$ and $v \in V \setminus (W \cup U)$ with $u \sim v$ such that

$$w(u,v)|f_{p_j}(u) - f_{p_j}(v)| \to w(u,v)|\tilde{f}(u) - \tilde{f}(v)| = \mathrm{lex}_{k+1}(\tilde{f})$$

so that for sufficiently large j,

$$w(u,v)|f_{p_j}(u) - f_{p_j}(v)| \geq \mathrm{lex}_{k+1}(\tilde{f}) - \frac{\delta}{4}.$$

Then we conclude

$$E_{p_j}(f_{p_j}) - E_{p_j}(h_{p_j})$$

$$= \sum_{u \in V} \sum_{v \sim u} w(u,v)^{p_j} \left(|f_{p_j}(v) - f_{p_j}(u)|^{p_j} - |h_{p_j}(v) - h_{p_j}(u)|^{p_j} \right)$$

$$= \sum_{u \in W \cup U} \sum_{\substack{v \sim u \\ v \notin (W \cup U)}} w(u,v)^{p_j} |f_{p_j}(v) - f_{p_j}(u)|^{p_j}$$

$$+ \sum_{u \in V \setminus (W \cup U)} \sum_{v \sim u} w(u,v)^{p_j} |f_{p_j}(v) - f_{p_j}(u)|^{p_j}$$

$$- \sum_{u \in W \cup U} \sum_{\substack{v \sim u \\ v \notin (W \cup U)}} w(u,v)^{p_j} |h_{p_j}(v) - h_{p_j}(u)|^{p_j}$$

$$- \sum_{u \in V \setminus (W \cup U)} \sum_{v \sim u} w(u,v)^{p_j} |h_{p_j}(v) - h_{p_j}(u)|^{p_j},$$

and further

$$E_{p_j}(f_{p_j}) - E_{p_j}(h_{p_j}) \geq \left(\text{lex}_{k+1}(\tilde{f}) - \frac{\delta}{4} \right)^{p_j} - C \left(\text{lex}_{k+1}(f) + \frac{\delta}{4} \right)^{p_j}$$

$$= \left(\text{lex}_{k+1}(\tilde{f}) - \frac{\delta}{4} \right)^{p_j} \left(1 - C \left(\frac{\text{lex}_{k+1}(f) + \frac{\delta}{4}}{\text{lex}_{k+1}(\tilde{f}) - \frac{\delta}{4}} \right)^{p_j} \right).$$

By definition of δ the quotient is smaller than 1, so that we obtain for sufficiently large j that $E_{p_j}(f_{p_j}) - E_{p_j}(h_{p_j}) > 0$. This contradicts that f_{p_j} is the minimizer of E_{p_j} and we are done. □

3 Midrange Filters and ∞-Harmonic Extensions

We consider the following minimization problem. For $x = (x_1, \ldots, x_n)^{\mathrm{T}}$, $x_i \in \mathbb{R}^m$ and $\omega = (\omega_1, \ldots, \omega_n)^{\mathrm{T}} \in (0,1]^n$ with $\omega_1 \geq \ldots \geq \omega_n$, we consider the *weighted midrange filter* $\text{midr}_\omega \colon \mathbb{R}^{mN} \to \mathbb{R}^m$ given by

$$\text{midr}_\omega \, x = \arg\min_{a \in \mathbb{R}^m} \left\{ \max_i \left(\omega_i |x_i - a| \right) \right\}. \tag{2}$$

If $\omega_i |x_i - a|$ is replaced by $\omega_i^2 |x_i - a|^2$, $i = 1, \ldots, n$, the minimizer remains the same. As the pointwise maximum of strongly convex functions with modulus $2\omega_i^2$, the functional $\max_i \left(\omega_i^2 |x_i - a|^2 \right)$ is strongly convex with modulus $2\omega_n^2$. Hence $\text{midr}_\omega \, x$ is uniquely determined.

In the non-weighted, scalar-valued case $m = 1$, the midrange filter simplifies to

$$\text{midr} \, x = \tfrac{1}{2} \left(\min_i x_i + \max_i x_i \right).$$

For the weighted filter, there is no analytical expression, even in the scalar-valued case.

Indeed what we call midrange filter from the signal processing point of view, is quite classical and can be found under various names in the literature as, e.g., *smallest circle/bounding sphere* or *weighted Chebyshev center/circumcenter*. The problem is known to be solvable in $\mathcal{O}(n)$ by linear programming, where the factor in the $\mathcal{O}(n)$ term depends sub-exponentially on the dimension m [16]. For a comparison of several algorithms see [26]. The minimization problem (2) can be generalized from $\mathbb{R}^m$, e.g., to on Hadamard spaces [1]. The case of symmetric positive definite matrices was considered in [25].

Lemma 2. *For real-valued functions, i.e., $m = 1$, the operator midr_ω is Lipschitz continuous with constant $L \leq 1$ with respect to the ∞-norm.*

Proof. First, we have that midr_ω is given by

$$\mathrm{midr}_\omega\, x = \frac{\omega_i x_i + \omega_j x_j}{\omega_i + \omega_j}, \qquad (i,j) \in \arg\max_{k,l} \frac{|x_k - x_l|}{1/\omega_k + 1/\omega_l}, \qquad (3)$$

see [17, Theorem 5].

For $x, y \in \mathbb{R}^n$ we consider the vector-valued function $x(t) := (1-t)x + ty$, $t \in [0,1]$. Then, $L_{i,j} \colon [0,1] \to \mathbb{R}$, $i,j = 1, \ldots, n$ defined by

$$L_{i,j}(t) = \frac{|x_i(t) - x_j(t)|}{1/\omega_i + 1/\omega_j}$$

are piecewise linear functions and hence the same holds for $L(t) := \sup_{i,j} L_{i,j}(t)$. Therefore, $[0,1]$ can be split into a finite number of intervals $[t_{k-1}, t_k]$, $k = 1, \ldots, K$ with $t_0 := 0$ and $t_K := 1$ with corresponding maximizing indices (i_k, j_k) on $[t_{k-1}, t_k]$, $k = 1, \ldots, K$. By the triangle inequality and (3), we get

$$\left| \mathrm{midr}_\omega\, x - \mathrm{midr}_\omega\, y \right|$$

$$\leq \sum_{k=1}^K \left| \mathrm{midr}_\omega\, x(t_{k-1}) - \mathrm{midr}_\omega\, x(t_k) \right|$$

$$= \sum_{k=1}^n \left| \frac{\omega_{i_k} x_{i_k}(t_{k-1}) + \omega_{j_k} x_{j_k}(t_{k-1})}{\omega_{i_k} + \omega_{j_k}} - \frac{\omega_{i_k} x_{i_k}(t_k) + \omega_{j_k} x_{j_k}(t_k)}{\omega_{i_k} + \omega_{j_k}} \right|$$

$$= \sum_{k=1}^n \left| \frac{\omega_{i_k}\left(x_{i_k}(t_{k-1}) - x_{i_k}(t_k)\right) + \omega_{j_k}\left(x_{j_k}(t_{k-1}) - x_{j_k}(t_k)\right)}{\omega_{i_k} + \omega_{j_k}} \right|.$$

Finally, the definition of $x(t)$ implies

$$\left| \mathrm{midr}_\omega\, x - \mathrm{midr}_\omega\, y \right| \leq \sum_{k=1}^n \frac{\omega_{i_k}(t_k - t_{k-1})}{\omega_{i_k} + \omega_{j_k}} \|x - y\|_\infty + \frac{\omega_{j_k}(t_k - t_{k-1})}{\omega_{i_k} + \omega_{j_k}} \|x - y\|_\infty$$

$$= \|x - y\|_\infty. \qquad \square$$

Unfortunately midr_ω is in general not Lipschitz in dimensions $m \geq 2$. Here we can only show that midr_ω is locally $\frac{1}{2}$-Hölder continuous.

Next, we are interested in applying the midrange filter for given functions $f\colon V \to \mathbb{R}^m$, i.e., for every $u \in V \setminus U$ we compute midr_ω for $x(u) := (f(v))_{v \sim u}$ with weights $\omega(u) := (w(u, v))_{v \sim u}$. A function $f \in \mathcal{H}_g(V)$ is called an ∞-*harmonic extension* of g if

$$f(u) = \mathrm{midr}_{\omega(u)}\, x(u) = \arg\min_{a \in \mathbb{R}^m} \left\{ \max_{v \sim u} w(u, v) | f(v) - a| \right\}$$

for all $u \in V \setminus U$. In other words, $f\colon \mathcal{H}_g(V)$ is an ∞-harmonic extension of g if it is a fixed point of the operator $\mathrm{Midr}_w\colon \mathcal{H}(V) \to \mathcal{H}(V)$ given by

$$\mathrm{Midr}_w\, f(u) := \begin{cases} f(u) & \text{if } u \in U, \\ \mathrm{midr}_{\omega(u)}\, x(u) & \text{otherwise.} \end{cases}$$

The relation between ∞-harmonic extension and lex minimal ones is given as follows.

Lemma 3. *The Lipschitz minimal extension $f \in \mathcal{H}_g(V)$ is an ∞-harmonic extension. For $m = 1$, an ∞-harmonic extension $f \in \mathcal{H}_g(V)$ is lex minimal and therefore uniquely determined.*

For non-weighted graphs the results follow from [23], see [12] for weighted graphs. The second statement of the lemma is in general not true for the setting $m \geq 2$. In the vector-valued case, an ∞-harmonic extension $f \in \mathcal{H}_g(V)$ is not necessarily a minimizer of E_∞. An example is given in Fig. 1, and a more sophisticated one in [23]. Also, if an ∞-harmonic extension is a minimizer of E_∞ it must not be lex minimal.

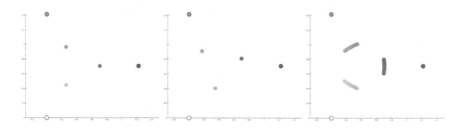

Fig. 1. Discrete ∞-harmonic extensions of the points $(0, 0, 0)$ (white), $(1, 0, 0)$ (red) and $(0, 1, 0)$ (green) in the RGB color cube visualized in the triangle plane. Left: lex-minimal extension, Middle: some ∞-harmonic extension, Right: set of all ∞-harmonic extensions. (Color figure online)

For the scalar-valued case $m = 1$, we can use that Midr_w is nonexpansive to deduce an iterative algorithm for computing the lex minimal extension of g based on the Krasnoselskii–Mann method for finite dimensional normed spaces [5, Corollaries 10, 11].

Theorem 4 (Krasnoselskii–Mann Iteration). *Let $T\colon \mathbb{R}^d \to \mathbb{R}^d$ be a nonexpansive mapping with nonempty fixed point set, where $\mathbb{R}^d$ is equipped with an arbitrary norm. Then, for every starting point $f^{(0)} \in \mathbb{R}^d$, the sequence of iterates $\{f^{(r)}\}_{r\in\mathbb{N}}$ generated by*

$$f^{(r+1)} := ((1 - \tau_r)I + \tau_r T)\, f^{(r)}, \qquad \tau_r \in (0,1)$$

converges to a fixed point of T provided that $\sum_{r=1}^{\infty} \tau_r = \infty$ and $\limsup_{r\to\infty} \tau_r < 1$.

For given $g\colon U \to \mathbb{R}$ and $f^{(0)} \in \mathcal{H}_g(V)$, the iteration scheme from Theorem 4 results in

$$f^{(r+1)}(u) := \begin{cases} g(u) & \text{if } u \in U, \\ f^{(r)}(u) + \tau\left(\mathrm{midr}_{\omega(u)}\, x^{(r)}(u) - f^{(r)}(u)\right) & \text{otherwise.} \end{cases} \tag{4}$$

Corollary 1. *Let $g\colon U \to \mathbb{R}$ be given. Then, for every $f^{(0)} \in \mathcal{H}_g(V)$ and $\tau \in (0,1)$, the sequence of iterates $\{f^{(r)}\}_{r\in\mathbb{N}}$ generated by (4) converges to the minimal Lipschitz extension of g.*

Proof. By Lemma 2, we have for $f_1, f_2 \in \mathcal{H}_g(V)$ that

$$\begin{aligned} \|\operatorname{Midr}_w f_1 - \operatorname{Midr}_w f_2\|_\infty &= \max_{u\in V} |\operatorname{Midr}_w f_1(u) - \operatorname{Midr}_w f_2(u)| \\ &= \max_{u\in V} |\mathrm{midr}_{\omega(u)}\, x_1(u) - \mathrm{midr}_{\omega(u)}\, x_2(u)| \\ &\le \max_{u\in V} \|x_1(u) - x_2(u)\|_\infty = \|f_1 - f_2\|_\infty. \end{aligned}$$

Hence $\operatorname{Midr}_w \colon \mathbb{R}^N \to \mathbb{R}^N$ is nonexpansive and the assertion follows by Theorem 4. $\qquad\square$

4   ∞-Laplacians on Scalar-Valued Functions

The minimizers of E_p, $p \in (1,\infty)$ in (1) are determined by the zero of the *anisotropic p-Laplacian operator*

$$\Delta_{w,p} f(u) := \sum_{v\in V} w(u,v)^p |f(u) - f(v)|^{p-2}\big(f(u) - f(v)\big) = 0$$

for $u \in V \setminus U$. So far we are not aware of a meaningful definition of an ∞-Laplacian for vector-valued functions. An attempt, which is unfortunately not well-defined, was done in [3]. Therefore, we restrict our attention to $m = 1$ in the following. For scalar valued-functions, the ∞-Laplacian on graphs was defined in [9,10] as $\Delta_{w,\infty}\colon \mathcal{H}_g(V) \to \mathcal{H}_g(V)$ with

$$\begin{aligned} \Delta_{w,\infty} f(u) &:= \frac{1}{2}\Big(\max_{v\sim u} w(u,v)\,(f(v) - f(u)) + \min_{v\sim u} w(u,v)\,(f(v) - f(u)) \Big) \\ &= \frac{1}{2}\Big(\max_{v\sim u} w(u,v)\,(f(v) - f(u)) - \max_{v\sim u} w(u,v)\,(f(u) - f(v)) \Big). \end{aligned}$$

For non-weighted graphs we have

$$\Delta_\infty f(u) = \frac{1}{2} \max_{v \sim u} \left(f(v) - f(u) \right) + \frac{1}{2} \min_{v \sim u} \left(f(v) - f(u) \right) \tag{5}$$
$$= \operatorname{midr} x(u) - f(u).$$

Hence, on $\mathcal{H}_g(V)$, the zero of Δ_∞ coincide with the fixed point of Midr, i.e. with the unique ∞-harmonic extension of g. For the weighted case, it was proved in [9] that there exists a unique extension $f \in \mathcal{H}_g(V)$ for which $\Delta_{w,\infty} f$ becomes zero. This can be also immediately concluded from the following theorem and Lemma 3. For a proof we refer to [12].

Theorem 5. *A function $f \in \mathcal{H}_g(V)$ is an ∞-harmonic extension if and only if $\Delta_{w,\infty} f(u) = 0$ for all $u \in V \setminus U$.*

For given $g \colon U \to \mathbb{R}$ and $f^{(0)} \in \mathcal{H}_g(V)$ we consider the iteration scheme

$$f^{(r+1)}(u) := \begin{cases} g(u) & \text{if } u \in U, \\ f^{(r)}(u) + \tau \Delta_{w,\infty} f^{(r)}(u) & \text{otherwise.} \end{cases} \tag{6}$$

This scheme was proposed in [9,10] without convergence proof. The authors only proved that in case of convergence the sequence converges to a zero of the ∞-Laplacian.

In the non-weighted case this can be rewritten by (5) as

$$f^{(r+1)}(u) = (1 - \tau) f^{(r)}(u) + \tau \operatorname{midr} x^{(r)}(u), \qquad u \in V \setminus U.$$

By Corollary 1 we see that the sequence of iterates (6) converges to the ∞-harmonic extension of g for $\tau \in (0,1)$.

For the weighted setting, this is also true by the following corollary.

Corollary 2. *Let $g \colon U \to \mathbb{R}$ be given. Then, for every $f^{(0)} \in \mathcal{H}_g(V)$ and $\tau \in (0,1)$, the sequence of iterates $\{f^{(r)}\}_{r \in \mathbb{N}}$ generated by (6) converges to the ∞-harmonic extension of g.*

Proof. Consider the operator $\Phi \colon \mathcal{H}(V) \to \mathcal{H}(V)$ defined by

$$\Phi f(u) := \begin{cases} g(u) & \text{if } u \in U, \\ f(u) + \Delta_{w,\infty} f(u) & \text{otherwise.} \end{cases}$$

By Theorem 5, Φ has a unique fixed point determined by the zero of the ∞-Laplacian. We show that Φ is nonexpansive. Then it follows immediately from Theorem 4 that the series $\{f^{(r)}\}_{r \in \mathbb{N}}$ converges to this fixed point. For $u \in V \setminus U$ we rewrite

$$\Delta_{w,\infty} f(u) = \frac{1}{2} \max_{v \sim u} w(u,v) \left(f(v) - f(u) \right) - \frac{1}{2} \max_{v \sim u} w(u,v) \left(f(u) - f(v) \right).$$

Let $f_i \in \mathcal{H}_g(V)$, $i = 1, 2$, and $y_i := \arg\max_{v \sim u} w(u, v)\,(f_i(v) - f_i(u))$, $z_i := \arg\max_{v \sim u} w(u, v)\,(f_i(u) - f_i(v))$. Then we obtain for $u \in V \setminus U$,

$$\Phi f_1(u) - \Phi f_2(u) = f_1(u) - f_2(u) + \Delta_{w,\infty} f_1(u) - \Delta_{w,\infty} f_2(u)$$

$$= f_1(u) - f_2(u) + \frac{w(u, y_1)}{2}\,(f_1(y_1) - f_1(u)) - \frac{w(u, z_1)}{2}\,(f_1(u) - f_1(z_1))$$

$$- \frac{w(u, y_2)}{2}\,(f_2(y_2) - f_2(u)) + \frac{w(u, z_2)}{2}\,(f_2(u) - f_2(z_2))$$

$$\leq f_1(u) - f_2(u) + \frac{w(u, y_1)}{2}\,(f_1(y_1) - f_1(u) - f_2(y_1) + f_2(u))$$

$$- \frac{w(u, z_2)}{2}\,(f_1(u) - f_1(z_2) - f_2(u) + f_2(z_2))$$

$$= \left(1 - \frac{w(u, y_1) + w(u, z_2)}{2}\right)(f_1(u) - f_2(u))$$

$$+ \frac{w(u, y_1)}{2}\,(f_1(y_1) - f_2(y_1)) + \frac{w(u, z_2)}{2}\,(f_1(z_2) - f_2(z_2))$$

and with $\|f_1 - f_2\|_\infty = \max_{u \in V} |f_1(u) - f_2(u)|$ further $\Phi f_1(u) - \Phi f_2(u) \leq \|f_1 - f_2\|_\infty$. Analogously, we get $\Phi f_1(u) - \Phi f_2(u) \geq -\|f_1 - f_2\|_\infty$ and thus $\|\Phi f_1 - \Phi f_2\|_\infty \leq \|f_1 - f_2\|_\infty$. Consequently, Φ is nonexpansive.   □

5   Conclusions

Starting with the proof that lex and L-lex minimal extensions of vector-valued functions on graphs are the same, we saw that for scalar-valued functions these extensions coincide with ∞-harmonics which are indeed the fixed points of midrange filters and also zeros of the ∞-Laplacian. Actually, we are dealing with numerical experiments for computing, resp. approximating vector-valued ∞-harmonic and minimal Lipschitz extensions. For the non-weighted case, a detailed discussion on continuity properties of the midrange operator can be found in [13]. In particular, it was proved that for pairwise different $x_i \in \mathbb{R}^m$, there exists a ball around $x \in \mathbb{R}^{mn}$ on which the midrange operator is Lipschitz continuous. It would be interesting if this results is also true in the weighted case. Finally, we are aware that topics of this paper were also considered for functions on continuous domains. However, this point of view is out of the scope of this conference paper.

References

1. Bačák, M.: Convex Analysis and Optimization in Hadamard Spaces. De Gruyter Series in Nonlinear Analysis and Applications, vol. 22. De Gruyter, Berlin (2014)
2. Belkin, M., Niyogi, P.: Laplacian eigenmaps for dimensionality reduction and data representation. Neural Comput. **15**, 1373–1396 (2003)

3. Bergmann, R., Tenbrinck, D.: Nonlocal inpainting of manifold-valued data on finite weighted graphs. In: Nielsen, F., Barbaresco, F. (eds.) GSI 2017. LNCS, vol. 10589, pp. 604–612. Springer, Cham (2017). https://doi.org/10.1007/978-3-319-68445-1_70

4. Bergmann, R., Tenbrinck, D.: A graph framework for manifold-valued data. SIAM J. Imaging Sci. **11**(1), 325–360 (2018)

5. Borwein, J., Reich, S., Shafrir, I.: Krasnoselski-Mann iterations in normed spaces. Can. Math. Bull. **35**(1), 21–28 (1992)

6. Caselles, V., Morel, J.-M., Sbert, C.: An axiomatic approach to image interpolation. IEEE Trans. Image Process. **7**(3), 376–386 (1998)

7. Descloux, J.: Approximations in l^p and Chebychev approximations. J. Soc. Ind. Appl. Math. **11**, 1017–1026 (1963)

8. Egger, A., Huotari, R.: Rate of convergence of the discrete Pólya algorithm. J. Approx. Theory **60**, 24–30 (1990)

9. Elmoataz, A., Desquesnes, X., Lakhdari, Z., Lézoray, O.: Nonlocal infinity Laplacian equation on graphs with applications in image processing and machine learning. Math. Comput. Simul. **102**, 153–163 (2014)

10. Elmoataz, A., Toutain, M., Tenbrinck, D.: On the p-Laplacian and ∞-Laplacian on graphs with applications in image and data processing. SIAM J. Imaging Sci. **8**(4), 2412–2451 (2015)

11. Gilboa, G., Osher, S.: Nonlocal operators with applications to image processing. SIAM Multiscale Model. Simul. **7**(3), 1005–1028 (2008)

12. Hertrich, J.: Infinity-Laplacians on scalar- and vector-valued functions and optimal Lipschitz extensions on graphs. Bachelor Thesis, TU Kaiserslautern (2018)

13. Ivanshin, P.N., Sosov, E.N.: Local Lipschitz property for the Chebyshev center mapping over N-nets. Matematichki Vesnik **60**(1), 9–22 (2008)

14. Kyng, R., Rao, A., Sachdeva, S., Spielman, D.A.: Algorithms for Lipschitz learning on graphs. In: Conference on Learning Theory, pp. 1190–1223 (2015)

15. Lazarus, A.J., Loeb, J., Propp, D.E., Stromquist, W.R., Ullman, H.D.: Combinatorial games under auction play. Games Economical Behav. **27**(2), 229–264 (1999)

16. Matoušek, J., Sharir, J., Welzl, E.: A subexponential bound for linear programming. Algorithmica **16**, 498–516 (1996)

17. Oberman, A.M.: A convergent difference scheme for the infinity Laplacian: construction of absolutely minimizing Lipschitz extensions. Math. Comput. **74**(251), 1217–1230 (2005)

18. Peres, Y., Schramm, O., Sheffield, S., Wilson, D.B.: Tug-of-war and infinity Laplacian. J. Am. Math. Soc. **22**(1), 167–210 (2005)

19. Pólya, G.: Sur un algorithm touours convergent pour obtenir les polynomes de meillure approximation de Tchebycheff pour une function continue quelconque. Comptes Rendus de l'Académie des Sciences Paris **157**, 840–843 (1913)

20. Quesada, J.M., Fernández-Ochoa, J., Martinez-Moreno, J., Bustamante, J.: The Pólya algorithm in sequence spaces. J. Approx. Theory **135**, 245–257 (2005)

21. Schmaltz, C., Peter, P., Mainberger, M., Ebel, F., Weickert, J., Bruhn, A.: Understanding, optimising, and extending data compression with anisotropic diffusion. Int. J. Comput. Vis. **108**(3), 222–240 (2014)

22. Serra, J.: Image Analysis and Mathematical Morphology. Academic Press, London (1982)

23. Sheffield, S., Smart, C.K.: Vector-valued optimal Lipschitz extensions. Commun. Pure Appl. Math. **65**(1), 128–154 (2012)

24. Shi, J., Malik, J.: Normalized cuts and image segmentation. IEEE Trans. Pattern Anal. Mach. Intell. **22**(8), 888–905 (2000)

25. Welk, M., Weickert, J., Becker, F., Schnörr, C., Feddern, C., Burgeth, B.: Median and related local filters for tensor-valued images. Signal Process. **87**(2), 291–308 (2007)
26. Xu, S., Freund, R.M., Sun, J.: Solution methodologies for the smallest enclosing circle problem. Computational Optimization and Applications. Int. J. **25**(1–3), 283–292 (2003)

PDEs and Level-Set Methods

The Convex-Hull-Stripping Median Approximates Affine Curvature Motion

Martin Welk[1](✉) and Michael Breuß[2]

[1] Institute of Biomedical Image Analysis, Private University of Health Sciences,
Medical Informatics and Technology, Eduard-Wallnöfer-Zentrum 1,
6060 Hall/Tyrol, Austria
martin.welk@umit.at
[2] Brandenburg University of Technology Cottbus-Senftenberg,
Platz der Deutschen Einheit 1, 03046 Cottbus, Germany
breuss@b-tu.de

Abstract. The median filter is one of the fundamental filters in image processing. Its standard realisation relies on a rank ordering of given data which is easy to perform if the given data are scalar values. However, the generalisation of the median filter to multivariate data is a delicate issue. One of the methods of potential interest for computing a multivariate median is the convex-hull-stripping median from the statistics literature. Its definition is of purely algorithmical nature, and it offers the advantageous property of affine equivariance.

While it is a classic result that the standard median filter approximates mean curvature motion, no corresponding assertion has been established up to now for the convex-hull-stripping median. The aim of our paper is to close this gap in the literature. In order to provide a theoretical foundation for the convex-hull-stripping median of multivariate images, we investigate its continuous-scale limit. It turns out that the resulting evolution is described by the well-known partial differential equation of affine curvature motion. Thus we have established in this paper a relation between two important models from image processing and statistics. We also present some experiments that support our theoretical findings.

Keywords: Median filter · Convex hull stripping ·
Partial differential equations · Curve evolution

1 Introduction

The median filter as introduced by Tukey [18] is a cornerstone of modern image processing, and it is used extensively in smoothing and denoising applications. The concept of the classic median filter relies on a rank ordering of input data, so that there exists a large range of algorithmic variations for realising its concept. However, while the median filtering of grey-value images is straightforward in terms of the natural total ordering of grey-value data within a filtering mask,

© Springer Nature Switzerland AG 2019
J. Lellmann et al. (Eds.): SSVM 2019, LNCS 11603, pp. 199–210, 2019.
https://doi.org/10.1007/978-3-030-22368-7_16

an extension of the median concept to multivariate data such as for instance in colour images is a delicate issue.

Generalisations of median filtering to multivariate data have been investigated by statisticians for more than a century [9]. Several strategies have been proposed and studied in the literature, see e.g. [16] for an useful overview. Let us elaborate on the possible generalisations and related work in some detail.

Earliest among the concepts that are still used today is the L^1 median that goes back to Weber's 1909 work in location theory [21] and was further studied, among others, by Gini [6], Weiszfeld [22], Austin [2], and Vardi [20]. The median value of scalar valued data may be considered as the minimiser of the sum of distances (i.e. absolute differences) of a variable parameter from the given data points [10]. The L^1 median generalises this observation by defining the median of data points in $\mathbb{R}^n$ as the location that minimises the sum of Euclidean distances from the given points. Whereas this minimiser can be computed efficiently [20], it lacks desirable invariance properties of the scalar-valued median. The scalar median is equivariant with respect to arbitrary monotonic transformations of the real line, meaning that the median operation $\mu : (\mathbb{R})^+ \to \mathbb{R}$ and any monotonic transformation $\tau : \mathbb{R} \to \mathbb{R}$ commute, $\mu \circ \tau^\otimes \equiv \tau \circ \mu$. Here, $(\mathbb{R})^+$ denotes the set of nonempty finite multisets of real numbers, and $\tau^\otimes$ the element-by-element application of τ to such a multiset. Let us note that this equivariance property is of some interest in the image processing context as it makes the standard median filter belong to the class of morphological filters. In contrast, the L^1 median is equivariant only with respect to the Euclidean and similarity groups.

As statisticians are often confronted with data from $\mathbb{R}^n$ that do not possess a natural Euclidean structure – i.e. the data belong to the vector or affine space $\mathbb{R}^n$ but not to a Euclidean space – , they have been interested since long time in alternatives to the L^1 median that would at least offer affine equivariance. Several such concepts have been proposed, including Oja's simplex median [11], the transformation–retransformation L^1 median [4,12], the half-space median [19] and the convex-hull-stripping median [3,14]. The latter will be in the focus of the present work.

Regarding the filtering of multivariate images, let us recall first that local image filters combine a selection step that selects at each image location a certain set of image values, with an aggregation step that computes from these values the filtered value at that location. In the simplest case, the selection step uses a sliding window so that an equally shaped neighbourhood is used at each location. The standard median filter [18] then uses the scalar median as aggregation step. In the case of a multivariate image, it is straightforward to combine the same selection step with any multivariate median filter to obtain a filtered multivariate image. Especially the L^1 median has been used for this purpose in the last decades, see [17,26]. More recently, multivariate image median filters based on the Oja as well as the transformation–retransformation L^1 median have been proposed [24].

Since images may be considered as discrete samplings of continuous-scale functions, it is desirable that a discretely operating filtering procedure should be related to a continuous-scale model. This may also give additional insight into important filtering properties, potential generalisations, or alternative implementations. For the scalar-valued median filter, such a connection was established by Guichard and Morel [7,8] who proved that in a continuous-scale limit, median filtering of 2D images approximates the partial differential equation (PDE) of mean curvature motion, a result which can straightforwardly be generalised to higher-dimensional scalar-valued images. In the case of multivariate images, similar PDE approximation results have been obtained in [24,25] for the L^1 median and in [24] for the Oja and transformation–retransformation L^1 median filter. In order to embed also the convex-hull-stripping median into such a framework, it is crucial to derive a continuous formulation of this originally entirely discrete procedure. On this basis, space-continuous analysis of e.g. convex-hull-stripping median filtering of multivariate images will be possible.

Our Contribution. The goal of our paper is to go the aforementioned first step in a continuous analysis of the convex-hull-stripping median, namely to equip it with a continuous formulation. In order to provide a theoretical foundation for the filtering of multivariate images by the convex-hull-stripping median, we investigate the continuous-scale limit of this median filter. We describe the continuous-scale counterpart of the discrete process defining the convex-hull-stripping median of finite sets in $\mathbb{R}^2$ by a PDE evolution which turns out to be in essence the well-known affine curvature motion [1,13]. We show experiments using bivariate images in $\mathbb{R}^2$ that support this finding.

Structure of the Paper. The paper is organised in accordance to the outlined contributions. After briefly recalling the algorithmical definition of the convex-hull-stripping median, we investigate its continuous-scale limit. The most important step, which forms the basis of all main results, is the proof of a technical lemma that we present in detail. After that we provide some experiments along with relevant remarks. The paper is finished by conclusive remarks.

2   Theory of Convex-Hull-Stripping Median Filtering

The convex-hull stripping median for multivariate data goes back to work by Barnett and Seheult [3,14]. After recalling its definition, we derive its continuous-scale limit in the two-dimensional case and turn then to discuss its possible application for the filtering of bivariate images.

2.1   The Convex-Hull-Stripping Median of Finite Sets

Considering a finite multiset of points in $\mathbb{R}^n$ (which means that points may appear with multiplicities greater than one), its convex hull is a polygon. By deleting all vertices of this polygon, a smaller multiset remains. This convex-hull-stripping procedure is repeated until an empty set remains. The convex hull of the last non-empty multiset in the constructed sequence is defined as the median of the initial multiset.

It is clear that the convex-hull-stripping median as defined above is in general a multiset itself. A unique median point may be chosen by some additional step like taking the centre of gravity. Later on, we will be interested in the continuous-scale case which is theoretically obtained as the limit of infinitely refined sampling from an assumed continuous density. In this situation the additional step for uniqueness is no longer relevant.

As the convex hull of a data multiset is equivariant under arbitrary affine transformations, it is clear that the convex-hull-stripping median, too, is affine equivariant.

A caveat of the construction that should be noted is that the convex-hull-stripping median of finite multisets does not always depend continuously on the input data. This is inherent to the discrete procedure underlying its definition in which the vertex set of the convex hull is updated set-wise in discrete algorithmic steps when point coordinates undergo continuous variations. The high-dimensional space of input multisets contains therefore a network of discontinuity hypersurfaces. In particular, all multisets with coincident data points (multiplicities greater than one) are candidates for discontinuities.

2.2 Continuous-Scale Limit

To best of our knowledge, the convex-hull-stripping median of a continuous-scale density has not been investigated so far. We will do this for data in $\mathbb{R}^2$.

Unlike the L^1 median or Oja median which are defined as minimisers of some objective function in $\mathbb{R}^n$, the convex-hull-stripping median is defined via an iterative process. Therefore it is no surprise that, when translating this concept to the situation of a continuous-scale density as input data, a time-continuous dynamical process arises that is described best by a PDE. The general strategy behind the following derivations is to apply the discrete convex-hull-stripping process to stochastic samplings of a given continuous density. When the sampling density is sent to infinity, the continuous process is approximated asymptotically with probability one. In its derivation, the stochastic samplings can therefore be studied in terms of expectation values. The main result of this paper reads as follows.

Proposition 1. *Let a piecewise smooth density $\gamma : \mathbb{R}^2 \to \mathbb{R}$ with compact support $\Omega_0 \subset \mathbb{R}^2$ in the Euclidean plane $\mathbb{R}^2$ be given. Assume that the boundary of Ω_0 is regular, and γ is differentiable on Ω_0. Let the point set $\mathcal{X}$ be a stochastic sampling of this density with sampling density $1/h^2$, i.e., there is on average one sampling point in an area in which the density integrates to h^2. For $h \to 0$, the convex-hull-stripping median of the set $\mathcal{X}$ asymptotically coincides with the vanishing point of the curve evolution*

$$
c_t(p,t) = \begin{cases} \gamma(c(p,t))^{-2/3}\kappa(p,t)^{1/3}n(p,t)\,, & c(p,t) \in \partial\operatorname{conv}(c(\cdot,t))\,, \\ 0 & else \end{cases} \tag{1}
$$

where $c : [0,L] \times [0,T] \to \mathbb{R}^2$ is a curve evolution of closed curves with curve parameter $p \in [0,L]$ and evolution time $t \in [0,T]$, which is initialised at time

$t = 0$ *with the boundary of the support set,* $\mathbf{c}_0 := \partial\Omega_0$. *Furthermore,* $\kappa(p,t)$
denotes the curvature and $\mathbf{n}(p,t)$ *the inward normal vector of* $\mathbf{c}$ *at* (p,t). *At any
time* t, *the evolution acts only on the part of* $\mathbf{c}$ *that is on the boundary* $\partial\,\mathrm{conv}\mathbf{c}$
of the convex hull of $\mathbf{c}$.

Remark 1. For a convex shape Ω_0 with uniform density $\gamma \equiv 1$ on Ω_0, the evolution reduces to the well-known affine (mean) curvature motion PDE $\mathbf{c}_t = \kappa^{1/3}\mathbf{n}$
which is known to reduce any closed regular initial curve to a single point in finite
time. We call this point the vanishing point of the evolution. The more general
Eq. (1) accounts for non-uniform density by the factor $1/\gamma^{2/3}$, and restricts the
evolution to convex segments of the boundary.

The proof of this proposition relies essentially on the following lemma.

Lemma 1. *Let a uniform density* $\gamma \equiv 1$ *within the disc* D_ϱ *of radius* ϱ *with
center* $O = (0,0)$ *in the Euclidean plane* $\mathbb{R}^2$ *be given. Let the point set* $\mathcal{X}$ *be
a stochastic sampling of this density with sampling density* $1/h^2$, *i.e., there is
on average one sampling point per area* h^2. *For* $h \to 0$, *one step of convex hull
stripping of the set* $\mathcal{X}$ *then asymptotically approximates a shrinkage of the disc
by* $C\varrho^{-1/3}h^{4/3}$ *with a positive constant* C *that does not depend on* ϱ *and* h.

Proof. Assume first that a uniform density $\gamma \equiv 1$ within some shape $\Omega \subset \mathbb{R}^2$ is
stochastically sampled with sampling density $1/h^2$. Then the number of sample
points within an area of measure a is for $a \to 0$ asymptotically a/h^2. Thus, the
probability that *no* sampling point is found in an area of measure A amounts to

$$\lim_{a \to 0}\left(1 - \frac{a}{h^2}\right)^{A/a} = \mathrm{e}^{-A/h^2} . \tag{2}$$

Consider now specifically the case $\Omega = \mathrm{D}_\varrho$ as specified in the hypothesis of
the lemma. Let P be any point on the boundary circle $\partial\Omega$. We seek the next
sampling point in positive orientation near $\partial\Omega$ that is a vertex of the convex hull
of the sample points, i.e., a sample point Q such that $\angle QPO \in [0, \pi/2]$ (angles
being measured in positive orientation) for which there is no sample point R
with $0 < \angle POR < \angle POQ$ for which the segments OR and PQ intersect.

For a small angle $\alpha > 0$ denote by P_α the point on the circle $\partial\Omega$ with
$\angle POP_\alpha = \alpha$. Consider the circle segment A_α of $\partial\Omega$ enclosed between the chord
$\overline{PP_\alpha}$ and the arc $\widehat{PP_\alpha}$ Then the probability that no sample point is found in A_α
is $\exp(-|A_\alpha|/h^2)$ where $|A_\alpha|$ denotes the area measure of A_α. For small $\alpha > 0$
one has

$$|A_\alpha| = \frac{1}{2}\varrho^2\alpha - \frac{1}{2}\varrho^2\sin\alpha = \frac{1}{12}\varrho^2\alpha^3 + \mathcal{O}(\alpha^4) . \tag{3}$$

Let α_Q denote the value of α for which the line PP_α goes through the sought
point Q. Using (3), the probability for α_Q to be at least α is approximately
$\exp(-\varrho^2\alpha^3/(12\,h^2))$. Therefore the probability density $p(\alpha)$ of α_Q is given by

$$p(\alpha) = -\frac{\mathrm{d}}{\mathrm{d}\alpha}\mathrm{e}^{-\varrho^2\alpha^3/(12\,h^2)} = \frac{\varrho^2\alpha^2}{4\,h^2}\mathrm{e}^{-\varrho^2\alpha^3/(12\,h^2)} . \tag{4}$$

Using integration by parts, the expectation value of α_Q is

$$E(\alpha) = \int_0^\infty \alpha p(\alpha)\, d\alpha = \int_0^\infty \alpha \cdot \frac{\varrho^2 \alpha^2}{4\,h^2} e^{-\varrho^2 \alpha^3/(12\,h^2)}\, d\alpha$$

$$= \underbrace{\left[-\alpha e^{-\varrho^2 \alpha^3/(12\,h^2)}\right]_0^\infty}_{(a)} + \underbrace{\int_0^\infty e^{-\varrho^2 \alpha^3/(12\,h^2)}\, d\alpha}_{(b)}$$

$$= \int_0^\infty e^{-\varrho^2 \alpha^3/(12\,h^2)}\, d\alpha = \sqrt[3]{\frac{12\,h^2}{\varrho^2}}\, \Gamma\left(\frac{4}{3}\right) \tag{5}$$

where part (a) was found to be zero, and part (b) has been evaluated using the integral $\int_0^\infty \exp(-z^3)\, dz = \Gamma(1/3)/3 = \Gamma(4/3)$, where Γ denotes the Gamma function. As the angle $\beta := \angle POQ$ is on average proportional to α_Q with some constant factor, its expectation value, too, satisfies

$$E(\beta) \sim \varrho^{-2/3} h^{2/3}\ . \tag{6}$$

As a consequence, the vertex count K of the convex hull of the sampling points has the expectation value

$$E(K) \doteq \frac{2\pi}{E(\beta)} = 2\pi\, C \frac{\varrho^{2/3}}{h^{2/3}} \tag{7}$$

with some positive constant C, where $\doteq$ denotes equality up to higher order terms.

Stripping the convex hull from $\mathcal{X}$ corresponds to some inward movement of the contour of Ω so that the reduced shape $\tilde{\Omega}$ is sampled by the remaining point set $\tilde{\mathcal{X}}$. Due to the given sampling density the area of $\mathcal{X} \setminus \tilde{\mathcal{X}}$ has the expectation value $E(K)\, h^2$. For rotational symmetry, the inward movement shall be in expectation equal along the periphery of Ω, such that $\tilde{\Omega}$ approximates a disc with smaller radius $\tilde{\varrho}$ whose area $\pi\tilde{\varrho}^2$ has the expectation value $\pi\varrho^2 - 2\pi\, C\varrho^{2/3} h^{4/3}$. From this one calculates that the expectation value of $\varrho - \tilde{\varrho}$ is asymptotically equal to $C\varrho^{-1/3} h^{4/3}$. Let us remark that $h \to 0$ is employed to this end so that in the limit $\tilde{\varrho}$ approaches ϱ, while the factor $h^{4/3}$ shows how fast this works with $h \to 0$. This concludes the proof.   □

We can now turn to prove our main result.

Proof (of Proposition 1). Whereas the argument in the proof of Lemma 1 is written for the entire disc, the process of convex hull stripping acts strictly locally on the convex boundary of the disc D_ϱ. The result is therefore valid for each infinitesimal convex portion of a compact support set Ω of a uniform density if one inserts for ϱ the radius of the osculating circle, i.e. $1/\kappa$. Next, a non-uniform density corresponds to a modification of the sampling density parameter h by a factor $1/\sqrt{\gamma}$ at each location. Finally, the evolution cannot take effect in portions of the boundary curve that do not belong to the convex hull of Ω.

The convex-hull-stripping median of $\mathcal{X}$ is obtained by iterative application of the procedure described in the lemma. It is therefore clear that the vanishing point of the curve evolution (1) is the sought median.   □

Proposition 1 formulated a curve evolution that starts with the possibly non-convex boundary curve of Ω_0, the support set of γ. As a consequence, the curve evolution in Proposition 1 consists of affine curvature motion and stationary behaviour, and the switching between the two depends on a non-local criterion (namely, whether the evolving curve point is on the convex hull boundary). As this criterion is difficult to evaluate within the context of a PDE discretisation, we reformulate the result in the following by starting the evolution with a *convex* curve *enclosing* the support set Ω.

However, when doing so, the curve evolution partially takes place outside Ω. There one has $\gamma \equiv 0$, so that this would lead to infinite speed of the evolution due to the factor $1/\gamma^{2/3}$. Therefore a regularisation is introduced.

Corollary 1. *Let the density γ with support Ω_0, and the sampled point set $\mathcal{X}$ with sampling density $1/h^2$ be given as in Proposition 1. Let $\tilde{\Omega}_0 \supseteq \Omega_0$ be a convex compact set with regular boundary. For $\varepsilon > 0$, define a piecewise smooth density γ_ε on $\tilde{\Omega}_0$ by $\gamma_\varepsilon(\boldsymbol{x}) = \gamma(\boldsymbol{x})$ for $\boldsymbol{x} \in \Omega_0$, $\gamma_\varepsilon(\boldsymbol{x}) = \varepsilon$ for $\boldsymbol{x} \in \tilde{\Omega}_0 \setminus \Omega_0$.*

For $h \to 0$, the convex-hull-stripping median of the set $\mathcal{X}$ asymptotically coincides with the limit for $\varepsilon \to 0$ of the vanishing point of the curve evolution

$$\tilde{\boldsymbol{c}}_t(p,t) = \gamma_\varepsilon(\tilde{\boldsymbol{c}}(p,t))^{-2/3}\kappa(p,t)^{1/3}\boldsymbol{n}_{\tilde{c}}(p,t) \tag{8}$$

where $\tilde{\boldsymbol{c}}$ is initialised at $t = 0$ with the convex closed curve $\partial\tilde{\Omega}_0$.

Remark 2. Practically it is sufficient to numerically compute the vanishing point with a sufficient small ε. Still, a numerical implementation must take care of the large difference in the evolution speed of (8) between the regions within and outside Ω_0.

Proof (of Corollary 1). In the limit $\varepsilon \to 0$, the evolution (8) will move the initial contour to the convex hull of Ω_0 within time $\mathcal{O}(\varepsilon)$, after which the evolution continues evolving this convex hull boundary, mimicking (1) for the coincident curve segments of $\tilde{\boldsymbol{c}}$ and $\boldsymbol{c}$ and keeping the remaining segments of $\tilde{\boldsymbol{c}}$ spanned as straight lines connecting the coincident segments.   □

3   Experiments

In this section we present experiments for the median filtering of bivariate images by the convex-hull-stripping median and the median obtained from the affine curvature flow. We emphasise that these experiments are driven by theoretical interest. As median filtering of (univariate as well as multivariate) images has been studied in many earlier works, we consider the application of the convex-hull-stripping median and its PDE counterpart in the context of an image filter as a good way to illustrate the asymptotic equivalence of both procedures as established in Proposition 1.

Besides the limited practical interest of bivariate images, we do not advocate these filters for applications at the time being because of their high computational expense. The computation of the discrete convex-hull-stripping median is already significantly slower than that of other multivariate median filters. The computation of medians by evaluating a PDE in the data space for each single pixel is computationally so expensive that it is presently only interesting from the theoretical viewpoint.

Discrete Convex-Hull-Stripping Median. At first glance, applying the convex-hull-stripping median as a local filter for discrete multivariate images appears to be a straightforward procedure. Shifting a sliding window across the image, one collects for each image pixel the values (points in the data space $\mathbb{R}^n$) of the neighbouring pixels within the window, then computes their convex-hull-stripping median and assigns it as the filtered data value to the given pixel.

However, the inherent discontinuity of the discrete convex-hull-stripping process renders the results of this procedure highly instable, since in typical digital images the selection process frequently yields multisets with coincident data points, which are often discontinuity locations in the input data space. Considering digital images as quantised samplings from continuous functions, it is clear that the high prevalence of coincident data points is actually an artifact of the quantisation, and should be countered by a suitable regularisation. To this end we use a stochastic perturbation approach that efficiently removes multiplicities from the data sets: Each selected data point $u \in \mathbb{R}^n$ is replaced with a fixed number p of data points $\tilde{u}_i = u + \nu_i$ where the ν_i are i.i.d. random perturbations with Gaussian distribution of a fixed small standard deviation σ. In our experiments on colour images with intensity range $[0, 255]$, we use $\sigma = 2$ and an augmentation factor $p = 5$. The convex-hull-stripping median is then computed from the perturbed and augmented data set at each pixel.

With this procedure, convex-hull-stripping median filter results can be computed for bivariate images. Figure 1a shows an RGB test image from which we create a bivariate image by averaging the red and green channels into a yellow channel, see Frame b. Frames d and e show the result of one and three iterations, resp., of a median filter where the sliding window around each pixel included its neighbours up to Euclidean distance 2 (thus, a 13-neighbourhood), and reflecting boundary conditions were used. Frame f shows the result of a single median filtering step with a larger sliding window that includes neighbours up to Euclidean distance 5 (altogether, 81 pixels). It is evident that the convex-hull-stripping median filter behaves similar to the scalar-valued median filter as it smoothes small image details, preserves edges and rounds corners. This is also in agreement with the observations in [23] for L^1 and Oja simplex median filtering of bivariate images; because of space limitations, we refrain from showing

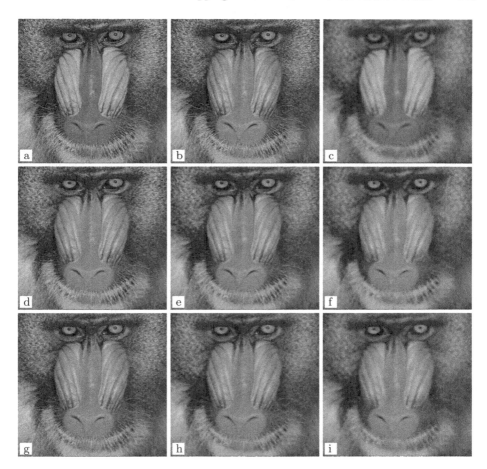

Fig. 1. Image filtering by convex-hull-stripping median and affine-curvature-flow median. **a** RGB image *baboon*, 512 × 512 pixels. – **b** Bivariate (yellow–blue) image obtained from a by averaging red and green channel. – **c** One iteration of convex-hull-stripping median filtering of the RGB image using a sliding window of radius 5. – **d–f** Convex-hull-stripping median filtering of the bivariate image: **d** Window radius 2, one iteration. – **e** Window radius 2, three iterations. – **f** Window radius 5, one iteration. – **g–i** Median filtering using affine curvature flow: **g** Window radius 2, one iteration. – **h** Window radius 2, three iterations. – **i** Window radius 5, one iteration. (Color figure online)

the (largely similar) results of these other median filters for our bivariate test image.

Whereas the implementation can easily be adapted to three-channel images, results are less favourable in this case. To demonstrate this, we show one result of convex-hull-stripping median filtering of the original RGB test image. As can be seen, the resulting image, shown in Fig. 1c, is much more blurred than the bivariate result in frame f which was obtained with the same sliding window and the same perturbation and augmentation parameters. This is mainly due to the fact that the data values in any local patch of a planar three-channel image tend to cluster around a hyperplane in the three-dimensional data space. Repeated deletion of the convex hull will therefore yield as the last non-empty set still a fairly extended point cloud near the hyperplane, from which then the average is computed. Thus, the result resembles much more a mean-value filter than a median filter. Similar problems occur also with other multivariate median filters for images where the dimension of the data space exceeds that of the image domain, see the discussion in [24]. To forge a sensible median filter for such images from the convex-hull-stripping approach therefore requires additional research which is beyond the scope of the present paper.

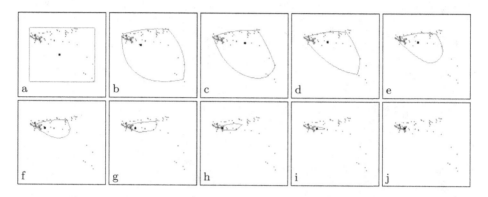

Fig. 2. Affine curvature flow for a set of bivariate data points. Blue: data points, light blue: smooth positive density (sum of Gaussians centred at data points), red: evolving contour (zero level-set of level-set function), black: current minimum of level-set function. **a** Initialisation with convex contour outside the bounding box of the data points. – **b–i** Intermediate states of progressing level-set evolution, shown every 2000 time steps (with adaptive step-size control). – **j** Final state (after approx. 17 600 iterations) when level-set function becomes entirely positive. The minimum of the level-set function defines the median. (Color figure online)

Affine Curvature Flow Median. To turn the PDE (8) into a median filter for bivariate images, it must be evaluated for each pixel. Moreover, the discrete set of data points obtained by the sliding-window selection needs to be turned into a smooth density γ. We do this by defining γ as sum of Gaussians with fixed standard deviation σ centred at the data points (for compact support, a cut-off is set sufficiently far outside the bounding box of the data points). Matching the perturbation procedure that was used in the convex-hull-stripping median filter above, we choose $\sigma = 2$ for our experiments.

Discretising the data space by a regular grid and initialising a convex contour encircling all data points, we compute (8) by a level-set method [15], using an explicit discretisation with upwinding and an adaptive step-size control ensuring a CFL-type condition. Spatial gradients in the denominator of the curvature expression are regularised by a small summand $\sim 10^{-6}$. Figure 2 shows an exemplary evolution for the data set of a single pixel.

Figure 1g–i show results for the full bivariate test image. Visual comparison with the discrete convex-hull-stripping median results in frame d–f confirms the largely similar smoothing behaviour. In a few pixels the computation suffered from inaccuracies introduced by the rather simple way of regularising the curvature expression, creating somewhat rigged edge structures. Refined numerics is needed to fix these problems.

As mentioned before, the per-pixel PDE evaluation comes at a high computational cost. With a parallel CUDA implementation on a powerful graphics card workstation, the computation of Fig. 1g–i took several days. Even with possible speed-up by narrow-band schemes, more advanced time-stepping and careful parameter tuning, a direct PDE evaluation as shown here remains prohibitive for practical applications.

4   Summary and Conclusions

We have validated theoretically that the convex-hull-stripping median approximates the PDE of affine curvature motion. Thus we have established a relation between two previously unconnected, important models from the literature, and at the same time we have bridged a gap between discrete and continuous-scale modelling. Let us also point out that our work implies that the convex-hull-stripping algorithm represents a non-standard discretisation of affine curvature motion. We conjecture that this may be an interesting aspect for future work.

As indicated, an efficient implementation of the convex-hull-stripping median is not trivial and a potential subject of future work, as might be its implementation within adaptive filtering methods. Let us also note that the multivariate L^1 median as a minimiser of a sum of Euclidean distances can be generalised to data on Riemannian manifolds [5]. In future work we plan to make use of the connection established in this paper to generalise convex-hull-stripping median filtering to data on Riemannian manifolds.

References

1. Alvarez, L., Guichard, F., Lions, P.L., Morel, J.M.: Axioms and fundamental equations in image processing. Arch. Ration. Mech. Anal. **123**, 199–257 (1993)
2. Austin, T.L.: An approximation to the point of minimum aggregate distance. Metron **19**, 10–21 (1959)
3. Barnett, V.: The ordering of multivariate data. J. Roy. Stat. Soc. A **139**(3), 318–355 (1976)
4. Chakraborty, B., Chaudhuri, P.: On a transformation and re-transformation technique for constructing an affine equivariant multivariate median. Proc. AMS **124**(6), 2539–2547 (1996)

5. Fletcher, P., Venkatasubramanian, S., Joshi, S.: The geometric median on Riemannian manifolds with applications to robust atlas estimation. NeuroImage **45**, S143–S152 (2009)
6. Gini, C., Galvani, L.: Di talune estensioni dei concetti di media ai caratteri qualitativi. Metron **8**, 3–209 (1929)
7. Guichard, F., Morel, J.M.: Partial differential equations and image iterative filtering. In: Duff, I.S., Watson, G.A. (eds.) The State of the Art in Numerical Analysis. IMA Conference Series (New Series), vol. 63, pp. 525–562. Clarendon Press, Oxford (1997)
8. Guichard, F., Morel, J.M.: Geometric partial differential equations and iterative filtering. In: Heijmans, H., Roerdink, J. (eds.) Mathematical Morphology and its Applications to Image and Signal Processing, pp. 127–138. Kluwer, Dordrecht (1998)
9. Hayford, J.F.: What is the center of an area, or the center of a population? J. Am. Stat. Assoc. **8**(58), 47–58 (1902)
10. Jackson, D.: Note on the median of a set of numbers. Bull. Am. Math. Soc. **27**, 160–164 (1921)
11. Oja, H.: Descriptive statistics for multivariate distributions. Stat. Probab. Lett. **1**, 327–332 (1983)
12. Rao, C.R.: Methodology based on the l_1-norm in statistical inference. Sankhyā A **50**, 289–313 (1988)
13. Sapiro, G., Tannenbaum, A.: Affine invariant scale-space. Int. J. Comput. Vis. **11**, 25–44 (1993)
14. Seheult, A.H., Diggle, P.J., Evans, D.A.: Discussion of Professor Barnett's paper. J. Roy. Stat. Soc. A **139**(3), 351–352 (1976)
15. Sethian, J.A.: Level Set and Fast Marching Methods. Cambridge University Press, Cambridge (1999)
16. Small, C.G.: A survey of multidimensional medians. Int. Stat. Rev. **58**(3), 263–277 (1990)
17. Spence, C., Fancourt, C.: An iterative method for vector median filtering. In: Proceedings of 2007 IEEE International Conference on Image Processing, vol. 5, pp. 265–268 (2007)
18. Tukey, J.W.: Exploratory Data Analysis. Addison-Wesley, Menlo Park (1971)
19. Tukey, J.W.: Mathematics and the picturing of data. In: Proceedings of the International Congress of Mathematics, pp. 523–532. Vancouver, Canada (1974)
20. Vardi, Y., Zhang, C.H.: A modified Weiszfeld algorithm for the Fermat-Weber location problem. Math. Program. A **90**, 559–566 (2001)
21. Weber, A.: Über den Standort der Industrien. Mohr, Tübingen (1909)
22. Weiszfeld, E.: Sur le point pour lequel la somme des distances de n points donnés est minimum. Tôhoku Math. J. **43**, 355–386 (1937)
23. Welk, M.: Partial differential equations of bivariate median filters. In: Aujol, J.-F., Nikolova, M., Papadakis, N. (eds.) SSVM 2015. LNCS, vol. 9087, pp. 53–65. Springer, Cham (2015). https://doi.org/10.1007/978-3-319-18461-6_5
24. Welk, M.: Multivariate median filters and partial differential equations. J. Math. Imaging Vis. **56**, 320–351 (2016)
25. Welk, M., Breuß, M.: Morphological amoebas and partial differential equations. In: Hawkes, P.W. (ed.) Advances in Imaging and Electron Physics, vol. 185, pp. 139–212. Elsevier Academic Press, Amsterdam (2014)
26. Welk, M., Weickert, J., Becker, F., Schnörr, C., Feddern, C., Burgeth, B.: Median and related local filters for tensor-valued images. Sig. Process. **87**, 291–308 (2007)

Total Variation and Mean Curvature PDEs on the Space of Positions and Orientations

Remco Duits[1(✉)], Etienne St-Onge[2], Jim Portegies[1], and Bart Smets[1]

[1] CASA, Eindhoven University of Technology, Eindhoven, The Netherlands
R.Duits@tue.nl
[2] SCIL, Sherbrooke Connectivity Imaging Lab, Sherbrooke, Canada

Abstract. Total variation regularization and total variation flows (TVF) have been widely applied for image enhancement and denoising. To include a generic preservation of crossing curvilinear structures in TVF we lift images to the homogeneous space $\mathbb{M} = \mathbb{R}^d \rtimes S^{d-1}$ of positions and orientations as a Lie group quotient in $SE(d)$. For $d = 2$ this is called 'total roto-translation variation' by Chambolle & Pock. We extend this to $d = 3$, by a PDE-approach with a limiting procedure for which we prove convergence. We also include a Mean Curvature Flow (MCF) in our PDE model on $\mathbb{M}$. This was first proposed for $d = 2$ by Citti et al. and we extend this to $d = 3$. Furthermore, for $d = 2$ we take advantage of locally optimal differential frames in invertible orientation scores (OS).

We apply our TVF and MCF in the denoising/enhancement of crossing fiber bundles in DW-MRI. In comparison to data-driven diffusions, we see a better preservation of bundle boundaries and angular sharpness in fiber orientation densities at crossings. We support this by error comparisons on a noisy DW-MRI phantom. We also apply our TVF and MCF in enhancement of crossing elongated structures in 2D images via OS, and compare the results to nonlinear diffusions (CED-OS) via OS.

Keywords: Total variation · Mean Curvature ·
Sub-Riemannian Geometry · Roto-translations · Denoising ·
Fiber enhancement

1 Introduction

In the last decade, many PDE-based image analysis techniques for tracking and enhancement of curvilinear structures took advantage of lifting the image data to the homogeneous space $\mathbb{M} = \mathbb{R}^d \rtimes S^{d-1}$ of d-dimensional positions and orientations, cf. [4,6,8,10,14,35]. The precise definition of this homogeneous space follows in the next subsection. Set-wise it can be seen as a Cartesian product $\mathbb{M} = \mathbb{R}^d \times S^{d-1}$. Geometrically it can be equipped with a roto-translation equivariant geometry and topology beyond the usual isotropic Riemannian setting.

© Springer Nature Switzerland AG 2019
J. Lellmann et al. (Eds.): SSVM 2019, LNCS 11603, pp. 211–223, 2019.
https://doi.org/10.1007/978-3-030-22368-7_17

Typically, these PDE-based image analysis techniques involve flows that implement morphological and (non)linear scale spaces or solve variational models. The key advantage of extending the image domain from $\mathbb{R}^d$ to the higher dimensional lifted space $\mathbb{M}$ is that the PDE-flows do not suffer from crossings as fronts can pass without collision. This idea was shown for image enhancement in [9,22]. In [22] the method of coherence enhancing diffusion (CED) [36], was lifted to $\mathbb{M}$ in a diffusion flow method called "coherence enhancing diffusion on invertible orientation scores" (CED-OS), that is recently generalized to 3D [25]. Also for geodesic tracking methods, it helps that crossing line structures are disentangled in the lifted data. Geodesic flows prior to steepest descent also rely on (related [5,8,33]) PDEs on $\mathbb{M}$ commuting with roto-translations. They can account for crossings/bifurcations/corners [4,8,18].

Nowadays PDE-flows on orientation lifts of *3D images* are indeed relevant for applications such as fiber enhancement [12,15,31,35] and fiber tracking [30] in Diffusion-Weighted Magnetic Resonance Imaging (DW-MRI), and in enhancement [25] and tracking [11] of blood vessels in 3D images.

As for PDE-based image denoising and enhancement, total variation flows (TVF) are more popular than nonlinear diffusion flows. Recently, Chambolle & Pock generalized TVF from $\mathbb{R}^2$ to $\mathbb{M} = \mathbb{R}^2 \times S^1$, via 'total roto-translation variation' (TV-RT) flows [8] of 2D images. They employ (a)symmetric Finslerian geodesic models on $\mathbb{M}$ cf. [18]. As TVF falls short on invariance w.r.t. monotonic co-domain transforms, we also consider a Mean Curvature Flow (MCF) variant in our PDE model on $\mathbb{M}$, as proposed for 2D (i.e. $d = 2$) by Citti et al. [9].

To get a visual impression of how such PDE-based image processing on lifted images (orientation scores) works, for the case of tracking and enhancement of curvilinear structures in images see Fig. 1. In the 3rd row of Fig. 1, and henceforth, we visualize a lifted image $U : \mathbb{R}^3 \rtimes S^2 \to \mathbb{R}^+$ by a grid of angular profiles $\{\,\mu\, U(\mathbf{x}, \mathbf{n})\,\mathbf{n} \mid \mathbf{x} \in \mathbb{Z}^3, \mathbf{n} \in S^2\,\}$, with fixed $\mu > 0$.

The main contributions of this article are:

- We set up a geometric PDE flow framework on $\mathbb{M}$ including TVF, MCF, and diffusion. We tackle the PDEs by a basic limiting procedure. For TVF we prove convergence via Gradient flow theory by Brezis-Komura [1,7].
- We extend TVF on $\mathbb{M}$ [8], and MCF on $\mathbb{M}$ [9] to the 2D and 3D setting.
- We apply TVF and MCF in the denoising and enhancement of crossing fiber bundles in fiber orientation density functions (FODF) of DW-MRI data. In comparison to data-driven diffusions, we show a better preservation of bundle boundaries and angular sharpness with TVF and MCF. We support this observation by error comparisons on a noisy DW-MRI phantom.
- We include locally optimal differential frames (LAD) [17] in invertible orientation scores (OS), and propose crossing-preserving denoising methods TVF-OS, MCF-OS. We show benefits of LAD inclusion on 2D data.
- We compare TVF-OS, MCF-OS to CED-OS on 2D images.

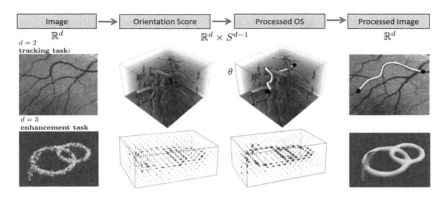

Fig. 1. Instead of direct PDE-based processing of an image, we apply PDE-based processing on a lifted image $U : \mathbb{R}^d \rtimes S^{d-1} \to \mathbb{R}$ (e.g. an orientation score: OS). The OS is obtained by convolving the image with a set of rotated wavelets allowing for stable reconstruction [4,14,25]. 2nd row: Vessel-tracking in a 2D image via geodesic PDE-flows in OS that underly TVF: [4,8,18], with $\mathbf{n} = (\cos\theta, \sin\theta)^T \in S^1$. 3rd row: CED-OS diffusion of a 3D image [17,25] visualized as a field of angular profiles. In this article we study image enhancement and denoising via TVF and MCF on $\mathbb{M} = \mathbb{R}^d \rtimes S^{d-1}$ and compare to nonlinear diffusion methods on $\mathbb{M}$ (like CED-OS).

2   Theory

2.1   The Homogeneous Space $\mathbb{M}$ of Positions and Orientations

Set $d \in \{2, 3\}$. Consider the rigid body motion group, $SE(d) = \mathbb{R}^d \rtimes SO(d)$. It acts transitively on $\mathbb{R}^d \times S^{d-1}$ by $(\mathbf{x}', \mathbf{n}') \mapsto g \odot (\mathbf{x}', \mathbf{n}')$ where

$$g \odot (\mathbf{x}', \mathbf{n}') := (\mathbf{R}\mathbf{x}' + \mathbf{x}, \mathbf{R}\mathbf{n}'), \text{ for all } g = (\mathbf{x}, \mathbf{R}) \in SE(d), \ (\mathbf{x}', \mathbf{n}') \in \mathbb{R}^d \times S^{d-1},$$

where $\mathbf{x} \in \mathbb{R}^d$ denotes a translation vector, $\mathbf{R} \in SO(d)$ denotes a rotation, and where $(\mathbf{x}', \mathbf{n}') \in \mathbb{R}^d \times S^{d-1}$. Now set $\mathbf{a} \in S^{d-1}$ as an a priori reference axis, say $\mathbf{a} = (1, 0)$ if $d = 2$ and $\mathbf{a} = \mathbf{e}_z = (0, 0, 1)^T$ if $d = 3$. The homogeneous space of positions and orientations is the partition of left-cosets:

$$\mathbb{M} := \mathbb{R}^d \rtimes S^{d-1} := SE(d)/H,$$

in $SE(d)$ and $H = \{g \in SE(d) \mid g \odot (\mathbf{0}, \mathbf{a}) = (\mathbf{0}, \mathbf{a})\}$. For $d = 2$ we have $\mathbb{M} \equiv SE(2)$. For $d = 3$ we have that

$$H = \{h_\alpha := (\mathbf{0}, \mathbf{R}_{\mathbf{a},\alpha}) \mid \alpha \in [0, 2\pi)\}, \tag{1}$$

where $\mathbf{R}_{\mathbf{a},\alpha}$ denotes a (counter-clockwise) rotation around $\mathbf{a} = \mathbf{e}_z$. Recall that by the definition of the left-cosets one has $H = \{\mathbf{0}\} \times SO(2)$, and $g_1 \sim g_2 \Leftrightarrow g_1^{-1}g_2 \in H$. This means that for $g_1 = (\mathbf{x}_1, \mathbf{R}_1)$, $g_2 = (\mathbf{x}_2, \mathbf{R}_2)$ one has

$$g_1 \sim g_2 \Leftrightarrow \mathbf{x}_1 = \mathbf{x}_2 \text{ and } \exists_{\alpha \in [0, 2\pi)} : \mathbf{R}_1 = \mathbf{R}_2 \mathbf{R}_{\mathbf{a},\alpha}.$$

The equivalence classes $[g] = \{g' \in SE(3) \mid g' \sim g\}$ are usually just denoted by $\mathbf{p} = (\mathbf{x}, \mathbf{n})$ as they consist of all rigid body motions $g = (\mathbf{x}, \mathbf{R_n})$ that map reference point $(\mathbf{0}, \mathbf{a})$ onto $(\mathbf{x}, \mathbf{n}) \in \mathbb{R}^3 \rtimes S^2$:

$$g \odot (\mathbf{0}, \mathbf{a}) = (\mathbf{x}, \mathbf{n}). \tag{2}$$

On tangent bundle $T(\mathbb{M}) = \{(\mathbf{p}, \dot{\mathbf{p}}) \mid \mathbf{p} \in \mathbb{M}, \dot{\mathbf{p}} \in T_\mathbf{p}(\mathbb{M})\}$ we set metric tensor:

$$\mathcal{G}_\epsilon|_\mathbf{p} (\dot{\mathbf{p}}, \dot{\mathbf{p}}) = D_S^{-1}|\dot{\mathbf{x}} \cdot \mathbf{n}|^2 + D_A^{-1}\|\dot{\mathbf{n}}\|^2 + \epsilon^{-2}D_S^{-1}\|\dot{\mathbf{x}} \wedge \mathbf{n}\|^2,$$
$$\text{for all } \mathbf{p} = (\mathbf{x}, \mathbf{n}) \in \mathbb{M}, \dot{\mathbf{p}} = (\dot{\mathbf{x}}, \dot{\mathbf{n}}) \in T_\mathbf{p}(\mathbb{M}) \tag{3}$$

with $0 < \epsilon \ll 1$ fixed, and with constant $D_s > 0$ costs for spatial motions and constant $D_A > 0$ costs for angular motions. For the sub-Riemannian setting ($\epsilon = 0$) we set $\mathcal{G}_0|_\mathbf{p} (\dot{\mathbf{p}}, \dot{\mathbf{p}}) = D_S^{-1}|\dot{\mathbf{x}} \cdot \mathbf{n}|^2 + D_A^{-1}\|\dot{\mathbf{n}}\|^2$ and constrain $\mathcal{G}_0$ to the sub-tangent bundle given by $\{((\mathbf{x}, \mathbf{n}), (\dot{\mathbf{x}}, \dot{\mathbf{n}})) \mid \dot{\mathbf{x}} \wedge \mathbf{n} = \mathbf{0}\}$.

Then (3) sets the Riemannian (resp. sub-Riemannian) gradient:

$$\nabla_\epsilon U(\mathbf{p}) = (\mathcal{G}_\epsilon^{-1}dU)(\mathbf{p}) \equiv$$
$$\left(D_S \, \mathbf{n}(\mathbf{n} \cdot \nabla_{\mathbb{R}^d}U(\mathbf{p})) + \epsilon^2 \, D_S(I - \mathbf{n} \otimes \mathbf{n})\nabla_{\mathbb{R}^d}U(\mathbf{p}) \, , \, D_A \, \nabla_{S^{d-1}}U(\mathbf{p})\right)^T, \tag{4}$$
$$\nabla_0 U(\mathbf{p}) = (\, D_S \, \mathbf{n}(\mathbf{n} \cdot \nabla_{\mathbb{R}^d}U(\mathbf{p})) \, , \, D_A \, \nabla_{S^{d-1}}U(\mathbf{p}) \,)^T,$$

for all differential functions $U \in C^1(\mathbb{M}, \mathbb{R})$.

We have particular interest for $U \in C^1(\mathbb{M}, \mathbb{R})$ that are 'orientation lifts' of input image $f : \Omega_f \to \mathbb{R}^+$. Such U are compactly supported within

$$\Omega = \Omega_f \times S^{d-1} \subset \mathbb{M}. \tag{5}$$

Such a lift may be (the real part of) an invertible orientation score [16] (cf. Fig. 1), a channel-representation [21], a lift by Gabor wavelets [2], or a fiber orientation density [29], where in general the absolute value $|U(\mathbf{x}, \mathbf{n})|$ is a probability density of finding a fiber structure at position $\mathbf{x} \in \mathbb{R}^d$ with local orientation $\mathbf{n} \in S^{d-1}$.

The corresponding norm of the gradient equals

$$\|\nabla_\epsilon U(\mathbf{p})\|_\epsilon = \sqrt{\mathcal{G}_\epsilon|_\mathbf{p} (\nabla_\epsilon U(\mathbf{p}), \nabla_\epsilon U(\mathbf{p}))}. \tag{6}$$

We set the following volume form on $\mathbb{M}$:

$$d\mu = D_S^{-1}d\mathbf{x} \wedge D_A^{-1}d\sigma_{S^{d-1}}. \tag{7}$$

This induces the following (sub-)Riemannian divergence

$$\text{div } \mathbf{v} = \begin{cases} \text{div}_{\mathbb{R}^3}\mathbf{v} + \text{div}_{S^2}\mathbf{v} & \text{for } \epsilon > 0, \\ \mathbf{n} \cdot \nabla_{\mathbb{R}^3}(\mathbf{n} \cdot \mathbf{v}) + \text{div}_{S^2}\mathbf{v} & \text{for } \epsilon = 0. \end{cases} \tag{8}$$

as the Lie derivative of $d\mu$ along $\mathbf{v}$ is $\mathcal{L}_\mathbf{v}d\mu = (\text{div } \mathbf{v})\,\mu$. TV on $\mathbb{R}^n$ is mainly built on the identity $\nabla \cdot (f\mathbf{v}) = f \, \nabla \cdot \mathbf{v} + \nabla f \cdot \mathbf{v}$. Similarly on $\mathbb{M}$ one has:

$$\text{div}(U\mathbf{v})(\mathbf{p}) = U(\mathbf{p}) \, \text{div } \mathbf{v}(\mathbf{p}) + \mathcal{G}_\epsilon|_\mathbf{p} (\nabla_\epsilon U(\mathbf{p}), \mathbf{v}(\mathbf{p})), \tag{9}$$

for all $\mathbf{p} \in \mathbb{M}$, from which we deduce the following integration by parts formula:

$$\int_\Omega U(\mathbf{p}) \operatorname{div} \mathbf{v}(\mathbf{p}) \, d\mu(\mathbf{p}) = \int_\Omega \mathcal{G}_{\mathfrak{e}}|_{\mathbf{p}} (\nabla_{\mathfrak{e}} U(\mathbf{p}), \mathbf{v}(\mathbf{p})) \, d\mu(\mathbf{p}), \tag{10}$$

for all $U \in C^1(\Omega)$ and all smooth vector fields $\mathbf{v}$ vanishing at the boundary $\partial\Omega$. This formula allows us to build a weak formulation of TVF on $\mathbb{M}$.

Definition 1 *(weak-formulation of TVF on $\mathbb{M}$).*
Let $U \in BV(\Omega)$ a function of bounded variation. Let $\chi_0(\Omega)$ denote the vector space of smooth vector fields that vanish at the boundary $\partial\Omega$. Then we define

$$TV_\varepsilon(U) := \sup_{\substack{\psi \in C_c^\infty(\Omega) \\ \mathbf{v} \in \chi_0(\Omega) \\ \|\mathbf{v}(\mathbf{p})\|_{\mathfrak{e}}^2 + |\psi(\mathbf{p})|^2 \le 1}} \int_\Omega \begin{pmatrix} \varepsilon \\ U(\mathbf{p}) \end{pmatrix} \cdot \begin{pmatrix} \psi(\mathbf{p}) \\ div\,\mathbf{v}(\mathbf{p}) \end{pmatrix} d\mu(\mathbf{p}) \tag{11}$$

For all $U \in BV(\Omega)$ we have $TV_0(U) \le TV_\varepsilon(U) \le TV_0(U) + \varepsilon|\Omega|$.

Lemma 1. *Let $\varepsilon, \mathfrak{e} \ge 0$. For $U \in C^1(\Omega, \mathbb{R})$ we have*

$$TV_\varepsilon(U) = \int_\Omega \sqrt{\|\nabla_{\mathfrak{e}} U(\mathbf{p})\|_{\mathfrak{e}}^2 + \varepsilon^2} \, d\mu(\mathbf{p}). \tag{12}$$

For $U \in C^2(\mathbb{M}, \mathbb{R})$ and $\mathfrak{e}, \varepsilon > 0$ we have $\partial TV_\varepsilon(U) = \operatorname{div} \circ \left(\frac{\nabla_\varepsilon(U)}{\|\nabla_\varepsilon(U)\|_{\mathfrak{e}}} \right)$.

Proof. First we substitute (10) into (11), then we apply Gauss theorem and use $U\mathbf{v}|_{\partial\Omega} = 0$. Then we apply Cauchy-Schwarz on $V_{\mathbf{p}} := \mathbb{R} \times T_{\mathbf{p}}(\mathbb{M})$ for each $\mathbf{p} \in \mathbb{M}$, with inner product $(\epsilon_1, \mathbf{v}_1) \cdot (\epsilon_2, \mathbf{v}_2) = \epsilon_1\epsilon_2 + \mathcal{G}_{\mathbf{p}}(\mathbf{v}_1, \mathbf{v}_2)$, which holds with equality iff the vectors are linearly dependent. Therefore we smoothly approximate $\frac{1}{\sqrt{\varepsilon^2 + \|\nabla_{\mathfrak{e}} U\|_{\mathfrak{e}}^2}}(\varepsilon, \nabla_{\mathfrak{e}} U)$ by $(\psi, \mathbf{v})$ to get (12). For $U \in C^2(\Omega, \mathbb{R})$, $\delta \in C_c^\infty(\Omega, \mathbb{R})$ we get $(\partial TV_\varepsilon(U), \delta)_{\mathbb{L}_2(\Omega)} = \lim_{h \downarrow 0} \frac{TV_\varepsilon(U + h\,\delta) - TV_\varepsilon(U)}{h} \overset{(10)}{=} (\operatorname{div} \circ \left(\frac{\nabla_\varepsilon(U)}{\|\nabla_\varepsilon(U)\|_{\mathfrak{e}}} \right), \delta)_{\mathbb{L}_2(\Omega)}$.

2.2 Total-Roto Translation Variation, Mean Curvature Flows on $\mathbb{M}$

Henceforth, we fix $\mathfrak{e} \ge 0$ and write $\nabla = \nabla_{\mathfrak{e}}$. We propose the following roto-translation equivariant enhancement PDEs on $\Omega \subset \mathbb{M}$, recall (5):

$$\begin{cases} \frac{\partial W^\varepsilon}{\partial t}(\mathbf{p}, t) = \left(\|\nabla W^\varepsilon(\mathbf{p}, t)\|^2 + \varepsilon^2 \right)^{\frac{a}{2}} \left(\operatorname{div} \circ \frac{\nabla W^\varepsilon(\cdot, t)}{(\|\nabla W^\varepsilon(\cdot, t)\|_{\mathfrak{e}}^2 + \varepsilon^2)^{\frac{b}{2}}} \right)(\mathbf{p}), & \\ 0 = \mathbf{N}(\mathbf{x}) \cdot \nabla_{\mathbb{R}^d} W^\varepsilon(\mathbf{x}, \mathbf{n}, 0) & \mathbf{p} = (\mathbf{x}, \mathbf{n}) \in \partial\Omega, \\ W^\varepsilon(\mathbf{p}, 0) = U(\mathbf{p}) & \mathbf{p} = (\mathbf{x}, \mathbf{n}) \in \Omega, \end{cases} \tag{13}$$

with evolution time $t \ge 0$, $0 < \varepsilon \ll 1$, and with parameters $a, b \in \{0, 1\}$. Regarding the boundary of Ω we note that $\mathbf{p} = (\mathbf{x}, \mathbf{n}) \in \partial\Omega \Leftrightarrow \mathbf{x} \in \partial\Omega_f, \mathbf{n} \in S^2$. We use Neumann boundary conditions as $\mathbf{N}(\mathbf{x})$ denotes the normal at $\mathbf{x} \in \partial\Omega_f$.

For $\{a, b\} = \{1, 1\}$ we have a geometric Mean Curvature Flow (MCF) PDE. For $\{a, b\} = \{0, 1\}$ we have a Total Variation Flow (TVF) [8]. For $\{a, b\} = \{0, 0\}$ we obtain a linear diffusion for which exact smooth solutions exist [28].

Remark 1. By the product rule (9) the right-hand side of (13) for $\varepsilon \downarrow 0$ becomes

$$\frac{\partial W^0}{\partial t} = \|\nabla W^0\|^{a-b} \Delta W^0 + 2b\,\overline{\kappa}_I \, \|\nabla W^0\|^a, \tag{14}$$

with the mean curvature $\overline{\kappa}_I(\mathbf{p}, t)$ of level set $\{\mathbf{q} \in \mathbb{M} \mid W^0(\mathbf{q}, t) = W^0(\mathbf{p}, t)\}$, akin to [24, ch;3.2], and with (possibly hypo-elliptic) Laplacian $\Delta = \mathrm{div} \circ \nabla$.

Remark 2. For MCF and TVF smooth solutions to the PDE (13) exist only under special circumstances. This lack of regularity is an advantage in image processing to preserve step-edges and plateaus in images, yet it forces us to define a concept of weak solutions. Here, we distinguish between MCF and TVF:

For MCF one relies on viscosity solution theory developed by Evans-Spruck [20], see also [23,32] for the case of MCF with Neumann boundary conditions. In [9, Thm 3.6] existence of C^1-viscosity solutions is shown for $d = 2$.

For TVF we will rely on gradient flow theory by Brezis-Komura [1,7].

Remark 3. Convergence of the solutions w.r.t. $\varepsilon \downarrow 0$ is clear from the exact solutions for $\{a, b\} = \{0, 0\}$, see [28, ch:2.7], and is also addressed for MCF [3,9]. For TVF one can rely on [18]. Next we focus on convergence results for $\varepsilon \downarrow 0$.

2.3   Gradient-Flow Formulations and Convergence

The total variation flow can be seen as a gradient flow of a lower-semicontinuous, convex functional in a Hilbert space, as we explain next.

If $F : H \to [0, \infty]$ is a proper (i.e. not everywhere equal to infinity), lower semicontinuous, convex functional on a Hilbert space H, the subdifferential of F in a point u in the finiteness domain of F is defined as

$$\partial F(u) := \{z \in H \mid (z, v - u) \leq F(v) - F(u) \text{ for all } v \in H\}.$$

The subdifferential is closed and convex, and thereby it has an element of minimal norm, called "the gradient of F in u" denoted by $\mathrm{grad}F(u)$. Let u_0 be in the closure of the finiteness domain of F. By Brezis-Komura theory, [7], [1, Thm 2.4.15] there is a unique locally absolutely continuous curve $u : [0, \infty) \to H$ s.t.

$$-u'(t) = \mathrm{grad}F(u(t)) \text{ for a.e. } t > 0 \text{ and } \lim_{t \downarrow 0} u(t) = u_0.$$

We call $u : [0, \infty) \to H$ the gradient flow of F starting at u_0.

The function $TV_\epsilon : L^2(\Omega) \to [0, \infty]$ is lower-semicontinuous and convex for every $\epsilon \geq 0$. This allows us to generalize solutions to the PDE (13) as follows:

Definition 2. *Let $U \in \Xi := BV(\Omega) \cap \mathbb{L}_2(\Omega)$. We define by $t \mapsto W^\epsilon(\cdot, t)$ the gradient flow of TV_ϵ starting at U.*

Remark 4. A smooth solution W^ϵ to (13) with $\{a, b\} = \{0, 1\}$ is a gradient flow.

Theorem 1 *(strong $\mathbb{L}_2$-convergence, stability and accuracy of TV-flows).*
Let $U \in \mathbb{L}_2(\Omega)$ and let W^ε be the gradient flow of TV_ε starting at U and $\varepsilon \geq 0, \mathfrak{e} \geq 0$. Let $t \geq 0$. Then

$$\lim_{\varepsilon \downarrow 0} W^\varepsilon(\cdot, t) = W^0(\cdot, t) \ in \ \mathbb{L}_2(\Omega).$$

More precisely, for $U \in BV(\Omega)$, we have for all $t \geq 0$:

$$\|W^\varepsilon(\cdot, t) - W^0(\cdot, t)\|_{\mathbb{L}_2(\Omega)} \leq 8\left(\|U\|_{L^2(\Omega)}(TV_0(U) + \delta)\delta t^2 \right)^{1/5} \ with \ \delta = \varepsilon|\Omega|$$

Theorem 1 follows from the following general result, if we take $F = TV_0$, $G = TV_\epsilon$, $\delta = \epsilon|\Omega|$.

Theorem 2. *Let $F : H \to [0, \infty]$ and $G : H \to [0, \infty]$ be two proper, (i.e. not everywhere equal to infinity), lower semi-continuous, convex functionals on a Hilbert space H, such that for some $\delta \geq 0$ one has*

$$F(u) - \delta \leq G(u) \leq F(u) + \delta$$

for all $u \in H$. Let $u_0, v_0 \in H$ and $E, M \geq 0$ be such that $F(u_0) \leq E$ and $G(v_0) \leq E$ and $\|u_0\| \leq M$, $\|v_0\| \leq M$. The gradient flow $u : [0, \infty) \to H$ of F starting at u_0, and the gradient flow $v : [0, \infty) \to H$ of G starting at v_0 satisfy

$$\|u(t) - v(t)\|_H \leq 16(ME\delta t^2)^{1/5} + \|u_0 - v_0\|_H$$

for all $0 \leq t \leq E^6 M^6/\delta^9$.

The proof can be found on the arxiv [19].

2.4   Numerics

We implemented the PDE system (13) by Euler forward time discretization, relying on standard B-spline or linear interpolation techniques for derivatives in the underlying tools of the gradient on $\mathbb{M}$ given by (4) and the divergence on $\mathbb{M}$ given by (8). For details see [12,22]. Also, the explicit upperbounds for stable choices of stepsizes can be derived by the Gershgorin circle theorem, [12,22].

The PDE system (13) can be re-expressed by a left-invariant PDE on $SE(d)$ as done in related previous works by several researchers [6,9,10,12,16,22]. For $d = 2$ this is straightforward as $SE(2) \equiv \mathbb{R}^2 \rtimes S^1$. For $d = 3$ and $\mathfrak{e} = 0$ one has

$$\text{div } \mathbf{v} \leftrightarrow \mathcal{A}_3 \tilde{v}^3 + \mathcal{A}_3 \tilde{v}^4 + \mathcal{A}_5 \tilde{v}^5, \ \nabla_0 W \leftrightarrow (D_S \mathcal{A}_3 \tilde{W}, D_A \mathcal{A}_4 \tilde{W}, D_A \mathcal{A}_5 \tilde{W})^T \quad (15)$$

where $\{\mathcal{A}_i\}$ is a basis of vector fields on $SE(3)$ given by $(\mathcal{A}_i f)(g) = \lim_{t \downarrow 0} \frac{f(ge^{tA_i}) - f(g)}{t}$ with a Lie algebra basis $\{A_i\}$ for $T_e(SE(3))$ as in [12,14], and with $\tilde{W}(\mathbf{x}, \mathbf{R}, t) = W(\mathbf{x}, \mathbf{R}a, t)$, $\tilde{v}^i(\mathbf{x}, \mathbf{R}, t) = v^i(\mathbf{x}, \mathbf{R}a, t)$. We used (15) to apply discretization on $SE(3)$ [12] in the software developed by Martin et al. [26], to our PDEs of interest (13) on $\mathbb{M}$ for $d = 3$.

Remark 5. The Euler-forward discretizations are not unconditionally stable. For $a = b = 0$, the Gerhsgorin circle theorem [12, ch.4.2] gives the stability bound

$$\Delta t \leq (\Delta t)_{crit} := \left(\frac{(d-1)D_A + D_S}{2h^2} + \frac{(d-1)D_A}{2h_a^2} \right)^{-1},$$

when using linear interpolation with spatial stepsize h and angular stepsize h_a. In our experiments, for $d = 2$ we set $h = 1$ and for $d = 3$ we took $h_a = \frac{\pi}{25}$ using an almost uniform spherical sampling from a tessellated icosahedron with $N_A = 162$ points. TVF required smaller times steps when ε decreases. Keeping in mind (14) but then applying the product rule (9) to the case $0 < \varepsilon \ll 1$, we concentrate on the first term as it is of order ε^{-1} when the gradient vanishes. Then we find $\Delta t \leq \varepsilon \cdot (\Delta t)_{crit}$ for TVF. For MCF we do not have this limitation.

3 Experiments

In our experiments, we aim to enhance contour and fiber trajectories in medical images and to remove noise. Lifting the image $f : \mathbb{R}^d \to \mathbb{R}$ towards its orientation lift $U : \mathbb{M} \to \mathbb{R}$ defined on the space of positions and orientations $\mathbb{M} = \mathbb{R}^d \rtimes S^{d-1}$ preserves crossings [22] and avoids leakage of wavefronts [18].

For our experiments for $d = 3$ the initial condition $U : \mathbb{M} \to \mathbb{R}^+$ is a fiber orientation density function (FODF) obtained from DW-MRI data [29].

For our experiments for $d = 2$ the initial condition U is an invertible orientation score (OS) and we adopt the flows in (13) via locally adaptive frames [17]. For both $d = 3$ (Subsect. 3.1) and $d = 2$ (Subsect. 3.2), we show advantages of TVF and MCF over crossing-preserving diffusion flows [12,22] on $\mathbb{M}$. We set $\mathfrak{e} = 0$ in all presented experiments as it gave better results than $\mathfrak{e} > 0$.

3.1 TVF and MCF on $\mathbb{R}^3 \rtimes S^2$ for Denoising FODFs in DW-MRI

In DW-MRI image processing one obtains a field of angular diffusivity profiles (orientation density function) of water-molecules. A high diffusivity in particular orientation correlates to biological fibers structure, in brain white matter, along that same direction. Crossing-preserving enhancement of FODF fields $U : \mathbb{M} \to \mathbb{R}^+$ helps to better identify structural pathways in brain white matter, which is relevant for surgery planning, see for example [27,29].

For a quantitative comparison we applied TVF, MCF and PM diffusion [12] to denoise a popular synthetic FODF $U : \mathbb{M} \to \mathbb{R}^+$ from the ISBI-HARDI 2013 challenge [13], with realistic noise profiles. In Fig. 2, we can observe the many crossing fibers in the dataset. Furthermore, we depicted the absolute $\mathbb{L}_2$-error $t \mapsto \|U - \Phi_t(U)(\cdot)\|_{\mathbb{L}_2(\mathbb{M})}$ as a function of the evolution parameter t, where $\Phi_t(U) = W_\varepsilon(\cdot, t)$ with optimized $\varepsilon = 0.02$ in the case of TVF (in green), and MCF (in blue), and where Φ_t is the PM diffusion evolution [12] on $\mathbb{M}$ with optimized PM parameter $K = 0.2$ (in red). We also depict results for $K = 0.1, 0.4$ (with the dashed lines) and $\varepsilon = 0.01, 0.04$. We see that the other parameter settings provide on average worse results, justifying our optimized parameter settings. We set $D_S = 1.0$, $D_A = 0.001$, $\Delta t = 0.01$. We observe that:

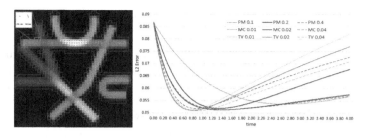

Fig. 2. Quantitative comparison of denoising a fiber orientation density function (FODF) obtained by (CSD) [34] from a benchmark DW-MRI dataset [13].

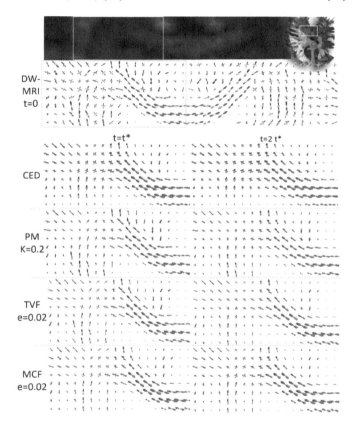

Fig. 3. Qualitative comparison of denoising a FODF obtained by (CSD) [34] from a standard DW-MRI dataset (with $b = 1000$ s/mm^2 and 54 gradient directions). For the CSD we used up to 8th order spherical harmonics, and the FODF is then spherically sampled on a tessellation of the icosahedron with 162 orientations.

- TVF can reach lower error values than MC-flow with adequate $\Delta t = 0.01$,
- MCF provides more stable errors for all $t > 0$, than TV-flow w.r.t. $\epsilon > 0$,

– TVF and MCF produce lower error values than PM-diffusion,
– PM-diffusion provides the most variable results for all $t > 0$.

For a qualitative comparison we applied TVF, MCF, PM diffusion and linear diffusion to a FODF field $U : \mathbb{M} \to \mathbb{R}^+$ obtained from a standard DW-MRI dataset (with $b = 1000$ s/mm^2, 54 gradient directions) via constrained spherical deconvolution (CSD) [34]. See Fig. 3, where for each method, we used the optimal parameter settings with the artificial data-set. We see that

– all methods perform well on the real datasets. Contextual alignment of the angular profiles better reflects the anatomical fiber bundles,
– MCF and TVF better preserve boundaries and angular sharpness,
– MCF better preserves the amplitude at crossings at longer times.

3.2   TVF and MCF on $\mathbb{R}^2 \rtimes S^1$ for 2D Image Enhancement/Denoising

The initial condition for our TVF/MCF-PDE (13) is set by an orientation score [14] of image $f : \mathbb{R}^2 \to \mathbb{R}$ given by $\mathcal{W}_\psi f(\mathbf{x}, \mathbf{n}) = (\psi_\mathbf{n} \star f)(\mathbf{x})$ where $\star$ denotes correlation and $\psi_\mathbf{n}$ is the rotated wavelet aligned with $\mathbf{n} \in S^1$. For ψ we use a cake-wavelet [14, ch:4.6] ψ with standard settings [26]. Then we compute:

$$f \mapsto \mathcal{W}_\psi f \mapsto \varPhi_t(\mathcal{W}_\psi f)(\cdot, \cdot) \mapsto f_t^a(\cdot) := \int_{S^1} \varPhi_t^a(\mathcal{W}_\psi f)(\cdot, \mathbf{n}) \, d\mu_{S^1}(\mathbf{n}). \qquad (16)$$

for $t \geq 0$. The cake-wavelets allow us to reconstruct by integration over S^1 only [14]. By the invertibility of the orientation score one thereby has $f = f_0^a$ so all flows depart from the original image. Furthermore, $U \mapsto W(\cdot, t) = \varPhi_t(U)$ denotes the flow operator on $\mathbb{M}$ (13), but then the PDE in (13) is re-expressed in the locally adaptive frame (LAD) $\{\mathcal{B}_i\}_{i=1}^3$ obtained by the method in [17,22,26]. The PDE then becomes

$$\frac{\partial \tilde{W}}{\partial t} = (\|\nabla_0 \tilde{W}\|_0^2 + \varepsilon^2)^{\frac{a}{2}} \sum_{i=1}^3 \tilde{D}_{ii} \, \mathcal{B}_i \circ (\|\nabla_0\|_0^2 + \varepsilon^2)^{-\frac{1}{2}} \mathcal{B}_i \tilde{W}, \qquad a \in \{0, 1\},$$

with $\tilde{D}_{11} = 1$, $\tilde{D}_{22} = \tilde{D}_{33}$ as in CED-OS [22, Eq. 72].

For $a = 0$ we call $f \mapsto f_t^a$ given by (16) a 'TVF-OS flow', for $a = 1$ we call it a 'MCF-OS flow'. In Fig. 4 we show how errors progress with $t \geq 0$. We see that inclusion of LAD is beneficial on the real image. In Fig. 5 we give a qualitative comparison to CED-OS [22]. Lines and plateaus are best preserved by TVF-OS.

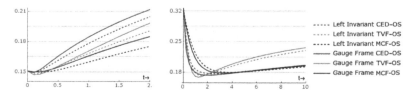

Fig. 4. Relative $\mathbb{L}_1$ errors of the spirals test image (left) and the collagen image (right) for the CED-OS, MCF-OS and TVF-OS methods.

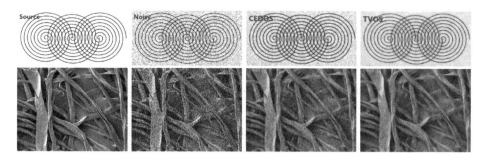

Fig. 5. From left to right; original image; noisy input image f, CED-OS output image [22], TVF-OS output image f_t^0. We have set $D_A = 0.01$, $D_S = 1$, and took $t = 10 \cdot t^*$ for the spirals image and $t = 2 \cdot t^*$ for the collagen image where t^* minimizes the relative $\mathbb{L}_1$-error to stress different qualitative behavior.

4   Conclusion

We have proposed a PDE system on the homogeneous space $\mathbb{M} = \mathbb{R}^d \rtimes S^{d-1}$ of positions and orientations, for crossing-preserving denoising and enhancement of (lifted) images containing both complex elongated structures and plateaus.

It includes TVF, MCF and diffusion flows as special cases, and includes (sub-) Riemannian geometry. Thereby we generalized recent related works by Citti et al. [9] and Chambolle and Pock [8] from 2D to 3D using a different numerical scheme with new convergence results (Theorem 1) and stability bounds. We used the divergence and intrinsic gradient on a (sub)-Riemannian manifold above $\mathbb{M}$ for a formal weak-formulation of total variation flows, which simplifies if the lifted images are differentiable (Lemma 1).

Compared to previous nonlinear crossing-preserving diffusion methods on $\mathbb{M}$, we showed improvements (Figs. 4, 5) over CED-OS methods [22] (for $d = 2$) and improvements over contextual fiber enhancement methods in DW-MRI processing (for $d = 3$) [12,15] on real medical image data. We observe that crossings and boundaries (of bundles and plateaus) are better preserved over time. We support this quantitatively by a denoising experiment on a benchmark DW-MRI dataset, where MCF performs better than TVF and both perform better than Perona-Malik diffusions, in view of error reduction and stability.

References

1. Ambrosio, L., Gigli, N., Savaré, G.: Gradient Flows in Metric Spaces and in the Space of Probability Measures, Bikhäuser (2005)
2. Baspinar, E., Citti, G., Sarti, A.: A geometric model of multi-scale orientation preference maps via gabor functions. JMIV **60**(6), 900–912 (2018)
3. Baspinar, E.: Minimal surfaces in Sub-Riemannian structures and functional geometry of the visual cortex. Ph.D. thesis, University of Bologna (2018)
4. Bekkers, E.: Retinal image analysis using Sub-Riemannian geometry in $SE(2)$. Ph.D. thesis, TU/e Eindhoven (2017)

5. Bekkers, E., Duits, R., Mashatkov, A., Sanguinetti, G.: A PDE approach to data-driven Sub-Riemannian geodesics in $SE(2)$. SIIMS **8**(4), 2740–2770 (2015)
6. Boscain, U., Chertovskih, R., Gauthier, J.P., Prandi, D., Remizov, A.: Highly corrupted image inpainting by hypoelliptic diffusion. JMIV **60**(8), 1231–1245 (2018)
7. Brézis, H.: Operateurs maximeaux monotones et semi-gropes de contractions dans les espaces de Hilbert, vol. 50. North-Holland Publishing Co., Amsterdam (1973)
8. Chambolle, A., Pock, T.: Total roto-translation variation. arXiv:17009.099532v2, pp. 1–47, July 2018
9. Citti, G., Franceschiello, B., Sanguinetti, G., Sarti, A.: Sub-riemannian mean curvature flow for image processing. SIIMS **9**(1), 212–237 (2016)
10. Citti, G., Sarti, A.: A cortical based model of perceptual completion in the roto-translation space. JMIV **24**(3), 307–326 (2006)
11. Cohen, E., Deffieux, T., Demené, C., Cohen, L.D., Tanter, M.: 3D vessel extraction in the rat brain from ultrasensitive doppler images. In: Gefen, A., Weihs, D. (eds.) Computer Methods in Biomechanics and Biomedical Engineering. LNB, pp. 81–91. Springer, Cham (2018). https://doi.org/10.1007/978-3-319-59764-5_10
12. Creusen, E.J., Duits, R., Florack, L., Vilanova, A.: Numerical schemes for linear and non-linear enhancement of DW-MRI. NM-TMA **6**(3), 138–168 (2013)
13. Daducci, A., Caruyer, E., Descoteaux, M., Thiran, J.P.: HARDI Reconstruction Challenge (2013). Published at IEEE ISBI 2013
14. Duits, R.: Perceptual organization in image analysis. Ph.D. thesis, TU/e (2005)
15. Duits, R., Creusen, E., Ghosh, A., Dela Haije, T.: Morphological and linear scale spaces for fiber enhancement in DW-MRI. JMIV **46**(3), 326–368 (2013)
16. Duits, R., Franken, E.M.: Left invariant parabolic evolution equations on $SE(2)$ and contour enhancement via invertible orientation scores, part I: linear left-invariant diffusion equations on $SE(2)$. QAM-AMS **68**, 255–292 (2010)
17. Duits, R., Janssen, M., Hannink, J., Sanguinetti, G.: Locally adaptive frames in the roto-translation group and their applications in medical image processing. JMIV **56**(3), 367–402 (2016)
18. Duits, R., Meesters, S., Mirebeau, J., Portegies, J.: Optimal paths for variants of the 2D and 3D reeds-shepp car with applications in image analysis. JMIV **60**, 816–848 (2018)
19. Duits, R., St.-Onge, E., Portegies, J., Smets, B.: Total variation and mean curvature PDEs on $\mathbb{R}^d \rtimes s^{d-1}$. Technical report https://arxiv.org/abs/1902.08145 (2019)
20. Evans, L.C., Spruck, J.: Motion of level sets by mean curvature. I. J. Differential Geom. **33**(3), 635–681 (1991)
21. Felsberg, M., Forssen, P.E., Scharr, H.: Channel smoothing: efficient robust smoothing of low-level signal features. IEEE PAMI **28**, 209–222 (2006)
22. Franken, E.M., Duits, R.: Crossing preserving coherence-enhancing diffusion on invertible orientation scores. IJCV **85**(3), 253–278 (2009)
23. Giga, Y., Sato, M.H.: Generalized interface evolution with the Neumann boundary condition. Proc. Japan Acad. Ser. A Math. Sci. **67**(8), 263–266 (1991)
24. Sapiro, G.: Geometric Partial Differential Equations and Image Analysis. CUP, Cambridge (2006)
25. Janssen, M.H.J., Janssen, A.J.E.M., Bekkers, E.J., Bescós, J.O., Duits, R.: Processing of invertible orientation scores of 3D images. JMIV **60**(9), 1427–1458 (2018)
26. Martin, F., Bekkers, E., Duits, R.: Lie analysis package (2017). www.lieanalysis.nl/
27. Meesters, S., et al.: Stability metrics for optic radiation tractography: towards damage prediction after resective surgery. J. Neurosci. Methods **288**, 34–44 (2017)

28. Portegies, J.M., Duits, R.: New exact and numerical solutions of the (convection-) diffusion kernels on SE(3). DGA **53**, 182–219 (2017)
29. Portegies, J.M., Fick, R., Sanguinetti, G.R., Meesters, S.P.L., Girard, G., Duits, R.: Improving fiber alignment in HARDI by combining contextual PDE flow with constrained spherical deconvolution. PLoS ONE **10**(10), e0138122 (2015)
30. Portegies, J.: PDEs on the Lie Group SE(3) and their applications in diffusion-weighted MRI. Ph.D. thesis, Dep. Math. TU/e (2018)
31. Reisert, M., Burkhardt, H.: Efficient tensor voting with 3D tensorial harmonics. In: IEEE Conference on CVPRW 2008, pp. 1–7 (2008)
32. Sato, M.H.: Interface evolution with Neumann boundary condition. Adv. Math. Sci. Appl. **4**(1), 249–264 (1994)
33. Schmidt, M., Weickert, J.: Morphological counterparts of linear shift-invariant scale-spaces. J. Math. Imaging Vis. **56**(2), 352–366 (2016)
34. Tournier, J.D., Calamante, F., Connelly, A.: MRtrix: diffusion tractography in crossing fiber regions. Int. J. Imaging Syst. Technol. **22**(1), 53–66 (2012)
35. Vogt, T., Lellmann, J.: Measure-valued variational models with applications to diffusion-weighted imaging. JMIV **60**(9), 1482–1502 (2018)
36. Weickert, J.A.: Coherence-enhancing diffusion filtering. Int. J. Comput. Vis. **31**(2/3), 111–127 (1999)

A Variational Convex Hull Algorithm

Lingfeng Li[1,2], Shousheng Luo[3,4], Xue-Cheng Tai[2(✉)], and Jiang Yang[1]

[1] Department of Mathematics, Southern University of Science and Technology,
Shenzhen, China
`yangj7@sustc.edu.cn`
[2] Department of Mathematics, Hong Kong Baptist University,
Kowlong Tong, Hong Kong
`lingfengli@life.hkbu.edu.hk, xuechengtai@hkbu.edu.hk`
[3] Beijing Computational Science Research Center, Beijing, China
`sluo1983@csrc.ac.cn`
[4] School of Mathematics and Statistics, Data Analysis Technology Lab,
Henan University, Kaifeng, China
`sluo@henu.edu.cn`

Abstract. Finding the convex hull of a given object or a point set is a very important problem. In this paper, we propose a variational convex hull model and numerical algorithms to solve it. Our model is based on level set representation. Efficient numerical algorithms and implementations based on splitting ideas are given. To test our proposed model, we conduct many experiments for objects represented by binary images, and the results suggest that our model can identify the convex hull accurately. Even more, our model can be easily modified to handle the outliers, and this is also demonstrated by numerical examples.

Keywords: Convex hull · Level set method · Image processing · Outliers detection

1 Introduction

Convex hull algorithm has been widely studied in the past three decades. It is an important problem in computational geometry and has been used in various areas, such as data clustering [1], robot motion planning [2], collision detection [3], image segmentation [4], etc.

Various algorithms have been proposed for finding the convex hull of a given point set, and all these methods fall into two different categories: exact algorithms and inexact algorithms. Some widely used exact convex hull algorithms are gift wrapping [5], Graham scan [6], quick hull [7], divide and conquer [8], and monotone chain [9]. These exact algorithms can identify the convex hull of a given point set efficiently, but they only return the vertices of the convex hull polygon.

In some cases, we may just want an approximate convex hull instead of the exact one. For example, when the given set of points contains some noise, finding

© Springer Nature Switzerland AG 2019
J. Lellmann et al. (Eds.): SSVM 2019, LNCS 11603, pp. 224–235, 2019.
https://doi.org/10.1007/978-3-030-22368-7_18

the exact convex hull of all given data would not be very helpful. In addition, finding approximate convex hulls can save computational time and storage space. Many researchers have proposed algorithms for approximating convex hulls, such as [10–12]. These algorithms can yield an approximate convex hull that contains a large majority of the given points very efficiently.

In some applications on images [13], we may need to find out all pixels that belong to the convex hull of a given set of points. In this case, all the algorithms we mentioned above cannot give the solution directly, and extra steps are needed to draw the regions of convex hulls, which is much more computationally expensive than just finding the vertices. Moreover, outliers in the given set of points pose a strong challenge for all the existing methods. However, outliers are often inevitable in the images acquired by cameras and scanners.

In this paper, we propose a variational method for convex hull problems of objects represented binary images. We use level set representations, which can identify all the points belonging to the convex hull. Level set function is a widespread tool in image processing, especially for image segmentation [14,15]. In our method, the convex hull region is denoted by the sub-level set of the signed distance function, and the convexity is guaranteed by the positivity of its Laplacian. This idea was first introduced in [16] and [17] for image segmentation with convex prior. In this work, we use the idea to design a variational model for convex hull problem. One thing we want to emphasize is that our method can be modified to handle the outliers easily. Alternating direction method of multiplier (ADMM) is employed to solve the proposed model efficiently. Many experiments on binary images were conducted to verify the effectiveness of the proposed method and algorithms, but only one example is shown in this paper due to page limitation.

This paper is organized as follows. In Sect. 2, we will give a brief introduction to level set representation of convex object, and then describe our proposed model and algorithm. In Sect. 3, we will present the numerical implementation of our model. In Sect. 4, some numerical experiments on binary images are illustrated.

2   Model Description

In this section, we first review some mathematical preliminaries. We then present the proposed method in detail.

2.1   Preliminaries

Assume the considered binary image representing an object is defined on a rectangle domain $\Omega \subset \mathbb{R}^2$, and the object is denoted by X. The binary image takes value 1 on the object X and 0 elsewhere. The convex hull of X, which we want to find, is usually defined in different forms. Here we present one of them as follows [18]:

Definition 1. *A set S is convex if and only if for any two points x_1 and x_2 in S, we have*

$$\theta x_1 + (1 - \theta)x_2 \in S,$$

where $0 \leq \theta \leq 1$.

Definition 2. *The **convex hull** of a set X, denoted by $\text{Conv}(X)$, is the intersection of all convex sets containing X.*

In other words, $\text{Conv}(X)$ is the smallest convex set containing X. Evidently, it is almost impossible to get $\text{Conv}(X)$ for binary image via the Definition (2), and various algorithms were proposed in the literature. We propose a new method to compute $\text{Conv}(X)$ via level set function representation technique.

For a given region $D \subset \mathbb{R}^2$, its corresponding signed distance function (SDF), a special level set function, is defined as

$$\phi(x) = \begin{cases} -\text{dist}(x, \partial D) & x \text{ is inside } \partial D \\ \text{dist}(x, \partial D) & x \text{ is outside } \partial D , \\ 0 & x \in \partial D \end{cases} \tag{1}$$

where ∂D is the boundary of D and $\text{dist}(x, \partial D)$ is the distance between $x \in \mathbb{R}^2$ and ∂D, i.e, $\text{dist}(x, \partial D) = \min\limits_{y \in \partial D} \|y - x\|_2^2$. To guarantee the convexity of ϕ, we need to introduce the following theorem from [17]:

Theorem 1. *Let ϕ be a signed distance function of a region D, and $\phi \in C^2(\mathbb{R}^2)$ almost everywhere. Then, D is convex if and only if*

$$\triangle\phi(x) \geq 0 \quad a.e. \tag{2}$$

In addition, the region D can be identified by the zero sub-level set, i.e., $D = \{x|\ \phi(x) \leq 0\}$.

2.2   Exact Convex Hull Model

Based on the discussions above, the aim to find the convex hull $\text{Conv}(X)$ of X can be converted to find the SDF of it. We assume that the convex hull of the given object lies in a subregion of Ω, denoted by Ω_1. Then, the level set function based method for convex hull problem can be simply formulated as

$$\phi^* = \arg\min_{\phi} \quad \int_{\Omega} [1 - H(\phi)]dx \tag{3}$$

$$\text{s.t.} \quad |\nabla\phi(x)|_2 = 1 \text{ in } \Omega_1, \triangle\phi(x) \geq 0 \text{ in } \Omega_1, \phi(x) \leq 0, \ x \in X,$$

where H is the Heaviside function and $\int(1 - H(\phi))dx$ measures the area of the object. The first two constraints are imposed to make ϕ be a convex SDF in Ω_1, and the last constraint makes X be enclosed in the zero sub-level set of ϕ. By solving this minimization problem, we can find the smallest convex set that contains X. Since we will pose extra constraints for ϕ on the boundary of Ω, we only require $\triangle\phi \geq 0$ in Ω_1 here. After having ϕ^*, the convex hull can be given by the zero sub-level set of ϕ^*.

To make this model more easier to solve, we develop the proposed model (3) in three aspects as follows. Firstly, since the derivative of $H(\phi)$ is nonzero only at the boundary of the region, we replace the region area $\int_\Omega (1 - H(\phi))dx$ by $\int_\Omega -\phi dx$. This term doesn't equal to the region area, but it has a similar effect to shrink the region size. Secondly, in order to regularize the shape of boundary during iterations, we can add the Total variation term without changing the optimizer. For some applications, it is also possible to identify some landmarks on the convex hull boundary, i.e. we know that the boundary of the convex hull must go through some points. We will present some techniques for identifying boundary landmarks in Subsect. 3.2. These landmarks information can also be Incorporated into the model. Let

$$F(\phi) = -\lambda_1\phi(x) + \lambda_2|\nabla H(\phi)|_2 + \sum_{k=1}^{K} \frac{\lambda_3}{2}\phi^2(x_k), \tag{4}$$

where x_k, $k = 1, 2, \ldots, K$ is the boundary landmarks. Then, the proposed model can be written as

$$\min_{\phi} \int_\Omega F(\phi)dx \tag{5}$$

$$\text{s.t. } |\nabla\phi(x)|_2 = 1, \ \triangle\phi(x) \geq 0, \ m(x)\phi(x) \leq 0 \text{ in } \Omega_1,$$

where $m(x)$ is the indicator function of X, and $\lambda_i > 0$, $i = 1, 2, 3$, are some constant parameters.

To solve (5), we introduce three auxiliary variables, $p = \triangle\phi$, $q = \nabla\phi$ and $z = m\phi$ with $p \geq 0, |q| = 1, z \leq 0$ in Ω_1. We also require ϕ to satisfy the following boundary conditions for the sake of computational simplicity.

$$\frac{\partial\phi}{\partial\mathbf{n}} = \frac{\partial\triangle\phi}{\partial\mathbf{n}} = 0 \quad \text{on} \quad \partial\Omega,$$

with $\mathbf{n}$ being the unit out-normal vector of $\partial\Omega$. Let's define three spaces

$$V = \left\{\phi \in H^2(\Omega) | \partial\phi/\partial\mathbf{n} = \partial\triangle\phi/\partial\mathbf{n} = 0 \quad \text{on } \partial\Omega\right\}, \tag{6}$$

$$V_1 = \left\{p \in H^1(\Omega) | \partial p/\partial\mathbf{n} = 0 \quad \text{on } \partial\Omega\right\}, \tag{7}$$

$$V_2 = \left\{q \in H^1(\Omega) \times H^1(\Omega) | \langle q, \mathbf{n}\rangle = 0 \quad \text{on } \partial\Omega\right\}. \tag{8}$$

Then we can obtain the augmented Lagrangian functional of (5) as

$$L(\phi, p, q, z, \gamma_1, \gamma_2, \gamma_3) = \int_\Omega F(\phi)dx + \langle\gamma_1, p - \triangle\phi\rangle + \langle\gamma_2, q - \nabla\phi\rangle$$

$$+ \langle\gamma_3, z - m\phi\rangle + \frac{\rho_1}{2}\|p - \triangle\phi\|_2^2 + \frac{\rho_2}{2}\|q - \nabla\phi\|_2^2 + \frac{\rho_3}{2}\|z - m\phi\|_2^2 \tag{9}$$

$$\text{s.t. } p \geq 0, |q| = 1, z \leq 0 \text{ in } \Omega_1,$$

where $\phi \in V$, $\gamma_1, p \in V_1$, $\gamma_2, q \in V_2$ and $\gamma_3, z \in L^2(\Omega)$.

2.3   The Algorithms for Proposed Models

To solve (9), we can use alternative direction method of multipliers (ADMM), see Algorithm 1. At each iteration of ADMM, where the number of iteration is denoted by t, we split the original problem into several sub-problems and update each variable separately.

Algorithm 1. Exact Convex Hull Algorithm

Input: A set of points X, a set of boundary landmarks, $\lambda_1, \lambda_2, \lambda_3, \rho_0$ and ρ_1. Maximum number of iteration M, and a threshold $\epsilon > 0$.

Output: ϕ

  Initialization : Let ϕ^1 to be the SDF of a polygon and $\phi^0 = 0$. Set $t = 1$.

1: **while** $t < M$ and $\frac{1}{|\Omega|} \int_\Omega |\phi^t(x) - \phi^{t-1}(x)| dx > \epsilon$ **do**

2:   $p^{t+1} = \arg \min_{p \in V_1} L(\phi^t, p, q^t, z^t, \gamma_1^t, \gamma_2^t, \gamma_3^t) = \max(0, \frac{\gamma_1^t}{\rho_1} + \triangle\phi^t)$ in Ω_1.

3:   $q^{t+1} = \arg \min_{q \in V_2} L(\phi^t, p^{t+1}, q, z^t, \gamma_1^t, \gamma_2^t, \gamma_3^t) = (\gamma_2^t + \rho_2\nabla\phi^t)/|\gamma_2^t + \rho_2\nabla\phi^t|$ in Ω_1.

4:   $z^{t+1} = \arg \min_{z \in L^2(\Omega)} L(\phi^t, p^{t+1}, q^{t+1}, z, \gamma_1^t, \gamma_2^t, \gamma_3^t) = \min(0, \gamma_3 + \frac{\rho_3}{\gamma_3})$.

5:   update ϕ by solving $\phi^{t+1} = \arg \min_{\phi \in V} L(\phi, p^{t+1}, q^{t+1}, z^{t+1}, \gamma_1^t, \gamma_2^t, \gamma_3^t)$.

6:   update γ_1, γ_2 and γ_3 by

$$\gamma_1^{t+1} = \gamma_1^t + \rho_1(\triangle\phi^{t+1} - p^{t+1}),$$
$$\gamma_2^{t+1} = \gamma_2^t + \rho_2(\nabla\phi^{t+1} - q^{t+1}),$$
$$\gamma_3^{t+1} = \gamma_3^t + \rho_3(m\phi^{t+1} - z^{t+1}).$$

7: **end while**

8: **return** $\{x|\phi(x) \leq 0\}$

The updates of $p, q, z, \gamma_1, \gamma_2$ and γ_3 are straightforward and easy to derive, so we only show the derivation of ϕ update here. By fixing other parameters, the ϕ update at the tth iteration is:

$$\phi^{t+1} = \arg \min_{\phi \in V} L(\phi, p^{t+1}, q^{t+1}, z^{t+1}, \gamma_1^t, \gamma_2^t, \gamma_3^t)$$

$$= \arg \min_{\phi \in V} \langle\gamma_1^t, \triangle\phi - p^{t+1}\rangle + \langle\gamma_2^t, \nabla\phi - q^{t+1}\rangle + \langle\gamma_3^t, m\phi - z^{t+1}\rangle$$

$$+ \frac{\rho_1}{2}\|\triangle\phi - p^{t+1}\|^2 + \frac{\rho_2}{2}\|\nabla\phi - q^{t+1}\|^2 + \frac{\rho_3}{2}\|m\phi - z^{t+1}\|^2 + \int_\Omega F(\phi)dx. \quad (10)$$

Using calculus of variations, we can derive

$$\triangle(\gamma_1^t + \rho_1(\triangle\phi^{t+1} - p^{t+1})) - \text{div}(\gamma_2^t + \rho_2(\nabla\phi^{t+1} - q^{t+1}))$$
$$+ m(\gamma_3 + \rho_3(\phi^{t+1} - z^{t+1})) + F'(\phi^{t+1}) = 0. \quad (11)$$

Approximately, (11) can be converted to solve the following biharmonic equations,

$$\rho_1 \triangle^2 \phi^{t+1} - \rho_2 \triangle \phi^{t+1}$$
$$= -F'(\phi^t) - \triangle(\gamma_1^t - \rho_1 p^{t+1}) + \mathrm{div}(\gamma_2^t - \rho_2 q^{t+1}) - m(\gamma_3 + \rho_3(\phi^t - z^{t+1})).$$
(12)

We denote right hand side of (12) as $rhs(\phi^t)$. Notice that it is a fourth order partial differential equation (PDE), so it is difficult to solve (12) efficiently. To make the computation easier, we add a proximity term $\frac{\rho_0}{2}\|\phi^{t+1} - \phi^t\|$ to (9), then (12) can be converted to

$$\begin{cases} \rho_1 \triangle^2 \phi^{t+1} - \rho_2 \triangle \phi^{t+1} + \rho_0 \phi^{t+1} = rhs(\phi^t) + \rho_0 \phi^t \\ \frac{\partial \phi}{\partial \mathbf{n}} = \frac{\partial \triangle \phi}{\partial \mathbf{n}} = 0 \text{ on } \partial\Omega \end{cases}, \quad (13)$$

where the right hand side of the first equation is denoted as $RHS(\phi^t)$ in the rest of this paper. By choosing $\rho_2 = 2\sqrt{\rho_0 \rho_1}$, we can separate the fourth order PDE (13) into two second order PDEs as follows:

$$\begin{cases} (\sqrt{\rho_1} \triangle - \sqrt{\rho_0})\psi^{t+1} \\ \frac{\partial \psi^{t+1}}{\partial \mathbf{n}} = 0 \text{ on } \partial\Omega, \end{cases} \quad (14)$$

$$\begin{cases} (\sqrt{\rho_1} \triangle - \sqrt{\rho_0})\phi^{t+1} = \psi^{t+1} \\ \frac{\partial \phi^{t+1}}{\partial \mathbf{n}} = 0 \text{ on } \partial\Omega. \end{cases} \quad (15)$$

We will discuss how to solve (14) and (15) efficiently in next section.

In (3), we require that ϕ must be non-positive at every point in X. However, when the set X contains noise or outliers, this constraint will not be appropriate. Instead, we can replace this constraint by a penalty term in the objective function, which will be able to handle the convex hull problem with outliers. The concrete model is omitted here due to page limitation, and only experiment results are present in Sect. 4.

3   Numerical Implementation

3.1   Discretization Scheme

In this section, we will give the efficient numerical implementation for Algorithm 1. Suppose our given binary image is of size $M \times N$, and then we denote the whole image domain as $\Omega_h = \{(m,n)|m = 1,\ldots,M, n = 1,\ldots,N\}$. In this model, we use the same discretization scheme with [17]. For the sake of completeness, we still present it here. For any feasible solution to (9) defined on Ω, we require that $\frac{\partial \phi}{\partial \mathbf{n}} = 0$ on $\partial\Omega$, so we extend Ω_h by 1 pixel on the boundary:

$$\phi(0,n) = \phi(1,n), \phi(M+1,n) = \phi(M,n), \quad (16)$$
$$\phi(m,0) = \phi(m,1), \phi(m,N+1) = \phi(m,N), \quad (17)$$

where $m = 1, \ldots, M$ and $n = 1, \ldots, N$. With this extension, we can define our gradient operator using forward difference:

$$\partial\phi(m, n) = \begin{bmatrix} \partial_x^+\phi(m, n) \\ \partial_y^+\phi(m, n) \end{bmatrix} = \begin{bmatrix} \phi(m + 1, n) - \phi(m, n) \\ \phi(m, n + 1) - \phi(m, n) \end{bmatrix}, \tag{18}$$

where $m = 1, \ldots, M$ and $n = 1, \ldots, N$. For the Laplacian operator, we can use central difference to approximate it:

$$\triangle\phi(m, n) = \partial_{x^2}^2\phi(m, n) + \partial_{y^2}^2\phi(m, n)$$
$$= \phi(m - 1, n) + \phi(m + 1, n) + \phi(m, n - 1) + \phi(m, n + 1) - 4\phi(m, n), \tag{19}$$

where $m = 1, \ldots, M$ and $n = 1, \ldots, N$. The divergence is approximated by

$$\text{div}(q)(m, n) = \partial_x^- q_1(m, n) + \partial_y^- q_2(m, n), \tag{20}$$

where $\partial_x^- q$ is defined as

$$\partial_x^- q(m, n) = \begin{cases} q(m, n) - q(m - 1, n) & 1 < m < M \\ q(1, n) & m = 1 \\ -q(M, n) & m = M, \end{cases} \tag{21}$$

for all $1 \le n \le N$, and $\partial_y^- q(m, n)$ is defined as

$$\partial_y^- q(m, n) = \begin{cases} q(m, n) - q(m, n - 1) & 1 < n < N \\ q(m, 1) & n = 1 \\ -q(m, N) & n = N, \end{cases} \tag{22}$$

for all $1 \le m \le M$. Notice that these discretization scheme satisfy $\text{div}(\partial\phi) = \triangle\phi$ in the discrete setting. Now the p, q, z and γ_i updates can be computed directly. Another thing we need to mention is the approximation of the Heaviside function $H(s)$ and its derivative $\delta(s)$. For numerical purposes, we use

$$H_a(s) = \frac{1}{2} + \frac{1}{\pi} \arctan(\frac{s}{a}) \quad \text{and} \quad \delta_a(s) = \frac{a}{a^2 + s^2}, \tag{23}$$

where a is a small positive number.

To solve the Laplacian equations (14) and (15), we can use the discrete cosine transform (DCT) [19]. Let's denote the DCT of sequence v by $\hat{v}$. Then, by some calculation, we can get

$$\widehat{\nabla_x^2\phi}(m, n) = 2\big(\cos\big(\frac{\pi(m - 1)}{M}\big) - 1\big)\hat{\phi}(m, n), \tag{24}$$

$$\widehat{\nabla_y^2\phi}(m, n) = 2\big(\cos\big(\frac{\pi(n - 1)}{N}\big) - 1\big)\hat{\phi}(m, n). \tag{25}$$

After discretizing (14) and (15), we can do DCT on both sides of them, and we have

$$r(m, n)\widehat{\psi^{t+1}}(m, n) = \widehat{RHS}(\phi^t)(m, n) \tag{26}$$

$$r(m, n)\widehat{\phi^{t+1}}(m, n) = \widehat{\psi^{t+1}}(m, n). \tag{27}$$

where $r(m,n) = [2\sqrt{\rho_1}(\cos(\frac{\pi(m-1)}{M}) + \cos(\frac{\pi(n-1)}{N}) - 2) - \sqrt{\rho_0}]$. Then $\widehat{\phi^{t+1}}(m,n)$ can be computed by

$$\widehat{\phi^{t+1}}(m,n) = \widehat{RHS}(\phi^t)(m,n)/r^2(m,n). \qquad (28)$$

After having $\widehat{\phi^{t+1}}$, we just need to do an inverse transform to get ϕ^{t+1}.

3.2   Initialization and Determination of Landmarks

A good initial value can make the algorithm converge faster. For our proposed model (5), We initialize ϕ to be the SDF of a polygon determined by the boundary landmarks.

Given a set of points X in $xy-$plane we can easily identify the points with the largest and the smallest x value and y value, denoted by r_1, l_1, t_1, b_1. Then we can rotate the $xy-$ axes by a small angle $\theta_0 = \pi/k$ for some positive integer k, and do the previous step to get r_2, l_2, t_2, b_2. We repeat this operation k times and get a set of landmarks V_L. In fact, the points in V_L must lie on the boundary of Conv(X), i.e., we can use the points in V_L as the boundary landmarks. To see this, we need to prove Theorem 2 first.

Theorem 2. *Given a set of 2-D points X with finite number of points, the point on the top, i.e, the point with largest y value, must lie on the boundary of the convex hull of X.*

Proof. Let's prove this by contradiction. Suppose the top point $v_t = (x_t, y_t)$ does not lie on the boundary of Conv(X), then it must be an interior point of Conv(X). Therefore, we can find a ball $B(v_t, \epsilon)$ centered at v_t with radius equals to $\epsilon > 0$ such that $B(v_t, \epsilon) \subset$ Conv(X). Consider the half-plane

$$\mathbb{H} = \{(x,y)|y \le y_t\}.$$

We know $X \subset \mathbb{H}$, since v_t is the top point in X. Then we take the intersection

$$C = \mathbb{H} \cap \text{Conv}(X).$$

Since Conv(X) and $\mathbb{H}$ are convex, C is also convex. Since X is a subset of Conv(X) and $\mathbb{H}$, C also contains all the points in X.

For the ball $B(v_t, \epsilon)$, the top half of it is in Conv(X) but not in $\mathbb{H}$, so we have

$$\text{Area}(C) \le \text{Area}(\text{Conv}(X)) - \pi\epsilon^2/2 < \text{Area}(\text{Conv}(X)).$$

Therefore, we find a convex region containing X with smaller area than Conv(X), which is a contradiction.

For any point v_i in V_L, we can always rotate the coordinates by some degree to make that point lie on the top of X, so v_i must lie on the boundary of Conv(X). Since some points may occur several times in V_L, we take the SDF of Conv(V_L) as the initial ϕ.

4   Numerical Experiment

4.1   Exact Convex Hull Model

In this part, we choose 9 pictures from [20] and use thresholding to obtain binary images. In this set of experiment, we use the same parameters for all images (Fig. 2). We set $\rho_1 = 15$, $\rho_3 = 1$, $\rho_0 = 1$, $\lambda_1 = 0.015$, $\lambda_2 = 5$, $\lambda_3 = 10$, $a = 0.5$ and $\epsilon = 0.005$. To initialize ϕ, we find 8 vertices for V_L and let ϕ^1 be the SDF of $\text{Conv}(V_L)$. The convex hulls found by Algorithm 1 are shown in Fig. 2.

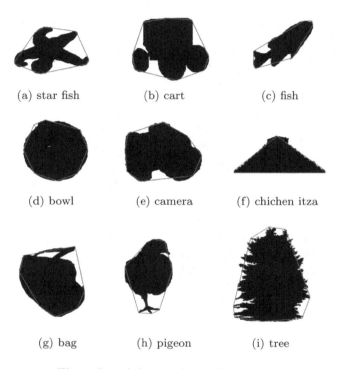

(a) star fish   (b) cart   (c) fish

(d) bowl   (e) camera   (f) chichen itza

(g) bag   (h) pigeon   (i) tree

Fig. 1. Initial shapes of ϕ in Algorithm 1

From the results, we can see the convex hulls given by our proposed algorithm can enclose the whole object accurately and the convexity of regions are also guaranteed. To measure the accuracy of our convex hulls, we use the relative distance measure: [21]

$$\text{err}(C_2) = \frac{\text{dist}_H(C_1, C_2)}{D(C_1)} \quad \text{and} \quad D(C_1) = 2\sqrt{\frac{area(C_1)}{\pi}}, \tag{29}$$

where C_2 is the convex hull found by our proposed algorithm, C_1 is the convex hull found by the benchmark algorithm and dist_H is the Hausdorff distance:

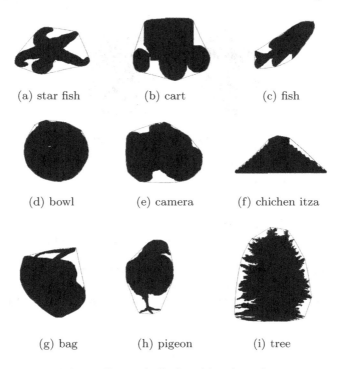

(a) star fish   (b) cart   (c) fish

(d) bowl   (e) camera   (f) chichen itza

(g) bag   (h) pigeon   (i) tree

Fig. 2. Convex hulls found by algorithm

$$\text{dist}_H(C_1, C_2) = \max\{\sup\{\text{dist}(x, C_2)|x \in C_1\}, \sup\{\text{dist}(y, C_1)|y \in C_2\}\}. \quad (30)$$

In this experiment, we used the quick hull algorithm [7] as bench mark. $D(C)$ is the equivalent diameters of C. The relative distance errors are shown in Table 1. From the results, we can see that our proposed algorithm can find the convex hull of given objects with very small error.

4.2 Convex Hull Model for Outliers

As we mentioned before, when the given data contains outliers, we don't require the curve to enclose all the given points. Instead, we allow our model to exclude some outliers when it is too "expensive" to enclose them. To see whether our modified model work, we test it on the tree image with simulated outliers.

Table 1. The relative errors of Algorithm 1

Name	star_fish	cart	fish	bowl	camera	chichen_itza	bag	pigeon	tree
Error	1.75%	1%	1.68%	0.52%	1.45%	2.74%	0.44%	1.13%	0.68%

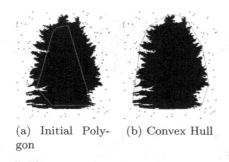

(a) Initial Polygon   (b) Convex Hull

Fig. 3. The results of the modified model

The noisy binary image with the initial curve and the convex hull are shown in Fig. 3. We also compute the relative error using (29), and the error is only 1.37%. From the results, we see that our proposed model can correctly filter out most of outliers and find the convex hull with small error. Though the error is larger than Table 1, it is still acceptable. What's more, we didn't find any other convex hull algorithm can achieve this good approximation under the existence of outliers.

5   Conclusion

Convex hull problems have been widely studied for many years and various efficient algorithms have been proposed. However, no existing method can handle the cases with outliers. In this paper, we proposed new convex hull models based on level set representation, which can either find the exact convex hull or convex hull without outliers. When using the modified algorithm to approximate the convex hull, by choosing appropriate parameters, we can filter out the outliers accurately. However, one limitation of our proposed algorithms is that we can only find the convex hull of one object. When the image contains several objects, we may want to identify the convex hulls for each of the objects. In our future work, we will try to modify our model in order to find several convex hulls simultaneously.

Acknowledgments. XC Tai would like to acknowledge the support from HKBU startup grant, RG(R)- RC/17-18/02-MATH, and FRG2/17-18/033.

References

1. Liparulo, L., Proietti, A., Panella, M.: Fuzzy clustering using the convex hull as geometrical model. Adv. Fuzzy Syst. **2015**, 6 (2015)
2. Hert, S., Lumelsky, V.: Motion planning in $\mathbf{R}^3$ for multiple tethered robots. IEEE Trans. Robot. Autom. **15**(4), 623–639 (1999)
3. Tomic, T., Ott, C., Haddadin, S.: External wrench estimation, collision detection, and reflex reaction for flying robots. IEEE Trans. Robot. **33**(6), 1467–1482 (2017)

4. Condat, L.: A convex approach to k-means clustering and image segmentation. In: Pelillo, M., Hancock, E. (eds.) EMMCVPR 2017. LNCS, vol. 10746, pp. 220–234. Springer, Cham (2018). https://doi.org/10.1007/978-3-319-78199-0_15

5. Jarvis, R.A.: On the identification of the convex hull of a finite set of points in the plane. Inf. Process. Lett. **2**, 18–21 (1973)

6. Graham, R.: An efficient algorithm for determining the convex hull of a finite planar set. Inf. Process. Lett. **26**, 132–133 (1972)

7. Barber, C.B., Dobkin, D.P., Huhdanpaa, H.: The quickhull algorithm for convex hulls. ACM Trans. Math. Softw. **22**(4), 469–483 (1996)

8. Preparata, F.P., Hong, S.J.: Convex hulls of finite sets of points in two and three dimensions. Commun. ACM **20**(2), 87–93 (1977)

9. Andrew, A.M.: Another efficient algorithm for convex hulls in two dimensions. Inf. Process. Lett. **9**(5), 216–219 (1979)

10. Bentley, J., Preparata, F., Faust, M.: Approximation algorithms for convex hulls. Commun. ACM **25**(1), 64–68 (1982)

11. Kavan, L., Kolingerova, I., Zara, J.: Fast approximation of convex hull. ACST **6**, 101–104 (2006)

12. Krvr, C.E., Ivan, S.: Sequential and parallel approximate convex hull algorithms. Comput. Artif. Intell. **14**(6), 597–610 (1995)

13. Hazra, A., Deb, K., Kundu, S., Hazra, P.: Shape oriented feature selection for tomato plant identification. Int. J. Comput. Appl. Technol. Res. **2**(4), 449–454 (2013)

14. Chan, T., Vese, L.: An active contour model without edges. In: Nielsen, M., Johansen, P., Olsen, O.F., Weickert, J. (eds.) Scale-Space 1999. LNCS, vol. 1682, pp. 141–151. Springer, Heidelberg (1999). https://doi.org/10.1007/3-540-48236-9_13

15. Tai, X.-C., Duan, J.: A simple fast algorithm for minimization of the elastica energy combining binary and level set representations. Int. J. Numer. Anal. Model. **14**(6), 809–821 (2017)

16. Yan, S., Tai, X.-C, Liu, J., Huang, H.-Y: Convexity shape prior for level set based image segmentation method. arXiv preprint arXiv:1805.08676 (2018)

17. Luo, S., Tai, X.-C.: Convex shape priors for level set representation. arXiv preprint arXiv:1811.04715 (2018)

18. o'Rourke, J.: Computational Geometry in C. Cambridge University Press, Cambridge (1998)

19. Strang, G.: The discrete cosine transform. SIAM Rev. **41**(1), 135–147 (1999)

20. Alpert, S., Galun, M., Basri, R., Brandt, A.: Image segmentation by probabilistic bottom-up aggregation and cue integration. In: Proceedings of the IEEE Conference on Computer Vision and Pattern Recognition, June 2007

21. Rufai, R.A.: Convex Hull Problems. Ph.D. thesis (2015)

PDE Evolutions for M-Smoothers: From Common Myths to Robust Numerics

Martin Welk[1][✉] and Joachim Weickert[2]

[1] Institute of Biomedical Image Analysis,
Private University of Health Sciences, Medical Informatics and Technology,
Eduard-Wallnöfer-Zentrum 1, 6060 Hall/Tyrol, Austria
martin.welk@umit.at
[2] Mathematical Image Analysis Group, Campus E1.7, Saarland University,
66041 Saarbrücken, Germany
weickert@mia.uni-saarland.de

Abstract. Local M-smoothers constitute an interesting and important class of image processing techniques with many connections to other methods. In our paper we derive a family of partial differential equations (PDEs) that result as limiting processes from M-smoothers which are based on local order-p means within a disc the radius of which tends to zero. The order p may take any nonzero value > -1. Thus, we also allow negative values which have never been considered before. In contrast to results from the literature, we show in the space-continuous case that mode filtering does not arise for $p \to 0$, but for $p \to -1$. Extending our filter class to p-values smaller than -1 allows to include e.g. the classical image sharpening flow of Gabor. Since our PDE class is highly anisotropic and may contain backward parabolic operators, designing adequate numerical methods is difficult. We present an L^∞-stable explicit finite difference scheme that satisfies a discrete maximum–minimum principle, is fairly efficient, and offers excellent rotation invariance. Although it solves parabolic PDEs, it makes consequent use of stabilisation concepts from the numerics of hyperbolic PDEs. Our experiments show that the PDEs for $p < 1$ are of specific interest: Their backward parabolic term creates favourable sharpening properties, while they appear to maintain the strong shape simplification properties of mean curvature motion.

Keywords: M-smoother · Partial differential equation · Mode filter · Backward parabolic operator · Anisotropy · Finite difference methods · Shape analysis

1 Introduction

M-Estimators. It has been observed long ago by Legendre [15] and Gauß [8] that the mean of a finite multiset $\mathcal{X} = \{a_1, a_2, \ldots, a_n\}$ of real numbers can be described as the minimiser of the sum of squared distances to the given numbers:

$$\text{mean}(\mathcal{X}) = \operatorname*{argmin}_{\mu \in \mathbb{R}} \sum\nolimits_{i=1}^{n} (\mu - a_i)^2 \,. \tag{1}$$

© Springer Nature Switzerland AG 2019
J. Lellmann et al. (Eds.): SSVM 2019, LNCS 11603, pp. 236–248, 2019.
https://doi.org/10.1007/978-3-030-22368-7_19

Likewise it has been noted by Jackson [13] that the median of $\mathcal{X}$ minimises the sum of absolute distances:

$$\text{median}(\mathcal{X}) = \underset{\mu \in \mathbb{R}}{\text{argmin}} \sum_{i=1}^{n} |\mu - a_i| \, . \tag{2}$$

This can be generalised to the notion of *order-p means* given by

$$\text{mean}_p(\mathcal{X}) := \underset{\mu \in \mathbb{R}}{\text{argmin}} \sum_{i=1}^{n} |\mu - a_i|^p \tag{3}$$

for any $p > 0$, with $\text{mean}_2 \equiv \text{mean}$, $\text{mean}_1 \equiv \text{median}$. This concept is introduced by Jackson [13] for $p > 1$ (as a means to disambiguate the median by $p \to 1^+$) whereas Barral Souto [2] is interested in the order-p means in their own right for general $p > 0$. In robust statistics, order-p means belong to the class of *M-estimators* [12].

Including the limiting case of the monomials as $|z|^0 = 0$ for $z = 0$, and 1 otherwise, [2] also extends the definition (3) to the case $p = 0$ for which the *mode* of $\mathcal{X}$, i.e. its most frequent value, is obtained. As also noted in [2], the limit $p \to \infty$ yields what is also called the mid-range value, i.e. the arithmetic mean of the extremal values of $\mathcal{X}$.

It is straightforward to rewrite the definition of order-p means for continuous distributions (densities) on $\mathbb{R}$ just by replacing sums with integrals: Let $\gamma : \mathbb{R} \to \mathbb{R}_0^+$ be a density (integrable in a suitable sense), then one defines

$$\text{mean}_p(\gamma) := \underset{\mu \in \mathbb{R}}{\text{argmin}} \int_{-\infty}^{\infty} \gamma(z) |\mu - z|^p \, \mathrm{d}z \, . \tag{4}$$

This notion of continuous order-p means has been considered by Fréchet [6] for $p > 1$.

M-Smoothers. In image processing, M-estimators are commonly used to build local image filters, see [22] for the median filter (in signal processing) and [21] for order-p means with $p > 0$. In a local filter, one takes at each location the grey-values from a neighbourhood (selection step) and computes some common value of these (aggregation step) that is assigned to the location in the filtered signal, see e.g. [5,9]. These filters can be iterated to generate a series of progressively processed images.

It has been noticed since long that these filters behave similar to certain image filters based on partial differential equations (PDEs). Mean filters are a spatial discretisation of linear diffusion. As proven in [11], iterated median filtering approximates mean curvature motion [1]. In [9], also the case of the mode filter (associated with $p = 0$) is considered.

Our Contribution. We derive a family of PDEs associated with M-smoothers based on order-p means with variable p and vanishing disc radius. In contrast to results from the literature, we also permit negative p-values with $p > -1$. Reconsidering the relation of order-p means and their corresponding PDEs to existing image filters, we show that in the space-continuous setting the mode filter does *not* arise for $p \to 0$, as is commonly assumed [9], but for $p \to -1$. Extending our PDE family to values $p < -1$ allows to cover also the sharpening Gabor flow [7,16], for which no M-smoothing counterpart is known. In spite of the fact that our PDE family is anisotropic and may even involve backward parabolic operators, we design an L^∞-stable numerical scheme with very good rotation invariance. Our experiments show that the PDEs for $p < 1$ are particularly attractive since they simultaneously allow image sharpening and shape simplification.

Structure of the Paper. In Sect. 2 we present our theory that allows us to derive PDE evolutions from M-smoothers. Our numerical algorithm is discussed in Sect. 3, and Sect. 4 is devoted to an experimental evaluation. Our paper is concluded with a summary in Sect. 5.

2   M-Smoothers, Mode and Partial Differential Equations

In this section, we derive PDEs for M-smoothers and the mode filter.

Generalised Order-p Means. In the following, M-smoothers are based on order-p means with $p > -1$, $p \neq 0$. As this range for p goes beyond the usual $p > 0$, let us first extend the definition of order-p means of continuous-scale distributions accordingly.

Definition 1. *Let z be a real random variable with the bounded, piecewise continuous density $\gamma : \mathbb{R} \to \mathbb{R}$. For $p \in (-1, +\infty) \setminus \{0\}$, define the order-$p$ mean of γ as*

$$\mathrm{mean}_p(\gamma) = \underset{\mu \in \mathbb{R}}{\mathrm{argmin}} \int_{\mathbb{R}} \gamma(\mu) \, \mathrm{sgn}(p) |\mu - z|^p \, \mathrm{d}z \ . \tag{5}$$

As $|z|^p$ is monotonically increasing on $\mathbb{R}_0^+$ for $p > 0$, but monotonically decreasing on $\mathbb{R}^+$ for $p < 0$, the $\mathrm{sgn}(p)$ factor in (5) ensures that in both cases an increasing penalty function is used.

For $p > 0$ the requirement of continuity of γ in Definition 1 can be relaxed; by modelling a discrete density as a weighted sum of delta peaks, the discrete order-p means as in [2] can be included in this definition.

The continuity is, however, essential for $p < 0$: In this case, the penalty function has a pole at $z = 0$ such that an improper integral is obtained; for $p > -1$ this integral exists provided that γ is continuous, i.e. no delta peaks are allowed. In particular, we cannot define an order-p mean with $-1 < p < 0$ for discrete distributions as considered in [2].

PDE Approximation Results. The proofs of the following propositions are given at the end of this section. The first proposition contains our first main result.

Proposition 1. *Let a smooth image $u : \mathbb{R}^2 \to \mathbb{R}$ be given, and let $\boldsymbol{x}_0 = (x_0, y_0)$ be a regular point, $|\boldsymbol{\nabla} u(\boldsymbol{x}_0)| > 0$. One step of order-p mean filtering of u with a disc-shaped window $\mathrm{D}_\varrho(\boldsymbol{x})$ and $p > -1$, $p \neq 0$ approximates for $\varrho \to 0$ a time step of size $\tau = \varrho^2 / (2p + 4)$ of an explicit time discretisation of the PDE*

$$u_t = u_{\xi\xi} + (p - 1)\, u_{\eta\eta} \tag{6}$$

where η and ξ are geometric coordinates referring at each image location to the direction of the positive gradient, and the level-line direction, respectively:

$$\mathrm{mean}_p \{ u(x, y) \mid (x, y) \in \mathrm{D}_\varrho(x_0, y_0) \} - u(x_0, y_0)$$

$$= \frac{\varrho^2}{2(p + 2)} \big(u_{\xi\xi}(x_0, y_0) + (p - 1) u_{\eta\eta}(x_0, y_0) \big) + \mathcal{O}(\varrho^{(\min\{p,0\}+5)/2}) \,. \tag{7}$$

At a local minimum (maximum) of u, i.e. $\boldsymbol{x}_0$ with $|\boldsymbol{\nabla} u(\boldsymbol{x}_0)| = 0$ where the Hessian $\mathrm{D}^2 u(\boldsymbol{x}_0)$ is positive (negative) semidefinite, the same filtering step fulfils for $\varrho \to 0$ the inequality $\mathrm{mean}_p \{ u(x, y) \mid (x, y) \in \mathrm{D}_\varrho(x_0, y_0) \} - u(x_0, y_0) \geq 0$ (≤ 0), thus approximates an evolution $u_t \geq 0$ ($u_t \leq 0$).

The approximation order in (7) is $\mathcal{O}(\varrho^{1/2})$ for positive p but reduces to $\mathcal{O}(\varrho^{(p+1)/2})$ for negative p.

For $p = 2$ and $p = 1$ the proposition yields the same PDEs as [9] except for a time rescaling which is due to the choice of a Gaussian window in [9].

Under analogous assumptions as in Proposition 1, one can also derive the PDE limit for the mode filter, where the mode is not obtained by a minimisation in the sense of (4) but directly as the maximum of the density of values in $\{ u(x, y) \mid (x, y) \in \mathrm{D}_\varrho(x_0, y_0) \}$.

Proposition 2. *Let u and $\boldsymbol{x}$ be as in Proposition 1. One step of mode filtering of u with a disc-shaped window $\mathrm{D}_\varrho(\boldsymbol{x})$ approximates for $\varrho \to 0$ a time step of size $\tau = \varrho^2 / 2$ of an explicit time discretisation of the PDE $u_t = u_{\xi\xi} - 2u_{\eta\eta}$ with ξ, η as in Proposition 1. At a local minimum (maximum), mode filtering approximates $u_t \geq 0$ ($u_t \leq 0$).*

The PDE for mode filtering coincides with the one given in [9], again up to time rescaling. We see, however, that (7) for $p \to 0$ does not yield the PDE from Proposition 2 but $u_t = u_{\xi\xi} - u_{\eta\eta}$. Instead, the mode filtering PDE is obtained for $p \to -1$. Inserting $p = -2$ into (7) yields $u_t = u_{\xi\xi} - 3u_{\eta\eta}$ which was stated as an image sharpening PDE in [7,16].

Proof (of Proposition 1). Assume w.l.o.g. that the regular location is $\boldsymbol{x}_0 = \boldsymbol{0} = (0, 0)$ with $u(0, 0) = 0$, and that the gradient of u at $(0, 0)$ is in the positive x direction, i.e. $u_x > 0$, $u_y = 0$. Setting $\alpha = u_x > 0$ and substituting $x = \varrho\xi$,

$y = \varrho\eta$, $u(x,y) = \varrho\alpha\omega(\xi,\eta)$, we can use Taylor expansion of ω up to third order to write for $(\xi,\eta) \in \mathrm{D}_1 \equiv \mathrm{D}_1(\mathbf{0})$

$$\omega(\xi,\eta) = \xi + \beta\xi^2\varrho + \gamma\xi\eta\varrho + \delta\eta^2\varrho + \varepsilon_0\xi^3\varrho^2 + \varepsilon_1\xi^2\eta\varrho^2 + \varepsilon_2\xi\eta^2\varrho^2 + \varepsilon_3\eta^3\varrho^3 + \mathcal{O}(\varrho^3) \ . \tag{8}$$

The sought M-smoother value $\mu = \varrho^2\alpha\kappa$ then corresponds to an extremum (minimum for $p > 0$, maximum for $p < 0$) of an integral in the unit disc D_1,

$$E(\kappa) = \iint_{\mathrm{D}_1} |\omega - \kappa\varrho|^p \, \mathrm{d}\eta \, \mathrm{d}\xi \tag{9}$$

For sufficiently small ϱ, this integral can be rewritten as

$$E(\kappa) = \int_{-1}^{1} W(\xi) \, |\omega(\xi,0) - \kappa\varrho|^p \, \mathrm{d}\xi + \mathcal{O}(\varrho^3) \ . \tag{10}$$

Herein, $W(\xi)$ essentially consists for each ξ of an integral along the level line of ω going through $(\xi,0)$. This integral measures the density of the value $\omega(\xi,0)$ in the overall distribution of ω values within D_1. Describing the level line by a function $\tilde{\xi}(\eta)$ on $[\eta_-^*(\xi), \eta_+^*(\xi)]$ where $\eta_\pm^*$ correspond to its end points on the boundary of D_1, $W(\xi)$ reads as

$$W(\xi) = \frac{\partial\omega}{\partial\xi}(\xi,0) \, V(\xi) \ , \qquad V(\xi) = \int_{\eta_-^*(\xi)}^{\eta_+^*(\xi)} \left(\frac{\partial\omega}{\partial\xi}(\tilde{\xi}(\eta),\eta) \right)^{-1} \mathrm{d}\eta \ . \tag{11}$$

By straightforward but lengthy calculations one obtains

$$\tilde{\xi}(\eta) = \xi - (\gamma\xi + \delta\eta)\eta\varrho$$
$$+ \left((2\beta\xi + \gamma\eta)(\gamma\xi + \delta\eta) - \varepsilon_1\xi^2 - \varepsilon_2\xi\eta - \varepsilon_3\eta^2 \right)\eta\varrho^2 + \mathcal{O}(\varrho^3) \ , \tag{12}$$

$$\eta_\pm^* = \pm\sqrt{1 - \xi^2} + \left(\gamma\xi^2 \pm \delta\xi\sqrt{1 - \xi^2} \right)\varrho + \left(\chi(\xi) \pm \psi(\xi)\sqrt{1 - \xi^2} \right)\varrho^2 + \mathcal{O}(\varrho^3) \tag{13}$$

where $\psi(\xi) = \psi_0 + \psi_1\xi + \psi_2\xi^2 + \psi_3\xi^3$ and $\chi(\xi)$ are polynomials in ξ of order 3 and 4, respectively, the exact coefficients of which are not further needed. Putting things together, one finds

$$W(\xi) = \left((w_{0,0} + w_{0,2}\varrho^2) + w_1\varrho\xi + w_2\varrho^2\xi^2 + w_3\varrho^2\xi^3 \right)\sqrt{1 - \xi^2} + \mathcal{O}(\varrho^3) \tag{14}$$

with

$$w_{0,0} = 2 \ , \qquad w_{0,2} = \frac{4}{3}\beta\delta + \frac{2}{3}\gamma^2 - \frac{2}{3}\varepsilon_2 + 2\psi_0 \ , \qquad w_1 = 2\delta + 2\psi_1\varrho \ ,$$
$$w_2 = -2\gamma^2 - \frac{4}{3}\beta\delta - \frac{2}{3}\gamma^2 + \frac{2}{3}\varepsilon_2 + 2\psi_2 \ , \qquad w_3 = 2\psi_3 \ . \tag{15}$$

To evaluate the outer integral (10), its integration interval $[-1,1]$ is split into four parts

$$E(\kappa) = \left(\int_{-1}^{-\sqrt{\varrho}} + \int_{-\sqrt{\varrho}}^{\nu\varrho} + \int_{\nu\varrho}^{\sqrt{\varrho}} + \int_{\sqrt{\varrho}}^{1} \right) W(\xi) \, |\omega(\xi,0) - \kappa\varrho|^p \, \mathrm{d}\xi + \mathcal{O}(\varrho^3) \tag{16}$$

where $\omega(\nu\varrho) = \kappa\varrho$. Herein, the modulus in the integrand can be resolved to $\pm(\omega(\xi, 0) - \kappa\varrho)$ in each part. Substituting further ξ to $-\xi$, $-\xi + \nu\varrho$, $\xi - \nu\varrho$, ξ, respectively, in the four intervals, all integrals can be reduced up to higher order terms to combinations of standard integrals $\int \xi^q \, d\xi$ (in the inner two intervals) or $\int \xi^q \sqrt{1 - \xi^2} \, d\xi$ (in the outer two intervals) with several q, and by combining all these one arrives at

$$E(\kappa) = \mathrm{const}(\kappa) + \mathcal{O}\left(\varrho^{\min\{(p+5)/2, 5/2\}}\right)$$
$$+ \left(-\frac{-4p(p-1)}{p+1}\beta\varrho^2 S - \frac{2p}{p+1}\delta\varrho^2 S\right)\kappa + \left(\frac{2(p+2)p}{p+1}\varrho^2 S\right)\kappa^2 \quad (17)$$

where the common constant S is defined by the definite integral

$$S = \int_{\arcsin\sqrt{\varrho}}^{\pi/2} \sin^{p+2}\varphi \, d\varphi. \quad (18)$$

The apex of the quadratic function in (17) yields the desired extremum of E, which is indeed a minimum for $p > 0$ and a maximum for $p < 0$:

$$\kappa = \frac{p-1}{p+1}\beta + \frac{1}{p+1}\delta + \mathcal{O}(\varrho^{(\min\{p,0\}+1)/2}). \quad (19)$$

Reverting our initial substitutions yields the claim of the proposition for regular points.

The inequalities for local minima (maxima) are obvious consequences of the fact that for any $\varrho > 0$ the mean-p filter value is in the convex hull of values $u(\boldsymbol{x})$, $\boldsymbol{x} \in \mathrm{D}_\varrho(\boldsymbol{x}_0)$. $\qquad\square$

Proof (of Proposition 2). With the same substitutions as in the previous proof, the mode of ω is given by the maximiser of $V(\xi)$. By a slight modification of the calculations of the previous proof one finds

$$V(\xi) = 2\left(1 + \delta\xi\varrho - 2\beta\xi\varrho\right)\sqrt{1 - \xi^2} + \mathcal{O}(\xi^2\varrho^2). \quad (20)$$

Equating $V'(\xi)$ to zero yields $\omega(\xi) = (\delta - 2\beta)\varrho + \mathcal{O}(\varrho^2)$ for the mode. For local minima (maxima), the same reasoning as in the previous proof applies. $\qquad\square$

PDE Evolutions. Propositions 1 and 2 state PDEs approximated by the respective M-smoothers at regular points, and inequalities that are valid at local minima and maxima. Let us briefly discuss how these results determine uniquely the evolutions of the entire image u (including critical points) approximated by the M-smoothers. To this end, notice first that in a connected critical region (a closed set in $\mathbb{R}^2$ consisting entirely of critical points, with nonempty interior), the inequalities for minima and maxima together imply $u_t = 0$.

An isolated critical point or a curve consisting of critical points, in contrast, is under the influence of the regular points surrounding it, and as long as the sign of the evolution speed u_t at the critical point and the surrounding regular points is of the same sign, the critical point will by a "sliding regime" be forced to adapt to the evolution speed of its regular surrounds. (A more detailed analysis will be given in a forthcoming paper.) If, however, the speed at the critical point is of opposite sign compared to all its surrounding regular points, they will coalesce to create a growing plateau which by the above remarks locks in at $u_t = 0$.

Without a detailed analysis which will be given in a forthcoming paper let us remark that the delimiter that forces u_t to 0 at some extrema never takes effect for $p \geq 1$ as in this case one has $u_t > 0$ ($u_t < 0$) in the surrounding regular points of minima (maxima). For $p < 1$ the lock-in at $u_t = 0$ first affects extrema where the Hessian of u is highly anisotropic (including curves consisting of extrema). As p decreases toward 0, this behaviour gradually extends to less anisotropic extrema. For $p < 0$, even isotropic critical points enter this regime such that now all local extrema develop into plateaus.

3   An L^∞-stable Numerical Scheme for the PDE Limit

Our PDE limit (6) creates two major numerical difficulties:

- It involves the anisotropic expressions $u_{\xi\xi}$ and $u_{\eta\eta}$. To reproduce their qualitative properties adequately, one has to take care that the discretisation approximates rotationally invariant behaviour well and that it satisfies a discrete maximum–minimum principle which prevents over- und undershoots.
- For $p < 1$, the sign in front of the operator $u_{\eta\eta}$ becomes negative, which results in a backward parabolic operator. Such operators are known to be ill-posed. They require additional stabilisation in the model and the numerics.

These challenges show that great care must be invested in the design of appropriate numerical algorithms. Thus, let us have a deeper look into our efforts along these lines.

Using $u_{\eta\eta} = \Delta u - u_{\xi\xi}$ and $u_{\xi\xi} = \mathrm{curv}(u)|\nabla u|$ with the isophote curvature $\mathrm{curv}(u)$ we rewrite (6) in a numerically more convenient form:

$$u_t = (2-p)\,\mathrm{curv}(u)|\nabla u| + (p-1)\,\Delta u. \tag{21}$$

If $p \geq 1$, we apply this equation in all locations, including extrema.

For $p < 1$, the second term describes backward diffusion, which we stabilise by freezing its action in extrema where $|\nabla u|$ vanishes:

$$u_t = (2-p)\,\mathrm{curv}(u)|\nabla u| + (p-1)\,\mathrm{sgn}(|\nabla u|)\,\Delta u. \tag{22}$$

In practice, our image domain is finite and of rectangular size. This motivates us to equip the Eqs. (21) and (22) with reflecting boundary conditions.

Both evolutions are replaced by explicit finite difference schemes on a regular grid of size h in x- and y-dimension and time step size τ. By $u_{i,j}$ we denote an approximation of u in pixel (i, j).

If $p \geq 1$, we discretise Δu in (21) with a nine-point stencil that averages an approximation aligned along the x- and y-axis with one aligned along the diagonal directions:

$$\frac{1}{2}\frac{1}{h^2}\begin{array}{|c|c|c|}\hline 0 & 1 & 0 \\\hline 1 & -4 & 1 \\\hline 0 & 1 & 0 \\\hline\end{array} + \frac{1}{2}\frac{1}{(\sqrt{2}\,h)^2}\begin{array}{|c|c|c|}\hline 1 & 0 & 1 \\\hline 0 & -4 & 0 \\\hline 1 & 0 & 1 \\\hline\end{array} = \frac{1}{4h^2}\begin{array}{|c|c|c|}\hline 1 & 2 & 1 \\\hline 2 & -12 & 2 \\\hline 1 & 2 & 1 \\\hline\end{array}. \tag{23}$$

This guarantees that all four principal grid directions are treated equally. Our experiments will also show that in this way, rotation invariance is approximated well.

For $p < 1$, the term $(p-1)\,\text{sgn}(|\nabla u|)\,\Delta u$ in (22) creates stabilised backward diffusion. Here we base our finite difference approximation on a minmod discretisation of Osher and Rudin [19], but average it again with its counterpart along the diagonal directions to guarantee equal treatment of the four principal grid directions. We denote the forward differences in x-, y-, and the diagonal directions $\mathbf{d} = (1,1)$ and $\mathbf{e} = (1,-1)$ by

$$u^x_{i,j} := \frac{u_{i+1,j} - u_{i,j}}{h}, \qquad u^y_{i,j} := \frac{u_{i,j+1} - u_{i,j}}{h}, \tag{24}$$

$$u^d_{i,j} := \frac{u_{i+1,j+1} - u_{i,j}}{\sqrt{2}\,h}, \qquad u^e_{i,j} := \frac{u_{i+1,j-1} - u_{i,j}}{\sqrt{2}\,h}, \tag{25}$$

and abbreviate the minmod function by

$$\text{M}(a,b,c) := \begin{cases} a & \text{if } |a| = \min\{|a|, |b|, |c|\}, \\ b & \text{if } |b| = \min\{|a|, |b|, |c|\}, \\ c & \text{if } |c| = \min\{|a|, |b|, |c|\}. \end{cases} \tag{26}$$

With these notations we approximate $\text{sgn}(|\nabla u|)\,\Delta u$ in pixel (i,j) by

$$\begin{aligned} &\tfrac{1}{2}\tfrac{1}{h}\big(\text{M}(u^x_{i+1,j}, u^x_{i,j}, u^x_{i-1,j}) - \text{M}(u^x_{i,j}, u^x_{i-1,j}, u^x_{i-2,j}) \\ &\quad + \text{M}(u^y_{i,j+1}, u^y_{i,j}, u^y_{i,j-1}) - \text{M}(u^y_{i,j}, u^y_{i,j-1}, u^y_{i,j-2})\big) \\ &+ \tfrac{1}{2}\tfrac{1}{\sqrt{2}h}\big(\text{M}(u^d_{i+1,j+1}, u^d_{i,j}, u^d_{i-1,j-1}) - \text{M}(u^d_{i,j}, u^d_{i-1,j-1}, u^d_{i-2,j-2}) \\ &\quad + \text{M}(u^e_{i+1,j-1}, u^e_{i,j}, u^e_{i-1,j+1}) - \text{M}(u^e_{i,j}, u^e_{i-1,j+1}, u^e_{i-2,j+2})\big). \end{aligned} \tag{27}$$

Let us now discuss our approximation of $(2-p)\,\text{curv}(u)|\nabla u|$. The isophote curvature

$$\text{curv}(u) = \frac{u_x^2 u_{yy} - 2u_x u_y u_{xy} + u_y^2 u_{xx}}{(u_x^2 + u_y^2)^{3/2}} \tag{28}$$

can be discretised in a straightforward way with central differences. To avoid a potential singularity in the denominator, we regularise by adding $\epsilon = 10^{-10}$ to $u_x^2 + u_y^2$. Moreover, note that the isophote curvature $\text{curv}(u)$ describes the inverse radius of the osculating circle to the level line. Since a discrete image does not have structures that are smaller than a single pixel, the smallest practically relevant radius is $\frac{h}{2}$. Thus, we impose a curvature limiter that restricts the computed result to the range $[-\frac{2}{h}, \frac{2}{h}]$.

Depending on the sign of $(2 - p)\,\mathrm{curv}(u)$, we may interpret $(2 - p)\,\mathrm{curv}(u)|\nabla u|$ either as a dilation term (for positive sign) or an erosion term (for negative sign) with a disc-shaped structuring element of radius $|(2-p)\,\mathrm{curv}(u)|$; see e.g. [1]. For a stable discretisation of $|\nabla u|$, we use the Rouy-Tourin upwind scheme [20]. In the dilation case, this comes down to

$$|\nabla u|_{i,j} \approx \sqrt{\left(\max\left(-u^x_{i-1,j},\, u^x_{i,j},\, 0\right)\right)^2 + \left(\max\left(-u^y_{i,j-1},\, u^y_{i,j},\, 0\right)\right)^2}, \qquad (29)$$

and in the erosion case to

$$|\nabla u|_{i,j} \approx \sqrt{\left(\max\left(-u^x_{i,j},\, u^x_{i-1,j},\, 0\right)\right)^2 + \left(\max\left(-u^y_{i,j},\, u^y_{i,j-1},\, 0\right)\right)^2}. \qquad (30)$$

Consistency. Since our resulting explicit scheme uses various one-sided – and thus first order – finite difference approximations within its upwind and minmod strategies, if follows that its general consistency order outside extrema is $\mathcal{O}(h+\tau)$. For the case $p = 2$, however, the second order stencil (23) gives $\mathcal{O}(h^2+\tau)$.

Stability. All components of our explicit scheme are specifically selected to create a nonnegative appoximation for an admissible time step size. This allows us to prove its L^∞ stability and a discrete maximum–minimum principle. Since the details are somewhat cumbersome and do not give more general insights, we sketch only the basic ideas by briefly analysing the contributions of the individual terms.

Let $\boldsymbol{u}^{k+1} = (\boldsymbol{I}+\tau\boldsymbol{A})\boldsymbol{u}^k$ be the matrix-vector notation of the explicit scheme for the diffusion evolution $u_t = \Delta u$. Here the vector $\boldsymbol{u}^k$ contains the values of u in all pixels at time $k\tau$, $\boldsymbol{I}$ is the unit matrix, and the matrix $\boldsymbol{A}$ represents the discretisation of Δu with the stencil (23) and reflecting boundary conditions that are implemented by mirroring. Then all entries of the iteration matrix $\boldsymbol{I}+\tau\boldsymbol{A}$ are nonnegative, if $1 - \frac{12\tau}{4h^2} \geq 0$. This leads to the stability condition $\tau \leq \frac{h^2}{3}$ for the forward diffusion process. Following the arguments in [19], the same stability limit can also be derived for an explicit scheme for the stabilised backward diffusion process $u_t = -\mathrm{sgn}(|\nabla u|)\Delta u$ that involves the minmod discretisation (27).

The explicit Rouy-Tourin scheme for the dilation/erosion evolutions $u_t = \pm|\nabla u|$ can be shown to be L^∞-stable for $\tau \leq \frac{h}{2}\sqrt{2}$.

By combining all these restrictions in a worst case scenario, it follows that our explicit scheme for the full evolution equation (21) or (22) must satisfy the stability condition

$$\tau \leq \frac{h^2}{2\sqrt{2}\,|2-p| + 3\,|p-1|}. \qquad (31)$$

It guarantees L^∞ stability and a discrete maximum–minimum principle. In practice this restriction is not very severe: With $h := 1$, it comes down to $\tau \leq \frac{1}{3}$

for the diffusion evolution ($p = 2$), to $\tau \leq \frac{1}{4}\sqrt{2} \approx 0.354$ for mean curvature motion ($p = 1$), and to $\tau \leq 0.069$ for the mode equation ($p = -1$). Thus, we can compute the PDE evolutions for M-smoothers not only in a stable way, but also fairly efficiently.

4 Experiments

In our experiments, we evaluate the PDE (6) with five different settings for p: a temporally rescaled midrange evolution ($p \to \infty$) using $u_t = u_{\eta\eta}$ with $\tau = 0.1$, the mean evolution leading to homogeneous diffusion ($p = 2$, $\tau = 0.25$), the median evolution yielding mean curvature motion ($p = 1$, $\tau = 0.25$), the mode evolution ($p = -1$, $\tau = 0.05$), and the Gabor flow ($p = -2$, $\tau = 0.04$).

Figure 1 illustrates the effect of these equations on the real-world test image *trui*. The CPU times for computing each of these results on a contemporary laptop are in the order of half a second. We observe that the midrange filter produces fairly jagged results, although it has a clear smoothing effect. Homogeneous diffusion does not suffer from jagged artifacts, but blurs also important structures such as edges. The median evolution smoothes only along isolines which results in a smaller deterioration of edge-like structures. The mode and the Gabor evolutions are very similar and may even enhance edges due to their backward parabolic term $-u_{\eta\eta}$.

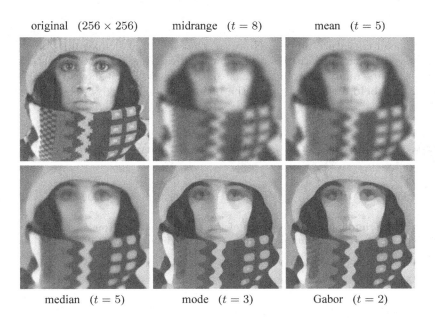

original (256 × 256)   midrange ($t = 8$)   mean ($t = 5$)

median ($t = 5$)   mode ($t = 3$)   Gabor ($t = 2$)

Fig. 1. Smoothing effect of the different evolution equations on the test image *trui*.

Figure 2 allows to judge if our numerical algorithm is capable of reproducing the rotationally invariant behaviour of its underlying PDE (6). We observe excellent rotation invariance. Moreover, we see that the mode and Gabor evolutions have comparable shrinkage properties as mean curvature motion. However, they differ from mean curvature motion by their backward term $-u_{\eta\eta}$, which can compensate dissipative artifacts that are caused by the discretisations of the forward parabolic term.

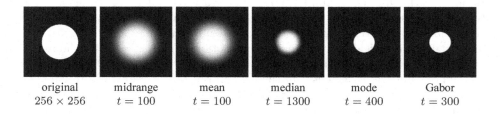

| original | midrange | mean | median | mode | Gabor |
| 256×256 | $t = 100$ | $t = 100$ | $t = 1300$ | $t = 400$ | $t = 300$ |

Fig. 2. Effect of the different evolution equations on a rotationally invariant test image.

In Fig. 3, we study the shape simplification properties of the mode evolution: It shrinks the binary shape of the kiwi in such a way that highly curved structures evolve faster than less curved ones, resulting in an evolution where nonconvex shapes become convex and vanish in finite time by shrinking to a circular point. Thus, the mode evolution appears to enjoy the same shape simplification qualities as mean curvature motion. However, that fact that it does not suffer from dissipative artifacts constitutes a distinctive advantage and makes it attractive for many shape analysis problems. Last but not least, Fig. 3 shows that the bird existed long before the egg, giving the ultimative answer to a deep problem in philosophy.

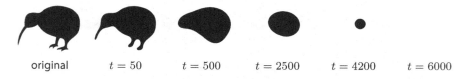

| original | $t = 50$ | $t = 500$ | $t = 2500$ | $t = 4200$ | $t = 6000$ |

Fig. 3. Shape simplification properties of the mode evolution. Image size: 579×449. Source of original image: https://commons.wikimedia.org/wiki/File:Kiwi_silhouette-by-flomar.svg.

5   Summary and Conclusions

We have established a comprehensive analysis that identifies the PDE limit for the full class of iterated M-smoothers with order-p means. It does not only reproduce known results for mean and median filtering, but also corrects a common misconception in the literature: We have shown the surprising fact that in the continuous limit, mode filtering does not correspond to $p = 0$, but results from the limit $p \to -1$. Moreover, our filter class $u_t = u_{\xi\xi} + (p-1)u_{\eta\eta}$ can also be extended to models that have no interpretation within the setting of M-smoothers, e.g. Gabor's classical method for $p = -2$.

Since literal implementations of some M-smoothers such as mode filtering can be highly nontrivial when using small local histograms [10,14], we have proposed a novel numerical algorithm that can handle the PDE evolution for arbitrary p-values. Although these evolutions can be highly anisotropic and may even exhibit backward parabolic behaviour, we managed to come up with an L^∞-stable finite difference scheme that is fairly efficient, satisfies a maximum–minimum principle and shows very good rotation invariance. This has been partly achieved by employing and adapting powerful stabilisation concepts from the numerics of hyperbolic PDEs, such as upwinding, minmod functions, and curvature limiters. Our numerical algorithm is applicable to any stable evolution of type $u_t = a\,u_{\xi\xi} + b\,u_{\eta\eta}$, where a and b may have arbitrary sign. Thus, it is of very general nature and covers also numerous applications beyond M-smoothing, including image interpolation [4], adaptive filter design [3], and many level set methods [18]. Exploring some of these applications is part of our ongoing work.

Our experiments indicate that the PDEs for $p < 1$, such as the mode evolution, are particularly appealing: They combine strong shape simplification properties with excellent sharpening qualities. They clearly deserve more research.

Connecting the class of M-smoothers to the family of PDE-based methods contributes one more mosaic stone to the mathematical foundations of image analysis. Since M-smoothers themselves are related to many other approaches [17,23,24], including W-smoothers, bilateral filters, mean-shift and robust estimation, our results can help to gain a broader and more coherent view on the entire field.

Acknowledgement. This project has received funding from the European Research Council (ERC) under the European Union's Horizon 2020 research and innovation programme (grant agreement No. 741215, ERC Advanced Grant INCOVID).

References

1. Alvarez, L., Guichard, F., Lions, P.L., Morel, J.M.: Axioms and fundamental equations in image processing. Arch. Ration. Mech. Anal. **123**, 199–257 (1993)
2. Barral Souto, J.: El modo y otras medias, casos particulares de una misma expresión matemática. Report No. 3, Cuadernos de Trabajo, Instituto de Biometria, Universidad Nacional de Buenos Aires, Argentina (1938)
3. Carmona, R., Zhong, S.: Adaptive smoothing respecting feature directions. IEEE Trans. Image Process. **7**(3), 353–358 (1998)
4. Caselles, V., Morel, J.M., Sbert, C.: An axiomatic approach to image interpolation. IEEE Trans. Image Process. **7**(3), 376–386 (1998)
5. Chu, C.K., Glad, I., Godtliebsen, F., Marron, J.S.: Edge-preserving smoothers for image processing. J. Am. Stat. Assoc. **93**(442), 526–556 (1998)

6. Fréchet, M.: Les élements aléatoires de nature quelconque dans un espace distancié. Ann. de l'Inst. Henri Poincaré **10**, 215–310 (1948)
7. Gabor, D.: Information theory in electron microscopy. Lab. Invest. **14**, 801–807 (1965)
8. Gauss, C.F.: Theoria motus corporum coelestium in sectionibus conicis solem ambientium. Perthes & Besser, Hamburg (1809)
9. Griffin, L.D.: Mean, median and mode filtering of images. Proc. Roy. Soc. London Ser. A **456**(2004), 2995–3004 (2000)
10. Griffin, L.D., Lillholm, M.: Mode estimation using pessimistic scale space tracking. In: Griffin, L.D., Lillholm, M. (eds.) Scale-Space 2003. LNCS, vol. 2695, pp. 266–280. Springer, Heidelberg (2003). https://doi.org/10.1007/3-540-44935-3_19
11. Guichard, F., Morel, J.M.: Partial differential equations and image iterative filtering. In: Duff, I.S., Watson, G.A. (eds.) The State of the Art in Numerical Analysis. IMA Conference Series (New Series), vol. 63, pp. 525–562. Clarendon Press, Oxford (1997)
12. Huber, P.J.: Robust Statistics. Wiley, New York (1981)
13. Jackson, D.: Note on the median of a set of numbers. Bull. Am. Math. Soc. **27**, 160–164 (1921)
14. Kass, M., Solomon, J.: Smoothed local histogram filters. ACM Trans. Graph. **29**(4), Article ID 100 (2010)
15. Legendre, A.M.: Nouvelles Méthodes pour la détermination des Orbites des Comètes. Firmin Didot, Paris (1805)
16. Lindenbaum, M., Fischer, M., Bruckstein, A.: On Gabor's contribution to image enhancement. Pattern Recogn. **27**, 1–8 (1994)
17. Mrázek, P., Weickert, J., Bruhn, A.: On robust estimation and smoothing with spatial and tonal kernels. In: Klette, R., Kozera, R., Noakes, L., Weickert, J. (eds.) Geometric Properties from Incomplete Data, Computational Imaging and Vision, vol. 31, pp. 335–352. Springer, Dordrecht (2006). https://doi.org/10.1007/1-4020-3858-8_18
18. Osher, S., Paragios, N. (eds.): Geometric Level Set Methods Imaging, Vision and Graphics. Springer, New York (2003). https://doi.org/10.1007/b97541
19. Osher, S., Rudin, L.: Shocks and other nonlinear filtering applied to image processing. In: Tescher, A.G. (ed.) Applications of Digital Image Processing XIV. Proceedings of SPIE, vol. 1567, pp. 414–431. SPIE Press, Bellingham (1991)
20. Rouy, E., Tourin, A.: A viscosity solutions approach to shape-from-shading. SIAM J. Numer. Anal. **29**, 867–884 (1992)
21. Torroba, P.L., Cap, N.L., Rabal, H.J., Furlan, W.D.: Fractional order mean in image processing. Opt. Eng. **33**(2), 528–534 (1994)
22. Tukey, J.W.: Exploratory Data Analysis. Addison-Wesley, Menlo Park (1971)
23. van den Boomgaard, R., van de Weijer, J.: On the equivalence of local-mode finding, robust estimation and mean-shift analysis as used in early vision tasks. In: Proceedings of the 16th International Conference on Pattern Recognition, Quebec City, Canada, vol. 3, pp. 927–930, August 2002
24. Winkler, G., Aurich, V., Hahn, K., Martin, A., Rodenacker, K.: Noise reduction in images: some recent edge-preserving methods. Pattern Recogn. Image Anal. **9**(4), 749–766 (1999)

Registration and Reconstruction

Variational Registration of Multiple Images with the SVD Based SqN Distance Measure

Kai Brehmer[1(✉)], Hari Om Aggrawal[1], Stefan Heldmann[2], and Jan Modersitzki[1,2]

[1] Institute of Mathematics and Image Computing, University of Lübeck, Lübeck, Germany
brehmer@mic.uni-luebeck.de
[2] Fraunhofer Institute for Digital Medicine MEVIS, Lübeck, Germany

Abstract. Image registration, especially the quantification of image similarity, is an important task in image processing. Various approaches for the comparison of two images are discussed in the literature. However, although most of these approaches perform very well in a two image scenario, an extension to a multiple images scenario deserves attention. In this article, we discuss and compare registration methods for multiple images. Our key assumption is, that information about the singular values of a feature matrix of images can be used for alignment. We introduce, discuss and relate three recent approaches from the literature: the Schatten q-norm based SqN distance measure, a rank based approach, and a feature volume based approach. We also present results for typical applications such as dynamic image sequences or stacks of histological sections. Our results indicate that the SqN approach is in fact a suitable distance measure for image registration. Moreover, our examples also indicate that the results obtained by SqN are superior to those obtained by its competitors.

Keywords: Groupwise registration · Dynamic imaging · 3D reconstruction

1 Introduction

Typical applications in medical imaging are to analyze spatio-temporal variations of bio-medical images. A prerequisite for such analysis is that images are aligned and in many cases joint registration of multiple images is required. Examples are, e.g., analysis of images from different time points and/or different complimentary modalities, atlas registration, longitudinal normalization, motion correction or image reconstruction [1,4,7,8,12,13,18,19,21].

A number of registration models are already available to register a pair of two images [15,20,22], but their simple extension to register a group of images might suffer from various problems. Generally, these pair-wise methods assume

© Springer Nature Switzerland AG 2019
J. Lellmann et al. (Eds.): SSVM 2019, LNCS 11603, pp. 251–262, 2019.
https://doi.org/10.1007/978-3-030-22368-7_20

one of the images as a reference image, and therefore registrations are implicitly biased towards the reference image. Moreover, the selection of a reference image from the given image sequence is not always a very straight forward process. Most importantly, these registration models are primarily influenced by features shared by the image pair and less affected by the features other images have in the image sequence. Therefore, this approach does not account the global information available in the image sequence. It has also been shown that these methods have slow convergence rate compared to the groupwise methods [2,3].

To avoid the selection of a reference image and the related bias, Joshi et al. [13] proposed the registration of each image from the image sequence with respect to the group mean of the registered image sequence. This approach does not need to define the reference image explicitly, moreover accounts the global information through the group mean. This approach inherits the assumption that every image in the image sequence is almost similar to the group mean.

Recently, Guyader [8] and Brehmer [2,3] proposed groupwise registration methods for a sequence of images. The underlying assumption is that images are linearly dependent if they are aligned. The linear dependency idea completely circumvents the need of defining a group mean image. Both of these methods construct an image matrix where each column is corresponding to an image from the sequence. Brehmer [2,3] estimates transformation fields by minimizing the rank of the matrix and implicitly forcing columns of the matrix to become linear dependent to each other. Guyader [8] utilizes the multivariate version of mutual information, called total correlation, to define a groupwise registration model.

The paper is structured as follows: In Sect. 2, we discuss mathematical formulations of SVD based image registration approaches. More precise, we discuss a general framework for groupwise registration models based on correlation maximization. In Sect. 3 we briefly discuss the used numerical setting. After that, in Sect. 4, we demonstrate the performance of some of the proposed methods on two datasets and compare them with other state-of-the-art methods.

2   Registration Approaches for Multiple Images

In this section, we describe our Schatten q-norm based distance measure SqN for multiple images. We start by briefly outlining a standard variational registration framework for two images [15]. We then present a straightforward extension for multiple images and discuss the drawbacks of the naive approach drawbacks. The main drawbacks are its sequential and thus ordering dependent assessment of the image frames and the weak coupling of image information over the frames.

We then present the setting of the SqN distance measure. The main idea is to make use of the singular values of an image feature array. Finally, we relate the Schatten q-norm based distance measure to work of Friedman et al. [6] and Guyader et al. [8].

2.1 Variational Registration Approach for Two Images

We start the discussion with a standard approach to image registration; see e.g. [15] for details. To simplify discussion, an image $\mathcal{T}$ is assumed to be a real valued intensity function $\mathcal{T} : \mathbb{R}^d \to \mathbb{R}$ with compact support in a domain $\Omega \subset \mathbb{R}^d$. Given two images $\mathcal{T}_0, \mathcal{T}_1$, the goal of image registration is to find a transformation $y : \mathbb{R}^d \to \mathbb{R}^d$ such that ideally $\mathcal{T}_1 \circ y \approx \mathcal{T}_0$, where $\mathcal{T} \circ y(x) := \mathcal{T}(y(x))$. To achieve this goal, we choose a variational framework where a joined functional

$$J^{\text{two}}(y; \mathcal{T}_0, \mathcal{T}_1) := D(\mathcal{T}_0, \mathcal{T}_1 \circ y) + S(y), \tag{1}$$

is to be minimized over an admissible set of transformations. Various choices for distance measures D and regularizers S are discussed in the literature; see e.g. [15] and references therein. A thorough discussion is beyond the scope of this paper. Here, we only briefly recall the L_2-norm (sum of squared distances, SSD), the normalized gradient field (NGF) [10], and the elastic potential [5]:

$$D^{\text{SSD}}(\mathcal{T}_0, \mathcal{T}_1 \circ y) := \tfrac{1}{2} \|\mathcal{T}_1 \circ y - \mathcal{T}_0\|^2_{L_2(\Omega)}, \tag{2}$$

$$D^{\text{NGF}}(\mathcal{T}_0, \mathcal{T}_1 \circ y) := \tfrac{1}{2} \int_\Omega \left[1 - \left\langle \frac{\nabla \mathcal{T}_1 \circ y}{\|\nabla \mathcal{T}_1 \circ y\|_\eta}, \frac{\nabla \mathcal{T}_0}{\|\nabla \mathcal{T}_0\|_\eta} \right\rangle^2 \right] dx \tag{3}$$

$$S^{\text{elas}}(y) := \tfrac{1}{2} \|\mu \operatorname{tr}(E^2) + \lambda \operatorname{tr}(E)^2\|^2_{L_2(\Omega)} \tag{4}$$

with $\|a\|_\eta := \sqrt{\langle a, a \rangle + \eta}$, $\eta > 0$ and strain $E := \nabla y + \nabla y^\top - I$ where I is the identity matrix.

Derivations of image intensities are also commonly used to quantify image similarity. For a unified conceptual framework, we introduce a feature map F that maps an image to a Hilbert space of features. Any metrics μ on the feature space can then be used for registration: $D(\mathcal{T}_0, \mathcal{T}_1) := \mu(F(\mathcal{T}_0), F(\mathcal{T}_1))$. Examples of such feature maps are e.g. intensity normalization $F^{\text{IN}}(\mathcal{T}) = \mathcal{T}/\|\mathcal{T}\|_{L_2}$ or the normalized gradient field, $F^{\text{NGF}}(\mathcal{T}) = \nabla \mathcal{T}/\|\nabla \mathcal{T}\|_\eta$, to name a few. Note that the NGF distance measure is based on $\nabla \mathcal{T}(x)/\|\nabla \mathcal{T}(x)\|_\eta$ whereas the feature map is based on $\nabla \mathcal{T}(x)/\|\nabla \mathcal{T}\|_{L_2}$.

2.2 Sequential Registration Approach for Multiple Images

Our goal is to extend the standard registration to sequences of images $T = (\mathcal{T}_1, \ldots, \mathcal{T}_K)$. Note that the images might be given as a time series such as our DCE-MRI example, a structured process such as the HISTO application, or even an unstructured ensemble of images such as an atlas generation.

The first approach is to simply apply the above framework sequentially. With transformations $Y = (y_1, \ldots, y_K)$ the corresponding energy to be minimized with respect to Y reads

$$J^{\text{seq}}(Y; T) := \sum_{k=2}^{K} \{ D(\mathcal{T}_{k-1} \circ y_{k-1}, \mathcal{T}_k \circ y_k) + S(y_k) \}. \tag{5}$$

Note that typically, one of the deformations is fixed, e.g., $y_1(x) := x$ for well-posedness. However, as the problem is usually too big to be solved straightforwardly, a non-linear Gauss-Seidel type iteration is usually applied. Here, one assumes that Y is a good starting guess and sequentially improves component by component for $\ell = 1, \ldots, K$ by determining optimizers

$$z^* \in \arg\min_z J^{\text{seq}}(y_1, \ldots, y_{\ell-1}, z, y_{\ell+1}, \ldots, y_K; T), \tag{6}$$

setting $y_\ell := z^*$ and iterates until convergence. This process is generally rather expensive and therefore slow. A problem is that the coupling of the different components of Y is weak. An update of y_ℓ has impact only every K-th step in the procedure. Therefore, potentially a high number of iterations is required.

2.3   Global Registration Approach for Multiple Images

Here, we propose a registration approach that provides a full coupling of all image frames. Our objective is to find a minimizer Y of the energy J^{glo},

$$J^{\text{glo}}(Y; T) := D^{\text{glo}}(T \circ Y) + S^{\text{glo}}(Y), \tag{7}$$

where we use the suggestive abbreviation $T \circ Y := (\mathcal{T}_0 \circ y_0, \ldots, \mathcal{T}_K \circ y_K)$ and for sake of simplicity let be $S^{\text{glo}}(Y) := \sum_{k=1}^{K} S(y_k)$ with S any of the regularizers discussed in Sect. 2.1. Clearly, one could debate for a more general or even stronger regularization of Y. However, this is not in the scope of the paper and we leave the discussion for future work. The essential contribution is thus the global distance measure that is based on the feature array $F(T) := [F(\mathcal{T}_1), \,\ldots,\, F(\mathcal{T}_K)]$ which comprises the features of the image sequence and its symmetric, positive semi-definite correlation matrix $C = \langle F, F \rangle \in \mathbb{R}^{K \times K}$ where C_{ij} assembles the correlations of $F(\mathcal{T}_i)$ and $F(\mathcal{T}_j)$. Note that we assumed F maps into a Hilbert space such that the correlation is well defined according to the corresponding inner product. Our key assumption is that the rank of the feature array is minimal if the image frames are aligned. Note that we actually aim to exclude the trivial situation $\text{rank}\, F = 0$ as this implies that all features are zero. We also note that the assumption may not hold for multi-modal images, if the feature map does not compensate intensity variations. Therefore, a plain image intensity based feature map may not be successful. If we expect that intensity changes will occur at similar positions in space, e.g., the NGF feature map is a valid choice.

2.4   Schatten q-norm Based Image Similarity Measure $D_{S,q}$

The above considerations suggest to choose $\text{rank}\, F$ as a distance measure. In [2,3], Brehmer et al. proposed to reformulate the rank minimization problem in terms of a relaxation of the rank function based on a so-called Schatten q-norm. Roughly speaking, the Schatten q-norm of an operator is the q-norm of the vector of its singular values. Thus

$$D_{S,q}(T) := \|F(T)\|_{S,q} := \left(\sum_{k=1}^{K} \sigma_k(F(T))^q \right)^{1/q} \tag{8}$$

where σ_k, $k = 1, \ldots, K$, denote the non-zero singular values of $F(T)$. Before we discuss numerical details, we relate this measure to other rank based similarity measures for image stacks. Particularly we address volume minimization of the feature parallelotope and correlation maximization of normalized features.

2.5   Volume Minimization of the Feature Parallelotope

The above approach can be linked to work of Guyader et al. [8]. To this end, we consider the minimization of the volume of the parallelotope spanned by the columns of $F(T)$. Equivalently, we can consider the determinant of C or, exploring the monotonicity of the logarithm, set

$$D(T) := \log(\det(C(T))) = \log(\textstyle\prod_{k=1}^{K} \sigma_k^2(F(T))) = 2\sum_{k=1}^{K} \log(\sigma_k(F(T))). \quad (9)$$

This expression is related to the volume of a normalized covariance matrix which is the total correlation in [8] and used as a similarity measure for group-wise registration.

However, a volume based approach has a severe drawback; see also the discussion in [11]. To illustrate this, we consider two feature vectors $f_1 \neq 0$ and f_2 with angle α. Hence, volume$(f_1, f_2) = \|f_1\|\|f_2\|\sin\alpha$. This value is minimal if the vectors are linearly dependent. Unfortunately, this also happens if $f_2 = 0$. In a registration context, this implies that a translation of one of the images, say, about the diameter of Ω yields a global optimizer. In [11] it is therefore suggested to replace the minimization of volume by a maximization of correlation $|\cos\alpha|$. This value is maximal iff and only iff $f_2 = \pm f_1$ and is in fact minimal if $f_2 = 0$. This subtle difference is very important in a registration context.

2.6   Correlation Maximization of Normalized Features

In this section we focus on correlation maximization and do not discuss the corresponding minimization formulation. We also assume that feature vectors are normalized, i.e. $\|F(\mathcal{T}_k)\| = 1$. For the correlation matrix $C(T) \in \mathbb{R}^{K,K}$ holds

$$C_{kk} = 1, \quad C_{jk} = \langle F(\mathcal{T}_j), F(\mathcal{T}_k)\rangle = \cos\gamma_{jk}, \quad (10)$$

where γ_{jk} denotes the angle between the j-th and k-th feature. In the two image setting it is therefore natural to maximize $|C_{1,2}|$ if we account both, for positive and negative correlation. This is the underlying idea of normalized cross correlation. Note that the NGF approach is still different as the correlation is computed point wise and finally averaged.

For the multiple image setting, the best scenario is $C \in \{\pm 1\}^{K,K}$. If only non-negative correlation is considered, the ideal case is $C(T) = 1 \cdot 1^{\top}$. On the opposite, the worst case scenario for registration is that $C(T) = I$ meaning all features are fully uncorrelated. Therefore, a suitable distance measure is to maximize the difference

$$D(T) := \|C(T) - I\|_M, \quad (11)$$

where $\|\cdot\|_M$ denotes a suitable matrix norm.

2.7 Correlation Maximization and Schatten q-norms

Specifically, choosing $\| \cdot \|_M = \| \cdot \|_{S,q}$ a Schatten q-norm in (11) we obtain

$$D(T) = \|C(T) - I\|_{S,q} = \left(\sum_{k=1}^{K} (\sigma_k^2(F(T)) - 1)^q \right)^{1/q}. \tag{12}$$

We investigate the special cases $q = 2$ and $q = \infty$. Note that

$$\|A\|_{S,\infty} = \sigma_{\max}(A), \quad \text{the largest singular value of } A, \text{ and}$$
$$\|A\|_{S,2}^2 = \sum_k \sigma_k^2 = \text{trace}(A^\top A) = \sum_{j,k} |a_{j,k}|^2 = \|A\|_{\text{Fro}}^2.$$

Thus, choosing the Schatten ∞-norm yields maximizing $\sigma_{\max}^2(F(T)) - 1$. This is equivalent to maximizing the largest singular value of $F(T)$, see also [6]:

$$\arg \max \|C(T) - I\|_{S,\infty} = \arg \max \sigma_{\max}(F(T)).$$

For the Schatten 2-norm we have $D(T) = \|C(T) - I\|_{S,2}^2 = \sum_{i \neq j} |C_{ij}|^2$ which shows that the distance is quadratic mean of the correlation among the image features. Furthermore, a direct computation shows

$$D(T) = \|C(T) - I\|_{S,2}^2 = \|F\|_{S,4}^4 - K.$$

Here, we exploit the special structure of correlation matrix C, i.e., $\text{trace}(C) = K$.

To this end, we define the two SqN distance measures for NGF features as follows:

$$\text{SqN}_4(T) := K - \|F^{\text{NGF}}(T)\|_{S,4}^4 \tag{13}$$
$$\text{SqN}_\infty(T) := -\sigma_{\max}(F^{\text{NGF}}(T)) \tag{14}$$

3 Numerical Methods

For the optimization of the functional J^{seq} (cf. (5)) we use the discretize-then-optimize framework introduced in [9]. The basic concept is to use a sequence of discretized finite dimensional optimization problems. A smooth approximation of the problem is represented with few degrees of freedom. It is expected that the optimization is fast as the problem is low dimensional and smooth. Its numerical solution is prolongated and then serves as a starting guess for the finer resolved problem. It is expected that a numerical solution can be computed fast, as the starting point is expected to be close to the solution. The process is generally terminated when reaching the resolution of the given data. Note that the images are only smoothed in the spatial domain.

To solve the discrete problem on a fixed resolution we use a quasi-Newton type approach. More precisely, we use L-BFGS with the Hessian of the regularizer as an initial approximation of the metric and a Wolfe linesearch; see, e.g. [16] for optimization and [15] for details.

For the optimization of J^{SqN},

$$J^{\mathrm{SqN}}(Y;T) := \mathrm{SqN}(T \circ Y) + S^{\mathrm{glo}}(Y), \tag{15}$$

we use similar concepts as above for the regularization term.

For the SqN distance, we remark that the distance is a rather simple algebraic expression of the singular values of the feature matrix. The challenging part is thus the derivative of the singular values. Here, we follow [17]. A singular value decomposition of the feature matrix $F \in \mathbb{R}^{n \times K}$ is denoted by $F = U\Sigma V^{\top}$, where the matrices $U = (u_{i,k}) \in \mathbb{R}^{n,n}$ and $V = (v_{j,k}) \in \mathbb{R}^{K,K}$ are orthogonal and $\Sigma \in \mathbb{R}^{n,K}$ is a non-negative diagonal matrix with the singular values $\sigma_k(F)$ as diagonal entries. From [17] we have the surprisingly simple relation $\frac{\partial \sigma_k(F)}{\partial F_{i,j}} = u_{i,k}v_{j,k}$ that is used in our implementation.

4   Results

We now present results for the registration of histological serial sectioning of a marmoset monkey brain as well as for DCE-MRI sequences of a human kidney. For the given datasets, we will compare the registration results of SqN_4, SqN_∞ in comparison to a total correlation based approach like in [8] and sequential NGF. We start with registrations of a serial sectioning of a marmoset monkey brain; data courtesy of Harald Möller, Max Planck Institute for Human Cognitive and Brain Sciences, Leipzig, Germany [14]. The dataset consists of every 4th slice of the original serial sectioning of the brain, in total 69 slices of sizes from 2252×3957 pixels up to 7655×9965 pixels. For proof of concept we reduced the number of pixels per slice to reduce computation time to a reasonable level. The objective of the registration of histological slices is to align them in order to reconstruct the $3D$ volume of the tissue.

Slice 5   Slice 30   Slice 46

Fig. 1. Three representative axial slices of a marmoset monkey brain dataset; data courtesy of Harald Möller [14]

Figure 1 shows three representative axial slices of the data set. The main difficulties of registering this particular dataset are the different sizes of the slices on the one hand and the translation of whole parts of the imagestack within the domain on the other hand. Furthermore we didn't use a pre-segmentation of the dataset to show robustness of the registration approaches against artifacts in the background region. The background region of the slices contains several markings of the examiners like white rectangles as well as dust and dirt from the object slide captured during the high resolution scanning process; see Fig. 1.

Figure 2 shows two sagittal slices (top and bottom row) through the image stack from the reduced, unregistered monkey brain dataset besides the registration results to illustrate the alignment of the slices. As expected the results of SqN_4 are quite similar to the results of SqN_∞. The computation for the groupwise approaches using SqN as well as the total correlation approach from [8] took about 45 to 50 min for a resolution of 128×158 pixels for each of the 69 slices. Compared to this, the sequential NGF approach with just one sweep needed about 2.2 times the computation time (ca. 110 min). However, from visual comparison it is obvious that many more sweeps are needed to achieve results comparable to those of the groupwise approaches; see Fig. 2. Everything was implemented in Python using Numpy and Scipy for optimization.

Moreover, we used a random permutation of the stack of histological serial sections to demonstrate invariance to the order of images of the singular value based groupwise registration approaches. We randomly permuted the order of images, registered the stack in random order using SqN_4 and reordered it afterwards; see Fig. 3, center column. As expected, the results are the same as for registration using SqN_4 without random permutation; cf. Figs. 2 and 3 for comparison.

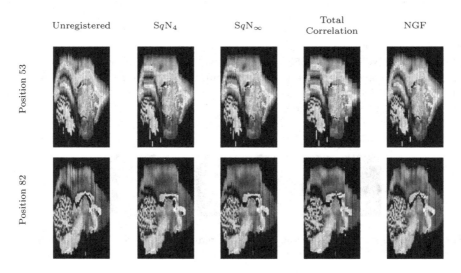

Fig. 2. Registration results for 3D reconstruction of the monkey brain datasets. For illustration, we show only 2D slices that are sagittal cuts at two positions, i.e., 53 and 82.

Next we present registration results for a DCE-MRI sequence of a human kidney; data courtesy of Jarle Rørvik, Haukeland University Hospital Bergen, Norway. Here, 3D images are taken at 45 time points. For ease of presentation and to have a reasonable level of computation time we show results for a 2D slice over time. More precisely, we use 178-by-95 coronal slices of a 178-by-95-by-30-by-45 volume for z-slice 18; see Fig. 4 for representative slices. All time points are used for registration. The objective here is to register the slices while maintaining the dynamics. Figure 5 illustrates the stack of slices for the different registration approaches using a sagittal cut through the stack, analog to the results for the histological serial sections shown in Fig. 2. The illustrated results were achieved using three different levels of spatial resolution up to half the original resolution in about 8 min per groupwise approach. The result of the sequential approach was achieved in about twice the time using just one sweep. For the alignment using the approach from [8], we couldn't find a parameter setting to achieve results comparable to the SqN-approaches.

5 Discussion and Conclusions

The registration of multiple images is an important task in image processing. Conventional approaches often use an extension of a pairwise approach for two images. In this paper, we demonstrate that this approach may come with numerous disadvantages and may be time consuming. We also describe and analyze a recently proposed alternative. The Schatten q-norm based SqN [2,3] distance measure is a reference for our investigations on different singular value based measures such as the maximization of correlation between different images as well as minimization of spanned volumes. For this purpose we have introduced a general formulation using feature maps that map images into Hilbert spaces.

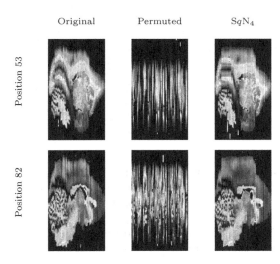

Fig. 3. Registration results after random permutation of the axial slices. As expected, the results are the same as for the non-permuted image stack; also see Fig. 2 for comparison.

Time point 5   Time point 11   Time point 21

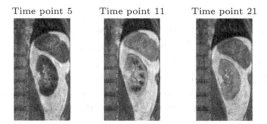

Fig. 4. Three representative 2D coronal slices of the 4D DCE-MRI dataset of a human kidney; data courtesy of Jarle Rørvik, Haukeland University Hospital, Bergen, Norway. The slices are shown at three different time points. The dataset is a 178-by-95-by-30-by-45 volume, the shown slices are 178-by-95.

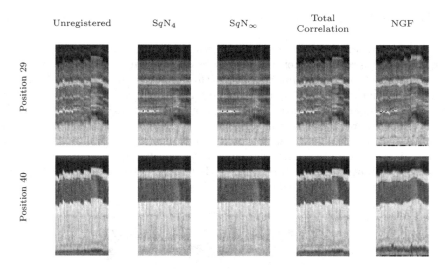

Fig. 5. Illustrated are sagittal cuts through the stack of 2D slices from a 4D DCE-MRI dataset of a human kidney at positions 29 and 40. The first column shows the unregistered stack. Right next to this the results of the different registration approaches are illustrated.

This opens a door for even further investigation on image registration methods for multiple images. With our numerical results we demonstrate that SqN based motion compensation is applicable in dynamic imaging as well as for the alignment of histological serial sections. Moreover, the results clearly show that SqN performs at least as good as standard approaches from the literature. In our experiments both the alignment and the computation time of the groupwise approaches were closer to a desirable solution than the sequential approach using pairwise NGF.

Furthermore, we outlined that a singular value based approach exploits the global information of a dataset, which cannot be achieved by using two-neighbourhoods in registration. In some specific applications, such as dynamic imaging or reconstruction of histological volumes from serial sections, this can avoid unwanted effects like the so-called banana-effect. Future work will address the optimal choice of the parameter q and investigations of different variants of feature maps. Finally, different regularization strategies will be investigated.

Acknowledgement. The authors acknowledge the financial support by the Federal Ministry of Education and Research of Germany in the framework of MED4D (project number 05M16FLA).

References

1. Bhatia, K.K., Hajnal, J.V., Puri, B.K., Edwards, A.D., Rueckert, D.: Consistent groupwise non-rigid registration for atlas construction. In: 2004 2nd IEEE International Symposium on Biomedical Imaging: Nano to Macro (IEEE Cat No. 04EX821), vol. 1, pp. 908–911 (2004)
2. Brehmer, K., Wacker, B., Modersitzki, J.: A novel similarity measure for image sequences. In: Klein, S., Staring, M., Durrleman, S., Sommer, S. (eds.) WBIR 2018. LNCS, vol. 10883, pp. 47–56. Springer, Cham (2018). https://doi.org/10. 1007/978-3-319-92258-4_5
3. Brehmer, K., Wacker, B., Modersitzki, J.: Simultaneous registration of image sequences - a novel singular value based images similarity measure. PAMM **18**(1), e201800370 (2018)
4. Cootes, T.F., Marsland, S., Twining, C.J., Smith, K., Taylor, C.J.: Groupwise diffeomorphic non-rigid registration for automatic model building. In: Pajdla, T., Matas, J. (eds.) ECCV 2004. LNCS, vol. 3024, pp. 316–327. Springer, Heidelberg (2004). https://doi.org/10.1007/978-3-540-24673-2_26
5. Fischler, M.A., Elschlager, R.A.: The representation and matching of pictorial structures. IEEE Tran. Comput. **22**(1), 67–92 (1973)
6. Friedman, S., Weisberg, H.F.: Interpreting the first eigenvalue of a correlation matrix. Educ. Psychol. Measure. **41**(1), 11–21 (1981)
7. Geng, X., Christensen, G.E., Gu, H., Ross, T.J., Yang, Y.: Implicit reference-based group-wise image registration and its application to structural and functional MRI. NeuroImage **47**(4), 1341–1351 (2009)
8. Guyader, J.M., et al.: Groupwise image registration based on a total correlation dissimilarity measure for quantitative MRI and dynamic imaging data. Sci. Rep. **8**(1), 13112 (2018)
9. Haber, E., Modersitzki, J.: A multilevel method for image registration. SIAM J. Sci. Comput. **27**(5), 1594–1607 (2006)
10. Haber, E., Modersitzki, J.: Intensity gradient based registration and fusion of multi-modal images. In: Larsen, R., Nielsen, M., Sporring, J. (eds.) MICCAI 2006. LNCS, vol. 4191, pp. 726–733. Springer, Heidelberg (2006). https://doi.org/10. 1007/11866763_89
11. Haber, E., Modersitzki, J.: Intensity gradient based registration and fusion of multi-modal images. In: Larsen, R., Nielsen, M., Sporring, J. (eds.) MICCAI 2006. LNCS, vol. 4191, pp. 726–733. Springer, Heidelberg (2006). https://doi.org/10. 1007/11866763_89

12. Huizinga, W., et al.: PCA-based groupwise image registration for quantitative mri. Med. Image Anal. **29**, 65–78 (2016)
13. Joshi, S., Davis, B., Jomier, M., Gerig, G.: Unbiased diffeomorphic atlas construction for computational anatomy. NeuroImage **23**, S151–S160 (2004)
14. Marschner, H., et al.: High-resolution quantitative magnetization transfer imaging of post-mortem marmoset brain. In: 22nd Annual Meeting of the International Society for Magnetic Resonance in Medicine (2014)
15. Modersitzki, J.: FAIR: Flexible Algorithms for Image Registration. SIAM, Philadelphia (2009)
16. Nocedal, J., Wright, S.J.: Numerical Optimization. Springer Series in Operations Research, 2nd edn. Springer, Heidelberg (2006). https://doi.org/10.1007/978-3-540-35447-5
17. Papadopoulo, T., Lourakis, M.I.A.: Estimating the jacobian of the singular value decomposition: theory and applications. In: Vernon, D. (ed.) ECCV 2000. LNCS, vol. 1842, pp. 554–570. Springer, Heidelberg (2000). https://doi.org/10.1007/3-540-45054-8_36
18. Polfliet, M., Klein, S., Huizinga, W., Paulides, M.M., Niessen, W.J., Vandemeulebroucke, J.: Intrasubject multimodal groupwise registration with the conditional template entropy. Med. Image Anal. **46**, 15–25 (2018)
19. Schmitt, O., Modersitzki, J., Heldmann, S., Wirtz, S., Fischer, B.: Image registration of sectioned brains. Int. J. Comput. Vis. **73**(1), 5–39 (2006)
20. Sotiras, A., Davatzikos, C., Paragios, N.: Deformable medical image registration: A survey. IEEE Trans. Med. Imaging **32**(7), 1153–1190 (2013)
21. Yigitsoy, M., Wachinger, C., Navab, N.: Temporal groupwise registration for motion modeling. Inf. Process. Med. Imaging **22**, 648–59 (2011)
22. Zitová, B., Flusser, J.: Image registartion methods: a survey. Image Vis. Comput. **21**, 977–1000 (2003)

Multi-tasking to Correct: Motion-Compensated MRI via Joint Reconstruction and Registration

Veronica Corona[1]([✉]), Angelica I. Aviles-Rivero[2], Noémie Debroux[1],
Martin Graves[3], Carole Le Guyader[4], Carola-Bibiane Schönlieb[1],
and Guy Williams[5]

[1] DAMTP, University of Cambridge, Cambridge, UK
{vc324,nd448,cbs31}@cam.ac.uk
[2] DPMMS, University of Cambridge, Cambridge, UK
ai323@cam.ac.uk
[3] Department of Radiology, University of Cambridge, Cambridge, UK
mjg40@cam.ac.uk
[4] LMI, Normandie Université, INSA de Rouen, Rouen, France
carole.le-guyader@insa-rouen.fr
[5] Department of Clinical Neurosciences, University of Cambridge, Cambridge, UK
gbw1000@cam.ac.uk

Abstract. This work addresses a central topic in Magnetic Resonance Imaging (MRI) which is the motion-correction problem in a joint reconstruction and registration framework. From a set of multiple MR acquisitions corrupted by motion, we aim at - *jointly* - reconstructing a single motion-free corrected image and retrieving the physiological dynamics through the deformation maps. To this purpose, we propose a novel variational model. First, we introduce an L^2 fidelity term, which intertwines reconstruction and registration along with the weighted total variation. Second, we introduce an additional regulariser which is based on the hyperelasticity principles to allow large and smooth deformations. We demonstrate through numerical results that this combination creates synergies in our complex variational approach resulting in higher quality reconstructions and a good estimate of the breathing dynamics. We also show that our joint model outperforms in terms of contrast, detail and blurring artefacts, a sequential approach.

Keywords: 2D registration · Reconstruction · Joint model ·
Motion correction · Magnetic Resonance Imaging ·
Nonlinear elasticity · Weighted total variation

1 Introduction

Magnetic Resonance Imaging (MRI) is a well-established modality that allows to capture details of almost any organ or anatomical structure of the human body. However, the lengthy period of time needed to acquire the necessary

© Springer Nature Switzerland AG 2019
J. Lellmann et al. (Eds.): SSVM 2019, LNCS 11603, pp. 263–274, 2019.
https://doi.org/10.1007/978-3-030-22368-7_21

measurements to form an MR image continues to be a major contributor to image degradation, which compromises clinical interpretation [27]. This prolonged time makes MRI highly sensitive to motion mainly because the timescale of physiological motion -including involuntary motion- is generally shorter than the acquisition time [27]. This motion is manifested as undesirable artefacts including geometric distortions and blurring, which causes a significant degradation of the image quality and affects the clinical relevance for diagnosis. Therefore, the question of *how to do motion correction in MRI?* is of great interest at both theoretical and clinical levels, and this is the problem addressed in this paper.

Although motion reduction strategies such as faster imaging (e.g. [18]) have shown potential results, *we focus on post-acquisition correction of MRI sequences*, which involves explicit estimation of the motion. In this context, there have been different attempts for motion correction in MRI, whose predominant scheme is based on image registration after reconstructing the images (i.e. to estimate a mapping between a pair of images). A set of algorithmic approaches use rigid registration including [1,14,16]. However, the intrinsic nature of the problem yields those techniques to fail to capture complex transformations - e.g. cardiac and breathing motion - which, in many clinical cases, are inevitable. To mitigate this limitation, different non-rigid image registration methods have been proposed in this domain - e.g. [9,13,15,17].

On the other side of the aforementioned methods that separately compute the image reconstruction and the motion estimation, there is another type of algorithmic approaches that computes those sub-tasks jointly. This philosophy was early explored in [24]. In that work, authors proposed a factorisation method, which uses the singular-value decomposition for recovering shape and motion under orthography. Particularly in the medical domain, and following a variational perspective [8], different works have been reported including for SPECT imaging [19,22], PET [6] and MRI [2]- to name a few. The works with a closer aim to ours are discussed next.

Schumacher et al. [22] presented an algorithmic approach that combines reconstruction and motion correction for SPECT imaging. The authors proposed a novel variational approach using a regulariser that penalises an offset of motion parameter to favour a mean location of the target object. However, the major limitation is that they only consider rigid motions. In the same spirit, Fessler [11] proposed a generic joint reconstruction/registration framework with a model based on a penalised-likelihood functional using a weighted least square fidelity term along with a spatial and a motion regulariser. Authors in [6] propose a joint model composed of a motion-aware likelihood function and a smoothing term for a simultaneous image reconstruction and motion estimation for PET data. More recently, Odille et al [21] proposed a joint model for MRI image reconstruction and motion estimation. This approach allows for an estimate of both intra and inter-image motion, meaning that, not only the misalignment problem is addressed but also it allows correcting for blurring/ghosting artefacts.

Most recently and motivated by the deep learning (DL) performance breakthrough, a set of algorithmic approaches has been devoted to investigate the benefits of DL for image registration - e.g. [25,26]. Although, certainly, those approaches deserve attention, their review goes beyond the scope of this paper.

Contributions. In this work, we follow the joint model philosophy, in which we seek to compute - simultaneously and jointly - the MR image reconstruction representing the true underlying anatomy and the registration tasks. Whilst the use of joint models (multi-tasking) has been widely explored for modalities such as PET and SPECT, the application for MRI is a largely untouched area. This is of great interest as MRIs highly sensitive to motion. With this motivation in mind we propose a variational model that allows for intra-scan motion correction in MRI. The significance of our approach is that by computing the reconstruction and registration tasks jointly one can exploit their strong correlation thus reducing error propagation and resulting in a significant motion correction. Whilst this is an important part of our solution, our main contributions are:

- A novel theoretically well-motivated mathematical framework for motion correction in MRI. It relies on multi-tasking to address reconstruction and registration jointly, and guarantees the preservation of anatomical structures and topology by ensuring the positivity of the Jacobian determinant.
- Evaluation of our approach on a set of numerical and visual results using MRI datasets. We also show that our joint model leads to a better output than a sequential approach using the FAIR toolbox with similar characteristics.

2   Joint Model for Motion Correction in MRI

In this section, we describe our approach for motion correction using MRI data. Firstly, we theoretically motivate our joint reconstruction and registration model and secondly, we formulate our joint variational model for motion correction. Finally, we describe an optimisation scheme to numerically solve our proposed joint model.

2.1   Mathematical Model and Notation

In an MRI setting, a target image u representing a part of the patient body is acquired in spatial-frequency space. Let us denote by $x_i \in L^2(\mathbb{R}^2)$ the i-th measurement acquired by an MRI scanner for $i = 1, \cdots, T$, where T is the number of acquisitions. The measured x_i can be modelled as $x_i = \mathcal{A}u_i + \varepsilon$, where $\mathcal{A} : L^2(\mathbb{R}^2) \rightarrow L^2(\mathbb{R}^2)$ is the forward Fourier transform and by $\mathcal{A}^* : L^2(\mathbb{R}^2) \rightarrow L^2(\mathbb{R}^2)$ the backward inverse Fourier transform. $\mathcal{A}$ is a linear and continuous operator according to Plancherel's theorem. In this work, we assume to have real-valued images, therefore, our MRI forward operator takes only the real part of complex-valued measurements.

Let Ω be the image domain, a connected bounded open subset of $\mathbb{R}^2$ of class $\mathcal{C}^1$, and $u : \Omega \rightarrow \mathbb{R}$ be the sought single reconstructed image depicting the true underlying anatomy. For theoretical purposes, we assume that it is equal to 0 on the boundary of Ω, so that, we can extend it by 0 and apply the operator $\mathcal{A}$.

We introduce the unknown deformation as $\varphi_i : \bar{\Omega} \rightarrow \mathbb{R}^2$, between the i-th acquisition and the image u. Moreover, let z_i be the associated displacements such that $\varphi_i = \text{Id} + z_i$, where Id is the identity function. At the practical level, these deformations should be with values in $\bar{\Omega}$. Yet, at the theoretical level, working with such spaces of functions result in losing the structure of vector space. Nonetheless, we can show that our model retrieves deformations with values in $\bar{\Omega}$ - based on Ball's results [4]. A deformation is a smooth mapping that is topology-preserving and injective, except possibly on the boundary if self-contact is allowed. We consider $\nabla\varphi_i : \Omega \to M_2(\mathbb{R})$ to be the gradient of the deformation, where $M_2(\mathbb{R})$ is the set of real square matrices of order two. After introducing the set of notations for this work, we now turn to to motivate the introduction of our mathematical model.

2.2   Joint Variational Model

In this work, we seek to extract - *jointly* - from a set of MR acquisitions x_i which are corrupted by motion, a clean static free motion-corrected reconstructed image u, along with the physiological dynamics through the deformation maps φ_i. With this purpose in mind, we consider the motion correction issue as a registration problem. In a unified variational framework, u and all φ_i's for $i = 1, \cdots, T$, are seen as optimal solutions of a specifically designed cost function composed of: (1) a regularisation on the deformations φ_i which prescribes the nature of the deformation, (2) a regularisation on u that allows for removing artefacts whilst keeping fine details and (3) a fidelity term measuring the discrepancy between the deformed mean - to which the forward operator has been applied - and the acquisitions, intertwining then the reconstruction and the registration tasks.

In order to allow for large and smooth deformations, we propose viewing the shapes to be matched as isotropic, homogeneous and hyperelastic materials. This outlook dictates the regularisation on the deformations φ_i, which is based on the stored energy function of an Ogden material. We introduce the following notations : $A : B = tr(A^T B)$ is the matrix inner product whilst $\|A\| = \sqrt{A : A}$ is the associated Frobenius matrix norm. In two dimensions, the stored energy function of an Ogden material, in its general form, is given by the following expression.

$$W_O(F) = \sum_{i=1}^{M} a_i \|F\|^{\gamma_i} + \Gamma(\det F), \text{ with } a_i > 0, \gamma_i \geq 1 \text{ for all } i = 1, \cdots, M \text{ and}$$

$\Gamma :]0; \infty[\to \mathbb{R}$, is a convex function satisfying $\lim_{\delta \to 0^+} \Gamma(\delta) = \lim_{\delta \to +\infty} \Gamma(\delta) = +\infty$.

In this work, we focus on the particular energy, which reads:

$$W_{Op}(F) = \begin{cases} a_1 \|F\|^4 + a_2 \left(\det F - \frac{1}{\det F}\right)^4 & \text{if } \det F > 0 \\ +\infty & \text{otherwise,} \end{cases}$$

with $a_1 > 0$, and $a_2 > 0$. The first term penalises the changes in length, whilst the second term enforces small changes in area. We check that this function falls within the general formulation of the stored energy function of an Ogden material: $W_{Op}(F) = \tilde{W}(\xi, \delta) = \begin{cases} a_1\|\xi\|^4 + a_2 \left(\delta - \frac{1}{\delta}\right)^4 & \text{if } \delta > 0, \\ +\infty & \text{otherwise,} \end{cases}$ $\tilde{W} : M_2(\mathbb{R}) \times \mathbb{R} \to$

$\mathbb{R}$ is continuous since $\lim_{\delta \to 0^+} \tilde{W}(\xi, \delta) = \lim_{\delta \to +\infty} \tilde{W}(\xi, \delta) = +\infty$, and is convex with

respect to $\delta \left(g : \begin{vmatrix} \mathbb{R}_*^+ \to \mathbb{R} \\ x \mapsto (x - \frac{1}{x})^4 \end{vmatrix} \text{ and } g'' : \begin{vmatrix} \mathbb{R}_*^+ \to \mathbb{R} \\ x \mapsto 4(x - \frac{1}{x})^2 \left(\frac{3x^4 + 4x^2 + 5}{x^4}\right) \geq 0 \end{vmatrix}\right)$.

The design of the function Γ is driven by Ball's results [4] guaranteeing the deformation to be a bi-Hölder homeomorphism, and therefore, preserving the topology. It also controls that the Jacobian determinant remains close to one to avoid expansions or contractions that are too large.

The aforementioned regulariser is then applied along a discrepancy measure, which allows to interconnect the reconstruction and the registration tasks, and a regularisation of our free motion-corrected slice based on the weighted total variation. With this purpose, let $g : \mathbb{R}^+ \rightarrow \mathbb{R}^+$ be an edge detector function satisfying $g(0) = 1$, g strictly decreasing and $\lim_{r \to +\infty} g(r) = 0$. For the sake of simplicity, we set $g_i = g(\|\nabla G_\sigma * \mathcal{A}^* x_i\|)$, with G_σ being a Gaussian kernel of variance σ and for theoretical purposes, we assume that there exists $c > 0$ such that $0 < c \le g_i \le 1$, and g_i is Lipschitz continuous for each $i = 1, \cdots, T$. In practice, we choose g to be the Canny edge detector. We follow Baldi's arguments [3] to introduce the weighted BV-space and the associated weighted total variation related to the weight g_i, for each $i = 1, \cdots, T$.

Definition 1 ([3, Definition 2]). *Let w be a weight function satisfying some properties (defined in [3] and fulfilled here by g_i). We denote by $BV_w(\Omega)$ the set of functions $u \in L^1(\Omega, w)$, which are integrable with respect to the measure $w(x)dx$, such that:*

$$\sup \left\{ \int_\Omega u\, div(\varphi)\, dx \; : \; |\varphi| \le w \text{ everywhere} \;, \; \varphi \in Lip_0(\Omega, \mathbb{R}^2) \right\} < +\infty,$$

with $Lip_0(\Omega, \mathbb{R}^2)$ the space of Lipschitz continuous functions with compact support. We denote by TV_w the previous quantity.

Given E a bounded open set in $\mathbb{R}^2$, with boundary of class $\mathcal{C}^2$, then we have $TV_{g_i}(\xi_E) = |\partial E|(\Omega, g_i) = \int_{\Omega \cap \partial E} g_i\, dH^1$, where ξ_E is the characteristic function of the set E. This quantity can be viewed as a new definition of the curve length-with a metric depending on the observations x_i. Minimising it is equivalent to locating the curve on the edges of $\mathcal{A}^* x_i$ where g_i is close to 0. We thus introduce the following fidelity term and regulariser for our single reconstruction:

$$F(u, (\varphi_i)_{i=1,\cdots,T}) = \frac{1}{T} \sum_{i=1}^T \delta TV_{g_i}(u \circ \varphi_i^{-1}) + \frac{1}{2} \|\mathcal{A}((u \circ \varphi_i^{-1})_e) - x_i\|_{L^2(\mathbb{R}^2)}^2,$$

where φ_i^{-1} is the inverse deformation and $(u \circ \varphi_i^{-1})_e$ is the extension by 0 of $(u \circ \varphi_i^{-1})$ outside Ω. Indeed, we assume that $u = 0$ and $\varphi_i = \varphi_i^{-1} = \text{Id}$ so that $u \circ \varphi_i^{-1} = 0$ on the boundary $\partial\Omega$. Moreover, if $u \circ \varphi_i^{-1} \in BV_{g_i,0}(\Omega)$, with $BV_{g_i,0}(\Omega) = BV_0(\Omega) \cap BV_{g_i}(\Omega)$ (see [12, Chap. 6.3.3] for more details on the trace operator on the space of functions of bounded variations), then $u \circ \varphi_i^{-1}$ is null on the boundary $\partial\Omega$ and can be extended by 0 outside the domain Ω. Due to the embedding theorem ([12, Chap. 6.3.3]), we have that $(u \circ \varphi_i^{-1})_e \in L^2(\mathbb{R}^2)$ and then Plancherel's theorem gives us $\mathcal{A}((u \circ \varphi_i^{-1})_e) \in L^2(\mathbb{R}^2)$ leading to the well-definedness of the fidelity term. The first term of F aims at aligning the edges of the deformed u ($u \circ \varphi_i^{-1}$) with the ones of the different acquisitions, whilst regularising the reconstructed image u. The second quantity targets to make the Fourier transform of the deformed u ($u \circ \varphi_i^{-1}$) as close as possible to the acquisitions x_i (in the L^2 sense) - inspired by the classical fidelity term in MRI reconstruction models.

By combining both terms, we finally get the following minimisation problem

$$\inf I(u, (\varphi_i)_{i=1,\cdots,T}) = F(u, (\varphi_i)_{i=1,\cdots,T}) + \frac{1}{T}\sum_{i=1}^{T}\int_{\Omega} W_{Op}(\nabla\varphi_i)\,dx. \qquad (P)$$

We now describe the numerical resolution of our joint model (P).

2.3 Numerical Method: Optimisation Scheme

In order to tackle the nonlinearity of the problem, we introduce three auxiliary variables v_i, w_i, f_i to mimic $\nabla\varphi_i$, $u \circ \varphi_i^{-1}$ and w_i, respectively, through quadratic penalty terms. This corresponds to reformulating the equality constraints $v_i = \nabla\varphi_i$, $w_i = u \circ \varphi_i^{-1}$ and $w_i = f_i$, into an unconstrained minimisation problem. We then propose solving the following decoupled (discretised) problem

$$\min_{u,\varphi_i,v_i,w_i,f_i} \frac{1}{T}\sum_{i=1}^{T}\sum_{x\in\Omega} W_{Op}(v_i(x)) + \frac{\gamma_1}{2}\|v_i - \nabla\varphi_i\|_2^2$$
$$+ \frac{\gamma_3}{2}\|Aw_i - x_i\|_2^2 + \frac{\gamma_2}{2}\|(w_i - u \circ \varphi_i^{-1})\sqrt{\det \nabla\varphi_i^{-1}}\|_2^2 \qquad (1)$$
$$+ \frac{1}{2\theta}\|f_i - w_i\|_2^2 + \mathrm{TV}_{g_i}(f_i).$$

We can now minimise (1) with an alternating splitting approach, where we solve the subproblem according to each individual variable whilst keeping the remaining variables fixed. We iterate the following updates for $k = 0, 1, \ldots$.

Sub-problem 1: Optimisation over v_i. In practice, v_i simulates the gradient of the displacements $z_i = (z_{i,1}, z_{i,2})$ associated to the deformations φ_i. For every v_i, we have $v_i = \begin{pmatrix} v_{11} & v_{12} \\ v_{21} & v_{22} \end{pmatrix}$. For the sake of readability, we drop here the dependency on i. We solve the Euler-Lagrange equation with a semi-implicit finite difference scheme (for N iterations and all i) and update v_i as:

$$\begin{cases} v_{11}^{k+1} = \dfrac{1}{1+dt\gamma}\left(v_{11}^k + dt(-4a_1\|v_i^k\|_F^2(v_{11}^k+1) - 4a_2(1+v_{22}^k)c_0c_1 + \gamma\dfrac{\partial z_1^k}{\partial x}\right), \\[2ex] v_{12}^{k+1} = \dfrac{1}{1+dt\gamma}\left(v_{12}^k + dt(-4a_1\|v_i^k\|_F^2 v_{12}^k + 4a_2 v_{21}^k c_0c_1 + \gamma\dfrac{\partial z_1^k}{\partial y}\right), \\[2ex] v_{21}^{k+1} = \dfrac{1}{1+dt\gamma}\left(v_{21}^k + dt(-4a_1\|v_i^k\|_F^2 v_{21}^k + 4a_2 v_{12}^k c_0c_1 + \gamma\dfrac{\partial z_2^k}{\partial x}\right), \\[2ex] v_{22}^{k+1} = \dfrac{1}{1+dt\gamma}\left(v_{22}^k + dt(-4a_1\|v_i^k\|_F^2(v_{22}^k+1) - 4a_2(1+v_{11}^k)c_0c_1 + \gamma\dfrac{\partial z_2^k}{\partial y}\right), \end{cases}$$

with $c_0 = \left(\det v_i^k - \frac{1}{\det v_i^k}\right)^3$ and $c_1 = 1 + \frac{1}{(\det v_i^k)^2}$.

Sub-problem 2: Optimisation over φ_i. We solve the Euler-Lagrange equation in φ_i, after making the change of variable $y = \varphi_i^{-1}(x)$ in the L^2 penalty term, for all i, using an L^2 gradient flow scheme with a semi-implicit Euler time stepping (for N iterations in the same inner loop as v_i):

$$0 = \gamma_1 \Delta \varphi_i^{k+1} + \gamma_1 \begin{pmatrix} \operatorname{div} v_{i,1}^{k+1} \\ \operatorname{div} v_{i,2}^{k+1} \end{pmatrix} + \gamma_2 (w_i^k \circ \varphi_i^k - u^k) \nabla w_i^k(\varphi_i^k).$$

Sub-problem 3: Optimisation over w_i. The update in w_i, for all i, reads:

$$w_i^{k+1} = \mathcal{A}^* \left\{ \frac{\mathcal{A}\left(\gamma_2 \det \nabla(\varphi_i^{-1})^{k+1} u \circ (\varphi_i^{-1})^{k+1} + \frac{f_i^k}{\theta}\right) + \gamma_3 x_i}{\gamma_2 \det \nabla(\varphi_i^{-1})^{k+1} + \gamma_3 + \frac{1}{\theta}} \right\}.$$

Sub-problem 4: Optimisation over f_i. For each k and i, this is solved via Chambolle projection algorithm [10], as in [7], for an inner loop over $n = 0, 1, \ldots$

$$f_i^{n+1} = w_i^{k+1} - \theta \operatorname{div} p_i^n,$$

$$p_i^{n+1} = \frac{p_i^n + \delta_t \nabla(\operatorname{div} p_i^n - w_i^{k+1}/\theta)}{1 + \frac{\delta_t}{g_i}|\nabla(\operatorname{div} p_i^n - w_i^{k+1}/\theta)|}.$$

After enough iterations, we set $f_i^{k+1} = f_i^{n+1}$.

Sub-problem 5: Optimisation over u. Finally, using the same change of variables as in sub-problem 2, the problem in u is simply

$$u^{k+1} = \frac{1}{T} \sum_{i=1}^{T} w_i^{k+1} \circ \varphi_i^{k+1}.$$

In the following section, we present the results obtained with our algorithmic approach, where the registration problem in v_i and φ_i, $i = 1, \ldots, T$, has been solved in a multi-scale framework, from coarser to finer grids. It is worth noticing that - even though in W_{Op}, we have a term controlling that the Jacobian determinant remains positive, we introduce an additional step in our implementation, to monitor that the determinant does not become negative. To do so, we use the following regridding technique: if at iteration N, $\det(\nabla \varphi^N)$ becomes too small or negative, we save $\varphi^{k,N-1}$ and re-initialise $\varphi^{k,N} = \operatorname{Id}$ and $v^{k,N} = 0$, and $w^k = w^k \circ \varphi^{k,N-1}$, and we then continue the loop on N. Finally, $\varphi^{k,final}$ is the composition of all the saved deformations.

3 Numerical Experiments

This sections describes the set of experimental results that are performed to validate our proposed joint model.

Data Description. We evaluate our framework on two datasets. The *Dataset* I^1 is a 4DMRI data acquired during free-breathing of the right liver lobe [23]. It was acquired on a 1.5T Philips Achieva system, TR = 3.1 ms, coils = 4, slices = 25, matrix size = 195 × 166. *Dataset* II^2 is a 2D T1-weighted dataset [5] acquired during free breathing of the entire thorax. It was acquired with a 3T Philips Achieva system with matrix size = 215 × 173, slice thickness = 8 mm, TR = 3.1 ms and TE = 1.9 ms. The experiments reported in this section were run under the same condition in a CPU-based Matlab implementation. We used an Intel core i7 with 4 GHz and a 16 GB RAM.

Table 1. Parameter values.

	a_1	a_2	γ_1	γ_2	γ_3	θ	σ	k	N	n
Dataset I	1	50	5	10^5	30	5	2	2	500	500
Dataset II	1	50	5	10^5	15	5	1.5	2	500	500

Parameter Selection. We present a discussion on the role of parameters in our model. The ranges of these parameters are rather stable for both experiments as shown in Table 1. The parameters in the regularisation term for the registration W_{Op} are a_1 and a_2. The former controls the smoothness of the deformation whilst the latter is more sensitive and can be viewed as a measure of rigidity allowing for a trade-off between topology preservation and the capacity of handling large deformations. If a_2 is chosen too small, the deformation may lose its injectivity property resulting in the loss of topology preservation, while if chosen too big, the registration accuracy may be impaired. The parameters γ_1, γ_2 and θ, balance the L^2 penalty terms. Finally, γ_3 weights the data-fitting term for the reconstruction. For the FAIR approach - which is described next - we set the regularisation parameter for Dataset I = 0.1 and Dataset II = 1.

Comparison with FAIR. We begin by evaluating our joint model against a sequential approach using the FAIR toolbox [20], which is a well-established framework for image registration. For a fair comparison, we set the characteristics of the compared approach closer to ours, as follows. Firstly, we choose the sum of squared differences as a distance measure on A^*x_i, and secondly, we set the hyperelastic regulariser - as it allows for similar modeling of the deformations.

The computations of FAIR and our approach were performed under the same conditions. We ran the same multi-scale approach. Moreover, we set the same reference image that we selected for the initialisation of our alternating scheme. This is important because of the non-convexity of our joint problem. After registering each frame to the reference, we average the results to obtain a mean image. In Fig. 1, we show the outputs - for both datasets - *uncorrected* being the coarse Euclidean mean, FAIR referring to the results of the process described above; and *ours* being the reconstructed image u obtained with our model. By visual inspection, we observe that the coarse Euclidean mean displays highly blurry effects along with loss of details whilst our approach improves in this regard. We also observed that - although FAIR improves over the coarse mean - outputs

---

[1] http://www.vision.ee.ethz.ch/~organmot/chapter_download.shtml.
[2] https://zenodo.org/record/55345#.XBOkvi2cbUZ.

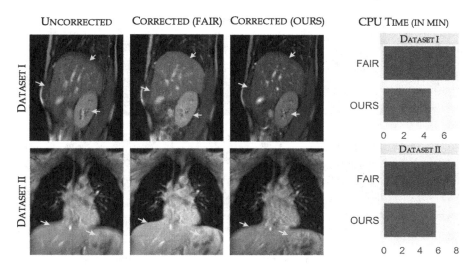

Fig. 1. Free-motion corrected comparisons: (from left to right) coarse Euclidean mean reconstructed image, FAIR and our joint model. Visual assessment in terms of blurring artefacts, loss of details and contrast are pointed out with yellow arrows. Elapsed time, in min, comparison between FAIR and ours approaches. (Color figure online)

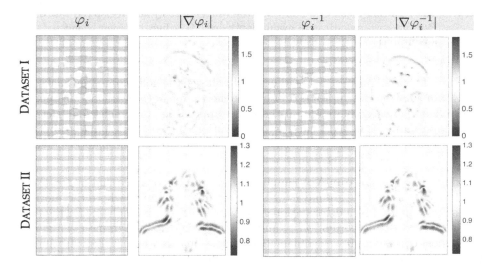

Fig. 2. Estimated motion and determinant maps of the deformation Jacobian. This is shown for the transformation φ_i and its inverse φ_i^{-1} for our two datasets.

generated with FAIR have a remarkable loss of contrast and detail. In addition, we point out that even though our method needs to solve 5 subproblems, the final computational time is less than the sequential FAIR approach.

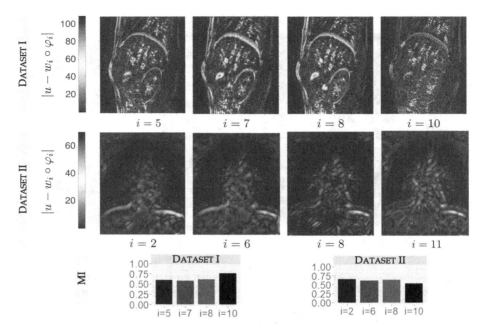

Fig. 3. Top and middle rows: visual assessment of the difference map, at specific times, between the resulting reconstruction u (in the range $[0, 255]$) and registered raw data $w_i \circ \varphi_i$. Bottom row: quantitative evaluation using MI values between u and $w_i \circ \varphi_i$ for the same frames.

Overall, visual inspection of the results shows the following drawbacks in which our joint model improves: (1) blurry effects: for both datasets, our model gives clearer images compared to the other approaches in which small details and shapes are lost, and (2) contrast preservation: our method is able to better preserve the contrast than other approaches, especially the one based on FAIR in which loss of contrast is clearly visible.

Further Analysis of our Approach. In Fig. 2, we can observe that our method produces a reasonable estimation of the breathing dynamics, where the motion fields are visualised by a deformation grid. More precisely, it displays - for the two datasets - the estimated motion φ_i and its inverse φ_i^{-1} for the frame $i = 5$ along with their corresponding Jacobian determinant maps, whose values are interpreted as follows: small deformations when values are closer to 1, big expansions when values are greater than 1, and big contractions when values are smaller than 1. Moreover, one can observe that the determinants remain positive in all cases, that is to say, our estimated deformations are physically meaningful and preserve the topology as required in a registration framework.

In Fig. 3, we show further results for the registration. We display a difference map (top) between our (adaptive) reference u, and the registered data $w_i \circ \varphi_i$ for sample frames for Dataset I and II. Moreover, we show the mutual information

(MI) between the same images as a quantitative measure of performance (bottom), where larger MI values indicates better alignment. Since the L^2 fidelity term drives the registration process in our optimisation framework, it sounds relevant to use the difference map to illustrate the accuracy of our model. However, one should also be aware that a small change in contrast - which often happens between two frames and thus between u and $w_i \circ \varphi_i$ - is clearly visible on the difference map, even if the overall anatomies are well aligned. Keeping in mind that, for the sake of tractability, the parameters are set globally and not adjusted for each frame, we can see that the intensities and the structures are well aligned for all the examples. This is corroborated by large MI values.

4 Conclusion

In this work we presented a new mathematical model to tackle motion correction in MRI in a joint reconstruction/registration framework. The underlying idea is to exploit redundancy in the temporal resolution in the data to compensate for motion artefacts due to breathing. The proposed multi-tasking approach solves simultaneously the problems of reconstruction and registration to achieve higher image quality and accurate estimation of the breathing dynamics. We mathematically motivated our joint model and presented our optimisation scheme. Furthermore, our experimental results show the potential of our approach for clinical applications.

Acknowledgements. VC acknowledges the financial support of the Cambridge Cancer Centre. Support from the CMIH and CCIMI University of Cambridge is greatly acknowledged.

References

1. Adluru, G., DiBella, E.V., Schabel, M.C.: Model-based registration for dynamic cardiac perfusion MRI. J. Magn. Reson. Imaging **24**(5), 1062–70 (2006)
2. Aviles-Rivero, A.I., Williams, G., Graves, M.J., Schönlieb, C.B.: Compressed sensing plus motion (CS+M): a new perspective for improving undersampled MR image reconstruction. arXiv preprint arXiv:1810.10828 (2018)
3. Baldi, A.: Weighted BV functions. Houston J. Math. **27**(3), 683–705 (2001)
4. Ball, J.M.: Global invertibility of Sobolev functions and the interpenetration of matter. Proc. Roy. Soc. Edinb. A **88**(3–4), 315–328 (1981)
5. Baumgartner, C.F., Kolbitsch, C., McClelland, J.R., Rueckert, D., King, A.P.: Autoadaptive motion modelling for MR-based respiratory motion estimation. Med. Image Anal. **35**, 83–100 (2017)
6. Blume, M., Martinez-Moller, A., Keil, A., Navab, N., Rafecas, M.: Joint reconstruction of image and motion in gated positron emission tomography. IEEE Trans. Med. Imaging **29**(11), 1892–1906 (2010)
7. Bresson, X., Esedoḡlu, S., Vandergheynst, P., Thiran, J.P., Osher, S.: Fast global minimization of the active contour/snake model. J. Math. Imaging Vis. **28**(2), 151–167 (2007)

8. Burger, M., Dirks, H., Schönlieb, C.B.: A variational model for joint motion estimation and image reconstruction. SIAM J. Imaging Sci. **11**(1), 94–128 (2018)
9. Burger, M., Modersitzki, J., Ruthotto, L.: A hyperelastic regularization energy for image registration. SIAM J. Sci. Comput. **35**(1), B132–B148 (2013)
10. Chambolle, A.: An algorithm for total variation minimization and applications. J. Math. Imaging Vis. **20**(1), 89–97 (2004)
11. Chun, S., Fessler, J.: Joint image reconstruction and nonrigid motion estimation with a simple penalty that encourages local invertibility. In: Proceedings of the SPIE, vol. 7258
12. Demengel, F., Demengel, G., Erné, R.: Functional Spaces for the Theory of Elliptic Partial Differential Equations. Universitext. Springer, London (2012). https://doi.org/10.1007/978-1-4471-2807-6
13. Droske, M., Rumpf, M.: A variational approach to nonrigid morphological image registration. SIAM J. Appl. Math. **64**(2), 668–687 (2004)
14. Gupta, S.N., Solaiyappan, M., Beache, G.M., Arai, A.E., Foo, T.K.: Fast method for correcting image misregistration due to organ motion in time-series MRI data. Magn. Reson. Med. **49**(3), 506–514 (2003)
15. Jansen, M., Kuijf, H., Veldhuis, W., Wessels, F., Van Leeuwen, M., Pluim, J.: Evaluation of motion correction for clinical dynamic contrast enhanced MRI of the liver. Phys. Med. Biol. **62**(19), 7556 (2017)
16. Johansson, A., Balter, J., Cao, Y.: Rigid-body motion correction of the liver in image reconstruction for golden-angle stack-of-stars DCE MRI. Magn. Reson. Med. **79**(3), 1345–1353 (2018)
17. Ledesma-Carbayo, M.J., Kellman, P., Arai, A.E., McVeigh, E.R.: Motion corrected free-breathing delayed-enhancement imaging of myocardial infarction using nonrigid registration. J. Magn. Reson. Imaging **26**(1), 184–90 (2007)
18. Lustig, M., Donoho, D., Pauly, J.M.: Sparse MRI: the application of compressed sensing for rapid MR imaging. Magn. Reson. Med. **58**, 1182–1195 (2007)
19. Mair, B.A., Gilland, D.R., Sun, J.: Estimation of images and nonrigid deformations in gated emission CT. IEEE Trans. Med. Imaging **25**(9), 1130–1144 (2006)
20. Modersitzki, J.: FAIR: Flexible Algorithms for Image Registration. SIAM, Philadelphia (2009)
21. Odille, F., et al.: Joint reconstruction of multiple images and motion in MRI: application to free-breathing myocardial T2 quantification. IEEE Trans. Med. Imaging **35**(1), 197–207 (2016)
22. Schumacher, H., Modersitzki, J., Fischer, B.: Combined reconstruction and motion correction in SPECT imaging. ITNS **56**(1), 73–80 (2009)
23. von Siebenthal, M., Szekely, G., Gamper, U., Boesiger, P., Lomax, A., Cattin, P.: 4D MR imaging of respiratory organ motion and its variability. Phys. Med. Biol. **52**(6), 1547 (2007)
24. Tomasi, C., Kanade, T.: Shape and motion from image streams under orthography: a factorization method. Int. J. Comput. Vis. **9**, 137–154 (1992)
25. de Vos, B.D., Berendsen, F.F., Viergever, M.A., Staring, M., Išgum, I.: End-to-end unsupervised deformable image registration with a convolutional neural network. In: Cardoso, M.J., et al. (eds.) DLMIA/ML-CDS -2017. LNCS, vol. 10553, pp. 204–212. Springer, Cham (2017). https://doi.org/10.1007/978-3-319-67558-9_24
26. Yang, X., Kwitt, R., Styner, M., Niethammer, M.: Quicksilver: fast predictive image registration-a deep learning approach. NeuroImage **158**, 378–396 (2017)
27. Zaitsev, M., Maclaren, J., Herbst, M.: Motion artifacts in MRI: a complex problem with many partial solutions. J. Magn. Reson. Imaging **42**(4), 887–901 (2015)

Variational Image Registration for Inhomogeneous-Resolution Pairs

Kento Hosoya[1] and Atsushi Imiya[2(✉)]

[1] School of Science and Engineering, Chiba University,
Yayoi-cho 1-33, Inage-ku, Chiba 263-8522, Japan
[2] Institute of Management and Information Technologies, Chiba University,
Yayoi-cho 1-33, Inage-ku, Chiba 263-8522, Japan
`imiya@faculty.chiba-u.jp`

Abstract. We propose a variational image registration method for a pair of images with different resolutions. Traditional image registration methods match images assuming that the resolutions of the reference and target images are homogeneous. For the registration of inhomogeneous-resolution image pairs, we first introduce a resolution-conversion method to harmonise the resolution of a pair of images using the rational-order pyramid transform. Then, we develop a variational method for image registration using this resolution-conversion method.

1 Introduction

We develop a variatinal method for the registration of reference and target images with different resolutions. For longitudinal analysis [1–4], we are required to register a temporal sequence of images observed by different modalities. The resolution of medical images depends on the modality of observation. If the mechanism of imaging systems is completely modelled, the inverse operation of blur yields clear images with the same resolution. The resolution of medical images depends on the modality of observation. Even if physically the same observation modality is used for measuring each image in a sequence, images with different resolutions are measured. For instance, the width of x-rays used for computerised tomography affects to the resolution of slice images. The same slice images measured by using x-rays with different energies possess the different resolutions [5], since the energy of x-rays mathematically defines the width of the x-ray beam.

If super-resolution is pre-processed to the low-resolution images of the registration pair for harmonisation of the resolutions of the pair of images, traditional techniques for image registration [6,7] with the same resolution can be used as the second step. In this two-step method, the quality of the super-resolution technique in the first step affects the image registration result. Therefore, a direct method to compute the displacement for image registration with different resolutions is demanded. The resolution of a pair of images for the registration is normalised by the rational-order pyramid transform. The rational-order pyramid transform, which possess similar mathematical properties with the generalised inverse [8]. is a generalisation of the traditional pyramid transform [9–13].

© Springer Nature Switzerland AG 2019
J. Lellmann et al. (Eds.): SSVM 2019, LNCS 11603, pp. 275–287, 2019.
https://doi.org/10.1007/978-3-030-22368-7_22

2   Mathematical Preliminaries

Setting $\boldsymbol{y} = (u, v)^\top$ and $\boldsymbol{x} = (x, y)^\top$ to be orthogonal coordinate systems on coarse- and fine-image planes, respectively, we deal with linear transforms between coarse- and fine-image planes. We assume that our images are elements of the Sobolev space $H^2(\mathbf{R}^2)$. The downsampling and upsampling operations with factor $\sigma > 0$ are defined as

$$S_\sigma f(x, y) = f(\sigma x, \sigma y), \qquad U_\sigma g(x, y) = \sigma^{-2} g(\sigma^{-1} x, \sigma^{-1} y). \qquad (1)$$

Since $S_\sigma U_\sigma f = f$ and $\int_{-\infty}^{\infty} \int_{-\infty}^{\infty} |S_\sigma|^2 dx dy = \int_{-\infty}^{\infty} \int_{-\infty}^{\infty} |f|^2 dx dy$, $U_\sigma = S_\sigma^*$ and U_σ is a partial isometric operator. For the function

$$w_\sigma(s) = \begin{cases} \frac{1}{\sigma} \left(1 - \frac{1}{\sigma} |s| \right) & |s| \leq \sigma, \\ 0 & |s| > \sigma, \end{cases} \qquad (2)$$

we define the linear transforms

$$f_\sigma(u, v) = R_\sigma f(u, v) = \int_{-\infty}^{\infty} \int_{-\infty}^{\infty} w_\sigma(x) w_\sigma(y) f(\sigma u - x, \sigma v - y) dx dy, \qquad (3)$$

$$g^\sigma(x, y) = E_\sigma g(x, y) = \frac{1}{\sigma^2} \int_{-\infty}^{\infty} \int_{-\infty}^{\infty} w_\sigma(u) w_\sigma(v) g\left(\frac{x - u}{\sigma}, \frac{y - v}{\sigma} \right) du dv. \qquad (4)$$

These transforms perform downsampling with factor σ after shift-invariant filtering and linear interpolation after upsampling with factor σ, respectively. Since the ratio of the numbers of pixels in images defines the resolution ratio, hereafter we assume that σ in Eqs. (3) and (4) is a rational number.

Definition 1. *In both the defined domain and the range space of the transformation R, the inner products of functions are defined as*

$$(f, g)_D = \int_{-\infty}^{\infty} \int_{-\infty}^{\infty} f(x, y) g(x, y) dx dy, \qquad (5)$$

$$(R_\sigma f, R_\sigma g)_R = \int_{-\infty}^{\infty} \int_{-\infty}^{\infty} R_\sigma f(u, v) R_\sigma g(u, v) du dv. \qquad (6)$$

Since the relation

$$\int_{-\infty}^{\infty} \int_{-\infty}^{\infty} R_\sigma f(u, v) g(u, v) du dv = \int_{-\infty}^{\infty} \int_{-\infty}^{\infty} f(x, y) E_\sigma g(x, y) dx dy \qquad (7)$$

is satisfied, the operator E_σ is the dual operator of R_σ.

We deal with the registration of $f(\boldsymbol{y})$ and $g(\boldsymbol{x})$, which are defined on the coarse- and fine-image planes, respectively. Using the pyramid transform $R_\sigma \boldsymbol{x} = (R_\sigma x, R_\sigma y)^\top$ and its dual transform $E_\sigma \boldsymbol{y} = (E_\sigma u, E_\sigma v)^\top$ for vector-valued functions, we define the minimisation problem

$$J(\boldsymbol{u}, \boldsymbol{v}) = \int_{-\infty}^{\infty} \int_{-\infty}^{\infty} \{ (f(\boldsymbol{y} + R_\sigma \boldsymbol{u}) - R_\sigma g(\boldsymbol{y} - \boldsymbol{v}))^2 + \lambda |\nabla \boldsymbol{v}|_F^2 \} du dv$$

$$+ \int_{-\infty}^{\infty} \int_{-\infty}^{\infty} \{ (E_\sigma f(\boldsymbol{x} + E_\sigma \boldsymbol{v}) - g(\boldsymbol{x} - \boldsymbol{u}))^2 + \lambda |\nabla \boldsymbol{u}|_F^2 \} dx dy, \qquad (8)$$

for the Frobenius norms $|\nabla u|_F$ and $|\nabla v|_F$ of Jacobian matrices ∇u and ∇v of the deformation vectors u and v on the fine- and coarse-image planes, respectively, with the constraints $\frac{\partial}{\partial m} v = 0$ and $\frac{\partial}{\partial n} u = 0$, where $\frac{\partial}{\partial m} v$ and $\frac{\partial}{\partial n} u$ denote directional differentiation on the boundaries of the domains of u and v, respectively. For the deformation filed vectors $u = (u_1(x, y), u_2(x, y))^\top$ and $v = (v_1(u, v), v_2(u, v))^\top$, the Frobenius norms of the Jacobian matrices ∇u and ∇v are

$$|\nabla u|_F^2 = tr \nabla u \nabla u^\top = u_{1x}^2 + u_{1y}^2 + u_{2x}^2 + u_{2y}^2, \tag{9}$$

$$|\nabla u|_F^2 = tr \nabla v \nabla v^\top = v_{1u}^2 + v_{1v}^2 + v_{2u}^2 + v_{2v}^2. \tag{10}$$

The minimisation of Eq. (8) matches a pair of images with different resolutions without involving any super-resolution process. The variation of Eq. (8) with respect to u and v derives the following system of the Euler-Lagrange equations:

$$(f(y + R_\sigma u) - R_\sigma g(y - v)) \nabla_v R_\sigma g(y - v) - \lambda \Delta v = 0, \qquad \frac{\partial}{\partial m} v = 0, \tag{11}$$

$$(E_\sigma f(x + E_\sigma v) - g(x - u)) \nabla_u g(x - u) - \lambda \Delta u = 0, \qquad \frac{\partial}{\partial n} u = 0. \tag{12}$$

The solutions of these equations are derived as the solutions for $t = \infty$ of the system of diffusion equations

$$\frac{\partial v}{\partial t} = \Delta v - \frac{1}{\lambda}(f(y + R_\sigma u) - R_\sigma g(y - v)) \nabla_v R_\sigma g(y - v), \qquad \frac{\partial}{\partial m} v = 0, \tag{13}$$

$$\frac{\partial u}{\partial t} = \Delta u - \frac{1}{\lambda}(E_\sigma f(x + E_\sigma v) - g(x - u)) \nabla_u g(x - u), \qquad \frac{\partial}{\partial n} u = 0. \tag{14}$$

3 Numerical Method for Variational Image Registration

3.1 Rational-Order Pyramid Transform

Setting $f_{ij} = f(\Delta i, \Delta j)$, the two-dimensional pyramid transform of order p and its dual transform are

$$R_p f_{mn} = h_{pm\,pn}, \quad h_{mn} = \sum_{\alpha,\beta=-(p-1)}^{(p-1)} \frac{p - |\alpha|}{p^2} \cdot \frac{p - |\beta|}{p^2} f_{m+\alpha\,n+\beta}, \tag{15}$$

$$E_p g_{pm+\alpha\,pn+\beta} = \frac{1}{p^2} \left(\frac{p - \alpha}{p} \cdot \frac{p - \beta}{p} g_{pm\,pn} + \frac{\alpha}{p} \cdot \frac{\beta}{p} g_{p(m+1)\,p(n+1)} \right), \tag{16}$$

for $\alpha, \beta = 0, 1, \cdots, (p-1)$. Using these transforms, we define the rational-order pyramid transform.

Definition 2. *The q/p-pyramid transform first constructs the upsampled images with order p by using linear interpolation. For the upsampled data, the pyramid transform of order q is applied.*

Definition 3. *The dual transform of the q/p-pyramid transform is achieved by downsampling to the result of the dual transform of the pyramid transform.*

Setting f to be the vectorisation of image array F, we have the following theorem.

Theorem 1. *The q/p-pyramid transform and its dual transform are expressed as*

$$f_{q/p} = (R_{q/p} \otimes R_{q/p})f, \qquad g^{q/p} = (E_{q/p} \otimes E_{q/p})g, \qquad (17)$$

where $R_{q/p}$ and $E_{q/p}$ are the matrix expression of the q/p-pyramid transform and its dual transform, respectively, for vectors.

For the $n \times n$ second-order difference matrix

$$D_n = \begin{pmatrix} -1 & 1 & 0 & 0 & \cdots & 0 & 0 \\ 1 & -2 & 1 & 0 & \cdots & 0 & 0 \\ 0 & 1 & -2 & 1 & \cdots & 0 & 0 \\ \vdots & \vdots & \vdots & \vdots & \ddots & \vdots & \vdots \\ 0 & 0 & 0 & \cdots & 0 & 1 & -1 \end{pmatrix} \qquad (18)$$

with the Neumann boundary condition, the eigendecomposition of D_n is

$$D_n \Phi_n = \Phi_n \Lambda_n, \quad \Phi_n^\top \Phi_n = I_n, \quad \Lambda_n = ((\lambda_k^{(m)} \delta_{kl})), \quad \lambda_k^{(m)} = 4 \sin^2 \frac{\pi k}{2n}. \quad (19)$$

The matrix Φ_n is the discrete cosine transform (DCT) matrix of type II for vectors in $\mathbf{R}^n$.

Setting $\Phi_n = (\varphi_0, \varphi_1, \cdots, \varphi_{n-1})$, Eq. (17) derives the following theorems. For the proofs of these theorems, see the appendix.

Theorem 2. *With the Neumann boundary condition, the q/p-pyramid transform $R_{q/p}$ is a linear transform from $\mathcal{L}\{\varphi_i \varphi_j^\top\}_{i=0}^{n-1}$ to $\mathcal{L}\{\varphi_i \varphi_j^\top\}_{i,j=0}^{\frac{p}{q}n-1}$ for $n = kq$.*

Theorem 3. *With the Neumann boundary condition, the dual transform $E_{q/p}$ of the q/p-pyramid transform $R_{q/p}$ is a linear transform from $\mathcal{L}\{\varphi_i \varphi_j^\top\}_{i=0}^{\frac{p}{q}n-1}$ to $\mathcal{L}\{\varphi_i \varphi_j^\top\}_{i,j=0}^{n-1}$ for $n = kq$.*

3.2   Discretisation of Variational Method

We assume that the resolution ration between a pair of images f and g is $q : p$. Semi-implicit discretisation of Eqs. (13) and (14) derives the system of iterations

$$(I - \tau L_n)v^{(m+1)} = v^{(m)} - \frac{\tau}{\lambda}(f(y + R_{q/p}u^{(m)}) - R_{q/p}g(y - v^{(m)}))\nabla_v R_{q/p}g(y - v^{(m)}), \quad (20)$$

$$(I - \tau L_{np/q})u^{(m+1)} = u^{(m)} - \frac{1}{\lambda}(E_{q/p}f(x + E_{q/p}v^{(m)}) - g(x - u^{(m)}))\nabla_u g(x - u^{(m)}). \quad (21)$$

For the Neumann boundary condition, the Laplacian matrix is expressed as

$$\boldsymbol{L}_k = \boldsymbol{D}_k \otimes \boldsymbol{I}_k + \boldsymbol{I}_k \otimes \boldsymbol{D}_k = \boldsymbol{U}_k^\top (\boldsymbol{\Lambda}_k \oplus \boldsymbol{\Lambda}_k) \boldsymbol{U}_k, \tag{22}$$

where

$$\boldsymbol{\Lambda}_k \oplus \boldsymbol{\Lambda}_k = \boldsymbol{\Lambda}_k \otimes \boldsymbol{I}_k + \boldsymbol{I}_k \otimes \boldsymbol{\Lambda}_k, \qquad \boldsymbol{U}_k = \boldsymbol{\Phi}_k \otimes \boldsymbol{\Phi}_k. \tag{23}$$

Therefore, for

$$F(\boldsymbol{u}^{(m)}, \boldsymbol{v}^{(m)}) = \frac{\tau}{\lambda}(f(\boldsymbol{y} + \boldsymbol{R}_{q/p}\boldsymbol{u}^{(m)})$$
$$-\boldsymbol{R}_{q/p}g(\boldsymbol{y} - \boldsymbol{v}^{(m)}))\nabla_v \boldsymbol{R}_{q/p}g(\boldsymbol{y} - \boldsymbol{v}^{(m)}), \tag{24}$$

$$G(\boldsymbol{u}^{(m)}, \boldsymbol{v}^{(m)}) = \frac{1}{\lambda}(\boldsymbol{E}_{q/p}f(\boldsymbol{x} + \boldsymbol{E}_{q/p}\boldsymbol{v}^{(m)})$$
$$-g(\boldsymbol{x} - \boldsymbol{u}^{(m)}))\nabla_u g(\boldsymbol{x} - \boldsymbol{u}^{(m)}), \tag{25}$$

Eqs. (20) and (21) become

$$\boldsymbol{v}^{(m+1)} = \boldsymbol{U}_{pn/q}^\top \boldsymbol{\Sigma}_{pn/q} \boldsymbol{U}_{pn/q} \boldsymbol{b}^{(m)}, \qquad \boldsymbol{b}^{(m)} = \boldsymbol{v}^{(m)} - F(\boldsymbol{u}^{(m)}, \boldsymbol{v}^{(m)}), \tag{26}$$

$$\boldsymbol{u}^{(m+1)} = \boldsymbol{U}_n^\top \boldsymbol{\Sigma}_n \boldsymbol{U}_n \boldsymbol{a}^{(m)}, \qquad \boldsymbol{a}^{(m)} = \boldsymbol{u}^{(m)} - G(\boldsymbol{u}^{(m)}, \boldsymbol{v}^{(m)}), \tag{27}$$

where

$$\boldsymbol{\Sigma}_{pn/q} = (\boldsymbol{I}_{pn/q} - \tau \boldsymbol{\Lambda}_{pn/q})^{-1} = (((1 - \tau(\lambda_i^{(pn/q)} + \lambda_j^{(pn/q)}))^{-1})), \tag{28}$$

$$\boldsymbol{\Sigma}_n = (\boldsymbol{I}_n - \tau \boldsymbol{\Lambda}_n)^{-1} = (((1 - \tau(\lambda_i^{(n)} + \lambda_j^{(n)}))^{-1})). \tag{29}$$

These relations imply that the displacement vectors $\boldsymbol{u}^{(m+1)}$ and $\boldsymbol{v}^{(m+1)}$, which establish registration of a pair of images with different resolutions, are computed using the two-dimensional DCT of type II. and filtering operations from $\boldsymbol{u}^{(m)}$ and $\boldsymbol{v}^{(m)}$, respectively.

4   Resolution Interpolation

Next, we develop a method for the generation of an average image h with an intermediate resolution from a pair of inhomogeneous-resolution images f and g. We assume that the resolution ratio of f and g is $q : p$ and that the resolution ratio of f and h is $r : s$. Assuming that the image h is an element in $\mathcal{L}(\{\boldsymbol{\varphi}_i\boldsymbol{\varphi}_j^\top\}_{i,j=0}^{n-1})$, f and g are elements in $\mathcal{L}(\{\boldsymbol{\varphi}_i\boldsymbol{\varphi}_j^\top\}_{i,j=0}^{\frac{q}{p}n-1})$ and $\mathcal{L}(\{\boldsymbol{\varphi}_i\boldsymbol{\varphi}_j^\top\}_{i,j=0}^{\frac{s}{t}n-1})$, respectively.

Setting $f_{q/p} = R_{q/p}f$ and $g^{r/s} = E_{r/s}g$, we define the minimisation problem

$$J(h, \boldsymbol{u}, \boldsymbol{v}) = \int_{\mathbf{R}^2} (h(\boldsymbol{x} + \boldsymbol{u}) - f_{q/p}(\boldsymbol{x}))^2 dxdy + \int_{\mathbf{R}^2} (h(\boldsymbol{x} + \boldsymbol{v}) - g^{r/s}(\boldsymbol{x}))^2 dxdy$$
$$+\mu \int_{\mathbf{R}^2} |\nabla h|^2 dxdy + \lambda \int_{\mathbf{R}^2} |\nabla \boldsymbol{u}|_F^2 dxdy + \lambda \int_{\mathbf{R}^2} |\nabla \boldsymbol{v}|_F^2 dxdy \tag{30}$$

for $h(x, y)$, u and v with the Neumann boundary condition.

From Eq. (30), we have the system of differential equations

$$\Delta h = \frac{1}{\mu} \left\{ (h(\boldsymbol{x} + \boldsymbol{u}) - f_{q/p}(\boldsymbol{x})) + (h(\boldsymbol{x} + \boldsymbol{v}) - g^{r/s}(\boldsymbol{x})) \right\}, \tag{31}$$

$$\Delta \boldsymbol{u} = \frac{1}{\lambda} \left\{ (h(\boldsymbol{x} + \boldsymbol{u}) - f_{q/p}(\boldsymbol{x})) \nabla_u h(\boldsymbol{x} + \boldsymbol{u}) \right\}, \tag{32}$$

$$\Delta \boldsymbol{v} = \frac{1}{\lambda} \left\{ (h(\boldsymbol{x} + \boldsymbol{v}) - g^{r/s}(\boldsymbol{x})) \nabla_v h(\boldsymbol{x} + \boldsymbol{v}) \right\}. \tag{33}$$

Then, the semi-implicit discretisation of the system of these diffusion equations

$$\frac{\partial h}{\partial t} = \Delta h - a(\boldsymbol{x}), \qquad \frac{\partial \boldsymbol{u}}{\partial t} = \Delta \boldsymbol{u} - \boldsymbol{b}(\boldsymbol{x}), \qquad \frac{\partial \boldsymbol{v}}{\partial t} = \Delta \boldsymbol{v} - \boldsymbol{c}(\boldsymbol{x}), \tag{34}$$

for

$$a(\boldsymbol{x}) = \frac{1}{\mu} \left\{ (h(\boldsymbol{x} + \boldsymbol{u}) - f_{q/p}(\boldsymbol{x})) + (h(\boldsymbol{x} + \boldsymbol{v}) - g^{r/s}(\boldsymbol{x})) \right\}, \tag{35}$$

$$\boldsymbol{b}(\boldsymbol{x}) = \frac{1}{\lambda} \left\{ (h(\boldsymbol{x} + \boldsymbol{u}) - f_{q/p}(\boldsymbol{x})) \nabla_u h(\boldsymbol{x} + \boldsymbol{u}) \right\}, \tag{36}$$

$$\boldsymbol{c}(\boldsymbol{x}) = \frac{1}{\lambda} \left\{ (h(\boldsymbol{x} + \boldsymbol{v}) - g^{r/s}(\boldsymbol{x})) \nabla_v h(\boldsymbol{x} + \boldsymbol{v}) \right\} \tag{37}$$

derives the system of iterations

$$\frac{h^{(m+1)} - h^{(m)}}{\tau} = \boldsymbol{L}_n h^{(m+1)} - \frac{1}{\mu} a^{(m)}, \tag{38}$$

$$\frac{\boldsymbol{u}^{(m+1)} - \boldsymbol{u}^{(m)}}{\tau} = \boldsymbol{L}_n \boldsymbol{u}^{(m+1)} - \frac{1}{\lambda} \boldsymbol{b}^{(m)}, \tag{39}$$

$$\frac{\boldsymbol{v}^{(m+1)} - \boldsymbol{v}^{(m)}}{\tau} = \boldsymbol{L}_n \boldsymbol{v}^{(m+1)} - \frac{1}{\lambda} \boldsymbol{c}^{(m)}, \tag{40}$$

where

$$a^{(m)} = (h(\boldsymbol{x} + \boldsymbol{u}^{(m)})^{(m)} - f_{q/p}(\boldsymbol{x})) + (h(\boldsymbol{x} + \boldsymbol{v}^{(m)})^{(m)} - g^{s/r}(\boldsymbol{x})), \tag{41}$$

$$\boldsymbol{b}^{(m)} = (h(\boldsymbol{x} + \boldsymbol{u}^{(m)})^{(m)} - f_{q/p}(\boldsymbol{x}))^{(m)} \nabla_u h(\boldsymbol{x} + \boldsymbol{u}^{(m)})^{(m)}, \tag{42}$$

$$\boldsymbol{c}^{(m)} = (h(\boldsymbol{x} + \boldsymbol{v}^{(m)})^{(m)} - g_{s/r}(\boldsymbol{x})) \nabla_v h(\boldsymbol{x} + \boldsymbol{v}^{(m)})^{(m)}. \tag{43}$$

Since all of these equations are expressed in the form

$$\frac{\boldsymbol{x}^{(m+1)} - \boldsymbol{x}^{(m)}}{\tau} = (\boldsymbol{D}_n \otimes \boldsymbol{I}_n + \boldsymbol{I}_n \otimes \boldsymbol{D}_n) \boldsymbol{x}^{(m+1)} - \frac{1}{\kappa} \boldsymbol{d}^{(m)}, \tag{44}$$

the DCT-based iteration method computes h_{ij}, $u_{1\,ij}$, $u_{2\,ij}$, $v_{1\,ij}$ and $v_{2\,ij}$ on the $n \times n$ grid.

5   Numerical Examples

5.1   Registration of Inhomogeneous-Resolution Pair

The resolutions of the two images of the dried sardine head in Fig. 1 are 600×450 and 200×150 pixels. The resolutions of the two images of the chicken wing bone in Fig. 2 are 650×650 and 390×390 pixels. Therefore, the resolution ratios of these image pairs are $1 : 1/3$ and $1 : 3/5$. We set $\tau = 10^{-4}$, $\lambda = 10^4$, and $m_{\max} = 2 \times 10^4$.

In Figs. 1 and 2, Figs. (a) and (e) show the original fine and coarse images, respectively. Since the pair of images are displayed with the same landscape, the coarse images are blurred. Figures (b) and (f) are the results for the registration of the coarse-resolution image to the fine-resolution image and the fine-resolution images to the coarse-resolution image, respectively. Figures (c) and (g) are the determinant of the Jacobian of the deformation vector at each point of the coarse-resolution image to the fine-resolution image and the determinant of the Jacobian of the deformation vector at each point the fine-resolution images to the coarse-resolution image, respectively. Figures (d) and (h) are the deformation vector fields on the fine- and coarse-resolution planes, respectively. Figures (i) and (m) are the original fine and coarse images expressed with the same landscape, respectively. Figures (j) and (n) show the resulting registrations on fine and coarse resolution spaces, respectively, with the same landscape. Figures (k) and (o) are heat charts of the determinant of Jacobian of the deformation vector at each point on the fine- and coarse-resolution planes, respectively, with the same landscape. Figures (n) and (p) are heat charts of the length of the deformation-field vector at each point on the fine-and coarse-resolution planes, respectively, with the same landscape. These results show that our method achieves the registration of inhomogeneous-resolution image pairs.

5.2   Resolution Interpolation

Figure 3 shows the results for resolution interpolation. From top to bottom, results for images of a dried sardine head, a chicken wing bone, which are measured by micro-CT, and a brain. The low-resolution image of the brain is generated by downsampling of the original image in Simulated Brain Data Set.

In Fig. 3 from left to right high-resolution images, initial images, resulting images and low-resolution images are shown. For the dried sardine, the resolution ratios of high-resolution, low-resolution and resulting images are 1, 1/3 and 1/2, respectively. For the chicken wing bone, the resolution ratios of high-resolution, low-resolution and resulting images are 1, 3/5 and 4/5, respectively. For the brains, the resolution ratios of high-resolution, low-resolution and generated images are 1, 1/2 and 2/3, respectively. Therefore, the resolutions of the resulting images of the dry-fish head, the chicken-wing bone and the brain are 400×300, 520×520 and 290×242, respectively.

Since the initial values are computed as

$$h^{(0)}(x, y) = \frac{pr}{pr + qs} f_{q/p}(x, y) + \frac{qs}{pr + qs} g^{s/r}(x, y).$$

the weighted average $h^{(0)}(x,y)$ of the inhomogeneous-resolution images yields blurred regions. We set $\tau = 10^{-3}$, $\mu = 10^{-2}$ and $\lambda = 10^4$. Furthermore, for dried sardine, chicken wing bone and brain, we set $m_{\max} = 5 \times 10^5$, $m_{\max} = 5 \times 10^5$ and $m_{\max} = 2 \times 10^6$.

Our resolution interpolation produces clear intermediate-resolution images as shown in Figs. 3(c), (k) and (o) for dried sardine, chicken wing bone, brain example 1 and brain example 2, respectively, by removing blurred regions. These results show that our method achieves acceptable performances for resolution interpolation.

6   Conclusions

We developed a variational image registration method for a pair of images with different resolutions measured by the same modality. In ref. [12], optical flow was computed from an image sequence if the resolution of a pair of successive images was increasing or decreasing. We extended this multiresolution optical flow computation based on the pyramid transform to the registration of a pair of images with different resolutions.

For the linear equation $Ax = y$, $x = A^\dagger y$ is the least-mean-square (LMS) solution. Although the generalised inverse $A^\dagger$ is the orthogonal projection to the range of A^*, A^* is a linear transform to the range of A^* [8]. Therefore, A^* approximates $A^\dagger$, although $x^* = A^*y$ is not the LMS solution to the linear equation $Ax = y$. These properties of the dual and generalised-inverse transforms imply that the pyramid transform approximates the transform from high-resolution images to low-resolution ones if the conversion of the resolution of an image is the scaling of the image sizes.

This research was supported by the Multidisciplinary Computational Anatomy and Its Application to Highly Intelligent Diagnosis and Therapy project funded by a Grant-in-Aid for Scientific Research on Innovative Areas from MEXT, Japan, and by Grants-in-Aid for Scientific Research funded by the Japan Society for the Promotion of Science. The discussions with Martin Welk in the early stage of this study is gratefully acknowledged.

Appendix: Proofs of Theorems 1, 2 and 3

Using the matrix expression of downsampling for vectors $\boldsymbol{S}_q = \boldsymbol{I}_n \otimes \boldsymbol{e}_1^q$, where $\boldsymbol{e}_1^q = (1, 0, \cdots, 0)^\top \in \mathbf{R}^q$, the two-dimensional downsampling is expressed as $\boldsymbol{G} = \boldsymbol{S}_q \boldsymbol{F} \boldsymbol{S}_q^\top$. This expression derives the following lemma.

Lemma 1. *Assuming that the domain of images is* $\mathcal{L}\{\boldsymbol{\varphi}_i \boldsymbol{\varphi}_i^\top\}_{i,j=0}^{n-1}$, *the range of images downsampled by factor* q *is* $\mathcal{L}\{\boldsymbol{\varphi}_i \boldsymbol{\varphi}_j^\top\}_{i,j=0}^{\frac{1}{q}n-1}$.

The $2p + 1$-dimensional diagonal matrix $\boldsymbol{N}_p = \left((n_{|i-j|})\right)$, where $n_k = \frac{p-k}{p}$ for $0 \leq p \leq k$. is expressed as $\boldsymbol{N}_p = \sum_{k=0}^{p} a_k \boldsymbol{D}_n^k$, where $\boldsymbol{D}_n^0 = \boldsymbol{I}_n$, for an

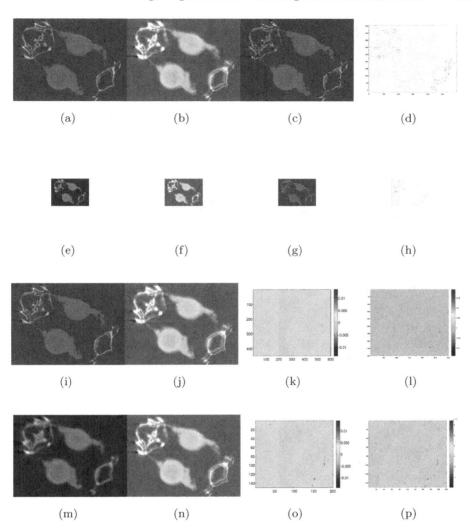

Fig. 1. Registration of inhomogeneous-resolution image pair. The resolutions of the image in (a) and (e) are 600×450 and 200×150 pixels, respectively. Therefore, the resolution ratio of this image pair is $1 : 1/3$. (a) Original fine-resolution image. (b) Registration on the fine-resolution plane. (c) Determinant of Jacobians of the deformation field vectors on the fine-resolution plane. (d) Deformation vector field on the fine-resolution plane. (e) Original coarse-resolution image. (f) Registration on the coarse-resolution plane. (g) Determinant of Jacobian of the deformation vector on the coarse-resolution plane. (h) Deformation vector field on the coarse-resolution plane. (i) Original fine-resolution image. (j) Registration on the fine-resolution plane. (k) Determinant of the Jacobians of the deformation vectors on the fine-resolution plane. (l) Length of deformation-field vector at each point on the fine-resolution plane. (m) Original coarse-resolution image expressed with the same landscape as in (i). (n) Registrations on the coarse-resolution plane. (o) Determinant of Jacobians of the deformation vectors on the coarse-resolution plane. (p) Length of deformation vector at each point on the coarse-resolution plane.

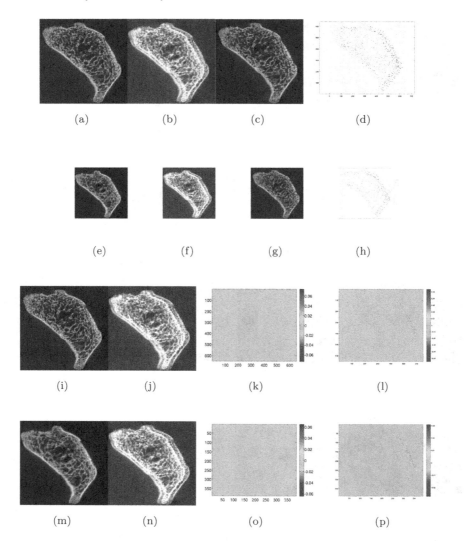

Fig. 2. Registration of inhomogeneous image pair. The resolutions of image in (a) and (e) are 650 × 650 and 390 × 390 pixels. Therefore, the resolution ratio this image pair is 1 : 3/5. (a) Original fine-resolution image. (b) Registration on the fine-resolution plane. (c) Determinant of Jacobians of the deformation field vectors on the fine-resolution plane. (d) Deformation vector field on the fine-resolution plane. (e) Original coarse-resolution image. (f) Registration on the coarse-resolution plane. (g) Determinant of Jacobian of the deformation vector on the coarse-resolution plane. (h) Deformation vector field on the coarse-resolution plane. (i) Original fine-resolution image. (j) Registration on the fine-resolution plane. (k) Determinant of the Jacobians of the deformation vectors on the fine-resolution plane. (l) Length of deformation-field vector at each point on the fine-resolution plane. (m) Original coarse-resolution image expressed with the same landscape as in (i). (n) Registrations on the coarse-resolution plane. (o) Determinant of Jacobians of the deformation vectors on the coarse-resolution plane. (p) Length of deformation vector at each point on the coarse-resolution plane.

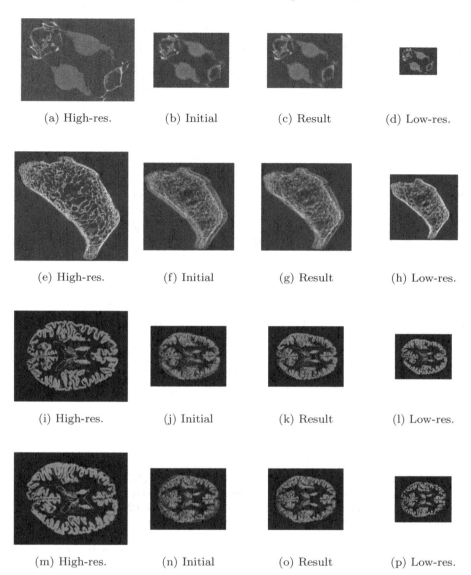

(a) High-res.   (b) Initial   (c) Result   (d) Low-res.

(e) High-res.   (f) Initial   (g) Result   (h) Low-res.

(i) High-res.   (j) Initial   (k) Result   (l) Low-res.

(m) High-res.   (n) Initial   (o) Result   (p) Low-res.

Fig. 3. Resolution interpolation. From top to bottom, results for images of the head of a dry-fish head and a chicken-wing bone, which are measured by micro CT, and two brain examples are shown. From left to right, high-resolution images, the initial images, the result images and low-resolution images are shown.

appropriate collection of coefficients $\{a_k\}_{k=0}^{p}$. The linear interpolation for the two-dimensional image $\boldsymbol{F}$ is $\boldsymbol{N}_p\boldsymbol{S}_p\boldsymbol{F}(\boldsymbol{N}_p\boldsymbol{S}_p)^{\top} = \boldsymbol{N}_p\boldsymbol{S}_p\boldsymbol{F}\boldsymbol{S}_p^{\top}\boldsymbol{N}_p^{\top}$. This expression implies the following lemma.

Lemma 2. *Assuming that the domain of images is* $\mathcal{L}\{\varphi_i\varphi_j^\top\}_{i,j=0}^{n-1}$ *the range of images interpolated by order* p *is* $\mathcal{L}\{\varphi_i\varphi_j^\top\}_{i,j=0}^{pn-1}$.

Furthermore, the pyramid transform of factor q is $\frac{1}{q^2}\boldsymbol{S}_q\boldsymbol{N}_q\boldsymbol{F}\boldsymbol{N}_q^\top\boldsymbol{S}_q^\top$, since the pyramid transform is achieved by downsampling after shift-invariant smoothing. This expression of the pyramid transform implies the following lemma.

Lemma 3. *With the Neumann boundary condition, the pyramid transform of order* q *is a linear transform from* $\mathcal{L}\{\varphi_i\varphi_j^\top\}_{i,j=0}^{n-1}$ *to* $\mathcal{L}\{\varphi_i\varphi_i^\top\}_{i,j=0}^{\frac{1}{q}n-1}$, *assuming* $n = kq$.

Moreover, the matrix form of the q/p-pyramid transform for the two-dimensional image $\boldsymbol{F}$ is

$$\boldsymbol{R}_{q/p}\boldsymbol{F}\boldsymbol{R}_{q/p}^\top = \frac{1}{q^2}\boldsymbol{S}_q\boldsymbol{N}_{q/p}\boldsymbol{S}_p\boldsymbol{F}(\boldsymbol{S}_q\boldsymbol{N}_{q/p}\boldsymbol{S}_p)^\top, \qquad \boldsymbol{N}_{q/p} = \boldsymbol{N}_q\boldsymbol{N}_p. \tag{45}$$

This expression implies the following lemma.

Lemma 4. *With the Neumann boundary condition, the* q/p *pyramid transform is a linear transform from* $\mathcal{L}\{\varphi_i\varphi_j^\top\}_{i,j=0}^{n-1}$ *to* $\mathcal{L}\{\varphi_i\varphi_j^\top\}_{i,j=0}^{\frac{p}{q}n-1}$.

References

1. Henn, S., Witsch, K.: Multimodal image registration using a variational approach. SIAM J. Sci. Comput. **25**, 1429–1447 (2004)
2. Hermosillo, G., Chefd'Hotel, C., Faugeras, O.: Variational methods for multimodal image matching. IJCV **50**, 329–343 (2002)
3. Hermosillo, G., Faugeras, O.: Well-posedness of two nonrigid multimodal image registration methods. SIAM J. Appl. Math. **64**, 1550–1587 (2002)
4. Durrleman, S., Pennec, X., Trouvé, A., Gerig, G., Ayache, N.: Spatiotemporal atlas estimation for developmental delay detection in longitudinal datasets. In: Yang, G.-Z., Hawkes, D., Rueckert, D., Noble, A., Taylor, C. (eds.) MICCAI 2009. LNCS, vol. 5761, pp. 297–304. Springer, Heidelberg (2009). https://doi.org/10.1007/978-3-642-04268-3_37
5. Shepp, L.A., Kruskal, J.: Computerized tomography: the new medical x-ray technology. Amer. Math. Monthly **85**, 420–439 (1978)
6. Modersitzki, J.: Numerical Methods for Image Registration. OUP, Oxford (2004)
7. Rumpf, M., Wirth, B.: A nonlinear elastic shape averaging approach. SIAM J. Imaging Sci. **2**, 800–833 (2009)
8. Campbell, S.L., Meyer Jr., C.D.: Generalized Inverses of Linear Transformations. Pitman, London (1979)
9. Burt, P.J., Adelson, E.H.: The Laplacian pyramid as a compact image code. IEEE Trans. Commun. **31**, 532–540 (1983)
10. Burt, P.J., Adelson, E.H.: A multiresolution spline with application to image mosaics. ACM Trans. Graph. **2**, 217–236 (1983)
11. Thevenaz, P., Unser, M.: Optimization of mutual information for multiresolution image registration. IEEE Trans. Image Process. **9**, 2083–2099 (2000)

12. Ohnishi, N., Kameda, Y., Imiya, A., Dorst, L., Klette, R.: Dynamic multiresolution optical flow computation. In: Sommer, G., Klette, R. (eds.) RobVis 2008. LNCS, vol. 4931, pp. 1–15. Springer, Heidelberg (2008). https://doi.org/10.1007/978-3-540-78157-8_1
13. Kropatsch, W.G.: A pyramid that grows by powers of 2. Pattern Recogn. Lett. **3**, 315–322 (1985)

Scale-Space Methods

Computing Nonlinear Eigenfunctions via Gradient Flow Extinction

Leon Bungert[✉], Martin Burger, and Daniel Tenbrinck

Department Mathematik, Universität Erlangen-Nürnberg,
Cauerstrasse 11, 91058 Erlangen, Germany
{leon.bungert,martin.burger,daniel.tenbrinck}@fau.de

Abstract. In this work we investigate the computation of nonlinear eigenfunctions via the extinction profiles of gradient flows. We analyze a scheme that recursively subtracts such eigenfunctions from given data and show that this procedure yields a decomposition of the data into eigenfunctions in some cases as the 1-dimensional total variation, for instance. We discuss results of numerical experiments in which we use extinction profiles and the gradient flow for the task of spectral graph clustering as used, e.g., in machine learning applications.

Keywords: Nonlinear eigenfunctions · Spectral decompositions · Gradient flows · Extinction profiles · Graph clustering

1 Introduction

Linear eigenvalue problems are of utter importance and a classical tool in signal and image processing. A frequently used tool here is the Fourier transform which basically decomposes a given signal into eigenfunctions of the Laplacian operator and makes frequency-based filtering possible. In addition, such problems also find their applications in machine learning and the treatment of large data sets [13, 16]. However, for some applications – as for example certain graph clustering tasks – linear theory does not suffice to achieve satisfactory results. Therefore, nonlinear eigenproblems, which involve a nonlinear operator, have gained in popularity over the last years since they can be successfully applied in far more complex and interesting application scenarios. However, solving such nonlinear eigenproblems is a challenging task and the techniques heavily depend on the structure of the involved operator. The setting we adopt is the following: we consider a Hilbert space $\mathcal{H}$ and study the eigenvalue problem related to the subdifferential ∂J of an absolutely one-homogeneous convex functional $J : \mathcal{H} \to \mathbb{R} \cup \{+\infty\}$. A prototypical example for such a functional is the total variation. So called *nonlinear eigenfunctions* are characterized by the inclusion

$$\lambda p \in \partial J(p), \tag{1}$$

where usually the normalization $\|p\| = 1$ is demanded to have a interpretable eigenvalue λ. Note that the operator ∂J is nonlinear and multivalued, in particular eigenfunctions do not form linear subspaces. Important properties and

© Springer Nature Switzerland AG 2019
J. Lellmann et al. (Eds.): SSVM 2019, LNCS 11603, pp. 291–302, 2019.
https://doi.org/10.1007/978-3-030-22368-7_23

characterizations of the nonlinear eigenfunctions are collected in [5]. Of particular interest in applications is the decomposition of some data $f \in \mathcal{H}$ into a linear combination of eigenfunctions which, for instance, allows for scale-based filtering in image processing [6,10,11] – analogously to linear Fourier methods. In other applications, as spectral graph clustering, one is rather interested in finding a specific eigenfunction that is in some way related to the data or captures topological properties of the domain. An important tool for this nonlinear spectral analysis is the so called *gradient flow* of the functional J

$$\begin{cases} u'(t) = -p(t), \quad p(t) \in \partial J(u(t)), \\ u(0) = f, \end{cases} \tag{GF}$$

whose connection to nonlinear eigenfunctions has been analysed in finite dimensions in [7] and in infinite dimensions in [5]. In particular, the authors proved that the gradient flow is able to achieve the above-mentioned decomposition task in some situations. Furthermore, it always generates one specific eigenfunction, called the *extinction profile* or *asymptotic profile* of f (cf. [1] for the special case of total variation flow). This profile is given by the subsequential limit of $p(t)$ as t tends to the extinction time of the flow. A different flow which also generates an eigenfunction was introduced and analyzed in [2,14]. A third way for obtaining eigenfunctions, being less rigorous and reliable, consists in computing the gradient flow (GF) of f and checking for subgradients $p(t)$ to be eigenfunctions.

The rest of this work is organized as follows: After recapping some notation and important results regarding gradient flows and associated eigenfunctions in Sect. 2, we analyze an iterative scheme in Sect. 3 which is based on extinction profiles and constitutes an alternative to the already existent decomposition into nonlinear eigenfunctions through subgradients of the gradient flow. Finally, in Sect. 4 we present some applications of extinction profiles, mainly to spectral clustering.

2   Gradient Flows and Eigenfunctions

Without loss of generality, we will assume that the data f is orthogonal to the null-space of the functional J which is denoted by $\mathcal{N}(J)$. For the example of total variation, this corresponds to calculating with data functions of zero mean. Furthermore, we will only be confronted with eigenvectors of eigenvalue 1, which follows naturally from the gradient flow structure. Normalizing them to have unit norm, shows that a suitable eigenvalue for an element p which meets $p \in \partial J(p)$ is given by $\|p\|$. A complete picture of the theory of gradient flows and nonlinear eigenproblems is given in [5], from where the following statements are taken.

An important property of the gradient flow (GF) is that it decomposes the data f into subgradients of the functional J, i.e., it holds

$$f = \int_0^\infty p(s)\,ds, \tag{2}$$

where the subgradients $p(s)$ enjoy the regularity of being elements in $\mathcal{H}$ and, furthermore, have minimal norm in the subdifferentials $\partial J(u(s))$, i.e. $\|p(s)\| \leq \|p\|$ for all $p \in \partial J(u(s))$. This naturally qualifies them for being eigenfunctions as it was shown in [5]. Furthermore, the solution u of (GF) extincts to zero in finite time under generic conditions on the functional J. More precisely, it has to satisfy a Poincaré-type inequality, namely that there is $C > 0$ such that

$$\|u\| \leq CJ(u) \tag{3}$$

holds for all u which are orthogonal to the null-space of J (cf. [5, Rem. 6.3]). Let us in the following assume that the solution u of (GF) extincts at time $0 < T < \infty$, in other words $u(t) = 0$ for all $t \geq T$. In that case, there is an increasing sequence of times (t_n) converging to T such that

$$p^* := \lim_{n \to \infty} \frac{1}{T - t_n} \int_{t_n}^{T} p(s)\, ds \tag{4}$$

is a non-trivial eigenfunction of ∂J, i.e., $p^* \neq 0$ and $p^* \in \partial J(p^*)$. The element p^* is refererred to as an extinction profile of f.

In the following, the term *spectral case* refers to the scenario that the subgradients $p(t)$ in (GF) are eigenfunctions themselves, i.e., $p(t) \in \partial J(p(t))$ for all $t > 0$. Using this together with the fact that $\|p(t)\|$ is decreasing in t implies that (2) becomes a decomposition of the datum into eigenfunctions with decreasing eigenvalues. Several scenarios and geometric conditions for this to happen were investigated in [5]. For instance, if the functional J is the total variation in one space dimension, a divergence and rotation sparsity term, or special finite dimensional ℓ^1-sparsity terms, one has this spectral case. Also for general J, a special structure of the data (cf. [4,5,15]) can yield the spectral case. Let us conclude the nomenclature by introducing the quantity

$$\|f\|_* = \sup\left\{\langle f, p \rangle \,:\, J(p) = 1,\ p \in \mathcal{N}(J)^{\perp}\right\}, \quad f \in \mathcal{H}, \tag{5}$$

and noting that (3) implies

$$\|f\|_* \leq C\|f\|, \quad \forall f \in \mathcal{H}. \tag{6}$$

3   An Iterative Scheme to Compute Nonlinear Spectral Decompositions

If one is not in the above-explained spectral case, the decomposition of an arbitrary data into nonlinear eigenfunctions is a hard task. A very intuitive approach into this direction is to compute an eigenfunction, subtract it from the data, and start again. As already mentioned there are several approaches to get hold of a nonlinear eigenfunction, one of which consists in the computation of extinction profiles. We will use these to define and analyze a recursive scheme for the decomposition of data into nonlinear eigenfunctions. We consider

$$\begin{cases} f_0 & := f, \\ f_{n+1} & := f_n - c_n p_n^*, \quad n \geq 0, \end{cases} \tag{S}$$

where p_n^* denotes the extinction profile of f_n and $c_n := \langle f_n, p_n^* \rangle / \|p_n^*\|^2$. Note that despite being explicit, the scheme still requires the non-trivial computation of the extinction profiles p_n^* of f_n. A numerical approach for this subprocedure is given in Sect. 4.1. The scheme can be rewritten as

$$f_{n+1} = f - \sum_{i=0}^{n} c_i p_i^*, \quad n \geq 0. \tag{S'}$$

Hence, if there is $N \in \mathbb{N}$ such that $f_{N+1} = 0$, it holds $f = \sum_{i=0}^{N} c_i p_i^*$, which means that f can be written as linear combination of finitely many eigenfunctions of ∂J. More generally, if there is some $g \in \mathcal{H}$ such that $f_n \to g$ as $n \to \infty$, one has $f = g + \sum_{i=0}^{\infty} c_i p_i^*$, which corresponds to the decomposition of f into a linear combination of countably many eigenfunctions and a rest g.

Let us start by collecting some essential properties of the iterative scheme (S).

Proposition 1. *One has the following statements:*

1. *The scheme* (S) *terminates (i.e. $f_{n+1} = f_n$) if and only if p_n^* is orthogonal to f_n. In the spectral case, this happens if and only if $f_n = 0$.*
2. *$\|f_n\|$ is strictly decreasing at a maximal rate until termination.*

In more detail, it holds

$$\|f_{n+1}\|^2 = \|f_n\|^2 - \frac{\langle f_n, p_n^* \rangle^2}{\|p_n^*\|^2}, \quad n \geq 0. \tag{7}$$

Proof. Ad 1.: Obviously the scheme terminates if $c_n = 0$ which is the case if and only if $\langle f_n, p_n^* \rangle = 0$. In the spectral case, c_n equals the extinction time of the gradient flow with initial data f_n (cf. [5]). Due to continuity of the solution of the gradient flow, this is zero if and only if $f_n = 0$.

Ad 2.: One has $\|f_n - c p_n^*\|^2 = \|f_n\|^2 - [2c \langle f_n, p_n^* \rangle - c^2 \|p_n^*\|^2]$ for any $n \geq 0$ and $c \in \mathbb{R}$. The term in square brackets is quadratic in c and zero for $c \in \{0, 2\langle f_n, p_n^* \rangle / \|p_n^*\|^2\}$. Hence, it is maximal for $c = c_n = \langle f_n, p_n^* \rangle / \|p_n^*\|^2$. This concludes the proof.

Remark 1 (Well-definedness). *Note that the scheme* (S) *is well-defined due to the Poincaré inequality* (3). *It can be used to bound the extinction time $T(f)$ of the gradient flow with datum f. In more detail, it holds $T(f) \leq C\|f\|$, as was shown in [5]. Consequently, the flows with datum f_n for $n \geq 1$ all have finite extinction time since, $T(f_n) \leq C\|f_n\| \leq C\|f\| < \infty$.*

We start with a Lemma that will be useful for proving convergence of (S) in the spectral case.

Lemma 1. *It holds $\sum_{n=0}^{\infty} \frac{\langle f_n, p_n^* \rangle^2}{\|p_n^*\|^2} < \infty$ and, in particular, $\lim_{n \to \infty} \frac{\langle f_n, p_n \rangle}{\|p_n^*\|} = 0$.*

Proof. Summing (7) for $n = 0, \ldots, K$ and using $f_0 = f$ yields

$$\|f\|^2 = \|f_{K+1}\|^2 + \sum_{n=0}^{K} \frac{\langle f_n, p_n^* \rangle^2}{\|p_n^*\|^2} \geq \sum_{n=0}^{K} \frac{\langle f_n, p_n^* \rangle^2}{\|p_n^*\|^2}.$$

Letting K tend to infinity concludes the proof.

Corollary 1. *It holds* $\sum_{n=0}^{\infty} \|f_{n+1} - f_n\|^2 = \sum_{n=0}^{\infty} \frac{\langle f_n, p_n^* \rangle^2}{\|p_n^*\|^2} < \infty$.

3.1   The Spectral Case

In the spectral case, one already has the decomposition of f as integral over $p(t)$ for $t > 0$, where $p(t)$ is given by (GF). Still, we study scheme (S) and prove that it provides an alternative decomposition into a discrete sum of eigenfunctions.

Theorem 1. *In the spectral case the sequence* (f_n) *generated by* (S) *converges weakly to* 0.

Proof. It holds according to [5] that $\|f_n\|_* = \langle f_n, p_n^* \rangle / J(p_n^*)$. Using $J(p_n^*) = \|p_n^*\|^2$ we find

$$\lim_{n \to \infty} \|f_n\|_* \|p_n^*\| = \lim_{n \to \infty} \frac{\langle f_n, p_n \rangle}{\|p_n^*\|} = 0$$

by Lemma 1. However, $\|p_n^*\|$ cannot be a null sequence since (6) implies $1 = \|p_n^*\|_* \leq C \|p_n^*\|^2$. Therefore, $\|f_n\|_*$ converges to zero which immediately implies that $f_n \rightharpoonup 0$ in $\mathcal{H}$.

Corollary 2 (Parseval identity). *In the spectral case one has* $f = \sum_{n=0}^{\infty} c_n p_n^*$, *where the equality is to be understood in the weak sense, i.e., tested against any* $v \in \mathcal{H}$. *In particular, for* $v = f$ *this implies the* Parseval identity

$$\|f\|^2 = \sum_{i=0}^{\infty} c_i \langle p_i^*, f \rangle,$$

which is a perfect analogy to the linear Fourier transform.

3.2   The General Case

In the general case, one only has much weaker statements about the iterates of the scheme (S). Indeed, one can only prove weak convergence of a subsequence of (f_n). Furthermore, the limit is not zero, in general, meaning that there remains a rest which cannot be decomposed into eigenfunctions by the scheme.

Theorem 2. *The sequence* (f_n) *generated by* (S) *admits a subsequence* (f_{n_k}) *that converges weakly to some* $g \in \mathcal{H}$.

Proof. By (7), the sequence (f_n) is bounded in $\mathcal{H}$ and, therefore, admits a convergent subsequence.

To conclude this section, we mention that scheme (S) provides an alternative decomposition into eigenfunctions in the spectral case. In the general case, however, the scheme is of limited use since it fails to decompose the whole datum, in general. Furthermore, the indecomposable rest can still contain a large amount of information as Fig. 1 shows.

  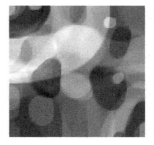

Fig. 1. From left to right: data f, indecomposable rest g, isolated TV-eigenfunctions

4   Applications

In the following, we discuss different applications for the proposed spectral decomposition scheme (S) and extinction profiles of the gradient flow (GF). After giving a short insight into the numerical computation of extinction profiles, we use the scheme and the gradient flow to compute and compare two different decompositions of a 1-dimensional signal into eigenfunctions of the total variation. Thereafter, we illustrate the use of the gradient flow and its extinction profiles for spectral graph clustering.

4.1   Numerical Computation of Extinction Profiles

Here we describe how to calculate the extinction profiles p_n^* of f_n in scheme (S). Given a time step size $\delta > 0$ the solution u_k of the gradient flow (GF) at time $t_k = \delta k$ with datum f is recursively approximated via the implicit scheme

$$u_k = \arg\min_{u \in \mathcal{H}} \frac{1}{2}\|u - u_{k-1}\|^2 + \delta J(u), \quad k \in \mathbb{N},$$

where $u_0 = f$. These minimization problems can be solved efficiently with a primal dual optimization algorithm [8]. Defining $p_k := (u_{k-1} - u_k)/\delta$ yields a sequence of pairs (u_k, p_k) where $p_k \in \partial J(u_k)$ for every $k \in \mathbb{N}$. Hence, in order to compute an extinction profile of f, we keep track of the quantities $\hat{p}_k := (p_k + p_{k-1})/2$ (cf. the definition of p^* in (4)), and define the extinction profile p^* of f as the element $\hat{p}_k$ which has the highest quotient $R(\hat{p}_k) := \|\hat{p}_k\|^2/J(\hat{p}_k)$ before extinction of the flow. Note that since $\partial J(u) \subset \partial J(0)$ holds for all $u \in \mathcal{H}$

(cf. [5]), both p_k and, by convexity, also $\hat{p}_k$ are in particular elements of $\partial J(0)$ and, thus, have a quotient smaller or equal than one. Hence, an extinction profile, being even an eigenfunction, can be identified by having a quotient of one. To design a criterion for extinction of the flow we make use of the fact that the subgradients $p(t)$ in the gradient flow have monotonously decreasing norms and define the extinction time as the number δk such that $\|p_k\|$ is below a certain threshold.

4.2   1D Total Variation Example

As already mentioned in Sect. 3, choosing J to be the one-dimensional total variation yields a spectral case, i.e., the sequence (f_n) in (S) weakly converges to zero which implies that $f = \sum_{i=0}^{\infty} c_i p_i^*$ is a decomposition into eigenfunctions. The rightmost images in Fig. 2 show the data signal in red, whereas the other four images depict the approximation of f by eigenfunctions of the gradient flow and the iterative scheme, respectively. The individual eigenfunctions (up to multiplicative constants) as computed by the gradient flow (GF) and the scheme (S) are given in Fig. 3. Hence, the sum of the top four or the bottom 18 eigenfunctions, respectively, yields back the red signal from Fig. 2.

Note that the gradient flow only needs four eigenfunctions to generate the data and hence gives a very sparse representation. However, the individual eigenfunctions have decreasing spatial complexity. This is a fundamental difference to the system of eigenfunctions generated by (S) which – being extinction profiles – all have low complexity. This qualitative difference can also be observed in Fig. 2 where the first eigenfunction of the gradient flow (top left) already contains all the structural information of the red signal whereas the approximation in the bottom row successively adds structure.

4.3   Spectral Clustering with Extinction Profiles

Spectral clustering arises in various real world applications, e.g., in discriminant analysis, machine learning, or computer vision. The aim in this task is to partition a given data set according to the spectral characteristics of an operator that captures the pairwise relationships between each data point. Based on the spectral decomposition of this operator one tries to find a partitioning of the data into sets of strongly related entities, called clusters, that should be clearly distinguishable with respect to a chosen feature. Within each cluster the belonging data points should be homogeneous with respect to this feature.

In order to model relationships between entities without further knowledge about the underlying data topology one may use finite weighted graphs. In this model each data point is represented by a vertex of the graph while the similarity between two data points is represented by a weighted edge connecting the respective vertices. For details on data analysis using finite weighted graphs we

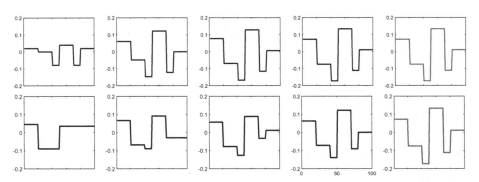

Fig. 2. Reconstruction of a 1D signal (red) based on TV eigenfunctions computed with the gradient flow scheme (GF) (top row) and the proposed extinction profile scheme (S) (bottom row). From left to right: sum of the first $1, 2, 3, 4$ (gradient flow), respectively $3, 6, 9, 18$ (recursive scheme) computed eigenfunctions, original data signal, see Fig. 3 for the individual eigenfunctions (Color figure online)

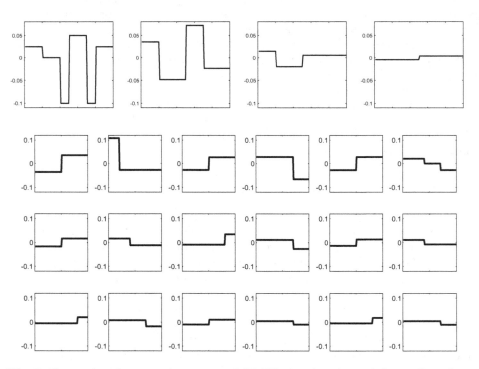

Fig. 3. Comparison between the computed 1D TV eigenfunctions of the gradient flow scheme (GF) and the proposed extinction profile scheme (S). Top: All four eigenfunctions computed with (GF). Bottom: first 18 eigenfunctions computed with (S).

refer to [9, 12]. In the literature it is well-known that there exists a strong mathematical relationship between spectral clustering and various minimum graph cut problems. For details we refer to [17].

For the task of spectral clustering one is typically interested in the eigenvectors of a discrete linear operator known as the *weighted graph Laplacian* Δ_w, which can be represented as a matrix L of the form $L = D - W$, for which D is a diagonal matrix consisting of the degree of each graph vertex and W is the adjacency matrix capturing the edge weights between vertices. Determining the discrete spectral decomposition of the graph Laplacian L is a common problem in mathematics and thus easy to compute. After determining the eigenvectors of L one performs the actual clustering, e.g., via a standard k-means algorithm or simple thresholding. Note that in various applications a spectral clustering based on solely one eigenvector, i.e., the corresponding eigenvector of the second-smallest eigenvalue, already yields interesting results, e.g., for image segmentation [12]. On the other hand, due to the linear nature of the graph Laplacian this approach is rather restricted in many real world applications. For this reason one aims to perform spectral clustering based on eigenfunctions of a nonlinear, possibly more suitable, operator. Bühler and Hein proposed in [3] an iterative scheme to compute eigenfunctions of a nonlinear operator known as the *weighted graph p-Laplacian*

$$\Delta_{w,p}f(x) = \sum_{y \sim x} w(x,y)^{\frac{p}{2}}|f(y) - f(x)|^{p-2}(f(y) - f(x)). \tag{8}$$

Note that this operator is a direct generalization of the standard graph Laplacian $\Delta_w = \Delta_{w,2}$ for $p = 2$. Their idea consists in computing an eigenfunction of the linear Laplacian Δ_w and use this as initialization for a non-convex minimization problem of a Rayleigh quotient which leads to an eigenfunction of $\Delta_{w,p}$ with $p < 2$. This procedure is repeated iteratively for decreasing $p \to 1$ by using the intermediate solutions as initialization for the next step. Their method already leads to satisfying results in situations in which a linear partitioning of the given data is not sufficient. However, as the authors state themselves, this approach often converges to unwanted local minima and is restricted to eigenfunctions corresponding to the second-smallest eigenvalue.

In Fig. 4 we compare the spectral clustering approach from [3] based on the nonlinear graph p-Laplace operator with the extinction profiles introduced in Sect. 2. For this we consider the following one-homogeneous, convex functional defined on vertex functions of a finite weighted graph

$$J(f) = \frac{1}{2}\sum_{x \in V}||\nabla_w f(x)||_1. \tag{9}$$

Here, $\nabla_w f(x)$ denotes the weighted gradient of the vertex function f in a vertex x. Note that this mimics a strong formulation of TV in the continuous case. The subgradient ∂J corresponds to the graph 1-Laplacian as special case of the graph p-Laplacian in (8) for $p = 1$. We test the different approaches on the "Two Moon" dataset with low noise variance (top row) and a slightly increased noise variance (bottom row). The second column shows the computed nonlinear eigenfunction by the Bühler and Hein approach. In case of low noise variance (top) the eigenfunction takes only two values and is piece-wise constant, partitioning the data well. However, the eigenfunction takes more values in the noisy case (bottom) and a subsequent k-means-based clustering with $k = 2$ does not yield a good partitioning of the data. In the third column we depict the extinction profiles computed with (GF), initialized with random values on the graph vertices. Note that both eigenfunctions are piece-wise constant and take only two values, thus inducing a binary partitioning directly. However, similar to the Bühler-Hein eigenfunction, the found eigenfunction for the noisy case is not suitable to partition the dataset correctly. Hence, we performed a third experiment in the right column in which we initialized 5% of the nodes per cluster with the values ± 1, respectively and set the others to zero. Thus, we enforced the computation of eigenfunctions that correctly partition the data. This can be interpreted as a semi-supervised spectral clustering approach.

Fig. 4. Spectral clustering results on the "Two Moon" dataset based on the *nonlinear graph 1-Laplace operator* for two different levels of noise variance (top: $\sigma^2 = 0.015$, bottom: $\sigma^2 = 0.02$). Eigenfunctions computed from left to right: k-nearest neighbor graph for $k = 10$, Bühler and Hein approach [3], extinction profile with random initialization, extinction profile with 5% manually labeled data points

In conclusion we state that spectral clustering based on nonlinear eigenfunctions is a potentially powerful tool for applications in data analysis and machine learning. However, we note that neither the Bühler and Hein approach discussed above nor extinction profiles guarantee a correct partioning of the data, in general. We could mitigate this drawback by using the fact that the chosen initialization of the gradient flow influences its extinction profile.

4.4 Outlook: Advanced Clustering with Higher-Order Eigenfunctions

Finally, we demonstrate some preliminary results of our numerical experiments on a more challenging data set known as "Three Moons". In this case one requires for spectral clustering an eigenfunction that is constant on each of the three half-moons. Thus, we aim to find eigenfunctions of the graph 1-Laplacian that correspond to a *higher* eigenvalue than the second-smallest one. For this reason it is apparent that Bühler and Hein's method in its simplest variant (without subsequent splitting) always fails in this scenario (see top-right image in Fig. 5). Also extinction profiles, having the lowest possible eigenvalue of all subgradients $p(t)$ of the gradient flow (GF) lead to unreasonable results. However, one can still make use of the other subgradients and select those that are close to an eigenfunction, which can be measured by the Rayleigh quotient. The second row in Fig. 5 shows three subgradients with a Rayleigh quotient of more than 90%, as they occur in the gradient flow. Note that the last one coincides with the extinction profile and obviously fails in separating all three moons since it only computes a binary clustering. Similarly, the first subgradient finds four clusters. The correct clustering into three moons is achieved by the subgradient in the center which was the last eigenfunction to appear before the extinction profile. This underlines the need of higher-order eigenfunctions for accurate multi-class spectral graph clustering.

Fig. 5. Comparison between the computed eigenfunction of the Bühler and Hein approach (top) and three subgradients of the gradient flow with decreasing norm and Rayleigh quotient $\geq 90\%$ (bottom) on the "Three moons" dataset

Acknowledgments. This work was supported by the European Union's Horizon 2020 research and innovation programme under the Marie Skłodowska-Curie grant agreement No 777826 (NoMADS). LB and MB acknowledge further support by ERC via Grant EU FP7 – ERC Consolidator Grant 615216 LifeInverse.

References

1. Andreu, F., Caselles, V., Diaz, J.I., Mazón, J.M.: Some qualitative properties for the total variation flow. J. Funct. Anal. **188**(2), 516–547 (2002)
2. Aujol, J.-F., Gilboa, G., Papadakis, N.: Theoretical analysis of flows estimating eigenfunctions of one-homogeneous functionals for segmentation and clustering. SIAM J. Imaging Sci. **11**, 1416–1440 (2018)
3. Bühler, T., Hein, M.: Spectral clustering based on the graph p-laplacian. In: International Conference on Machine Learning, pp. 81–88 (2009)

4. Bungert, L., Burger, M.: Solution paths of variational regularization methods for inverse problems. Inverse Prob. (2019). IOP Publishing
5. Bungert, L., Burger, M., Chambolle, A., Novaga,M.: Nonlinear spectral decompositions by gradient flows of one-homogeneous functionals. arXiv preprint arXiv:1901.06979 (2019)
6. Burger, M., Eckardt, L., Gilboa, G., Moeller, M.: Spectral representations of one-homogeneous functionals. In: Aujol, J.-F., Nikolova, M., Papadakis, N. (eds.) SSVM 2015. LNCS, vol. 9087, pp. 16–27. Springer, Cham (2015). https://doi.org/10.1007/978-3-319-18461-6_2
7. Burger, M., Gilboa, G., Moeller, M., Eckardt, L., Cremers, D.: Spectral decompositions using one-homogeneous functionals. SIAM J. Imaging Sci. **9**(3), 1374–1408 (2016)
8. Chambolle, A., Pock, T.: A first-order primal-dual algorithm for convex problems with applications to imaging. J. Math. Imaging Vis. **40**(1), 120–145 (2011)
9. Elmoataz, A., Toutain, M., Tenbrinck, D.: On the p-laplacian and ∞-laplacian on graphs with applications in image and data processing. SIAM J. Imaging Sci. **8**(4), 2412–2451 (2015)
10. Gilboa, G.: A total variation spectral framework for scale and texture analysis. SIAM J. Imaging Sci. **7**(4), 1937–1961 (2014)
11. Gilboa, G.: Nonlinear Eigenproblems in Image Processing and Computer Vision. ACVPR. Springer, Cham (2018). https://doi.org/10.1007/978-3-319-75847-3
12. Meng, Z., Merkurjev, E., Koniges, A., Bertozzi, A.L.: Hyperspectral image classification using graph clustering methods. Image Process. On Line **7**, 218–245 (2017)
13. Ng, A.Y., Jordan, M.I., Weiss, Y.: On spectral clustering: analysis and an algorithm. In: Advances in neural information processing systems, pp. 849–856 (2002)
14. Nossek, R.Z., Gilboa, G.: Flows generating nonlinear eigenfunctions. J. Sci. Comput. **75**(2), 859–888 (2018)
15. Schmidt, M.F., Benning, M., Schönlieb, C.-B.: Inverse scale space decomposition. Inverse Prob. **34**(4), 045008 (2018)
16. Shi, J., Malik, J.: Normalized cuts and image segmentation. IEEE Trans. Pattern Anal. Mach. Intell. **22**, 107 (2000). Departmental Papers (CIS)
17. von Luxburg, U.: A tutorial on spectral clustering. Stat. Comput. **17**(4), 395–416 (2007)

Sparsification Scale-Spaces

Marcelo Cárdenas$^{(\boxtimes)}$, Pascal Peter, and Joachim Weickert

Mathematical Image Analysis Group, Faculty of Mathematics and Computer Science, Campus E1.7, Saarland University, 66041 Saarbrücken, Germany
{cardenas,peter,weickert}@mia.uni-saarland.de

Abstract. We introduce a novel scale-space concept that is inspired by inpainting-based lossy image compression and the recent denoising by inpainting method of Adam et al. (2017). In the discrete setting, the main idea behind these so-called sparsification scale-spaces is as follows: Starting with the original image, one subsequently removes a pixel until a single pixel is left. In each removal step the missing data are interpolated with an inpainting method based on a partial differential equation. We demonstrate that under fairly mild assumptions on the inpainting operator this general concept indeed satisfies crucial scale-space properties such as gradual image simplification, a discrete semigroup property or invariances. Moreover, our experiments show that it can be tailored towards specific needs by selecting the inpainting operator and the pixel sparsification strategy in an appropriate way. This may lead either to uncommitted scale-spaces or to highly committed, image-adapted ones.

Keywords: Scale-space · Sparsification · Inpainting · Diffusion · Image compression

1 Introduction

Lossy image compression methods and scale-spaces share similar philosophies in a number of aspects: The former ones are based on information reduction, while the latter ones aim at image simplification. Both the amount of information reduction as well as the degree of image simplification can be steered by a free parameter: the quality parameter which influences the compression rate, and the scale parameter. Moreover, both concepts reveal naturally certain denoising properties: For instance, wavelet shrinkage can be used both for compression as well as for denoising, and nonlinear diffusion methods create scale-spaces that offer structure-preserving denoising.

In view of these similarities, it is surprising that attempts of both communities to fertilise each other are still fairly limited. One of the reasons lies in the fact that both paradigms traditionally rely on different techniques: Typical lossy compression methods are based on orthogonal or unitary transforms such as the discrete cosine or the discrete wavelet transforms [13,23], whereas scale-spaces usually involve partial differential equations (PDEs) [2,9,19,25] or pseudodifferential evolutions [6,21].

© Springer Nature Switzerland AG 2019
J. Lellmann et al. (Eds.): SSVM 2019, LNCS 11603, pp. 303–314, 2019.
https://doi.org/10.1007/978-3-030-22368-7_24

One of the few notable attempts to connect both worlds goes back to Chambolle and Lucier [4], who interpreted iterated shift-invariant wavelet shrinkage as a nonlinear smoothing scale-space. Although these authors emphasised that it is not given by a PDE, this statement may be questioned in view of several equivalence results between iterated wavelet shrinkage and nonlinear diffusion processes [26].

Inpainting-based compression [7] constitutes an alternative to classical transform-based codecs: One stores only a carefully selected subset of all pixels and reconstructs the missing image structures by inpainting which usually involves PDEs. The less pixels are stored, the more information is discarded. This suggests that also such approaches reveal scale-space properties. This claim is also supported by a recent paper by Adam et al. [1], who use closely related inpainting ideas and data sparsification concepts to design novel denoising methods.

Our Goals. In the present paper we use these ideas to introduce a novel class of scale-space processes which we name sparsification scale-spaces. For the sake of simplicity, we restrict ourselves to the discrete setting. The basic idea of constructing such a scale-space is as follows: One starts with the original image and gradually removes all pixels until only a single pixel remains. At those locations where no information is kept, we interpolate the missing image structure with some PDE-based inpainting. While this framework is strongly inspired by inpainting-based image compression which uses similar sparsification strategies [11], several questions have to be answered in this context, e.g.

- Can one prove that this scale-space representation creates indeed a simplifying transformation? Does it also satisfy other typical scale-space properties such as a semigroup structure or invariances?
- How powerful and flexible is this family of scale-spaces? Can it be made uncommited or can one adapt it in order to reward important structures with a larger lifetime over scales? Is it strongly dependent on the inpainting operator?

Our paper will give answers to all these questions.

Structure of the Paper. Our publication is organised as follows. In Sect. 2 we formalise the concept of discrete sparsification scale-spaces and establish a number of theoretical properties which show that our processes under consideration indeed qualify as scale-spaces under fairly mild assumptions on the inpainting operator. Section 3 is devoted to additional aspects of specific interest, and it presents several experiments which illustrate the flexibility of the sparsification scale-space framework. The paper is concluded with a summary and an outlook in Sect. 4.

2   Theoretical Results

In this section we provide a formalisation of discrete sparsification scale-spaces, and we show that they satisfy all essential properties of a space- and time-discrete scale-space.

2.1   Formalisation of Discrete Sparsification Scale-Spaces

To define our sparsification scale-space in a discrete setting, we have to select its two components first: a sparsification strategy, and an inpainting method that involves a scale-space operator. Thus, we will first elaborate on these two concepts and then present our formal definition.

Sparsification Strategy. We represent a discrete greyscale image with N pixels by a vector $f \in \mathbb{R}^N$. Let its corresponding index set be denoted by $J = \{1, ..., N\}$. We define a *known data sequence* $(K^\ell)_{\ell=0}^{N-1}$ of N nested subsets of J by imposing two conditions:

- The first set K^0 satisfies $K^0 = J$.
- For $\ell = 1, ..., N-1$, every set K^ℓ is a subset of $K^{\ell-1}$ and contains $N - \ell$ elements.

Thus, K^ℓ is created by removing a single index m_ℓ from $K^{\ell-1}$. Different strategies for pixel removal will be discussed in Sect. 3.

Instead of specifying a known data sequence $(K^\ell)_{\ell=0}^{N-1}$, we can equivalently specify its *sparsification path* $m = (m_1, ..., m_{N-1})$. Obviously the sparsification path offers a very compact representation of the known data sequence, which is useful when we want to store this information.

Another equivalent representation of the known data sequence is the *mask sequence*. It consists of a sequence $(C^\ell)_{\ell=0}^{N-1}$ of diagonal matrices $C^\ell \in \mathbb{R}^{N \times N}$ with the following property: The diagonal elements $c_{i,i}^\ell$ are 1 if $i \in K^\ell$, and 0 elsewhere.

Scale-Space Induced Inpainting. Apart from these sparsification concepts, we also need a discrete inpainting method. To this end, we consider a linear operator $A \in \mathbb{R}^{N \times N}$ or a nonlinear operator $A(u) : \mathbb{R}^N \to \mathbb{R}^{N \times N}$ that generates a space-discrete scale-space evolution $\{u(t) \mid t \geq 0\}$ via

$$u(0) = f, \tag{1}$$

$$\frac{du}{dt} = A(u)\, u. \tag{2}$$

Here we have chosen the nonlinear formulation, which includes the linear one. We assume that this scale-space satisfies classical requirements that can be classified into three categories [2]:

- architectural properties (e.g. the semigroup property),
- simplification properties (such as energy minimisation properties, the existence of a Lyapunov functional, or causality in terms of a maximum–minimum principle),
- invariances (e.g. w.r.t. additive shifts in the grey values).

This space-discrete scale-space can be seen as a dynamical system. Typically it arises from a space discretisation of a continuous scale-space which evolves an image under a partial differential equation (PDE), e.g. a diffusion equation [25].

Moreover, the structure of A also involves discretisations of boundary conditions, such as reflecting (homogeneous Neumann) boundary conditions.

To use the operator $A(u)$ for inpainting purposes, let us assume that we are given some image $f \in \mathbb{R}^N$, and that its grey values are only reliable in a subset K of its pixel set J. Then one can inpaint the missing information in $J \setminus K$ by solving the following problem for u:

$$C(u - f) - (I - C) A(u) u = 0 \tag{3}$$

where $I \in \mathbb{R}^{N \times N}$ is the identity matrix, and the diagonal matrix C denotes the inpainting mask associated to the known pixel set K. Thus, our restoration u is identical to f on K, and it is inpainted with the operator $A(u)$ in $J \setminus K$. This operator uses the data f in K as additional Dirichlet boundary data.

One may wonder if it is natural to base an inpainting process on a scale-space operator. Actually this makes a lot of sense: It is not difficult to see that basically all PDE-based inpainting operators enjoy scale-space properties, although some representatives such as higher-order operators may not have been used in this context. For more details on PDE-based inpainting we refer to [22].

Sparsification Scale-Spaces. Having a sparsification strategy and an inpainting process at our disposal, it is not difficult to define a sparsification scale-space. Let $f = (f_i)_{i=1}^N \in \mathbb{R}^N$ be our image, $(K^\ell)_{\ell=0}^{N-1}$ a known data sequence, $(C^\ell)_{\ell=0}^{N-1}$ its associated mask sequence, and let $A(u)$ denote a scale-space induced inpainting operator. Then we define a (space- and time-discrete) *sparsification scale-space* $(u^\ell)_{\ell=0}^{N-1}$ of f as the solution set of the following sequence of inpainting problems:

$$C^\ell (u^\ell - f) - (I - C^\ell) A(u^\ell) u^\ell = 0 \qquad (\ell = 0, ..., N-1). \tag{4}$$

If we interpret this equation as a space-discrete elliptic PDE, our time-discrete scale-space solves a sequence of elliptic problems. In this sense, it has some structural similarities with scale-spaces created by the iterative regularisation methods of Radmoser et al. [17]. The central remaining question, however, is the following: In which sense does a sparsification scale-space satisfy the typical scale-space properties? It shall be answered next.

2.2   Scale-Space Properties

The good news after the somewhat tedious formalisation in the previous subsection is that it makes a subsequent analysis of the scale-space properties of sparsification scale-spaces surprisingly simple and intuitive. Let us now verify six important properties.

Property 1 (Original Image as Initial State).
For $\ell = 0$, Eq. (4) comes down to

$$C^0 (u^0 - f) - (I - C^0) A(u^0) u^0 = 0, \tag{5}$$

where C^0 denotes the mask associated to K^0. However, K^0 is defined as the full index set J, which implies that C^0 is identical to the identity matrix I. In this case, Eq. (5) simplifies to $u^0 - f = 0$.

Property 2 (Semigroup Property). The semigroup property states that one can construct a scale-space in a cascadic manner. This property follows directly from the fact that u^ℓ depends only on f and the nested known data sequence $(K^\ell)_{\ell=0}^{N-1}$. As long as one keeps f and this sequence, the sparsification scale-space has an obvious semigroup structure: A result $u^{\ell+n}$ can be obtained in $\ell+n$ steps by starting from $u^0 = f$, or in n steps by starting from u^ℓ.

Property 3 (Maximum–Minimum Principle). Traditionally, maximum–minimum principles are related to causality in the sense of Hummel [8]. They state that during the scale-space evolution, the greyscale range of the gradually simplified images lies in the range of the original image. For sparsification scale-spaces this property is always satisfied whenever the scale-space induced inpainting operator $A(u)$ fulfils a maximum-minimum principle: Then the inpainting solution $u^\ell = (u_i^\ell)_{i=1}^N$ satisfies

$$\min_{j\in J} f_j \ \le\ \min_{j\in K^\ell} f_j \ \le\ u_i^\ell \ \le\ \max_{j\in K^\ell} f_j \ \le\ \max_{j\in J} f_j \qquad \text{for all } i \in J, \qquad (6)$$

since the known data sequence $(K^\ell)_{\ell=0}^{N-1}$ is nested and starts with $K^0 = J$.

Property 4 (Lyapunov Sequences). Often the scale-space evolution (2) can be derived as the gradient descent evolution of some energy function $E(u)$. For instance, in a linear setting with operator A, one minimises the quadratic form

$$E(u) = \tfrac{1}{2}\, u^\top A u. \qquad (7)$$

In our inpainting setting, we have additional interpolation constraints on the set K^ℓ, which obviously results in a larger energy. By increasing ℓ, we reduce the nested constraint set K^ℓ, which means that we also reduce the energy gradually. This discussion shows that energy minimisation properties of the scale-space operator $A(u)$ allow us to interpret the energy as a Lyapunov sequence for the sparsification scale-space, i.e.

$$E(u^\ell) \le E(u^{\ell-1}) \qquad \text{for } \ell = 1,...,N-1. \qquad (8)$$

Lyapunov sequences provide important criteria for the simplification properties of scale-spaces [25]. Interestingly, they may even exist when no energy function is known: For instance, it is sufficient that the operator $A(u)$ satisfies a maximum–minimum principle. Then the nested known data sequence $(K^\ell)_{\ell=0}^{N-1}$ implies that the total image contrast

$$\Phi(u^\ell) \ :=\ \max_{i\in J} u_i^\ell - \min_{i\in J} u_i^\ell \ =\ \max_{i\in K^\ell} f_i - \min_{i\in K^\ell} f_i \qquad (9)$$

decreases in ℓ. Thus, $(\Phi(u^\ell))_{\ell=0}^{N-1}$ can be regarded as a Lyapunov sequence. Recently such sequences based on the total image contrast have been introduced by Welk et al. [27] to establish scale-space properties for FAB diffusion.

Property 5 (Invariances). Any invariance of the scale-space induced inpainting operator $A(u)$ is directly inherited to its sparsification scale-space.

Property 6 (Convergence to a Flat Steady-State). A discrete sparsification scale-space consists of N images $\{u^0, ..., u^{N-1}\}$. The final image u^{N-1} is always obtained by inpainting with a single mask pixel. In this case, a scale-space induced inpainting operator $A(u)$ that implements reflecting boundary conditions will create a flat image.

This completes our discussion of the essential properties that are satisfied by sparsification scale-spaces. We observe that all properties are either directly inherited from the scale-space induced inpainting operator or follow from the nested structure of the known data sequence.

3   Specific Aspects and Experiments

Our theoretical framework from the previous section allows to define sparsification scale-spaces for any PDE-based scale-space operator that can be expressed by the minimisation of an energy. Therefore, many well-known filters are viable for this task. The simplest one is the harmonic operator (also known as homogeneous diffusion) [9], but our model also covers higher-order linear operators such as biharmonic inpainting [5]. Moreover, a variety of nonlinear filters are also applicable. For instance, all operators investigated by Peter et al. [16] fit into the framework. This includes the nonlinear isotropic Perona–Malik model [14] as well as anisotropic filters, such as the approach of Tschumperlé and Deriche [24] or the method of Roussos and Maragos [18].

The experiments in Fig. 1 illustrate the influence of different inpainting methods on the behaviour of the corresponding sparsification scale-space. With the same randomly sparsified masks, both scale-space simplify the image in a similar fashion. In both cases, we start with the original image at 100% mask density. The density acts as the scale parameter: With decreasing number of mask pixels, the reconstructions gradually loose detail until only a single pixel remains and we reach a constant steady-state. Note that this implies that for an image with $m \times n$ pixels, we reach a finite extinction time after $m \cdot n - 1$ discrete time steps. Visually, the inpainting results at intermediate time steps differ significantly due to the individual properties of the operators: Harmonic inpainting creates visible singularities at the locations of known data. They are caused by the logarithmic singularity of the Green's function of the 2D Laplacian. Biharmonic inpainting yields much smoother results, since its Green's function is continuously differentiable. For more information on Green's functions we refer the reader to [12].

In addition to the influence of the inpainting approach, sparsification scale-spaces depend significantly on the order in which mask points are removed. The framework from Sect. 2 only requires that a single pixel is removed in each discrete time step, but it does not prescribe which one. For a given inpainting operator, this order of removal, the sparsification path, uniquely defines the scale-space.

On one hand, sparsification strategies can be uncommitted, such as the random removal of one mask point per step in Fig. 1. It does not depend on the original image in any way. By randomising the removal, we ensure that the lifetime expectation of each pixel is identical. Thus, the notion of uncommittedness must be understood in a statistical sense. On the other hand, our framework also allows adaptive sparsification that takes into account the image structure. Figure 2 shows results of a global adaptive sparsification strategy that has been inspired by the denoising by inpainting approach of Adam et al. [1]. In each step, we remove the single mask point that yields the smallest mean squared error (MSE) of the reconstruction. To speed up this algorithm, we consider 200 randomly selected mask points as candidates for removal in every iteration instead of an exhaustive search. Both the homogeneous and biharmonic inpainting yield considerably different results compared to the uncommitted sparsification path of Fig. 2. The adaptive sparsification assigns different life times to different pixels according to their importance for the reconstruction. This preserves important features of the original image over a longer period of time. Although the inpainting operators are still linear, this adaptive behaviour resembles nonlinear scale-spaces such as the Perona-Malik model, where image edges selected by a contrast parameter have a longer survival time in scale-space [14]. However, it should be noted that the adaptive sparsification is parameter-free.

Apart from denoising by inpainting, adaptive inpainting scale-spaces are of significant importance for image compression. Stochastic sparsification approaches similar to the one in our experiment have been proposed to identify known data to be stored in inpainting-based codecs [11,15]. In particular, there are variants of sparsification [15] that allow to store the corresponding masks efficiently as a binary tree. This concept is the foundation of the most successful inpainting-based codecs [7,20]. Thus, sparsification scale-spaces have a direct impact on practical applications in the field of image compression.

The distinction between uncommitted and committed scale-spaces sheds some light on an interesting additional aspect: Traditional uncommitted scale-spaces such as the Gaussian scale-space try to remove information as quickly and completely as possible. Sometimes this criterion is even stated explicitly, e.g. in Iijima's principle of maximum loss of figure impression [10]. On the other hand, a committed scale-space that aims at representations which are useful for image compression tries to achieve the exact opposite: It simplifies an image while aiming at a maximal preservation of the figure impression. It is nice and useful that sparsification scale-spaces allow to accommodate both philosophies.

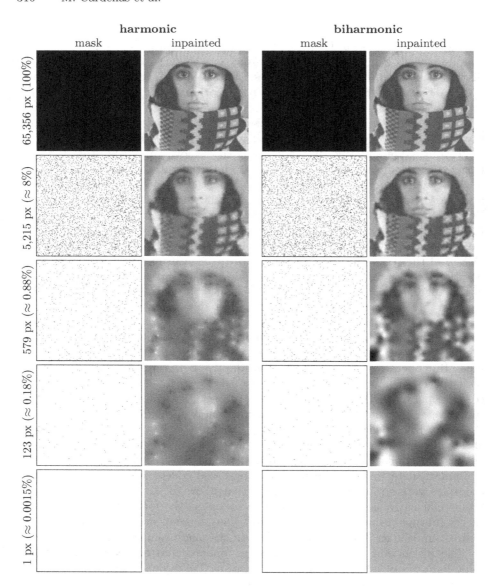

Fig. 1. Randomly Sparsified Scale-Spaces: A sparsification with purely random selection of the removed pixels provides the same mask for harmonic and biharmonic inpainting. These uncommitted scale-spaces differ only due to the inpainting operator.

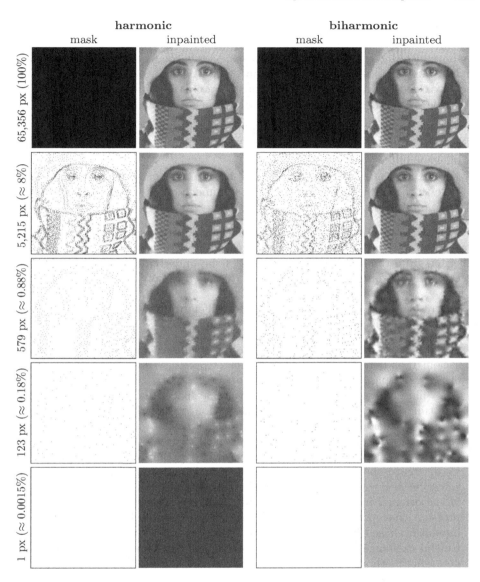

Fig. 2. Adaptively Sparsified Scale-Spaces: An image-adaptive stochastic sparsification leads to different masks for harmonic and biharmonic inpainting. Important image structures such as edges are better preserved than for an uncommitted sparsification scale-spaces.

4 Conclusions and Future Work

We have introduced and analysed a new scale-space concept: sparsification scale-spaces. They simplify a digital image via a gradual removal of pixels and inpaint the discarded image structures. By construction, they offer a great flexibility which lies in the freedom how their two ingredients are chosen:

- The sparsification path can be uncommitted or adaptive. In the latter case one follows a specific task-driven strategy, which allows e.g. structure-preserving image simplification.
- For the inpainting operator, any PDE-based scale-space operator is permissible that can be derived from energy minimisation or satisfies a maximum–minimum principle.

Our theoretical analysis has shown that sparsification scale-spaces satisfy all reasonable assumptions on a space- and time-discrete scale-space. They are either inherited from the scale-space properties of the underlying inpainting operator or they follow from the nested structure of the gradually sparsified inpainting data.

Interestingly, sparsification scale-spaces differ from most other scale-spaces in a number of aspects:

- The space discretisation of the image directly implies a natural time discretisation: It is determined by the number of pixels. We are not aware of any other scale-space which shares this property.
- After finitely many time steps, adaptive scale-spaces reach their finite extinction time. Apart from a few exceptions such as the total variation (TV) flow [3] and evolutions based on mean curvature motion or its affine invariant variant [2], this behaviour is not often met in the scale-space literature.
- Sparsification scale-spaces are perfectly suited for compression applications: If we disregard specific coding aspects such as quantisation, an adaptive sparsification scale-space can contain the entire family of compressed versions of the original image. This natural link between scale-space ideas and compression concepts is new and has been made possible by the paradigm of inpainting-based image compression. It may lay the foundations of a more fruitful exchange of ideas between both fields, and it may also inspire novel denoising methods.

In our future work, we are going to analyse and evaluate a broader spectrum of representatives within the large family of sparsification scale-spaces, covering e.g. also nonlinear methods and approaches that are not derived from a variational formulation.

It should be noted that we have restricted ourselves to discrete scale-spaces so far. Interestingly, the entire framework can also be extended to continuous sparsification scale-spaces. However, since their theory involves more technical challenges, it will be treated in a journal paper.

Acknowledgements. This project has received funding from the European Research Council (ERC) under the European Union's Horizon 2020 research and innovation programme (grant agreement no. 741215, ERC Advanced Grant INCOVID).

References

1. Adam, R.D., Peter, P., Weickert, J.: Denoising by inpainting. In: Lauze, F., Dong, Y., Dahl, A.B. (eds.) SSVM 2017. LNCS, vol. 10302, pp. 121–132. Springer, Cham (2017). https://doi.org/10.1007/978-3-319-58771-4_10
2. Alvarez, L., Guichard, F., Lions, P.L., Morel, J.M.: Axioms and fundamental equations in image processing. Arch. Ration. Mech. Anal. **123**, 199–257 (1993)
3. Andreu-Vaillo, F., Caselles, V., Mazon, J.M.: Parabolic Quasilinaer Equations Minimizing Linear Growth Functionals, Progress in Mathematics, vol. 223. Birkhäuser, Basel (2004)
4. Chambolle, A., Lucier, B.L.: Interpreting translationally-invariant wavelet shrinkage as a new image smoothing scale space. IEEE Trans. Image Process. **10**(7), 993–1000 (2001)
5. Duchon, J.: Interpolation des fonctions de deux variables suivant le principe de la flexion des plaques minces. RAIRO Anal. Numérique **10**, 5–12 (1976)
6. Duits, R., Florack, L., de Graaf, J., ter Haar Romeny, B.: On the axioms of scale space theory. J. Math. Imaging Vis. **20**, 267–298 (2004)
7. Galić, I., Weickert, J., Welk, M., Bruhn, A., Belyaev, A., Seidel, H.P.: Image compression with anisotropic diffusion. J. Math. Imaging Vis. **31**(2–3), 255–269 (2008)
8. Hummel, R.A.: Representations based on zero-crossings in scale space. In: Proceedings of 1986 IEEE Computer Society Conference on Computer Vision and Pattern Recognition, pp. 204–209. IEEE Computer Society Press, Miami Beach, June 1986
9. Iijima, T.: Basic theory on normalization of pattern (in case of typical one-dimensional pattern). Bull. Electrotechnical Lab. **26**, 368–388 (1962). in Japanese
10. Iijima, T.: Basic equation of figure and observational transformation. Syst. Comput. Controls **2**(4), 70–77 (1971). in English
11. Mainberger, M., Hoffmann, S., Weickert, J., Tang, C.H., Johannsen, D., Neumann, F., Doerr, B.: Optimising spatial and tonal data for homogeneous diffusion inpainting. In: Bruckstein, A.M., ter Haar Romeny, B.M., Bronstein, A.M., Bronstein, M.M. (eds.) SSVM 2011. LNCS, vol. 6667, pp. 26–37. Springer, Heidelberg (2012). https://doi.org/10.1007/978-3-642-24785-9_3
12. Melnikov, Y.A., Melnikov, M.Y.: Green's Functions: Construction and Applications. De Gruyter, Berlin (2012)
13. Pennebaker, W.B., Mitchell, J.L.: JPEG: Still Image Data Compression Standard. Springer, New York (1992)
14. Perona, P., Malik, J.: Scale space and edge detection using anisotropic diffusion. IEEE Trans. Pattern Anal. Mach. Intell. **12**, 629–639 (1990)
15. Peter, P., Hoffmann, S., Nedwed, F., Hoeltgen, L., Weickert, J.: Evaluating the true potential of diffusion-based inpainting in a compression context. Signal Process. Image Commun. **46**, 40–53 (2016)
16. Peter, P., Weickert, J., Munk, A., Krivobokova, T., Li, H.: Justifying tensor-driven diffusion from structure-adaptive statistics of natural images. In: Tai, X.-C., Bae, E., Chan, T.F., Lysaker, M. (eds.) EMMCVPR 2015. LNCS, vol. 8932, pp. 263–277. Springer, Cham (2015). https://doi.org/10.1007/978-3-319-14612-6_20

17. Radmoser, E., Scherzer, O., Weickert, J.: Scale-space properties of nonstationary iterative regularization methods. J. Visual Commun. Image Represent. **11**(2), 96–114 (2000)
18. Roussos, A., Maragos, P.: Tensor-based image diffusions derived from generalizations of the total variation and Beltrami functionals. In: Proceedings of 17th IEEE International Conference on Image Processing, Hong Kong, pp. 4141–4144, September 2010
19. Scherzer, O., Weickert, J.: Relations between regularization and diffusion filtering. J. Math. Imaging Vis. **12**(1), 43–63 (2000)
20. Schmaltz, C., Peter, P., Mainberger, M., Ebel, F., Weickert, J., Bruhn, A.: Understanding, optimising, and extending data compression with anisotropic diffusion. Int. J. Comput. Vis. **108**(3), 222–240 (2014)
21. Schmidt, M., Weickert, J.: Morphological counterparts of linear shift-invariant scale-spaces. J. Math. Imaging Vis. **56**(2), 352–366 (2016)
22. Schönlieb, C.B.: Partial Differential Equation Methods for Image Inpainting. Cambridge University Press, New York (2015)
23. Taubman, D.S., Marcellin, M.W. (eds.): JPEG 2000: Image Compression Fundamentals, Standards and Practice. Kluwer, Boston (2002)
24. Tschumperlé, D., Deriche, R.: Vector-valued image regularization with PDEs: a common framework for different applications. IEEE Trans. Pattern Anal. Mach. Intell. **27**(4), 506–516 (2005)
25. Weickert, J.: Anisotropic Diffusion in Image Processing. Teubner, Stuttgart (1998)
26. Weickert, J., Steidl, G., Mrázek, P., Welk, M., Brox, T.: Diffusion filters and wavelets: what can they learn from each other? In: Paragios, N., Chen, Y., Faugeras, O. (eds.) Handbook of Mathematical Models in Computer Vision, pp. 3–16. Springer, New York (2006). https://doi.org/10.1007/0-387-28831-7_1
27. Welk, M., Weickert, J., Gilboa, G.: A discrete theory and efficient algorithms for forward-and-backward diffusion filtering. J. Math. Imaging Vis. **60**(9), 1399–1426 (2018)

Stable Explicit p-Laplacian Flows Based on Nonlinear Eigenvalue Analysis

Ido Cohen$^{(\boxtimes)}$, Adi Falik, and Guy Gilboa

Technion - Israel Institute of Technology, 32000 Haifa, Israel
{idoc,adifalik}@campus.technion.ac.il, guy.gilboa@ee.technion.ac.il

Abstract. Implementation of nonlinear flows by explicit schemes can be very convenient, due to their simplicity and low-computational cost per time step. A well known drawback is the small time step bound, referred to as the CFL condition, which ensures a stable flow. For p-Laplacian flows, with $1 < p < 2$, explicit schemes without gradient regularization require, in principle, a time step approaching zero. However, numerical implementations show explicit flows with small time-steps are well behaved. We can now explain and quantify this phenomenon.

In this paper we examine explicit p-Laplacian flows by analyzing the evolution of nonlinear eigenfunctions, with respect to the p-Laplacian operator. For these cases analytic solutions can be formulated, allowing for a comprehensive analysis. A generalized CFL condition is presented, relating the time step to the inverse of the nonlinear eigenvalue. Moreover, we show that the flow converges and formulate a bound on the error of the discrete scheme. Finally, we examine general initial conditions and propose a dynamic time-step bound, which is based on a nonlinear Rayleigh quotient.

1 Introduction

Nonlinear diffusion equations are known as an effective method for creating a systematic series of simplifications of a signal. This has numerous applications in signal and image processing [25]. Recently, there is a growing interest in the p-Laplacian flow, we refer to as the p-flow, defined by

$$u_t = \Delta_p u, \quad u(t = 0) = f, \tag{1}$$

where f is an initial condition and $\Delta_p u$ is the p-Laplacian defined as

$$\Delta_p u = \operatorname{div}\left(|\nabla u|^{p-2}\nabla u\right). \tag{2}$$

The effectiveness of the p-Laplacian is demonstrated in [2,5,6,16,22] for $1 \leq p \leq 2$ with p dependent on the gradient of the image. For the case of constant p, properties of the p-flow in the context of image filtering are discussed in [18–20,24], establishing the relations to the Gauge coordinates. For $p \to 1$, the evolution of (1) approximates the total variation (TV) flow [1].

© Springer Nature Switzerland AG 2019
J. Lellmann et al. (Eds.): SSVM 2019, LNCS 11603, pp. 315–327, 2019.
https://doi.org/10.1007/978-3-030-22368-7_25

One can model the p-flow as a nonlinear diffusion process [23]:

$$u_t = \text{div}\left(c(|\nabla u|)\nabla u\right), \quad u\left(t = 0\right) = f, \tag{3}$$

where $c(|\nabla u|) = |\nabla u|^{p-2}$ is the gradient dependent diffusion coefficient. For such flows there is no analytic solution in the general case and a numerical approximation should be used. One of the most popular methods in numerical analysis of PDEs is the finite difference (FD) method. It is based on discretizing the spatial and temporal domains and substituting the derivatives by finite difference approximations. There are several ways to model the p-flow by FD. The most simple one is the explicit scheme, in which the flow (1) is approximated by [18]

$$u_{k+1} = u_k + dt \cdot \Delta_p\left(u_k\right), \quad u_0 = f, \tag{4}$$

where Δ_p is the discretized p-Laplace operator and k is the iteration index. This scheme is the most intuitive and convenient to solve, yet it implies a stability limitation on the step size dt as we discuss later on. Another common method is the semi-implicit scheme which approximates the spatial derivatives by involving both u_k and u_{k+1}. This scheme offers an unconditional stability which is independent of dt yet it is more intensive computationally than the explicit method as it requires solving a system of linear equations at each time step.

When applying an explicit FD scheme one needs to resolve the bound on dt for which the flow is stable. A well known stability condition, is the *Courant-Friedrichs-Lewy* (CFL) condition [9]. Applying the CFL theory to nonlinear flows (3) in the d-dimensional case with equal spatial intervals Δh, yields that convergence of the numerical solution is guaranteed for time steps bounded by,

$$dt \leq \frac{(\Delta h)^2}{2d \max_{x,t} c(|\nabla u|)}. \tag{5}$$

This bound introduces a major problem for p-flows with $1 \leq p < 2$. As $|\nabla u| \to 0$ it can be seen that the time-step bound in (5) approaches zero. A common remedy is to use a regularization of c by $c(|\nabla u|) = \left(|\nabla u|^2 + \epsilon^2\right)^{\frac{p-2}{2}}$, where ϵ is proportional to the time step size [2]. However, this regularization raises an issue regarding the size of ϵ. Large values lead to a coarse approximation of the actual flow, whereas small values require small time steps.

A more general approach for stability analysis, leading also to (5), was introduced by von-Neumann, based on Fourier and eigenvalue analysis of the linearized equation [10,17,21]. A general explicit scheme can be expressed by:

$$u_{k+1} = u_k + dt \cdot A^{(p)}\left(u_k\right) \cdot u_k, \tag{6}$$

where the matrix $A^{(p)}\left(u_k\right)$ is the operator of the flow evaluated at iteration k. For the linear case, where $p = 2$, the matrix $A^{(p)} = A^{(2)}$ is constant and does not depend on u_k. In this case, a sufficient condition of numerical stability can be determined by the spectral radius, λ_{max}, of the matrix $A^{(2)}$,

$$dt \leq \frac{2}{\lambda_{max}}, \tag{7}$$

where λ_{max} is the maximum absolute eigenvalue of the matrix $A^{(2)}$. In the nonlinear case the radius of A^p should be evaluated every iteration. This stability condition is widely used in various numerical methods [12, 26].

In this work we analyze explicit numerical schemes of the p-Laplacian flow for $1 < p < 2$, where no ϵ approximation is used. We gain insight by analytically solving for nonlinear eigenfunctions as initial conditions. Two time step policies are compared with respect to their stability and accuracy.

2   Preliminary

We recall the discrete Hilbert space $\mathbb{R}^n$ with standard discrete Euclidean inner product and norm definitions, i.e. $\langle u, v \rangle = \sum_i u_i \cdot v_i$ and $\|u\|^2 = \langle u, u \rangle$, respectively. The p-Dirichlet energy for a function $u \in \mathbb{R}^n$ is defined by,

$$J_p(u) := \frac{1}{p} \langle |\nabla u|^p, 1 \rangle = \frac{1}{p} \langle |\nabla u|^{p-2} \nabla u, \nabla u \rangle = \frac{1}{p} \langle -\text{div} \left(|\nabla u|^{p-2} \nabla u \right), u \rangle$$
$$= \frac{1}{p} \langle -\Delta_p u, u \rangle,$$

(8)

where $\Delta_p u$ is defined in Eq. (2) and div and ∇ are discrete divergence and gradient operators, respectively. We note that $-\Delta_p$ is a positive semi-definite operator and that $J_p(u) \geq 0$, $\forall u \in \mathbb{R}^n$. The gradient descent with respect to J_p is the p-flow, defined in Eq. (1). The p-Laplacian operator is $p-1$ homogeneous i.e.

$$\Delta_p(a \cdot u) = a|a|^{p-2} \cdot \Delta_p(u), \quad \forall a \in \mathbb{R}.$$

(9)

In the analysis below, we also use the following known relation. Every concave function $f : \mathbb{R} \rightarrow \mathbb{R}$ in C^1 admits

$$f(t) \leq f(a) + f'(a)(t - a), \quad \forall t, a \in \mathbb{R}.$$

(10)

The inequality becomes strict if f is strictly concave.

2.1   Eigenfunctions of the p-Laplacian

A comprehensive analysis of 1-Laplacian eigenfunctions in the spatial continuous setting was conducted in [1], as part of the investigation of the *total-variation flow*. An analytic solution of the TV-flow was formulated for initial conditions which are eigenfunctions, where the shape is preserved and the contrast decays linearly. Recently, an active research is conducted on nonlinear spectral representations and decompositions based on TV [13] and on one-homogeneous functionals in general [3]. Lately, this approach was applied to p-homogeneous functionals when $1 < p < 2$ [7].

Let us define a nonlinear eigenfunction of the p-Laplacian as

$$\Delta_p(f) = \lambda \cdot f, \quad \lambda \in \mathbb{R}, \quad \textbf{(EF)}$$

where the corresponding eigenvalue λ admits

$$\lambda = \langle \Delta_p(f), f \rangle / \|f\|^2 = -pJ_p(f)/\|f\|^2 \leq 0. \tag{11}$$

The existence of eigenfunctions was established in [11]. It was shown in [7] that when f admits (**EF**) the solution of (1) is of the form

$$u(t) = a(t) \cdot f, \tag{12a}$$

$$a(t) = \left([(2-p)\lambda t + 1]^+ \right)^{\frac{1}{2-p}} \quad \text{or} \quad \frac{d}{dt}a(t) = a(t)|a(t)|^{p-2}\lambda, \; a(0) = 1, \tag{12b}$$

where $q^+ = \max\{q, 0\}$. This solution vanishes in finite extinction time

$$T_{ext} = -\left[(2-p)\lambda \right]^{-1}. \tag{13}$$

These results coincide with [1] when $p = 1$. A solution of the form (12a) is referred to as a shape preserving flow since the shape of f is preserved and only its contrast changes throughout the flow. We now turn to analyze explicit time discrete evolutions of (1).

3   Analysis of Explicit Schemes for Shape Preserving Flows

To gain insight on the flow, implemented by explicit schemes, we turn to analyze initial conditions which are eigenfunctions, as they have analytic solutions. The time domain of the flow is discretized by time steps dt_k, where k denotes the iteration number. Note that, as oppose to (4), dt_k can change with iterations. The explicit scheme of Eq. (1) is therefore given by:

$$u_{k+1} = u_k + dt_k \cdot \Delta_p(u_k), \qquad u_0 = f. \qquad (\mathbf{pFlow})$$

We say that the solution of (**pFlow**) is shape preserving if it can be written as:

$$u_k = a_k \cdot f, \qquad (\mathbf{ShP})$$

where $a_k \in \mathbb{R}$. The next Lemma discusses the conditions to obtain such solutions and formulates the recurrence relation of a_{k+1} as a function of a_k.

Lemma 1. *The solution of* (**pFlow**) *is* (**ShP**) *for all* $dt_k > 0$ *iff* f *admits* (**EF**). *In this case the recurrence relating* a_{k+1} *to* a_k *is:*

$$a_{k+1} = a_k \left(1 + |a_k|^{p-2}\lambda dt_k \right), \qquad a_0 = 1. \tag{14}$$

Proof. $\Rightarrow$ We assume f is an eigenfunction and prove (**ShP**). The proof is based on induction. From (**pFlow**) and (**EF**) we have, $u_1 = u_0 + \Delta_p(u_0) \cdot dt_0 =$

$f(1 + \lambda dt_0)$. Thus, the induction assumption (**ShP**) holds for $k = 1$. Now let us assume (**ShP**). Then,

$$
\underbrace{u_{k+1} \quad = \quad u_k + dt_k \cdot \Delta_p u_k}_{Eq.(\textbf{pFlow})} \quad \underbrace{= \quad a_k f + dt_k \cdot \Delta_p (a_k f)}_{\text{Induction assumption}}
$$

$$
\underbrace{= a_k f + a_k |a_k|^{p-2} dt_k \cdot \Delta_p f}_{Eq.(9)} \underbrace{= (a_k + \lambda a_k |a_k|^{p-2} dt_k) \cdot f = a_{k+1} f.}_{Eq.(\textbf{EF})}
$$

(15)

$\Leftarrow$ Now, we assume (**ShP**) and prove f admits (**EF**). According to (**pFlow**) and (**ShP**) we can write $u_1 = u_0 + \Delta_p(u_0) \cdot dt_0$ and therefore $a_1 \cdot f = f + \Delta_p(f) \cdot dt_0$. Then, $\Delta_p(f) = \frac{a_1 - 1}{dt_0} \cdot f = \lambda f$, therefore, f is an eigenfunction. Moreover, from (15) we readily have (14). $\qquad\square$

Note, the recurrence relation (14) can be viewed as an explicit first order approximation of (12b). In the rest of this section we discuss the advantages and limitations of adaptive versus fixed step sizes.

3.1   Convergence Criterion for Adaptive Step Size

We are looking for dt_k that guarantees convergence of a_k to zero.

Theorem 1 (Convergence for adaptive step size policy). *If the solution of* (**pFlow**) *admits* (**ShP**) *and the step size is*

$$
dt_k = -\delta \lambda^{-1} |a_k|^{2-p}, \quad \delta \in (0, 1], \tag{16}
$$

then $\lim_{k \to \infty} a_k = 0$, *and the extinction time is*

$$
\hat{T}_{ext}(\delta) = -\delta \lambda^{-1} \left[1 - (1 - \delta)^{2-p} \right]^{-1}. \tag{17}
$$

In addition, the following relation holds

$$
-\lambda^{-1} \leq \hat{T}_{ext}(\delta) \leq T_{ext} \tag{18}
$$

where T_{ext} *is defined in Eq.* (13).

Proof. Substituting the step size (16) in (14) yields $a_{k+1} = a_k - a_k |a_k|^{p-2} \lambda \cdot \delta \cdot \lambda^{-1} |a_k|^{2-p} = a_k (1 - \delta)$. Thus, $\{a_k\}$ is simply a geometrical series,

$$
a_k = (1 - \delta)^k, \quad k = 0, 1, .. \infty. \tag{19}
$$

For $\delta \in (0, 1]$ this series converges to zero. The total time for convergence is

$$
\hat{T}_{ext}(\delta) = \sum_{k=0}^{\infty} dt_k = \begin{cases} -\delta \lambda^{-1} \sum_{k=0}^{\infty} (1 - \delta)^{k(2-p)} = \frac{-\delta}{\lambda[1-(1-\delta)^{2-p}]} & \delta \neq 1 \\ -\lambda^{-1} & \delta = 1 \end{cases}. \tag{20}
$$

Now, let us show $\hat{T}_{ext}(\delta)$ is monotonically strictly decreasing with $\delta \in (0,1]$. The derivative of $\hat{T}_{ext}(\delta)$ is

$$\frac{d}{d\delta}\hat{T}_{ext}(\delta) = -\underbrace{\frac{(1-\delta)^{(1-p)}}{\lambda\left[1-(1-\delta)^{(2-p)}\right]^2}}_{>0} \underbrace{\left[(1-\delta)^{(p-1)} - 1 + (p-1)\delta\right]}_{<0 \text{ as } (1-\delta)^{(p-1)} \text{ is strictly concave (Eq. (10))}} < 0.$$

Equation (18) is obtained by, $-\lambda^{-1} = \hat{T}_{ext}(1) \leq \hat{T}_{ext}(\delta) \leq \underbrace{\hat{T}_{ext}(\delta \to 0)}_{\text{L'Hôpital}} = T_{ext}$.   □

Remark 1. Lemma 1 yields that u_k is an eigenfunction with eigenvalue $\lambda_k = \lambda|a_k|^{p-2}$. Consequently, from (16) we deduce that $\{a_k\}$ converges when $dt_k \leq -1/\lambda_k$. But, according to (19) $\{a_k\}$ converges for all $\delta \in (0,2)$ i.e.:

$$dt_k < -2/\lambda_k. \tag{21}$$

This conclusion coincides with the stability condition of von-Neumann, Eq. (7). However, we limit ourselves to $\delta \in (0,1]$ to refrain from negative a_k since it contradicts (12b).

Error Bound. Let us compare between the semi-discrete solution (12) and the discrete solution (19) of shape preserving flows. These solutions can be compared only at the discrete times t_k,

$$t_k = \sum_{n=0}^{k-1} dt_n = -\frac{\delta}{\lambda}\left[\frac{1-(1-\delta)^{k(2-p)}}{1-(1-\delta)^{2-p}}\right]. \tag{22}$$

The error is defined by

$$E_k = \|u(t_k) - u_k\|^2. \tag{23}$$

From (12a), **(ShP)** and (23) we have $E_k = (a(t_k) - a_k)^2 \|f\|^2$. The following theorem provides upper and lower bounds on the error.

Theorem 2 (Error bound). *For the explicit flow* **(pFlow)** *with adaptive time steps defined in* (16) *the error* (23) *is bounded by*

$$E_k \leq \lambda^2 \left(\hat{T}_{ext}(\delta) - T_{ext}\right)^2 \|f\|^2. \tag{24}$$

Moreover, let us denote $e_k = a(t_k) - a_k$, $a'(t)$ *as the time derivative of* (12b) *and define* $\tilde{t}_k$ *as the time for which* $a(\tilde{t}_k) = a_k$, *with* $d_k = t_k - \tilde{t}_k$. *Then,* $\forall k \in \mathbb{N}$, e_k *is upper- and lower-bounded by,*

$$0 \leq a'(\tilde{t}_k)d_k \leq e_k \leq a'(t_k)d_k \leq \lambda\left(\hat{T}_{ext}(\delta) - T_{ext}\right). \tag{25}$$

Proof. We show that $a_k \leq a(t_k)$. Since $a(t)$ is strictly monotonically decreasing, it is equivalent to show that if $a(\tilde{t}_k) = a_k$ then $t_k \leq \tilde{t}_k$. Let us compute $\tilde{t}_k$,

$$a(\tilde{t}_k) = a_k \quad \Rightarrow \quad \left[(2-p)\lambda\tilde{t}_k + 1\right]^{\frac{1}{2-p}} = (1-\delta)^k \quad \Rightarrow \quad \tilde{t}_k = \frac{(1-\delta)^{k(2-p)} - 1}{(2-p)\lambda}.$$

Then d_k can be expressed as

$$d_k = \underbrace{\tilde{t}_k - t_k}_{\text{Eq. (22)}} = \frac{(1-\delta)^{k(2-p)} - 1}{(2-p)\lambda} + \frac{\delta}{\lambda} \cdot \frac{1 - (1-\delta)^{k(2-p)}}{1 - (1-\delta)^{2-p}}$$

$$= \underbrace{\left[1 - (1-\delta)^{k(2-p)}\right]}_{\substack{\text{Eqs. (13), (17)} \\ \geq 0}} \underbrace{\left[\overbrace{\delta\lambda^{-1}\left[1 - (1-\delta)^{2-p}\right]^{-1}}^{-\hat{T}_{ext}} \overbrace{-\lambda^{-1}(2-p)^{-1}}^{T_{ext}}\right]}_{\geq 0 \text{ Theorem 1}} \geq 0.$$

Therefore, $a_k \leq a(t_k)$. Since $a(t)$ is convex we have

$$a'(\tilde{t}_k) \geq \frac{a(t_k) - a_k}{t_k - \tilde{t}_k} \geq a'(t_k) \Rightarrow \underbrace{a'(\tilde{t}_k)}_{\leq 0} \underbrace{\left(t_k - \tilde{t}_k\right)}_{\leq 0} \leq a(t_k) - a_k \leq \underbrace{-a'(t_k)}_{\leq -\lambda} \underbrace{\left(\tilde{t}_k - t_k\right)}_{\leq T_{ext} - \hat{T}_{ext}(\delta)}$$

and thus we obtain (24). □

From Theorem 1 we have $\hat{T}_{ext}(\delta \to 0) = T_{ext}$, thus (24) yields $E_k \to 0$, $\forall k$ as $\delta \to 0$. A relaxation of the CFL condition was suggested in [14], however, this method does not well approximate the decay profile. We note that in [15] the error of the approximated nonlocal p-Laplacian was studied for the continuous and the discrete settings.

3.2   Stability Criterion for Fixed Step Size

Common explicit formulations use a fixed step size. In the following theorem we show certain stability of the flow for fixed step sizes.

Theorem 3 (Lyapunov stability for fixed step size). *Let u_k be a solution of* **(pFlow)** *with a fixed step size $dt_k = dt > 0$, where f admits* **(EF)**. *We denote $\tau = (-\lambda dt)^{\frac{1}{2-p}}$. Then for every dt, $\exists K < \infty$ s.t.*

$$J_p(u_k) \leq \tau^p J_p(f), \quad \forall k > K. \tag{26}$$

Moreover, if the series $\{|a_k|\}$ converges, it converges either to 0 or to $\tau \left(\frac{1}{2}\right)^{\frac{1}{2-p}}$.

Proof. First, let us note that Lemma 1 is valid for all $dt_k > 0$ and therefore (14) holds. The assertion (26) is equivalent to the statement that for every $dt > 0$

there exists a finite K such that $|a_k| \leq \tau$, $\forall k > K$. Note, as long as $a_k > \tau$ the series a_k is monotonically strictly decreasing,

$$a_k - a_{k+1} = -\lambda |a_k|^{p-1} \cdot dt > -\lambda(-\lambda \cdot dt)^{\frac{p-1}{2-p}} dt = \tau > 0. \tag{27}$$

Moreover, if $a_k > \tau$ then $a_{k+1} = a_k + \lambda a_k |a_k|^{p-2} \cdot dt > \tau \left[1 + \lambda(-\lambda \cdot dt)^{-1} dt\right] = 0$. Therefore, there exists some k where $a_k \in [0, \tau]$ and from (27) we can bound the number of steps by $K \leq \tau^{-1}$. Now, we show that if $|a_k| < \tau$ then $|a_m| < \tau$, $\forall m > k$. Let us assume $0 < a_k < \tau$, then

$$\underbrace{0 < a_k < \tau}_{A} \Rightarrow \underbrace{0 < (a_k)^{p-1} < \tau^{p-1}}_{B} \Rightarrow \underbrace{-\tau < (a_k)^{p-1}\lambda dt < 0}_{C}.$$

The expression in C is obtained by multiplying the expression B by $(\lambda dt) < 0$ and using the definition of τ. Summation of the relations A and C yields $-\tau < a_k + (a_k)^{p-1}\lambda dt < \tau$, therefore $|a_{k+1}| < \tau$. If $-\tau < a_k < 0$ then in a similar manner we get $|a_{k+1}| < \tau$. The rest of this proof is by induction and trivial. If the series $|a_k|$ converges we have $a_k = a_{k+1}$ or $a_k = -a_{k+1}$. According to the recurrence (14) the first option is possible only if $a_k = 0$ and the second if $|a_k| = \left(-\frac{\lambda \cdot dt}{2}\right)^{\frac{1}{2-p}} = \tau\left(\frac{1}{2}\right)^{\frac{1}{2-p}}$. In this case J_p converges either to 0 or to

$$\lim_{k \to \infty} J_p(u_k) = \tau^p \left(\frac{1}{2}\right)^{\frac{p}{2-p}} J_p(f) = \left(-\frac{\lambda \cdot dt}{2}\right)^{\frac{p}{2-p}} J_p(f). \tag{28}$$

$\square$

Corollary 1. *To ensure $J_p(u_k) \leq C$ in finite k we require*

$$0 < dt < -\frac{1}{\lambda}\left(\frac{C}{J_p(f)}\right)^{\frac{2-p}{p}}. \tag{29}$$

Proof. Immediately from Theorem 3. $\square$

4   Arbitrary Initial Conditions

Now, we assume that f is an arbitrary function and examine the conditions for $J_p(u_k)$ to converge to zero with an adaptive step size.

Theorem 4. *Let $f \in \mathbb{R}^n$, $\tilde{\lambda}_k = \|\Delta_p u_k\|^2 / \langle \Delta_p u_k, u_k \rangle$, and $\delta \in (0, 2)$. Then $J_p(u_k) \to 0$ as $k \to \infty$ for*

$$dt_k = -\delta/\tilde{\lambda}_k. \tag{30}$$

Proof. Let us recall $pJ_p(u_k) = -\langle \Delta_p u_k, u_k \rangle \geq 0$. Using **(pFlow)**, we can express $\|u_{k+1}\|^2$ by $\|u_{k+1}\|^2 = \|u_k\|^2 + 2dt_k \langle u_k, \Delta_p u_k \rangle + dt_k^2 \|\Delta_p(u_k)\|^2$. Using (30) we can write the difference between two successive elements as

$$\|u_{k+1}\|^2 - \|u_k\|^2 = \underbrace{(-2 + \delta)}_{<0} \delta p^2 \underbrace{\frac{J_p^2(u_k)}{\|\Delta_p u_k\|^2}}_{\geq 0} \leq 0 \tag{31}$$

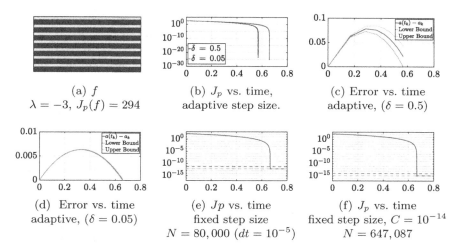

Fig. 1. Behaviour of J_p as a function of time with different step size policies when f is an eigenfunction. (a) The initial condition f. (b) Theorem 1, convergence with adaptive step size, iterations: $N = 59$, $(\delta = 0.5)$ (blue), $N = 762$, $(\delta = 0.05)$ (red), the theoretical extinction time is $T_{ext} = 2/3$, Eq. (13). (c), (d) Theorem 2, the actual errors (blue) and the theoretical bounds (red), Eq. (25). (e) Theorem 3 fixed time step for fixed dt $(r = 2.874 \, \text{s})$ and, (f) Corollary 1, fixed time step for fixed C $(dt = 1.2363 \cdot 10^{-06}, r = 24.4 \, \text{s})$. In (e) and (f) the dashed dot yellow line is the convergence value Eq. (28), the dashed red line is the upper bound Eq. (26). (Color figure online)

Thus, $\{\|u_k\|^2\}$ is monotonically decreasing and bounded from below by zero, therefore, converges. Therefore, the difference between two successive elements converges to zero. Since $(-2 + \delta)\delta p^2$ is a fixed number, the ratio $J_p^2(u_k)/\|\Delta_p u_k\|^2 \to 0$ as $k \to \infty$. As $\|\Delta_p u_k\|$ is bounded from above in the discrete case, $J_p(u_k)$ converges to zero. $\quad\square$

Discussion. From the analysis of eigenfunctions and Remark 1, we deduce that $0 < dt_k < -2/\lambda_k$ guarantees convergence of the flow. Theorem 4 implies a very similar restriction for an arbitrary initial condition, $0 < dt_k < -2/\tilde{\lambda}_k$. $\tilde{\lambda}_k$ can be interpreted as a variant of a nonlinear Rayleigh quotient. For eigenfunctions $\tilde{\lambda}_k = \lambda_k$. From (31), the fastest decrease is when $\delta = 1$, i.e. when $dt_k = -\tilde{\lambda}_k$.

Fixed Step Size. For arbitrary initial conditions and fixed step size we do yet have a full theory. In this case we allow an error tolerance in u of approximately $\pm h$. Thus we would like to avoid oscillations above an amplitude of h. The most oscillatory eigenfunction (highest absolute eigenvalue) of amplitude of at least h is a "chessboard"-like signal with amplitude h. Its eigenvalue can be computed by $\lambda_{max} = -p\frac{J_p(u)}{\|u\|^2} = -8^{\frac{p}{2}} h^{p-2}$. We require

$$dt = -\lambda_{max}^{-1}. \tag{32}$$

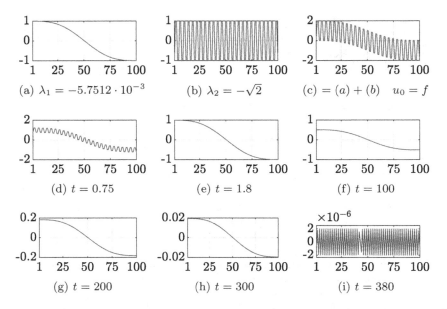

Fig. 2. Two combined eigenfunctions $(dt = 10^{-3})$

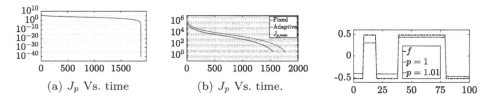

(a) J_p Vs. time   (b) J_p Vs. time.

Fig. 3. Behaviour of J_p as a function of time with different step size policies when f is a natural image (Lena). (a) Theorem 4, adaptive step size, $r = 338$ s $N = 265826$. (b) Eq. (32), $J_{p,min}$ (4.011) and dt (0.0066) are computed for $h = 10^{-3}$. The fixed policy reaches $J_{p,min}$ in $N = 233734$ steps and the adaptive policy in $N = 20635$ steps.

Fig. 4. TV-flow and p-flow $(p = 1.01)$. The p-flow runs with fixed time step $dt = 10^{-4}$. Edges are preserved with no significant oscillations.

Based on Theorem 3, we can expect that the minimum of J_p will reach $J_{p,min} \approx 8^{\frac{p}{2}} \cdot h^p \cdot |\Omega|/p$ (where $|\Omega|$ is the image size).

5   Results and Conclusion

We present several examples which demonstrate the theoretical results established in previous sections. We evolve the flow (**pFlow**) for different initial conditions f with adaptive and fixed step sizes. A common approximation of the

p-Laplacian is $\Delta_{p,\epsilon} u = \text{div}\left((|\nabla u|^2 + \epsilon^2)^{\frac{p-2}{2}} \nabla u\right)$. To avoid that we formulate the p-Laplacian differently, $\Delta_p u = \text{div}\left(|\nabla u|^{p-1} \nabla u/|\nabla u|\right)$. The term $\nabla u/|\nabla u|$ can be evaluated either with a $sign()$ function in $1D$ or with the cosine and sine for the angle between u_x and u_y (we use the Matlab function $\text{atan}\,2(u_x, u_y)$ which is implemented by a look-up table). Since $p - 1 > 0$ we avoid division by zero (note that for zero gradient the p-Laplacian is zero for any $p > 1$). The discrete divergence and gradient operators use standard forward and backward difference, as defined in [4]. We use $p = 1.5$, unless stated otherwise. The eigenfunctions were numerically calculated using the algorithm [8]. We denote by r the total running time and by N the number of iterations. All experiments were run on an i7-8700k CPU machine @ 3.70 GHz, 64 GB RAM.

Eigenfunctions as Initial Condition. Figure 1 depicts flows for adaptive and fixed step sizes, illustrating Theorems 1, 2, 3 and Corollary 1. One can clearly observe that when the initial condition is an eigenfunction the adaptive step size is highly efficient and accurate. In Fig. 2 we present an evolution of two combined eigenfunctions with fixed dt. The initial condition (Fig. 2c) is a sum of two eigenfunctions (Figs. 2a and b). The Figs. 2d–i show different stages in the evolution. Note, the extinction times of the eigenfunctions (separately) are $T_1 = 347$ and $T_2 = \sqrt{2}$ and Figs. 2e and i depict approximately these stages. Moreover, the residue (Fig. 2i) is compatible with Theorem 3 as if the initial condition is only the eigenfunction with larger eigenvalue (Fig. 2b), where $|a_\infty| = (-\lambda_2 \cdot dt/2)^{\frac{1}{2-p}} = 2 \cdot 10^{-6}$.

Arbitrary Initial Condition. Figure 3 depicts the flow for adaptive and fixed step size policies, illustrating Theorem 4 and the following discussion in Sect. 4. From the running time and the number of iterations, given within the figure's caption, it is easy to see that the adaptive step size policy is faster than the fixed one.

TV-Flow Approximation. We approximate the TV-flow with $p = 1.01$ with fixed $dt = 10^{-4}$ and compare it with an optimization-based solution using [4]. In Fig. 4 the results after a unit time are shown. It can be observed both methods yield comparable solutions. Further study is needed to quantify the differences.

Conclusion. In this study we analyze explicit schemes of the p-Laplacian flow via evolutions of eigenfunctions. We examine two step size policies – adaptive and fixed, providing analytic solutions and error bounds. For adaptive step size with arbitrary initial conditions we prove convergence of the flow. For fixed step size we prove Lyapunov stability for an eigenfunction as initial condition and empirically show similar behavior for arbitrary initializations. In general, we conclude that the adaptive step size policy is better in terms of both computational efficiency and accuracy. The conditions for convergence generalize the von-Neumann conditions to a nonlinear eigenvalue setting. This appears as a promising direction in the analysis of numerical schemes of nonlinear flows.

Acknowledgements. This project is supported by the European Union's Horizon 2020 research and innovation program under the Marie Skłodowska-Curie grant agreement No. 777826. We acknowledge support by the Israel Science Foundation (grant No. 718/15). This work was supported by the Technion Ollendorff Minerva Center.

References

1. Andreu, F., Ballester, C., Caselles, V., Mazón, J.M., et al.: Minimizing total variation flow. Differ. Integr. Eqn. **14**(3), 321–360 (2001)
2. Baravdish, G., Svensson, O., Åström, F.: On backward p (x)-parabolic equations for image enhancement. Numer. Funct. Anal. Optim. **36**(2), 147–168 (2015)
3. Burger, M., Gilboa, G., Moeller, M., Eckardt, L., Cremers, D.: Spectral decompositions using one-homogeneous functionals. SIAM J. Imaging Sci. **9**(3), 1374–1408 (2016)
4. Chambolle, A.: An algorithm for total variation minimization and applications. J. Math. Imaging Vis. **20**(1–2), 89–97 (2004)
5. Chen, J., Guo, J.: Image restoration based on adaptive p-Laplace diffusion. In: 3rd International Congress on Image and Signal Processing, vol. 1, pp. 143–146. IEEE (2010)
6. Chen, Y., Levine, S., Rao, M.: Variable exponent, linear growth functionals in image restoration. SIAM J. Appl. Math. **66**(4), 1383–1406 (2006)
7. Cohen, I., Gilboa, G.: Shape preserving flows and the p-Laplacian spectra, HAL preprint hal-01870019
8. Cohen, I., Gilboa, G.: Energy dissipating flows for solving nonlinear eigenpair problems. J. Comput. Phys. **375**, 1138–1158 (2018)
9. Courant, R., Friedrichs, K., Lewy, H.: Über die partiellen differenzengleichungen der mathematischen physik. Math. Ann. **100**(1), 32–74 (1928)
10. Crank, J., Nicolson, P.: A practical method for numerical evaluation of solutions of partial differential equations of the heat-conduction type. Adv. Comput. Math. **6**(1), 207–226 (1996)
11. García Azorero, J., Peral Alonso, I.: Existence and nonuniqueness for the p-Laplacian. Commun. Partial. Differ. Equ. **12**(12), 126–202 (1987)
12. Gawronska, E., Sczygiol, N.: Relationship between eigenvalues and size of time step in computer simulation of thermomechanics phenomena. In: Proceedings of the International MultiConference of Engineers and Computer Scientists, vol. 2 (2014)
13. Gilboa, G.: A total variation spectral framework for scale and texture analysis. SIAM J. Imaging Sci. **7**(4), 1937–1961 (2014)
14. Grewenig, S., Weickert, J., Bruhn, A.: From box filtering to fast explicit diffusion. In: Goesele, M., Roth, S., Kuijper, A., Schiele, B., Schindler, K. (eds.) DAGM 2010. LNCS, vol. 6376, pp. 533–542. Springer, Heidelberg (2010). https://doi.org/10.1007/978-3-642-15986-2_54
15. Hafiene, Y., Fadili, J., Elmoataz, A.: Nonlocal p-Laplacian evolution problems on graphs. SIAM J. Numer. Anal. **56**(2), 1064–1090 (2018)
16. Huang, C., Zeng, L.: Level set evolution model for image segmentation based on variable exponent p-Laplace equation. Appl. Math. Model. **40**(17–18), 7739–7750 (2016)
17. Iserles, A.: A First Course in the Numerical Analysis of Differential Equations. No. 44. Cambridge University Press, Cambridge (2009)
18. Kuijper, A.: Image processing by minimising l^p norms. Pattern Recogn. Image Anal. **23**(2), 226–235 (2013)

19. Kuijper, A.: p-Laplacian driven image processing. In: IEEE International Conference on Image Processing, ICIP 2007, vol. 5, p. V-257. IEEE (2007)
20. Kuijper, A.: Geometrical PDEs based on second-order derivatives of gauge coordinates in image processing. Image Vis. Comput. **27**(8), 1023–1034 (2009)
21. Lax, P.D., Richtmyer, R.D.: Survey of the stability of linear finite difference equations. Commun. Pure Appl. Math. **9**(2), 267–293 (1956)
22. Liu, Q., Guo, Z., Wang, C.: Renormalized solutions to a reaction-diffusion system applied to image denoising. DCDS-B **21**(6), 1839–1858 (2016)
23. Perona, P., Malik, J.: Scale-space and edge detection using anisotropic diffusion. PAMI **12**(7), 629–639 (1990)
24. Wei, W., Zhou, B.: A p-Laplace equation model for image denoising. Inform. Technol. J. **11**, 632–636 (2012)
25. Weickert, J.: Anisotropic Diffusion in Image Processing, vol. 1. Teubner, Stuttgart (1998)
26. Widrow, B., McCool, J.M., Larimore, M.G., Johnson, C.R.: Stationary and non-stationary learning characteristics of the LMS adaptive filter. Proc. IEEE **64**(8), 1151–1162 (1976)

Provably Scale-Covariant Networks from Oriented Quasi Quadrature Measures in Cascade

Tony Lindeberg[✉]

Computational Brain Science Lab,
Division of Computational Science and Technology,
KTH Royal Institute of Technology, Stockholm, Sweden
tony@kth.se

Abstract. This article presents a continuous model for hierarchical networks based on a combination of mathematically derived models of receptive fields and biologically inspired computations. Based on a functional model of complex cells in terms of an oriented quasi quadrature combination of first- and second-order directional Gaussian derivatives, we couple such primitive computations in cascade over combinatorial expansions over image orientations. Scale-space properties of the computational primitives are analysed and it is shown that the resulting representation allows for provable scale and rotation covariance. A prototype application to texture analysis is developed and it is demonstrated that a simplified mean-reduced representation of the resulting QuasiQuadNet leads to promising experimental results on three texture datasets.

1 Introduction

The recent progress with deep learning architectures has demonstrated that hierarchical feature representations over multiple layers have higher potential compared to approaches based on single layers of receptive fields. A limitation of current deep nets, however, is that they are not truly scale covariant. A deep network constructed by repeated application of compact 3×3 or 5×5 kernels, such as AlexNet [1], VGG-Net [2] or ResNet [3], implies an implicit assumption of a preferred size in the image domain as induced by the discretization in terms of local 3×3 or 5×5 kernels of a fixed size. Thereby, due to the non-linearities in the deep net, the output from the network may be qualitatively different depending on the specific size of the object in the image domain, as varying because of *e.g.* different distances between the object and the observer. To handle this lack of scale covariance, approaches have been developed such as spatial transformer networks [4], using sets of subnetworks in a multi-scale fashion [5] or by combining deep nets with image pyramids [6]. Since the size normalization performed

The support from the Swedish Research Council (contract 2018-03586) is gratefully acknowledged.

J. Lellmann et al. (Eds.): SSVM 2019, LNCS 11603, pp. 328–340, 2019.
https://doi.org/10.1007/978-3-030-22368-7_26

by a spatial transformer network is not guaranteed to be truly scale covariant, and since traditional image pyramids imply a loss of image information that can be interpreted as corresponding to undersampling, it is of interest to develop continuous approaches for deep networks that guarantee true scale covariance or better approximations thereof.

The subject of this article is to develop a continuous model for capturing non-linear hierarchical relations between features over multiple scales in such a way that the resulting feature representation is provably scale covariant. Building upon axiomatic modelling of visual receptive fields in terms of Gaussian derivatives and affine extensions thereof, which can serve as idealized models of simple cells in the primary visual cortex [7–9], we will propose a functional model for complex cells in terms of an oriented quasi quadrature measure. Then, we will combine such oriented quasi quadrature measures in cascade, building upon the early idea of Fukushima [10] of using Hubel and Wiesel's findings regarding receptive fields in the primary visual cortex [11] to build a hierarchical neural network from repeated application of models of simple and complex cells.

We will show how the scale-space properties of the quasi quadrature primitive in this representation can be theoretically analyzed and how the resulting hand-crafted network becomes provably scale and rotation covariant, in such a way that the multi-scale and multi-orientation network commutes with scaling transformations and rotations over the spatial image domain. Experimentally, we will investigate a prototype application to texture classification based on a substantially mean-reduced representation of the resulting QuasiQuadNet.

2 The Quasi Quadrature Measure over a 1-D Signal

Consider the scale-space representation $L(x; s)$ of a 1-D signal $f(x)$ defined by convolution with Gaussian kernels $g(x; s) = \exp(-x^2/2s)/\sqrt{2\pi s}$ and with scale-normalized derivatives according to $\partial_{\xi^n} = \partial_{x^n,\gamma-norm} = s^{n\gamma/2} \partial_x^n$ [12].

Quasi Quadrature in 1-D. Motivated by the fact that the first-order derivatives primarily respond to the locally odd component of the signal, whereas the second-order derivatives primarily respond to the locally even component of a signal, it is natural to aim at a differential feature detector that combines locally odd and even components in a complementary manner. By specifically combining the first- and second-order scale-normalized derivative responses in a Euclidean way, we obtain a quasi quadrature measure of the form

$$\mathcal{Q}_{x,norm} L = \sqrt{\frac{s\, L_x^2 + C\, s^2\, L_{xx}^2}{s^\Gamma}} \tag{1}$$

as a modification of the quasi quadrature measures previously proposed and studied in [12,13], with the scale normalization parameters γ_1 and γ_2 of the first- and second-order derivatives coupled according to $\gamma_1 = 1 - \Gamma$ and $\gamma_2 = 1 - \Gamma/2$ to enable scale covariance by adding derivative expressions of different orders only for the scale-invariant choice of $\gamma = 1$. This differential entity can be seen as an approximation of the notion of a quadrature pair of an odd and even filter

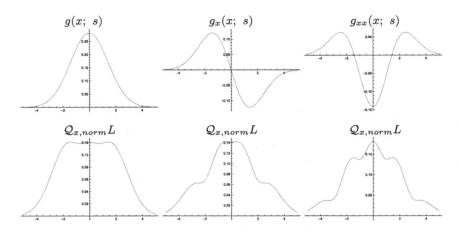

Fig. 1. 1-D Gaussian derivatives up to orders 0, 1 and 2 for $s_0 = 1$ with the corresponding 1-D quasi quadrature measures computed from them at scale $s = 1$ for $C = 8/11$. (Horizontal axis: $x \in [-5, 5]$.)

as more traditionally formulated based on a Hilbert transform, while confined within the family of differential expressions based on Gaussian derivatives.

Figure 1 shows the result of computing this quasi quadrature measure for a Gaussian peak as well as its first- and second-order derivatives. As can be seen, the quasi quadrature measure is much less sensitive to the position of the peak compared to *e.g.* the first- or second-order derivatives. Additionally, the quasi quadrature also has some degree of spatial insensitivity for a first-order derivative (a local edge model) and a second-order derivative.

Determination of C. To determine the weighting parameter C between local second-order and first-order information, let us consider a Gaussian blob $f(x) = g(x; s_0)$ with spatial extent given by s_0 as input model signal. By using the semi-group property of the Gaussian kernel $g(\cdot; s_1) * g(\cdot; s_2) = g(\cdot; s_1 + s_2)$, the quasi quadrature measure can be computed in closed form

$$\mathcal{Q}_{x,norm}L = \frac{s^{\frac{1-\Gamma}{2}} e^{-\frac{x^2}{2(s+s_0)}} \sqrt{x^2(s+s_0)^2 + Cs(s+s_0-x^2)^2 + 2}}{\sqrt{2\pi}(s+s_0)^{5/2}}. \tag{2}$$

By determining the weighting parameter C such that it minimizes the overall ripple in the squared quasi quadrature measure for a Gaussian input

$$\hat{C} = \mathrm{argmin}_{C \geq 0} \int_{x=-\infty}^{\infty} \left(\partial_x (\mathcal{Q}_{x,norm}^2 L) \right)^2 dx, \tag{3}$$

we obtain

$$\hat{C} = \frac{4(s+s_0)}{11s}, \tag{4}$$

which in the special case of choosing $s = s_0$ corresponds to $C = 8/11 \approx 0.727$. This value is very close to the value $C = 1/\sqrt{2} \approx 0.707$ derived from an equal contribution condition in [13, Eq. (27)] for the special case of choosing $\Gamma = 0$.

Scale Selection Properties. To analyze the scale selection properties of the quasi quadrature measure, let us consider the result of using Gaussian derivatives of orders 0, 1 and 2 as input signals, *i.e.*, $f(x) = g_{x^n}(x; s_0)$ for $n \in \{0, 1, 2\}$.

For the zero-order Gaussian kernel, the scale-normalized quasi quadrature measure at the origin is given by

$$\mathcal{Q}_{x,norm}L\big|_{x=0,n=0} = \frac{\sqrt{C}s^{1-\Gamma/2}}{2\pi(s + s_0)^2}. \tag{5}$$

For the first-order Gaussian derivative kernel, the scale-normalized quasi quadrature measure at the origin is

$$\mathcal{Q}_{x,norm}L\big|_{x=0,n=1} = \frac{s_0^{1/2}s^{(1-\Gamma)/2}}{2\pi(s + s_0)^2}, \tag{6}$$

whereas for the second-order Gaussian derivative kernel, the scale-normalized quasi quadrature measure at the origin is

$$\mathcal{Q}_{x,norm}L\big|_{x=0,n=2} = \frac{3\sqrt{C}s_0 s^{1-\Gamma/2}}{2\pi(s + s_0)^3}. \tag{7}$$

By differentiating these expressions with respect to scale, we find that for a zero-order Gaussian kernel the maximum response over scale is assumed at

$$\hat{s}\big|_{n=0} = \frac{s_0 (2 - \Gamma)}{2 + \Gamma}, \tag{8}$$

whereas for first- and second-order derivatives, respectively, the maximum response over scale is assumed at

$$\hat{s}\big|_{n=1} = \frac{s_0 (1 - \Gamma)}{3 + \Gamma}, \qquad \hat{s}\big|_{n=2} = \frac{s_0 (2 - \Gamma)}{4 + \Gamma}. \tag{9}$$

In the special case of choosing $\Gamma = 0$, these scale estimates correspond to

$$\hat{s}\big|_{n=0} = s_0, \qquad \hat{s}\big|_{n=1} = \frac{s_0}{3}, \qquad \hat{s}\big|_{n=2} = \frac{s_0}{2}. \tag{10a-c}$$

Thus, for a Gaussian input signal, the selected scale level will for the most scale-invariant choice of using $\Gamma = 0$ reflect the spatial extent $\hat{s} = s_0$ of the blob, whereas if we would like the scale estimate to reflect the scale parameter of first- and second-order derivatives, we would have to choose $\Gamma = -1$. An alternative motivation for using finer scale levels for the Gaussian derivative kernels is to regard the positive and negative lobes of the Gaussian derivative kernels as substructures of a more complex signal, which would then warrant the use of finer scale levels to reflect the substructures of the signal ((10b) and (10c)).

3   Oriented Quasi Quadrature Modelling of Complex Cells

In this section, we will consider an extension of the 1-D quasi quadrature measure (1) into an oriented quasi quadrature measure of the form

$$Q_{\varphi,norm} L = \sqrt{\frac{\lambda_\varphi L_\varphi^2 + C \lambda_\varphi^2 L_{\varphi\varphi}^2}{s^\Gamma}},\tag{11}$$

where L_φ and $L_{\varphi\varphi}$ denote directional derivatives of an affine Gaussian scale-space representation [14, ch. 15] of the form $L_\varphi = \cos\varphi\, L_{x_1} + \sin\varphi\, L_{x_2}$ and $L_{\varphi\varphi} = \cos^2\varphi\, L_{x_1 x_1} + 2\cos\varphi\, \sin\varphi\, L_{x_1 x_2} + \sin^2\varphi\, L_{x_2 x_2}$, and with λ_φ denoting the variance of the affine Gaussian kernel (with $x = (x_1, x_2)^T$)

$$g(x;\ s, \Sigma) = \frac{1}{2\pi s\sqrt{\det\Sigma}} e^{-x^T \Sigma^{-1} x/2s}\tag{12}$$

in direction φ, preferably with the orientation φ aligned with the direction α of either of the eigenvectors of the composed spatial covariance matrix $s\,\Sigma$, with

$$\Sigma = \frac{1}{\max(\lambda_1, \lambda_2)} \begin{pmatrix} \lambda_1 \cos^2\alpha + \lambda_2 \sin^2\alpha & (\lambda_1 - \lambda_2)\cos\alpha\,\sin\alpha \\ (\lambda_1 - \lambda_2)\cos\alpha\,\sin\alpha & \lambda_1 \sin^2\alpha + \lambda_2 \cos^2\alpha \end{pmatrix}\tag{13}$$

normalized such that the main eigenvalue is equal to one.

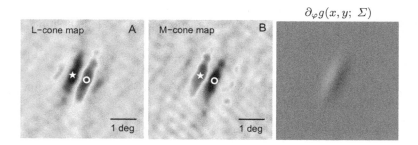

Fig. 2. Example of a colour-opponent receptive field profile for a double-opponent simple cell in the primary visual cortex (V1) as measured by Johnson *et al.* [15] (Fig. 1(a–b) Copyright Society for Neuroscience with permission): (left) Responses to L-cones corresponding to long wavelength red cones, with positive weights represented by red and negative weights by blue. (middle) Responses to M-cones corresponding to medium wavelength green cones, with positive weights represented by red and negative weights by blue. (right) Idealized model of the receptive field from a first-order directional derivative of an affine Gaussian kernel $\partial_\varphi g(x, y;\ \Sigma)$ according to (14) for $\sigma_1 = \sqrt{\lambda_1} = 0.6$, $\sigma_2 = \sqrt{\lambda_2} = 0.2$ in units of degrees of visual angle, $\alpha = 157°$ and with positive weights for the red-green colour-opponent channel $U = R - G$ with positive values represented by red and negative values represented by green. (Color figure online)

Affine Gaussian derivative model for linear receptive fields. According to the normative theory for visual receptive fields in Lindeberg [8,9], directional derivatives of affine Gaussian kernels constitute a canonical model for visual receptive fields over a 2-D spatial domain. Specifically, it was proposed that simple cells in the primary visual cortex (V1) can be modelled by directional derivatives of affine Gaussian kernels, termed *affine Gaussian derivatives*, of the form

$$T_{\varphi^m}(x_1, x_2;\ s, \Sigma) = \partial_\varphi^m \left(g(x_1, x_2;\ s, \Sigma) \right). \qquad (14)$$

Figure 2 shows an example of the spatial dependency of a colour-opponent simple cell that can be well modelled by a first-order affine Gaussian derivative over an R-G colour-opponent channel over image intensities. Corresponding modelling results for non-chromatic receptive fields can be found in [8,9].

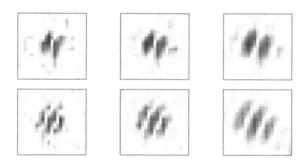

Fig. 3. Significant eigenvectors of a complex cell in the cat primary visual cortex, as determined by Touryan *et al.* [16] (Fig. 5(b) Copyright Elsevier with permission) from the response properties of the cell to a set of natural image stimuli, using a spike-triggered covariance method (STC) that computes the eigenvalues and the eigenvectors of a second-order Wiener kernel using three different parameter settings (cutoff frequencies) in the system identification method (from left to right). Qualitatively, these kernel shapes agree well with the shapes of first- and second-order affine Gaussian derivatives.

Affine Quasi Quadrature Modelling of Complex Cells. Figure 3 shows functional properties of a complex cell as determined from its response properties to natural images, using a spike-triggered covariance method (STC), which computes the eigenvalues and the eigenvectors of a second-order Wiener kernel (Touryan *et al.* [16]). As can be seen from this figure, the shapes of the eigenvectors determined from the non-linear Wiener kernel model of the complex cell do qualitatively agree very well with the shapes of corresponding affine Gaussian derivative kernels of orders 1 and 2. Motivated by this property and theoretical and experimental motivations for modelling receptive field profiles of simple cells by affine Gaussian derivatives, we propose to model complex cells by a possibly post-smoothed (spatially pooled) oriented quasi quadrature measure of the form (11)

$$(\overline{\mathcal{Q}}_{\varphi,norm}L)(\cdot;\ s_{loc}, s_{int}, \Sigma_\varphi) = \sqrt{g(\cdot;\ s_{int}, \Sigma_\varphi) * (\mathcal{Q}_{\varphi,norm}^2 L)(\cdot;\ s_{loc}, \Sigma_\varphi)} \quad (15)$$

where $s_{loc} \Sigma_\varphi$ represents an affine covariance matrix in direction φ for computing directional derivatives and $s_{int} \Sigma_\varphi$ represents an affine covariance matrix in the same direction for integrating pointwise affine quasi quadrature measures over a region in image space.

The pointwise affine quasi quadrature measure $(\mathcal{Q}_{\varphi,norm}L)(\cdot; s_{loc}, \Sigma_\varphi)$ can be seen as a Gaussian derivative based analogue of the energy model for complex cells as proposed by Adelson and Bergen [17] and Heeger [18]. It is closely related to a proposal by Koenderink and van Doorn [19] of summing up the squares of first- and second-order derivative responses and nicely compatible with results by De Valois et al. [20], who showed that first- and second-order receptive fields typically occur in pairs that can be modelled as approximate Hilbert pairs.

The addition of a complementary post-smoothing stage as determined by the affine Gaussian weighting function $g(\cdot; s_{int}, \Sigma_\varphi)$ is closely related to recent results by Westö and May [21], who have shown that complex cells are better modelled as a combination of two spatial integration steps.

By choosing these spatial smoothing and weighting functions as affine Gaussian kernels, we ensure an affine covariant model of the complex cells, to enable the computation of affine invariants at higher levels in the visual hierarchy.

The use of multiple affine receptive fields over different shapes of the affine covariance matrices $\Sigma_{\varphi,loc}$ and $\Sigma_{\varphi,int}$ can be motivated by results by Goris et al. [22], who show that there is a large variability in the orientation selectivity of simple and complex cells. With respect to this model, this means that we can think of affine covariance matrices of different eccentricity as being present from isotropic to highly eccentric. By considering the full family of positive definite affine covariance matrices, we obtain a fully affine covariant image representation able to handle local linearizations of the perspective mapping for all possible views of any smooth local surface patch.

4   Hierarchies of Oriented Quasi Quadrature Measures

Let us in this first study disregard the variability due to different shapes of the affine receptive fields for different eccentricities and assume that $\Sigma = I$. This restriction enables covariance to scaling transformations and rotations, whereas a full treatment of affine quasi quadrature measures over all positive definite covariance matrices would have the potential to enable full affine covariance.

An approach that we shall pursue is to build feature hierarchies by coupling oriented quasi quadrature measures (11) or (15) in cascade

$$F_1(x, \varphi_1) = (\mathcal{Q}_{\varphi_1,norm} L)(x) \tag{16}$$

$$F_k(x, \varphi_1, ..., \varphi_{k-1}, \varphi_k) = (\mathcal{Q}_{\varphi_k,norm} F_{k-1})(x, \varphi_1, ..., \varphi_{k-1}), \tag{17}$$

where we have suppressed the notation for the scale levels assumed to be distributed such that the scale parameter at level k is $s_k = s_0 \, r^{2(k-1)}$ for some $r > 1$, e.g., $r = 2$. Assuming that the initial scale-space representation L is computed at scale s_0, such a network can in turn be initiated for different values of s_0, also distributed according to a geometric distribution.

This construction builds upon an early proposal by Fukushima [10] of building a hierarchical neural network from repeated application of models of simple and complex cells [11], which has later been explored in a hand-crafted network based on Gabor functions by Serre *et al.* [23] and in the scattering convolution networks by Bruno and Mallat [24]. This idea is also consistent with a proposal by Yamins and DiCarlo [25] of using repeated application of a single hierarchical convolution layer for explaining the computations in the mammalian cortex. With this construction, we obtain a way to define continuous networks that express a corresponding hierarchical architecture based on Gaussian derivative based models of simple and complex cells within the scale-space framework.

Each new layer in this model implies an expansion of combinations of angles over the different layers in the hierarchy. For example, if we in a discrete implementation discretize the angles $\varphi \in [0, \pi[$ into M discrete spatial orientations, we will then obtain M^k different features at level k in the hierarchy. To keep the complexity down at higher levels, we will for $k \geq K$ in a corresponding way as done by Hadji and Wildes [26] introduce a pooling stage over orientations

$$(\mathcal{P}_k F_k)(x, \varphi_1, ..., \varphi_{K-1}) = \sum_{\varphi_k} F_k(x, \varphi_1, ..., \varphi_{K-1}, \varphi_k), \qquad (18)$$

and instead define the next successive layer as

$$F_k(x, \varphi_1, ..., \varphi_{k-2}, \varphi_{K-1}, \varphi_k) = (\mathcal{Q}_{\varphi_k, norm} \mathcal{P}_{k-1} F_{k-1})(x, \varphi_1, ..., \varphi_{K-1}) \qquad (19)$$

to limit the number of features at any level to maximally M^{K-1}. The proposed hierarchical feature representation is termed QuasiQuadNet.

Scale Covariance. A theoretically attractive property of this family of networks is that the networks are provably scale covariant. Given two images f and f' that are related by a uniform scaling transformation $f(x) = f'(Sx)$ for some $S > 0$, their corresponding scale-space representations L and L' will be equal $L'(x'; s') = L(x; s)$ and so will the scale-normalized derivatives $s'^{n/2} L'_{x'^n}(x'; s') = s^{n/2} L_{x_i^n}(x; s)$ based on $\gamma = 1$ if the spatial positions are related according to $x' = Sx$ and the scale levels according to $s' = S^2 s$ [12, Eqns. (16) and (20)]. This implies that if the initial scale levels s_0 and s'_0 underlying the construction in (16) and (17) are related according to $s'_0 = S^2 s_0$, then the first layers of the feature hierarchy will be related according to $F'_1(x', \varphi_1) = S^{-\Gamma} F_1(x, \varphi_1)$ [13, Eqns. (55) and (63)]. Higher layers in the feature hierarchy are in turn related according to

$$F'_k(x', \varphi_1, ..., \varphi_{k-1}, \varphi_k) = S^{-k\Gamma} F_k(x, \varphi_1, ..., \varphi_{k-1}, \varphi_k) \qquad (20)$$

and are specifically equal if $\Gamma = 0$. This means that it will be possible to perfectly match such hierarchical representations under uniform scaling transformations.

Rotation Covariance. Under a rotation of image space by an angle α, $f'(x') = f(x)$ for $x' = R_\alpha x$, the corresponding feature hierarchies are in turn equal if the orientation angles are related according to $\varphi'_i = \varphi_i + \alpha$ $(i = 1..k)$

$$F'_k(x', \varphi'_1, ..., \varphi'_{k-1}, \varphi'_k) = F_k(x, \varphi_1, ..., \varphi_{k-1}, \varphi_k). \qquad (21)$$

5   Application to Texture Analysis

In the following, we will use a substantially reduced version of the proposed quasi quadrature network for building an application to texture analysis.

If we make the assumption that a spatial texture should obey certain stationarity properties over image space, we may regard it as reasonable to construct texture descriptors by accumulating statistics of feature responses over the image domain, in terms of e.g mean values or histograms. Inspired by the way the SURF descriptor [27] accumulates mean values and mean absolute values of derivative responses and the way Bruno and Mallat [24] and Hadji and Wildes [26] compute mean values of their hierarchical feature representations, we will initially explore reducing the QuasiQuadNet to just the mean values over the image domain of the following 5 features

$$\{\partial_\varphi F_k, |\partial_\varphi F_k|, \partial_{\varphi\varphi} F_k, |\partial_{\varphi\varphi} F_k|, \mathcal{Q}_\varphi F_k\}. \tag{22}$$

These types of features are computed for all layers in the feature hierarchy (with $F_0 = L$), which leads to a 4000-D descriptor based on $M = 8$ uniformly distributed orientations in $[0, \pi[$, 4 layers in the hierarchy delimited in complexity by directional pooling for $K = 3$ with 4 initial scale levels $\sigma_0 = \sqrt{s_0} \in \{1, 2, 4, 8\}$.

Table 1. Performance results of the mean-reduced QuasiQuadNet in comparison with a selection of among the better methods in the extensive performance evaluation by Liu *et al.* [34] (our results in slanted font).

	KTH-TIPS2b	CUReT	UMD
FV-VGGVD [28] (SVM)	88.2	99.0	99.9
FV-VGGM [28] (SVM)	79.9	98.7	99.9
MRELBP [29] (SVM)	77.9	99.0	99.4
FV-AlexNet [28] (SVM)	77.9	98.4	99.7
mean-reduced QuasiQuadNet LUV (SVM)	78.3	98.6	
mean-reduced QuasiQuadNet grey (SVM)	75.3	98.3	97.1
ScatNet [24] (PCA)	68.9	99.7	98.4
MRELBP [29]	69.0	97.1	98.7
BRINT [30]	66.7	97.0	97.4
MDLBP [31]	66.5	96.9	97.3
mean-reduced QuasiQuadNet LUV (NNC)	72.1	94.9	
mean-reduced QuasiQuadNet grey (NNC)	70.2	93.0	93.3
LBP [32]	62.7	97.0	96.2
ScatNet [24] (NNC)	63.7	95.5	93.4
PCANet [33] (NNC)	59.4	92.0	90.5
RandNet [33] (NNC)	56.9	90.9	90.9

The second column in Table 1 shows the result of applying this approach to the KTH-TIPS2b dataset [35] for texture classification, consisting of 11 classes ("aluminum foil", "cork", "wool", "lettuce leaf", "corduroy", "linen", "cotton", "brown bread", "white bread", "wood" and "cracker") with 4 physical samples from each class and photos of each sample taken from 9 distances leading to 9 relative scales labelled "2", ..., "10" over a factor of 4 in scaling transformations and additionally 12 different pose and illumination conditions for each scale, leading to a total number of $11 \times 4 \times 9 \times 12 = 4752$ images. The regular benchmark setup implies that the images from 3 samples in each class are used for training and the remaining sample in each class is used for testing over 4 permutations. Since several of the samples from the same class are quite different from each other in appearance, this implies a non-trivial benchmark which has not yet been saturated.

When using nearest-neighbour classification on the mean-reduced grey-level descriptor, we get 70.2% accuracy, and 72.1% accuracy when computing corresponding features from the LUV channels of a colour-opponent representation. When using SVM classification, the accuracy becomes 75.3% and 78.3%, respectively. Comparing with the results of an extensive set of other methods in Liu et al. [34], out of which a selection of the better results are listed in Table 1, the results of the mean-reduced QuasiQuadNet are better than classical texture classification methods such as locally binary patterns (LBP) [32], binary rotation invariant noise tolerant texture descriptors [30] and multi-dimensional local binary patterns (MDLBP) [31] and also better than other handcrafted networks, such as ScatNet [24], PCANet [33] and RandNet [33]. The performance of the mean-reduced QuasiQuadNet descriptor does, however, not reach the performance of applying SVM classification to Fischer vectors of the filter output in learned convolutional networks (FV-VGGVD, FV-VGGM [28]).

By instead performing the training on every second scale in the dataset (scales 2, 4, 6, 8, 10) and the testing on the other scales (3, 5, 7, 9), such that the benchmark does not primarily test the generalization properties between the different very few samples in each class, the classification performance is 98.8% for the grey-level descriptor and 99.6% for the LUV descriptor.

The third and fourth columns in Table 1 show corresponding results of texture classification on the CUReT [36] and UMD [37] texture datasets, with random equally sized partitionings of the images into training and testing data. Also for these datasets, the performance of the mean-reduced descriptor is reasonable compared to other methods.

6 Summary and Discussion

We have presented a theory for defining hand-crafted hierarchical networks by applying quasi quadrature responses of first- and second-order directional Gaussian derivatives in cascade. The purpose behind this study has been to investigate if we could start building a bridge between the well-founded theory of scale-space representation and the recent empirical developments in deep learning, while at the same time being inspired by biological vision. The present work is intended as an initial work in this direction, where we propose the family of quasi quadrature networks as a new baseline for hand-crafted networks with associated provable covariance properties under scaling and rotation transformations.

By early experiments with a substantially mean-reduced representation of the resulting QuasiQuadNet, we have demonstrated that it is possible to get quite promising performance on texture classification, and comparable or better than other hand-crafted networks, although not reaching the performance of learned CNNs. By inspection of the full non-reduced feature maps, which could not be shown here because of the space limitations, we have also observed that some representations in higher layers may respond to irregularities in regular textures (defect detection) or corners or end-stoppings in regular scenes.

Concerning extensions of the approach, we propose to: (i) complement the computation of quasi quadrature responses by divisive normalization [38] to enforce a competition between multiple feature responses, (ii) explore the spatial relationships in the full feature maps that are suppressed in the mean-reduced representation and (iii) incorporate learning mechanisms.

References

1. Krizhevsky, A., Sutskever, I., Hinton, G.E.: ImageNet classification with deep convolutional neural networks. In: NIPS, pp. 1097–1105 (2012)
2. Simonyan, K., Zisserman, A.: Very deep convolutional networks for large-scale image recognition. In: ICLR (2015). arXiv:1409.1556
3. He, K., Zhang, X., Ren, S., Sun, J.: Deep residual learning for image recognition. In: Proceedings of Computer Vision and Pattern Recognition (CVPR 2016), pp. 770–778 (2016)
4. Jaderberg, M., Simonyan, K., Zisserman, A., Kavukcuoglu, K.: Spatial transformer networks. In: NIPS, pp. 2017–2025 (2015)
5. Cai, Z., Fan, Q., Feris, R.S., Vasconcelos, N.: A unified multi-scale deep convolutional neural network for fast object detection. In: Leibe, B., Matas, J., Sebe, N., Welling, M. (eds.) ECCV 2016. LNCS, vol. 9908, pp. 354–370. Springer, Cham (2016). https://doi.org/10.1007/978-3-319-46493-0_22
6. Lin, T.Y., Dollár, P., Girshick, R., He, K., Hariharan, B., Belongie, S.: Feature pyramid networks for object detection. In: CVPR (2017)
7. Koenderink, J.J., van Doorn, A.J.: Generic neighborhood operators. IEEE-TPAMI 14, 597–605 (1992)
8. Lindeberg, T.: Generalized Gaussian scale-space axiomatics comprising linear scale-space, affine scale-space and spatio-temporal scale-space. J. Math. Imaging Vis. 40, 36–81 (2011)
9. Lindeberg, T.: A computational theory of visual receptive fields. Biol. Cybern. 107, 589–635 (2013)
10. Fukushima, K.: Neocognitron: a self-organizing neural network model for a mechanism of pattern recognition unaffected by shift in position. Biol. Cybern. 36, 193–202 (1980)
11. Hubel, D.H., Wiesel, T.N.: Brain and Visual Perception. Oxford University Press, New York (2005)

12. Lindeberg, T.: Feature detection with automatic scale selection. Int. J. Comput. Vis. **30**, 77–116 (1998)
13. Lindeberg, T.: Dense scale selection over space, time and space-time. SIAM J. Imaging Sci. **11**, 407–441 (2018)
14. Lindeberg, T.: Scale-Space Theory in Computer Vision. Springer, Dordrecht (1993). https://doi.org/10.1007/978-1-4757-6465-9
15. Johnson, E.N., Hawken, M.J., Shapley, R.: The orientation selectivity of color-responsive neurons in Macaque V1. J. Neurosci. **28**, 8096–8106 (2008)
16. Touryan, J., Felsen, G., Dan, Y.: Spatial structure of complex cell receptive fields measured with natural images. Neuron **45**, 781–791 (2005)
17. Adelson, E., Bergen, J.: Spatiotemporal energy models for the perception of motion. JOSA A **2**, 284–299 (1985)
18. Heeger, D.J.: Normalization of cell responses in cat striate cortex. Vis. Neurosci. **9**, 181–197 (1992)
19. Koenderink, J.J., van Doorn, A.J.: Receptive field families. Biol. Cybern. **63**, 291–298 (1990)
20. De Valois, R.L., Cottaris, N.P., Mahon, L.E., Elfer, S.D., Wilson, J.A.: Spatial and temporal receptive fields of geniculate and cortical cells and directional selectivity. Vis. Res. **40**, 3685–3702 (2000)
21. Westö, J., May, P.J.C.: Describing complex cells in primary visual cortex: a comparison of context and multi-filter LN models. J. Neurophys. **120**, 703–719 (2018)
22. Goris, R.L.T., Simoncelli, E.P., Movshon, J.A.: Origin and function of tuning diversity in Macaque visual cortex. Neuron **88**, 819–831 (2015)
23. Serre, T., Wolf, L., Bileschi, S., Riesenhuber, M., Poggio, T.: Robust object recognition with cortex-like mechanisms. IEEE-TPAMI **29**, 411–426 (2007)
24. Bruna, J., Mallat, S.: Invariant scattering convolution networks. IEEE-TPAMI **35**, 1872–1886 (2013)
25. Yamins, D.L.K., DiCarlo, J.J.: Using goal-driven deep learning models to understand sensory cortex. Nat. Neurosci. **19**, 356–365 (2016)
26. Hadji, I., Wildes, R.P.: A spatiotemporal oriented energy network for dynamic texture recognition. In: ICCV, pp. 3066–3074 (2017)
27. Bay, H., Ess, A., Tuytelaars, T., van Gool, L.: Speeded Up Robust Features (SURF). CVIU **110**, 346–359 (2008)
28. Cimpoi, M., Maji, S., Vedaldi, A.: Deep filter banks for texture recognition and segmentation. In: CVPR, pp. 3828–3836 (2015)
29. Liu, L., Lao, S., Fieguth, P.W., Guo, Y., Wang, X., Pietikäinen, M.: Median robust extended local binary pattern for texture classification. IEEE-TIP **25**, 1368–1381 (2016)
30. Liu, L., Long, Y., Fieguth, P.W., Lao, S., Zhao, G.: BRINT: binary rotation invariant and noise tolerant texture classification. IEEE-TIP **23**, 3071–3084 (2014)
31. Schaefer, G., Doshi, N.P.: Multi-dimensional local binary pattern descriptors for improved texture analysis. In: ICPR, pp. 2500–2503 (2012)
32. Ojala, T., Pietikäinen, M., Maenpaa, T.: Multiresolution gray-scale and rotation invariant texture classification with local binary patterns. IEEE-TPAMI **24**, 971–987 (2002)
33. Chan, T.H., Jia, K., Gao, S., Lu, J., Zeng, Z., Ma, Y.: PCANet: A simple deep learning baseline for image classification? IEEE-TIP **24**, 5017–5032 (2015)
34. Liu, L., Fieguth, P., Guo, Y., Wang, Z., Pietikäinen, M.: Local binary features for texture classification: taxonomy and experimental study. Pattern Recogn. **62**, 135–160 (2017)

35. Mallikarjuna, P., Targhi, A.T., Fritz, M., Hayman, E., Caputo, B., Eklundh, J.O.: The KTH-TIPS2 database. KTH Royal Institute of Technology (2006)
36. Varma, M., Zisserman, A.: A statistical approach to material classification using image patch exemplars. IEEE-TPAMI **31**, 2032–2047 (2009)
37. Xu, Y., Yang, X., Ling, H., Ji, H.: A new texture descriptor using multifractal analysis in multi-orientation wavelet pyramid. In: CVPR, pp. 161–168 (2010)
38. Carandini, M., Heeger, D.J.: Normalization as a canonical neural computation. Nat. Rev. Neurosci. **13**, 51–62 (2012)

A Fast Multi-layer Approximation
to Semi-discrete Optimal Transport

Arthur Leclaire[1(✉)] and Julien Rabin[2(✉)]

[1] Univ. Bordeaux, IMB, Bordeaux INP, CNRS, UMR 5251, 33400 Talence, France
arthur.leclaire@math.u-bordeaux.fr
[2] Normandie Univ., UNICAEN, ENSICAEN, CNRS, GREYC, Caen, France
julien.rabin@unicaen.fr
http://www.math.u-bordeaux.fr/~aleclaire
http://sites.google.com/site/rabinjulien

Abstract. The optimal transport (OT) framework has been largely used in inverse imaging and computer vision problems, as an interesting way to incorporate statistical constraints or priors. In recent years, OT has also been used in machine learning, mostly as a metric to compare probability distributions. This work addresses the semi-discrete OT problem where a continuous source distribution is matched to a discrete target distribution. We introduce a fast stochastic algorithm to approximate such a semi-discrete OT problem using a hierarchical multi-layer transport plan. This method allows for tractable computation in high-dimensional case and for large point-clouds, both during training and synthesis time. Experiments demonstrate its numerical advantage over multi-scale (or multi-level) methods. Applications to fast exemplar-based texture synthesis based on patch matching with two layers, also show stunning improvements over previous single layer approaches. This *shallow* model achieves comparable results with state-of-the-art deep learning methods, while being very compact, faster to train, and using a single image during training instead of a large dataset.

Keywords: Optimal transport · Texture synthesis · Patch matching

1 Introduction

Optimal transport (OT) [16,21] provides a powerful tool to measure the distance between two probability distributions. It has found several applications in image processing, for example feature matching [15] which requires comparing histograms of gradient orientations, color transfer [13] where OT distances are used to compare color distributions, texture mixing [14,22] which formulates as computing barycenters of texture models for the OT distance, shape interpolation [18] which uses OT distance on shapes identified on probability measures. Here we will focus on exemplar-based texture synthesis, which consists in producing a (possibly very large) image that has the same aspect as a given texture sample while not being a verbatim copy.

© Springer Nature Switzerland AG 2019
J. Lellmann et al. (Eds.): SSVM 2019, LNCS 11603, pp. 341–353, 2019.
https://doi.org/10.1007/978-3-030-22368-7_27

A common way to formulate texture synthesis is to ask for an image which is as random as possible while respecting a certain number of statistical constraints. Thus, texture synthesis can be addressed by exploiting OT distances to compare distributions of linear or nonlinear filter responses [19], or comparing directly patch distributions [3,5]. In this paper, we build on the model proposed in [3] which performs texture synthesis by applying a well-chosen OT map on 3×3 patches in a coarse to fine manner (starting from a white noise at the lowest resolution). One limitation of this work is that the restriction to 3×3 patches prevents the proper reproduction of geometric structures at larger scales. Complementary to the multiresolution strategy, a common way of dealing with large geometric structures is to work with larger patches as is usually done in popular patch-based methods, as [8] (8×8 to 32×32 patches) or [2] (7×7 patches). But then we must face the fact that OT maps are very difficult to compute in such a high-dimensional setting.

Following [3] we are particularly interested in the semi-discrete OT case, where one wishes to send an absolutely continuous probability measure μ onto a discrete probability measure ν while minimizing a transportation cost. For the semi-discrete case, as shown in [1,7], the OT map can be obtained as a weighted nearest-neighbor (NN) assignment, which is parameterized by a finite-dimensional vector v that solves a concave optimization problem. Several first and second order schemes have been proposed to solve this problem, based on exact gradient computations [6,7,9,11]. Unfortunately, these exact methods are very difficult to apply in high dimension (or when the point set supporting ν is large) where exact gradients computations are intractable. To cope with that, one can use stochastic optimization techniques that rely on Monte-Carlo estimations of the gradient, as proposed in [3,4]. But such a stochastic method has a very slow convergence in high dimension, especially when the support of target measure has a large cardinal. One can improve convergence speed by working in a multi-scale framework, meaning that one precomputes a decomposition of the target measure *i.e.* several simplified versions of the target discrete distribution (with a clustering algorithm, for example K-means), and computes the OT map by progressively refining the target distribution. It has been noted in [9,11] that the approximate OT map obtained at one scale could be used to initialize the map at the next finer scale.

Here we propose an alternative way of exploiting this multi-scale framework. Indeed, we suggest to optimize a transport map having a parametric form which combines several layers of weighted NN assignments. To distinguish from previous works, such a map will be called a *multi-layer* transport map. Thanks to its inherent structure, during the iterative optimization algorithm, the multi-layer map attains a near-optimal transport cost in a much faster way than previous algorithm. But the price to pay is that we deviate from the truly optimal solution, meaning that the precomputed decomposition of the target measure introduces a bias that cannot be canceled (even at infinite time). As a byproduct, once estimated, such a multi-layer map can be applied much faster because it trades one NN projection on a large finite set by hierarchical NN projections on much smaller sets. Such properties are very profitable for imaging applications, where satisfying the marginal constraints is certainly more important than the true optimality in terms of transportation cost.

In Sect. 2 we first recall the framework of semi-discrete optimal transport. In Sect. 3 we detail the construction of multi-layer transport maps and their stochastic optimization. In Sect. 4, we propose numerical experiments that illustrate the benefits of these multi-layer maps to realize approximate OT: first, we illustrate on a one-dimensional example the gain in computational time to optimize the parameters; and next, we use these multi-layer maps to tackle difficult examples of OT in the space of large patches, which leads to remarkable results in texture synthesis.

2   Semi-discrete Optimal Transport

Given two probability measures μ, ν on $\mathbb{R}^D$, the Monge formulation of OT writes

$$\inf_T \int_{\mathbb{R}^D} \|x - T(x)\|^2 d\mu(x), \tag{OT-M}$$

where the infimum is taken over all measurable functions $T : \mathbb{R}^D \to \mathbb{R}^D$ for which the push-forward-measure $T_\sharp \mu$ equals ν. General conditions for existence and unicity of solutions can be found in [16,21]. In this paper we focus on the semi-discrete case, meaning that μ is an absolutely continuous probability distribution on $\mathbb{R}^D$ with density ρ and that ν is supported on $Y = \{y_j, \ j \in J\}$ finite with

$$\nu = \sum_{j \in J} \nu_j \delta_{y_j} \quad \text{s.t.} \quad \sum_{j \in J} \nu_j = 1 \text{ and } \nu_i \geq 0. \tag{1}$$

As proved in [1,7,9], the solution of (OT-M) is a weighted NN assignment

$$T_{Y,v}(x) := \operatorname*{argmin}_{y_j \in Y} \|x - y_j\|^2 - v(j). \tag{2}$$

This map is defined almost everywhere, and its preimages define a partition of $\mathbb{R}^D$ (up to a negligible set), called the *Laguerre cells*. The solution $T_{Y,v}$ is entirely parameterized by a vector v that maximizes the concave function $H_{Y,\nu}(v) := \mathbb{E}_{X \sim \mu}[h_{Y,\nu}(X, v)]$ where

$$h_{Y,\nu}(x, v) = \left(\min_{j \in J} \|x - y_j\|^2 - v(j) \right) + \sum_{j \in J} v(j)\nu_j, \tag{3}$$

so that we have for all $v \in \mathbb{R}^J$ and for almost all $x \in \mathbb{R}^D$,

$$\partial_{v(j)} h_{Y,\nu}(x, v) = -\mathbb{1}_{T_{Y,v}(x)=y_j} + \nu_j, \quad \text{and} \quad \partial_{v(j)} H_{Y,\nu}(v) = -\mu(T_{Y,v}^{-1}(y_j)) + \nu_j. \tag{4}$$

This important result has been used in [1,7,9,11] to optimize weights v when distributions μ and ν are defined on $\mathbb{R}^D$ for $D = 2$ or $D = 3$ dimensions, making use of gradient descent or L-BFGS algorithm. But in higher dimension, exact gradient computation is challenging, and thus following [3,4], one may turn to the Average Stochastic Gradient Descent (ASGD) Algorithm 1 to solve it.

Algorithm 1. ASGD to estimate OT map $T_{Y,v}$ that solves (OT-M)

1: **Inputs:** source density μ, target discrete distribution ν,
   initial assignment weight (*e.g.* $\tilde{v} = 0$ and $v = 0$), and gradient step (*e.g.* $C = 1$)
2: **for** $t = 1, \dots, T$ **do**
3:    Draw a sample $x \sim \mu$
4:    Compute the gradient $g \leftarrow \nabla_v h_{Y,\nu}(x, \tilde{v})$ (see Eq. (4))
5:    Gradient ascent of weights: $\tilde{v} \leftarrow \tilde{v} + \frac{C}{\sqrt{t}} g$
6:    Average of updates: $v \leftarrow \frac{t-1}{t} v + \frac{1}{t} \tilde{v}$
7: **return** v

---

3 Multi-layer Approximation of the Transport Map

A classical approach in signal processing is to use multi-scale representation to accelerate or handle large data. This strategy has recently gained some interest to compute approximate or exact optimal transport maps in large discrete or semi-discrete problems [9,11,12,17] to overcome some convergence speed issues.

3.1 Multi-scale Semi-discrete Optimal Transport

The discrete target measure ν can be decomposed from fine ($\ell = 0$) to coarse ($\ell = L - 1$) scales by setting

$$\nu^\ell = \sum_{j \in J^\ell} \nu_j^\ell \delta_{y_j^\ell}, \quad \text{supported by } Y^\ell = \{y_j^\ell, \ j \in J^\ell\}. \tag{5}$$

Starting from $\nu^0 = \nu$ (and $Y^0 = Y$), the measure $\nu^{\ell+1}$ should be a close approximation of ν^ℓ with a decreasing budget support composed of $|J^{\ell+1}|$ points.

In practice, following [11], we first extract (for example with the K-means algorithm) a clustering $Y^\ell = \bigsqcup_{j \in J^{\ell+1}} C_j^\ell$ of the support Y^ℓ of ν^ℓ, and we denote by $Y^{\ell+1} = \{y_j^{\ell+1}, j \in J^{\ell+1}\}$ the corresponding cluster centroids. Next, we compute the weights by gathering the mass in each cluster $\nu_j^{\ell+1} = \nu^\ell(C_j^\ell)$. An illustration of this multi-scale clustering is given in Fig. 1.

The authors of [9,11] propose to estimate the optimal transportation map $T_{Y,v}$ from μ towards ν at the finest scale ($\ell = 0$) using a bottom-up strategy (from $\ell = L - 1$ to $\ell = 0$) that sequentially estimates the maps T_{Y^ℓ, v^ℓ} that realizes the OT between μ and ν^ℓ. The idea is to accelerate the optimization of v^ℓ by starting from an initial estimate that is extrapolated from the solution $v^{\ell+1}$ computed at the previous scale. While this method computes an exact solution, some heuristics is needed to extrapolate v^ℓ from $v^{\ell+1}$ (for example in [11], propagating the value $v^{\ell+1}(j)$ associated to centroid $y_j^{\ell+1}$ to the points in the corresponding cluster C_j^ℓ).

Let us also briefly mention that multi-scale strategies have been proposed to deal with *discrete* OT problems (*i.e.* when both distributions are discrete)

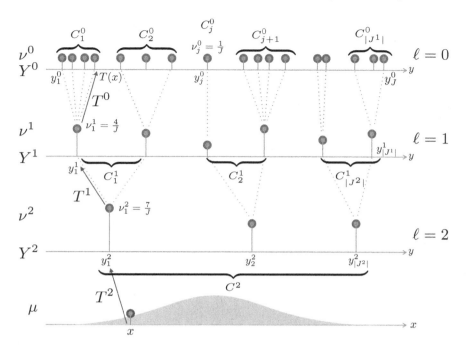

Fig. 1. Illustration of the multi-scale approximation of the discrete target distribution ν in the 1D case, and the multi-layer transport map $T(x)$ applied sequentially with $L = 3$ layers. See the text for more details about notation.

with very large data sets (yet in small dimension): an exact multiscale approach from [17], an approximation method in [12], and multi-level OT with Wasserstein-1 [10].

3.2   Multi-layer Transport Map

In this work, we consider a different approach that *approximates* the OT map with multi-layer transport maps to accelerate computation time for both the estimation of the mapping (off-line) and the mapping itself (on-line). The main difference with [11] is that the multi-scale representation of the target distribution (as illustrated in Fig. 1) is used to sequentially estimate a *hierarchical* Laguerre cell partitioning of the source distribution. This hierarchical clustering induces a tree structure, and thus applying the multi-layer map consists in a tree search which is faster than a direct global NN projection. More precisely, the proposed approach is as follows:

– **Training:** We estimate simultaneously along the tree structure, a weight vector $v_{j^\ell}^\ell$ defining the transport $T_{C_{j^\ell}^\ell, v_{j^\ell}^\ell}$, which maps the corresponding Laguerre cell indexed by $j^{\ell+1}$ of the source distribution μ (defined itself at the previous layer $\ell+1$ by a centroid $y_{j^{\ell+1}}^{\ell+1}$) to the cluster $C_{j^\ell}^\ell$ of points from the multi-scale target distribution ν^ℓ. The relationship between indices j^ℓ and $j^{\ell+1}$ depends entirely on the fixed hierarchical clustering that is performed beforehand, and is detailed in the paragraph hereafter.

– **Synthesis:** To find the image $T(x) \in Y$ of a point x, one should trace back the hierarchy of Laguerre cells to which x belongs and then apply the finest estimated transport map.

Eventually, the multi-layer map takes the form

$$T^\ell(x) = T_{C_{j^\ell}^\ell, v_{j^\ell}^\ell}(x) \tag{6}$$

where j^ℓ indicates the index of the cluster the point x is assigned to at layer ℓ. These indices are *recursively* defined (starting from $\ell = L-1$ where there is only one root cluster $C^{L-1} = Y^{L-1}$, as illustrated in Fig. 1), namely $j^\ell(x)$ (denoted simply j^ℓ for the sake of brevity) is computed from the weighted NN for the previous layer $\ell+1$

$$j^\ell = \operatorname*{argmin}_{j \mid y_j^{\ell+1} \in C_{j^{\ell+1}}^{\ell+1}} \|x - y_j^{\ell+1}\|^2 - v_{j^{\ell+1}}^{\ell+1}(j) . \tag{7}$$

Notice that the final value of the multi-layer map is $T(x) = T^0(x)$, but its evaluation requires to compute L transport maps $T^\ell(x)$ (actually the sequence of indices $j^{\ell+1}(x)$) from $\ell = L-1$ to $\ell = 0$. In the experiments shown in this paper, we restrict to the case of only two layers, which is illustrated in Fig. 2.

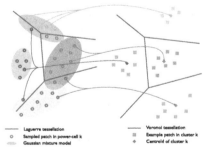 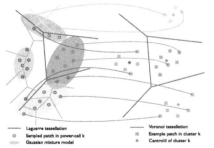

First layer $\ell = 1$: the coarse transport map T^1 maps every points of a Laguerre cell j of μ to the centroid of the corresponding cluster C_j^0 in ν.

Second layer $\ell = 0$: the fine mapping $T = T^0$ is defined from the transport maps $T_{C_j^0, v_j^\ell}$.

Fig. 2. Illustration of a multi-layer map (for $L = 2$ layers). Here the source distribution μ is chosen to be a Gaussian mixture model with 4 components (in graylevels). For each layer ℓ, the arrows illustrate $T^\ell(x)$ (arrows) the multi-layer mapping of samples x drawn from μ (circle points on the left) to the points of the discrete distribution ν^ℓ (diamonds for layer $\ell = 1$ and square for layer $\ell = 0$).

The weights $v_{j^\ell}^\ell$ (for all $\ell \in \{0, \ldots, L-1\}$ and $j^\ell \in J^\ell$) are optimized so that it realizes the semi-discrete OT between the restriction of μ to the Laguerre cell $\mathcal{L}_{j^\ell}^\ell = (j^\ell)^{-1}(j^\ell(x))$ and the restriction $\nu_{j^\ell}^\ell$ of ν^ℓ to $C_{j^\ell}^\ell$. In other words, assuming that $T^{\ell+1}$ is fixed (*i.e.* for a given $j^{\ell+1}$), we seek to maximize

$$H^\ell(v^\ell) = \sum_{j^\ell \in J^\ell} \int_{\mathcal{L}_{j^\ell}^\ell} \min_{\substack{j \\ y_j^\ell \in C_{j^\ell}^\ell}} \left(\|x - y_j^\ell\|^2 - v_{j^\ell}^\ell(j) \right) d\mu(x) + \sum_{\substack{j \\ y_j^\ell \in C_{j^\ell}^\ell}} v_{j^\ell}^\ell(j)\nu_j^\ell. \quad (8)$$

Notice that this cost is composed of $|J^{\ell+1}|$ separate semi-discrete OT problems with restrictions of μ and ν^ℓ. Indeed, the problem (8) is the dual convex of the Kantorovich formulation of OT between those restricted measures. Writing instead the Monge formulation of these separates sub-problems, and denoting by S^ℓ the map that gathers the restrictions $T_{C_{j^\ell}^\ell, v_{j^\ell}^\ell}$ on the cells $\mathcal{L}_{j^\ell}^\ell$, we get that S^ℓ minimizes the usual cost

$$\int_{\mathbb{R}^D} \|S^\ell(x) - x\|^2 d\mu(x) = \sum_{j^\ell \in J^\ell} \int_{\mathcal{L}_{j^\ell}^\ell} \|S^\ell(x) - x\|^2 d\mu(x) \quad (9)$$

but with marginal constraints on each of the Laguerre cells $\mathcal{L}_{j^\ell}^\ell$. We can apply ASGD to treat each of these sub-problems. It requires to sample the restricted measure μ. But in practice, we treat simultaneously all the sub-problems in all layers: at each new sample we update only the gradient relative to the current active Laguerre cell of the hierarchy, which in the end is equivalent to sample all restricted measures. This optimization procedure is summarized in Algorithm 2.

Algorithm 2. ASGD for the estimation of the multi-layer map $T_{Y,v}$.

Inputs: source density μ, target distribution ν, gradient step C, number of layers L and number of iterations T

1: Hierarchical clustering $\{\nu^\ell, \ell = 0, \ldots, L-1\}$ of ν
2: Set $v_j^\ell \leftarrow 0, \forall \ell, j$ (weights initialization)
3: Set $n_j^\ell \leftarrow 0, \forall \ell, j$ (number of visits in cluster C_j^ℓ)
4: **for** $t = 1, \ldots, T$ **do**
5:     Draw a sample $x \sim \mu$
6:     **for** $\ell = L-1, \ldots, 0$ **do**
7:         Compute the corresponding cluster index: $j \leftarrow j^\ell(x)$
8:         $n_j^\ell \leftarrow n_j^\ell + 1$
9:         $g \leftarrow \nabla_v h_{C_j^\ell, \nu_j^\ell}(x, \tilde{v}_j^\ell)$ (in Eq. (4))
10:        $\tilde{v}_j^\ell \leftarrow \tilde{v}_j^\ell + \frac{C}{\sqrt{n_j^\ell}} g$
11:        $v_j^\ell \leftarrow v_j^\ell + \frac{1}{n_j^\ell}\left(\tilde{v}_j^\ell - v_j^\ell\right)$

Outputs: $\{\nu^\ell\}_{\ell \leq L}$ and $\{v^\ell\}_{\ell \leq L}$

---

4   Numerical Experiments

In this section we illustrate that multi-layer transport maps can be used to approximate semi-discrete optimal transport, and that they can be computed more efficiently. We start with a simple one-dimensional example (where the true solution of optimal transport is explicit) and next turn to some imaging applications in higher dimension.

4.1   One-Dimensional Example

Here we consider the optimal transport problem when $D = 1$, between the normalized Gauss distribution $\mu \sim \mathcal{N}(0, 1)$ and the discrete uniform distribution ν on $J = 10^3$ (then $J = 10^4$) equally spaced points between -1 and 1 (see illustration in Fig. 1). We will compare the simple ASGD method (Algorithm 1) to the proposed multi-layer scheme (Algorithm 2), and also to the multi-scale approach of [11]. For the multi-layer scheme, we only consider two layers. For that we use only one clustering of the target points, with 10 clusters (computed with the K-means algorithm). For the multi-scale scheme of [11], except for the finest scale, the transported measure is obtained by extrapolating the weights to the corresponding cluster.

The comparison will focus on the distance between the transported measure $T_{\sharp}\mu$ and the target measure ν (which does not entirely reflect the optimality in terms of transportation cost). We will use the Kolmogorov distance between those two discrete measures, that is, the L^{∞} distance between the cumulative distribution functions (cdf). The computation of the cdf for the mono-layer map has already been explained in [3], and it can be extended to the multi-layer map. Notice that the cost of one iteration is not the same for those three iterative algorithms. Thus we will compare the evolution of the distance with respect to the true computation time (in seconds) and not the number of iterations (approximately 10^5 iterations for $J = 1000$, and 10^6 for $J = 10\,000$). The results are reported in Fig. 3.

One can observe that the multi-layer scheme quickly realizes a good approximation of the target distribution, much faster than the mono-scale. This advantage is greater as the number of points increases, making this approach interesting to estimate and approximate transport for large set of points. However, one can observe that the limiting value for the multi-layer scheme is larger than the one with the mono-scale version, which reflects the bias that is introduced by imposing the multi-layer form of the transportation plan (associated to one fixed decomposition of ν). The multi-scale approach does not perform well in this scenario, where the error can increase when changing scale. One explanation may be that the initialization from the previous scale should be handled more carefully.

4.2   Texture Synthesis

To illustrate the interest of the proposed multi-layer approach, we use the multi-scale patch-based texture synthesis framework proposed in [3], where the

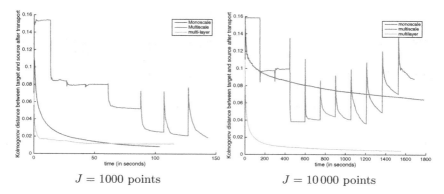

$J = 1000$ points $J = 10\,000$ points

Fig. 3. We monitor the evolution of the distance between $T_\sharp \mu$ and ν during the three iterative algorithms explained in the text. The horizontal axis represents the computational time (in seconds) and the vertical axis the Kolmogorov distance between the current transported measure $T_\sharp \mu$ and the target measure ν. The number of clusters for the multi-layer approach is $|J^1| = \frac{J}{10}$. See the text for comments.

semi-discrete OT is now estimated with Algorithm 2 instead of the single scale Algorithm 1. This enables to use larger patches (7×7 color patches instead of 3×3 in [3], so that dimension is $D = 147$), as well as a larger number of points (up to $J = 16\,000$ instead of $J = 1000$ in [3]). Additionally, as the patches are larger, we use Gaussian weighting for aggregation to avoid excessive blurring when averaging patch, and we also use a final optimal transportation at the finest resolution with 3×3 patches to restore fine details that are lost for the same reason. See Fig. 4 for an illustration.

Parameters for both methods are fixed through experiments: $S = 4$ resolutions and a Gaussian mixture model with 10 components are used for the texture model (see [3] for details); the patch size is $w \times w$ ($w = 3$ for 1-layer OT and $w = 7$ for bi-layer OT); $J = 1000$ patches are randomly sampled from the example image when using 1-layer OT, and $J = 1000, 2000, 4000, 16\,000$ (resp.) patches (from coarse to fine resolution) when using bi-layer OT, using $|J^1| = 10, 10, 20, 40$ (resp.) clusters that are estimated by a K-means algorithm at the first layer); a centered Gaussian weight is used for patch averaging of patch with bi-level OT, with a standard deviation $\sigma = \frac{w}{2}$; Algorithm 1 and Algorithm 2 are used with 10^6 iterations.

Figure 5 illustrates the interest of the multi-layer patch transportation over the single-layer approach originally proposed in [3]. As already mentioned, the latter is restricted to small patch dimension due to convergence issues. The use of large patch makes it possible to capture and synthesize larger scale structures from the example image.

Figure 6 shows a small comparison with the state-of-the-art Texture Networks [20] approach. This method first trains a convolutional network generator using an example image and a deep neural network feature extractor (VGG-19)

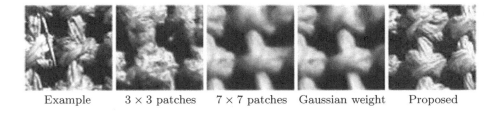

Example   3 × 3 patches   7 × 7 patches   Gaussian weight   Proposed

Fig. 4. Limitations of the 1-layer OT model of [3] for texture synthesis when using large patches. Images are cropped to display small details, full images are shown in Fig. 5. From left to right: example image, result from [3] (with 3 × 3 patches), result when using 7 × 7 patches without/with Gaussian weighting, and the proposed approach.

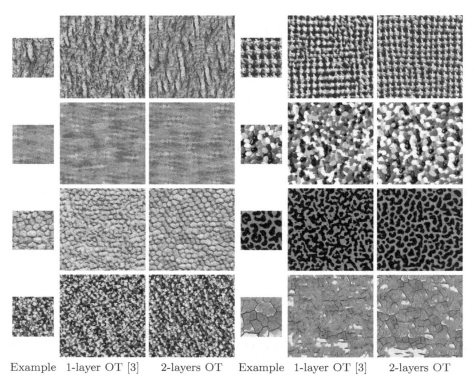

Example   1-layer OT [3]   2-layers OT   Example   1-layer OT [3]   2-layers OT

Fig. 5. Texture synthesis comparison between the 1-layer OT model of [3] and our 2-layers OT model.

that is trained separately on a very large classification dataset of natural images (ImageNet), and then performs fast texture synthesis by feeding the generator with a random input. While both methods produce similar results on regular pattern (first row of Fig. 6), it is not the case for other type of textures (second

Examples   2-layer OT   [20]   Examples   2-layer OT   [20]

Fig. 6. Texture synthesis comparison between the proposed approach (multi-layers Optimal Transport of 7×7 patches) and Texture Networks [20] (a feed-forward convolutional neural network). Note that for the *pumpkins* example, $S = 6$ resolutions has been used instead of $S = 4$ in the rest of experiments to capture large scale object information.

row), where the method of [20] tends to generate pseudo-periodic images. Note that, while our approach does not requires the use of a GPU for the training stage, both methods are completely parallel and achieve similar computation time during synthesis (within a second on a GPU for 1024×1024 images).

5   Discussion and Conclusion

We have described a new strategy to approximate the semi-discrete optimal transport problem based on hierarchical multi-layer transport maps. A simple stochastic algorithm has been proposed and shown to be effective with only two layers. It allows for faster training and synthesis than previous multi-scale approaches, and makes it also possible to deal with larger point sets in higher dimension. Its application to texture synthesis demonstrates its practical interest, making it possible to compete with recent machine learning techniques (based on deep convolutional neural networks).

In this preliminary work, we have only experimented with bi-layer transport maps, but we expect the proposed model to be even more efficient when exploiting its full multi-layer potential, as done by deep learning techniques. However, as shown in experiments for the multi-scale approach, this raises the problem of the optimal setting of the hierarchical structure (similarly to the importance of network design in machine learning) to avoid strong bias in the approximation of the optimal transportation map. Besides, the convergence of the stochastic algorithm for estimating multi-layer transport maps remains to be investigated, since it does not correspond to a convex problem anymore.

Acknowledgments. This project has been carried out with support from the French State, managed by the French National Research Agency (ANR-16-CE33-0010-01).

References

1. Aurenhammer, F., Hoffmann, F., Aronov, B.: Minkowski-type theorems and least-squares clustering. Algorithmica **20**(1), 61–76 (1998)
2. Barnes, C., Shechtman, E., Finkelstein, A., Goldman, D.B.: PatchMatch: a randomized correspondence algorithm for structural image editing. ACM Trans. Graph. TOG **28**(3), 24 (2009)
3. Galerne, B., Leclaire, A., Rabin, J.: A texture synthesis model based on semi-discrete optimal transport in patch space. SIAM J. Imaging Sci. **11**(4), 2456–2493 (2018)
4. Genevay, A., Cuturi, M., Peyré, G., Bach, F.: Stochastic optimization for large-scale optimal transport. In: Proceedings of NIPS, pp. 3432–3440 (2016)
5. Gutierrez, J., Rabin, J., Galerne, B., Hurtut, T.: Optimal patch assignment for statistically constrained texture synthesis. In: Lauze, F., Dong, Y., Dahl, A.B. (eds.) SSVM 2017. LNCS, vol. 10302, pp. 172–183. Springer, Cham (2017). https://doi.org/10.1007/978-3-319-58771-4_14
6. Kitagawa, J.: An iterative scheme for solving the optimal transportation problem. Calc. Var. Partial Differ. Equ. **51**(1–2), 243–263 (2014)
7. Kitagawa, J., Mérigot, Q., Thibert, B.: A Newton algorithm for semi-discrete optimal transport. J. Eur. Math Soc. (2017)
8. Kwatra, V., Essa, I., Bobick, A., Kwatra, N.: Texture optimization for example-based synthesis. ACM TOG **24**(3), 795–802 (2005)
9. Lévy, B.: A numerical algorithm for L2 semi-discrete optimal transport in 3D. ESAIM: M2AN **49**(6), 1693–1715 (2015)
10. Liu, J., Yin, W., Li, W., Chow, Y.T.: Multilevel optimal transport: a fast approximation of wasserstein-1 distances. arXiv preprint arXiv:1810.00118 (2018)
11. Mérigot, Q.: A multiscale approach to optimal transport. Comput. Graph. Forum **30**(5), 1583–1592 (2011)
12. Oberman, A.M., Ruan, Y.: An efficient linear programming method for optimal transportation. arXiv preprint arXiv:1509.03668 (2015)
13. Rabin, J., Peyré, G.: Wasserstein regularization of imaging problems. In: 2011 IEEE International Conference on Image Processing, ICIP 2011 (2011)
14. Rabin, J., Peyré, G., Delon, J., Bernot, M.: Wasserstein barycenter and its application to texture mixing. In: Bruckstein, A.M., ter Haar Romeny, B.M., Bronstein, A.M., Bronstein, M.M. (eds.) SSVM 2011. LNCS, vol. 6667, pp. 435–446. Springer, Heidelberg (2012). https://doi.org/10.1007/978-3-642-24785-9_37
15. Rubner, Y., Tomasi, C., Guibas, L.J.: A metric for distributions with applications to image databases. In: Sixth International Conference on Computer Vision, pp. 59–66. IEEE (1998)
16. Santambrogio, F.: Optimal Transport for Applied Mathematicians. Birkäuser (2015)
17. Schmitzer, B.: A sparse multiscale algorithm for dense optimal transport. J. Math. Imaging Vis. **56**(2), 238–259 (2016)
18. Solomon, J., et al.: Convolutional wasserstein distances: efficient optimal transportation on geometric domains. ACM Trans. Graph. (TOG) **34**(4), 66 (2015)
19. Tartavel, G., Gousseau, Y., Peyré, G.: Variational texture synthesis with sparsity and spectrum constraints. J. Math. Imaging Vis. **52**(1), 124–144 (2015)
20. Ulyanov, D., Lebedev, V., Vedaldi, A., Lempitsky, V.: Texture networks: feed-forward synthesis of textures and stylized images. In: Proceedings of the International Conference on Machine Learning, vol. 48, pp. 1349–1357 (2016)

21. Villani, C.: Topics in Optimal Transportation. American Mathematical Society (2003)
22. Xia, G., Ferradans, S., Peyré, G., Aujol, J.: Synthesizing and mixing stationary gaussian texture models. SIAM J. Imaging Sci. **7**(1), 476–508 (2014)

Segmentation and Labeling

Global Similarity with Additive Smoothness for Spectral Segmentation

Vedrana Andersen Dahl[(✉)] [iD] and Anders Bjorholm Dahl[iD]

Department of Applied Mathematics and Computer Science,
Technical University of Denmark, Kongens Lyngby, Denmark
{vand,abda}@dtu.com

Abstract. Faithful representation of pairwise pixel affinities is crucial
for the outcome of spectral segmentation methods. In conventional affinity models only close-range pixels interact, and a variety of subsequent
techniques aims at faster propagation of local grouping cues across long-
range connections. In this paper we propose a general framework for
constructing a full-range affinity matrix. Our affinity matrix consists of
a global similarity matrix and an additive proximity matrix. The similarity in appearance, including intensity and texture, is encoded for each
pair of image pixels. Despite being full-range, our similarity matrix has
a simple decomposition, which exploits an assignment of image pixels to
dictionary elements. The additive proximity enforces smoothness to the
segmentation by imposing interactions between near-by pixels. Our approach allows us to assess the advantages of using a full-range affinity for
various spectral segmentation problems. Within our general framework
we develop a few variants of full affinity for experimental validation. The
performance we accomplish on composite textured images is excellent,
and the results on natural images are promising.

Keywords: Image segmentation · Spectral methods · Affinity matrix

1 Introduction

Spectral clustering is fundamental in several popular methods for unsupervised
image segmentation [2,7,13]. The outcome of spectral segmentation depends
largely on pairwise pixel affinities. In general, there are two terms contributing
to affinities: a similarity term (how alike are the pixels values) and a proximity
term (how spatially close are the pixels).

In the original formulation of normalized cuts [13] similarity and proximity
are both modelled using a Gaussian similarity function spanning a certain scale.
The two contributions are multiplied, so only pairs of pixels that are within the
distance range and within the intensity range get assigned non-zero affinity. The
reasoning behind is that close-by pixels with similar intensity value are likely
to belong to one object. An argument in favor of short-range interactions is the
desire to reduce computational cost of the eigenvalue problem by using sparse
matrices.

© Springer Nature Switzerland AG 2019
J. Lellmann et al. (Eds.): SSVM 2019, LNCS 11603, pp. 357–368, 2019.
https://doi.org/10.1007/978-3-030-22368-7_28

Fig. 1. The left column shows spectral segmentation of an image of size 2400×3200 using full-range affinity, middle column is an image from the Berkeley Segmentation Dataset [1] and right is a composed texture from [12]. All images are segmented using no training data.

Since longer range interactions generally make segmentation better, multi-scale model [2] combines both coarse and fine-level details, but still on a fixed neighborhood radius. Another popular approach is to use superpixels to propagate grouping cues across larger image regions [7,16,17]. Such aggregating methods depend on the quality of initial oversegmentation.

Instead of purely merging regions, the method proposed in [6] utilizes initial oversegmentation to propagate local affinities in the affinity matrix. In a single optimization step all pairwise affinities are estimated, yielding impressive segmentation results. The approach is however still based on propagating, and not directly modelling full-range affinities.

Further challenge involves modelling pairwise similarity for images with regions characterized by cues like boundary and texture. In the intervening contour approach [2,10] the pairwise pixel affinity is extended with a term which measures the magnitude of image edges on the straight line connecting two pixels. Furthermore, textons are extracted from the image to define texture cue, but still only close-by pixels are connected.

In our approach, pairwise pixel similarity and pairwise pixel proximity are treated more independently. We model the full-range pairwise similarity, so pairs of similar pixels have positive affinity regardless of their distance. Pixel similarity is defined in terms of a dictionary assignment: two pixels are similar it they link to the same dictionary element (or two similar dictionary elements, or if they link to the dictionary elements in the similar way). This allows for a compact representation of the full-range similarity using a biadjacency matrix, which also makes the computation of eigenvectors efficient. As for the proximity, we incorporate it as an additive smoothness term modelled by the Gaussian similarity function. The reasoning behind is that pixels likely to belong to one region are either similar or close-by. This construction has great advantages when segmenting images containing textured regions. Close-by pixels in highly textured regions need not be similar, but, due to the repetitive nature of the texture, similar pixels will be found in a larger neighborhood, creating many unconnected networks within the textured region. Additive smoothness contributes with additional links, connecting those networks. The resulting general framework is capable of handling both textured and not-textured images.

The main contribution of our paper is an efficient and versatile method for approximating a full affinity matrix. This is useful for segmenting an image into global regions of homogeneous appearance, and may be included in more specialized segmentation engines. Our approach is rather different from other spectral methods, both in the problem statement and in the proposed solution. The aim of this paper is to introduce this novel affinity matrix and to provide insight in its properties and use.

The key contributions of our paper are:

1. For modelling full-range similarity we propose an efficient representation of a similarity matrix based on dictionary assignment.
2. For spectral segmentation we propose using a full-range similarity matrix and additive proximity. Our method has linear time complexity and can handle images with millions of pixels, see Fig. 1, left column.
3. For segmenting natural images we propose similarity matrix based on dictionary containing image patches and normalized image patches, see Fig. 1, middle column.
4. For segmenting textured images we propose similarity matrix based on dictionary containing image patches and SIFT features, see Fig. 1, right column.

2   Method

Our approach follows the common framework of spectral segmentation. To set the stage, we start this section with a brief review of a spectral segmentation using a conventional affinity matrix. After that, we cover the basic principles of constructing our novel full-range similarity matrix, describe few of its variants, and discuss its use.

2.1   Conventional Affinity Matrix

For an image I containing n pixels, an affinity matrix $\mathbf{W}$ is an $n \times n$ matrix representing a graph build upon the image. Matrix elements $w_{ii'}$ represent a weight of the edge between pixels i and i'. Dividing nodes of this graph in k partitions, using some optimality criteria, provides image segmentation. A popular objective function is normalized cut [13], which can be conveniently rewritten using the graph Laplacian $\mathbf{L} = \mathbf{D} - \mathbf{W}$, where $\mathbf{D} = \mathrm{diag}(\mathbf{W}1_{n \times 1})$ is a diagonal degree matrix. Relaxation of this discrete NP hard optimization problem yields the solution as the smallest k generalized eigenvectors of the generalized eigenproblem $\mathbf{L}\mathbf{u} = \lambda\mathbf{D}\mathbf{u}$. Real-valued eigenvectors lead to a new k-dimensional representation of the image pixels which enhances the clustering properties in the data. Typically k-means algorithm is used for detecting clusters of pixels in this new representation. For a detailed coverage of spectral segmentation we refer to [15]. It also shows that the solution equivalent to normalized cut may be found as largest k eigenvectors of the matrix $\mathbf{D}^{-1/2}\mathbf{W}\mathbf{D}^{-1/2}$ multiplied by $\mathbf{D}^{-1/2}$, which is the approach we choose in practice.

Constructing affinity matrix is a crucial step in spectral segmentation. Conventional approach is to define pixels affinities based on their intensities and positions in the image grid. In [2,13] pairwise affinity is defined as a product of intensity similarity term and spatial proximity term

$$w_{ii'} = \begin{cases} \exp\left(-\frac{d_I(i,i')^2}{2\sigma_I^2} - \frac{d_X(i,i')^2}{2\sigma_X^2}\right) & \text{if } d_X(i,i') < r \\ 0 & \text{otherwise} \end{cases} \tag{1}$$

where $d_I(i,i') = |I_i - I'_i|$ is the absolute intensity difference and $d_X(i,i') = \|X_i - X'_i\|_2$ is the spatial distance between pixels, r is the radius of the interaction and σ_I and σ_X are parameters of the Gaussian similarity functions. Approach is that close-by pixels with similar intensity value are likely to belong to one object.

2.2   Full-Range Affinity Matrix

Instead of defining affinity as a multiplication of the similarity with the short-range proximity we will define full-range similarity and add the proximity. By replacing multiplication with addition we replace a similar-*and*-close assumption with the similar-*or*-close assumption.

Dictionary Assignment. The initial step of our method involves constructing a dictionary and assigning every image pixel to a dictionary element. In the simplest version a dictionary is constructed by collecting pixel intensities from the image and clustering those in m clusters using k-means algorithm. All image pixels are then assigned to the nearest intensity and this assignment is denoted $D(i)$, where $i \in \{1, \ldots, n\}$, $D(i) \in \{1, \ldots, m\}$.

Biadjacency Matrix. Dictionary assignment can be represented as a biadjacency matrix $\mathbf{B}$, where an element b_{ij} takes a value 1 if $D(i) = j$ and 0 otherwise. An $n \times n$ matrix $\mathbf{S} = \mathbf{B}\mathbf{B}^\mathsf{T}$ is the simplest variant of our full-range similarity matrix. An important property of $\mathbf{S}$ is

$$s_{ii'} = \sum_{j=1}^{m} b_{ij}b_{i'j} = \begin{cases} 1 \text{ if } D(i) = D(i') \\ 0 \text{ otherwise} \end{cases} \tag{2}$$

We say that $\mathbf{S}$ encodes global binary similarity on the basis of dictionary assignment. This full-range similarity and its extensions are fundamental for our method.

A Few Variants of a Full-Range Similarity. For the simplest segmentation problems the similarity matrix given by (2) might suffice, but generally we want a better description of pixel similarity, and here we describe a few possibilities. Since all variants of full-range similarity may be used in the final segmentation model, we will denote them all by $\mathbf{S}$.

Pixel Similarity Weighted by Dictionary Similarity. One extension of the full affinity incorporates an $m \times m$ matrix $\mathbf{A}$ containing pairwise similarity between dictionary elements, computed in a conventional way using Gaussian similarity function as explained in connection with similarity term from Eq. (1). In a weighted similarity matrix $\mathbf{S} = \mathbf{B}\mathbf{A}\mathbf{B}^{\mathsf{T}}$ an element

$$s_{ii'} = \sum_{j=1}^{m} \sum_{j'=1}^{m} a_{jj'} b_{ij} b_{i'j'} = a_{D(i)D(i')} \tag{3}$$

is a similarity between dictionary element $D(i)$, where pixel i is assigned to, and a dictionary element $D(i')$, where pixel i' is assigned to. As such, this version of $\mathbf{S}$ approximates the conventional full pairwise similarity matrix, and the quality of the approximation depends on how well the dictionary represents regions present in the image. To our knowledge this type of a full affinity approximation has not been used in spectral image segmentation.

Dictionary of Features and Patches. Another extension concerns the dictionary construction. When image regions are characterized by the texture, we want a better representation of pixel similarity. Instead of clustering pixel intensities, we have an option of creating clusters according to any local feature, for example a SIFT feature. In another choice, we extract a quadratic $M \times M$ patch around each image pixel, and use the collected pixel intensities as a feature vector. The dictionary is constructed by clustering patches from the image using k-means algorithm with Euclidian distance.

Dictionary of Patch Pixels. Using image patches as features opens for another possibility. Instead of linking each pixel to one of m dictionary elements, we can, in an approach similar to [3,4], define links between n image pixels and mM^2 dictionary pixels. Each assignment $D(i)$ now contributes with M^2 links between the image and the dictionary. Since every pixel takes one of M^2 positions in an patch, rows of $\mathbf{B}$ now sum to M^2. The motivation for this approach is a desire for a better localization of the boundaries between segments.

A Final Similarity-Proximity Model. Full-range similarity matrix is a key ingredient to our affinity model for spectral segmentation. However, using similarity alone we risk a very fragmented segmentation, as it ignores the spatial position of image pixels.

Proximity Matrix. To enforce the smoothness of the segmentation we define a proximity term which links the close-by pixels according to a Gaussian similarity function of spatial distances. Proximity matrix **P** has elements

$$p_{ii'} = \frac{1}{2\pi\sigma^2} \exp\left(-\frac{d_X(i,i')^2}{2\sigma^2}\right),$$

where σ is a smoothing radius. For σ small compared to image dimensions, matrix **P** will in effect be a sparse matrix.

Affinity Matrix. We want to incorporate the proximity matrix to our final model without reducing the range of interactions between pixels. Therefore we add the proximity links to full-range similarity matrix, leading to the affinity matrix

$$\mathbf{W} = \mathbf{S} + \alpha\mathbf{P}$$

where **S** is one of the variants of the similarity matrix and **P** is the proximity matrix weighted by α.

Normalization. In spectral clustering, normalizing the affinity matrix will favor more balanced segmentation. This is important if degrees of the graph nodes vary considerably. In case of our proximity matrix **P**, the degree of the nodes is constant except at the boundary, so we only need to normalize **S**. In similarity matrix **S**, a pixel is linked to all other pixels belonging to the same dictionary element. This property allows us to define a normalized similarity matrix

$$\mathbf{S} = \mathbf{B}\, \mathrm{diag}(\mathbf{B}\mathbf{1}_{m\times 1})^{-1}\mathbf{B}^{\mathsf{T}}.$$

Eigendecomposition. To calculate leading eigenvectors of **W** using power methods such as Lanczos solver [5] we do not need an explicit representation of **W**. It is sufficient to define the result of the matrix-vector computation $\mathbf{y} = \mathbf{W}\mathbf{x}$ for every **x**. With minor modifications, depending on the variant of the similarity matrix, we have

$$\mathbf{y} = \mathbf{B}\, \mathrm{diag}(\mathbf{B}\mathbf{1}_{m\times 1})^{-1}\mathbf{B}^{\mathsf{T}}\mathbf{x} + \alpha\mathbf{P}\mathbf{x}.$$

Computational Efficiency. The similarity term can be efficiently computed by multiplying from right to left. Since **B** is $n \times m$ binary and sparse matrix with n non-zero entries, the first and the last multiplication is $\mathcal{O}(n)$. The time complexity of the multiplication in the middle does not depend on n, yielding time complexity of $\mathcal{O}(m)$ for diagonal matrix $\mathrm{diag}(\mathbf{B}\mathbf{1})^{-1}$. Had we used a formulation from Eq. (3) multiplication with **A** would give complexity of $\mathcal{O}(m^2)$.

The proximity term can also be efficiently computed by arranging values of **x** in an image grid and filtering this image-like construction with a one-dimensional Gaussian kernel in each direction. The time complexity of this operation is $\mathcal{O}(nr)$, where r is the size of the kernel. Assuming that dictionary size m and kernel size r do not depend on the image size n, the running time for computing **Wx** is linear with respect to n.

Fig. 2. Example demonstrating the global property of the spectral segmentation when employing our full-range affinity. All instances of grass, sky, skin and orange tulips belong to the same regions. Left: the original image with segments found using our method outlined in red. Middle left: average colors within segments. Middle right: segments in false colors. Right: manual segmentation from BSDB shown with false colors [1] lacks the global property. (Color figure online)

The time complexity of the Lanczos method is $O(ln)+O(lf(n))$ where l is the maximal number of matrix-vector computations required and $f(n)$ is the cost of each matrix-vector computation [13]. Consequently, the total running time for the eigendecomposition of our full-range affinity matrix is $O(ln)$.

The maximal number of matrix-vector computations depends on the ability of the affinity matrix to propagate clustering cues. In our experience, a few hundred iterations is usually enough, regardless of the size of the image. We conclude that in practice our algorithm has linear time complexity.

3   Results

We have carried out a number of experiments in order to investigate the properties of our novel full range affinity matrix.

Since the self-similarity is most evident in composite images, we validate the quality of our method on segmenting Brodatz texture mosaics from [12]. To demonstrate the potential of our method in a more realistic setting, we include segmentations examples of natural images from the Berkeley Segmentation Dataset [1,11]. In order to retain a focus on our contribution, we show those results without incorporating additional cues e.g. an edge term or a shape prior.

Global Property. As a consequence of a full-range similarity, all image regions with similar texture and color belong to the same segment despite being disconnected. An example of this is shown in Fig. 2, where an image of the girl is segmented using 14 labels. The flowers, grass and bushes in the background, despite being separated, are labeled together due to the similar color and texture. The same is seen for the skin, sky, stockings and the paved curb.

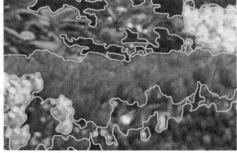

Fig. 3. Segmentation of a very large image. Left is a 4800 × 6400 image with segmentation boundaries in cyan. Right is a 750 × 1125 cutout indicated by a rectangle in the left image. (Color figure online)

Efficiency. Our affinity matrix allows for an efficient decomposition, so we can process large images. In Fig. 3 we show an image containing more than 30 million pixels segmented in 5 segments. The dictionary used for this segmentation is based on 3 × 3 patches grouped in 200 clusters. Smoothness parameters are $\sigma = 1$ and $\alpha = 15$.

Textured Images. We have made a quantitative evaluation of our method on a Randen texture dataset [12], with images composed of 5–16 different textures, allowing us to evaluate our method against ground truth. In this experiment, we base the dictionary on dense SIFT features grouped into 500 clusters. The smoothness parameters are $\sigma = 8$ and $\alpha = 4$. The segmentation on subset of images from the dataset is shown in the Fig. 4.

Quantitative results for the texture segmentation images are shown in Table 1, compared against other results reported for the dataset. It is important to note that the methods we compare against are all *trained* on examples of textured images, which gives them a great advantage. Our method takes a number of classes as input, but no other information about the textures. In other words, our method performs pixel clustering (unsupervised learning) while methods we compare to perform pixel classification (supervised learning). Still we obtain comparable results for most of the images. We compare to supervised methods because no unsupervised results have been reported for the Randen dataset.

Natural Images. Segmentation of natural images from the Berkeley Segmentation Dataset [1,11] in Fig. 5 is included to illustrate additional properties and the potentials of our method. These experiments have been carried out using 3 × 3 image patches clustered into dictionaries of 1000 elements. Segmenting into 2 to 6 regions was attempted, and we show the most meaningful result. Since global property is not part of the manually segmented reference data in the Berkeley Segmentation Dataset (see Fig. 2) we did not quantify our method against reference data.

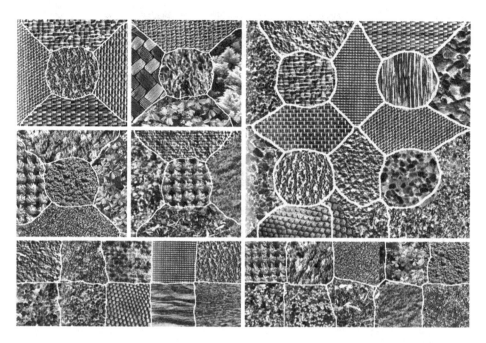

Fig. 4. Texture segmentation examples. Images 1 (top left), 2 (top middle), 3 (middle left), 4 (middle), 6 (top right), 8 (bottom left) and 9 (bottom right) from the Randen dataset [12].

Table 1. Comparison with Randen et al. [12], Lillo et al. [8], Mairal et al. [9], and Skretting and Engang [14] (best reported method). Reported number is the error rate. Unlike our method, the methods we compare to use training data.

Image no.	[12] 1999	[8] 2007	[9] 2008	[14] 2014	Our (without training)	Image no.	[12] 1999	[8] 2007	[9] 2008	[14] 2014	Our (without training)
1	7.20	3.37	1.61	2.00	1.68	7	41.70	21.67	8.80	4.14	31.57
2	18.90	16.05	16.42	3.24	4.43	8	32.30	21.96	2.24	4.80	2.55
3	20.60	13.03	4.15	4.01	6.74	9	27.80	9.61	2.04	3.90	7.99
4	16.80	6.62	3.67	2.55	6.58	10	0.70	0.36	0.17	0.42	0.24
5	17.20	8.15	4.58	1.26	9.29	11	0.20	1.33	0.60	0.61	0.26
6	34.70	18.66	9.04	6.72	10.06	12	2.50	1.14	0.78	0.70	0.75

Fig. 5. Segmentation results for a selection of images from the Berkeley Segmentation Dataset.

Smoothness Parameters. Influence of the model parameters on the segmentation results is shown in Fig. 6. We show the effect of changing the smoothing radius for the proximity matrix σ and the weight for proximity matrix α. For this experiment we use textured images and the dictionary based only on image patches without using SIFT features. This is because we also want to demonstrate the influence of the patch size M on the results.

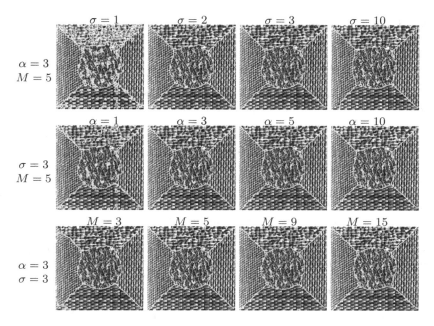

Fig. 6. The influence of the model parameters on the segmentation results. Parameters are: smoothing radius for proximity σ, proximity weight α and patch size M.

4 Conclusion

The contribution of this paper is a general framework for the construction of the global affinity matrix using full-range similarity and additive smoothness. The full-range affinity matrix is constructed employing a dictionary and a bipartite graph representation linking pixels in the image to elements/pixels in the dictionary.

Our approach is extremely versatile. Within the same general framework we apply our method for unsupervised spectral segmentation of the grayscale textured images and the rgb natural images. Our approach is also very efficient. Unlike other spectral methods, we obtain a linear time complexity for segmentation using our affinity matrix. This is because we do not represent the affinity matrix explicitly, and therefore it does not need to be sparse. We employ Lanczos algorithm that iteratively updates eigenvectors using an implicit matrix vector multiplication.

Our empirical investigation shows how the method partitions an image into regions with similar texture and color despite these being spatially separated. In addition we obtain impressive results on images composed of Brodatz textures, where no training data is used and only the number of segments is given to the algorithm.

In conclusion, we see our presented approach for spectral segmentation as an interesting alternative to conventional methods, appropriate for segmenting large images using a full-range affinity.

References

1. Arbelaez, P., Maire, M., Fowlkes, C., Malik, J.: Contour detection and hierarchical image segmentation. IEEE Trans. Pattern Anal. Mach. Intell. **33**(5), 898–916 (2011)
2. Cour, T., Benezit, F., Shi, J.: Spectral segmentation with multiscale graph decomposition. In: Conference on Computer Vision and Pattern Recognition, vol. 2, pp. 1124–1131. IEEE (2005)
3. Dahl, A.B., Dahl, V.A.: Dictionary based image segmentation. In: Paulsen, R.R., Pedersen, K.S. (eds.) SCIA 2015. LNCS, vol. 9127, pp. 26–37. Springer, Cham (2015). https://doi.org/10.1007/978-3-319-19665-7_3
4. Dahl, V.A., Dahl, A.B.: A probabilistic framework for curve evolution. In: Lauze, F., Dong, Y., Dahl, A.B. (eds.) SSVM 2017. LNCS, vol. 10302, pp. 421–432. Springer, Cham (2017). https://doi.org/10.1007/978-3-319-58771-4_34
5. Golub, G., Van Loan, C.F.: Matrix Computations, p. 642. Johns Hopkins University Press, Baltimore (1996)
6. Kim, T.H., Lee, K.M., Lee, S.U.: Learning full pairwise affinities for spectral segmentation. IEEE Trans. Pattern Anal. Mach. Intell. **35**(7), 1690–1703 (2013)
7. Li, Z., Wu, X.M., Chang, S.F.: Segmentation using superpixels: a bipartite graph partitioning approach. In: Conference on Computer Vision and Pattern Recognition, pp. 789–796. IEEE (2012)
8. Lillo, A.D., Motta, G., Storer, J., et al.: Texture classification based on discriminative features extracted in the frequency domain. In: International Conference on Image Processing, vol. 2, pp. II–53. IEEE (2007)
9. Mairal, J., Bach, F., Ponce, J., Sapiro, G., Zisserman, A.: Discriminative learned dictionaries for local image analysis. In: Conference on Computer Vision and Pattern Recognition, pp. 1–8. IEEE (2008)
10. Malik, J., Belongie, S., Leung, T., Shi, J.: Contour and texture analysis for image segmentation. Int. J. Comput. Vis. **43**(1), 7–27 (2001)
11. Martin, D., Fowlkes, C., Tal, D., Malik, J.: A database of human segmented natural images and its application to evaluating segmentation algorithms and measuring ecological statistics. In: International Conference on Computer Vision, vol. 2, pp. 416–423. IEEE (2001)
12. Randen, T., Husoy, J.H.: Filtering for texture classification: a comparative study. IEEE Trans. Pattern Anal. Mach. Intell. **21**(4), 291–310 (1999)
13. Shi, J., Malik, J.: Normalized cuts and image segmentation. IEEE Trans. Pattern Anal. Mach. Intell. **22**(8), 888–905 (2000)
14. Skretting, K., Engan, K.: Energy minimization by α-erosion for supervised texture segmentation. In: Campilho, A., Kamel, M. (eds.) ICIAR 2014. LNCS, vol. 8814, pp. 207–214. Springer, Cham (2014). https://doi.org/10.1007/978-3-319-11758-4_23
15. Von Luxburg, U.: A tutorial on spectral clustering. Stat. Comput. **17**(4), 395–416 (2007)
16. Wang, X., Li, H., Bichot, C.E., Masnou, S., Chen, L.: A graph-cut approach to image segmentation using an affinity graph based on ℓ_0-sparse representation of features. In: International Conference on Image Processing, pp. 4019–4023. IEEE (2013)
17. Wang, X., Tang, Y., Masnou, S., Chen, L.: A global/local affinity graph for image segmentation. IEEE Trans. Image Process. **24**(4), 1399–1411 (2015)

Segmentation of 2D and 3D Objects with Intrinsically Similarity Invariant Shape Regularisers

Jacob Daniel Kirstejn Hansen and François Lauze[✉]

Department of Computer Science, University of Copenhagen, Copenhagen, Denmark
{jdkh,francois}@di.ku.dk

Abstract. This paper presents a 2D and 3D variational segmentation approach based on a similarity invariant, i.e., translation, scaling, and rotation invariant shape regulariser. Indeed, shape moments of order up to 2 for shapes with limited symmetries can be combined to provide a shape normalisation for the group of similarities. In order to obtain a segmentation objective function, a two-means or two-local-means data term is added to it. Segmentation is then obtained by standard gradient descent on it. We demonstrate the capabilities of the approach on a series of experiments, of different complexity levels. We specifically target rat brain shapes in MR scans, where the setting is complex, because of bias field and complex anatomical structures. Our last experiments show that our approach is indeed capable of recovering brain shapes automatically.

1 Introduction

With the advent of 3D imaging devices, especially X-ray computerised tomography and magnetic resonance imaging, there has been a growing need for 3D segmentation methods that can handle a large variety of signals. Images produced from these modalities can show content with various degrees of complexity and structures, from almost fully random phases to highly structured data, for instance in medical imaging when imaging different types of tissues and organs. In this work, we are interested in the latter, where segmentation targets specific structures characterised by a shape distribution.

In this paper, we develop a variational segmentation–matching approach which incorporates shape priors and is at the same time robust to specific acquisition problems. In addition to noise, MR images are in general corrupted by a bias field coming from combinations of dropoff effects, as seen in Fig. 1 (a). For this, we will use a data fidelity term robust to noise and bias field. We will discuss it in Sect. 2. Our emphasis in this paper is, however, more on shape. We will first briefly review related ideas of shape priors in the next paragraph.

In a series of works [10], Kendall defined shape spaces based on equivalence classes of finite dimensional collections of points, so as to factor out scale, position, and pose. For segmentation, Cootes *et al.* Active Shapes Models [2] is the seminal work on the representation of shape distributions and shape priors, and

© Springer Nature Switzerland AG 2019
J. Lellmann et al. (Eds.): SSVM 2019, LNCS 11603, pp. 369–380, 2019.
https://doi.org/10.1007/978-3-030-22368-7_29

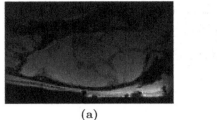

(a)   (b)

Fig. 1. Slice of an MRI scan of a rat cranium (a), 3D brain segmentation (b).

since then a large body of literature has been generated. A complete review is, however, out-of-scope here. An operational representation of shape distribution requires in general two main ingredients: representation of geometric objects in $\mathbb{R}^2$ or $\mathbb{R}^3$ and identifications of classes of such objects as a unique shape, in general via specific group actions. Of course, these two ingredients are rarely independent of each other. Typically, shape priors and regularisers are built from similarity criteria between forms or some of their extracted features, which are invariant by the group operation. One general way is to introduce explicit minimisation over the transformation group in the prior definition. In this line of work, Chan and Zhu [1], in 2D, explicitly minimise a shape term over the similarity group $\mathbf{S}(2)$ of scaling, rotation and translation in $\mathbb{R}^2$. A similar idea, within a levelset framework, and competing priors, was used by Fusseneger et al. in [7]. Mezhgich, M'Hiri and Ghorbel [12] use properties of the Fourier-Mellin transform cross-spectrum w.r.t. rotation and scaling to segment-match shapes from one or more training shapes. Wang, Yeung and Chan [14] define an affine matching and segmentation, based on the action of the affine group on a class of shape representing functions called interior-points-to-shape relations.

Another consists of finding *canonical forms* i.e., special representatives or features. On special interest are the representations or features which exhibit invariance by group action. Such representatives might be complex to define, however that is our approach in this work. 2D rotations were already present in the Diffusion Snakes [4], which incorporates elements of active shapes. Our work goes in a slightly different direction and extends the work of Cremers and Soatto [3] to incorporate rotations. Foulonneau et al. [6] extract affine canonical moments, i.e., normalised moments which are invariant by an affine transformation in 2D, and define invariant shape dissimilarity from them.

The rest of this paper is organised as follows. In Sect. 2, we start with mathematical preliminaries. We look at invariant shape terms and deduce a shape prior/regulariser term. As mentioned above, we will also discuss the chosen data fidelity term. Together they form the cost function of the method. Our algorithm is essentially a gradient descent on the cost function and we derive its first variation and gradient in Sect. 3. Validation of the capabilities and inherent transform invariance of our proposed model is done experimentally in Sect. 4. Finally, Sect. 5 discusses and concludes the proposed model.

2   Derivation of the Model

In this section, we start by introducing notations and elementary points about shapes and some types of canonical forms, that allows us to define invariant shape priors. We then introduce our objective, containing both data fidelity terms and shape regularisers.

Normalisers and Canonical Forms. If G is a group and $\mathcal{T}$ a set with a (left) G-action, a G-*normaliser* is a mapping $\alpha : \mathcal{T} \to G$ such that $\alpha(g \cdot A) = \alpha(A)g^{-1}$. It clearly satisfies $\alpha(\alpha(A) \cdot A) = \mathrm{id}_G$, thus the name. Its *group*-inverse (not to confuse with inverse mapping) $\alpha(-)^{-1} : A \mapsto \tau(A) = \alpha(A)^{-1} \in G$ is just a G-covariant mapping. Given a mapping $F : \mathcal{T} \to \mathcal{U}$, the mapping $\tilde{F} : A \mapsto F(\alpha(A) \cdot A)$ is G-invariant. Indeed, if $B = g \cdot A$, clearly $\alpha(B) \cdot B = \alpha(A) \cdot A$. For this reason, normalisation provides *canonical forms* for the G-set $\mathcal{T}$. In order to be able to use calculus methods, we need some differentiable structure on $\mathcal{T}$ with differentiability of normalizers. We assume that we have them in this paper.

In the sequel, we exhibit normalisers for shape related sets and some Lie subgroups G of the affine group, $\mathrm{Aff}(n)$ of invertible linear transforms and translations, with $n = 2$ or 3. The main subgroup of interest is the subgroup of similarities, $\mathbf{S}(n) \simeq (\mathbb{R}_+^* \times SO(n)) \ltimes \mathbb{R}^n$, of (positive) scalings, rotations and translations. To us, an *object* is a compact set of $\mathbb{R}^n$, with non-empty interior and regular boundary so as to be able to compute shape derivatives (see [5]). We will denote this set by $\mathcal{S}$. The affine group acts naturally on $\mathcal{S}$ and a G-shape is the orbit of an object under the action of G. More generally, we will consider the set $\mathcal{F}$ of integrable functions with compact support, non-empty interior, and values in $[0, 1]$, so that an object A is naturally represented by its characteristic function, χ_A. G acts on $\mathcal{F}$ by $g \cdot v = x \mapsto v(g^{-1}x)$.

For the group of translation and/or scaling, $G = \mathbb{R}_+^* \ltimes \mathbb{R}^n$, normalisers are easy to construct. For instance, set $\mu_0(A) = \int_A dx$ its volume and $\mu_1(A) = \mu_0(A)^{-1} \int_A x\, dx$ and define the transformation $\tau(A) = (\mu_0(A), \mu_1(A)) : x \mapsto \mu_0(A)x + \mu_1(A)$. Then $v \mapsto \tau(A)^{-1}$ is a normaliser. Scale normalisation can also be performed via the form's standard deviation

$$\sigma(A) = \left(\frac{1}{\mu_0(A)} \int_A |x - \mu_1(A)|^2\, dx \right)^{\frac{1}{2}}. \tag{2.1}$$

Setting $\gamma(A) = (\sigma(A), \mu_1(A))$, γ^{-1} is the normaliser used in [3]. Extending this low-order moments approach to similarities necessitates a few restrictions. We use the object's centred second order moment, i.e., the mapping

$$A \in \mathcal{S} \mapsto \Sigma(A) = \frac{1}{\mu_0(A)} \int_A (x - \mu_1(A))(x - \mu_1(A))^T\, dx \in SPD(n) \tag{2.2}$$

with $SPD(n)$ the space of symmetric positive-definite matrices of order n, and $\sigma(A)^2 = \mathrm{Tr}(\Sigma(A))$. A general eigenvalue/eigenvector decomposition $\Sigma(A) = R_A \Lambda_A R_A^T$ with $R_A \in O(n)$ and Λ_A diagonal, is the solution of a system of analytic equations. Restricted to the open subset $U \subset SPD(n)$ with *distinct eigenvalues*, there are finitely many solutions for R_A $(n!2^n)$. The choice of one of these solutions, still denoted by R_A, extends to U via the implicit function theorem [11]. Choose one of the branches for which $R_A \in SO(n)$ (there are $n!2^{n-1}$ such choices, and can be reduced to 2^{n-1} if eigenvalues are sorted). Restrict also the object space to $\mathcal{S}_d = \Sigma^{-1}(U)$, it is a $\mathbf{S}(n)$-invariant subspace of $\mathcal{S}$ and $A \mapsto \tau_A = (\sigma(A), R_A, \mu_1(A))$ is well-defined from what precedes, and $\mathbf{S}(n)$-covariant; it is straightforward to check that for $g = (s, S, t) \in \mathbf{S}(n)$ $\tau_{g \cdot A} = g\tau_A$ as

$$\Sigma(g \cdot A) = s^2 S\Sigma(A)S^T = s^2 SR_A \Lambda_A R_A^T S^T, \quad \mu_1(g \cdot A) = g\mu_1(A).$$

Thus $\alpha(A) = \tau_A^{-1}$ is a $\mathbf{S}(n)$-normaliser on $\mathcal{S}_d$. The same construction, with same restrictions, carries out for $\mathcal{F}$, over a subset $\mathcal{F}_d$ defined in a similar way to $\mathcal{S}_d$.

Extending this construction to $\text{Aff}(n)$ by factorizing $\Sigma(A)$ is however not possible. As $\Sigma(g.A) = U\Sigma U^T$ for $g = (U, t)$ in $\text{Aff}(n)$, a normalizer should satisfy $\alpha(A)^{-1} = (r(\Sigma(A)), \mu_1(A))$ for a *smooth* square-root $r : SPD(n) \to GL(n)$ *and satisfy* $r(U\Sigma U^T) = Ur(\Sigma), U \in GL(n)$. But the image of $U \mapsto Ur(\Sigma)$ is $GL(n)$, and r being smooth, $r(SPD(n)) \subsetneq GL(n)$ as $\dim SPD(n) < \dim GL(n), n > 1$.

Shape Invariant Dissimilarities and Shape Regulariser. Given a *model object* A_0, which we assume normalised, we choose dissimilarities of the form

$$d^2(A, A_0) = \frac{1}{2} \int_{\mathbb{R}^n} \left(\chi_{\alpha(A) \cdot A}(x) - \chi_{A_0}(x)\right)^2 dx = \frac{1}{2} \int_{\mathbb{R}^n} \left(\chi_A(\tau_A x) - \chi_{A_0}(x)\right)^2 dx \tag{2.3}$$

This is clearly a $\mathbf{S}(n)$-invariant measure. We then follow [3] and build a kernel density estimator from training data made of normalised model shapes $A_1, \ldots, A_N$, $p(A|A_1, \ldots, A_N) \propto F(A, A_1, \ldots, A_N) = \sum_{i=1}^{N} e^{-\frac{d^2(A, A_i)}{2\rho^2}}$, for a $\rho > 0$. This density is well defined if we restrict it to a finite dimensional subspace of $\mathcal{S}_d$, otherwise the normalising constant $c(A_1, \ldots, A_N) = \int_{A_d} F(A, A_1, \ldots, A_N) dA$ is in general undefined. However we do not need a definite prior, we need the well defined regulariser

$$\mathcal{E}_S(A) = -\log F(A, A_1, \ldots, A_N) = -\log \left(\sum_{i=1}^{N} e^{-\frac{d^2(A, A_i)}{2\rho^2}}\right). \tag{2.4}$$

Data Fidelity Term and Proposed Formulation. Different data fidelity terms can be used in conjunction with the regulariser. Two data terms are used in this work. The first one is a classical global 2-means term, while the second is a local means term, better suited to MR data. In both cases, the signal is represented by a function $u : \Omega \in \mathbb{R}^n \to \mathbb{R}$ and we want to partition the image domain Ω into two regions, A and $\Omega \backslash A$. The two-means term is

$$\mathcal{E}_{D_g}(A, c_1, c_2) = \frac{1}{2} \int_{\Omega} \left((u - c_1)^2 \chi_A + (u - c_2)^2 \chi_{\Omega \backslash A}\right) dx \tag{2.5}$$

with $c_1, c_2 \in \mathbb{R}$. MR image data suffers among others from bias fields, that are only partially corrected for by standard techniques. We use ideas from [8] and propose the two-local-means term

$$\mathcal{E}_{D_l}(A, c_1, c_2) = \frac{1}{2} \int_{\Omega} g * \left[(u - c_1(x))^2 \chi_A + (u - c_2(x))^2 \chi_{\Omega \setminus A} \right](x)\, dx \qquad (2.6)$$

with c_1 a smoothed version of u on A, c_2 a smoothed version of u on $\Omega \setminus A$ and g a smoothing kernel, e.g. a Gaussian or nearest neighbours (NN) kernel, which we assume to be even symmetric. Finally we propose to minimise the following criterion

$$\mathcal{E}(A, c_1, c_2) = \mathcal{E}_D(A, c_1, c_2) + \kappa \mathcal{E}_S(A) \qquad (2.7)$$

with $\kappa > 0$ a trade-off parameter between data fidelity and shape.

From Shapes to Label Fields. In all the previous constructions, A can be replaced by χ_A or a v in $\mathcal{F}$. Moments are extended trivially to $\mu_0(v)$, $\mu_1(v)$, $\Sigma(v)$, and $\sigma(v)$. This case and $\mathcal{F}_d$ represent the subspace of functions in $\mathcal{F}_d$ for which $\Sigma(v)$ has distinct eigenvalues.

3   First Variations and Optimisation

In this section, we compute the first variation for (2.7) with a bit more focus for the term (2.4) as we follow [8] for the first term.

Derivative of Shape Regulariser. We start by stating moment derivatives $d_v \mathbf{m}.w$ of the moment $\mathbf{m}(v)$, $v \in \mathcal{F}$ for a "variation" w of v, i.e., functions $\mathbf{v}(t)$ for which $\mathbf{v}(0) = v$ and $\dot{\mathbf{v}}(0) = w$. The moment derivatives are defined as $\frac{d}{dt}\big|_{t=0} \mathbf{m}(\mathbf{v}(t))$, for the moments $\mu_0(v)$, $\mu_1(v)$ and $\Sigma(v)$. Then we compute $d_v \tau.w$ instead of the derivative of the normaliser $d_v \alpha.w$. We use the notation $\langle f, g \rangle$ for the integral $\int_{\mathbb{R}^n} f(x)g(x)\, dx$. Straightforward calculations give

$$d_v \mu_0.w = \langle 1, w \rangle, \quad d_v \mu_1.w = \left\langle \frac{\cdot - \mu_1(v)}{\mu_0(v)}, w \right\rangle, \qquad (3.1)$$

$$d_v \Sigma.w = \left\langle \frac{(\cdot - \mu_1(v))(\cdot - \mu_1(v))^T - \Sigma(v)}{\mu_0(v)}, w \right\rangle, d_v \sigma.w = \left\langle \frac{|\cdot - \mu_1(v)|^2 - \sigma^2(v)}{2\sigma(v)\mu_0(v)}, w \right\rangle \qquad (3.2)$$

To compute $d_v R.w$, we use classical formulas for the derivatives of eigenvectors and eigenvalues of a symmetric matrix. If $e_1, \ldots, e_n$ and $\lambda_1 \ldots, \lambda_n$ are the eigenvectors and associated eigenvalues of $\Sigma(v)$,

$$d_v R.w = \frac{1}{\mu_0(v)} \left\langle \left[\sum_{j \neq i} \frac{e_i^T(\cdot - \mu_1(v)) e_j^T (\cdot - \mu_1(v)) e_j}{\lambda_i - \lambda_j} \right]_{i=1}^n, w \right\rangle \qquad (3.3)$$

Putting it together, one get $(d_v \tau.w)(x) = ((d_v \sigma.w)R(v) + \sigma(v)d_v R.w)\, x + d_v \mu_1.w$. We set $v_i = \chi_{A_i}$ and provide the first variation of the dissimilarity measure $\mathcal{L}^i(v) = \frac{1}{2} \int (v(\tau_v x) - v_i(x))^2 \, dx$. A complete computation is not feasible within the page limit and of limited interest anyway.

$$d_v \mathcal{L}^i . w = \int (v(\tau_v x) - v_i(x)) \nabla_{\tau_v x} v \cdot (d_v \tau . w)(x) \, dx + \int (v(\tau_v x) - v_i(x)) w(\tau_v x) \, dx \tag{3.4}$$

The first integral can be rewritten as

$$\frac{1}{\mu_0(v)} \left[\int_{\mathbb{R}^n} \left\{ \left\langle \int S(y, v) y^T \, dy, \frac{|x - \mu_1(v)|^2 - \sigma(v)^2}{2\sigma(v)} R(v) + \sigma(v) \Xi(x, v) \right\rangle_F + \left(\int S(y, v) \, dy \right)^T (x - \mu_1(v)) \right\} w(x) \, dx \right] \tag{3.5}$$

where we have set

$$S(x, v) = (v(\tau_v x) - v_i(x)) \nabla_{\tau_v x} v, \quad \Xi(x, v) = \left[\sum_{j \neq i} \frac{e_i^T (x - \mu_1(v)) e_j^T (x - \mu_1(v)) e_j}{\lambda_i - \lambda_j} \right]_{i=1}^n .$$

The second integral in (3.4) becomes, after a mere change of variables,

$$\frac{1}{\sigma(v)} \int (v(x) - v_i(\tau_v^{-1} x)) w(x) \, dx. \tag{3.6}$$

From (3.5) and (3.6), one obtain $d_v \mathcal{L}^i . w = \langle G_S^i(v), w \rangle$ where

$$G_S^i(v) = \frac{1}{\mu_0(v)} \left\langle \int S(y, v) y^T \, dy, \frac{|x - \mu_1(v)|^2 - \sigma(v)^2}{2\sigma(v)} R(v) + \sigma(v) \Xi(x, v) \right\rangle_F$$
$$+ \frac{1}{\mu_0(v)} \left(\int S(y, v) \, dy \right)^T (x - \mu_1(v)) + \frac{1}{\sigma(v)} \left(v(x) - v_i \left(\frac{R(v)^T (x - \mu_1(v))}{\sigma(v)} \right) \right) . \tag{3.7}$$

Finally, using standard differentiations rules, one get that

$$d_v \mathcal{E}_S . w = -\frac{1}{2\rho^2} \left\langle \frac{\sum_{i=1}^N e^{-\frac{\mathcal{L}^i(v)}{2\rho^2}} G_S^i(v)}{\sum_{i=1}^N e^{-\frac{\mathcal{L}^i(v)}{2\rho^2}}}, w \right\rangle = \langle G_S(v), w \rangle \tag{3.8}$$

Derivatives of Data Fidelity Terms. Using a variation w over Ω, a trivial calculation gives, for (2.5)

$$d_v \mathcal{E}_{D_g} . w = \langle 2u(c_2 - c_1) + (c_1 - c_2)^2, w \rangle_{L^2(\Omega)} = \langle G_{D_g}, w \rangle_{L^2(\Omega)} \tag{3.9}$$

and is a bit more complex for (2.6)

$$d_v \mathcal{E}_{D_l} . w = \langle u^2 g * \chi_\Omega - 2u * g(c_2 - c_1) + g * (c_1^2 - c_2^2), w \rangle_{L^2(\Omega)} = \langle G_{D_l}, w \rangle_{L^2(\Omega)} . \tag{3.10}$$

Optimisation. We can be tempted to identify the gradient as $-\kappa G_S(v) - G_D(v)$ with $G_D = G_{D_g}$ or G_{D_l}. Unfortunately, G_S contains non local terms, this means that constrains on the support of w must be enforced. When v is a characteristic function and shape derivatives [5] are used, these variations are distributions supported along ∂A and have the form $\mathbf{v}\delta_{\partial A}$ or its normal component along ∂A, $(\mathbf{v} \cdot \mathbf{n}_{\partial A})\mathbf{n}_{\partial A}\delta_{\partial A}$ with $\mathbf{v}$ a vector field defined at least in a neighbourhood of ∂A. Here, instead of imposing a special form to our w we use a simple narrow band approach: we restrict their support to $(\text{supp}v)_\varepsilon = \text{supp}v \oplus B(0,\varepsilon)$, the dilation of $\text{supp}v$ by a ball of radius ε. In practice ε is one or two pixels (voxels) and we implement the projected gradient descent step. The algorithm is described in Algorithm 1. Note that because of the narrow band implementation, the initial estimate (or its dilation) should overlap the object we want to segment. This may limit this type of approach, however, for the application that we have in mind, rat brain segmentation, we usually start with a very large overlap due to the MR measurement setting.

Algorithm 1. Sketch of the full algorithm.

Input: $A_1, \ldots, A_n$ training shapes, $\rho > 0$ the deviation parameter, $\kappa > 0$ the data vs. shape trade-off parameter, $\delta t > 0$ the descent step parameter, $\eta > 0$ the convergence threshold, $\zeta \in (0,1)$ the binarisation threshold, v^0 initial shape function.

Output: Segmentation function / object closed to the training shapes.

    repeat

$$v^{n+1} = \mathcal{P}_{\mathcal{F}}\left(v^n - \delta t\left((G_D(v^n) + \kappa G_S(v^n))|_{(\text{supp }v)_\varepsilon}\right)\right)$$

    until $\frac{\|v^{n+1} - v^n\|^2}{|\Omega|} < \eta$

    return $\mathbf{v} := (v^n > \zeta)$.

---

4 Experimental Validation

We target MR scans of rats, but there is no ground truth available for these volumes. Therefore we first present two synthetic experiments where only one shape is used, then two experiments on an MR scan where we have segmented five rat brains by a combination of classical variational methods and postprocessing so as to extract and learn brain shapes. We use them to build a shape regulariser and proceed to segment an unseen MR volume. To evaluate segmentation when we have ground truth, we have used the Dice-Sørensen Coefficient (DSC) score: $DSC(X,Y) = \frac{2|X \cap Y|}{|X| + |Y|}$ where $|\cdot|$ is the cardinality of a set. The binarisation threshold in Algorithm 1 has been fixed to 0.5. There are classical techniques for computing a good value for ρ in (2.4), though, in this work we fixed it to $\rho = 1.5$ and 1 for the experimental and synthetic respectively, as they always provided satisfactory results.

A Simple 2D Experiment. In this experiment, as a simple validation of the framework, we are given one training shape and an image obtained by a similarity transform, with added noise and some small level of occlusion. The data term used is the two-means term (2.5). We illustrate our prior, input data, initial estimate and final segmentation in Fig. 2. The values for κ, η and δt in Algorithm 1 are respectively 5.0, 3×10^{-5} and 0.1. The reported DSC is 98.02%. We observe a spurious small contour in (d), which would have disappeared with a slightly lower threshold. Scaling and rotations necessitate interpolation, which is also a source of inaccuracies. This is likewise present in the other experiments.

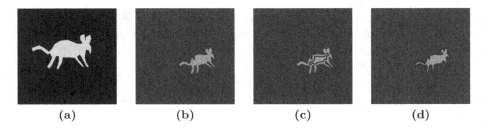

(a)   (b)   (c)   (d)

Fig. 2. 2D experiment. (a): Shape prior. (b): Data to be segmented. (c): Initial shape estimate. (d): Final segmentation.

A 3D Experiment. In this experiment, we use a simple training shape, a binarised 3D rendering of the letter "F", while we generated a more complex 3D image and added noise and complex structures to it. Here too, the data term is the two-means term (2.5). This is illustrated in Fig. 3. The values for κ, η and δt in Algorithm 1 are respectively 10.0, 1.5×10^{-6} and 0.1. The reported DSC score is 90.7%, which may seem a bit low. Among other things, this is due to the relatively important deviation between the training shape (very flat) and the somewhat crenellated ground truth, as can be observed from Fig. 3.

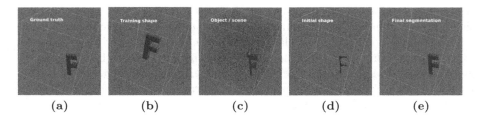

(a)   (b)   (c)   (d)   (e)

Fig. 3. Synthetic 3D example. The ground truth (a) is a crenellated 3D 'F' shape. The training shape (b) is a simple 3D 'F' shape with flat faces. (c) is a noisy and cluttered 3D scene (at one of the corners and side of the letter), (d) is the initial shape for segmentation and (e) the final shape.

Rat Brain Segmentation. First, we start by a 2D experiment, mainly because it makes visualisation easier. Medically annotated data is not available for the specific dataset series we worked on, and to our knowledge, no automatic segmentation exists for MR brain scans of rodents. Validation is therefore based on qualitative assessment of the achieved segmentation. Without annotated data, we have used the work in [9] to obtain the brain segmentations.

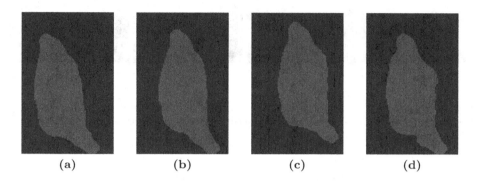

(a)   (b)   (c)   (d)

Fig. 4. Training shape masks used in 2D segmentation of rotated rat cranium scan.

Figure 5 shows the effect of the shape regularisation in 2D. The rotated MR slice in Fig. 5b is after the bias field correction procedure of [13]. It is clear that the bias field has not been fully eliminated. We therefore use the local means data term (2.6). In the next subfigure we illustrate the different terms in the objective function. Evolving only the regulariser term should result in a sort of mean training shape. This is illustrated in Fig. 5a. We use data in the form of a rotated slice from a MR brain, which is represented in Fig. 5b. The red square contour inside represent the initial segmentation guess. Optimising using only the local two-means data term results in an over-segmentation where part of the skull and other anatomical elements are segmented: this can be seen in Fig. 5c. Finally, Fig. 5d shows the obtained segmentation with our approach.

We now illustrate the segmentation in 3D. Using the pipeline of variational and image cleaning methods of [9], we segmented five rat brains from MR scans. With careful postprocessing, three of the segmentations were refined and added to our set of training shapes, providing ten final training shapes. No initial alignment of the training data was performed. As initial guess, we use one of the training shapes, slightly eroded, with its original pose and scale parameters. The physical setup indeed guarantees that pose and scale parameters for the different scans should vary moderately from scan to scan. Two-dimensional slices for 4 of the original shapes have already been shown in Fig. 4. We tested our approach on an MR volume which could not be segmented by the approach of [9] (it would either incorporate part of the skull or severely under-segment the brain). The values for κ, η and δt in Algorithm 1 are respectively 2.0, 1×10^{-5} and 2.0. Moreover, the two-local-means term (2.6) was used with a very simple $3 \times 3 \times 3$

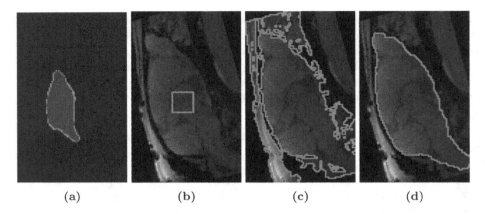

Fig. 5. (a): "Mean" training shape computed by minimisation of regulariser. (b): Rotated slice. (c): Local two-means clustering. (d): Segmentation with both terms. (Color figure online)

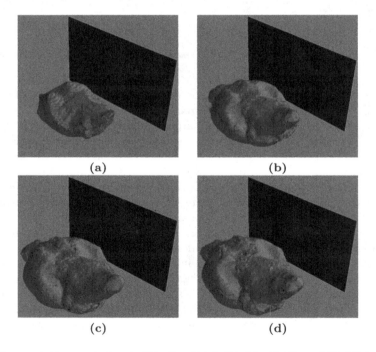

Fig. 6. Segmentation evolution in 3D after 0 (a), 24 (b), 49 (c), and 75 (d) iterations.

local mean kernel g. Some evolution steps are displayed in Fig. 6. Small angular variations can be seen by observing the relative position of the shape and the reference black plane. In Sect. 4 we visualise the results for different viewing angles (Fig. 7).

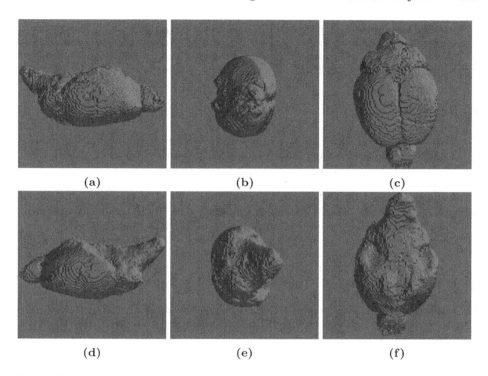

(a)   (b)   (c)

(d)   (e)   (f)

Fig. 7. Final segmentation results of our proposed model for different viewing angles. (a) and (d) shows the x and $-x$ direction respectively, y and $-y$ directions in (b) and (e) and z and $-z$ in (c) and (f).

5   Conclusion

In this paper, we have proposed a 2D and 3D approach for the segmentation of objects whose distribution of shapes can be approximated via training shapes. This allowed us to construct a regularisation term, within a variational frame-work, which is invariant under similarity transforms. We coupled the regulariser with a classic or robust data attachment term, to obtain segmentation objective functions and computed their gradients analytically. We have demonstrated this approach in several scenarios, particularly on MRI scans of rat crania. We are working on a more objective evaluation of our approach for rat brains, which is somewhat complicated in the absence of ground truth or accepted gold standard. We have used projected gradient descent on simple shape representations via relaxed characteristic functions. More efficient optimisation methods could be considered, and other types of shape representation could be used efficiently, especially for shape distributions where topology does not change.

Acknowledgements. J. Hansen and F. Lauze thank S. Darkner[1], K. N. Mortensen[2], S. Sanggaard (see footnote 2), H. Benveniste[3], and M. Nedergaard[4] (see footnote 2) for the data. F. Lauze acknowledges the support of the Center for Stochastic Geometry and Advanced Bioimaging (CSGB).

References

1. Chan, T., Zhu, W.: Level set based shape prior segmentation. In: IEEE Computer Society Conference on Computer Vision and Pattern Recognition, CVPR 2005, vol. 2, pp. 1164–1170. IEEE (2005)
2. Cootes, T., Taylor, C., Cooper, D., Graham, J.: Active shape models-their training and application. CVIU **61**(1), 38–59 (1995)
3. Cremers, D., Kohlberger, T., Schnörr, C.: Shape statistics in kernel space for variational image segmentation. Pattern Recogn. **36**(9), 1929–1943 (2003)
4. Cremers, D., Tischhäuser, F., Weickert, J., Schnörr, C.: Diffusion snakes: introducing statistical shape knowledge into the mumford-shah functional. Int. J. Comput. Vis. **50**(3), 295–313 (2002)
5. Delfour, M., Zolésio, J.-P.: Shapes and Geometries. Advances in Design and Control, Siam (2001)
6. Foulonneau, A., Charbonnier, P., Heitz, F.: Affine-invariant geometric shape priors for region-based active contours. IEEE Trans. Pattern Anal. Mach. Intell. **28**(8), 1352–1357 (2006)
7. Fussenegger, M., Deriche, R., Pinz, A.: A multiphase level set based segmentation framework with pose invariant shape priors. In: Narayanan, P.J., Nayar, S.K., Shum, H.-Y. (eds.) ACCV 2006. LNCS, vol. 3852, pp. 395–404. Springer, Heidelberg (2006). https://doi.org/10.1007/11612704_40
8. Hansen, J.D.K., Lauze, F.: Local mean multiphase segmentation with HMMF models. In: Lauze, F., Dong, Y., Dahl, A.B. (eds.) SSVM 2017. LNCS, vol. 10302, pp. 396–407. Springer, Cham (2017). https://doi.org/10.1007/978-3-319-58771-4_32
9. Hansen, J.D.K., et al.: Brain extraction and segmentation framework for bias field rich cranial MRI scans of rats. In: Annual meeting ISMRM, Paris, France, June 2018, p. 3249 (2018). Book of Abstracts
10. Kendall, D.: Shape manifolds, procrustean metrics and complex projective spaces. Bull. London Math. Soc. **16**, 81–121 (1984)
11. Magnus, J.: On differentiating eigenvalues and eigenvectors. Econometric Theor. **1**, 179–191 (1985)
12. Mezghich, M.A., Slim, M., Ghorbel, F.: Invariant shape prior knowledge for an edge-based active contours invariant shape prior for active contours. In: 2014 International Conference on Computer Vision Theory and Applications (VISAPP) (2014), pp. 454–461. IEEE (2014)
13. Tustison, N.J., et al.: N4ITK: improved N3 bias correction. IEEE Trans. Med. Imaging **29**(6), 1310–1320 (2010)
14. Wang, J., Yeung, S.-K., Chan, K.L.: Matching-constrained active contours with affine-invariant shape prior. Comput. Vis. Image Underst. **132**, 39–55 (2015)

[1] DIKU, University of Copenhagen.
[2] Center for Translational Neuromedicine, University of Copenhagen.
[3] Anesthesiology, Yale School of Medicine, Yale University.
[4] Center for Translational Neuromedicine, University of Rochester.

Lattice Metric Space Application to Grain Defect Detection

Yuchen He[(✉)] and Sung Ha Kang

School of Mathematics, Georgia Institute of Technology, Atlanta, GA, USA
yhe306@gatech.edu, kang@math.gatech.edu

Abstract. We propose a new model for grain defect detection based on the theory of lattice metric space [7]. The lattice metric space $(\mathcal{L}, d_{\mathcal{L}})$ shows outstanding advantages in representing lattices. Utilizing this advantage, we propose a new algorithm, Lattice clustering algorithm (LCA). After over-segmentation using regularized k-means, the merging stage is built upon the lattice equivalence relation. Since LCA is built upon $(\mathcal{L}, d_{\mathcal{L}})$, it is robust against missing particles, deficient hexagonal cells, and can handle non-hexagonal lattices without any modification. We present various numerical experiments to validate our method and investigate interesting properties.

1 Introduction

In crystalline materials, a grain is a homogeneous region that is composed of a single layer of crystal [11]. The presence of crystal defects such as particle dislocation, grain deformation, and grain boundary has unfavorable influences on macro-scale properties of the materials. To automatically detect these structures in atomic-scale 2D crystal images, many methods are developed in the literature. Variational model-based methods [2–4,8,16,22] extract and classify the grains by minimizing certain functional energy; wavelet type methods [5,14,19,21] measure the local properties of wave-like components to reveal crystal defects; and Voronoi type methods [12,13,20] detect the problematic particles by comparing the Voronoi cells with hexagonal polygons. Different from these methods, we approach the grain defect detection problem by clustering particles in an abstract metric space directly, where lattice representation is unique, and comparison between lattice patterns is systematic.

In this paper, we apply the lattice metric space $(\mathcal{L}, d_{\mathcal{L}})$ developed in [7] to grain defect detection problem, and explore its further properties. As a general framework to describe and compare arbitrary lattice patterns, the lattice metric space is advantageous. First, $\mathcal{L}$ identifies each grain using a pair of complex numbers, the descriptors $(\beta, \rho) \in \mathbb{C}^2$ which encapsulates full structural information about the lattice. Second, the lattice metric space $(\mathcal{L}, d_{\mathcal{L}})$ can detect lattice inconsistencies such as grain boundaries in non-hexagonal crystalline materials without any particular modification. Voronoi type methods are limited to hexagonal lattices, and other techniques are involved to analyze non-hexagonal

© Springer Nature Switzerland AG 2019
J. Lellmann et al. (Eds.): SSVM 2019, LNCS 11603, pp. 381–392, 2019.
https://doi.org/10.1007/978-3-030-22368-7_30

ones [12]. In 2D, the other types of lattices are oblique (e.g., orthoclase), square (e.g., halite), primitive rectangular (e.g., epsomite) and centered rectangular (e.g., hemimorphite). These classes are encoded in (β, ρ) and further refined by considering the metric structure on $\mathcal{L}$. This property makes our method robust against missing particles and deficient hexagonal cells. Third, the metric $d_{\mathcal{L}}$ provides a single-valued yet comprehensive measurement of the dissimilarities between any grains. It is robust and sensitive enough to expose the grain boundaries and to reveal particle dislocations as well as continuous deformations.

Based on the lattice metric space theory, we propose an efficient algorithm to extract grains and detect grain defects. The main idea of our method is to classify particles into visually distinct grains by checking their vicinities. For each particle, by comparing a 9-point stencil with nearby points, a lattice is extracted and mapped to $\mathcal{L}$. In $\mathcal{L}$, Riemannian center of mass is non-unique, which poses an obstacle for application of clustering methods requiring the notion of centers, such as k-means [1]. To tackle this challenge, we over-segment the lattices using regularized k-means [10], and merge them in $\mathcal{L}$ with an agglomerative hierarchical clustering method [1]. We define a function considering the sizes of clusters to assist the cut-off selection for the merging.

This paper is organized as follows. In Sect. 2, we review the lattice metric space theory, and present new properties. We describe our grain defect detection algorithm, LCA, in Sect. 3, which is followed by various numerical experiments in Sect. 4. We conclude this paper in Sect. 5.

2   Lattice Metric Space: Review and New Properties

We first review the most relevant definitions in the lattice metric space theory [7], then discuss its new properties of $(\mathcal{L}, d_{\mathcal{L}})$ in relation to grain defect detection.

2.1   Review on Lattice Metric Space

Any 2D lattice is denoted by $\Lambda(b_1, b_2)$, where the pair $(b_1, b_2) \in \mathbb{C}^2$ forms a positive minimal basis for the given lattice. This means that all the lattice points in $\Lambda(b_1, b_2)$ can be located by taking linear combinations of b_1 and b_2 using integer coefficients. Among a set of basis representations (b_1, b_2), the lattice comparison is difficult. To address the problem of infinite equivalent basis representations, the *descriptors* are introduced:

$$(\beta, \rho) \in \mathbb{C}^2 : \ \beta := b_1, \ \rho := b_2/b_1, \tag{1}$$

where β is referred to as a *scale descriptor*, and ρ is a *shape descriptor*. The notation $\Lambda\langle\beta, \rho\rangle$ denotes the same lattice $\Lambda(\beta, \beta\rho)$, but using descriptors. Examples of this correspondence are shown in Fig. 1. The domain for β is $\mathcal{K} := \mathbb{C} \setminus \{0\}$, and that for ρ is $\mathcal{P}$:

$$\mathcal{P} := \{z \in \mathbb{C} \mid |z| \geq 1, |\mathrm{Re}(z)| \leq \frac{1}{2}, \mathrm{Im}(z) > 0\} \subset \mathbb{C}, \tag{2}$$

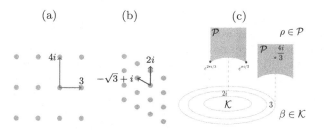

Fig. 1. [From (b_1, b_2) to (β, ρ)](a) Lattice $\Lambda(3, 4i)$ whose basis is colored red. (b) Lattice $\Lambda(2i, -\sqrt{3}+i)$ whose basis is colored blue. (c) The red dot represents the lattice in (a), and the blue dots represent the same lattice in (b). (Color figure online)

which is closely related to the fundamental regions of modular group actions. In [7], the authors show that equivalent positive minimal bases induces equivalence relations among descriptors using the modular group theory. Based on which, they define the lattice space $\mathcal{L}$ as follows:

Definition 1. *The lattice space $\mathcal{L}$ is a topological space constructed by:*

$$\mathcal{L} := \left(\mathcal{K}/\sim_1 \times \mathcal{P}/\sim_2\right)/\sim_3, \tag{3}$$

where the three equivalence relations are:

1. $\beta \sim_1 -\beta$, $\forall \beta \in \mathcal{K}$, *i.e.*, $\Lambda\langle\beta, \rho\rangle = \Lambda\langle-\beta, \rho\rangle$
2. $\rho \sim_2 \rho'$, $\forall \rho, \rho' \in \mathcal{P}$, *for* $Im(\rho) = Im(\rho')$ *and* $|Re(\rho)| = |Re(\rho')| = 1/2$, *i.e.*, $\Lambda\langle\beta, \rho\rangle = \Lambda\langle\beta, \rho'\rangle$,
3. $\langle[\beta]_1, [\rho]_2\rangle \sim_3 \langle[\beta\rho]_1, [-1/\rho]_2\rangle$, $\forall \beta \in \mathcal{K}$, $\forall \rho \in \mathcal{P}$, *for* $|\rho| = 1$, *i.e.*, $\Lambda\langle\beta, \rho\rangle = \Lambda\langle\beta\rho, -1/\rho\rangle$.

Considering all the equivalence relations in Definition 1, lattice metric $d_{\mathcal{L}}$ is deduced. Given any two descriptor pairs $(\beta, \rho), (\beta', \rho') \in \mathcal{K} \times \mathcal{P}$, a product metric D on $\mathcal{K}/\sim_1 \times \mathcal{P}/\sim_2$ is defined as:

$$D((\beta, \rho), (\beta', \rho')) = \sqrt{d_{\mathcal{K}}(\beta, \beta')^2 + d_{\mathcal{P}}(\rho, \rho')^2}. \tag{4}$$

The equivalence relations $\sim_1$ and $\sim_2$ (See Definition 1) are incorporated into $d_{\mathcal{K}}$ and $d_{\mathcal{P}}$ respectively as follows:

$$d_{\mathcal{K}}(\beta, \beta') = \min\{D_{\mathcal{K}}(\beta, \beta'), D_{\mathcal{K}}(-\beta, \beta')\}, \tag{5}$$

$$D_{\mathcal{K}}(\beta, \beta') = \sqrt{w(|\beta| - |\beta'|)^2 + (1-w)(\cos^{-1}\frac{Re(\beta\overline{\beta'})}{|\beta||\beta'|})^2}; \tag{6}$$

$$d_{\mathcal{P}}(\rho, \rho') = \min\{D_{\mathcal{P}}(\rho, \rho'), D_{\mathcal{P}}(\rho-1, \rho'), D_{\mathcal{P}}(\rho+1, \rho')\}, \tag{7}$$

$$D_{\mathcal{P}}(\rho, \rho') = 2\ln\frac{|\rho - \rho'| + |\rho - \overline{\rho'}|}{2\sqrt{Im(\rho)Im(\rho')}}. \tag{8}$$

Note that $w \in [0,1]$ is a parameter which adjusts the sensitivity between angle and length, and $D_{\mathcal{P}}$ is the well-known Poincaré metric [6] on upper-half plane restricted to $\mathcal{P}$. In practice, no extra care is needed if either $\rho + 1$ or $\rho - 1$ is outside of $\mathcal{P}$, since (8) remains well-defined, and the minimization in (7) eliminates these extra values. Upon D, the third equivalence relation $\sim_3$ is embodied by considering the shortest path in $(\mathcal{K}/\sim_1 \times, \mathcal{P}/\sim_2, D)$ connecting any pair of given lattices. In fact, there are only eight extra types of paths to compare $\{D_j\}_{j=1}^8$ besides the direct path in the product space. Explicit expressions of these paths are stated in [7]. The formal definition of $d_{\mathcal{L}}$ is:

$$d_{\mathcal{L}} = \min\{D, D_1, \cdots, D_8\}. \tag{9}$$

2.2   Properties in Relation to Grain Defect Detection

The representation $\Lambda\langle\beta,\rho\rangle$ provides a universal framework to characterize lattices. For example, $\Lambda\langle 10, i\rangle$ represents a cubic lattice; $\Lambda\langle 10, e^{i\pi/3}\rangle$ denotes a hexagonal lattice; and the notation $\Lambda\langle e^{-2\pi i/9}, e^{4\pi i/9}\rangle$ represents a centered rectangular lattice. This property allows our algorithm in Sect. 3 to be independent from the type of lattice.

A useful feature of $d_{\mathcal{L}}$ is that it can adjust sensitivity to orientation versus scale. Depending on the application, we can change the weight parameter w in (6) to emphasize the inconsistency in angles or lengths. When $w = 0$, only the lattice orientation is considered, which is similar to [4]. Figure 2(a) shows $d_{\mathcal{L}}(\Lambda\langle 10, e^{\pi i/3}\rangle, \Lambda\langle(10 + \Delta|\beta|)e^{i\pi/9}, e^{i\pi/3}\rangle)$ when $w = 0.5, 0.05$ and 0.005, where the scale variation $\Delta|\beta| \in [-5, 5]$ and orientation difference is fixed at $e^{i\pi/9}$.

Typical configuration of atoms present hexagonal patterns, and to $d_{\mathcal{L}}$, the misorientation between hexagonal lattices is more distinguishable. Figure 2(b) shows $d_{\mathcal{L}}(\Lambda\langle 10, e^{\pi i/3}\rangle, \Lambda\langle\beta,\rho\rangle)$ for a set of different β with unit ρ, whose arguments varies within $[\pi/3, 2\pi/3]$. Notice that $d_{\mathcal{L}}$ has the biggest differences when $\Delta\mathrm{Arg}\,\rho = 0$ and $\Delta\mathrm{Arg}\,\rho = 60°$ corresponding to the left and right boundary respectively, which produces hexagonal lattices. This shows that $d_{\mathcal{L}}$ is most sensitive to the misorientation between hexagonal lattices. Since rotating a hexagonal by 30° clockwise and counter-clockwise result in an identical lattice, we see the red curve ($\mathrm{Arg}\,\beta = 30°$) is mirror-symmetric. The two global minima of this red curve bring up an important property that the Riemannian center in $(\mathcal{L}, d_{\mathcal{L}})$ is not unique. As a consequence, we can not directly apply centroid-based analysis.

From a different perspective, in Fig. 2(c), we compute $d_{\mathcal{L}}(\Lambda\langle 10, \rho\rangle, \Lambda\langle\beta,\rho\rangle)$ for a set of different ρ with β satisfying $|\beta| = 10$ and $\mathrm{Arg}\,\beta$ varies within $[0, \pi]$. Since $\Lambda\langle 10, e^{\pi i/3}\rangle$, $\Lambda\langle 10e^{\pi i/3}, e^{\pi i/3}\rangle$ and $\Lambda\langle 10e^{2\pi i/3}, e^{\pi i/3}\rangle$ are equivalent (hexagonal) lattices, and $\Lambda\langle 10, e^{\pi i/2}\rangle$ and $\Lambda\langle 10e^{\pi i/2}, e^{\pi i/2}\rangle$ are equivalent (rectangular) lattices, their distances are 0 respectively. Figure 2(b) and (c) also show that lattice metric $d_{\mathcal{L}}$ is able to measure the differences between lattices of any Bravais lattice types. The comparison using $d_{\mathcal{L}}$ considers differences between equivalence classes of lattice representations, which is different from [4] where lattice differences are measured via intensities between image patches.

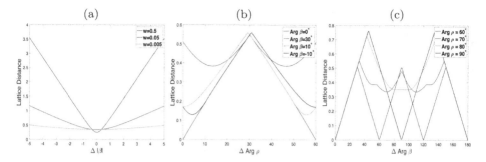

Fig. 2. [Properties of $d_\mathcal{L}$] (a) Effect of changing w. (b) Misorientation of hexagonal lattices is emphasized using $d_\mathcal{L}$, which corresponds to the left and right edges. (c) High symmetry of the hexagonal lattice is reflected by the symmetry of the blue curve. Lattices to be compared are not necessarily of the same type, and $d_\mathcal{L}$ considers the lattice equivalence relations.

3   Lattice Clustering Algorithm

We propose an efficient algorithm to capture structural defects in multigrain from a gray-scale image. Our approach is based on clustering, and particles with visually similar vicinity are grouped. The algorithm is summarized in Table 1.

In Step 1, each particle is located by identifying local maxima refined by fitting a narrow Gaussian. For each particle p_j, k-nearest ($k = 5$ in our method) particles are found, denoted as $q_1, q_2 \cdots, q_k$. For each pair of (q_s, q_t), $s \neq t$, we construct a 9-point stencil with Gaussian weight located at $p_j + a(q_s - p_j) + b(q_t - p_j)$, $a, b = 0, \pm 1$. We choose the pair with the highest response and compute its corresponding descriptors (β_j, ρ_j). Then p_j is assigned with the lattice $\Lambda\langle\beta_j, \rho_j\rangle$.

In Step 2, we first cluster the lattices $\mathcal{A} = \{\Lambda\langle\beta_j, \rho_j\rangle\}_{j=1}^N$ by clustering the descriptors $\{(\beta_j, \rho_j)\}_{j=1}^N$ in $\mathbb{R}^4$ equipped with the Euclidean norm. As discussed in Sect. 2, the quotient geometry of $\mathcal{L}$ is non-trivial, which complicates directly clustering the descriptors in $\mathcal{L}$. To overcome this difficulty, we over-segment the lattices using the regularized k-means [10] in $\mathbb{R}^4$ first. The reason is that if (β_j, ρ_j) and (β_k, ρ_k) are close in $\mathbb{R}^4$, then $\Lambda\langle\beta_j, \rho_j\rangle$ and $\Lambda\langle\beta_k, \rho_k\rangle$ are close in $\mathcal{L}$, but the inverse is not true. Consequently, there might be two clusters whose members are similar lattices. We denote the resulting clusters by $\{\mathcal{C}_l\}_{l=1}^K$ with centers $\{\Lambda\langle\overline{\beta}_l, \overline{\rho}_l\rangle\}_{l=1}^K$, where $(\overline{\beta}_l, \overline{\rho}_l)$ is the Euclidean center of pairs of descriptors in $\mathcal{C}_l$.

In Step 3, we merge the clusters $\{\mathcal{C}_l\}_{l=1}^K$ considering the equivalence relations among lattices. We compute the pair-wise lattice distances $d_\mathcal{L}$ among $\{\Lambda\langle\overline{\beta}_l, \overline{\rho}_l\rangle\}_{l=1}^K$ and employ the standard agglomerative hierarchical clustering method [18]. This method requires a cut-off $t > 0$, and cluster centers closer than t will be merged. To assist choosing the optimal threshold, we consider the following function:

Table 1. Lattice Clustering Algorithm

Lattice Clustering Algorithm (LCA algorithm)

Inputs:

1. U: given gray-scale image;
2. λ: the parameter in regularized k-means;
3. T: threshold for merging.

Step 1. Particle and local lattice identification. Each local maxima is refined by fitting a narrow Gaussian to find each particle p_j, $j = 1, 2, \cdots, N$. Among the k-nearest $(k = 5)$ neighbors of each particle p_j, two vectors which gives the best match of a 9-point stencil is picked for p_j. Let $\mathcal{A}$ be the collection of such vectors in $\mathbb{R}^4$.
Step 2. Apply the regularized k-means to $\mathcal{A}$, and obtain K clusters $\{\mathcal{C}_l\}_{l=1}^K$ with Euclidean centers $\{(\overline{\beta}_l, \overline{\rho}_l)\}_{l=1}^K$.
Step 3. Considering the equivalence relations among descriptors, merge $\{\mathcal{C}_l\}_{l=1}^K$ by clustering the lattices $\{\Lambda\langle\overline{\beta}_l, \overline{\rho}_l\rangle\}_{l=1}^K$ in $\mathcal{L}$ with hierarchical clustering using the threshold T.

$$g(t) = \max_{m=1,2,\cdots,s} \left\{ \frac{\sum_{j=J_{m-1}+1}^{J_m} |\mathcal{C}_j|}{\max_{j=J_{m-1}+1,\cdots,J_m}\{|\mathcal{C}_j|\}} \right\}, \tag{10}$$

which measures the energy of the current stage of clustering using cut-off t; here, without loss of generality, $J_0 = 0$, $\{\mathcal{C}_j\}_{j=1}^{J_1}$, $\{\mathcal{C}_j\}_{j=J_1+1}^{J_2}, \cdots$, and $\{\mathcal{C}_j\}_{j=J_{s-1}+1}^{J_s}$, represent the clusters of particles to be merged respectively when the cut-off is t; and $|\mathcal{C}_l|$ denotes the counting measure $l = 1, 2, \cdots, K$. In general, $g(t)$ is a staircase function of $t > 0$. In order for the merging to be *stable* and *substantial*, we pick optimal thresholds T in the intervals upon which the graph of g is flat and the previous discontinuity has high jump.

4   Numerical Experiments and Discussion

We apply LCA to images from the literature [2,4,9,15,17]. Throughout this paper, we fix the weighting parameter $w = 0.05$, and the suitable range for λ in regularized k-means is found to be $0.2 \sim 0.5$. The particles belonging to a common grain region are colored with identical color.

Lattice Representation: (b_1, b_2) versus (β, ρ). Compared to basis representation (b_1, b_2), descriptor representation (β, ρ) is stable. Arg b_2 ranges from 0 to 2π, yet Arg ρ takes value in $[\pi/3, 2\pi/3]$, which is more restrictive. In practice, grains in polycrystalline materials generally have similar lattice patterns, thus the shape descriptors ρ in the data are concentrated. The benefit of $\mathcal{L}$ is exemplified in Fig. 3. Fixing a particle, colored red, as the reference, we compare the lattice representations (b_1, b_2) and (β, ρ) to those at the reference. Darker color indicates closer in the chosen distance. If we compare (b_1, b_2) in Euclidean metric for $\mathbb{R}^4$, then (d) and (f) display the results for two distinct reference points which are nearly random. Using the same reference points, if we apply the lattice metric $d_{\mathcal{L}}$ on (β, ρ), (e) and (g) show the advantage of using descriptor representation compared to (d) and (f) respectively. The color distributions in (e) and (g) comply with our perception of homogeneity. Figure 3(e) and (g) also reveal grain defects and continuous deformation within homogeneous regions.

We note this experiment is done by comparing every particle to the reference (red point) without using LCA. Every lattice label is compared to that of the reference particle directly in $\mathcal{L}$, and the distance is represented by the color. This approach shows excellent results, yet the computation is very slow. In the following, we apply LCA, which is more computationally efficient.

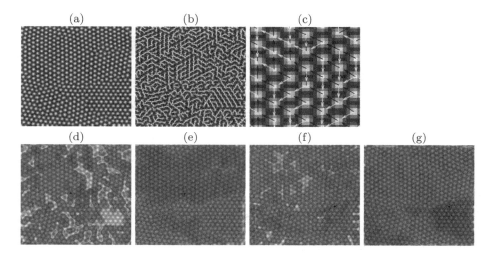

Fig. 3. [Direct classification using $d_{\mathcal{L}}$] (a) PFC image from [4] Fig. 4. (b) Lattice labels obtained in Step 1 of LCA. (c) zoomed-in partial region from (b). (d) and (f) show Euclidean distance function of (b_1, b_2) representation with respect to that of the red points. (e) and (g) show the lattice distance function $d_{\mathcal{L}}$ of (β, ρ) representations with respect to that of the red points, which are the same as those in (d) and (f) respectively. Linear interpolation is applied to fill the color in (d)–(g). (Color figure online)

General Example of LCA. There are two parameters in LCA: λ and T. λ is a penalty parameter in the regularized k-means, which implicitly controls the number of clusters K. In general, choosing λ is straightforward [10], and the optimal ones are found within wide intervals. Large λ penalizes clusters with small elements, hence fewer clusters with even sizes are obtained; small λ produces more clusters with even sizes. Applying LCA to the image in Fig. 3(a), we plot the function $g(t)$ in Fig. 4(a) when t ranges from 0.1 to 1. Notice that $g(t)$

is a staircase function of t, and the figure shows clear plateau within intervals, e.g. $[0.25, 0.35]$, $[0.35, 0.45]$, $[0.47, 0.7]$. One should avoid choosing the threshold T near the jump-discontinuities. Any perturbation near these points will produce substantially different results. The results with T from the flat regions $T = 0.4$, $T = 0.5$ and $T = 0.8$ present different levels of details, as shown in (b)–(d) in order. Compared to (c), (b) distinguishes minor disorientation such as those between red and green regions. While large scale boundaries remain noticeable in (c), only particles inconsistent with ambient patterns are accentuated in (d). Typical particle defects are not identified in region based methods, e.g., compare Fig. 3 with Fig. 5 of [4]. Otherwise, both methods agree on large scale grain boundaries. We also note that regions labeled with the same color have similar lattice pattern, and they can be separated in different locations, for example, the green regions in (b).

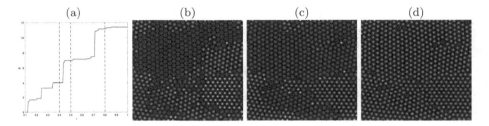

Fig. 4. Apply LCA to the image Fig. 3(a). Here (a) shows the curve $g(t)$. Results when (b) $T = 0.4$, (c) $T = 0.5$ and (d) $T = 0.8$ show the effect of T. (Color figure online)

Over-Segmentation Using k-means Versus Regularized k-means. In LCA, we choose the regularized k-means to over-segment. Using regularized k-means, the number of clusters K is implicitly controlled by the parameter λ, and K dynamically evolves depending on the data. This is desirable, especially when we do not know a suitable K a priori. Moreover, regularized k-means is stable. Every step during the clustering is deterministic, hence the results from regularized k-means are reproducible. To justify our choice, we experiment and compare the regularized k-means with the most common clustering method, k-means [1]. First, K has a direct impact on the results, yet it is not straightforward to choose. If we apply the k-means instead of the regularized k-means in LCA Step 1, fix $T = 0.5$, and rerun the algorithm for 100 times for different choices of the number of clusters K, then Fig. 5(a) shows that the number of identified grains clearly depends on K. Second, the random initialization in k-means causes difficulties in finding a stable threshold T. Even with the same parameters, e.g., $K = 30, T = 0.5$, it exhibits different results each time we run the algorithm, as shown in (b) and (c). Notice that the blue region in (b) is connected, yet the corresponding purple region in (c) is not. Also, the point defects in the upper right within the grains have different features in these results. In (d), using $K = 50$

and the same threshold $T = 0.5$ produces a result similar to Fig. 5(c), yet this is not reproducible. However, all the results in Fig. 4 are deterministic in the sense that, for any given image, the same combination of λ and T yields an identical result.

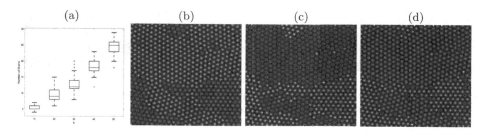

Fig. 5. [Instability of k-means] (a) Box-plot of the number of grains against parameter K in k-means. (b)–(d) use k-means with different initializations. (b) and (c) set $K = 30$ and $T = 0.5$ and (d) uses $K = 50$ and $T = 0.5$. (Color figure online)

Graph of g and Grain Features. The domain for the function g defined in (10) is closely related to the lattice metric $d_{\mathcal{L}}$, and the graph of g reveals geometrical features about the multigrain. In Fig. 6(a), we apply LCA, and g is plotted in (b). Observe that there are 3 major jumps in the graph of g, which exactly correspond to the grain regions in (a). The grain in the bottom of (c) is merged with the top one when $T = 0.5$ in the plateau after the first jump is changed to $T = 0.8$ after the second jump, as shown in (d). It is also clear that the height of each jump is proportionally related to the grain area. See Figs. 7 and 8 for images with two grain regions. Notice that the threshold T are chosen on the plateau of the function g respectively.

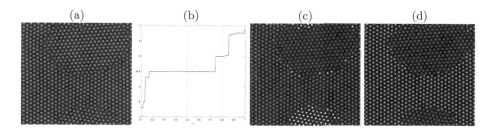

Fig. 6. (a) There are 3 grains: one on the top, one in the middle, and one in the bottom. Image from [2] Fig. 1. (b) The curve $g(t)$ with 3 major jump-discontinuities. (c) $T = 0.5$. (d) $T = 0.8$.

Behavior of LCA Near Grain Boundaries. In Figs. 6(c), 7(c) and 8(c), LCA assigns the boundary particles to either of the neighboring lattice patterns; in

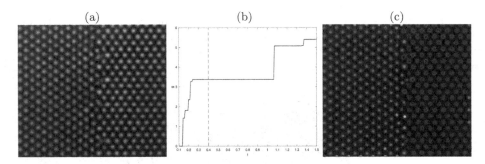

Fig. 7. (a) There are 2 grains with a regular boundary. Image adapted from [17] Fig. 6(a). (b) The curve $g(t)$ with 2 major jump-discontinuities. (c) Result with $T = 0.4$.

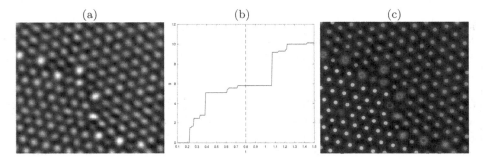

Fig. 8. (a) There are 2 grains presented and the grain boundary is irregular. Image adapted from [9] Fig. 6. (b) The curve $g(t)$ with 2 major jump-discontinuities and the jump is rough. (c) Result with $T = 0.8$.

other cases, LCA creates new classes. This differs from grain boundary identification using variational approaches, where spatial constraints such as length minimization are added. In the framework of lattice metric space, each particle is classified only based on k-nearest points, thus it is free to create new clusters.

Grain Boundary Detection in Non-hexagonal Crystalline Materials. As we discussed in Sect. 2, the generality of the lattice metric space theory allows analysis applicable to any types of Bravais lattices. Since LCA is established on this framework, we can directly employ LCA to images composed by arbitrary types of grains. Figure 9(a) is a HAADF-STEM image presenting a grain boundary in well-annealed, body-centered cubic (BCC) Fe. The graph of g in (b) implies choices for $T \gtrsim 0.6$, and the result with $T = 0.7$ is shown in (c). Due to fewer symmetries than the hexagonal lattices, the boundary between two cubic lattices has a relatively larger gap; hence it is challenging for methods that depend on hexagonal cells. LCA, with the flexibility supported by the lattice metric space, classifies the particles on the edges of the gap correctly.

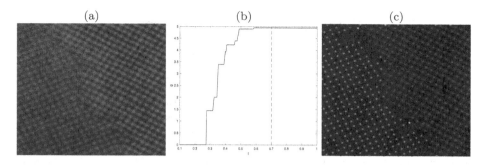

Fig. 9. (a) Grain boundary between non-hexagonal grains. Image adapted from [15] Fig. 3. (b) The curve $g(t)$. (c) Result with $T = 0.7$.

5   Conclusion

In this paper, we apply the lattice metric space theory to grain defect detection problems. Most methods such as [4,16] are free from particle position estimation, since they focus on homogeneous regions. We detect inconsistencies in the local patterns based on positions of k-nearest neighbors of each particle. More sophisticated techniques can be applied during this preprocessing stage. Theoretically, this approach is easily linked with the lattice metric space theory, which provides a uniform framework to classify particles in materials of any type of 2D Bravais lattice. We propose a lattice clustering algorithm, LCA. Different from variational methods, we emphasize the neighboring similarity of each particle rather than large scale homogeneity, thus LCA is superior at identifying various grain defects including grain boundaries. Since length minimization is not applicable, isolated particle deviating from the lattice point of the neighboring pattern will create a new cluster. For example, besides the grain boundaries, the dots in the left bottom region of Fig. 9 are recognized as new clusters, since the adjacent lattice patterns are not exactly regular.

References

1. Arthur, D., Vassilvitskii, S.: k-means++: the advantages of careful seeding. In: Proceedings of the Eighteenth Annual ACM-SIAM Symposium on Discrete Algorithms, pp. 1027–1035. Society for Industrial and Applied Mathematics (2007)
2. Berkels, B., Rätz, A., Rumpf, M., Voigt, A.: Identification of grain boundary contours at atomic scale. In: Sgallari, F., Murli, A., Paragios, N. (eds.) SSVM 2007. LNCS, vol. 4485, pp. 765–776. Springer, Heidelberg (2007). https://doi.org/10.1007/978-3-540-72823-8_66
3. Berkels, B., Rätz, A., Rumpf, M., Voigt, A.: Extracting grain boundaries and macroscopic deformations from images on atomic scale. J. Sci. Comput. **35**(1), 1–23 (2008)
4. Boerdgen, M., Berkels, B., Rumpf, M., Cremers, D.: Convex relaxation for grain segmentation at atomic scale. In: VMV, pp. 179–186 (2010)

5. Daubechies, I.: A nonlinear squeezing of the continuous wavelet transform based on auditory nerve models. In: Wavelets in Medicine and Biology, pp. 527–546 (1996)
6. Farkas, H.M., Kra, I.: Riemann Surfaces, pp. 9–31. Springer, New York (1992). https://doi.org/10.1007/978-1-4612-2034-3
7. He, Y., Kang, S.H.: Lattice identification and separation: theory and algorithm. arXiv:1901.02520 (2018)
8. Hirvonen, P., et al.: Grain extraction and microstructural analysis method for two-dimensional poly and quasicrystalline solids. arXiv:1806.00700 (2018)
9. Huang, P.Y., et al.: Grains and grain boundaries in single-layer graphene atomic patchwork quilts. Nature 469(7330), 389 (2011)
10. Kang, S.H., Sandberg, B., Yip, A.M.: A regularized k-means and multiphase scale segmentation. Inverse Prob. Imaging 5(2), 407–429 (2011)
11. Kittel, C., McEuen, P., McEuen, P.: Introduction to Solid State Physics, vol. 8. Wiley, New York (1996)
12. La Boissoniere, G.M., Choksi, R.: Atom based grain extraction and measurement of geometric properties. Model Simul. Mater. Sci. Eng. 26(3), 035001 (2018)
13. Lazar, E.A., Han, J., Srolovitz, D.J.: Topological framework for local structure analysis in condensed matter. Proc. Nat. Acad. Sci. 112(43), E5769–E5776 (2015)
14. Lu, J., Yang, H.: Phase-space sketching for crystal image analysis based on synchrosqueezed transforms. SIAM J. Imaging Sci. 11(3), 1954–1978 (2018)
15. Medlin, D., Hattar, K., Zimmerman, J., Abdeljawad, F., Foiles, S.: Defect character at grain boundary facet junctions: analysis of an asymmetric $\sigma = 5$ grain boundary in Fe. Acta Mater. 124, 383–396 (2017)
16. Mevenkamp, N., Berkels, B.: Variational multi-phase segmentation using high-dimensional local features. In: 2016 IEEE Winter Conference on Applications of Computer Vision (WACV), pp. 1–9. IEEE (2016)
17. Radetic, T., Lancon, F., Dahmen, U.: Chevron defect at the intersection of grain boundaries with free surfaces in Au. Phys. Rev. Lett. 89(8), 085502 (2002)
18. Rokach, L., Maimon, O.: Clustering methods. In: Maimon, O., Rokach, L. (eds.) Data Mining and Knowledge Discovery Handbook, pp. 321–352. Springer, Boston (2005). https://doi.org/10.1007/0-387-25465-X_15
19. Singer, H., Singer, I.: Analysis and visualization of multiply oriented lattice structures by a two-dimensional continuous wavelet transform. Phys. Rev. E 74(3), 031103 (2006)
20. Stukowski, A.: Structure identification methods for atomistic simulations of crystalline materials. Model. Simul. Mater. Sci. Eng. 20(4), 045021 (2012)
21. Yang, H., Lu, J., Ying, L.: Crystal image analysis using 2D synchrosqueezed transforms. Multiscale Model Simul. 13(4), 1542–1572 (2015)
22. Zosso, D., Dragomiretskiy, K., Bertozzi, A.L., Weiss, P.S.: Two-dimensional compact variational mode decomposition. J. Math. Imaging Vis. 58(2), 294–320 (2017)

Learning Adaptive Regularization for Image Labeling Using Geometric Assignment

Ruben Hühnerbein[1](✉), Fabrizio Savarino[1], Stefania Petra[2], and Christoph Schnörr[1]

[1] Image and Pattern Analysis Group, Heidelberg University, Heidelberg, Germany
`ruben.huehnerbein@iwr.uni-heidelberg.de`
[2] Mathematical Imaging Group, Heidelberg University, Heidelberg, Germany

Abstract. We introduce and study the *inverse problem of model parameter learning for image labeling*, based on the linear assignment flow. This flow parametrizes the assignment of labels to feature data on the assignment manifold through a linear ODE on the tangent space. We show that both common approaches are *equivalent*: either differentiating the continuous system and numerical integration of the state and the adjoint system, or discretizing the problem followed by constrained parameter optimization. Experiments demonstrate how a parameter prediction map based on kernel regression and optimal parameter values, enables the assignment flow to perform *adaptive regularization* that can be directly applied to novel data.

Keywords: Image labeling · Assignment manifold ·
Linear assignment flows · Parameter learning · Geometric integration ·
Adaptive regularization

1 Introduction

The *image labeling problem*, i.e. the problem to classify images depending on the spatial context, has been thoroughly investigated during the last two decades. While the evaluation (inference) of such models is well understood [10], *learning the parameters* of such models has remained elusive, in particular for models with higher connectivity of the underlying graph. Various sampling-based and other approximation methods exist (cf., e.g. [17] and references therein), but the *relation* between approximations of the *learning problem* on the one hand, and approximations of the subordinate *inference problem* on the other hand, is less well understood [14].

This paper is based on the *assignment flow for image labeling* introduced in [2] and on the *linear* assignment flow introduced and studied in [16]. These flows are induced by dynamical systems that evolve on an elementary statistical manifold, the so-called *assignment manifold*. Regarding the *inference task*, it has been shown recently that the assignment flow can evaluate discrete graphical models [3].

© Springer Nature Switzerland AG 2019
J. Lellmann et al. (Eds.): SSVM 2019, LNCS 11603, pp. 393–405, 2019.
https://doi.org/10.1007/978-3-030-22368-7_31

Contribution. In this paper, we take a first step towards *learning the model parameters* using the linear assignment flow. These parameters control local geometric averaging as part of the vector field which drives the flow. The problem to *learn* these parameters was raised in [2, Section 5 and Fig. 14]. See Fig. 1 below for an illustration. Our contribution can be characterized as follows.

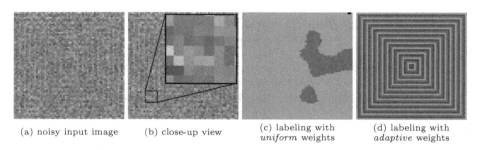

| (a) noisy input image | (b) close-up view | (c) labeling with *uniform* weights | (d) labeling with *adaptive* weights |

Fig. 1. Running the linear assignment flow with *uniform* weights for geometric averaging corresponds – roughly speaking – to labeling with a graphical model that involves a Potts prior (uniform label distances) for regularization. As it is well known, this does not work, e.g., for fine spatial structures contaminated with noise (cf. panels (a), (b) and the result (c)). This paper presents a *general* method for *learning how to regularize image labeling* using linear assignment flows, by Riemannian gradient flows on the manifold of regularization parameters, so as to recognize 'familiar' image structure (panel (d)).

Exact Inference. The inference problem is solved *exactly* during learning, unlike in work based on discrete graphical models, as discussed above. As a consequence, the 'predictive power' of learned parameters can be expected to be larger.

Reproducible Results. Our results are based on a simple algorithm for numerical geometric integration and hence are *reproducible*. This contrasts with current research on deep networks and complex software tools for training, that are more powerful in applications but are also less well understood [13].

Networks as Dynamical Systems. Our work ties in with research on networks from a *dynamical systems* point of view, that emanated from [9] in computer science and has also been promoted recently in mathematics [4]. The recent work [7], for example, studied stability issues of discrete-time network dynamics using techniques of numerical ODE integration. The authors adopted the *discretize-then-differentiate* viewpoint on the parameter estimation problem and suggested symplectic numerical integration in order to achieve better stability.

Commuting Modelling Diagram. Our work contrasts in that inference is always *exact* during learning, unlike [7] where learning is based on *approximate* inference. Furthermore, symplectic numerical integration is a *consequence*, in our case, of resolving the dilemma of *what path* to choose in the 'modelling diagram' of Fig. 2: the fact that both paths *commute* qualifies our approach as a proper (though rudimentary) method of *optimal control* (cf. [12]).

Organization of the Paper. Section 2 summarizes the assignment flow, the linear assignment flow and related concepts. Section 3 details our approach: both paths of the diagram of Fig. 2 are worked out and shown to be *equivalent*. Experiments in Sect. 4 demonstrate that a basic kernel-based regression estimator based on optimal parameter values and corresponding features makes the assignment flow *adaptive* and directly applicable to *novel* data.

$$\frac{d}{d\Omega} E\big(V(T,\Omega)\big)$$
$$\text{s.t. } \dot{V} = f(V,\Omega)$$
$$(1)$$

$$E\big(V(T,\Omega)\big) \text{ s.t. } \dot{V} = f(V,\Omega) \xrightarrow{\text{differentiate}} \textbf{adjoint system}$$

$$\text{discretize}\Big\downarrow \qquad\qquad\qquad\qquad \Big\downarrow \text{discretize}$$

$$\textbf{nonlinear program} \xrightarrow[\text{differentiate}]{} \textbf{sensitivity}$$

Fig. 2. Illustration of the methodological part of this paper. LEFT: The *sensitivity* of an objective function E with respect to parameters Ω to be estimated, defines a gradient flow on the parameter manifold for adapting initial parameter values. The parameter dependency of E is implicitly given through the linear assignment flow $V(t)$ at the terminal point of time $t = T$, whose state equation depends on Ω. RIGHT: Our approach satisfies the commuting diagram, i.e. *identical* results are obtained either if the continuous problem is differentiated first and than discretized (blue path), or the other way around (violet path). (Color figure online)

2   Preliminaries

This section summarizes the *assignment flow* and its approximation, the *linear assignment flow*, introduced in [2] and [16], respectively. The latter provides the basis for our approach to parameter estimation developed in Sect. 3.

2.1   Assignment Flow

Let $G = (I, E)$ be a given undirected graph with vertices $i \in I$ indexing data $\mathcal{F}_I = \{f_i : i \in I\} \subset \mathcal{F}$ given in a metric space $(\mathcal{F}, d)$. The edge set E specifies neighborhoods $\mathcal{N}_i = \{k \in I : ik = ki \in E\}$ for every pixel $i \in I$ along with positive weight vectors $w_i \in \text{rint}\,\Delta_{|\mathcal{N}_i|}$, where $\text{rint}\,\Delta_n = \Delta_n \cap \mathbb{R}^n_{++}$ denotes the relative interior of the probability simplex Δ_n.

Along with $\mathcal{F}_I$, *prototypical data (labels)* $\mathcal{L}_J = \{l_j \in \mathcal{F} : j \in J\}$ are given that represent classes $j = 1, \ldots, |J|$. *Supervised image labeling* denotes the task to assign precisely one prototype l_j to each datum f_i in a spatially coherent way. These assignments are represented at each pixel i by probability vectors

$$W_i \in \mathcal{S} := (\text{rint}\,\Delta_{|J|}, g_{FR}), \quad i \in I \tag{2}$$

on the relative interior of the simplex $\Delta_{|J|}$, that together with the Fisher-Rao metric g_{FR} becomes a Riemannian manifold denoted by $\mathcal{S}$. Collecting all assignment vectors into a strictly positive, row-stochastic matrix

$$W = (W_1, \ldots, W_{|I|})^\top \in \mathcal{W} = \mathcal{S} \times \cdots \times \mathcal{S} \subset \mathbb{R}^{|I| \times |J|} \tag{3}$$

defines a point on the *assignment manifold* $\mathcal{W}$. Image labeling is accomplished by geometrically integrating the *assignment flow* (the r.h.s. is defined below)

$$\dot{W} = \Pi_W\big(S(W)\big), \qquad W(0) = \mathbb{1}_{\mathcal{W}} := \frac{1}{|J|}\mathbb{1}_{|I|}\mathbb{1}_{|J|}^\top \qquad \text{(barycenter)}, \qquad (4)$$

that evolves from the barycenter $W(0)$ towards *pure* assignment vectors, i.e. each vector W_i approaches the ε-neighborhood of some unit vector at some vertex of $\mathcal{S}$ and hence a *labeling* after trivial rounding.

In order to explain the rationale behind (4), we need the following maps based on the affine e-connection of information geometry [1] in place of the Levi-Civita connection on the tangent bundle of the manifolds $\mathcal{S}$ and $\mathcal{W}$: With tangent space $T_0 = T_p\mathcal{S}$ independent of the base point $p \in \mathcal{S}$, we define

$$\mathbb{R}^{|J|} \ni z \mapsto \Pi_p(z) = \big(\operatorname{Diag}(p) - pp^\top\big)z \in T_0, \qquad (5a)$$

$$\mathcal{S} \times T_0 \ni (p,v) \mapsto \operatorname{Exp}_p(v) = \frac{e^{\frac{v}{p}}}{\langle p, e^{\frac{v}{p}}\rangle}p \in \mathcal{S}, \qquad (5b)$$

$$\mathcal{S} \times \mathcal{S} \ni (p,q) \mapsto \operatorname{Exp}_p^{-1}(q) = \Pi_p \log\frac{q}{p} \in T_0, \qquad (5c)$$

$$\mathcal{S} \times \mathbb{R}^{|J|} \ni (p,z) \mapsto \exp_p(z) = \operatorname{Exp}_p \circ \Pi_p(z) = \frac{pe^z}{\langle p, e^z\rangle} \in \mathcal{S}, \qquad (5d)$$

where multiplication, subdivision and the exponential function $e^{(\cdot)}$ apply *componentwise* to strictly positive vectors in $\mathcal{S}$. Corresponding maps $\Pi_W, \operatorname{Exp}_W, \exp_W$ in connection with the product manifold (3) are defined analogously, as is the tangent space $\mathcal{T}_0 = T_0 \times \cdots \times T_0$.

The vector field defining the assignment flow on the right-hand side of (4) is defined as follows. Given a metric d, data $\mathcal{F}_I$ and labels $\mathcal{L}_J$, distance vectors $D_i = (d(f_i, l_1), \ldots, d(f_i, l_{|J|}))^\top$ are defined at each pixel $i \in I$ and mapped to the assignment manifold by

$$L(W) = \exp_W(-\tfrac{1}{\rho}D) \in \mathcal{W}, \quad L_i(W_i) = \exp_{W_i}(-\tfrac{1}{\rho}D_i) = \frac{W_i e^{-\frac{1}{\rho}D_i}}{\langle W_i, e^{-\frac{1}{\rho}D_i}\rangle}, \qquad (6)$$

where $\rho > 0$ is a user parameter for normalizing the scale of the data. These *likelihood vectors* represent 'data terms' in conventional variational approaches, and they are *spatially regularized* in a way conforming to the geometry of $\mathcal{S}$, to obtain

$$S(W) = \mathcal{G}^\omega(L(W)) \in \mathcal{W}, \quad \mathcal{G}_i^\omega(W) := \operatorname{Exp}_{W_i}\Big(\sum_{k\in\mathcal{N}_i} w_{ik}\operatorname{Exp}_{W_i}^{-1}(W_k)\Big). \qquad (7)$$

Note that (7) is *parametrized* by the 'weight patches' $(w_{ik})_{k\in\mathcal{N}_i}$, $i \in I$. **Learning these parameters from data is the subject of this paper.**

The assignment flow (4) now simply *approximates* the *Riemannian gradient ascent flow* $\dot{W} = \nabla_{\mathcal{W}}J(W)$ with respect to the correlation functional $J(W)$,

$$\nabla_{\mathcal{W}}J(W) = \Pi_W(\nabla J(W)), \qquad J(W) = \langle W, S(W)\rangle, \qquad (8)$$

based on the approximation of the Euclidean gradient $\nabla J(W) \approx S(W)$, which is justified by the slow dynamics of $S(W(t))$ due to averaging (7), relative to the fast dynamics of $W(t)$.

2.2 Linear Assignment Flow

The *linear assignment flow*, introduced by [16], approximates the mapping (7) as part of the assignment flow (4) by

$$\dot{W} = \Pi_W\Big(S(W_0) + dS_{W_0}\Pi_{W_0}\log\frac{W}{W_0}\Big), \quad W(0) = W_0 = \mathbb{1}_{\mathcal{W}} \in \mathcal{W}. \quad (9)$$

This flow is still *nonlinear* but admits the following parametrization [16, Prop. 4.2]

$$W(t) = \mathrm{Exp}_{W_0}\big(V(t)\big), \quad \dot{V} = \Pi_{W_0}\big(S(W_0) + dS_{W_0}V\big), \quad V(0) = 0, \quad (10)$$

where the latter ODE is *linear* and defined on the tangent space $\mathcal{T}_0$. Assuming that V results from stacking row-wise the tangent vectors V_i for each pixel $i \in I$, the Jacobian dS_{W_0} is given by the block matrix

$$dS_{W_0} = \big(A_{ik}(W_0)\big)_{i,k\in I}, \quad A_{ik}(W_0)(V_k) = \begin{cases} w_{ik}\Pi_{S_i(W_0)}\big(\frac{V_k}{W_{0k}}\big), & k \in \mathcal{N}_i, \\ 0, & k \notin \mathcal{N}_i. \end{cases} \quad (11)$$

It is the *linearity* of (10) with respect to both the tangent vector V and the parameters w_{ik} (see (11)), that makes this approach attractive for parameter estimation.

3 Learning Adaptive Regularization Parameters

This section describes our contribution as illustrated by Fig. 2: an objective function for determining optimal weights – called *parameters* – that steer the linear assignment flow towards given ground-truth labelings (Sect. 3.1); a *continuous* approach to parameter estimation based on the adjoint dynamical system (Sect. 3.2), which is the method of choice when estimating *many* parameters [5]; alternatively, a *discrete* approach based on discretizing first the parameter estimation problem followed by nonlinear programming (Sect. 3.3); showing that either way yields the *same* result (Sect. 3.4); specifying the resulting algorithm (Sect. 3.5); finally, closing the loop by learning a simple predictor function that maps features extracted from *novel* data to proper weights (Sect. 3.6).

3.1 Parameter Estimation by Trajectory Optimization

We consider the following constrained optimization problem

$$\min_{\Omega \in \mathcal{P}} \quad E\big(V(T)\big) \tag{12a}$$

$$\text{s.t.} \quad \dot{V}(t) = f(V(t), \Omega), \quad t \in [0, T], \quad V(0) = 0_{|I|\times|J|}. \tag{12b}$$

where (12b) is given by (10) and the remaining symbols are defined as follows.

Ω parameters $(w_{ik})_{k \in \mathcal{N}_i}$, $i \in I$ of (11) to be estimated;
$\mathcal{P}$ manifold of parameters $\Omega \in \mathcal{P} := \text{rint}\, \Delta_{|\mathcal{N}|}^{|I|} = \text{rint}(\Delta_{|\mathcal{N}|} \times \cdots \times \Delta_{|\mathcal{N}|})$;
$|\mathcal{N}|$ equal size of each local neighborhood $|\mathcal{N}| = |\mathcal{N}_i|$ of pixel $i \in I$;
$V(T)$ tangent vectors solving (12b) at termination time $V(t = T)$;
$\quad\quad V(T) = V(T, \Omega)$ depends on Ω through (12b).

A concrete example for an objective (12a) is

$$E(V(T)) = D_{\text{KL}}\left(W^*, \exp_{\mathbb{1}_{\mathcal{W}}}(V(T))\right), \tag{13}$$

which evaluates the Kullback-Leibler distance of the labeling induced by $V(T)$ from a given ground-truth labeling $W^* \in \mathcal{W}$. In words, our objective is to estimate parameters Ω that control the linear assignment flow (9), by minimizing E. Since the dependency $\Omega \mapsto E$ is only *implicitly* given through (12b), the major task – illustrated by Fig. 2 – is to determine the *sensitivity* $\frac{d}{d\Omega}E(V(T))$, which in turn is used to adapt the parameters. Next, we detail both paths of Fig. 2 for evaluating Eq. (1).

3.2   Parameter Estimation: Continuous Approach

The following theorem makes precise the upper path to the right of Fig. 2.

Theorem 1 (parameter sensitivity: continuous case). *Let the objective E be defined by (12). Then*

$$\frac{dE}{d\Omega} = \int_0^T \left(\frac{\partial f}{\partial \Omega}\right)^\top \lambda(t) dt, \tag{14a}$$

where $\lambda(t)$ satisfies the adjoint differential equation

$$\dot{\lambda}(t) = -\left(\frac{\partial f}{\partial V}\right)^\top \lambda(t), \qquad \lambda(T) = \frac{\partial E}{\partial V}(V(T)), \tag{14b}$$

which has to be solved backwards in time.

Proof. Setting up the Lagrangian

$$L(V, \Omega) = E(V)\big|_{t=T} - \int_0^T \langle \lambda, F(\dot{V}, V, \Omega) \rangle dt \tag{15}$$

with multiplier $\lambda(t)$ and $F(\dot{V}, V, \Omega) := \dot{V}(t) - f(V(t), \Omega) \equiv 0$, we get

$$\frac{dE}{d\Omega} = \frac{dL}{d\Omega} = \left(\frac{\partial V}{\partial \Omega}\right)^\top \frac{\partial E}{\partial V}\bigg|_{t=T} - \int_0^T \left(\frac{\partial F}{\partial \dot{V}}\frac{\partial \dot{V}}{\partial \Omega} + \frac{\partial F}{\partial V}\frac{\partial V}{\partial \Omega} + \frac{\partial F}{\partial \Omega}\right)^\top \lambda\, dt \tag{16}$$

where integration applies componentwise. Using $\frac{\partial F}{\partial \dot{V}} = I$, we partially integrate the first term under the integral,

$$\int_0^T \left(\frac{\partial \dot{V}}{\partial \Omega}\right)^\top \lambda dt = \left(\frac{\partial V}{\partial \Omega}\right)^\top \lambda \bigg|_{t=0}^T - \int_0^T \left(\frac{\partial V}{\partial \Omega}\right)^\top \dot{\lambda}\, dt, \tag{17}$$

to obtain with $\frac{\partial F}{\partial \Omega} = -\frac{\partial f}{\partial \Omega}$, $\frac{\partial F}{\partial V} = -\frac{\partial f}{\partial V}$

$$\frac{dE}{d\Omega} = \left(\frac{\partial V}{\partial \Omega}\right)^{\top} \frac{\partial E}{\partial V}\bigg|_{t=T} - \left(\frac{\partial V}{\partial \Omega}\right)^{\top} \lambda\bigg|_{t=0}^{T} + \int_0^T \left(\frac{\partial V}{\partial \Omega}\right)^{\top} \dot{\lambda} \, dt \qquad (18a)$$

$$+ \int_0^T \left(\frac{\partial f}{\partial V}\frac{\partial V}{\partial \Omega} + \frac{\partial f}{\partial \Omega}\right)^{\top} \lambda \, dt. \qquad (18b)$$

Thus, with $\frac{\partial V}{\partial \Omega}(0) = 0$, factoring out the unknown Jacobian $\frac{\partial V}{\partial \Omega}$ and choosing $\lambda(t)$ such that the corresponding coefficient vanishes, we obtain

$$\frac{dE}{d\Omega} = \int_0^T \left(\frac{\partial f}{\partial \Omega}\right)^{\top} \lambda(t) \, dt, \quad \text{where } \lambda(t) \text{ solves} \qquad (19a)$$

$$\dot{\lambda}(t) = -\left(\frac{\partial f}{\partial V}\right)^{\top} \lambda(t), \qquad \lambda(T) = \frac{\partial E}{\partial V}(V(T)). \qquad (19b)$$

$\square$

Discretization. Due to lack of space, we only consider the simplest *geometric* scheme for numerically integrating the ODEs (12b) and (14b), namely the symplectic Euler method [8], and get

$$V_{k+1} = V_k + h_k f(V_k, \Omega), \qquad (20a)$$

$$\lambda_{k+1} = \lambda_k - h_k \big(\partial_V f(V_k, \Omega)\big)^{\top} \lambda_{k+1}, \qquad (20b)$$

with iteration index $k = 0, \ldots, N - 1$, step sizes h_k and $\sum_{k\in[N]} h_k = T$. This amounts to apply the *explicit* Euler scheme to the linear assignment flow and the *implicit* Euler scheme to the adjoint system. We can avoid the implicit step (20b) by reversing the order of integration $k = N - 1, \ldots, 0$, to obtain

$$\lambda_k = \lambda_{k+1} + h_k \big(\partial_V f(V_k, \Omega)\big)^{\top} \lambda_{k+1}, \qquad (21)$$

with initial condition $\lambda_N = \lambda(T)$ given by (14b). As a consequence, we first iterate (20a), store all iterates V_k and then iterate (21).

3.3 Parameter Estimation: Discrete Approach

In contrast to the previous section, we first *discretize* problem (12) and then *differentiate* the resulting *nonlinear program* (Fig. 2: violet path).

As mentioned in the previous section, we only consider the simplest integration scheme, the explicit Euler method. Using (20a), problem (12) becomes the *nonlinear optimization problem*

$$\min_{\Omega \in \mathcal{P}} \quad E(V_N) \quad \text{s.t.} \quad V_{k+1} = V_k + h_k f(V_k, \Omega), \qquad k = 0, \ldots, N - 1, \qquad (22)$$

with $V_0 = 0_{|I| \times |J|}$. The result analogous to Theorem 1 follows.

Theorem 2. (parameter sensitivity: discrete case). *The sensitivity of problem* (22) *with respect to the parameter Ω is given by*

$$\frac{dE}{d\Omega} = \sum_{k=1}^{N} h_{k-1} \left(\frac{\partial f(V_{k-1}, \Omega)}{\partial \Omega} \right)^{\top} \overline{\lambda}_k, \tag{23a}$$

where the discrete adjoint vectors $\overline{\lambda}_k$ are given by

$$\overline{\lambda}_k = \frac{\partial E(V_N)}{\partial V_k}, \quad k = 0, \dots, N. \tag{23b}$$

Proof. Skipped due to lack of space. □

We notice that (23a) corresponds to a discretization of (14a). Likewise, the discrete adjoints (23b) can be based on the theory of *automatic differentiation (AD)* [6]. Specifically, in view of the *geometric* numerical integration of the system (20), formula (23a) can be seen as a non-Euclidean version of the *reverse mode* of AD.

3.4 Differentiate or Discretize First?

We briefly indicate why either approach of Sect. 3.2 or Sect. 3.3 yields the same result, that is diagram of Fig. 2 *commutes* indeed.

We take a closer look at expression (23b) due to the *discrete* problem formulation and compute

$$\overline{\lambda}_k = \frac{\partial E(V_N)}{\partial V_k} = \left(\frac{\partial V_{k+1}}{\partial V_k} \right)^{\top} \frac{\partial E(V_N)}{\partial V_{k+1}} = \left(\frac{\partial V_{k+1}}{\partial V_k} \right)^{\top} \overline{\lambda}_{k+1} \tag{24a}$$

$$\overset{(22)}{=} \left(\frac{\partial (V_k + h_k f(V_k, \Omega))}{\partial V_k} \right)^{\top} \overline{\lambda}_{k+1} = \left(I + h_k \partial_V f(V_k, \Omega) \right)^{\top} \overline{\lambda}_{k+1} \tag{24b}$$

$$\overline{\lambda}_k = \overline{\lambda}_{k+1} + h_k \left(\partial_V f(V_k, \Omega) \right)^{\top} \overline{\lambda}_{k+1}, \tag{24c}$$

which agrees with (21) derived from the *continuous* problem formulation.

3.5 Parameter Estimation Algorithm

Our strategy for *parameter adaption* is to follow the *Riemannian gradient descent flow* on the parameter manifold $\mathcal{P}$,

$$\dot{\Omega} = -\nabla_{\mathcal{P}} E(V(T, \Omega)) = -\Pi_{\Omega} \left(\frac{d}{d\Omega} E(V(T, \Omega)) \right), \quad \Omega(0) = \mathbb{1}_{\mathcal{P}}, \tag{25}$$

with Π_{Ω} given by (5a) and with the *unbiased* initialization $\Omega(0) = \mathbb{1}_{\mathcal{P}}$, i.e. *uniform* weights at every patch $\mathcal{N}_i$ around pixel $i \in I$. We discretize flow (25) using the *geometric* explicit Euler scheme (cf. [16])

$$\Omega_{k+1} = \exp_{\Omega_k} \left(-h_k \nabla_{\mathcal{P}} E(V_N(\Omega_k)) \right), \quad \Omega_0 = \Omega(0), \quad k = 1, 2, \dots \tag{26}$$

We summarize the two main procedures for parameter learning.

Algorithm 1. Discretized *Riemannian flow.* (25)

Data: initial weights $\Omega_0 = \mathbb{1}_{\mathcal{P}}$, objective function $E\big(V(T)\big)$
Result: weight parameter estimates Ω^*
`// geometric Euler integration`
1 **for** $k = 0, \ldots, K$ **do**
2 $\quad$ compute $\frac{d}{d\Omega} E\big(V_N(\Omega_k)\big)$; $\qquad\qquad\qquad$ `// Algorithm 2`
3 $\quad$ $\Omega_{k+1} = \exp_{\Omega_k}\big(- h_k \Pi_\Omega\big(\frac{d}{d\Omega_k} E(V_N)\big)\big)$;

In practice, we terminate the iteration when $\|\Omega_{k+1} - \Omega_k\| \le \varepsilon$. This determines K.

Algorithm 2. Compute the sensitivity $\frac{dE(V_N)}{d\Omega}$ by (23a).

Data: current weights Ω_j
Result: objective value $E(V_N(\Omega_j))$, sensitivity $\frac{dE(V_N)}{d\Omega_j}$
`// forward Euler integration`
1 **for** $k = 0, \ldots, N - 1$ **do**
2 $\quad$ $V_{k+1} = V_k + h f(V_k, \Omega_j)$;
3 compute $\lambda_N = \frac{\partial E(V_N)}{\partial V_N}$;
4 set $\frac{dE(V_N)}{d\Omega} = 0$;
`// backward Euler integration`
5 **for** $k = N - 1, \ldots, 0$ **do**
6 $\quad$ $\lambda_k = \lambda_{k+1} + h_k \big(\frac{\partial f(V_k, \Omega_j)}{\partial V}\big)^\top \lambda_{k+1}$;
7 $\quad$ $\frac{dE(V_N)}{d\Omega} \mathrel{+}= h_{k-1}\big(\frac{\partial f(V_{k-1}, \Omega_j)}{\partial \Omega}\big)^\top \lambda_k$; $\qquad$ `// summand of (23a)`

3.6 Parameter Prediction

We apply kernel-based local regression in order to *generalize* the relation between features and optimal weight parameters for regularization to *novel* data.

Specifically, let $\Omega^* = (w_{ik}^*)_{k \in \mathcal{N}_i}$, $i \in I$ denote the output of Algorithm 1, and let $\mathcal{F}_i^* = (f_k^*)_{k \in \mathcal{N}_i}$, $i \in I$ denote the corresponding image 'feature patches' of the training data that were used to compute Ω^*. The goal of regression is to generalize the relation between $\mathcal{F}_i^*$, $i \in I$ and Ω^* to novel image data and features $\mathcal{F}_i$ in terms of a *weight map*

$$\widehat{w} \colon \mathcal{F}_i \to \mathcal{P}, \qquad (f_k)_{k \in \mathcal{N}_i} \to (w_{ik})_{k \in \mathcal{N}_i}, \quad i \in I, \tag{27}$$

where $\widehat{w}$ does *not* depend on the pixel location $i \in I$. In words, the *local spatial context* in terms of novel features f_k observed within an image patch $k \in \mathcal{N}_i$ around each pixel $i \in I$, is used to *predict* the regularization parameters $(w_{ik})_{k \in \mathcal{N}_i}$.

Using the basic Nadaraya–Watson kernel regression estimator [15, Section 5.4] based on the data

$$\left(\mathcal{F}_i^*, (w_{ik}^*)_{k\in\mathcal{N}_i}\right)_{i\in N} \tag{28}$$

which constitutes a *coreset* [11] of the data produced offline by Algorithm 1 from (possibly many) training images, we explore (see Sect. 4) the predictor map

$$\widehat{w}(\mathcal{F}_i) = \sum_{j\in[N]} \frac{K_h(\mathcal{F}_i, \mathcal{F}_j^*)}{\sum_{j'\in[N]} K_h(\mathcal{F}_i, \mathcal{F}_{j'}^*)} w_j^*, \qquad w_j^* = (w_{jk}^*)_{k\in\mathcal{N}_j}, \tag{29}$$

where $K_h(\cdot, \cdot)$ is a standard kernel function (Gaussian, Epanechnikov, etc.) applied to a suitable distance between the feature sets $\mathcal{F}_j$ and $\mathcal{F}_{j'}^*$. The *bandwidth parameter* h is determined by cross-validation using the data (28).

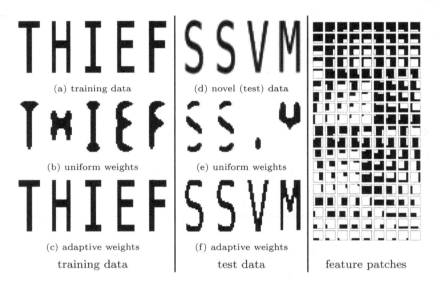

(a) training data   (d) novel (test) data

(b) uniform weights   (e) uniform weights

(c) adaptive weights   (f) adaptive weights

training data   test data   feature patches

Fig. 3. Patch-based adaptive regularization of binary letters. LEFT COLUMN: **(a)** Training data and corresponding 5×5 feature patches (RIGHTMOST COLUMN). **(b)** Uniform regularization fails even with perfect features. **(c)** Perfect adaptive reconstruction (sanity check). CENTER COLUMN: **(d)** Test data to be labelled using the features and regularizer trained on (a). **(e)** Uniform regularization fails. **(f)** Adaptive regularization predicts *curvilinear* structures of (d) using 'knowledge' based on (a) where *only vertical and horizontal* structures occur.

4   Experiments

We illustrate adaptive regularization for image labeling by two further experiments, based on the objective function (13), in addition to the experiment with non-binary data illustrated by Fig. 1.

Figure 3 shows patch-based labeling of *curvilinear* letters (center column) using an adaptive regularizer trained on letters with *vertical and horizontal* structures only (left column). See the figure caption for details. This result indicates adaptivity in a two-fold way: use of *non-uniform* weights that are predicted *online* as optimal parameters for *novel* image data *not* seen before.

Figure 4 shows results for *curvilinear line structures* contaminated by noise. The geometry of these scenes correspond to *random* Voronoi diagrams, that is training and test scenes *differ*. The raw grayvalue data together with the outputs of a ridge filter and a Laplacian-of-Gaussian filter were used as feature vectors

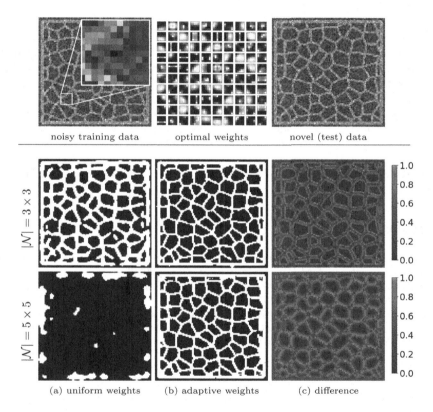

Fig. 4. Adaptive regularization and labeling of curvilinear noisy line structures. Training and test data were randomly generated and hence differ completely. Non-adaptive regularization (uniform weights) returns dilated incomplete structures at the smallest scale ($|\mathcal{N}| = 3 \times 3$) or fails completely ($|\mathcal{N}| = 5 \times 5$). Panel 'optimal weights' illustrates a sample of *non-uniform optimal* weight patches $(w_{ik}^*)_{k \in \mathcal{N}_i}$, for a couple of pixels $i \in I$, computed during the *training* phase. Panels 'difference' illustrate the deviation of weight patches from uniform weights during the *test* phase for *each* pixel. The corresponding labelings (panels (b)) are almost perfect except for minor boundary effects.

of dimension 3. The results illustrate the labeling with *uniform* regularization using the smallest scale $|\mathcal{N}| = 3 \times 3$ returns *dilated* and *incomplete* structures, and fails completely at the larger scale $|\mathcal{N}| = 5 \times 5$. The adaptive regularizer yields almost perfect results except for minor boundary effects. This illustrates that after the training phase, when using optimal parameters Ω^* determined for training data, then the linear assignment flow has enough predictive power in order to 'know' how to average geometrically when confronted with novel image data not seen before.

5   Conclusion

We introduced an novel approach to image labeling based on adaptive regularization of the linear assignment flow. Parameter estimation relies on a consistent discretization and geometric numerical integration that is easy to reproduce. The prediction component can be based on any state-of-the-art kernel method from machine learning.

Our future work will extend the numerics to more general symplectic integrators and the approach itself to state-dependent parameter prediction, in connection with spatial multiscale representations of image data.

Acknowledgement. Support from the German Science Foundation, grant GRK 1653, is gratefully acknowledged.

References

1. Amari, S.I., Nagaoka, H.: Methods of Information Geometry. American. Mathematical Society, Oxford University Press, Oxford (2000)
2. Aström, F., Petra, S., Schmitzer, B., Schnörr, C.: Image labeling by assignment. J. Math. Imaging Vis. **58**(2), 211–238 (2017)
3. Åström, F., Hühnerbein, R., Savarino, F., Recknagel, J., Schnörr, C.: MAP image labeling using Wasserstein messages and geometric assignment. In: Lauze, F., Dong, Y., Dahl, A.B. (eds.) SSVM 2017. LNCS, vol. 10302, pp. 373–385. Springer, Cham (2017). https://doi.org/10.1007/978-3-319-58771-4_30
4. Weinan, E.: A proposal on machine learning via dynamical systems. Commun. Math. Stat. **5**(1), 1–11 (2017)
5. Giles, M.B., Pierce, N.A.: An introduction to the adjoint approach to design. Flow Turbul. Combust. **65**(3), 393–415 (2000)
6. Griewank, A.: A mathematical view of automatic differentiation. Acta Numerica **12**, 321–398 (2003)
7. Haber, E., Ruthotto, L.: Stable architectures for deep neural networks. Inverse Prob. **34**(1), 014004 (2017)
8. Hairer, E., Lubich, C., Wanner, G.: Geometric Numerical Integration. Springer, Heidelberg (2006). https://doi.org/10.1007/3-540-30666-8
9. He, K., Zhang, X., Ren, S., Sun, J.: Deep residual learning for image recognition. In: Proceedings of CVPR (2016)

10. Kappes, J., et al.: A comparative study of modern inference techniques for structured discrete energy minimization problems. Int. J. Comput. Vis. **115**(2), 155–184 (2015)
11. Phillips, J.: Coresets and sketches. In: Handbook of Discrete and Computational Geometry. CRC Press (2016). Chapter 48
12. Ross, I.: A roadmap for optimal control: the right way to commute. Ann. N. Y. Acad. Sci. **1065**(1), 210–231 (2005)
13. Shalev-Shwartz, S., Shamir, O., Shammah, S.: Failures of gradient-based deep learning. CoRR abs/1703.07950 (2017)
14. Wainwright, M.J.: Estimating the "Wrong" graphical model: benefits in the computation-limited setting. J. Mach. Learn. Res. **7**, 1829–1859 (2006)
15. Wasserman, L.: All of Nonparametric Statistics. Springer, New York (2006). https://doi.org/10.1007/0-387-30623-4
16. Zeilmann, A., Savarino, F., Petra, S., Schnörr, C.: Geometric numerical integration of the assignment flow. CoRR abs/1810.06970 (2018)
17. Zhu, S.C., Liu, X.: Learning in Gibbsian fields: how accurate and how fast can it be? IEEE Trans. Patt. Anal. Mach. Intell. **24**(7), 1001–1006 (2002)

Direct MRI Segmentation from k-Space Data by Iterative Potts Minimization

Lukas Kiefer[1,2(✉)], Stefania Petra[1], Martin Storath[3],
and Andreas Weinmann[2,4]

[1] Mathematical Imaging Group, Heidelberg University, Heidelberg, Germany
[2] Department of Mathematics and Natural Sciences, Hochschule Darmstadt,
Darmstadt, Germany
`lukas.kiefer@h-da.de`
[3] Department of Applied Natural Sciences and Humanities,
Hochschule Würzburg-Schweinfurt, Würzburg, Germany
[4] Institute of Computational Biology, Helmholtz Zentrum München,
Munich, Germany

Abstract. We consider the problem of image segmentation in magnetic resonance imaging (MRI) directly from k-space measurements as opposed to more standard approaches that first reconstruct the image before segmenting it. The model we employ is the piecewise-constant Mumford-Shah model (the Potts model) in connection with the MRI reconstruction problem. The output of our proposed scheme is a piecewise-constant function which yields a segmentation of the image domain. To solve the involved non-convex and non-smooth optimization problem, we adopt an iterative minimization strategy based on surrogate functionals. Numerical experiments illustrate the potential of the approach when applied on undersampled MRI data.

Keywords: MRI segmentation ·
Joint reconstruction and segmentation · Potts model ·
Piecewise-constant Mumford-Shah model ·
Regularization of inverse problems

1 Introduction

Magnetic resonance imaging (MRI) is an established method for noninvasive medical imaging. A related method is nuclear magnetic resonance spectroscopy in organic chemistry. A special application is the detection of small particles [32] in the context of cellular imaging. In many medical-imaging applications segmentation is needed [27]. In this work, we propose a method for the segmentation of images directly from their MRI data or, more precisely, from their k-space

This work was supported by the German Research Foundation (DFG STO1126/2-1 & WE5886/4-1).

© Springer Nature Switzerland AG 2019
J. Lellmann et al. (Eds.): SSVM 2019, LNCS 11603, pp. 406–418, 2019.
https://doi.org/10.1007/978-3-030-22368-7_32

data. To this end, we consider the piecewise-constant Mumford-Shah model (the Potts model) for the MRI reconstruction problem.

The piecewise-constant Mumford-Shah model has been successfully applied for various imaging tasks. For pure image segmentation we refer to [25, 34, 36]. In the context of inverse problems, examples are [28] for incomplete Radon data, [19] for SPECT and Radon data and [35] for Radon and blurred data. The regularizing nature of the piecewise-constant Mumford-Shah model in the sense of inverse problems was shown in [29]. In this context, it was also shown that the boundaries of the discontinuity set of the underlying signal detected by (general) Mumford-Shah approaches are regularized as well [15]. The feasible candidates for the piecewise-constant Mumford-Shah/Potts MRI problem are piecewise-constant which means that they also induce segmentations. Due to the regularizing nature of the Potts model reasonable segmentations can be obtained from undersampled and noisy MRI data.

MRI Forward Model. Our goal is to segment images directly from their under-sampled (simulated) MRI data. Concerning the data acquisition process of an MR tomograph we adopt the forward model of [12] where the MRI imaging oper-ator is modeled as the Fourier transform $\mathcal{F}$ which was sampled by a sampling operator $\mathcal{S}$. Further, since the images under consideration are assumed to be real-valued, the imaging operator is given by

$$A : \mathbb{R}^N \to \mathbb{C}^m, \quad Au = \mathcal{S}\mathcal{F}\iota(u), \tag{1}$$

where $\iota : \mathbb{R}^N \to \mathbb{C}^N$, $\iota(u) = u + 0 \cdot i$ is the inclusion map and $\mathcal{F} : \mathbb{C}^N \to \mathbb{C}^N$ is the (discrete) Fourier transform. The sampling operator $\mathcal{S} : \mathbb{C}^N \to \mathbb{C}^m$ acts on a generic x by $(\mathcal{S}x)_j = x_{\pi(j)}$ with $\pi(j)$ being the location of the j-th sample in the Fourier domain. In practice, measurements are often corrupted by noise. Therefore, the measured data f is in general given by $f = Au + \nu$, where u is the unknown image and the entries of the (unknown) noise vector ν are often modeled as i.i.d. Gaussians with zero mean and standard deviation σ.

The algorithmic scheme used in this paper involves the adjoint of the MRI imaging operator. By the definition in (1), $A^* = \iota^* \circ \mathcal{F}^* \circ \mathcal{S}^*$, where ι^* is the restriction onto the real part, $\mathcal{F}^* = \mathcal{F}^{-1}$ is the inverse Fourier transform and $\mathcal{S}^*$ is given by $\mathcal{S}^*(y) = \sum_{j=1}^m e_{\pi(j)} y_j$ for the standard basis $(e_j)_{j=1}^N$.

In practice, we often have $m \ll N$ due to restrictions on the measurement process, e.g., by time constraints or by consideration for the patients [31]. Hence, the operator A cannot be directly inverted and regularization becomes necessary.

Related Work. In [22], the authors propose simultaneous reconstruction and segmentation by labeling in the context of computed tomography (CT) problems with limited field of view and missing data, where the labels evolve during the iterations. Zisler et al. [42] propose a method for joint image reconstruction and labeling for fixed labels and apply it to CT problems. Their regularization is based on the Kullback-Leibler-divergence (KL) and the objective is minimized via DC programming. In [5], a simultaneous reconstruction and segmentation method is applied to dynamic SPECT data. The authors' model uses a KL data fidelity term and a regularizer related to the Chan-Vese model for image segmentation. In [6], the authors perform spinal MRI segmentation after reconstruction by normalized cuts using Nyström approximation. Recently, Corona et al. [10] proposed joint reconstruction and segmentation with special emphasis on MRI. They obtain a reconstruction and a segmentation, where the segmentation is given w.r.t. a labeling of discrete input labels. They use total variation regularization for both the reconstruction and segmentation process using Bregman distances. The two processes are related via an extra coupling penalty and the convex subproblems are solved with primal-dual algorithms.

In the paper [16] already reconstructed brain MRI images are segmented by applying nonparametric density estimation using the mean shift algorithm in the joint spatial-range domain. Another class of post-reconstruction segmentation methods are learning-based approaches with convolutional neural networks, e.g., for brain tumor segmentation [26].

In [35], the authors propose an algorithmic approach to the inverse Potts model for inverse imaging problems and apply it to Radon data and blurred data. In [1], Bar et al. apply the Ambrosio-Tortorelli approximation of the piecewise smooth Mumford-Shah model to debluring problems (which gives near piecewise-constant solutions for large smoothness penalties). The method principally carries over to MRI problems. The authors of [38] propose an iterative algorithm based on surrogate functionals to solve the inverse Potts model for univariate signals and prove convergence to a local minimizer.

Contributions and Organization. In this paper, we propose a method for direct segmentation of images from their MRI data. In Sect. 2, we briefly explain the iterative minimization strategy of [17] which is based on the iterative minimization of surrogate functionals and apply it to the Potts MRI problem. We see that the involved (non-separable) subproblems of the algorithmic scheme can be computed exactly and efficiently. Further, the cluster points of the algorithm are local minimizers of the Potts MRI problem. In Sect. 3, we illustrate the potential of the proposed method for direct segmentation of MRI data by applying it to synthetic and simulated realistic undersampled k-space data.

2   Proposed Method

Potts-Regularized MRI Problem. To segment images directly from MRI measurements we adopt the Potts model (aka the piecewise-constant Mumford-Shah model). It is formulated in terms of the following minimization problem

$$\operatorname*{argmin}_{u:\Omega\subset\mathbb{R}^d\to\mathbb{R}} \|Au - f\|_2^2 + \gamma\|\nabla u\|_0, \tag{2}$$

where the first term measures the data fidelity of u to the data f (recall A is the MRI operator defined in (1)) and $\gamma > 0$ is a model parameter (the jump-penalty) which controls the level of regularization induced by the second term, that is, the coarseness of the segmentation. Regarding the regularization term, the gradient

∇u is given in the distributional sense and the jump-term $\|\nabla u\|_0$ as the boundary length of the discontinuity set in terms of the $(d-1)$-dimensional Hausdorff measure. Since a candidate u which is not piecewise-constant yields $\|\nabla u\|_0 = \infty$ [29], solutions of (2) are piecewise-constant. Segmentations of the domain Ω and piecewise-constant functions on Ω are in one-to-one correspondence and a solution of (2) provides both a segmentation and a piecewise-constant approximation of the image to be reconstructed.

In this paper, we consider discrete data and discrete two-dimensional images. A discretization of (2) based on finite differences leads to

$$\operatorname*{argmin}_{u:\Omega'\to\mathbb{R}} \|Au - f\|_2^2 + \gamma \sum_{s=1}^{S} \omega_s \|\nabla_{a_s} u\|_0, \tag{3}$$

where the image u is discretized on the image domain $\Omega' = \{1,\ldots,n_1\} \times \{1,\ldots,n_2\}$. Further, the integer vectors $a_s \in \mathbb{Z}^2$ form a finite set of directions, $\omega_s > 0$ are the corresponding weights and the term $\nabla_{a_s} u(i,j)$ denotes the directional difference $u_{(i,j)+a_s} - u_{i,j}$. Moreover, $\|\nabla_{a_s} u\|_0$ counts the number of nonzero entries of $\nabla_{a_s} u$. In order to obtain near isotropic results, we utilize the unit vectors a_1, a_2 together with the diagonal vectors a_3, a_4 as directions and use the weights $\omega_s = \sqrt{2} - 1$ for $s = 1, 2$ and $\omega_s = 1 - \sqrt{2}/2$ for $s = 3, 4$, see [8,34].

Derivation of the Algorithm. We adapt next the iterative Potts algorithm from [17] to solve the MRI Potts problem (3). To this end, we introduce splitting variables $u_1, \ldots, u_S$ corresponding to the offset vectors $a_1, \ldots, a_S$ under the constraint that they are equal. Thus, we reformulate (3) as

$$\operatorname*{argmin}_{u_1,\ldots,u_S} \sum_{s=1}^{S} \tfrac{1}{S}\|Au_s - f\|_2^2 + \gamma\omega_s\|\nabla_{a_s} u_s\|_0,$$

$$\text{subject to} \quad u_1 = u_2 = \ldots = u_S. \tag{4}$$

Note that solving (4) is equivalent to solving (3). Next we consider the quadratic penalty relaxation of (4) which corresponds to minimizing the functional

$$P_{\gamma,\rho}(u_1,\ldots,u_S) = \sum_{s=1}^{S} \tfrac{1}{S}\|Au_s - f\|_2^2 + \gamma\omega_s\|\nabla_{a_s} u_s\|_0 + \rho\|u_s - u_{(s+1)\bmod S}\|_2^2. \tag{5}$$

That is, we relax the hard constraints in (4) by replacing them with the squared Euclidean distances between consecutive splitting variables weighted by a positive parameter ρ. Thus, the coupling parameter ρ controls the similarity of the splitting variables, i.e., a large value of ρ enforces the u_s to be close to each other and in the limit case $\rho \to \infty$ they become equal and yield a solution to (4).

The basic principle of the algorithm is to solve successive instances of the quadratic penalty relaxation (5) for a (strictly) increasing sequence of coupling parameters ρ_k. We first consider a fixed ρ_k and recall the surrogate approach of [17] for the quadratic penalty relaxation. Concerning further details on surrogate functionals we refer to [11]. In order to derive the surrogate functional of (5), we rewrite $P_{\gamma,\rho}$ in terms of the (formal) block matrix B and the vector g which are given by

$$
B = \begin{pmatrix} S^{-1/2}A & & & & & \\ & S^{-1/2}A & & & & \\ & & \ddots & & & \\ & & & & S^{-1/2}A & 0 \\ \rho^{1/2}I & -\rho^{1/2}I & 0 & \cdots & & 0 \\ 0 & \rho^{1/2}I & -\rho^{1/2}I & 0 & \cdots & 0 \\ \vdots & & \ddots & & & \vdots \\ -\rho^{1/2}I & 0 & & \cdots & & \rho^{1/2}I \end{pmatrix} \quad \text{and} \quad g = \begin{pmatrix} S^{-1/2}f \\ S^{-1/2}f \\ \vdots \\ S^{-1/2}f \\ 0 \\ \vdots \\ \vdots \\ 0 \end{pmatrix}. \quad (6)
$$

Moreover, we define the difference operator D by

$$
D(u_1,\ldots,u_S) = \begin{pmatrix} \nabla_{a_1}u_1 \\ \vdots \\ \nabla_{a_S}u_S \end{pmatrix}, \quad \text{and} \quad \|D(u_1,\ldots,u_S)\|_{0,\omega} = \sum_{s=1}^{S} \omega_s \|\nabla_{a_s}u_s\|_0. \quad (7)
$$

Together, (6) and (7) yield

$$
P_{\gamma,\rho}(u_1,\ldots,u_S) = \|B(u_1,\ldots,u_S)^T - g\|_2^2 + \gamma\|D(u_1,\ldots,u_S)\|_{0,\omega} \quad (8)
$$

whose surrogate functional reads

$$
P_{\gamma,\rho}^{\mathrm{surr}}(u_1,\ldots,u_S,v_1,\ldots,v_S) = \tfrac{1}{L_\rho^2}\|B(u_1,\ldots,u_S)^T - g\|_2^2 + \tfrac{\gamma}{L_\rho^2}\|D(u_1,\ldots,u_S)\|_{0,\omega}
$$
$$
- \tfrac{1}{L_\rho^2}\|B(u_1,\ldots,u_S)^T - B(v_1,\ldots,v_S)^T\|_2^2 + \|(u_1,\ldots,u_S)^T - (v_1,\ldots,v_S)^T\|_2^2
$$

for a constant $L_\rho \geq 1$ that is chosen larger than the spectral norm of B so that B/L_ρ is a contraction. More precisely, we choose L_ρ such that

$$
L_\rho^2 > 1/S + 4\rho. \quad (9)
$$

Then, by [17, Lemma 8] and the next Lemma which estimates the spectral norm of the MRI imaging operator A as defined in (1), B/L_ρ is contractive.

Lemma 1. *The spectral norm of the MRI imaging operator A given by (1) fulfills $\|A\|_2 \leq 1$.*

Proof. Since A is defined by the composition $\mathcal{S} \circ \mathcal{F} \circ \iota$ we have the estimate $\|A\|_2 \leq \|\mathcal{S}\|_2\|\mathcal{F}\|_2\|\iota\|_2$, where all three factors on the right handside are equal to 1: indeed $\mathcal{F}$ is a unitary operator, ι the canonical injection and as at least one point in the frequency domain is sampled, $\mathcal{S} \neq 0$ and $\|\mathcal{S}\|_2 = 1$.

Recall that the surrogate iterations concerning the minimization of (5) read

$$
(u_1^{n+1},\ldots,u_S^{n+1}) \in \operatorname*{argmin}_{u_1,\ldots,u_S} P_{\gamma,\rho}^{\mathrm{surr}}(u_1,\ldots,u_S,u_1^n,\ldots,u_S^n). \quad (10)
$$

In the following, we derive a more explicit formulation of (10). By exploiting properties of the inner product (cf., e.g., [11]) we obtain

$$
\begin{aligned}
P_{\gamma,\rho}^{\mathrm{surr}}&(u_1,\dots,u_S,v_1,\dots,v_S) \\
&= \left\| (u_1,\dots,u_S)^T - \left((v_1,\dots,v_S)^T - \tfrac{1}{L_\rho^2} B^T \left(B(v_1,\dots,v_S)^T - g \right) \right) \right\|_2^2 \\
&\quad + \tfrac{\gamma}{L_\rho^2} \| D(u_1,\dots,u_S) \|_{0,\omega} + R(v_1,\dots,v_S),
\end{aligned}
\tag{11}
$$

where $R(v_1,\dots,v_S)$ is a term that only depends on the v_s, and so it can be omitted when minimizing $P_{\gamma,\rho}^{\mathrm{surr}}$ w.r.t. the u_s. Using the reformulation (11) of $P_{\gamma,\rho}^{\mathrm{surr}}$, the surrogate iterations (10) can be written as

$$
(u_1^{n+1},\dots,u_S^{n+1}) \in \operatorname*{argmin}_{u_1,\dots,u_S} \sum_{s=1}^{S} \| u_s - h_s^{(n)} \|_2^2 + \tfrac{\gamma \omega_s}{L_\rho^2} \| \nabla_{a_s} u_s \|_0,
\tag{12}
$$

where $h_s^{(n)}$ is given by

$$
h_s^{(n)} = u_s^{(n)} + \tfrac{1}{SL_\rho^2} A^* f - \tfrac{1}{SL_\rho^2} A^* A u_s^{(n)} - \tfrac{\rho}{L_\rho^2} \left(u_s^{(n)} - u_{(s+1)\bmod S}^{(n)} \right),
\tag{13}
$$

for all $s = 1,\dots,S$. Note that the forward step (13) corresponds to a Landweber update [21] for the least squares problem induced by B and g which essentially involves applying the MRI forward operator and its adjoint. The MRI forward operator only appears in the forward step (13). In contrast to [2,11], the backward step (12) does not decompose into pixelwise problems that could be solved by a thresholding technique. However, they are tractable as explained next.

The crucial point in the iterative scheme (12)–(13) is that the problem in the backward step (12) can be solved exactly and efficiently. To see this, we observe that (12) can be solved for each u_s separately. Subsequently, the emerging s problems decompose into univariate Potts problems for data along the paths in $h_s^{(n)}$ induced by the direction a_s. Univariate Potts problems can be solved exactly and efficiently by dynamic programming [7,14,23,24,40]. We prune the search space to improve runtimes [18,34]. For further details we refer to [14,33].

We gathered all necessary ingredients to formulate the algorithmic scheme to solve the MRI Potts problem (4).

Algorithm 1. *Define a strictly increasing sequence of coupling parameters by $\rho_k = \tau^k \rho_0$ for $\tau > 1, \rho_0 > 0$. Further define a stopping parameter t which obeys*

$$
t > 2(2 - 2\cos(2\pi/S))^{-1/2} S^{-1/2} \| f \|
$$

and a strictly decreasing null sequence δ_k. Initialize $u_s^{(0)} = u_s^{(0,0)} = A^* f$ for $s = 1,\dots,S$.

1. *While for any $s = 1,\dots,S$,*

$$
\| u_s^{(k,n)} - u_{(s+1)\bmod S}^{(k,n)} \|_2 > \tfrac{t}{\rho_k}, \quad or \quad \| u_s^{(k,n)} - u_s^{(k,n-1)} \|_2 > \tfrac{\delta_k}{L_{\rho_k}},
\tag{14}
$$

   do

a. $h_s^{(k,n)} = u_s^{(k,n)} + \frac{1}{SL_{\rho_k}^2} A^* f - \frac{1}{SL_{\rho_k}^2} A^* A u_s^{(k,n)} - \frac{\rho_k}{L_{\rho_k}^2} \left(u_s^{(k,n)} - u_{(s+1)\bmod S}^{(k,n)} \right)$,

b. $\left(u_1^{(k,n+1)}, \ldots, u_S^{(k,n+1)} \right) \in \underset{u_1,\ldots,u_S}{argmin} \sum_{s=1}^{S} \| u_s - h_s^{(k,n)} \|_2^2 + \frac{\gamma \omega_s}{L_{\rho_k}^2} \| \nabla_{a_s} u_s \|_0$,

2. Set $u_s^{(k+1)} = u_s^{(k+1,0)} = u_s^{(k,n)}$, $\rho_{k+1} = \tau \rho_k$, update $\delta_k \to \delta_{k+1}$, update $L_{\rho_k} \to L_{\rho_{k+1}}$ according to (9) and goto 1.

Convergence. The following result ensures that Algorithm 1 is well-defined and that its cluster points are local minimizers of (3).

Theorem 1. *Let $u^{(k,n)}$ be the sequence defined by Algorithm 1. Then, (i) for any $k \in \mathbb{N}$, there is an $n \in \mathbb{N}$ such that the inner stopping criterion (14) will be fulfilled. (ii) The sequence $u^{(k)}$ defined by Algorithm 1 has a cluster point and each cluster point is a local minimizer of the Potts MRI problem (3) (in particular $u_s^* = u_t^*$ for all s, t).*

Both statements are stated and proven in [17, Thms. 6 and 7] in a general setting.

3   Experiments

Here we show the applicability of the proposed method in a series of numerical experiments. First, we provide all necessary implementation details and specify the methods we compare with the proposed one. Then we apply the proposed method to synthetic and simulated realistic data. Throughout this section, we let the noise level η be given by $\eta = N\sigma/\|\hat{f}\|_2$, where $\hat{f}$ denotes the clean data and $N = n_1 n_2$ is the number of pixels of the groundtruth image.

Implementation Details. We implemented Algorithm 1 for the system of directions consisting of the unit vectors together with the diagonal vectors. Their weights were chosen as described in Sect. 2, $\omega_{1,2} = \sqrt{2} - 1$, $\omega_{3,4} = 1 - \sqrt{2}/2$. Concerning the coupling sequence, we set the update parameter to $\tau = 1.05$ and the starting value $\rho_0 = 10^{-3}$, i.e., $\rho_k = 1.05^k \rho_0$. Furthermore, we chose the zero sequence $\delta_k = 1/(0.98\rho_k)$. In our experiments, the results improved by applying the modified step-size rule $L_\rho^\lambda = L_\rho \left(\lambda + (1 - (n+1)^{-1/2})(1 - \lambda) \right)$ for $\lambda = 0.15$ in the inner iterations of Algorithm 1. We stopped the computations when the relative difference $\|u_1^{(k)} - u_2^{(k)}\|_2/(\|u_1^{(k)}\|_2 + \|u_2^{(k)}\|_2)$ became smaller than 10^{-6}.

Methods for Comparison. We give a short description of the methods we compare our method with. In the joint segmentation and reconstruction approach of [10], they consider the problem

$$\underset{u \in \mathbb{R}^N, v \in C}{argmin} \frac{1}{2}\|Au - f\|_2^2 + \alpha \, TV(u) + \delta \sum_{i=1}^{N} \sum_{j=1}^{\ell} v_{ij}(c_j - u_i)^2 + \beta \, TV(v), \quad (15)$$

where u denotes the reconstruction and $v \in C = \{ z \in [0,1]^{N \times \ell} : \sum_{j=1}^{\ell} z_{ij} = 1, \, \forall i \in N \}$ encodes the assignment of the ℓ fixed input labels $c_1, \ldots, c_\ell$ to the

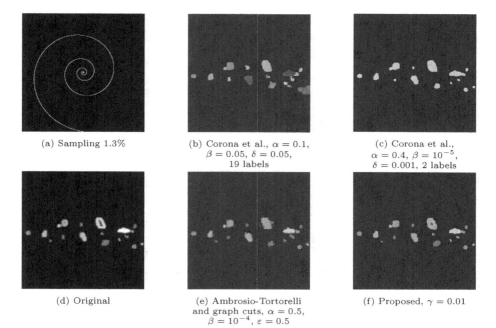

(a) Sampling 1.3%

(b) Corona et al., $\alpha = 0.1$, $\beta = 0.05$, $\delta = 0.05$, 19 labels

(c) Corona et al., $\alpha = 0.4$, $\beta = 10^{-5}$, $\delta = 0.001$, 2 labels

(d) Original

(e) Ambrosio-Tortorelli and graph cuts, $\alpha = 0.5$, $\beta = 10^{-4}$, $\varepsilon = 0.5$

(f) Proposed, $\gamma = 0.01$

Fig. 1. Direct segmentation of a synthetic particle image from spirally sampled k-space data corrupted by Gaussian noise of level $\eta = 0.15$. The proposed method detects all blobs and distinguishes them from each other.

pixels by discrete probability distributions for each pixel. Instead of solving (15), Bregman iterations are applied to u and v in an alternating fashion. The convex subproblems are solved by primal-dual algorithms. In our implementation, we chose the input labels by applying k-means to the groundtruth image.

We further compare with the Ambrosio-Tortorelli reconstruction model. Our implementation follows [1], where the Euler-Lagrange equations, given by

$$2\alpha v\|\nabla u\|_2^2 + \beta\frac{v - 1}{2\varepsilon} - 2\varepsilon\beta\nabla^2 v = 0, \qquad A^*(Au - f) - 2\alpha\mathrm{div}(v^2\nabla u) = 0,$$

are solved using a MINRES solver and using the method of conjugate gradients, respectively. The smoothness penalty is given by $\alpha > 0$, the edge penalty by $\beta > 0$ and $\varepsilon > 0$ is an edge smoothing parameter for the edge indicator v. Subsequently, we applied graph cuts [3,4,20] to the reconstruction to obtain a segmentation. We used 256 discrete input labels which were chosen by k-means.

In case of pure MRI reconstruction we further compare with (isotropic) total variation regularization [30] computed using the primal-dual algorithm of [9].

Synthetic Data. In Fig. 1, we segment a synthetic "blobs" image which mimics particles or bubbles in a liquid from spirally undersampled k-space data. The Ambrosio-Tortorelli model with subsequent graph cuts segmentation captures

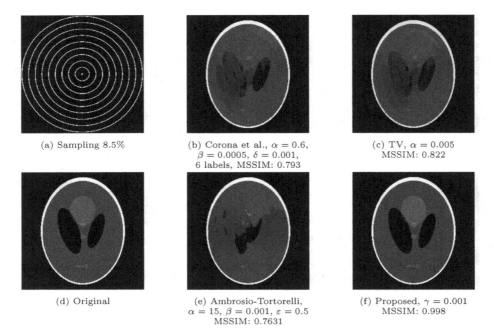

(a) Sampling 8.5%

(b) Corona et al., $\alpha = 0.6$,
$\beta = 0.0005$, $\delta = 0.001$,
6 labels, MSSIM: 0.793

(c) TV, $\alpha = 0.005$
MSSIM: 0.822

(d) Original

(e) Ambrosio-Tortorelli,
$\alpha = 15$, $\beta = 0.001$, $\varepsilon = 0.5$
MSSIM: 0.7631

(f) Proposed, $\gamma = 0.001$
MSSIM: 0.998

Fig. 2. Reconstruction of the Shepp-Logan phantom from undersampled concentric k-space data corrupted by Gaussian noise of level $\eta = 0.08$. The reconstructions by the TV model, the Ambrosio-Tortorelli model and the method of Corona et al. exhibit sampling artifacts. The proposed method provides a near-exact reconstruction.

all blobs, but suffers from clutter. The method of [10] detects all but one blob, yet creates some spurious segments. We observed an improved result for the method of [10] in terms of the latter if we set the number of input labels to two. Then, however, neighboring segments cannot be well-distinguished. The proposed method detects all blobs and also distinguishes them from each other.

Figure 2 shows reconstructions for the Shepp-Logan phantom from undersampled concentric k-space data [41]. Since the object to be reconstructed has a piecewise-constant structure, the proposed method provides a potential reconstruction as well. We measure the reconstruction quality by the mean structural similarity index (MSSIM) [37] whose range is $[0, 1]$ and higher values indicate better results. For each method, the input parameters were chosen empirically w.r.t. the MSSIM. The respective reconstructions by the Ambrosio-Tortorelli model, total variation regularization and the method of Corona et al. reconstruct the coarse structures, but suffer from sampling artifacts. The proposed method provides a near-exact reconstruction and achieves a higher MSSIM.

Simulated Realistic Data. To simulate realistic data, we applied the MRI imaging operator (1) to a slice of an originally fully-sampled human knee from

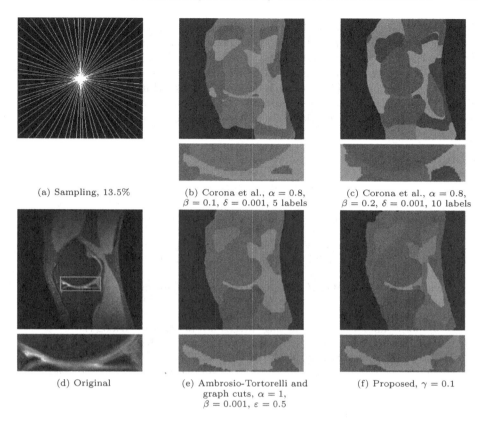

(a) Sampling, 13.5%

(b) Corona et al., $\alpha = 0.8$, $\beta = 0.1$, $\delta = 0.001$, 5 labels

(c) Corona et al., $\alpha = 0.8$, $\beta = 0.2$, $\delta = 0.001$, 10 labels

(d) Original

(e) Ambrosio-Tortorelli and graph cuts, $\alpha = 1$, $\beta = 0.001$, $\varepsilon = 0.5$

(f) Proposed, $\gamma = 0.1$

Fig. 3. Segmentation of a knee from simulated radially sampled k-space data corrupted by Gaussian noise of level $\eta = 0.1$ and detailed views of the highlighted area between the shinbone and the thighbone.

the dataset by Epperson et al. [13].[1] In Fig. 3, we segment this image from k-space data sampled by a radial profile whose line order is based on the golden ratio [39]. All methods segment the basic structures, e.g. they separate the shinbone from the thighbone. The detailed view of the region between shinbone and thighbone reveals that all segmentations produced by the comparison methods merge the right part of this region with the surrounding large segment (mainly consisting of muscle tissue). The proposed method creates distinct segments for the region and differentiates the region from the large neighboring segments.

4   Conclusion

In this paper, we have proposed a segmentation strategy using the Potts MRI model (or piecewise-constant Mumford-Shah MRI model) to segment an image

---

[1] Subject 10, layer 100; The dataset was downloaded from http://mridata.org.

directly from its MRI data. In the special case of (approximately) piecewise-constant images, the proposed method can be seen as MRI reconstruction as well. To solve the non-convex optimization problem we utilized an iterative strategy based on surrogate functionals whose non-separable subproblems can be solved exactly and efficiently. The cluster points of the algorithm are local minimizers of the Potts MRI model. Eventually, we demonstrated the applicability of the proposed method for synthetic and simulated realistic data, respectively.

References

1. Bar, L., Sochen, N., Kiryati, N.: Semi-blind image restoration via Mumford-Shah regularization. IEEE Trans. Image Process. **15**(2), 483–493 (2006)
2. Blumensath, T., Davies, M.: Iterative thresholding for sparse approximations. J. Fourier Anal. Appl. **14**(5–6), 629–654 (2008)
3. Boykov, Y., Kolmogorov, V.: An experimental comparison of min-cut/max-flow algorithms for energy minimization in vision. IEEE Tran. Pattern Anal. **26**(9), 1124–1137 (2004)
4. Boykov, Y., Veksler, O., Zabih, R.: Fast approximate energy minimization via graph cuts. IEEE Trans. Pattern Anal. **23**(11), 1222–1239 (2001)
5. Burger, M., Rossmanith, C., Zhang, X.: Simultaneous reconstruction and segmentation for dynamic SPECT imaging. Inverse Probl. **32**(10), 104002 (2016)
6. Carballido-Gamio, J., Belongie, S., Majumdar, S.: Normalized cuts in 3-D for spinal MRI segmentation. IEEE Trans. Med. Imaging **23**(1), 36–44 (2004)
7. Chambolle, A.: Image segmentation by variational methods: Mumford and Shah functional and the discrete approximations. SIAM J. Appl. Math. **55**(3), 827–863 (1995)
8. Chambolle, A.: Finite-differences discretizations of the Mumford-Shah functional. ESAIM Math. Model. Numer. Anal. **33**(02), 261–288 (1999)
9. Chambolle, A., Pock, T.: A first-order primal-dual algorithm for convex problems with applications to imaging. J. Math. Imaging Vis. **40**(1), 120–145 (2011)
10. Corona, V., et al.: Enhancing joint reconstruction and segmentation with non-convex Bregman iteration. Preprint, arXiv:1807.01660 (2018)
11. Daubechies, I., Defrise, M., De Mol, C.: An iterative thresholding algorithm for linear inverse problems with a sparsity constraint. Commun. Pure Appl. Math. **57**(11), 1413–1457 (2004)
12. Ehrhardt, M., Betcke, M.: Multicontrast MRI reconstruction with structure-guided total variation. SIAM J. Imaging Sci. **9**(3), 1084–1106 (2016)
13. Epperson, K., et al.: Creation of fully sampled MR data repository for compressed sensing of the knee. In: Proceedings of the 22nd Annual Meeting for Section for Magnetic Resonance Technologists, Salt Lake City, Utah, USA (2013)
14. Friedrich, F., Kempe, A., Liebscher, V., Winkler, G.: Complexity penalized M-estimation. J. Comput. Graph. Stat. **17**(1), 201–224 (2008)
15. Jiang, M., Maass, P., Page, T.: Regularizing properties of the Mumford-Shah functional for imaging applications. Inverse Probl. **30**(3), 035007 (2014)
16. Jiménez-Alaniz, J., Medina-Bañuelos, V., Yáñez-Suárez, O.: Data-driven brain MRI segmentation supported on edge confidence and a priori tissue information. IEEE Tran. Med. Imaging **25**(1), 74–83 (2006)

17. Kiefer, L., Storath, M., Weinmann, A.: Iterative Potts minimization for the recovery of signals with discontinuities from indirect measurements - the multivariate case. Preprint, arXiv:1812.00862 (2018)
18. Killick, R., Fearnhead, P., Eckley, I.: Optimal detection of changepoints with a linear computational cost. J. Am. Stat. Assoc. **107**(500), 1590–1598 (2012)
19. Klann, E., Ramlau, R., Ring, W.: A Mumford-Shah level-set approach for the inversion and segmentation of SPECT/CT data. Inverse Probl. Imaging **5**(1), 137–166 (2011)
20. Kolmogorov, V., Zabin, R.: What energy functions can be minimized via graph cuts? IEEE Trans. Pattern Anal. **26**(2), 147–159 (2004)
21. Landweber, L.: An iteration formula for Fredholm integral equations of the first kind. Am. J. Math. **73**(3), 615–624 (1951)
22. Lauze, F., Quéau, Y., Plenge, E.: Simultaneous reconstruction and segmentation of CT scans with shadowed data. In: Lauze, F., Dong, Y., Dahl, A.B. (eds.) SSVM 2017. LNCS, vol. 10302, pp. 308–319. Springer, Cham (2017). https://doi.org/10.1007/978-3-319-58771-4_25
23. Mumford, D., Shah, J.: Boundary detection by minimizing functionals. In: Proceedings of CVPR IEEE, vol. 17, pp. 137–154 (1985)
24. Mumford, D., Shah, J.: Optimal approximations by piecewise smooth functions and associated variational problems. Commun. Pure Appl. Math. **42**(5), 577–685 (1989)
25. Nguyen, R.M., Brown, M.S.: Fast and effective L_0 gradient minimization by region fusion. In: Proceedings of the IEEE ICCV, pp. 208–216 (2015)
26. Pereira, S., Pinto, A., Alves, V., Silva, C.: Brain tumor segmentation using convolutional neural networks in MRI images. IEEE Tran. Med. Imaging **35**(5), 1240–1251 (2016)
27. Pham, D., Xu, C., Prince, J.: Current methods in medical image segmentation. Annu. Rev. Biomed. Eng. **2**(1), 315–337 (2000)
28. Ramlau, R., Ring, W.: A Mumford-Shah level-set approach for the inversion and segmentation of X-ray tomography data. J. Comput. Phys. **221**(2), 539–557 (2007)
29. Ramlau, R., Ring, W.: Regularization of ill-posed Mumford-Shah models with perimeter penalization. Inverse Probl. **26**(11), 115001 (2010)
30. Rudin, L., Osher, S., Fatemi, E.: Nonlinear total variation based noise removal algorithms. Physica D **60**(1), 259–268 (1992)
31. Santos, J., Cunningham, C., Lustig, M., Hargreaves, B., Hu, B., Nishimura, D., Pauly, J.: Single breath-hold whole-heart MRA using variable-density spirals at 3T. Magn. Reson. Med. **55**(2), 371–379 (2006)
32. Shapiro, E., Skrtic, S., Sharer, K., Hill, J., Dunbar, C., Koretsky, A.: MRI detection of single particles for cellular imaging. PNAS **101**(30), 10901–10906 (2004)
33. Storath, M., Kiefer, L., Weinmann, A.: Smoothing for signals with discontinuities using higher order Mumford-Shah models. Preprint, arXiv:1803.06156 (2018)
34. Storath, M., Weinmann, A.: Fast partitioning of vector-valued images. SIAM J. Imaging Sci. **7**(3), 1826–1852 (2014)
35. Storath, M., Weinmann, A., Frikel, J., Unser, M.: Joint image reconstruction and segmentation using the Potts model. Inverse Probl. **31**(2), 025003 (2015)
36. Strekalovskiy, E., Cremers, D.: Real-time minimization of the piecewise smooth Mumford-Shah functional. In: Fleet, D., Pajdla, T., Schiele, B., Tuytelaars, T. (eds.) ECCV 2014. LNCS, vol. 8690, pp. 127–141. Springer, Cham (2014). https://doi.org/10.1007/978-3-319-10605-2_9
37. Wang, Z., Bovik, A., Sheikh, H., Simoncelli, E.: Image quality assessment: from error visibility to structural similarity. IEEE TIP **13**(4), 600–612 (2004)

38. Weinmann, A., Storath, M.: Iterative Potts and Blake-Zisserman minimization for the recovery of functions with discontinuities from indirect measurements. Proc. R. Soc. A **471**(2176), 20140638 (2015)
39. Winkelmann, S., Schaeffter, T., Koehler, T., Eggers, H., Doessel, O.: An optimal radial profile order based on the golden ratio for time-resolved MRI. IEEE TMI **26**(1), 68–76 (2007)
40. Winkler, G., Liebscher, V.: Smoothers for discontinuous signals. J. Nonparametr. Stat. **14**(1–2), 203–222 (2002)
41. Wu, H., Lee, J.H., Nishimura, D.: MRI using a concentric rings trajectory. Magn. Reson. Med. **59**(1), 102–112 (2008)
42. Zisler, M., Åström, F., Petra, S., Schnörr, C.: Image reconstruction by multilabel propagation. In: Lauze, F., Dong, Y., Dahl, A.B. (eds.) SSVM 2017. LNCS, vol. 10302, pp. 247–259. Springer, Cham (2017). https://doi.org/10.1007/978-3-319-58771-4_20

A Balanced Phase Field Model
for Active Contours

Jozsef Molnar[1], Ervin Tasnadi[1(✉)], and Peter Horvath[1,2]

[1] Biological Research Centre, Hungarian Academy of Sciences, Szeged, Hungary
{molnar.jozsef,tasnadi.ervin,horvath.peter}@brc.mta.hu
[2] Institute for Molecular Medicine Finland, University of Helsinki, Helsinki, Finland

Abstract. In this paper we present a balanced phase field model that eliminates the often undesired curvature-dependent shrinking of the zero level set, while maintaining the smooth interface necessary to calculate fundamental quantities such as the normal vector or the curvature of the represented contour. The proposed model extends the Ginzburg-Landau phase field energy with a higher order smoothness term. The relative weights are determined with the analysis of the level set motion in a curvilinear system adapted to the zero level set. The proposed level set framework exhibits strong shape maintaining capability without significant interference with the active (e.g. a segmentation) model.

1   Introduction

Active contours have become one of the most widely-used techniques for image segmentation [1]. Early parametric contours used a Lagrangian description of the discretized boundary to solve the Euler-Lagrange equation derived from an appropriate energy functional. Their most important difficulties are the need for periodic redistribution of the points and efficiently tracking topology changes.

The first problem was solved, and the second partially addressed, with the advent of geometric active contours [2,3]. These used an implicit representation of the contour as the zero level set of an appropriately constructed function, subsequently discretized on a fixed grid (Eulerian description). The level set method based on the Hamilton-Jacobi formulation was introduced in [4].

The level set function is usually initialized to signed distance function. During the contour evolution however, the distance property (required for the stability and accuracy) is not retained without specific handling. This is a major drawback. Different solutions were proposed to cope with this problem. The two main approaches are (a) reinitialization and (b) extension of the PDE associated with the original problem with a term that penalizes the deviations from the distance function. Beyond the theoretical incoherence with the Hamilton-Jacobi formulation [5], rebuilding the signed distance function in the whole domain is slow. The partial remedy for this problem can be the narrow band technique [6] for the

We acknowledge the LENDULET-BIOMAG Grant (2018-342) and the European Regional Funds (GINOP-2.3.2-15-2016-00001, GINOP-2.3.2-15-2016-00037).

© Springer Nature Switzerland AG 2019
J. Lellmann et al. (Eds.): SSVM 2019, LNCS 11603, pp. 419–431, 2019.
https://doi.org/10.1007/978-3-030-22368-7_33

price of higher complexity. The extension of the original PDE with a distance regularization term [7] may add instability (see [8]) and increase complexity [9]. More importantly, these approaches may move the zero level set away from the expected stopping location. We propose a solution to overcome this problem.

The Ginzburg-Landau phase field model was used in image segmentation in [10] as an alternative to the Hamilton-Jacobi formulation. It possesses interesting advantages as greater topological freedom, the possibility of a 'neutral' initialization. Here we stress another aspect: phase field models automatically form narrow band, a useful property that can only be achieved using additive regularization [9] in the case of Hamilton-Jacobi formulation. Moreover, unlike the reaction-diffusion model [8,11], it exhibits fast shape recovery due to the double well potential term incorporated in its functional. On the other hand, the Ginzburg-Landau phase field energy is proportional to the length of the contour that causes curvature dependent shrinking of the level sets. In some cases they are rather destructive and the Euler's elastica is used instead (e.g. [12]).

The calculation of the fundamental quantities requires a smooth transition across a certain neighbourhood of the zero level set. On the other hand, any method dedicated to this transitional shape maintaining should have the least possible interference with the segmentation PDE. Specifically, any curvature dependent behaviour should be an intentionally designed part of the segmentation model itself. The Ginzburg-Landau phase field obviously violates this 'least possible interference' requirement. In this paper we propose a *balanced phase field model* that eliminates the curvature driven shrinking, while maintains the smooth transition around the zero level set.

2   The Phase Field Model

2.1   Minimizing the Contour Energy Using Phase Field

In the level set framework, the representation of contours is given by a level set function of two variables $\phi(x, y)$. The quantities of the segmentation problem are extracted from this function, such as the unit normal vector $\mathbf{n} = \frac{\nabla\phi}{|\nabla\phi|}$ or the curvature $\kappa = -\nabla \cdot \left(\frac{\nabla\phi}{|\nabla\phi|}\right)$ where ∇ is the gradient operator and "$\cdot$" stands for the scalar (dot) product, *i.e.* $\nabla \cdot \mathbf{v}$ is the divergence of the vector field $\mathbf{v}$. The level set function is usually maintained on a uniform grid and its derivatives are approximated by finite differences. Such calculation requires the level set function to be approx. Linear around a small neighborhood of the zero level.

For its simplicity and the fast transition-developing property we chose the phase field model [10]. The Ginzburg-Landau functional is defined as

$$\iint_\Omega \frac{D_o}{2} |\nabla\phi|^2 + \lambda_o \left(\frac{\phi^4}{4} - \frac{\phi^2}{2}\right) dA, \quad dA = dxdy \tag{1}$$

D_o and λ_o are weights. Its solution is the scalar field ϕ with ± 1 stable values representing the local minimal energies of the field and - due to the first term - a localized transition between these values that naturally represents narrow band.

Albeit the energy (1) could be incorporated into any segmentation functional, this would extremely complicate the analysis of such a complex system. There is though another way using the phase field equation: that is using it in "shape maintaining" role, solving its associated Euler-Lagrange equation independently of and before the segmentation. In either case we wish the phase field equation ideally to maintain the shape of ϕ without moving its level sets. This idea is similar to the regularization of the level set by reinitialization or the diffusion phase of the reaction-diffusion model.

First, we assess the results of the Ginzburg-Landau phase field analysis using linear ansatz (see [10]). One can show that the width of the transition is

$$w_{o*} = \sqrt{\frac{15 D_o}{\lambda_o}}, \tag{2}$$

and the energy of the transitional band is approximately proportional to the perimeter of the innermost (zero) level set. These approximations are valid wherever $w_{o*} |\kappa| \ll 1$. The associated Euler-Lagrange equation is

$$- D_o \triangle \phi + \lambda_o \left(\phi^3 - \phi \right) = 0 . \tag{3}$$

3   Higher Order Smoothness Terms for Phase Field Model

In this section we examine a phase field $\phi(x, y)$ with Laplacian smoothness $(\triangle \phi)^2$- as a potential candidate for our purpose. Note that the origin of the energy can be chosen freely. If the phase field satisfies the condition of constancy almost everywhere except the regions of transitions, the origin is expediently chosen to be the energy level of $\phi = \pm 1$. In this case, the whole energy is equivalent to the energy of the transitions and can be written as

$$\iint_\Omega \frac{D}{2} (\triangle \phi)^2 + \lambda \left(\frac{\phi^4}{4} - \frac{\phi^2}{2} + \frac{1}{4} \right) dx dy . \tag{4}$$

The Euler-Lagrange equation associated with this functional is

$$D \triangle \triangle \phi + \lambda \left(\phi^3 - \phi \right) = 0 . \tag{5}$$

To estimate the energy (4) we use curvilinear coordinates.

3.1   Approximate System Energy

In the vicinity of the curve $\mathbf{r}(s)$ (s is the arc length parameter), the plane can be parameterized as $\mathbf{R}(s, p) = \mathbf{r}(s) + p \mathbf{n}(s)$, where $\mathbf{n}(s)$ is the unit normal vector of the curve at s and p is the coordinate in the normal direction. The metric tensor components are the scalar (dot) products of the covariant basis vectors $\mathbf{R}_s = \frac{\partial \mathbf{R}}{\partial s}$, $\mathbf{R}_p = \frac{\partial \mathbf{R}}{\partial p}$ and takes the form $[g_{ik}] = \text{diag} \left[(1 - p\kappa)^2, 1 \right]$, where $\kappa = \kappa(s)$ is the curvature of the curve $\mathbf{r}$ at s given by the Frenet-Serret formula $\mathbf{n}_s = -\kappa\mathbf{e}$ ($\mathbf{e}$ is the unit tangent vector). The invariant infinitesimal area is $dA = \sqrt{g}\,dsdt$ where $g = \det[g_{ik}]$. Using these, the Laplacian $\triangle\phi$ in the curved system $(u^1, u^2) = (s, p)$ is given by the Laplace-Beltrami operator $\triangle\phi = \frac{1}{\sqrt{g}}\frac{\partial\sqrt{g}g^{ik}\frac{\partial\phi}{\partial u^k}}{\partial u^i} = \frac{1}{1-p\kappa}\left[\frac{\partial(1-p\kappa)^{-1}\frac{\partial\phi}{\partial s}}{\partial s} + \frac{\partial(1-p\kappa)\frac{\partial\phi}{\partial p}}{\partial p}\right]$ (in the general expression the components of the inverse metric $[g^{ik}] = [g_{ik}]^{-1}$ and the Einstein summation convention are used). It can be rearranged as

$$\triangle\phi = \frac{1}{(1-p\kappa)^2}\frac{\partial^2\phi}{\partial s^2} + \frac{p}{(1-p\kappa)^3}\frac{d\kappa}{ds}\frac{\partial\phi}{\partial s} + \frac{\partial^2\phi}{\partial p^2} - \frac{\kappa}{1-p\kappa}\frac{\partial\phi}{\partial p}. \tag{6}$$

Now we choose $\mathbf{r}(s)$ to be the zero level set and use the following simplifications:

1. the constant level sets are equidistant to $\mathbf{r}(s)$ *i.e.* $\phi = const \rightarrow \frac{\partial^n\phi}{\partial s^k\partial p^{n-k}} = 0$, $k \in [1, n]$ along the parameter lines $p = const$
2. the transition is confined to a stripe $\left(-\frac{w}{2}, \frac{w}{2}\right)$ along the zero level set contour
3. the osculating circle is significantly bigger than the stripe width: $1 - p\kappa \approx 1$

then energy (4) expressed in the (s, p) system becomes

$$\oint\int_{-\frac{w}{2}}^{\frac{w}{2}}\frac{D}{2}(\phi'' - \kappa\phi')^2 + \lambda\left(\frac{\phi^4}{4} - \frac{\phi^2}{2} + \frac{1}{4}\right)dpds \tag{7}$$

$\phi(s, p) = \phi(p)$ and $\kappa = \kappa(s)$ is the curvature measured on the zero level set and prime notation is used for the derivatives wrt p.

3.2   Cubic Ansatz

In the presence of the second derivative ϕ'' in (7), the linear ansatz is not applicable. The next simplest choice is a cubic ansatz with boundary conditions $\phi\left(-\frac{w}{2}\right) = -1$, $\phi\left(\frac{w}{2}\right) = 1$ and $\phi'\left(-\frac{w}{2}\right) = \phi'\left(\frac{w}{2}\right) = 0$. The function satisfying these conditions is

$$\phi(p) = -\frac{4}{w^3}p^3 + \frac{3}{w}p. \tag{8}$$

Its derivatives are:

$$\phi' = -\frac{12}{w^3}p^2 + \frac{3}{w}, \quad \phi'' = -\frac{24}{w^3}p, \quad \phi''' = -\frac{24}{w^3}. \tag{9}$$

The square of the approximate Laplacian, obtained from (6) is $(\phi'' - \kappa\phi')^2 = \left[-\frac{24}{w^3}p - \kappa\left(-\frac{12}{w^3}p^2 + \frac{3}{w}\right)\right]^2$. The inner integral in (7) is symmetrical, therefore the terms having odd powers of p do not contribute to the energy. Integrating this smoothness term results $\frac{24D}{w}\left(\frac{1}{w^2} + \frac{\kappa^2}{10}\right)$. Now it can be seen that the appearance of the curvature in the energy violates the assumptions 1 and 2 given in Sect. 3.1. The contribution of the second term $\frac{\kappa^2}{10} = \frac{1}{10r_O^2} \ll \frac{1}{w^2}$ is however, very modest thus omitted in the subsequent calculations. Similarly, the inner integral of the phase field double well potential term $\lambda \left(\frac{\phi^4}{4} - \frac{\phi^2}{2} + \frac{1}{4} \right)$ is approximately $0.1\lambda w$, hence the approximate energy of (7) is $L \left(\frac{24D}{w^3} + \frac{\lambda w}{10} \right)$ (L is the contour length). Deriving it wrt w, the optimal width of the transitional region is

$$w_* = \sqrt[4]{\frac{720D}{\lambda}} . \tag{10}$$

3.3   The Motion of the Level Sets

The Euler-Lagrange equation (5) can be expressed in the curvilinear system aligned with the zero level set applying the Laplace-Beltrami operator once again to the Eq. (6) (and multiplying the result by $\sqrt{g} = 1 - p\kappa$). The Euler-Lagrange terms having the derivatives of ϕ by the contour parameter s can be omitted in the result. This approximate equation is

$$D \left(-A\phi' - 2\kappa\phi''' - \kappa^2\phi'' + \phi'''' \right) + \lambda \left(\phi^3 - \phi \right) = 0$$

$$A = 3p \left(\frac{d\kappa}{ds} \right)^2 + \frac{d^2\kappa}{ds^2} + \kappa^3 . \tag{11}$$

The shape of the numerical solution for (11) is close to the cubic ansatz (see Fig. 2). Due to the assumed symmetry of the zero level set, its motion is governed by $-D \left(\frac{d^2\kappa}{ds^2} + \kappa^3 \right) \phi' - 2D\kappa\phi''' = 0$ or using the cubic ansatz at $p = 0$ and (10):

$$\frac{48D}{w_*^3}\kappa - \left(\frac{d^2\kappa}{ds^2} + \kappa^3 \right) \frac{3D}{w_*} = 0 . \tag{12}$$

Equation (12) describes either a static state wherever the curvature is identically zero or shrinking proportional to the curvature where the radius r_O of the osculating circle is significantly bigger than the thickness of the transition and this width varies slowly (Fig. 1).

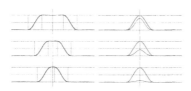

Fig. 1. Alteration of the phase field function in normal direction (thin blue line). Left column: small curvature, the cubic ansatz (thick violet line) is valid. Right column: high curvature, cubic approximation is invalid. (Color figure online)

For contours with constant curvature a static solution would be at radius $\frac{48\kappa}{w_*^3} = \frac{3\kappa^3}{w_*} \rightarrow r = \frac{w_*}{4}$, however this is not the case. Around this curvature value neither the assumption $1 - p\kappa \approx 1$ nor the cubic ansatz approximation are valid. Under these circumstances the phase field function can no longer be modeled with cubic ansatz (see Fig. 2 right column). The theoretical minimum value while the phase field is shape-retaining is $\frac{w_*}{2}$.

Therefore, we conclude that energy (4) does not fulfill our expectation stated in the beginning of this section because it still has a curvature dependent term.

3.4   Motion of the Level Sets of the Original Model

Assuming again that the conditions that led to the simplified energy expression (7) are valid, the Euler-Lagrange equation of the Ginzburg-Landau phase field (3) reduced to the normal direction using the Laplacian (6) at $1 - p\kappa \approx 1$ and $\frac{\partial^n \phi}{\partial s^n} = 0$ is: $-D_o \left(\phi'' - \kappa\phi'\right) + \lambda_o \left(\phi^3 - \phi\right) = 0$. From this expression, the motion of the zero level set is governed by

$$D_o \kappa \phi' = 0, \tag{13}$$

that is a pure curvature-driven motion.

4   Phase Field Model for Reinitialization Purpose

The motion of the zero level set is basically curvature driven both for the Ginzburg-Landau (13) and the higher order smoothness (12) models. This effect can be eliminated by the appropriate combination of the smoothness terms $(\nabla\phi)^2$ and $(\triangle\phi)^2$. First we calculate the optimal width for the functional

$$\iint_\Omega \frac{D}{2} |\triangle\phi|^2 - \frac{D_o}{2} |\nabla\phi|^2 + \lambda \left(\frac{\phi^4}{4} - \frac{\phi^2}{2} + \frac{1}{4}\right) dA. \tag{14}$$

The approximate energy - using simplifications 1.-3. introduced in Sect. 3.1 - is

$$L \int_{-\frac{w}{2}}^{\frac{w}{2}} \frac{D}{2} \left(\phi'' + \kappa\phi'\right)^2 - \frac{D_o}{2} \left(\phi'\right)^2 + \lambda \left(\frac{\phi^4}{4} - \frac{\phi^2}{2} + \frac{1}{4}\right) dp, \tag{15}$$

where the length of the contour is $L = \oint ds$ is independent of w. Substituting the cubic ansatz (8), the integral (15) (divided by L) becomes

$$24D \left(\frac{1}{w^3} + \frac{\kappa^2}{10w}\right) - \frac{12D_o}{5w} + \frac{\lambda w}{10}. \tag{16}$$

The term dependent on the square of the curvature is again negligible, hence omitted. From expression (16), the optimal width is given by derivation wrt w

$$\boxed{\lambda w^4 - 24D_o w^2 - 720D = 0} \tag{17}$$

that can be solved for the optimal width w_*. The solution is

$$w_* = \sqrt{\frac{12}{\lambda} \left(-D_o + \sqrt{D_o^2 + 5D\lambda}\right)}. \tag{18}$$

Now we use the approximate Euler-Lagrange equation associated with (15)

$$D\left(-A\phi' - 2\kappa\phi''' - \kappa^2\phi'' + \phi''''\right) + D_o\left(\phi'' - \kappa\phi'\right) + \lambda\left(\phi^3 - \phi\right) = 0 \qquad (19)$$

to derive condition for the curvature-independent solution (here A is defined in (11)). From (19) the curvature-dependent term is eliminated with the condition: $-D_o\phi' - 2D\phi''' \doteq 0$. Substituting the cubic ansatz (8) (at $p = 0$) we get:

$$\boxed{-D_o\frac{3}{w} + D\frac{48}{w^3} = 0} \;\rightarrow\; D = \frac{w^2}{16D_o}. \qquad (20)$$

The width (17) and the curvature (20) constraints determine the weights for the solution with curvature driven shrinking effect removed. There are other terms, e.g. terms included in factor A, but the influence of those is much weaker. In fact the impact of the term $D\left(\frac{d^2\kappa}{ds^2} + \kappa^3\right)$ is similar to that of the solution of the Euler's elastica $\oint \frac{D}{2}\kappa^2 ds$ with associated Euler-Lagrange equation: $D\left(\frac{d^2\kappa}{ds^2} + \frac{1}{2}\kappa^3\right) = 0$. The numerical tests confirm that the phase field used in this manner - satisfying Eqs. (17), (20) - essentially fulfills the "transitional shape maintenance" role while standing still.

4.1 Determining Weights

Given two constraints (17), (20) for the energy (14), one of the weights can be chosen freely (say $D_o = 1$). The calculation of the remaining weights are as follows. First determine the width: depending on the highest order of the derivatives n (occurring either in the segmentation model or the phase field itself), we need at least $n + 1$ grid points around the zero level set using finite central difference schemes. This suggests about twice as big (as a cautious choice) thickness of the phase field transition to remain within the range where it is approximately linear, *i.e.* $w \geqq 2(n + 1)$ is recommended. Second, solving (17) and (20), the weights the functions of the width parameter w such as:

$$D_o = 1, \; D = \frac{w^2}{16}, \; \lambda = \frac{21}{w^2}. \qquad (21)$$

The Euler-Lagrange equation associated with the proposed energy (14), using the calculated weights (21) dependent on the width parameter w is therefore

$$\boxed{\frac{w^2}{16}\triangle\triangle\phi + \triangle\phi + \frac{21}{w^2}\left(\phi^3 - \phi\right) = 0}. \qquad (22)$$

In (22) the Laplace operator can be expressed wrt the standard basis as $\triangle\Phi = \frac{\partial^2\Phi}{\partial x^2} + \frac{\partial^2\Phi}{\partial y^2}$, $\Phi \in \{\phi, \triangle\phi\}$ and discretized on a uniform grid using finite differences. Its gradient descent was used in the tests. The method can be efficiently implemented as a 5×5 linear filter plus a point-wise cubic term acting on the uniform grid used to discretize the level set function. The approximation $r_{min} \approx X(n + 1)$ (where X is the grid size that can be smaller or greater than a pixel) also determines the size of the segmentable smallest image-feature.

5   Experimental Evaluation

In this section, we show that our balanced phase field model (a) maintains a smooth transition of the level set in a narrow band during the evolution, while (b) it has minimal side effect on the contour at the same time. This section is organized as follows: first, we compare the Ginzburg-Landau phase field model to the proposed one to show that the latter has much better contour preserving performance. Next, we compare the proposed model, the reaction diffusion model [8] and a reinitialization method [6] on synthetic and real data.

5.1   Stability Tests: Comparing to the Ginzburg-Landau Model

The test environment is prepared to guarantee the synchronous snapshot production for the illustrations: the simulation space is splitted in the middle such that the phase field evolutions are governed by the proposed energy (14) on one side and the Ginzburg-Landau energy (1) on the other.

For the first test, the pure phase field equations were used. Figure 2 left shows the initial contour preserving capability of the balanced phase field model compared with the Ginzburg-Landau model on the right side.

Fig. 2. Alteration after the same iteration numbers of the phase fields when $w = 10$, left: the balanced phase field - interfaces are barely moved; right: the Ginzburg-Landau phase field: interfaces are displaced significantly. Weights are set according to (21) and (2) for the balanced model and the Ginzburg-Landau model respectively.

The proposed phase field model was also applied to real data segmentation, using a selective segmentation model [12]. The energy to be minimized is: $E = \alpha S + \beta P + \gamma D + \delta \mathcal{E}$, where $\mathcal{P} = \frac{1}{2} \left[\oint dA - q \left(\oint ds \right)^2 \right]^2$ is the "plasma shape" prior (q is the shape parameter, the ratio of the enclosed area and the square of the perimeter), the data term of the original model was replaced by the simplest anisotropic edge energy $\mathcal{D} = \oint \nabla I \cdot \mathbf{n} ds$ (see [13]); the $\mathcal{E}$ is the Euler elastica, while α, β, γ, δ are weights. This segmentation model was chosen, because of its sensitivity to any size decreasing effect due to the term $S = \frac{1}{3} \left(\oint dA - A_0 \right)^3$ which is used at its inflection point at the preferred size A_0. The initial contours were produced by simple thresholding. Segmentation steps and the sequence of the phase field sectional values along a horizontal line are shown in Fig. 3.

For the test $X = 1$ pixel grid size and $w = 10$ width values were used; the maximum speed of the evolution by the segmentation model was set such that its maximal value could not exceed the grid size. Preceding the segmentation step, the gradient descent equation of (22) is iterated and the phase field is

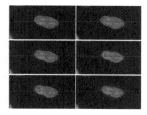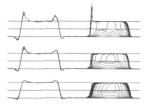

Fig. 3. Left: Evolution of the selective segmentation example (grid size: 1 pixel, w = 10). Segmentation result: First column: the balanced; Second column: the Ginzburg-Landau phase fields. Right: Details of evolution along the blue horizontal line using reinitializing iterations 2, 5, 15 for shape maintenance. First column: the balanced phase field; Second column: Ginzburg-Landau phase field. (Color figure online)

updated in a reinitialization loop to recover a reasonably smooth interface. Then the segmentation gradient descent moves the contour towards the solution (but deteriorates its shape).

The transitional-shape recovery and the segmentation results using the balanced and the Ginzburg-Landau models, depending on the number of the phase field iterations in the reinitialization step (denoted by n) are assessed here: At $n = 2$ neither the balanced nor the Ginzburg-Landau models can be considered stable, at $n = 5$ both models provide stable transition, however the Ginzburg-Landau model develops extremely steep slopes, while at $n = 15$, both models exhibit high degree of stability as well as widths close to the designed/predicted ones. Regardless the number of the phase field iterations used, the selective segmentation [12] combined with the Ginzburg-Landau model ends up in the collapse of the contour, whilst its combination with the balanced phase field model provides the expected solution.

5.2 Comparing to the Reaction-Diffusion Model

The reaction diffusion model (RD) [8] is also proposed to diminish the interference with the segmentation (active) model. The shape maintenance of the level set function is achieved by adding a diffusion term $\varepsilon\Delta\phi$ to the gradient descent of the active model, therefore $\phi_t = \varepsilon\Delta\phi + \frac{1}{\varepsilon}F|\nabla\phi|$, where ε is a small

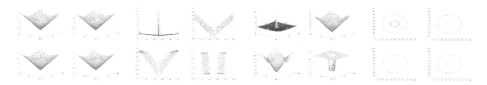

Fig. 4. First two columns: initial level sets; second: level set during evolution (intersection at $x = 50$); third: final level sets, fourth: final contours. In each group - top: DI, RM, bottom: RD, BPF.

constant and F represents the gradient descent equation of the active model. We first show that both the RD and the proposed balanced phase field model fulfill the shape maintenance role. Next, we show that RD moves the interface more significantly compared to the proposed model. In case of RD, this shrinking side-effect eventually leads to the disappearing of some objects. For the quantitative results, we borrowed a Jaccard-distance based metric similar to the one used in the 2018 Data Science Bowl (DSB2018) competition [14]. The only modification is that we used the threshold levels $t = 0.1$ to $t = 0.95$ with steps 0.5 (inclusive).

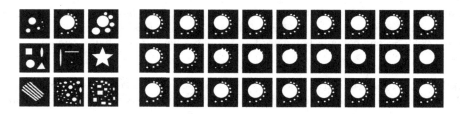

Fig. 5. Left: sample synthetic masks encoding initial contours; right: level set evolutions on a sample image from the left side with RM/RD/BPF from top to bottom, respectively.

Shape Maintenance Tests: We compared three different models to ours. We tested the reaction diffusion method (RD), the balanced phase field model (BPF), a reinitialization method (RM) [6] using the reinitialization equation: $\phi_t + S(\phi_0)(|\nabla\phi| - 1) = 0$, where $S(\phi) = \frac{\phi}{\sqrt{\phi^2 + (|\nabla\phi|\Delta x)^2}}$, (the same method as the one used in the RD paper and referred to as re-initialization. (For the implementation, see the online supplementary material of that paper), and lastly, no shape maintenance) (DI - direct implementation). The first test inherits from Fig. 5 of the RD paper [8], $\Delta t_1, \Delta t_2$ (used for the numerical solution of the RD equation, see the RD paper for details) are 0.1. For the RM, Δx and Δy is 1 and $\alpha = 0.5$, while $w = 8$ in the BPF model. The force term in the gradient descent of the active contour model is simply 1, grid dimensions: 100×100, number of iterations: 200. The results are shown in Fig. 4.

Synthetic Tests: The models are compared to each other by performing a level set evolution using 11 synthetic initial contours (subset of these masks are shown in Fig. 5) using 0 as a force term in the active model. In this setting, we would assume that the initial contours are not moving. A sample evolution is visualized in Fig. 5 using the RM, RD and the BPF methods. The same test performed with all of the synthetic initial contours. The quantitative results using the modified DSB2018 metric presented in Fig. 6 left. The ground truth is the initial contour and the accuracy is measured during the evolution on every image. Simple statistics summarizes the results in Table 1 left.

Table 1. Segmentation accuracy of the synthetic and the real tests. Peak: the best scores reached during the evolution. RD can not keep the accuracy of the active model in long term, it has massive side-effects while the BPF behaves similarly to the reinitialization method.

	Synthetic			Real		
	RM	RD	BPF	RM	RD	BPF
Result mean	0.943	0.339	0.882	0.365	0.093	0.343
Peak mean	0.948	0.983	0.975	0.608	0.802	0.794

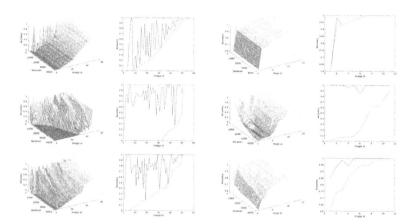

Fig. 6. Left: quantitative results for the synthetic test. From top to bottom: RM, RD, BPF. Columns: first: accuracy on each image during the evolution, second: peak (maximum acc. on an image, blue) and the final acc. for the images. Right: results on patches extracted from the DSB2018 training set, GAC model ($\nu = 0.5$). From top to bottom: RM, RD, BPF. (Color figure online)

Real Tests: We also compared the methods to each other using the geodesic active contour model (GAC) [15] on 51 real images containing nuclei extracted from the DSB2018 training set with random sampling. In the GAC model, the force term in the gradient descent equation is: $\nabla \cdot \left(g(I)\frac{\nabla\phi}{|\nabla\phi|} \right) + \nu g(I)$, where $g(I) = \frac{1}{1+(\nabla G(15,1.5)*I)^2}$ is the edge indicator function (the same as the one used in the RD paper for the tests). The quantitative results with this model are presented in Fig. 6 right. The parameters are unchanged. A sample test image used for this test is shown in Fig. 7. Simple statistics presented in Table 1 right. The parameters left unchanged since the last test, except the Δt_2 that is 0.001 in this case. In conclusion, the BPF outperformed RD both for the synthetic and real tests and produced results that are comparable to the RM method.

(a) The most accurate contours achievable with each of the methods.

(b) The contours after 40k iterations with different reinitialization methods.

Fig. 7. Active contour evolutions with different reinitialization methods on a patch from the DSB2018 dataset: red: BPF, green: RD, blue: RM. Even if the active model is able to achieve good accuracy (Fig. 7a) the contour vanishes by the time if we use the RD method (*serious side effect* on the active model). The proposed model has contour preserving ability comparable to the reference method (*marginal side effect* on the active model) (Fig. 7b). (Color figure online)

6   Discussion

In this paper we proposed and analyzed a *balanced phase field model* as an alternative to the Ginzburg-Landau level set framework. The proposed model exhibits very fast shape recovery (essentially) without moving the level sets *i.e.* its interference with the "active" (*e.g.* segmentation) PDE is negligible. This important property makes this level set formulation suitable for accurate segmentation. Similar balancing could be used for any model that includes Laplacian smoothness term in their gradient descent equation such as the reaction-diffusion model.

References

1. Kass, M., Witkin, A., Terzopoulos, D.: Snakes: active contour models. IJCV **1**(4), 321–331 (1988)
2. Caselles, V., Catté, F., Coll, T., Dibos, F.: A geometric model for active contours in image processing. Numer. Math. **66**(1), 1–31 (1993)
3. Malladi, R., Sethian, J.A., Vemuri, B.C.: Shape modeling with front propagation: a level set approach. IEEE TPAMI **17**(2), 158–175 (1995)
4. Osher, S., Sethian, J.A.: Fronts propagating with curvature-dependent speed: algorithms based on Hamilton-Jacobi formulations. J. Comput. Phys. **79**(1), 12–49 (1988)
5. Gomes, J., Faugeras, O.: Reconciling distance functions and level sets. J. Vis. Commun. Image Represent. **11**(2), 209–223 (2000)
6. Peng, D., Merriman, B., Osher, S., Zhao, H., Kang, M.: A PDE-based fast local level set method. J. Comput. Phys. **155**(2), 410–438 (1999)
7. Li, C., Xu, C., Gui, C., Fox, M.D.: Distance regularized level set evolution and its application to image segmentation. IEEE TIP **19**(12), 3243–3254 (2010)
8. Zhang, K., Zhang, L., Song, H., Zhang, D.: Reinitialization-free level set evolution via reaction diffusion. IEEE TIP **22**(1), 258–271 (2013)
9. Wang, X., Shan, J., Niu, Y., Tan, L., Zhang, S.-X.: Enhanced distance regularization for re-initialization free level set evolution with application to image segmentation. Neurocomputing **141**, 223–235 (2014)

10. Rochery, M., Jermyn, I., Zerubia, J.: Phase field models and higher-order active contours. In: Tenth IEEE International Conference on Computer Vision ICCV 2005, vol. 2, pp. 970–976. IEEE (2005)
11. Esedog, S., Tsai, Y.-H.R., et al.: Threshold dynamics for the piecewise constant mumford-shah functional. J. Comput. Phys. **211**(1), 367–384 (2006)
12. Molnar, J., Szucs, A.I., Molnar, C., Horvath, P.: Active contours for selective object segmentation. In: IEEE WACV, pp. 1–9. IEEE (2016)
13. Kimmel, R., Bruckstein, A.M.: Regularized Laplacian zero crossings as optimal edge integrators. IJCV **53**(3), 225–243 (2003)
14. 2018 Data Science Bowl. https://www.kaggle.com/c/data-science-bowl-2018
15. Caselles, V., Kimmel, R., Sapiro, G.: Geodesic active contours. IJCV **22**(1), 61–79 (1997)

Unsupervised Labeling by Geometric and Spatially Regularized Self-assignment

Matthias Zisler[1(✉)], Artjom Zern[1], Stefania Petra[2], and Christoph Schnörr[1]

[1] Image and Pattern Analysis Group, Heidelberg University, Heidelberg, Germany
zisler@math.uni-heidelberg.de
[2] Mathematical Imaging Group, Heidelberg University, Heidelberg, Germany

Abstract. We introduce and study the *unsupervised self-assignment flow* for labeling image data (euclidean or manifold-valued) without specifying any class prototypes (labels) beforehand, and without alternating between data assignment and prototype evolution, which is common in unsupervised learning. Rather, a *single* smooth flow evolving on an elementary statistical manifold is geometrically integrated which assigns given data to *itself*. Specifying the *scale* of spatial regularization by geometric averaging suffices to induce a *low-rank* data representation, the emergence of prototypes together with their number, and the data labeling. Connections to the literature on low-rank matrix factorization and on data representations based on discrete optimal mass transport are discussed.

1 Introduction

The assignment flow introduced by [3] is a smooth dynamical system for image labeling. Its state is given by discrete distributions (assignment vectors) assigned to each pixel, like in established discrete graphical models [13]. These states evolve on probability simplices equipped with the Fisher-Rao metric in order to minimize locally a distance to class prototypes, commonly called *labels*. Label assignments emerge gradually as the flow evolves and are spatially regularized by geometric averaging over local neighborhoods. Convergence analysis for a particular multiplicative update scheme was reported in [4].

Due to its *smoothness* and the wide range of *sparse* numerical updates that can be derived by geometrically integrating the assignment flow [22], this approach provides an attractive alternative to established discrete graphical models for image labeling [13], that rely on convex relaxations for large-scale problems, which do not scale well with increasing problem size or increasing numbers of labels. The evaluation of graphical models using the assignment flow has been demonstrated recently [12].

A common problem of *supervised* image labeling concerns the specification of labels beforehand, that is to determine prototypical features that properly represent different data categories into which given image data should be classified in a spatially coherent way. As a remedy, the *adaption* of labels during the

J. Lellmann et al. (Eds.): SSVM 2019, LNCS 11603, pp. 432–444, 2019.
https://doi.org/10.1007/978-3-030-22368-7_34

assignment process was proposed recently [23]: Starting with an 'overcomplete' dictionary of prototypes (labels) that is efficiently computed beforehand using conventional metric clustering, prototypes evolved on a corresponding feature manifold while being coupled to the assignment flow.

While this approach largely compensates the lack of prior knowledge of adequate labels, it does not directly address the fundamental question: How may prototypes emerge *directly* from the data during the assignment process *without any* prior coding using conventional clustering? This paper describes our first step towards a *completely unsupervised* approach to image labeling.

Contribution. Our approach is to apply the supervised assignment flow (Sect. 2) to the *self-assignment* of the given data (Sect. 3). This gives rise to a *factorized* affinity matrix that is parametrized by the assignments and converges to a *low-rank representation* of the data, *solely* induced by the *scale* of the spatial regularization performed by the assignment flow. This process also defines the formation of feature prototypes (labels) and a proper number of classes.

Related Work. Our work utilizes information geometry and relates to the current dynamically evolving literature on clustering using low-rank matrix factorizations, and on data representation using discrete optimal mass transport. These relations are discussed in Sect. 4 after presenting our novel approach.

We conclude with experiments in Sect. 5. In order to illustrate one-to-one the content of the preceding sections, our implementation does *not* involve any further changes of the assignment flow. In particular, we did not change and adapt the numerics in order to exploit the low-rank structure for large problem sizes right from the beginning. We leave such aspects for future work and used small and medium problem sizes for our experiments, to be able to apply the assignment flow *directly* to the self-assignment of given data, and to study how labels emerge in a completely unsupervised way.

2   Assignment Flow: Supervised Labeling

We summarize the assignment flow for *supervised* image labeling introduced by [3]. Let $G = (I, E)$ be a given undirected graph with vertices $i \in I$ indexing data $\mathcal{F}_I = \{f_i \colon i \in I\} \subset \mathcal{F}$ given in a metric space $(\mathcal{F}, d)$. The edge set E specifies neigborhoods $\mathcal{N}_i = \{k \in I \colon ik = ki \in E\} \cup \{i\}$ for every pixel $i \in I$, together with positive weight vectors $w_i \in \mathrm{rint}\,\Delta_{|\mathcal{N}_i|}$, where $\Delta_n \subset \mathbb{R}^n$ denotes the probability simplex.

Along with $\mathcal{F}_I$, *prototypical data (labels)* $\mathcal{G}_J = \{g_j \in \mathcal{F} \colon j \in J\}$ are given representing classes $j = 1, \ldots, |J|$. *Supervised image labeling* denotes the task to assign precisely one prototype g_j to each datum f_i in a spatially coherent way. These assignments are represented at each pixel i by probability vectors

$$W_i \in \mathcal{S} := \mathrm{rint}\,\Delta_{|J|}, \quad i \in I \tag{1}$$

on the relative interior of the simplex $\Delta_{|J|}$, that together with the Fisher-Rao metric g_{FR} becomes a Riemannian manifold denoted by $\mathcal{S}$. Collecting all assignment vectors into a strictly positive, row-stochastic matrix

$$W = (W_1, \ldots, W_{|I|})^\top \in \mathcal{W} = \mathcal{S} \times \cdots \times \mathcal{S} \subset \mathbb{R}^{|I| \times |J|} \tag{2}$$

defines a point on the *assignment manifold* $\mathcal{W}$. Image labeling is accomplished by geometrically integrating the *assignment flow* (the r.h.s. is defined below)

$$\dot{W} = \Pi_W(S(W)), \qquad W(0) = \mathbb{1}_{\mathcal{W}}, \tag{3}$$

that evolves from the barycenter $W(0)$ towards *pure* assignment vectors, i.e. each vector W_i approaches the ε-neighborhood of some unit vector at some vertex of $\mathcal{S}$ and hence a *labeling* after trivial rounding.

In order to explain the rationale behind (3), we need the following maps based on the affine e-connection of information geometry [2] in place of the Levi-Civita connection on the tangent bundle of the manifolds $\mathcal{S}$ and $\mathcal{W}$: With tangent space $T_0 = T_p\mathcal{S}$ independent of the base point $p \in \mathcal{S}$, we define

$$\mathbb{R}^{|J|} \ni z \mapsto \Pi_p(z) = \left(\text{Diag}(p) - pp^\top\right)z \in T_0, \tag{4a}$$

$$\mathcal{S} \times T_0 \ni (p, v) \mapsto \text{Exp}_p(v) = \frac{e^{\frac{v}{p}}}{\langle p, e^{\frac{v}{p}}\rangle}p \in \mathcal{S}, \tag{4b}$$

$$\mathcal{S} \times \mathcal{S} \ni (p, q) \mapsto \text{Exp}_p^{-1}(q) = \Pi_p \log \frac{q}{p} \in T_0, \tag{4c}$$

$$\mathcal{S} \times \mathbb{R}^{|J|} \ni (p, z) \mapsto \exp_p(z) = \text{Exp}_p \circ \Pi_p(z) = \frac{pe^z}{\langle p, e^z\rangle} \in \mathcal{S}, \tag{4d}$$

where multiplication, subdivision and the exponential function $e^{(\cdot)}$ apply *componentwise* to strictly positive vectors in $\mathcal{S}$. Corresponding maps $\Pi_W, \text{Exp}_W, \exp_W$ in connection with the product manifold (2) are defined analogously.

The vector field defining the assignment flow on the right-hand side of (3) is defined as follows. Given the metric d, data $\mathcal{F}_I$ and labels $\mathcal{G}_J$, distance vectors $D_i = (d(f_i, g_1), \ldots, d(f_i, g_{|J|}))^\top$ are defined at each pixel $i \in I$ and mapped to the assignment manifold by

$$L(W) = \exp_W\left(-\tfrac{1}{\rho}D\right) \in \mathcal{W}, \quad L_i(W_i) = \exp_{W_i}\left(-\tfrac{1}{\rho}D_i\right) = \frac{W_i e^{-\frac{1}{\rho}D_i}}{\langle W_i, e^{-\frac{1}{\rho}D_i}\rangle}, \tag{5}$$

where $\rho > 0$ is a user parameter for normalizing the scale of the data. These *likelihood vectors* represent 'data terms' in conventional variational approaches, and they are *spatially regularized* in a way conforming to the geometry of $\mathcal{S}$, to obtain

$$S(W) = \mathcal{R}^w(L(W)) \in \mathcal{W}, \quad \mathcal{R}_i^w(W) := \text{Exp}_{W_i}\left(\sum_{k \in \mathcal{N}_i} w_{ik} \text{Exp}_{W_i}^{-1}(W_k)\right). \tag{6}$$

The assignment flow (3) is well-defined based on (6). In addition, following [3], it *may* also be interpreted from a variational perspective as *approximate* Riemannian gradient ascent flow $\dot{W} = \nabla_{\mathcal{W}} J(W)$ with respect to the correlation functional $J(W)$,

$$\nabla_{\mathcal{W}} J(W) = \Pi_W(\nabla J(W)), \qquad J(W) = \langle W, S(W) \rangle, \tag{7}$$

based on the approximation of the Euclidean gradient $\nabla J(W) \approx S(W)$, which is justified by the slow dynamics of $S(W(t))$ due to averaging (6), relative to the fast dynamics of $W(t)$.

3   Approach: Label Learning Through Self-assignment

In this section we generalize the assignment flow to completely *unsupervised* scenarios. Specifically, we do *not* assume a set of prototypes $\mathcal{G}_J$ to be given. Rather, we initially set $\mathcal{G}_J = \mathcal{F}_I$ and consider each datum *both* as data point $f_i \in \mathcal{F}_I$ and (its copy) as label $f_i \in \mathcal{G}_J$. Consequently, the distance matrix of (5) is now defined as

$$D = \big(d(f_i, f_k)\big)_{i,k \in I}. \tag{8}$$

Integrating the assignment flow then performs a *spatially regularized self-assignment* of the data, based on which the set $\mathcal{G}_J$ evolves and forms *prototypes* in an unbiased and unsupervised way.

We regard these prototypes as *latent* variables denoted by $g_j \in \mathcal{G}_J$, to be distinguished from $f_i \in \mathcal{F}_I$ which are both *data points* and *labels*.

3.1   Rationale

Due to initially setting $\mathcal{G}_J = \mathcal{F}_I$, we have $J = I$ and the row-stochastic assignment matrix (2) is quadratic: $W \in \mathcal{W} \subset \mathbb{R}^{|I| \times |I|}$. Adopting from [3] the interpretation of the entry W_{ij} as posterior probability of assigning label f_j conditioned on the observation f_i,

$$W_{ij} = P(j|i), \quad j \in J, \ i \in I, \qquad P(i) = \frac{1}{|I|}, \quad i \in I \tag{9}$$

together with uniform prior probabilities $P(i)$ due to the absence of any supervision, Bayes' rule yields the probability of observing datum f_i conditioned on the label f_j,

$$P(i|j) = \frac{P(j|i)P(i)}{P(j)} = \frac{P(j|i)P(i)}{\sum_{l \in I} P(j|l)P(l)} \overset{(9)}{=} \frac{P(j|i)}{\sum_{l \in I} P(j|l)} \overset{(9)}{=} \frac{W_{ij}}{\sum_{l \in I} W_{lj}} \tag{10a}$$

$$= \big(WC(W)^{-1}\big)_{ij} \quad \text{with} \quad C(W) := \text{Diag}(W^\top \mathbb{1}_{|I|}), \tag{10b}$$

that is by *normalizing* the *columns* of W. Since the *rows* of W are normalized by definition, this *symmetry* reflects our ansatz to form prototypes from the *entire* given data set $\mathcal{F}_I$.

Next we introduce and compute the *probabilities of self-assignments* $f_i \leftrightarrow f_k$ by marginalizing over the labels f_j, $j \in J$,

$$A_{ki}(W) := \sum_{j \in J} P(k|j)P(j|i) \overset{(9),(10)}{=} \sum_{j \in J} (WC(W)^{-1})_{kj} W_{ij} = (WC(W)^{-1}W^\top)_{ki}. \tag{11}$$

The resulting *(self)-affinity matrix*

$$A(W) \in \mathbb{R}_+^{|I| \times |I|}, \qquad A(W) = A(W)^\top, \quad A(W)\mathbb{1}_{|I|} = A(W)^\top \mathbb{1}_{|I|} = \mathbb{1}_{|I|} \tag{12}$$

is nonnegative, symmetric and doubly stochastic. It represents the mutual influence of the features at all pixels, as a function of the assignment matrix W.

As a consequence, we propose to replace the objective (7) used in the *supervised* case which maximizes the correlation of assignments and a spatially regularized representation of the affinity between data and prototypes, by the objective function

$$\min_{W \in \mathcal{W}} E(W), \qquad E(W) = \langle D, A(W) \rangle, \tag{13}$$

which in the present *unsupervised* scenario minimizes the correlation between the data (distance) matrix D (8) and the self-affinity matrix $A(W)$: whenever the feature distance between pixel i and k is *large*, the affinity probability $A_{ik}(W)$ between this pair of pixels should *decrease*, subject to mass conservation (12).

The *latent* (hidden) prototypes $\mathcal{G}_J$ that emerge from the data $\mathcal{F}_I$ are *implicitly* determined by the column-normalized assignment matrix (10) that minimizes (13): entries $(P(i|j))_{i \in I}$ signal the relative contribution of each data point i to forming the prototype g_j. How g_j is actually computed depends on the nature of the feature space $\mathcal{F}$ whose properties only matter at this point: a corresponding weighted average has to be well-defined. In the simplest case, the space $\mathcal{F}$ is Euclidean and prototype g_j, $j \in J$ is defined as the convex combination of all data points f_i, $i \in I$ with the probabilities $P(i|j)$, $i \in I$ as coefficients, i.e.

$$g_j = \sum_{i \in I} (WC(W)^{-1})_{ij} f_i. \tag{14}$$

Of particular interest is the capability of this process to represent given data by *few* prototypes. Our approach accomplishes this in a natural way, solely depending on the *scale* at which spatial regularity is enforced in terms of the neighborhood size $|\mathcal{N}_i|$ for geometric averaging (6).

3.2   Computational Approach

We explain our approach as adaption of the *supervised* assignment flow (7) to *unsupervised* scenarios.

The vector field on the right-hand side of (7) involves the geometrically and spatially averaged likelihood vectors (5), which in turn result from mapping the feature distance matrix D to $L(W) \in \mathcal{W}$ on the assignment manifold. In view of the data terms of established variational segmentation approaches (see [6,15] for the binary and non-binary case, respectively), we regard $D = \nabla_W \langle D, W \rangle$ as Euclidean gradient of such a basic data term.

A natural way to adapt the assignment flow to the present unsupervised setting is to replace this gradient by the Euclidean gradient of the objective (13), that is we redefine (5) as

$$L(W) = \exp_W \left(-\tfrac{1}{\rho} \nabla E(W) \right) \in \mathcal{W}, \quad L_i(W) = \frac{W_i e^{-\frac{1}{\rho} \nabla E(W)_i}}{\langle W_i, e^{-\frac{1}{\rho} \nabla E(W)_i} \rangle}, \quad \rho > 0 \quad (15)$$

with the gradient of (13) given by

$$\nabla E(W) = 2 DWC(W)^{-1} - \mathbb{1}_{|I|} \operatorname{diag}\left(C(W)^{-1} W^\top DWC(W)^{-1} \right)^\top, \quad (16)$$

where $\operatorname{diag}(\cdot)$ denotes the vector of diagonal elements of a matrix.

Besides this modification of (5), the remaining formulas (6), (7) do not change. The *unsupervised self-assignment flow*, therefore, reads

$$\dot{W}(t) = \Pi_{W(t)}\big(S(W(t))\big), \quad W(0) = \exp_{\mathbb{1}_\mathcal{W}}(-\varepsilon D) \in \mathcal{W}, \quad 0 < \varepsilon \ll 1, \quad (17)$$

where the initial point is a small perturbation of the barycenter $\mathbb{1}_\mathcal{W}$, in order to break the symmetry of the expression defining the gradient (16) that would result from choosing $W = \mathbb{1}_\mathcal{W}$ as initial point. For numerical schemes that properly integrate flows of the form (17) evolving on the manifold $\mathcal{W}$, we refer to [22].

We point out that the perturbed initialization (17) is not required in the supervised case in which distances are not averaged: compare $L(W)$ of (15) with the supervised version (5). Indeed, comparing again (5) and (15), we may view the gradient (16) as a *time-varying* distance matrix

$$D(t) = D(W(t)) := \nabla E(W(t)), \qquad D(0) = \nabla E\big(\exp_{\mathbb{1}_\mathcal{W}}(-\varepsilon D)\big), \quad (18a)$$

$$D(t)_{ij} \overset{(16)}{=} 2\langle D_i, W_C^j \rangle - \langle W_C^j, DW_C^j \rangle \quad \text{with} \quad W_C^j := (WC(W)^{-1})_{*j}, \quad (18b)$$

that emanates from (8) and takes into account the formation of the latent prototypes g_j, $j \in J$, caused by the self-assignment process (17). The prototypes are *implicitly* represented by the normalized column vectors W_C^j of the assignment matrix W, where each component $(W_C^j)_i = P(i|j)$ (see (10)) represents the support (affinity) of pixel $i \in I$. Accordingly, distances $D_{ij} = d(f_i, f_j)$ due to (8) are replaced in (18) by the *time-varying* averaged distances

$$D(t)_{ij} = 2\mathbb{E}\big[d(f_i, \mathcal{F}_I)\big|g_j\big] - \mathbb{E}\big[d(\mathcal{F}_I, \mathcal{F}_I)\big|g_j\big], \quad (19a)$$

$$\text{where} \quad \mathbb{E}\big[d(f_i, \mathcal{F}_I)\big|g_j\big] = \sum_{k \in I} P(k|j) d(f_i, f_k), \quad (19b)$$

of feature f_i to all features f_k supporting prototype g_j.

3.3   Spatially Regularized Optimal Transport

Problem (13) may also be regarded as *discrete optimal transport* problem [17], with D as cost matrix, with a transportation plan $A(W)$ *parametrized* by the assignment matrix W, and with marginal constraints (12) implied by the constraint $W \in \mathcal{W}$ of (13).

At first glance, this problem looks uninteresting since a uniform measure $\mathbb{1}_{|I|}$ assigned to all data is transported to another uniform measure (see (12)). This, however, reflects the fact that our approach is *completely* unsupervised. Moreover, since $D_{ii} = d(f_i, f_i) = 0$, $i \in I$, the trivial solution $A^* = I$ would attain the lower bound $0 \leq E(W) = \langle D, A \rangle$ and hence be optimal, *if A were not* constrained through the parametrization $A = A(W)$. And this trivial solution would correspond to the (useless) situation in which *each* data point is kept as prototype, which can only happen if assignment to pixels do not interact at all.

This trivial situation is ruled out through the *spatial interaction* of pixel assignments in terms of the geometric averaging map (6), which is a key property of the assignment flow (17). This interaction induces the formation of a (depending on the spatial scale: much) smaller subset of prototypes as latent variables and in turn a *low-rank factorization* of the affinity matrix (12), because many components j of the diagonal matrix $C(W)$ in (11) converge to 0. Rewriting the objective (13) as

$$\langle D, A(W) \rangle \overset{(11)}{=} \langle D, WC(W)^{-1}W^\top \rangle = \langle DWC^{-1}, W \rangle \overset{(18b)}{=} \langle DW_C, W \rangle, \quad (20)$$

with column-normalized, *asymmetric* assignment matrix W_C, admits the following interpretation complementing the discussion of (18) and takes into account that *spatial regularization* is 'built in' the unsupervised assignment flow (17):

Minimizing (13) through the flow (17) amounts to *spatially regularized* optimal transport of the uniform measure $\mathbb{1}_{|I|}$ assigned to all data points f_i, $i \in I$, to the positive measure $w(\mathcal{G}_J(t))$,

$$\mathbb{1}_{|I|} = W(t)\mathbb{1}_{|I|} \quad \rightarrow \quad W(t)^\top \mathbb{1}_{|I|} =: w(\mathcal{G}_J(t)), \quad (21)$$

where the latter concentrates on the *effective* support $J(w) \subset I$ of the emerging prototypes $g_j(t)$, $j \in J$. The corresponding cost matrix $DW_C = DW_C(t)$ of (20) for determining the transport plan $W(t)$ is given by the initial distance matrix (8) after averaging these distances with respect to the probability distributions that correspond to the normalized columns of $W(t)$,

$$\left(DW_C(t)\right)_{ij} \overset{(10)}{=} \sum_{i' \in I} P(i'|j)(t)d(f_i, f_{i'}) \overset{(19)}{=} \mathbb{E}\big[d(f_i, \mathcal{F}_I)|g_j\big], \quad j \in J. \quad (22)$$

At each point of time t, the assignment matrix $W(t)$ concentrates measure on those prototypes g_j, $j \in J$ (implicitly represented by the support of (21)) for which the average ('within-cluster') distances (22) are small.

4   Related Work

The literature on unsupervised learning and clustering is vast. We briefly comment on relations of our new approach to few closely related works from two general viewpoints: *nonnegative matrix factorization* and *discrete optimal transport* – see [7] and [17,18] as general references.

4.1   Nonnegative Matrix Factorization (NMF)

NMF is concerned with representing a nonnegative input data matrix $F \in \mathbb{R}_+^{|I| \times d}$ (d is the feature dimension) in terms of a product of two nonnegative matrices $G, H \geq 0$. Non-negativity is key for interpreting the factors as weights H and as a dictionary G of prototypes. In addition, the rank constraint rank $(HG^\top) \leq k$ has to be supplied as user parameter.

Archetypal Analysis [8] is an early work where the representation of data by prototypes was proposed. The factorization approach reads

$$F \approx HG^\top F \tag{23}$$

where both nonnegative factors $G^\top$ and H are constrained to be row-stochastic, so that $G^\top F$ are prototypes formed by convex combinations of data (feature) vectors that, in turn, are combined in a convex way using the weights H. The factorization is determined as local minimum by alternating minimization. Alternating updates of data assignment and prototype formation is common to most algorithms for unsupervised learning.

A major shortcoming of this early work is that the algorithm directly operates on the data rather than abstracting from the data space through a distance or a similarity function, as is common nowadays in machine learning. Likewise, Eq. (8) shows that we abstract from the data space through distance functions which could simply be induces by Euclidean norms or – when using more involved data models – by Riemannian distances of manifold-valued features.

Zass and Shashua [21] studied the clustering problem in the form

$$\max_G \langle A(F), GG^\top \rangle \quad \text{subject to} \quad G \geq 0, \quad G^\top G = I, \quad GG^\top \mathbb{1} = \mathbb{1} \tag{24}$$

where the data F are represented by any positive-semidefinite, symmetric affinity matrix $A(F)$. In particular, they showed that all three constraints together imply *hard* clustering which is combinatorially difficult, and that the second orthogonality constraint which accounts for normalization as done with basic spectral relaxations [16,19], is the weakest one. Accordingly, they proposed a two-step procedure after dropping the second constraint: compute the closest symmetric doubly-stochastic nonnegative approximation of the data matrix $A(F)$, followed by a second step for determining a completely positive factorization $GG^\top$, $G \geq 0$. The same set-up was proposed by [20] except for determining a locally optimal solution in a single iterative process using DC-programming. Likewise, [14] explored symmetric nonnegative factorizations but ignored the constraint enforcing that $GG^\top$ is doubly-stochastic which is crucial for cluster normalization.

These works are close to our approach (13) (the distances D_{ij} to be minimized are turned into affinities by (15)), in that $A(W)$ given by (11), (12) is a nonnegative, symmetric, doubly-stochastic low-rank factorization whose factors are not constrained to be orthogonal. Key differences are that we determine $A(W) = WC(W)^{-1}W^\top$ by a single *smooth continuous* process (17) that may be turned into a discrete iterative process performing numerical integration techniques [22], and that a numerical low rank is induced by *spatial regularization*.

4.2 Discrete Optimal Mass Transport (DOMT)

Authors of [5] adopted an interesting viewpoint on the clustering problem: associating a sample distribution q with the given data and a distribution p with an unknown subset of the data regarded as prototypes, the problem of determining the latter is defined as the task to minimize the transport costs between q and p, subject to a cardinality (sparsity) constraint on p in order to obtain a *smaller* subset of representative prototypes. The combinatorially hard cardinality constraint is turned into a specific convex penalty, imposed on the transport map having p as marginal. Since the number of data points in our scenarios is already large, working in the 'lifted' space of transport mappings with quadratic dimension is computationally infeasible, however.

The recent work [11] provides a related and natural reformulation of the clustering problem based on discrete optimal transport. Prototypes are defined as Wasserstein barycenters [1,9] of the assignment distributions, and the squared Wasserstein distance is decomposed into a sum of within-cluster and between-cluster distances, analogous to the classical decomposition of the total scatter matrix associated with patterns in a Euclidean feature space [10]. The authors promote low-rank transport maps not only to cope with the curse of dimensionality but also as an effective method to achieve stability under sampling noise.

The relations to our work are not as direct as the relations to work on NMF discussed in Sect. 4.1. On the one hand, the objectives (13), (20) together with the constraints admit an interpretation as DOMT, and our approach involving spatial regularization also leads to low-rank transport maps. On the other hand, averaging for prototype formation is based on the geometry of the Fisher-Rao metric rather than on the Wasserstein distance. We leave a more detailed discussion of these interesting aspects for future work.

5 Experiments

We demonstrate the proposed **unsupervised self-assignment flow (USAF)**, Eq. (17) by depicting the self-assignments of various scenes.

Implementation. The USAF was integrated numerically using the geometric Euler scheme [22] with step-size $h = 1.0$, together with the renormalization strategy from [3] with $\varepsilon = 10^{-10}$, termination criterion (average entropy $\leq 10^{-3}$) which ensures almost unique assignments. Default values are $\rho = 0.1$ (scale normalization), a $|\mathcal{N}| = 3 \times 3$ neighborhood with uniform weights w_i for spatial averaging and the ℓ_1-norm for the distance matrix of (5). We restricted the problem size to 64×64 pixels, since exploiting numerically the low-rank structure for larger problem sizes is beyond the scope of this paper. Finally, for Euclidean data, the self-assignments are $u_i = \sum_{j \in J} W_{ij} g_j$, $i \in I$ with g_j due to (14).

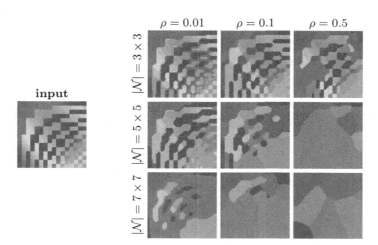

Fig. 1. Unsupervised image labeling through self-assignment of RGB data, depending on $|\mathcal{N}|$ and ρ. Increasing either value decreases the number of prototypes.

Parameter Influence. Figure 1 illustrates the influence of the only two user parameters on the self-assignment (labeling) of prototypes the emerge from the RGB input data. We observe that increasing the spatial scale (averaging) or the selectivity parameter reduces the number of prototypes.

Comparison with Supervised Assignment Flows (SAF) [3] **and Unsupervised Assignment Flows (UAF)** [23]**.** We adopted the implementation from [3,23] and determined first the *effective* number k of protoypes using USAF, given by $\mathrm{rank}(W)$, and then used k-means to determine also k prototypes for both (SAF: prototypes are fixed) and (UAF: prototypes may evolve). The comparison (Fig. 2) shows that *decoupling* prototype formation and spatial inference (SAF and UAF) leads to labelings that *mix spatial scales* in a way that is difficult to control. The self-assignment returned by (USAF: keeping the default value $\rho = 0.1$ fixed), on the other hand, clearly demonstrates that prototype formation is *solely determined by spatial scale*, i.e. by a single parameter.

Manifold Valued Data. Figure 3 shows $\mathcal{S}^1$-valued orientation data (panel 'angular') extracted from the fingerprint image using the structure tensor. The USAF returns a natural partition in terms of *prototypical orientations* extracted from the data itself, just based on the spatial scale at which the USAF operates.

This happens *without extra costs*, since USAF *separates* data from the assignment manifold and hence manifold-specific operations are *not* needed. Prototypes *on* the manifold *may* be computed, of course, as weighted Riemannian means using the probabilities of (10).

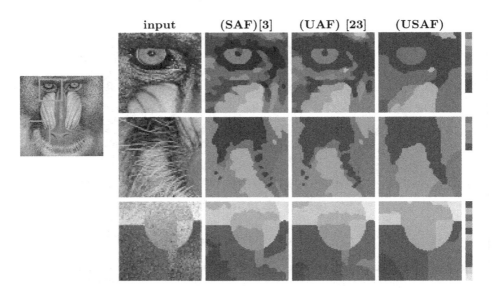

Fig. 2. Comparison of supervised (SAF), unsupervised (UAF) and self-assignment (USAF) flows. The right-most column depicts the prototypes $\mathcal{G}_J$ returned by the USAF, solely determined by the spatial scale, as the corresponding labeling (partition) reflects. By contrast, both SAF and UAF mix spatial scales due to prespecified prototypes.

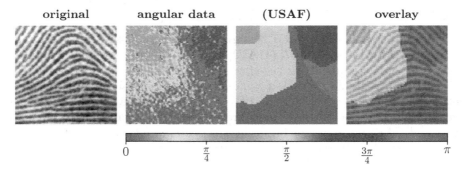

Fig. 3. Manifold valued data. $\mathcal{S}^1$-values angular data are extracted as input data from a fingerprint image. The USAF, operating with 5×5 neighborhoods, returns $|\mathcal{G}_J| = 6$ prototypes and a natural labeling (partition) on $\mathcal{W}$, *without* any operation *on* the data manifold $\mathcal{F}$, due to the separation of $\mathcal{F}$ and $\mathcal{W}$. Nevertheless, the partition *does* reflect the geometry of $\mathcal{F}$: orientations 0 and π are identified, for example.

6   Conclusion

We presented a novel geometric flow for completely unsupervised image labeling. A clear probabilistic interpretable stochastic factorization of the self-affinity matrix constitutes a low-rank representation of the input data, whereas the complexity (number of effective prototypes) is exclusively induced by spatial regularization, which is performed in an geometric and unbiased way. Experiments demonstrated the approach on Euclidean and manifold valued data.

In future work, we plan to exploit the low-rank structure by globally restricting the complexity of the solution, which immediately enables handling large problem instances.

Acknowledgement. Support from the German Science Foundation, grant GRK 1653, is gratefully acknowledged.

References

1. Agueh, M., Carlier, G.: Barycenters in the Wasserstein space. SIAM J. Math. Anal. **43**(2), 904–924 (2011)
2. Amari, S.I., Nagaoka, H.: Methods of Information Geometry. American Mathematical Society and Oxford University Press, Oxford (2000)
3. Åström, F., Petra, S., Schmitzer, B., Schnörr, C.: Image labeling by assignment. J. Math. Imaging Vis. **58**(2), 211–238 (2017)
4. Bergmann, R., Fitschen, J.H., Persch, J., Steidl, G.: Iterative multiplicative filters for data labeling. Int. J. Comput. Vis. **123**(3), 435–453 (2017)
5. Carli, F.P., Ning, L., Georgiou, T.T.: Convex clustering via optimal mass transport. arXiv preprint arXiv:1307.5459 (2013)
6. Chan, T.F., Esedoglu, S., Nikolova, M.: Algorithms for finding global minimizers of image segmentation and denoising models. SIAM J. Appl. Math. **66**(5), 1632–1648 (2006)
7. Cichocki, A., Zdunek, A., Phan, A.H., Amari, S.I.: Nonnegative Matrix and Tensor Factorizations. Wiley, Chichester (2009)
8. Cutler, A., Breiman, L.: Archetypal analysis. Technometrics **36**(4), 338–347 (1994)
9. Cuturi, M., Peyré, G.: Semidual regularized optimal transport. SIAM Rev. **60**(4), 941–965 (2018)
10. Devyver, P.A., Kittler, J.: Pattern Recognition: A Statistical Approach. Prentice Hall, London (1982)
11. Forrow, A., Hütter, J.C., Nitzan, M., Schiebinger, G., Rigollet, P., Weed, J.: Statistical optimal transport via geodesic hubs. arXiv preprint arXiv:1806.07348 (2018)
12. Hühnerbein, R., Savarino, F., Åström, F., Schnörr, C.: Image labeling based on graphical models using wasserstein messages and geometric assignment. SIAM J. Imaging Sci. **11**(2), 1317–1362 (2018)
13. Kappes, J., et al.: A comparative study of modern inference techniques for structured discrete energy minimization problems. Int. J. Comput. Vis. **115**(2), 155–184 (2015)
14. Kuang, D., Yun, S., Park, H.: SymNMF: nonnegative low-rank approximation of a similarity matrix for graph clustering. J. Glob. Optim. **62**(3), 545–574 (2015)

15. Lellmann, J., Schnörr, C.: Continuous multiclass labeling approaches and algorithms. SIAM J. Imag. Sci. **4**(4), 1049–1096 (2011)
16. von Luxburg, U.: A tutorial on spectral clustering. Stat. Comput. **17**(4), 395–416 (2007)
17. Peyré, G., Cuturi, M.: Computational Optimal Transport. CNRS (2018)
18. Santambrogio, F.: Optimal Transport for Applied Mathematicians. Birkhäuser, New York (2015)
19. Shi, J., Malik, J.: Normalized cuts and image segmentation. IEEE Trans. Patt. Anal. Mach. Intell. **22**, 888–905 (2000)
20. Yang, Z., Corander, J., Oja, E.: Low-rank doubly stochastic matrix decomposition for cluster analysis. J. Mach. Learn. Res. **17**, 1–25 (2016)
21. Zass, R., Shashua, A.: A unifying approach to hard and probabilistic clustering. In: Proceedings of ICCV (2005)
22. Zeilmann, A., Savarino, F., Petra, S., Schnörr, C.: Geometric Numerical Integration of the Assignment Flow. CoRR abs/1810.06970 (2018)
23. Zern, A., Zisler, M., Åström, F., Petra, S., Schnörr, C.: Unsupervised label learning on manifolds by spatially regularized geometric assignment. In: Brox, T., Bruhn, A., Fritz, M. (eds.) GCPR 2018. LNCS, vol. 11269, pp. 698–713. Springer, Cham (2019). https://doi.org/10.1007/978-3-030-12939-2_48

Variational Methods

Aorta Centerline Smoothing and Registration Using Variational Models

Luis Alvarez[1(✉)], Daniel Santana-Cedrés[1], Pablo G. Tahoces[2],
and José M. Carreira[3]

[1] CTIM, DIS, Universidad de Las Palmas de Gran Canaria,
Las Palmas de Gran Canaria, Spain
lalvarez@ulpgc.es, dsantana@ctim.es
[2] DEC, Universidad de Santiago de Compostela, Santiago de Compostela, Spain
pablo.tahoces@usc.es
[3] Complejo Hospitalario Universitario de Santiago (CHUS),
Santiago de Compostela, Spain
josemartin.carreira@usc.es

Abstract. In this work we present an application of variational techniques to the smoothing and registration of aorta centerlines. We assume that a $3D$ segmentation of the aorta lumen and an initial estimation of the aorta centerline are available. The centerline smoothing technique aims to maximize the distance of the centerline to the boundary of the aorta lumen segmentation but keeping the curve smooth. The proposed registration technique computes a rigid transformation by minimizing the squared Euclidean distance between the points of the curves, using landmarks and taking into account that the curves can be of different lengths. We present a variety of experiments on synthetic and real scenarios in order to show the performance of the methods.

Keywords: Aorta centerline · 3D curve smoothing ·
3D curve registration · Variational methods

1 Introduction

Aorta segmentation and centerline estimation from CT scans is an important issue in the diagnosis of cardiovascular diseases. In this paper, we present an application of variational techniques to the smoothing and registration of aorta centerlines. We will assume that we have previously estimated a segmentation of the aorta lumen, given by a 3D set A, and an initial estimation of the aorta centerline, $\mathcal{C}(s)$, given by a $3D$ curve $\mathcal{C} : [0, |\mathcal{C}|] \to R^3$, where s represents the arc-length parameter and $|\mathcal{C}|$ the length of the curve. As shown in Sect. 2, there are a variety of techniques to obtain the aorta segmentation and centerline.

One of the main motivations of this paper is that usually, the centerline estimation techniques include noise in the location of the aorta centerline points. This noise can produce significant errors in some important aorta measures as

© Springer Nature Switzerland AG 2019
J. Lellmann et al. (Eds.): SSVM 2019, LNCS 11603, pp. 447–458, 2019.
https://doi.org/10.1007/978-3-030-22368-7_35

the aorta length between 2 given points which is a critical measure in some medical treatments such as aortic stent implantations. To address this problem, we propose a centerline smoothing procedure inspired by the following variational model introduced in [7]:

$$\mathcal{C}_w = \underset{\mathcal{C}:\mathcal{C}(0)=p_0,\mathcal{C}(|\mathcal{C}|)=p_1}{\arg\min} \int_0^{|\mathcal{C}|} P(\mathcal{C}(s))ds + w|\mathcal{C}|, \tag{1}$$

where $P(x)$ is the potential, $|\mathcal{C}|$ is the curve length, and $w \geq 0$ is a parameter to balance both terms. In [7], the authors deal with $2D$ curves. However, in our approach, we extend the formulation to $3D$ curves.

The second topic we address in this paper is the registration of 2 centerlines. This is an important issue for patients follow-up, when different CT-scans are acquired in different periods of times. To facilitate the registration procedure, we take into account that given the way the CT scan is obtained, the point $\mathcal{C}(s_0)$ in the aortic arch with minimum z value (if the scan is performed from head to feet, which it is usually the case) can be used as an aortic landmark.

The main contribution of this paper is the application of variational techniques to aorta centerline smoothing and registration. In the case of smoothing, we use the model proposed in [7] but extended to $3D$ curves, using as potential the signed distance function to the aorta segmentation boundary. We propose a completely new numerical scheme adapted to 3D curves. In the case of aorta registration, we propose to compute the rigid transformation between 2 aorta centerlines using landmarks and minimizing the squared Euclidean distance between the corresponding points of both curves. Futhermore, the energy used allows the registration of curves of different lengths. In fact, the smoothing and the registration of centerlines are related problems and as the experiments carried out in this work show, smoothing the centerlines improves the accuracy of the registration.

The rest of the paper is organized as follows: in Sect. 2, we present some related works. In Sect. 3, we study in details the proposed variational method for aorta centerline smoothing. In Sect. 4, we describe the variational method for centerline registration. In Sect. 5, we show some experiments on synthetic and real centerlines. Finally, in Sect. 6, we present some conclusions.

2   Related Work

Automatic aorta segmentation algorithms from CT scans have been previously developed (see [10] for a survey). For instance, in [13], the authors proposed an iterative method based on building a 2-D region for segmenting the ascending aorta. In [2], a tracking procedure of the aorta centerline is presented. In [12], the authors introduce a method for the automatic estimation of the aorta segmentation and the centerline estimation. In [1], an active contour method for the aorta segmentation is proposed.

In [5] and [11], some energies for curvature penalized minimal path are proposed to regularize $2D$ curves. In [4], the authors propose a minimal path approach for tubular structures segmentation in $2D$ images with applications to retinal vessel segmentation.

Concerning 3D curve registration, the iterative closest point (ICP) (see [3]), is a general purpose method for the registration of 3D curves and surfaces which requires initialization of the expected rigid transform. In [9], the authors introduce a scale-space approach for registration of tree vessel structures.

3   Variational Methods for Aorta Centerline Smoothing

In order to perform the aorta centerline smoothing, we propose an extension to 3D curves of the variational model (1) introduced in [7] for $2D$ curves. We use as potential $P(x)$ the signed distance function $d_{\partial A}(x)$ to the boundary of the aorta segmentation A given by

$$P(x) = d_{\partial A}(x) = \begin{cases} d(x, \partial A) \ if \ x \notin A, \\ -d(x, \partial A) \ if \ x \in A. \end{cases} \tag{2}$$

Using this potential in (1) we aim to maximize the distance between the centerline and the boundary of the aorta lumen segmentation, but keeping the curve smooth. Thus, the final variational model for the aorta centerline smoothing is given by

$$\mathcal{C}_w = \underset{\mathcal{C}:\mathcal{C}(0)=p_0, \mathcal{C}(|\mathcal{C}|)=p_1}{\arg\min} E_S(\mathcal{C}) \equiv \int_0^{|\mathcal{C}|} d_{\partial A}(\mathcal{C}(s))ds + w|\mathcal{C}|, \tag{3}$$

where the parameter w balances both both energy terms. The larger the value of w, the more regular the curve is expected to be.

Numerical Scheme to Minimize Energy (3)

We propose a basic numerical scheme to minimize energy (3). In practice, the centerline $\mathcal{C}$ is given by a collection of 3D points $\{\mathcal{C}_i\}_{i=1,..,N_C}$, where $\mathcal{C}_1 = p_0$ and $\mathcal{C}_{N_C} = p_1$. We assume that

$$\|\mathcal{C}_i - \mathcal{C}_{i-1}\| = h \quad \text{for all } i = 2, .., N_C - 1, \tag{4}$$
$$\|\mathcal{C}_{N_C} - \mathcal{C}_{N_C-1}\| \leq h,$$

that is, we use a curve parameterization with constant arc-length h. Usually h depends on the image spacial resolution of the CT scan. Typically, h ranges from 0.5 mm to 1 mm.

For the left part of energy (3), we consider in each point $\mathcal{C}_i$ a gradient descent type scheme of the form

$$\mathcal{C}_i^{n+1} = \mathcal{C}_i^n - \delta \nabla d_{\partial A}(\mathcal{C}_i^n). \tag{5}$$

To simplify the numerical scheme of the right part of energy (3) we consider the curvature shortening flow described, for instance, in [6]. This flow tends to reduce the length of the curve and can be formulated as follows:

$$\mathcal{C}_i^{n+1} = \mathcal{C}_i^n + \delta k_i^n \mathcal{N}_i^n, \tag{6}$$

where k_i^n represents an approximation to the $2D$ curvature and $\mathcal{N}_i^n$ the unit normal direction in $\mathcal{C}_i^n$ of the curve restricted to the plane given by the points $\mathcal{C}_{i-1}^n, \mathcal{C}_i^n, \mathcal{C}_{i+1}^n$. By combining both schemes and adding the weight w, we obtain the following minimization scheme for (3)

$$\mathcal{C}_i^{n+1} = \mathcal{C}_i^n - \delta \nabla d_{\partial A}(\mathcal{C}_i^n) + \delta w k_i^n \mathcal{N}_i^n, \tag{7}$$

where $\delta > 0$ represents the discretization step. We point out that, after each iteration, we need to reparameterize $\{\mathcal{C}_i^{n+1}\}$ in order to preserve the constant arc-length condition (4). We compute the unit normal vector $\mathcal{N}_i^n$ as

$$\mathcal{N}_i^n = \begin{cases} \dfrac{\frac{\mathcal{C}_{i-1}^n + \mathcal{C}_{i+1}^n}{2} - \mathcal{C}_i^n}{\left\| \frac{\mathcal{C}_{i-1}^n + \mathcal{C}_{i+1}^n}{2} - \mathcal{C}_i^n \right\|} & \text{if } \frac{\mathcal{C}_{i-1}^n + \mathcal{C}_{i+1}^n}{2} \neq \mathcal{C}_i^n, \\[2ex] \overrightarrow{0} & \text{otherwise}, \end{cases} \tag{8}$$

and the curvature k_i^n is approximated as the quotient between the angle, $\theta_i^n = \angle \mathcal{C}_{i-1}^n \mathcal{C}_i^n \mathcal{C}_{i+1}^n$, of the vectors $\overrightarrow{\mathcal{C}_{i-1}^n \mathcal{C}_i^n}$ and $\overrightarrow{\mathcal{C}_i^n \mathcal{C}_{i+1}^n}$ and the arc-length h, that is

$$k_i^n = \frac{\theta_i^n}{h}. \tag{9}$$

We use as stopping criterion of the iterative scheme (7) the condition

$$\frac{|E_S(\mathcal{C}^n) - E_S(\mathcal{C}^{n-1})|}{|E_S(\mathcal{C}^{n-1})|} < \epsilon, \tag{10}$$

where $E_S(\mathcal{C})$ is the energy defined in (3). $\epsilon > 0$ is a parameter to fix the stopping criterion of the scheme. In the experiments presented in this paper we use $\epsilon = 10^{-8}$.

Scheme (7) is a basic approximation of a minimizer of energy (3) but it is not derived from the Euler-Lagrange equation of (3) due to the choice of the curvature shortening flow introduced in (6). Nevertheless, despite this theoretical limitation, in the experiments we show that scheme (7) behaves as a good minimizer of (3) (see Fig. 3).

Automatic Estimation of the Discretization Step δ

We can estimate δ automatically using a two-step process: on the one hand, using as potential $P(x)$ the normalized signed distance function, we have that $\|\nabla P(\mathcal{C}_i^n)\| \leq 1$. Therefore, by imposing in the scheme (5) that

$$\delta \leq \frac{h}{2}, \tag{11}$$

we obtain that the point C_i^{n+1} is closer to C_i^n than to C_{i-1}^n and C_{i+1}^n. On the other hand, with respect to the curvature part we impose that

$$\delta w k_i^n \leq \left\| \frac{C_{i-1}^n + C_{i+1}^n}{2} - C_i^n \right\|, \tag{12}$$

which means that the curvature flow never makes the point C_i^n to move to the other side of the segment $\overline{C_{i-1}^n C_{i+1}^n}$. Using a straightforward computation, we obtain that the above condition is equivalent to

$$\delta w \frac{\theta_i^n}{h} \leq h \cos\left(\frac{\pi - \theta_i^n}{2}\right). \tag{13}$$

We observe that the function

$$f(\theta) = \delta w \frac{\theta}{h} - h \cos\left(\frac{\pi - \theta}{2}\right), \tag{14}$$

satisfies that

$$f(\pi) = \delta w \frac{\pi}{h} - h \leq 0 \Leftrightarrow \delta \leq \frac{h^2}{w\pi},$$

and we can easily check that if

$$\delta \leq \frac{h^2}{w\pi}, \tag{15}$$

then $f(\theta) \leq 0$ for any $\theta \in [0, \pi]$ and then (13) is satisfied. Therefore, joining both estimations of δ for each part of the numerical scheme, we can fix automatically δ as

$$\delta = \frac{\min\{\frac{h}{2}, \frac{h^2}{w\pi}\}}{2}. \tag{16}$$

In practice, we experienced that using this choice for δ we obtain a stable numerical evolution of (7).

4   Variational Methods for Aorta Centerline Registration

As mentioned in the introduction, to facilitate the registration procedure, we use the aortic arch point with the minimum value of z, $C(s_0)$, as an aorta landmark. We can assume that, given two centerlines C and C' of the same patient and their corresponding landmarks s_0 and s_0', the corresponding point of $C(s_0)$ in the other centerline is in a vicinity of $C'(s_0')$. Notice that one landmark can not correspond exactly with the other one because a modification in the patient position during the CT scan acquisition can produce that both landmarks do not match exactly.

Given 2 corresponding positions s_0 and s' in both centerlines, that is $C(s_0)$ corresponds to $C'(s')$, we obtain the 3D rigid transformation (given by a 3×3 rotation matrix and a 3D translation vector t) which transforms C' into C by minimizing the following energy

$$E(s_0, s', R, t) = \frac{\int_{\max\{-s_0, -s'\}}^{\min\{-s_0+|\mathcal{C}|, -s'+|\mathcal{C}'|\}} ||(\mathcal{C}(s+s_0) - (R \cdot \mathcal{C}'(s+s') + t)||_2^2 ds}{\min\{-s_0 + |\mathcal{C}|, -s' + |\mathcal{C}'|\} - \max\{-s_0, -s'\}}.$$

$$(17)$$

As showed below, for a fixed value of s_0, s', the above minimization problem has a close-form solution. The energy (17) takes into account that the length of both centerlines can be different. Indeed, for instance, according to the way the CT acquisition is performed the size of the aortic centerline in the abdominal area can be different. Finally, we compute the best rigid transformation between both centerlines as

$$(R, t) = \underset{s' \in [s_0' - r, s_0' + r]}{\arg\min} E(s_0, s', R, t), \tag{18}$$

where $r \geq 0$ is a parameter which determines the vicinity of the landmark $\mathcal{C}'(s_0')$ used to look for the point $\mathcal{C}'(s')$ corresponding to the landmark $\mathcal{C}(s_0)$.

Close form Solution of the Minimization Problem (17)

We observe that, in practice, if s_0 and s' are fixed and the curves are discretized, then the minimization problem (18) is equivalent to minimize the energy

$$\tilde{E}(R, t) = \sum_i ||(\mathcal{C}_{i+i_0} - (R \cdot \mathcal{C}'_{i+i'} + t)||_2^2,$$

for a certain range of i values. In [8], it is showed that this minimization problem has a close-form solution based on a quaternion representation of the rotation matrix (see [8] for more details). We point out that the rigid transformation obtained with the proposed method can be used as an initial guess for iterative methods like ICP (see [3]).

5   Experimental Setup

To verify the accuracy of the proposed method for smoothing the centerline of the aorta, we build the aorta synthetic phantom illustrated in Fig. 1. This centerline is used as an approximation of a real one. The discretized arc-length h (see (4)) is taken equal to 1. Using this centerline we build an aorta segmentation by drawing spheres centered in the aorta centerline points. The radii of the spheres is taken in the range of values expected along real centerlines. To study the convergence of the numerical scheme (7), we add some noise to the position of the original centerline points $\{\mathcal{C}_i\}_{i=1,..,N_C}$ in the following way

$$\mathcal{C}_i^0 = \mathcal{C}_i + (6 + \mathcal{U}(-1, 1), 6 + \mathcal{U}(-1, 1), 6 + \mathcal{U}(-1, 1))^T, \tag{19}$$

where $\mathcal{U}(-1, 1)$ follows the uniform probability distribution in the interval $[-1, 1]$. We use $\mathcal{C}^0$ as the initial guess for the scheme (7). To evaluate the robustness and convergence of the algorithm $\mathcal{C}^0$ is chosen quite far from the ground truth. We denote by $\mathcal{C}_w^\infty$ the asymptotic state of the scheme (7) accordingly to the parameter w and the stopping criterion (10). In this case the original centerline $\mathcal{C}$ represents the ground-truth and the curve $\mathcal{C}_w^\infty$ represents an approximation of $\mathcal{C}$. In Fig. 1 we show the curves $\mathcal{C}$, $\mathcal{C}^0$ and $\mathcal{C}_w^\infty$ for $w = 0, 1, 10, 50$. We observe that for the values $w = 0, 1, 10$, the smoothed curve $\mathcal{C}_w^\infty$ is close to the original centerline $\mathcal{C}$. However, for $w = 50$, the smoothing effect is too strong and $\mathcal{C}_{50}^\infty$ is quite far from $\mathcal{C}$. In Fig. 2 we show a zoom of the same curves in the abdominal area, where we can appreciate the influence of the smoothing parameter w.

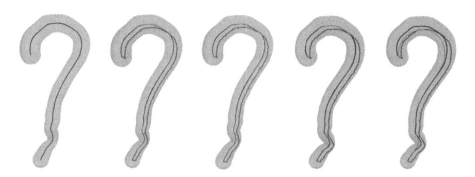

Fig. 1. From left to right we show the original synthetic centerline $\mathcal{C}$, in black, and the segmentation of the aorta phantom in grey (first image). Next we show the curves $\mathcal{C}$ (black), $\mathcal{C}^0$ (red), and $\mathcal{C}_w^\infty$ (green) for $w = 0, 1, 10, 50$. (Color figure online)

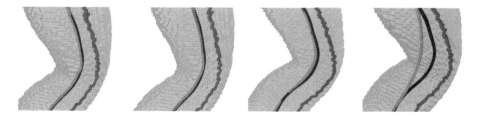

Fig. 2. From left to right we show a zoom of the abdominal area of the aorta phantom where we can see the original centerline, $\mathcal{C}$ (black), the segmentation of the aorta phantom (grey) and the curves $\mathcal{C}^0$ (red) and $\mathcal{C}_w^\infty$ (green) for $w = 0, 1, 10, 50$. (Color figure online)

To measure the quality of the approximation in a quantitative way, we use, as approximation error, the following distance between the ground-truth $\mathcal{C}$ and the approximation $\mathcal{C}^n$:

$$d(\mathcal{C}, \mathcal{C}^n) = \sqrt{\frac{1}{2N_{\mathcal{C}}} \sum_{i=1}^{N_{\mathcal{C}}} d^2(\mathcal{C}_i, \mathcal{C}^n) + \frac{1}{2N_{\mathcal{C}^n}} \sum_{i=1}^{N_{\mathcal{C}^n}} d^2(\mathcal{C}_i^n, \mathcal{C})}, \qquad (20)$$

where, for a given $3D$ point p and a curve $\mathcal{C}$, $d(p, \mathcal{C})$ is the Euclidean distance of p to the curve $\mathcal{C}$. In Table 1 we present some quantitative results about the number of iterations using the stopping criterion (10), the length $|\mathcal{C}_w^\infty|$, the energy $E_S(\mathcal{C}_w^\infty)$ defined in (3) and the error $d(\mathcal{C}, \mathcal{C}_w^\infty)$ defined by (20). We observe that the error $d(\mathcal{C}, \mathcal{C}_w^\infty)$ is similar for $w = 0$ and $w = 1$. However, there is an important difference with respect to the centerline length. The length of the original curve $\mathcal{C}$ is 457 mm, whereas the length of $\mathcal{C}_0^\infty$ is much larger (464.86 mm). The reason for this length discrepancy is that if we remove the smoothing term, the obtained 3D curve tends to zigzag, which produces an artificial increase of the centerline length. The length of $\mathcal{C}_1^\infty$ (458.05 mm) is much closer to the original one.

Table 1. Quantitative results obtained for the phantom centerline $\mathcal{C}_w^\infty$ for $w = 0, 1, 10, 50$.

w	N. Iterations (10)	Final length (mm)	Final energy $E_S(\mathcal{C}_w^\infty)$ (3)	Final error (mm) (20)
0	352	464.86	-6728.28	0.670057
1	267	458.05	-6184.34	0.676328
10	1721	453.39	-2031.11	0.733271
50	4556	428.74	16128.73	4.094052

In Figs. 3, 4 and 5 we show, for different values of w, the evolution of the length of $\mathcal{C}^n$, the energy $E_S(\mathcal{C}^n)$ and the error $d(\mathcal{C}, \mathcal{C}^n)$ for the first 1000 iterations of scheme (7). We observe a nice convergence behavior of $E_S(\mathcal{C}^n)$. As expected, the evolution of the length of $\mathcal{C}^n$ is strongly influenced by the value of w, and if the smoothing parameter w is too high, $\mathcal{C}^n$ does not converge towards $\mathcal{C}$.

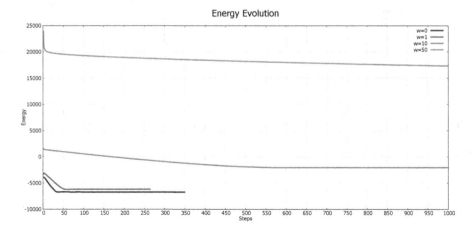

Fig. 3. Evolution of $E_S(\mathcal{C}^n)$ for the phantom centerline with different values of w.

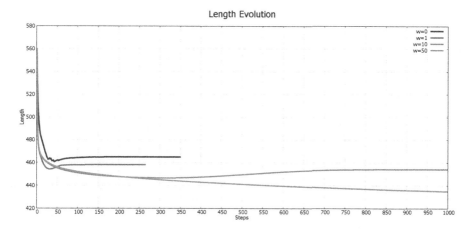

Fig. 4. Evolution of $|\mathcal{C}^n|$ for the phantom centerline with different values of w.

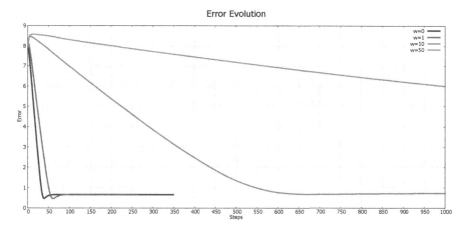

Fig. 5. Evolution of $d(\mathcal{C}, \mathcal{C}^n)$ for the phantom centerline with different values of w.

In Fig. 6 we illustrate the results of the proposed method for centerline smoothing using a real CT scan. We can observe in the figure the smoothing effect introduced by the curve regularization.

Next, we present some experiments to show the accuracy of the proposed 3D curve registration technique. We use 2 CT scans of the same patient provided to us by the Department of Radiology of the University Hospital of Santiago de Compostela, Spain. In Fig. 7, we compare the centerlines and segmentations obtained from both CT scans before and after registration using the original centerlines. The average squared distance error obtained in the centerline registration given by (18) is 7.54 mm. We also compute the distance between the

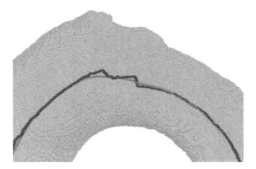

Fig. 6. We present a zoom in the aortic arch of one of the real aortas presented in Fig. 7. We show the segmentation (grey), the original centerline (red) and the centerline smoothed with the proposed method with $w = 1$ (green). (Color figure online)

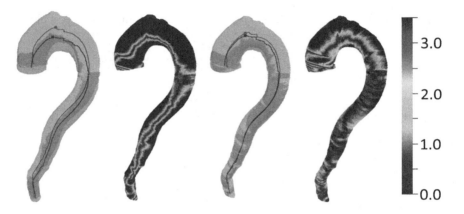

Fig. 7. We compare the centerlines and segmentations obtained from 2 CT scans of the same patient before and after registration. The landmarks used are marked with a sphere. We use the original centerlines and the radius r used in (18) to look in a vicinity of the landmark is $r = 25$. We use a colormap to illustrate the distance between the boundary of both segmentations (before and after registration).

boundaries of the segmentations. We use a colormap to illustrate such distance (blue means distance 0 and red distance 3.5 (in mm)).

In Fig. 8, we show the results of the same experiment but the centerlines have been previously smoothed. In this case, the average squared distance error obtained in (18) is 1.38 mm, which is much smaller than the one previously obtained without smoothing the centerlines. In particular, it means that by smoothing the centerlines we improve the accuracy of the registration procedure. This behaviour is expected due to the noise present in the original centerlines. This noise introduces disturbances in the centerline parameterizations that can produce significant errors in the registration procedure. By smoothing the centerlines, we strongly reduce these errors, achieving a greater precision in the

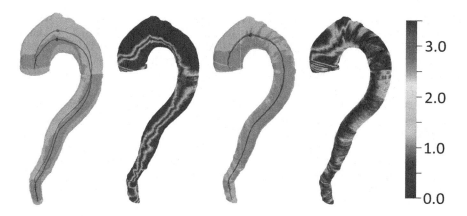

Fig. 8. We perform the same comparison that in Fig. 7 but in this case the centerlines have been previously smoothed with $w = 1$.

calculation of the distance between any two points of the centerline of the aorta. This result could be of great interest for the development of medical applications, where the precise measurement of distances is necessary for the diagnosis of diseases and follow-up of patients.

6   Conclusions

In this paper we have presented an application of variational models to the problem of aorta centerline smoothing and registration which are both relevant issues in medical imaging. For centerline smoothing we use an extension to $3D$ curves of a well-known variational model introduced in [7] for $2D$ curves and we propose a completely new numerical scheme. The proposed method for centerline registration is based on curve parameterization and the use of landmarks, providing close-form solutions for the estimation of the rigid transformation, even in the case of curves of different lengths. The experiments performed on synthetic and real centerlines for both methods provide promising results. In particular, we have managed to improve the accuracy of the registration technique by smoothing the centerlines.

Acknowledgement. This research has partially been supported by the MINECO projects references TIN2016-76373-P (AEI/FEDER, UE) and MTM2016-75339-P (AEI/FEDER, UE) (Ministerio de Economía y Competitividad, Spain).

References

1. Alemán-Flores, M., et al.: Segmentation of the aorta using active contours with histogram-based descriptors. In: Stoyanov, D., et al. (eds.) LABELS/CVII/STENT -2018. LNCS, vol. 11043, pp. 28–35. Springer, Cham (2018). https://doi.org/10.1007/978-3-030-01364-6_4

2. Alvarez, L., et al.: Tracking the aortic lumen geometry by optimizing the 3D orientation of its cross-sections. In: Descoteaux, M., Maier-Hein, L., Franz, A., Jannin, P., Collins, D.L., Duchesne, S. (eds.) MICCAI 2017. LNCS, vol. 10434, pp. 174–181. Springer, Cham (2017). https://doi.org/10.1007/978-3-319-66185-8_20

3. Besl, P.J., McKay, N.D.: A method for registration of 3-D shapes. IEEE Trans. Pattern Anal. Mach. Intell. **14**(2), 239–256 (1992)

4. Chen, D., Zhang, J., Cohen, L.D.: Minimal paths for tubular structure segmentation with coherence penalty and adaptive anisotropy. IEEE Trans. Image Process. **28**(3), 1271–1284 (2019)

5. Chen, D., Mirebeau, J.M., Cohen, L.D.: Global minimum for curvature penalized minimal path method. In: Proceedings of the British Machine Vision Conference (BMVC), pp. 86.1-86.12. BMVA Press, September 2015

6. Chou, K.S., Zhu, X.P.: The Curve Shortening Problem. Chapman and Hall/CRC, Boca Raton (2001)

7. Cohen, L.D., Kimmel, R.: Global minimum for active contour models: a minimal path approach. Int. J. Comput. Vis. **24**(1), 57–78 (1997)

8. Faugeras, O.D., Hebert, M.: The representation, recognition, and locating of 3-D objects. Int. J. Rob. Res. **5**(3), 27–52 (1986)

9. Heldmann, S., Papenberg, N.: A scale-space approach for image registration of vessel structures. In: Meinzer, H.P., Deserno, T.M., Handels, H., Tolxdorff, T. (eds) Bildverarbeitung für die Medizin 2009. Informatik aktuell, pp. 137–141. Springer, Heidelberg (2009). https://doi.org/10.1007/978-3-540-93860-6_28

10. Lesage, D., Angelini, E.D., Bloch, I., Funka-Lea, G.: A review of 3D vessel lumen segmentation techniques: models, features and extraction schemes. Med. Image Anal. **13**(6), 819–845 (2009). includes Special Section on Computational Biomechanics for Medicine

11. Mirebeau, J.M.: Fast-marching methods for curvature penalized shortest paths. J. Math. Imaging Vis. **60**(6), 784–815 (2018)

12. Tahoces, P.G., et al.: Automatic estimation of the aortic lumen geometry by ellipse tracking. Int. J. Comput. Assist. Radiol. Surg. **14**(2), 345–355 (2019)

13. Wang, S., Fu, L., Yue, Y., Kang, Y., Liu, J.: Fast and automatic segmentation of ascending aorta in MSCT volume data. In: 2009 2nd International Congress on Image and Signal Processing, pp. 1–5, October 2009

A Connection Between Image Processing and Artificial Neural Networks Layers Through a Geometric Model of Visual Perception

Thomas Batard[1]([envelope]), Eduard Ramon Maldonado[1,2], Gabriele Steidl[3], and Marcelo Bertalmío[4]

[1] Crisalix, Barcelona, Spain
{thomas.batard,eduard.ramon}@crisalix.com
[2] Universitat Politécnica de Catalunya, Barcelona, Spain
[3] TU Kaiserslautern, Kaiserslautern, Germany
steidl@mathematik.uni-kl.de
[4] Universitat Pompeu Fabra, Barcelona, Spain
marcelo.bertalmio@upf.edu

Abstract. In this paper, we establish a connection between image processing, visual perception, and deep learning by introducing a mathematical model inspired by visual perception from which neural network layers and image processing models for color correction can be derived. Our model is inspired by the geometry of visual perception and couples a geometric model for the organization of some neurons in the visual cortex with a geometric model of color perception. More precisely, the model is a combination of a Wilson-Cowan equation describing the activity of neurons responding to edges and textures in the area V1 of the visual cortex and a Retinex model of color vision. For some particular activation functions, this yields a color correction model which processes simultaneously edges/textures, encoded into a Riemannian metric, and the color contrast, encoded into a nonlocal covariant derivative. Then, we show that the proposed model can be assimilated to a residual layer provided that the activation function is nonlinear and to a convolutional layer for a linear activation function. Finally, we show the accuracy of the model for deep learning by testing it on the MNIST dataset for digit classification.

Keywords: Differential geometry · Variational model · Image processing · Vision · Neural network

1 Introduction

Deep learning techniques based on artificial neural networks (ANNs) provide state-of-the-arts results in many computer vision tasks [18]. The original goal of the ANN approach was to solve problems in the same way as a human brain. However, over time, attention moved to performing specific tasks, leading to

© Springer Nature Switzerland AG 2019
J. Lellmann et al. (Eds.): SSVM 2019, LNCS 11603, pp. 459–471, 2019.
https://doi.org/10.1007/978-3-030-22368-7_36

deviations from biology. Nonetheless, some layers like the convolutional layers [25] and the residual units [19,20], which are at the core of some efficient ANNs, combine linear and nonlinear operators mimicking the neuronal activity (see e.g. [27] and references therein).

The linear operator involved in ANNs is mainly the standard (Euclidean) convolution operator. One of the success of the convolution operator for deep learning tasks is, besides its simplicity which enables fast computation, its ability to generate different types of filters, like isotropic/anisotropic and low-pass/high-pass. Another key property satisfied by the convolution operator is the fact that it commutes with translations, as ANNs satisfying invariance/equivariance with respect to group transformations are desirable in many computer vision tasks like the ones involving recognition or classification. Starting with the seminal convolutional neural network (CNN) [25] involving the standard convolution operator, ANNs enlarging the symmetry group to the Euclidean transformations have been developed by applying the convolution on the special Euclidean Lie group SE(2) [2] or some of its subgroups [12,13].

Based on the following two properties of the convolution on SE(2), we are led to propose a new operator in the context of ANNs. First, the convolution on SE(2) is related to the interactions between neurons located at different orientation columns in the area V1 of the visual cortex. However, this convolution does not take into account the specificity of these interactions, with the presence of horizontal connections between these neurons and a function modeling neurons activation. Then, the convolution on SE(2) requires the lifting of the visual input from the 2D domain to SE(2). Nonetheless, no lifting is required to reach Euclidean invariance, as this latter can be obtained by means of well-chosen differential operators on the 2D domain. Hence, our proposal is to first, consider a model describing the activity of a population of neurons responding to spatial features in V1 called the Wilson-Cowan equations [7,10,14,29,30], then to derive an operator from a 2D version of the Wilson-Cowan equations.

The main contribution of this paper is two-folds. On one hand, we establish a connection between the proposed operator and ANNs by showing that a linear activation of the neurons yields a convolutional layer corresponding to a polynomial of order 1 of a differential operator as in [22,23], whereas a nonlinear activation yields a residual unit [19]. On the other hand, we establish a connection with image processing by showing that, for some nonlinear activation function, the proposed operator is related to a variational model refining the color correction models in [1,4], and a linear activation yields a quadratic relaxation of the variational model.

In Sect. 2, we first present a recent form of the Wilson-Cowan equations for neurons responding to edges and textures and its main properties [10,14]. Then, we introduce a 2D simplification of this model by making use of a cell selectivity principle [11]. Finally, we show for some particular activation functions, that the proposed 2D simplification corresponds to the gradient descent equations of variational models for image processing refining the one in [4].

In Sect. 3, we improve the model developed in Sect. 2 by coupling the simplified Wilson-Cowan equations with a model of color perception introduced in [1] and inspired by the Retinex theory of color vision [24]. For particular activation functions, we show that the coupling model represents the gradient descent equations of variational models refining the one in [1].

In Sect. 4, we first derive an operator from the variational models developed in Sect. 3 and establish a connection between this operator and ANNs. In particular, we show that a linear activation function yields an accurate convolution operator by showing the ability of this operator to generate different types of local filters and commute with isometries. Finally, we apply this operator in the context of deep learning by inserting it into a simple ANN and test this latter on the MNIST dataset for digit classification.

2 From Wilson-Cowan Equations to Image Processing

2.1 Wilson-Cowan Equations Encoding the Edges and Textures

It has been shown by Hubel and Wiesel that there exist neurons responding selectively to the local orientation of the visual input in the area V1 of the visual cortex [21]. The Wilson-Cowan equations [7,8,29,30] describe the temporal evolution of the mean activity of these neurons. More recently, the existence of populations of neurons in V1 encoding the structure tensor $\mathcal{T} \in \mathrm{SPD}(2,\mathbb{R})$, see [15], and organized as a column at each cortical position has been suggested by Chossat and Faugeras [10]. Here, $\mathrm{SPD}(2,\mathbb{R})$ denotes the cone of symmetric positive definite matrices equipped with the affine invariant metric and corresponding distance function $d_{\mathrm{SPD}(2,\mathbb{R})}$. Based on this and inspired by the Wilson-Cowan equations, an evolution equation describing the temporal activity of such neurons has been proposed in [10]. Treating V1 as a planar sheet Ω, the evolution of the mean activity $a\colon \Omega \times \mathrm{SPD}(2,\mathbb{R}) \times [0,\infty) \to \mathbb{R}$ of a population of cells with cortical coordinates $x \in \mathbb{R}^2$ and structure tensor preference $\mathcal{T}$ can be modeled (for low firing rates) with the following integro-differential equation (*generalized Wilson-Cowan equations*)

$$\frac{\partial a(x,\mathcal{T},t)}{\partial t} = -\,\alpha\, a(x,\mathcal{T},t) + h(x,\mathcal{T},t)$$
$$+ \int_{\Omega} \int_{\mathrm{SPD}(2,\mathbb{R})} w((x,\mathcal{T}),(y,\mathcal{T}'))\, \sigma(a(y,\mathcal{T}',t))\, d\mathcal{T}'\, dy, \quad (1)$$

where $\alpha \geq 0$, σ is an activation function, h the visual input and w the synaptic weights.

2.2 Properties of the Equations

On the Synaptic Weights. Following the decomposition of the synaptic weights in the original Wilson-Cowan equations into vertical and horizontal terms [9], Faye and Chossat [14] suggested that the synaptic weights in (1) are of the form

$$w((x,\mathcal{T}),(y,\mathcal{T}')) = w_{ver}(d_{SPD(2,\mathbb{R})}(\mathcal{T},\mathcal{T}'))\,\delta(x-y) + \lambda\, w_{hor}^{\kappa}((x,\mathcal{T}),(y,\mathcal{T}'))(1 - \delta(x-y)),$$
$$(2)$$

which are split into vertical connections w_{ver} between neurons in the same column and horizontal connections w_{hor}^κ between neurons in different columns. Here, λ is a trade-off parameter whose sign indicates whether the horizontal connections have a net excitatory or inhibitory effect. In particular, they propose the horizontal connections of the form

$$w_{hor}^\kappa((x,T),(y,T')) = K_\rho \left(\sqrt{(x-y)^T(\mathbb{I}_2 + \kappa T)(x-y)} \right) \times \mathcal{K}(d_{SPD(2,\mathbb{R})}(T,T')),$$
(3)

where K_ρ is a Gaussian kernel of variance ρ, $\kappa \geq 0$, and $\mathcal{K}$ is an even positive function of compact support.

Equivariance with Respect to Group Transformations. One of the key property of the original Wilson-Cowan equations is their equivariance with respect to $E(2)$ transformations (assuming that h is 0) [7]. Faye and Chossat [14] show that, under some action of $E(2)$ on $\mathbb{R}^2 \times SPD(2,\mathbb{R})$, the Wilson-Cowan equations (1) satisfy an $E(2)$-equivariance as well. Moreover, in the limit case where $\kappa = 0$, they point out that the equation admits a $GL(2,\mathbb{R})$ symmetry.

On the Activation Function. As mentioned by Wilson and Cowan [30], the qualitative properties of solutions of the original Wilson-Cowan equations do not depend on the particular form of the activation function σ and it is likely that σ will differ across the neocortex. Then, assuming that the same property holds for the Wilson-Cowan equations (1), and following the way Bertalmío and Cowan rewrote the activation function of the original Wilson-Cowan equation in [6], we replace the term $\sigma[a(y,T',t)]$ by $\sigma[f(a(x,T,t),a(y,T',t))]$ for some real-valued function f.

The activation function σ is often assumed to be nonlinear and having a specific form, i.e. it is taken as a non-decreasing function such that $\sigma(0) = 0$, saturating at the infinity, and shaped as a sigmoid. In this context, a linear function can be seen as the limit of such a function when increasing its stiffness, and linear activations have actually been considered as well (see e.g. [28]).

2.3   Connection to Image Processing Models

Simplifying the Wilson-Cowan Equations Through Cell Selectivity. In [11], the visual cortex is abstracted as $\mathbb{R}^2 \times S^1$, so that the columns only encode the orientations. Here, it is supposed that the simple cells sensitive to the orientation of the gradient of the visual input are maximally activated and selected. Assuming that the same property holds for the columns encoding edges and textures, the cell sensitives to the structure tensor T_h of the visual input h are maximally activated and selected at each cortical position.

Together with the rewriting of the activation function aforementioned, it yields the following *simplified Wilson-Cowan equation*

$$\frac{\partial a(x,t)}{\partial t} = -\alpha\, a(x,t) + h(x,t) + \lambda \int_\Omega w(x,y)\, \sigma[f(a(x,t),a(y,t))]\, dy, \qquad (4)$$

Fig. 1. Contrast modification model (9). From left to right: Result of the model for $\lambda < 0$ - Original image - Result of the model for $\lambda > 0$.

where the weight function w is of the form

$$w(x,y) = K_\rho \left(\sqrt{(x-y)^T \left(\mathbb{I}_2 + \kappa \mathcal{T}_h(x) \right)(x-y)} \right) \times \mathcal{K} \left(d_{SPD(2,\mathbb{R})}(\mathcal{T}_h(x), \mathcal{T}_h(y)) \right). \tag{5}$$

This simplified Wilson-Cowan equation differs from the one in [6]. Indeed, whereas the weight function (5) encodes the structure tensor of the visual input at each cortical position, the simplification in [6] produces weights of the form

$$w(x,y) = K_\rho(\|x-y\|). \tag{6}$$

where $\| \cdot \|$ stands for the Euclidean vector norm.

Image Restoration Model with a Nonlinear Activation Function.
Assuming in the following that the weight w is symmetric, the input h is constant in time, $\alpha = 1$, and the activation function is of the form

$$\sigma[f(a(x,t), a(y,t))] = s_\epsilon(w(x,y)[a(x,t) - a(y,t)]),$$

where s_ϵ is a differentiable approximation of the sign function, the evolution equation (4) can be seen as the gradient descent equation of a differentiable approximation of the variational model

$$\arg\min_a \frac{1}{2} \int_\Omega (a(x) - h(x))^2 \, dx - \frac{\lambda}{2} \int_{\Omega^2} w(x,y) |a(y) - a(x)| \, dx \, dy. \tag{7}$$

Note that the weight (5) can be symmetrized in a straightforward way if we replace $\mathcal{T}_h(x)$ by $(\mathcal{T}_h(x) + \mathcal{T}_h(y))/2$ in the first term.

For w of the form (6), we recover the variational model in [4], which realizes contrast enhancement for $\lambda > 0$ and contrast reduction for $\lambda < 0$ provided that the variance ρ is large.

Image Restoration Model with a Linear Activation Function. Assuming that the activation function is of the form

$$\sigma[f(a(x,t), a(y,t))] = w(x,y)[a(x,t) - a(y,t)],$$

the evolution equation (4) corresponds to the gradient descent of the following variational problem

$$\arg\min_a \frac{1}{2}\int_\Omega (a(x) - h(x))^2\, dx - \frac{\lambda}{2}\int_{\Omega^2} w^2(x,y)(a(y) - a(x))^2\, dx\, dy. \qquad (8)$$

In particular, for w being a normalized Gaussian kernel, the solution of the corresponding Euler-Lagrange equations is

$$a = \mathcal{F}^{-1}\left(\frac{\mathcal{F}(h)}{(1-\lambda) - \lambda\,\mathcal{F}(w^2)}\right) \qquad (9)$$

where $\mathcal{F}, \mathcal{F}^{-1}$ are the Fourier transform and its inverse, respectively.

Figure 1 shows the results of applying formula (9) to each channel of the input image h in Fig. 1 (center), where the variance of the Gaussian kernel w is taken large, and for two opposite signs of λ. As in the nonlinear case (7), the model performs contrast enhancement for $\lambda > 0$ (Fig. 1 (right)) and contrast reduction for $\lambda < 0$ (Fig. 1 (left)).

3   Wilson-Cowan Equations and Color Perception Models

3.1   A Geometric Model of Color Perception

Retinex models of color vision aim to reproduce the perception of the colors of a scene inspired by psychophysical/physiological knowledge about color vision [5,31]. They can be interpreted as the averaging of perceptual distances between image pixels, as pointed out in [1].

Given an RGB color image $u = (u^1, u^2, u^3)$, the perceived image $L = (L^1, L^2, L^3)$ is, according to Kernel-Based Retinex [5], given for $k = 1, 2, 3$, by

$$L^k(x) = \int_{y:u^k(y)\geq u^k(x)} w(x,y)\left[A\log\left(\frac{u^k(x)}{u^k(y)}\right) + 1\right] dy + \int_{y:u^k(y)<u^k(x)} w(x,y)\, dy,$$

where w is a Gaussian kernel and A is a constant, which can be rewritten as

$$L^k(x) = \int_{y\in\Omega} w(x,y)\,\zeta\,(\log[u^k(x)] - \log[u^k(y)])\, dy$$

$$= \int_{y\in\Omega} w(x,y)\,\zeta\left(\int_{\gamma_{y,x}} \nabla log(u^k)(\gamma_{y,x}(t))dt\right) dy \qquad (10)$$

for some nonlinear function ζ, and for any path $\gamma_{y,x}$ joining y to x, see [1].

The quantity $\nabla log(u^k)$ can be interpreted as the perceived gradient of the image according to Weber's law in vision. However, Weber's law suffers from several limitations, and formula (10) can be improved replacing $\nabla \log(u^k)$ by a more accurate representation of the perceptual gradient.

Based on the assumption that the color constancy property comes from an equivariance of the perceived gradient with respect to light changes, Georgiev

[16] suggested that a well-chosen covariant derivative ∇^E on a vector bundle E is a good candidate to describe the perceived gradient, due to the invariance of this differential operator with respect to moving frame changes.

Coupling formula (10) with Georgiev's approach yields the following expression of the perceived image

$$L(x) = \int_{y \in \Omega} w(x,y) \, \zeta \left(\int_{\gamma_{y,x}} \nabla^E_{\gamma'_{y,x}(t)} u(\gamma_{y,x}(t)) \, dt \right) dy$$

$$= \int_{y \in \Omega} w(x,y) \, \zeta \left(u(x) - \tau^E_{y,x,\gamma_{y,x}} u(y) \right) dy, \tag{11}$$

where $\tau^E_{y,x,\gamma_{y,x}}$ denotes the parallel transport map with respect to ∇^E along a path $\gamma_{y,x}$ joining y to x.

Finally, let us mention that the vector bundle framework is consistent with neurophysiological studies suggesting the existence of neurons in V1 responding to colors and organized as a column at each cortical coordinates [21,26], and with the presence of horizontal connections between color columns suggested in [3]. Indeed, the columns can be assimilated to the fibers of the bundle and the horizontal connections between the columns to the parallel transport map.

3.2   Connection to Image Processing Models

Section 3.1 revealed a close relation between the formula (11) describing the perceived colors of a scene and the integral term in (4). In what follows, we assume that the vector bundle is equipped with a metric $\| \cdot \|$ and a covariant derivative ∇^E compatible with this metric. Again, we assume that $\alpha = 1$, h constant in time and w symmetric.

Image Restoration Model with a Nonlinear Activation Function.
Assuming that the activation function in (4) is of the form

$$\sigma[f(a(x,t), a(y,t))] = s_\epsilon(w(x,y)[a(x,t) - \tau^E_{y,x,\gamma_{y,x}} a(y,t)]),$$

where $s_\epsilon = z/\sqrt{\|z\|^2 + \epsilon}$ is the differentiable approximation of $s(z) = z/\|z\|$, then the evolution equation (4) corresponds to the gradient descent equation of a differentiable approximation of the variational problem

$$\arg\min_a \frac{1}{2} \int_\Omega \|a(x) - h(x)\|^2 \, dx - \frac{\lambda}{2} \int_{\Omega^2} w(x,y) \| \tau^E_{y,x,\gamma_{y,x}} a(y) - a(x) \| \, dx \, dy. \tag{12}$$

Our variational model (12) is a refinement of the one for color image correction introduced in [1] as the weight function in this latter is of the form (6), whereas the proposed weight function, which is a symmetrization of the weight (5), encodes the structure tensor, such that our model can take the edges and textures of the image into account.

Fig. 2. Comparison of two contrast enhancement models. From left to right: Original image "Giocondarioca" - Result of the model (9) - Result of the model (14).

Image Restoration Model with a Linear Activation Function. Assuming that the activation function in (4) is of the form

$$\sigma[f(a(x,t), a(y,t))] = w(x,y)[a(x,t) - \tau^E_{y,x,\gamma_{y,x}} a(y,t)],$$

the evolution equation (4) corresponds to the gradient descent equation of the variational model

$$\arg\min_a \frac{1}{2}\int_\Omega \|a(x) - h(x)\|^2\,dx - \frac{\lambda}{2}\int_{\Omega^2} w^2(x,y)\| \tau^E_{y,x,\gamma_{y,x}} a(y) - a(x)\|^2\,dx\,dy. \tag{13}$$

In the particular case where the covariant derivative is flat and the weight w is a normalized Gaussian kernel, the solution of the Euler-Lagrange equations is

$$a = P\left(\mathcal{F}^{-1}\left(\frac{\mathcal{F}(P^{-1}h)}{(1-\lambda) - \lambda\,\mathcal{F}(w^2)}\right)\right), \tag{14}$$

where P is the moving frame in which the covariant derivative is trivial.

Figure 2 compares the model (14) for a flat covariant derivative encoding some brightness perception phenomenon, called the Helmholtz-Kohlrausch effect (see Sect. 5.5.2 in [1] for the expression of the moving frame P), to its Euclidean restriction (9). Both models are applied with the same parameter $\lambda > 0$ and the same large variance for the Gaussian kernel. We observe that the model (14) provides a better result as it preserves more the colors of the original image.

4   Connection to Artificial Neural Networks

4.1   A New Operator and Its Connections to Existing Layers

The Proposed Operator. An ANN can be described as a sequence of functions $\{a^l\}$, $l = 0, \cdots, L$, with $a^l \colon \Omega^l \longrightarrow \mathbb{R}^{n_l}$ and operators $\{H^l\}$, such that $a^{l+1} =$

$H^l(a^l)$. Hence, an operator in an ANN should be able to modify the dimension of its input. Then, together with the coupling between the simplified Wilson-Cowan equation (4) and the color perception model (11), we derive the following layer

$$H^l(a^l): x \longmapsto W^l\, a^l(x) + Q^l \int_{\Omega^l} w(x,y)\, \beta(w(x,y)[\tau^{E^l}_{y,x,\gamma_{y,x}} W^l a^l(y) - W^l a^l(x)])\, dy,$$
(15)

where β is antisymmetric, $W^l \in End(\mathbb{R}^{n_l}; \mathbb{R}^{n_{l+1}})$, $Q^l \in End(\mathbb{R}^{n_{l+1}})$ and τ^{E^l} is the parallel transport map associated to a covariant derivative ∇^{E^l}.

Connection to Convolutional Layers. It is common to call convolutional any layer where the operator H^l is linear. Then, assuming that $\beta = Id$, the operator (15) writes

$$H^l(a^l) = W^l a^l + Q^l\, \Delta^{E^l\,NL}_w W^l a^l,$$
(16)

where the operator $\Delta^{E^l\,NL}_w := \frac{1}{2} \nabla^{E^l\,NL*}_w \nabla^{E^l\,NL}_w$ corresponds to an extension of the (Euclidean) nonlocal Laplacian [17] to vector bundles, called generalized nonlocal Laplacian, with

$$\nabla^{E\,NL}_w u: (x,y) \longmapsto w(x,y)[\tau^E_{y,x,\gamma_{y,x}} u(y) - u(x)]$$

being the nonlocal covariant derivative induced by a covariant derivative ∇^E and a weight function w, and

$$\nabla^{E\,NL*}_w \eta: x \longmapsto \int_\Omega w(x,y)[\tau^E_{y,x,\gamma_{y,x}} \eta(y,x) - \eta(x,y)]dy$$

its adjoint (see [1] for more details).

The layer (16) is a polynomial of degree 1 of a differential operator, as in the graph convolutional networks [22] and surface networks [23] approaches.

Connection to Residual Layers. A residual layer is of the form

$$a^{l+1} = W^l a^l + F^l(a^l),$$
(17)

where the so-called residual function F^l combines linear and nonlinear operators.

Then, we can observe that, for β nonlinear, the operator (15) can be considered as a residual layer where

$$F^l = Q^l\, \nabla^{E^l\,NL*}_w \beta\, \nabla^{E^l\,NL}_w W^l.$$

4.2 Learning the Parameters of the Operator

In deep learning, the parameters of the key layers of an ANN (e.g. convolutional or residual layers) are learned on a well-chosen training data, this latter depending on the task to be performed (classification, recognition, denoising, etc). In what follows, we show that the parameters of the operator (15) have to satisfy some constraints in order for this operator to be consistent in the context of deep learning.

Avoid Overfitting. One of the main constraint when designing an ANN for deep learning is to avoid overfitting, as this latter makes the ANN memorize rather than learn the training data, and which greatly affects the results on the test data afterwards. Overfitting can occur if the number of trainable parameters is too big. In the CNN approach [25], this problem is addressed by convolving the signal with a spatially invariant template of small support. Then, in the context of the operator (15), this problem can be addressed in a two-folds way: first, by reducing the spatial support of the weight function; then, by making the structure tensor T^l and the connection 1-forms ω^{E^l} be constant on the whole domains Ω^l. This makes the weight function be layer dependent and of the form

$$w^l(x, y) = K_\rho \left(\sqrt{(x - y)^T \left(\mathbb{I}_2 + \kappa T^l \right) (x - y)} \right)$$

for some structure tensor T^l, where the inner term corresponds to the distance between x and y induced by the metric $g^l := \mathbb{I}_2 + \kappa T^l$.

Generate Different Types of Local Filters. It is crucial to make the key layers of an ANN be able to generate a large variety of filters. In the CNN approach, this is achieved by learning the template to which the signal is convolved. Moreover, the spatial invariance of the template makes the convolutional layer be equivariant with respect to translations, and consequently makes the neural network satisfy some invariance with respect to translations, which is a desirable property in many computer vision tasks.

We claim that the operator (15) can generate a great variety of filters as well by making the parameters $T^l, \omega^{E^l}, W_l, Q_l$ be trainable. For instance, the operator can be a low-pass or a high-pass filter, depending on the values of W^l and Q^l. Moreover, the metric g^l determines whether the filter is isotropic ($g^l \equiv \mathbb{I}_2$) or anisotropic ($g^l \not\equiv \mathbb{I}_2$), and invariance with respect to the isometries of (Ω^l, g^l) can be achieved for (15) of the form (16) and $\Delta_w^{E^l^{NL}}$ reduced to its local form. Finally, the parallel transport maps τ^{E^l} make the operators H^l mix the channels, yielding new types of filters.

4.3   Experiments

We test the operator (15) for the task of digit classification on the MNIST dataset. The experiment consists in, given an ANN well-designed for this task, replacing the key layer in this ANN by the operator (15), and test the subsequent ANN for three different configurations of the operator.

The ANN we consider here is inspired by the one available in https://github. com/keras-team/keras, and stacks the following layers: 1 convolution (32 features), 1 activation (LeakyReLU), 1 maxpooling, 1 convolution (64 features), 1 activation (LeakyReLU), 1 maxpooling, 1 convolution (64 features), 1 activation (LeakyReLU), 1 maxpooling, 1 dropout, 1 flatten, 1 dense, 1 activation (ReLU), 1 dropout, 1 dense. Here, we consider the form (16) of the operator.

Table 1 shows the mean scores after 50 epochs over 10 tests for each configuration of the operator, where the size of the spatial support of the weight w is 3×3. The chosen optimizer is Adam with learning rate parameter 10^{-3}. We observe that the best scores are obtained when both the metric and the connection 1-form are learned, which corresponds to the most accurate configuration in terms of visual perception.

Table 1. Mean scores of the proposed operator tested on the MNIST dataset.

Configuration of the operator	Score after 50 epochs	Highest score
Euclidean ($g \equiv Id$, $\omega^E \equiv 0$)	0.9927	0.9935 (after 39 epochs)
Vector bundle 1 ($g \equiv Id$, learning ω^E)	0.9929	0.9938 (after 38 epochs)
Vector bundle 2 (learning g and ω^E)	0.9933	0.9944 (after 39 epochs)

5   Conclusion

We showed that an operator inspired by vision neuroscience and psychophysics connects image processing models and some layers in artificial neural networks. In order to show the accuracy of this operator for deep learning tasks, we tested it on a simple dataset for classification. Further work will be devoted to insert this operator into deeper ANNs for more complex computer vision and image processing tasks like denoising and deblurring. Moreover, the proposed geometric framework enables to design ANNs on non-Euclidean domains in order to address tasks involving geometric data like 3D shapes and graphs.

References

1. Batard, T., Bertalmío, M.: A geometric model of brightness perception and its application to color images correction. J. Math. Imag. Vis. **60**(6), 849–881 (2018)
2. Bekkers, E.J., Lafarge, M.W., Veta, M., Eppenhof, K.A.J., Pluim, J.P.W., Duits, R.: Roto-Translation Covariant Convolutional Networks for Medical Image Analysis. In: Frangi, A.F., Schnabel, J.A., Davatzikos, C., Alberola-López, C., Fichtinger, G. (eds.) MICCAI 2018. LNCS, vol. 11070, pp. 440–448. Springer, Cham (2018). https://doi.org/10.1007/978-3-030-00928-1_50
3. Ben-Shahar, O., Zucker, S.W.: Hue geometry and horizontal connections. Neural Netw. **17**(5–6), 753–771 (2004). Special Issue Vision and Brain
4. Bertalmío, M., Caselles, V., Provenzi, E., Rizzi, A.: Perceptual color correction through variational techniques. IEEE Trans. Image. Process. **16**(4), 1058–1072 (2007)
5. Bertalmío, M., Caselles, V., Provenzi, E.: Issues about retinex theory and contrast enhancement. Int. J. Comput. Vis. **83**(1), 101–119 (2009)
6. Bertalmío, M., Cowan, J.D.: Implementing the retinex algorithm with wilson-cowan equations. J. Physiol. Paris **103**(1–2), 69–72 (2009)
7. Bressloff, P.C., Cowan, J.D., Golubitsky, M., Thomas, P.J., Wiener, M.C.: Geometric visual hallucinations, euclidean symmetry and the functional architecture of striate cortex. Phil. Trans. Roy Soc. Lond. B **356**, 299–330 (2001)

8. Bressloff, P.C., Cowan, J.D., Golubitsky, M., Thomas, P.J., Wiener, M.C.: What geometric visual hallucinations tell us about the visual cortex. Neural Comput. **14**(3), 473–491 (2002)
9. Bressloff, P.C., Cowan, J.D.: The functional geometry and local and horizontal connections in a model of V1. J. Physiol. Paris **97**(2–3), 221–236 (2003)
10. Chossat, P., Faugeras, O.: Hyperbolic planforms in relation to visual edges and textures perception. PLoS Comput. Biol. **5**(12), 1–16 (2009)
11. Citti, G., Sarti, A.: A gauge field of modal completion. J. Math. Imag. Vis. **52**(2), 267–284 (2015)
12. Cohen, T.S., Welling, M.: Group equivariant convolutional networks. In: Proceedings of International Conference on Machine Learning ICML (2016)
13. Dieleman, S., De Fauw, J., Kavukcuoglu, K.: Exploiting cyclic symmetry in convolutional neural networks. In: Proceedings of International Conference on Machine Learning ICML (2016)
14. Faye, G., Chossat, P.: A spatialized model of visual texture perception using the structure tensor formalism. Netw. Heterogen. Media **8**(1), 211–260 (2013)
15. Förstner, W., Gülch, E.: A fast operator for detection and precise location of distinct points, corners and centres of circular features. In: Proceedings ISPRS Intercommission Conference on Fast Processing of Photogrammetric Data, pp. 281–305 (1987)
16. Georgiev, T.: Relighting, retinex theory, and perceived gradients. In: Proceedings of Mirage (2005)
17. Gilboa, G., Osher, S.: Nonlocal operators with applications to image processing. Multiscale Model. Simul. **7**(3), 1005–1028 (2008)
18. Goodfellow, I., Bengio, Y., Courville, A.: Deep Learning. MIT Press, Cambridge (2016)
19. He, K., Zhang, X., Ren, S., Sun, J.: Deep residual learning for image recognition. In: Proceedings of IEEE Conference on Computer Vision Pattern Recognition CVPR (2016)
20. Huang, G., Zhuang, L., van der Maaten, L., Weinberger, K.Q.: Densely connected convolutional networks. In: Proceedings of IEEE Conference on Computer Vision Pattern Recognition CVPR (2017)
21. Hubel, D.H.: Eye, Brain and Vision. Scientific American Library, W.H. Freeman & Co, New York (1988)
22. Kipf, T.N., Welling, M.: Semi-supervised classification with graph convolutional networks. In: Proceedings of International Conference on Learning Representation ICLR (2017)
23. Kostrikov, I., Jiang, Z., Panozzo, D., Zorin, D., Bruna, J.: Surface networks. In: Proceedings of IEEE Conference on Computer Vision Pattern Recognition CVPR (2017)
24. Land, E., McCann, J.J.: Lightness and retinex theory. J. Opt. Soc. Am. **61**(1), 1–11 (1971)
25. LeCun, Y., Bottou, L., Bengio, Y., Haffner, P.: Gradient-based Learning applied to document recognition. Proc. IEEE **86**(11), 2278–2324 (1998)
26. Livingstone, M.S., Hubel, D.H.: Anatomy and physiology of a color system in the primate visual cortex. J. Neurosci. **4**(1), 309–356 (1984)
27. Martinez-Garcia, M., Cyriac, P., Batard, T., Bertalmío, M., Malo, J.: Derivatives and inverse of cascaded linear+nonlinear neural models. PLOS ONE **13**(10), e0201326 (2018)
28. Neves, L.L., Monteiro, L.H.A.: A linear analysis of coupled wilson-cowan neuronal populations. Comput. Intell. Neurosci. (2016). Article ID 8939218, 6 pages

29. Wilson, H.R., Cowan, J.D.: Excitatory and inhibitory interactions in localized populations of model neurons. Biophys. J. **12**(1), 1–24 (1972)
30. Wilson, H.R., Cowan, J.D.: A mathematical theory of the functional dynamics of cortical and thalamic nervous tissue. Kybernetik **13**(2), 55–80 (1973)
31. Yeonan-Kim, J., Bertalmío, M.: Analysis of retinal and cortical components of retinex algorithms. J. Electron. Imaging **26**(3), 031208 (2017)

A Cortical-Inspired Model for Orientation-Dependent Contrast Perception: A Link with Wilson-Cowan Equations

Marcelo Bertalmío[1], Luca Calatroni[2(✉)], Valentina Franceschi[3],
Benedetta Franceschiello[4], and Dario Prandi[5]

[1] DTIC, Universitat Pompeu Fabra, Barcelona, Spain
`marcelo.bertalmio@upf.edu`
[2] CMAP, École Polytechnique CNRS, Palaiseau, France
`luca.calatroni@polytechnique.edu`
[3] IMO, Université Paris-Sud, Orsay, France
`valentina.franceschi@math.u-psud.fr`
[4] Fondation Asile des Aveugles and Laboratory for Investigative Neurophysiology,
Lausanne, Switzerland
`benedetta.franceschiello@fa2.ch`
[5] CNRS, L2S, CentraleSupélec, Gif-sur-Yvette, France
`dario.prandi@l2s.centralesupelec.fr`

Abstract. We consider a differential model describing neuro-physiological contrast perception phenomena induced by surrounding orientations. The mathematical formulation relies on a cortical-inspired modelling [11] largely used over the last years to describe neuron interactions in the primary visual cortex (V1) and applied to several image processing problems [14,15,21]. Our model connects to Wilson-Cowan-type equations [26] and it is analogous to the one used in [3,4,16] to describe assimilation and contrast phenomena, the main novelty being its explicit dependence on local image orientation. To confirm the validity of the model, we report some numerical tests showing its ability to explain orientation-dependent phenomena (such as grating induction) and geometric-optical illusions [18,24] classically explained only by filtering-based techniques [7,20].

Keywords: Orientation-dependent modelling ·
Wilson-Cowan equations · Primary visual cortex ·
Contrast perception · Variational modelling

1 Introduction

Many, if not most, popular vision models consist in a cascade of linear and non-linear (L+NL) operations [17]. This happens for models describing both visual perception – e.g., the Oriented Difference Of Gaussians (ODOG) [7] or

© Springer Nature Switzerland AG 2019
J. Lellmann et al. (Eds.): SSVM 2019, LNCS 11603, pp. 472–484, 2019.
https://doi.org/10.1007/978-3-030-22368-7_37

the Brightness Induction Wavelet Model (BIWaM) [20] – and neural activity [10]. However, L+NL models, while suitable in many cases for retinal and thalamic activity, are not adequate to predict neural activity in the primary visual cortex area (V1). In fact, according to [10], such models have low predictive power (they can explain less than 40% of the variance of the data). On the other hand, several vision models are not in the form of a cascade of L+NL operations, such as those describing neural dynamics via Wilson-Cowan (WC) equations [9, 26]. These equations describe the state $a(x, \theta, t)$ of a population of neurons with V1 coordinates $x \in \mathbb{R}^2$ and orientation preference $\theta \in [0, \pi)$ at time $t > 0$ as

$$\frac{\partial}{\partial t}a(x, \theta, t) = -\alpha a(x, \theta, t) + \nu \int_0^\pi \int_{\mathbb{R}^2} \omega(x, \theta \| x', \theta') \sigma(a(x', \theta', t)) \, dx' \, d\theta' + h(x, \theta, t).$$

(1)

Here, $\alpha, \nu > 0$ are fixed parameters, $\omega(x, \theta \| x', \theta')$ is an interaction weight, $\sigma : \mathbb{R} \to \mathbb{R}$ is a sigmoid saturation function and h represents the external stimulus. In [3–5] the authors show how orientation independent WC-type equations admit a variational formulation through an associated energy functional which can be linked to histogram equalisation, visual adaptation and efficient coding [19].

In this paper, we consider a generalisation of this modelling and introduce explicit orientation dependence via a lifting procedure inspired by neurophysiological models of V1 [11, 14, 21]. Interestingly, the Euler-Lagrange equations associated with the proposed functional yield orientation-dependent WC-type equations analogous to (1). We remark that WC-type cortical-inspired models have been previously considered in [1, 22], though without any variational interpretation and with only very few applications to contour detection for spatio-temporal imaging. The numerical results of our enriched variational model show how it can effectively be applied to reproduce more elaborated visual perception phenomena in comparison to both previous orientation-independent WC-type models and state-of-the-art ODOG [7] and BIWaM [20] L+NL models.

We firstly test our model on orientation-dependent Grating Induction (GI) phenomena (generalising the ones presented in [7, Fig. 3], see also [18]) and show a direct dependence of the processed image on the orientation, which cannot be reproduced via orientation-independent models, but which is in accordance with the reference result provided by the ODOG model. We then test the proposed model on the Poggendorff illusion, a geometrical optical effect where a misalignment of two collinear segment is induced by the presence of a surface [24, 25], see Fig. 6. For this example our model is able to reconstruct the perceptual bias better than all the reference models considered.

2   Orientation-Independent Variational Models

Let $Q \subset \mathbb{R}^2$ be a rectangular image domain and let $f : Q \to [0, 1]$ be a normalised image on Q. Let further $\omega : Q \times Q \to \mathbb{R}_+$ be a given positive symmetric weighting function such that for any $x \in Q$, $\omega(x, \cdot) \in L^1(Q)$. For a fixed parameter $\alpha > 1$ we consider the piecewise affine sigmoid $\sigma_\alpha : \mathbb{R} \to [-1, 1]$ defined by $\sigma_\alpha(\rho) := \min\{1, \max\{\alpha\rho, -1\}\}$, and Σ_α to be any function such that $\Sigma'_\alpha = \sigma_\alpha$. Trivially, the function Σ_α is convex and even.

In [3,4,16] the authors consider the following energy as a variational model for contrast and assimilation phenomena, defined for a given initial image f_0:

$$\mathcal{E}(f) := \frac{1}{2} \int_Q (f(x) - \mu(x))^2 \, dx + \frac{\lambda}{2} \int_Q (f(x) - f_0(x))^2 \, dx$$

$$- \frac{1}{4M} \int_Q \int_Q \omega(x,y) \Sigma_\alpha \big(f(x) - f(y)\big) \, dx \, dy. \quad (2)$$

Here, $\mu : Q \to \mathbb{R}$ is the local mean of the initial image f_0. Such term can encode a global reference to the "Grey-World" principle [4] (in this case $\mu(x) = \frac{1}{2}$ for any $x \in Q$), or a filtering around $x \in Q$ computed either via a single-Gaussian convolution [3] or a sum of Gaussian filters [16], consistently with the modelling of the multiple inhibition effects happening at a retinal-level [27]. Finally, the parameter $M \in (0,1]$ stands for a normalisation constant, while $\lambda > 0$ represents a weighting parameter enforcing the attachment of the solution f to f_0.

The gradient descent associated with $\mathcal{E}$ corresponds to a Wilson-Cowan-type equation similar to (1), where the visual activation is assumed to be independent of the orientation, as discussed in [8]. The parameters α and ν are set $\alpha = 1$ and $\nu = \frac{1}{2M}$ and the time-invariant external stimulus h is equal to $\mu + \lambda f_0$:

$$\frac{\partial}{\partial t} f(x,t) = -(1+\lambda)f(x,t) + (\mu(x) + \lambda f_0(x)) + \frac{1}{2M} \int_Q \omega(x,y) \sigma_\alpha \big(f(x,t) - f(y,t)\big) \, dy.$$

$$(3)$$

Note that in [16] the authors consider in (2) a convolution kernel ω which is a convex combination of two 2D Gaussians with different standard deviations. While this variation of the model is effective in describing *assimilation* effects, the lack of dependence on the local perceived orientation in (3) makes such modelling intrinsically not adapted to orientation-induced contrast/colour perception effects as the ones observed in [7,20,23]. Up to our knowledge, the only perceptual models capable to explain the latter are the ones based on oriented Difference of Gaussian filtering coupled with some non-linear processing, such as the ODOG and the BIWaM models described in [6,7] and [20], respectively.

3   A Cortical-Inspired Model

Let us denote by $R > 0$ the size of the visual plane and let $D_R \subset \mathbb{R}^2$ be the disk $D_R := \{x_1^2 + x_2^2 \leq R^2\}$. Fix $R > 0$ such that $Q \subset D_R$. In order to exploit the properties of the roto-translation group $SE(2)$ on images, we now consider them to be elements in the set: $\mathcal{I} = \{f \in L^2(\mathbb{R}^2, [0,1]) \text{ such that } \operatorname{supp} f \subset D_R\}$. We remark that fixing $R > 0$ is necessary, since contrast perception is strongly dependent on the scale of the features under consideration w.r.t. the visual plane.

In the following, after introducing the functional lifting under consideration, which follows well-established ideas contained, e.g., in [11,14,21], we present the proposed cortical-inspired extension of the energy functional (2), and discuss the numerical implementation of the associated gradient descent. We remark that the resulting procedure thus consists of a linear lifting combined with WC-type evolution, which is non-linear due to the presence of the sigmoid σ_α.

3.1   Functional Lifting

Each neuron ξ in V1 is assumed to be associated with a receptive field (RF) $\psi_\xi \in L^2(\mathbb{R}^2)$ such that its response under a visual stimulus $f \in \mathcal{I}$ is given by

$$F(\xi) = \langle \psi_\xi, f \rangle_{L^2(\mathbb{R}^2)} = \int_{\mathbb{R}^2} \overline{\psi_\xi(x)} f(x) \, dx. \qquad (4)$$

Motivated by neuro-phyisiological evidence, we assume that each neuron is sensible to a preferred position and orientation in the visual plane, i.e., $\xi = (x, \theta) \in \mathcal{M} = \mathbb{R}^2 \times \mathbb{P}^1$. Here, $\mathbb{P}^1$ is the projective line that we represent as $[0, \pi]/ \sim$, with $0 \sim \pi$. Moreover, we assume that RFs of different neurons are "deducible" one from the other via a linear transformation. Let us explain this in detail.

The double covering of $\mathcal{M}$ is given by the Euclidean motion group $SE(2) = \mathbb{R}^2 \rtimes S^1$, that we consider endowed with its natural semi-direct product structure

$$(x, \theta) \star (y, \varphi) = (x + R_\theta y, \theta + \varphi), \quad \forall (x, \theta), (y, \varphi) \in SE(2), \quad R_\theta = \begin{pmatrix} \cos\theta & -\sin\theta \\ \sin\theta & \cos\theta \end{pmatrix}.$$

In particular, the above operation induces an action of $SE(2)$ on $\mathcal{M}$, which is thus an homogeneous space. Observe that $SE(2)$ is unimodular and that its Haar measure (the left and right-invariant measure up to scalar multiples) is $dx d\theta$.

We now denote by $\mathcal{U}(L^2(\mathbb{R}^2)) \subset \mathcal{L}(L^2(\mathbb{R}^2))$ the space of linear unitary operators on $L^2(\mathbb{R}^2)$ and let $\pi : SE(2) \to \mathcal{U}(L^2(\mathbb{R}^2))$ be the *quasi-regular representation* of $SE(2)$ which associates to any $(x, \theta) \in SE(2)$ the unitary operator $\pi(x, \theta) \in \mathcal{U}(L^2(\mathbb{R}^2))$, i.e., the action of the roto-translation (x, θ) on square-integrable functions on $\mathbb{R}^2$. Namely, $\pi(x, \theta)$ acts on $\psi \in L^2(\mathbb{R}^2)$ by

$$[\pi(x, \theta)\psi](y) = \psi((x, \theta)^{-1} y) = \psi(R_{-\theta}(y - x)), \qquad \forall y \in \mathbb{R}^2.$$

Moreover, we let $\Lambda : SE(2) \to \mathcal{U}(L^2(SE(2)))$ be the *left-regular representation*, which acts on functions $F \in L^2(SE(2))$ as

$$[\Lambda(x, \theta)F](y, \varphi) = F((x, \theta)^{-1} \star (y, \varphi)) = F(R_{-\theta}(y - x), \varphi - \theta), \quad \forall (y, \theta) \in SE(2).$$

Letting $L : L^2(\mathbb{R}^2) \to L^2(\mathcal{M})$ be the operator that transforms visual stimuli into cortical activations, it is natural to require that

$$L \circ \pi(x, \theta) = \Lambda(x, \theta) \circ L, \qquad \forall (x, \theta) \in SE(2).$$

Under mild continuity assumption on L, it has been shown in [21] that L is then a continuous wavelet transform. That is, there exists a *mother wavelet* $\Psi \in L^2(\mathbb{R}^2)$ satisfying $\pi(x, \theta)\Psi = \pi(x, \theta + \pi)\Psi$ for all $(x, \theta) \in SE(2)$, and such that

$$Lf(x, \theta) = \langle \pi(x, \theta)\Psi, f \rangle, \qquad \forall f \in L^2(\mathbb{R}^2), (x, \theta) \in \mathcal{M}. \qquad (5)$$

Observe that the operation $\pi(x, \theta)\Psi$ above is well defined for $(x, \theta) \in \mathcal{M}$ thanks to the assumption on Ψ. By (4), the above representation of L is equivalent to the fact that the RF associated with the neuron $(x, \theta) \in \mathcal{M}$ is the roto-translation of the mother wavelet, i.e., $\psi_{(x, \theta)} = \pi(x, \theta)\Psi$.

Remark 1. Letting $\Psi^*(x) := \overline{\Psi(-x)}$, the above formula can be rewritten as

$$Lf(x,\theta) = \int_{\mathbb{R}^2} \overline{\Psi(R_{-\theta}(y-x))} f(y)\,dy = \left[f * (\Psi^* \circ R_{-\theta}) \right](x),$$

where $f * g$ denotes the standard convolution on $L^2(\mathbb{R}^2)$.

Notice that, although images are functions of $L^2(\mathbb{R}^2)$ with values in $[0,1]$, it is in general not true that $Lf(x,\theta) \in [0,1]$. However, from (5) we deduce that Lf is a continuous function, satisfying $|Lf| \le \|\Psi\|_{L^2(\mathbb{R}^2)} \|f\|_{L^2(\mathbb{R}^2)}$.

Neuro-physiological evidence shows that a good fit for the RFs is given by Gabor filters, whose Fourier transform is the product of a Gaussian with an oriented plane wave [12]. However, these filters are quite challenging to invert, and are parametrised on a bigger space than $\mathcal{M}$, which takes into account also the frequency of the plane wave and not only its orientation. For this reason, in this work we chose to consider as wavelets the *cake wavelets* introduced in [13], see also [2]. These are obtained via a mother wavelet Ψ^{cake} whose support in the Fourier domain is concentrated on a fixed slice, depending on the number of orientations one aims to consider in the numerical implementation. For the sake of integrability, the Fourier transform of this mother wavelet is then smoothly cut off via a low-pass filtering, see [2, Sect. 2.3] for details. Observe, however, that in order to lift to $\mathcal{M}$ and not to $SE(2)$, we consider a non-oriented version of the mother wavelet, given by $\tilde{\psi}^{cake}(\omega) + \tilde{\psi}^{cake}(e^{i\pi}\omega)$, in the notations of [2].

An important feature of cake wavelets is that, in order to recover the original image, it suffices to define the so-called *projection operator* by

$$P : L^2(\mathcal{M}) \to L^2(\mathbb{R}^2), \qquad PF(x) := \int_{\mathbb{P}^1} F(x,\theta)\,d\theta, \qquad F \in L^2(\mathcal{M}) \qquad (6)$$

Indeed, by construction of cake wavelets it holds that $\int_{\mathbb{P}^1} R_\theta \Psi^{cake}\,d\theta = 1$ and thus a straightforward application of Fubini's Theorem shows that $P \circ L = \mathrm{Id}$.

3.2   WC-Type Mean Field Modelling

The natural extension of (2) to the cortical setting introduced in the previous section is the following energy on functions $F \in L^2(\mathcal{M})$:

$$\mathcal{E}(F) = \frac{1}{2}\|F - G_0\|^2_{L^2(\mathcal{M})} + \frac{\lambda}{2}\|F - F_0\|^2_{L^2(\mathcal{M})}$$
$$- \frac{1}{4M}\int_{\mathcal{M}}\int_{\mathcal{M}} \omega(x,\theta\|x',\theta')\Sigma_\alpha\big(F(x,\theta) - F(x',\theta')\big)\,dxd\theta\,dx'd\theta'. \quad (7)$$

Here, G_0 is the lift $L\mu$ of the local mean appearing in (2), and F_0 is the lift Lf_0 of the given initial image f_0. The constant $M \in (0,1]$ will be specified later. Furthermore, $\omega : \mathcal{M} \times \mathcal{M} \to \mathbb{R}_+$ is a positive symmetric weight function and, as in [4], we denote by $\omega(x,\theta\|x',\theta')$ its evaluation at $((x,\theta),(x',\theta'))$. A sensible choice for ω would be an approximation of the heat kernel of the anisotropic diffusion associated with the structure of V1, as studied in [11,14]. However, in this work we restrict to the case where $\omega(x,\theta\|x',\theta') = \omega(x-x',\theta-\theta')$ is a (normalised) 3D Gaussian in $\mathcal{M}$, since this is sufficient to describe many perceptual phenomena not explained by the previous 2D model (2). A possible improvement could be obtained by choosing ω as the sum of two such Gaussians at different scales in order to better describe, e.g., lightness assimilation phenomena.

Thanks to the symmetry of ω and the oddness of σ, computations similar to those of [4] yield that the gradient descent associated with (7) is

$$\frac{\partial}{\partial t}F(x,\theta,t) = -(1+\lambda)F(x,\theta,t) + G_0(x,\theta) + \lambda F_0(x,\theta)$$

$$+\frac{1}{2M}\int_{\mathcal{M}}\omega(x,\theta\|x',\theta')\sigma_\alpha\big(F(x,\theta,t) - F(x',\theta',t)\big)\,dx'd\theta'. \quad (8)$$

This is indeed a Wilson-Cowan-type equation, where the external cortical stimulus h in (1) is $h = G_0 + \lambda F_0$. We observe that in general there is no guarantee that rng L is invariant w.r.t. the evolution $t \mapsto F(\cdot,\cdot,t)$ given by (8). That is, in general, although $F_0 = Lf_0$, if $t > 0$ there is no image $f(\cdot,t) \in L^2(\mathbb{R}^2)$ such that $F(\cdot,\cdot,t) = Lf(\cdot,\cdot,t)$. We will nevertheless assume that the perceived image for any cortical activation is given by the projection operator (6).

3.3   Discretisation via Gradient Descent

In order to numerically implement the gradient descent (8), we discretise the initial (square) image f_0 as an $N\times N$ matrix. (Here, we assume periodic boundary conditions). We additionally consider $K \in \mathbb{N}$ orientations, parametrised by $k \in \{1,\dots,K\} \mapsto \theta_k := (k-1)\pi/K$.

The discretised lift operator, still denoted by L, then transforms $N\times N$ matrices into $N \times N \times K$ arrays. Its action on an $N \times N$ matrix f is defined by

$$(Lf)_{n,m,k} = \mathcal{F}^{-1}\left((\mathcal{F}f) \odot (R_{\theta_k}\mathcal{F}\Psi^{\text{cake}})\right)_{n,m} \quad n,m \in \{1,\dots,N\},\ k \in \{1,\dots,K\},$$

where $\odot$ is the Hadamard (i.e., element-wise) product of matrices, $\mathcal{F}$ denotes the discrete Fourier transform, R_{θ_k} is the rotation of angle θ_k, and Ψ^{cake} is the cake mother wavelet.

Denoting by $F^0 = Lf_0$, and by $G_0 = L\mu$ where μ is a Gaussian filtering of f_0, we find that the explicit time-discretisation of the gradient descent (8) is

$$\frac{F^{\ell+1} - F^\ell}{\Delta t} = -(1+\lambda)F^\ell + G_0 + \lambda F^0 + \frac{1}{2M}\mathcal{R}_{F^\ell}, \qquad \Delta t \ll 1,\ \ell \in \mathbb{N}.$$

Here, for a given 3D Gaussian matrix W encoding the weight ω, and an $N \times N \times M$ matrix F, we let, for any $n,m \in \{1,\dots,N\}$ and $k \in \{1,\dots,K\}$,

$$(\mathcal{R}_F)_{n,m,k} := \sum_{n',m'=1}^{N}\sum_{k'=1}^{K} W_{n-n',m-m',k-k'}\sigma(F_{n,m,k} - F_{n',m',k'}).$$

We refer to [4, Sect. IV.A] for the description of an efficient numerical approach used to compute the above quantity in the 2D case and that can be translated verbatim to the 3D case under consideration.

After a suitable number of iterations $\bar{\ell}$ of the above algorithm (measured by the criterion $\|F^{\ell+1} - F^\ell\|_2/\|F^\ell\|_2 \leq \tau$, for a fixed tolerance $\tau \ll 1$), the output image is then found via (6) as $\bar{f}_{n,m} = \sum_{k=1}^{K} F_{n,m,k}^{\bar{\ell}}$.

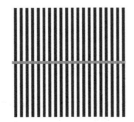

(a) Relative orientation $\theta = \pi/2$.　　　(b) Relative orientation $\theta = \pi/3$.

Fig. 1. Grating inductions with varying background orientation.

4　Numerical Results

In this section we present the results obtained by applying the cortical-inspired model presented in the previous section to a class of well-established phenomena where contrast perception is affected by local orientations.

We compare the results obtained by our model with the corresponding Wilson-Cowan-type 2D model (2)–(3) for contrast enhancement considered in [3,16]. For further reference, we also report comparisons with two standard reference models based on oriented Gaussian filtering. The former is the ODOG model [7] where the output is computed via a convolution of the input image with oriented difference of Gaussian filters in six orientations and seven spatial frequencies. The filtering outputs within the same orientation are then summed in a non-linear fashion privileging higher frequencies. The second model used for comparison is the BIWaM model, introduced in [20]. This is a variation of the ODOG model, the difference being the dependence on the local surround orientation of the contrast sensitivity function[1].

Parameters. All the considered images are 200×200 pixel. We always consider lifts with $K = 30$ orientations. The relevant cake wavelets are then computed following [2] for which the frequency band bw is set to bw $= 4$ for all experiments. In (8), we compute the local mean average μ and the integral term by Gaussian filtering with standard deviation $\sigma_\mu > \sigma_\omega$. The gradient descent algorithm stops when the relative stopping criterion defined in Sect. 3.3 with a tolerance $\tau = 10^{-2}$ is verified.

---

[1] For our comparisons we used the ODOG and BIWaM codes freely available at https://github.com/TUBvision/betz2015_noise.

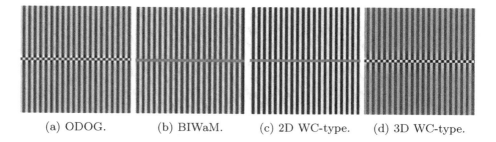

(a) ODOG.   (b) BIWaM.   (c) 2D WC-type.   (d) 3D WC-type.

Fig. 2. Model outputs of input Fig. 1a. Parameters for (d): $\sigma_\mu = 10$, $\sigma_\omega = 5$, $\lambda = 0.5$.

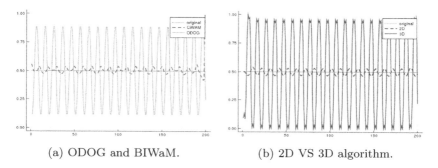

(a) ODOG and BIWaM.   (b) 2D VS 3D algorithm.

Fig. 3. Middle line-profiles of outputs in Fig. 2.

4.1   Grating Induction with Oriented Relative Background

Grating induction (GI) is a contrast effect first described in [18] and later studied, e.g., in [7]. In this section we describe our results about GI with *relative background orientation*, see Fig. 1. Here, when the background has different orientation from the central grey bar, a grating effect, i.e. an alternation of dark-grey/light-grey patterns within the central bar, is produced and perceived by the observer. This phenomenon is contrast dependent, as the intensity of the induced grey patterns (dark-grey/light-grey) is in opposition with the background grating. Moreover, it is also orientation-dependent as the magnitude of the phenomenon varies based on the background orientation, and is maximal when the background bars are orthogonal to the central grey one.

Discussion on Computational Results. We observe that model (8) predicts, in accordance with visual perception, the appearance of a counter-phase grating in the central grey bar, see Figs. 2d and 4d. The same result is obtained by the ODOG model, see Figs. 2a and 4a. In particular, Figs. 3 and 5 show higher intensity profile when the background gratings are orthogonal to the central line, with respect to the case of background angle equal to $\pi/3$, see orange and green dashed line. On the other hand, the BIWaM model and the WC-type 2D model for contrast enhancement do not appear suitable to describe this phenomenon. See for comparison the red and blue dashed lines in Figs. 3 and 5.

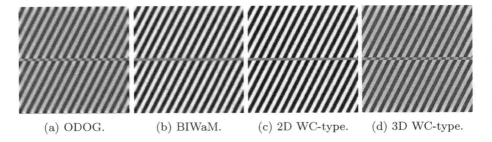

(a) ODOG.   (b) BIWaM.   (c) 2D WC-type.   (d) 3D WC-type.

Fig. 4. Model outputs of input Fig. 1b. Parameters for (d): $\sigma_\mu = 10$, $\sigma_\omega = 5$, $\lambda = 0.5$.

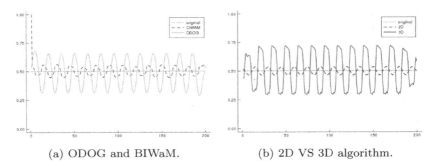

(a) ODOG and BIWaM.   (b) 2D VS 3D algorithm.

Fig. 5. Middle line-profiles of outputs in Fig. 4.

4.2   Poggendorff Illusion

The Poggendorff illusion (see Fig. 6c) consists in the super-position of a surface on top of a continuous line, which then induces a misalignment effect. This phenomenon has been deeply investigated [24,25] and studied via neuro-physical experiments, see, e.g., [24]. Here, we consider a variation of the Poggendorff illusion, where the background is constituted by a grating pattern, see Fig. 6a.

Discussion on Computational Results. The result obtained by applying (8) to Fig. 6a is presented in Figs. 6d, 7d. As in Sect. 4.1, we observe an induced counter-phase grating in the central grey bar. However, the question here is whether it is possible to reconstruct numerically the perceived misalignment between a fixed black stripe in the bottom part of Fig. 6a and its collinear prosecution lying in the upper part. Note that the perceived alignment differs from the actual geometrical one: for a fixed black stripe in the bottom part, the alignment of the corresponding collinear top stripe is perceived slightly flushed left, see Fig. 6c, where single stripes have been isolated for better visualisation. To answer the question, let us focus on Fig. 6b and mark by a continuous green line a fixed black stripe in the bottom part of the image. In order to find the corresponding perceived collinear stripe in the upper part, we follow how the model propagates the marked stripe across the central surface, by drawing a dashed line. As

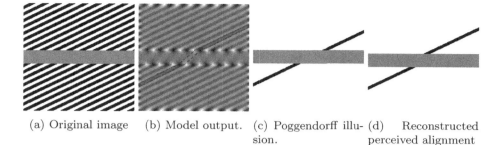

(a) Original image   (b) Model output.   (c) Poggendorff illu- (d)   Reconstructed
                                             sion.   perceived alignment

Fig. 6. A variation of the Poggendorff illusion. The presence of the grey central surface in (a) induces a misalignment of the background lines of the output (b). In (c), we present the original Poggendorff illusion, extracted from (a), and in (d) the reconstruction deduced from (b). Parameters: $\sigma_\mu = 10$, $\sigma_\omega = 3$, $\lambda = 0.5$.

expected, the prosecution provided by the algorithm following this procedure does not correspond to its actual collinear prosecution.

The proposed algorithm hence computes an output in agreement with our perception. Comparison with reference models is presented in Figs. 7 and 8. We observe that the results obtained via the proposed 3D-WC model cannot be reproduced by the BIWaM nor the WC-type 2D model, which moreover induce non-counter-phase grating in the central grey bar. On the other hand, the result obtained by the ODOG model is consistent with ours, but presents a much less evident alternating grating within the central grey bar. In particular, the induced oblique bands are not visibly connected across the whole grey bar, i.e. their induced contrast is very poor and, consequently, the induced edges are not as sharp as the ones reconstructed via our model, see Fig. 8 for an example on the middle-line profile.

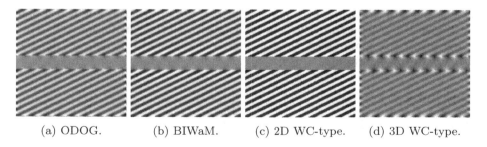

(a) ODOG.   (b) BIWaM.   (c) 2D WC-type.   (d) 3D WC-type.

Fig. 7. Reconstruction of the Poggendorff illusion Fig. 6a via reference models.

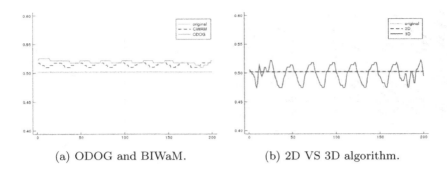

(a) ODOG and BIWaM.   (b) 2D VS 3D algorithm.

Fig. 8. Middle line-profiles of outputs in Fig. 7.

5   Conclusions

We presented a cortical-inspired setting extending the approach used in [3,4] to describe contrast phenomena in images. By mimicking the structure of V1, the model explicitly takes into account information on local image orientation and it relates naturally to Wilson-Cowan-type equations. The model can be efficiently implemented via convolution with appropriate kernels and discretised via standard explicit schemes. The information on the local orientation allows to describe contrast phenomena as well as orientation-dependent illusions, outperforming the models introduced in [3,4]. The introduced mean field model is competitive with the results obtained by applying some popular orientation-dependent filtering such as the ODOG and the BIWaM models [7,20].

Further investigations should address a more accurate modelling reflecting the actual structure of V1. In particular, this concerns the lift operation where the cake wavelet filters should be replaced by Gabor filtering, as well as the interaction weight ω which could be taken to be the anisotropic heat kernel of [11, 14]. Finally, extensive numerical experiments should be performed to assess the compatibility of the model with psycho-physical tests measuring the perceptual bias induced by these and other phenomena such as the tilt illusion [23]. This would provide insights about the robustness of the model in reproducing the visual pathway behaviour.

References

1. Barbieri, D., Citti, G., Cocci, G., Sarti, A.: A cortical-inspired geometry for contour perception and motion integration. JMIV **49**(3), 511–529 (2014)
2. Bekkers, E., Duits, R., Berendschot, T., ter Haar Romeny, B.: A multi-orientation analysis approach to retinal vessel tracking. JMIV **49**(3), 583–610 (2014)
3. Bertalmío, M.: From image processing to computational neuroscience: a neural model based on histogram equalization. Front. Comput. Neurosci. **8**, 71 (2014)
4. Bertalmío, M., Caselles, V., Provenzi, E., Rizzi, A.: Perceptual color correction through variational techniques. IEEE Trans. Image Process. **16**(4), 1058–1072 (2007)

5. Bertalmío, M., Cowan, J.D.: Implementing the retinex algorithm with Wilson-Cowan equations. J. Physiol. Paris **103**(1), 69–72 (2009)
6. Blakeslee, B., Cope, D., McCourt, M.E.: The oriented difference of Gaussians (ODOG) model of brightness perception: overview and executable mathematica notebooks. Behav. Res. Methods **48**(1), 306–312 (2016)
7. Blakeslee, B., McCourt, M.E.: A multiscale spatial filtering account of the White effect, simultaneous brightness contrast and grating induction. Vision Res. **39**(26), 4361–4377 (1999)
8. Bressloff, P.C., Cowan, J.D., Golubitsky, M., Thomas, P.J., Wiener, M.C.: Geometric visual hallucinations, Euclidean symmetry and the functional architecture of striate cortex. Philos. Trans. Roy. Soc. Lond. B Biol. Sci. **356**, 299–330 (2001)
9. Bressloff, P.C., Cowan, J.D.: An amplitude equation approach to contextual effects in visual cortex. Neural Comput. **14**(3), 493–525 (2002)
10. Carandini, M., et al.: Do we know what the early visual system does? J. Neurosci. **25**(46), 10577–10597 (2005)
11. Citti, G., Sarti, A.: A cortical based model of perceptual completion in the roto-translation space. JMIV **24**(3), 307–326 (2006)
12. Daugman, J.G.: Uncertainty relation for resolution in space, spatial frequency, and orientation optimized by two-dimensional visual cortical filters. J. Opt. Soc. Am. A **2**(7), 1160–1169 (1985)
13. Duits, R., Felsberg, M., Granlund, G., ter Haar Romeny, B.: Image analysis and reconstruction using a wavelet transform constructed from a reducible representation of the euclidean motion group. Int. J. Comput. Vis. **72**(1), 79–102 (2007)
14. Duits, R., Franken, E.: Left-invariant parabolic evolutions on $SE(2)$ and contour enhancement via invertible orientation scores. Part I: linear left-invariant diffusion equations on $SE(2)$. Quart. Appl. Math. **68**(2), 255–292 (2010)
15. Franceschiello, B., Sarti, A., Citti, G.: A neuromathematical model for geometrical optical illusions. JMIV **60**(1), 94–108 (2018)
16. Kim, J., Batard, T., Bertalmío, M.: Retinal processing optimizes contrast coding. J. Vis. **16**(12), 1151–1151 (2016)
17. Martinez-Garcia, M., Cyriac, P., Batard, T., Bertalmío, M., Malo, J.: Derivatives and inverse of cascaded linear+nonlinear neural models. PLOS One **13**(10), 1–49 (2018)
18. McCourt, M.E.: A spatial frequency dependent grating-induction effect. Vis. Res. **22**(1), 119–134 (1982)
19. Olshausen, B.A., Field, D.J.: Vision and the coding of natural images: the human brain may hold the secrets to the best image-compression algorithms. Am. Sci. **88**(3), 238–245 (2000)
20. Otazu, X., Vanrell, M., Parraga, C.A.: Multiresolution wavelet framework models brightness induction effects. Vis. Res. **48**(5), 733–751 (2008)
21. Prandi, D., Gauthier, J.P.: A Semidiscrete Version of the Petitot Model as a Plausible Model for Anthropomorphic Image Reconstruction and Pattern Recognition. SpringerBriefs in Mathematics. Springer, Cham (2017). https://doi.org/10.1007/978-3-319-78482-3
22. Sarti, A., Citti, G.: The constitution of visual perceptual units in the functional architecture of V1. J. comput. Neurosci. **38**(2), 285–300 (2015)
23. Self, M.W., et al.: Orientation-tuned surround suppression in mouse visual cortex. J. Neurosci. **34**(28), 9290–9304 (2014)
24. Weintraub, D.J., Krantz, D.H.: The Poggendorff illusion: amputations, rotations, and other perturbations. Attent. Percept. Psychol. **10**(4), 257–264 (1971)

25. Westheimer, G.: Illusions in the spatial sense of the eye: geometrical-optical illusions and the neural representation of space. Vis. Res. **48**(20), 212–2142 (2008)
26. Wilson, H.R., Cowan, J.D.: Excitatory and inhibitory interactions in localized populations of model neurons. BioPhys. J. **12**(1), 1–24 (1972)
27. Yeonan-Kim, J., Bertalmío, M.: Retinal lateral inhibition provides the biological basis of long-range spatial induction. PLOS One **11**(12), 1–23 (2016)

A Total Variation Based Regularizer Promoting Piecewise-Lipschitz Reconstructions

Martin Burger[1], Yury Korolev[2(✉)], Carola-Bibiane Schönlieb[2], and Christiane Stollenwerk[3]

[1] Department Mathematik, University of Erlangen-Nürnberg, Cauerstr. 11, 91058 Erlangen, Germany
martin.burger@fau.de
[2] Department of Applied Mathematics and Theoretical Physics, University of Cambridge, Wilberforce Road, Cambridge CB3 0WA, UK
{yk362,cbs31}@cam.ac.uk
[3] Institute for Analysis and Numerics, University of Münster, Einsteinstr. 62, 48149 Münster, Germany
ChristianeStollenwerk@web.de

Abstract. We introduce a new regularizer in the total variation family that promotes reconstructions with a given Lipschitz constant (which can also vary spatially). We prove regularizing properties of this functional and investigate its connections to total variation and infimal convolution type regularizers TVL^p and, in particular, establish topological equivalence. Our numerical experiments show that the proposed regularizer can achieve similar performance as total generalized variation while having the advantage of a very intuitive interpretation of its free parameter, which is just a local estimate of the norm of the gradient. It also provides a natural approach to spatially adaptive regularization.

Keywords: Total variation · Total generalized variation · First order regularization · Image denoising

1 Introduction

Since it has been introduced in [20], total variation (TV) has been popular in image processing due to its ability to preserve edges while imposing sufficient regularity on the reconstructions. There have been numerous works studying the geometric structure of TV-based reconstructions (e.g., [8,11,16,17,19]). A typical characteristic of these reconstructions is the so-called *staircasing* [17, 19], which refers to piecewise-constant reconstructions with jumps that are not present in the ground truth image. To overcome the issue of staircasing, many other TV-type regularizers have been proposed, perhaps the most successful of which being the Total Generalized Variation (TGV) [3,5]. TGV uses derivatives

© Springer Nature Switzerland AG 2019
J. Lellmann et al. (Eds.): SSVM 2019, LNCS 11603, pp. 485–497, 2019.
https://doi.org/10.1007/978-3-030-22368-7_38

of higher order and favours reconstructions that are piecewise-polynomial; in the most common case of TGV^2 these are piecewise-affine.

While TGV greatly improves the reconstruction quality compared to TV, the fact that it uses second order derivatives typically results in slower convergence of iterative optimization algorithms and therefore increases computational costs of the reconstruction. Therefore, there has been an effort to achieve a performance similar to that of TGV with a first-order method (i.e. a method that only uses derivatives of the first order). In [9,10], infimal convolution type regularizers TVL^p have been introduced that use an infimal convolution of the Radon norm and an L^p norm applied to the weak gradient of the image. For an $u \in L^1(\Omega)$, TVL^p is defined as follows

$$TVL^p{}_{\alpha,\beta}(u) := \min_{w \in L^p(\Omega;\mathbb{R}^d)} \alpha||Du - w||_{\mathfrak{M}} + \beta||w||_{L^p(\Omega;\mathbb{R}^d)}.$$

where D is the weak gradient, $\alpha, \beta > 0$ are constants and $1 < p \leqslant \infty$. It was shown that for $p = 2$ the reconstructions are piecewise-smooth, while for $p = \infty$ they somewhat resemble those obtained with TGV.

The regularizer we introduce in the current paper also aims at achieving a similar performance with second-order methods while only relying on first order derivatives. It can be seen either as a relaxiation of TV obtained by extending its kernel from constants to all functions with a given Lipschitz constant (for this reason, we call this new regularizer TV_{pwL}, with 'pwL' standing for 'piecewise-Lipschitz'), or as an infimal convolution type regularizer, where the Radon norm is convolved with the characteristic function of a certain convex set.

We start with the following motivation. Let $\Omega \subset \mathbb{R}^2$ be a bounded Lipschitz domain and $f \in L^2(\Omega)$ a noisy image. Recall the ROF [20] denoising model

$$\min_{u \in BV(\Omega)} \frac{1}{2}||u - f||^2_{L^2(\Omega)} + \alpha||Du||_{\mathfrak{M}},$$

where $D: L^1(\Omega) \to \mathfrak{M}(\Omega, \mathbb{R}^2)$ is the weak gradient, $\mathfrak{M}(\Omega, \mathbb{R}^2)$ is the space of vector-valued Radon measures and $\alpha > 0$ is the regularization parameter. Introducing an auxiliary variable $g \in \mathfrak{M}(\Omega, \mathbb{R}^2)$, we can rewrite this problem as follows

$$\min_{\substack{u \in BV(\Omega) \\ g \in \mathfrak{M}(\Omega, \mathbb{R}^2)}} \frac{1}{2}||u - f||^2_{L^2(\Omega)} + \alpha||g||_{\mathfrak{M}} \qquad s.t.\ Du = g.$$

Our idea is to relax the constraint on Du as follows

$$\min_{\substack{u \in BV(\Omega) \\ g \in \mathfrak{M}(\Omega, \mathbb{R}^2)}} \frac{1}{2}||u - f||^2_{L^2(\Omega)} + \alpha||g||_{\mathfrak{M}} \qquad s.t.\ |Du - g| \leqslant \gamma$$

for some positive constant, function or measure γ. Here $|Du - g|$ is the variation measure corresponding to $Du - g$ and the symbol "$\leqslant$" denotes a partial order in the space of signed (scalar valued) measures $\mathcal{M}(\Omega)$. This problem is equivalent to

$$\min_{\substack{u \in BV(\Omega) \\ g \in \mathfrak{M}(\Omega, \mathbb{R}^2)}} \frac{1}{2}\|u - f\|_{L^2(\Omega)}^2 + \alpha\|Du - g\|_{\mathfrak{M}} \quad s.t. \ |g| \leqslant \gamma, \tag{1}$$

which we take as the starting point of our approach.

This paper is organized as follows. In Sect. 2 we introduce the primal and dual formulations of the TV_{pwL} functional and prove their equivalence. In Sect. 3 we prove some basic properties of TV_{pwL} and study its relationship with other TV-type regularizers. Section 4 contains numerical experiments with the proposed regularizer.

2 Primal and Dual Formulations

Let us first clarify the notation of the inequality for signed measures in (1).

Definition 1. *We call a measure $\mu \in \mathcal{M}(\Omega)$ positive if for every subset $E \subseteq \Omega$ one has $\mu(E) \geqslant 0$. For two signed measures $\mu_1, \mu_2 \in \mathcal{M}(\Omega)$ we say that $\mu_1 \leqslant \mu_2$ if $\mu_2 - \mu_1$ is a positive measure.*

Now let us formally define the new regularizer.

Definition 2. *Let $\Omega \subset \mathbb{R}^2$ be a bounded Lipschitz domain, $\gamma \in \mathcal{M}(\Omega)$ be a finite positive measure. For any $u \in L^1(\Omega)$ we define*

$$\mathrm{TV}_{pwL}^{\gamma}(u) := \inf_{g \in \mathfrak{M}(\Omega, \mathbb{R}^2)} \|Du - g\|_{\mathfrak{M}} \quad s.t. \ |g| \leqslant \gamma$$

where $\|\cdot\|_{\mathfrak{M}}$ denotes the Radon norm and $|g|$ is the variation measure [4] corresponding to g, i.e. for any subset $E \subset \Omega$

$$|g|(E) := \sup\left\{\sum_{i=1}^{\infty} \|g(E_i)\|_2 \mid E = \bigcup_{i \in \mathbb{N}} E_i, \ E_i \ \text{pairwise disjoint}\right\}$$

(see also the polar decomposition of measures [1]).

The inf in Definition 2 can actually be replaced by a min since we are dealing with a metric projection onto a closed convex set in the dual of a separable Banach space. We also note that for $\gamma = 0$ we immediately recover total variation.

We defined the regularizer $\mathrm{TV}_{pwL}^{\gamma}$ in the general case of γ being a measure, but we can also choose γ to be a Lebesgue-measurable function or a constant. In this case the inequality is understood in the sense $|g| \leqslant \gamma \, d\mathcal{L}$ where $\mathcal{L}$ is the Lebesgue measure, resulting in $|g|$ being absolutely continuous with respect to $\mathcal{L}$. We will not distinguish between these cases in what follows and just write $|g| \leqslant \gamma$.

As with standard TV, there is also an equivalent dual formulation of TV_{pwL}.

Theorem 1. *Let* $\gamma \in \mathcal{M}(\Omega)$ *be a positive finite measure and* Ω *a bounded Lipschitz domain. Then for any* $u \in L^1(\Omega)$ *the* TV_{pwL}^γ *functional can be equivalently expressed as follows*

$$\mathrm{TV}_{pwL}^\gamma(u) = \sup_{\substack{\varphi \in \mathcal{C}_0^\infty(\Omega;\mathbb{R}^2) \\ |\varphi|_2 \leqslant 1}} \left\{ \int_\Omega u \ \mathrm{div} \ \varphi \ dx - \int_\Omega |\varphi|_2 d\gamma \right\},$$

where $|\varphi|_2$ *denotes the pointwise 2-norm of* φ.

Proof. Since by the Riesz-Markov-Kakutani representation theorem the space of vector valued Radon measures $\mathfrak{M}(\Omega, \mathbb{R}^2)$ is the dual of the space $\mathcal{C}_0(\Omega, \mathbb{R}^2)$, we can rewrite the expression in Definition 2 as follows

$$\mathrm{TV}_{pwL}^\gamma(u) = \inf_{\substack{g \in \mathfrak{M}(\Omega,\mathbb{R}^2) \\ |g| \leqslant \gamma}} \|Du - g\|_{\mathfrak{M}} = \inf_{\substack{g \in \mathfrak{M}(\Omega,\mathbb{R}^2) \\ |g| \leqslant \gamma}} \sup_{\substack{\varphi \in \mathcal{C}_0(\Omega;\mathbb{R}^2) \\ |\varphi|_2 \leqslant 1}} (Du - g, \varphi).$$

In order to exchange inf and sup, we need to apply a minimax theorem. In our setting we can use the Nonsymmetrical Minimax Theorem from [2, Th. 3.6.4]. Since the set $\{g \mid |g| \leqslant \gamma\} \subset \mathfrak{M}(\Omega, \mathbb{R}^2) = (\mathcal{C}_0(\Omega, \mathbb{R}^2))^*$ is bounded, convex and closed and the set $\{\varphi \mid \|\varphi\|_{2,\infty} \leqslant 1\} \subset \mathcal{C}_0(\Omega, \mathbb{R}^2)$ is convex, we can swap the infimum and the supremum and obtain the following representation

$$\mathrm{TV}_{pwL}^\gamma(u) = \sup_{\substack{\varphi \in \mathcal{C}_0(\Omega;\mathbb{R}^2) \\ |\varphi|_2 \leqslant 1}} \inf_{\substack{g \in \mathfrak{M}(\Omega,\mathbb{R}^2) \\ |g| \leqslant \gamma}} (Du - g, \varphi)$$

$$= \sup_{\substack{\varphi \in \mathcal{C}_0(\Omega;\mathbb{R}^2) \\ |\varphi|_2 \leqslant 1}} \left[(Du, \varphi) - \sup_{\substack{g \in \mathfrak{M}(\Omega,\mathbb{R}^2) \\ |g| \leqslant \gamma}} (g, \varphi) \right] = \sup_{\substack{\varphi \in \mathcal{C}_0(\Omega;\mathbb{R}^2) \\ |\varphi|_2 \leqslant 1}} \left[(Du, \varphi) - (\gamma, |\varphi|_2) \right].$$

Noting that the supremum can actually be taken over $\varphi \in \mathcal{C}_0^\infty(\Omega; \mathbb{R}^2)$, we obtain

$$\mathrm{TV}_{pwL}^\gamma(u) = \sup_{\substack{\varphi \in \mathcal{C}_0^\infty(\Omega;\mathbb{R}^2) \\ |\varphi|_2 \leqslant 1}} \left[(u, -\mathrm{div}\,\varphi) - (\gamma, |\varphi|_2) \right]$$

which yields the assertion upon replacing φ with $-\varphi$.

3   Basic Properties and Relationship with Other TV-type Regularizers

Influence of γ. It is evident from Definition 2 that a larger γ yields a larger feasible set and a smaller value of TV_{pwL}. Therefore, $\mathrm{TV}_{pwL}^\gamma \geqslant \mathrm{TV}_{pwL}^{\bar\gamma}$ whenever $0 \leqslant \gamma \leqslant \bar\gamma$. In particular, we get that $\mathrm{TV}_{pwL}^\gamma \leqslant \mathrm{TV}_{pwL}^0 = \mathrm{TV}$ for any $\gamma \geqslant 0$.

Lower-Semicontinuity and Convexity. Lower-semicontinuity is clear from Definition 2 if we recall that the infimum is actually a minimum. Convexity follows from the fact that TV_{pwL} is an infimal convolution of two convex functions.

Absolute One-Homogeneity. Noting that TV_{pwL} is the distance from the convex set $\{g \in \mathfrak{M}(\Omega; \mathbb{R}^2) \mid |g| \leqslant \gamma\}$, we conclude that it is absolute one-homogeneous if and only if this set consists of just zero, i.e. when $\gamma = 0$ and $\mathrm{TV}_{pwL} = \mathrm{TV}$.

Coercivity. We have seen that $\mathrm{TV}_{pwL}^{\gamma} \leqslant \mathrm{TV}$ for any $\gamma \geqslant 0$, i.e. TV_{pwL} is a lower bound for TV. If $\gamma(\Omega)$ is finite, the converse inequality (up to a constant) also holds and we obtain topological equivalence of TV_{pwL} and TV.

Theorem 2. *Let $\Omega \subset \mathbb{R}^2$ be a bounded Lipschitz domain and $\gamma \in \mathcal{M}(\Omega)$ a positive finite measure. For every $u \in L^1(\Omega)$ we obtain the following relation:*

$$TV(u) - \gamma(\Omega) \leqslant \mathrm{TV}_{pwL}^{\gamma}(u) \leqslant TV(u).$$

Proof. We already established the right inequality. For the left one we observe that for any $g \in \mathfrak{M}(\Omega, \mathbb{R}^2)$ such that $|g| \leqslant \gamma$ the following estimate holds

$$\|Du - g\|_{\mathfrak{M}} \geqslant \|Du\|_{\mathfrak{M}} - \|g\|_{\mathfrak{M}} \geqslant \|Du\|_{\mathfrak{M}} - \|\gamma\|_{\mathfrak{M}} = TV(u) - \gamma(\Omega),$$

which also holds for the infimum over g.

The left inequality in Theorem 2 ensures that TV_{pwL} is coercive on BV_0, since $\mathrm{TV}(u_n) \to \infty$ implies $\mathrm{TV}_{pwL}^{\gamma}(u_n) \geqslant \mathrm{TV}(u_n) - \gamma(\Omega) \to \infty$. Upon adding the L^1 norm, we also get coercivity on BV. This ensures that TV_{pwL} can be used for regularisation of inverse problems in the same scenarios as TV.

Topological equivalence between TV_{pwL} and TV is understood in the sense that if one is bounded then the other one is too. Being not absolute one-homogeneous, however, TV_{pwL} cannot be an equivalent norm on BV_0.

Null Space. We will study the null space in the case when $\gamma \in L_+^\infty(\Omega)$ is a Lebesgue measurable function and the inequality $|g| \leqslant \gamma$ in Definition 2 is understood in the sense that $|g| \leqslant \gamma d\mathcal{L}$ with $\mathcal{L}$ being the Lebesgue measure.

Proposition 1. *Let $u \in L^1(\Omega)$ and $\gamma \geqslant 0$ be an L^∞ function. Then $\mathrm{TV}_{pwL}^{\gamma}(u) = 0$ if and only if the weak derivative Du is absolutely continuous with respect to the Lebesgue measure and its 2-norm is bounded by γ a.e.*

Proof. We already noted that, since the space $\mathfrak{M}(\Omega; \mathbb{R}^2)$ is the dual of the separable Banach space $C_0(\Omega, \mathbb{R}^2)$, the infimum in Definition 2 is actually a minimum and there exists a $\tilde{g} \in \mathfrak{M}(\Omega; \mathbb{R}^2)$ with $|g| \leqslant \gamma d\mathcal{L}$ such that

$$\mathrm{TV}_{pwL}^{\gamma}(u) = \|Du - \tilde{g}\|_{\mathfrak{M}}.$$

If $\mathrm{TV}_{pwL}^{\gamma}(u) = 0$, then $Du = \tilde{g}$ and therefore $0 \leqslant |Du| \leqslant \gamma d\mathcal{L}$, which implies that $|Du|$ is absolutely continuous with respect to $\mathcal{L}$ and can be written as

$$|Du|(A) = \int_A f \, d\mathcal{L}$$

for any $A \subseteq \Omega$. The function $f \in L_+^1(\Omega)$ is the Radon-Nikodym derivative $\frac{d|Du|}{d\mathcal{L}}$. From the condition $|Du| \leqslant \gamma \, d\mathcal{L}$ we immediately get that $f \leqslant \gamma$ a.e.

Remark 1. We notice that the set $\{u \in L^1(\Omega): \text{TV}_{pwL}(u) = 0\}$ is not a linear subspace, therefore, we should rather speak of the null set than the null space.

Remark 2. Proposition 1 implies that all functions in the null set are Lipschitz continuous with (perhaps, spatially varying) Lipschitz constant γ; hence the name of the regularizer.

Luxemburg Norms. For a positive convex nondecreasing function $\varphi: \mathbb{R}_+ \to \mathbb{R}_+$ with $\varphi(0) = 0$ the Luxemburg norm $\| \cdot \|_\varphi$ is defined as follows [18]

$$\|u\|_\varphi = \sup \left\{ \lambda > 0: \int \varphi(|u|/\lambda)\, d\mu \leqslant 1 \right\}.$$

We point out a possible connection between TV_{pwL} and a Luxemburg norm corresponding to $\varphi(x) = (x - c)_+$ with a suitable constant $c > 0$. However, we do not investigate this connection in this paper.

Relationship with Infimal Convolution Type Regularizers. We would like to highlight a relationship to infimal convolution type regularizers TVL^P [9,10]. Indeed, as already noticed earlier, TV_{pwL} can be written as an infimal convolution

$$TV_{pwL}^\gamma(u) = \inf_{\substack{g \in \mathfrak{M}(\Omega;\mathbb{R}^2) \\ |g| \leqslant \gamma}} \|Du - g\|_{\mathcal{M}} = \inf_{g \in \mathfrak{M}(\Omega,\mathbb{R}^2)} \{\|Du - g\|_{\mathfrak{M}} + \chi_{C_\gamma}(g)\}, \quad (2)$$

where $C_\gamma := \{\eta \in \mathfrak{M}(\Omega; \mathbb{R}^2) \mid |\eta|_p \leqslant \gamma\}$. If $\gamma > 0$ is a constant, we obtain a bound on the $2, \infty$-norm of g. This highlights a connection to the TVL^∞ regularizer [10]: for any particular weight in front of the ∞-norm in TVL^∞ (and for a given u), the auxiliary variable g will have some value of the ∞-norm and if we use this value as γ in TV_{pwL}^γ, we will obtain the same reconstruction. The important difference is that with TVL^∞ we don't have direct control over this value and can only influence it in an indirect way through the weights in the regularizer. In TV_{pwL} this parameter is given explicitly and can be either obtained using additional a priori information about the ground truth or estimated from the noisy image.

Similar arguments can be made in the case of spatially variable γ if the weighting in TVL^∞ is also allowed to vary spatially.

4 Numerical Experiments

In this section we want to compare the performance of the proposed first order regularizer with a second order regularizer, TGV, in image denosing. We consider images (or 1D signals) corrupted by Gaussian noise with a known variance and use the residual method [14] to reconstruct the noise-free image, i.e. we solve (in the discrete setting)

$$\min_{u \in \mathbb{R}^N} \mathcal{J}(u) \quad \text{s.t. } \|u - f\|_2^2 \leqslant \sigma^2 \cdot N, \quad (3)$$

where f is the noisy image, $\mathcal{J}$ is the regularizer (TV_{pwL} or TGV), σ is the standard deviation of the Gaussian noise and N is the number of pixels in the image (or the number of entries in the 1D signal). We solve all problems in MATLAB using CVX [15]. For TGV we use the parameter $\beta = 1.25$, which is in the range $[1, 1.5]$ recommended in [12].

A characteristic feature of the proposed regularizer TV_{pwL} is its ability to efficiently encode the information about the gradient of the ground truth (away from jumps) if such information is available. Our experiments showed that the quality of the reconstruction significantly depends on the quality of the (local) estimate of the norm of the gradient of the ground truth.

The ideal application for TV_{pwL} would be one where we have a good estimate of the gradient of the ground truth away from jumps which, however, may occur at unknown locations and be of unknown magnitude. If such an estimate is not available, we can roughly estimate the gradient of the ground truth from the noisy signal, which is the approach we take.

4.1 1D Experiments

We consider the ground truth shown in Fig. 1a (green dashed line). The signal is discretized using $N = 1000$ points. We add Gaussian noise with variance $\sigma = 0.1$ and obtain the noisy signal shown in the same Figure (blue solid line).

To use TV_{pwL}, we need to estimate the derivative of the true signal away from the jumps. Therefore, we need to detect the jumps, but leave the signal intact away from them (up to a constant shift). This is exactly what happens if the image is overregularized with TV. We compute a TV reconstruction by solving the ROF model

$$\min_{u \in \mathbb{R}^N} \frac{1}{2}\|u - f\|_2^2 + \alpha\,\text{TV}(u) \tag{4}$$

with a large value of α (in this example we took $\alpha = 0.5$). The result is shown in Fig. 1a (red solid line). The residual, which we want to use to estimate the derivative of the ground truth, is shown in Fig. 1b. Although the jumps have not been removed entirely, this signal can be used to estimate the derivative using filtering.

The filtered residual (we used the build-in MATLAB function 'smooth' with option 'rlowess' (robust weighted linear least squares) and $\omega = 50$) is shown in Fig. 1c. This signal is sufficiently smooth to be differentiated. We use central differences; to suppress the remaining noise in the filtered residual we use a step size for differentiation that is 20 times the original step size. The result is shown in Fig. 1d (reg solid line) along with the true derivative (green dashed line). We use the absolute value of the so computed derivative as the parameter γ.

The reconstruction obtained using TV_{pwL}^{γ} is shown in Fig. 1e. We see that the reconstruction is best in areas where our estimate of the true derivative was most faithful (e.g., between 4 and 5). But also in other areas the reconstruction is good and preserves the structure of the ground truth rather well. We notice a

492   M. Burger et al.

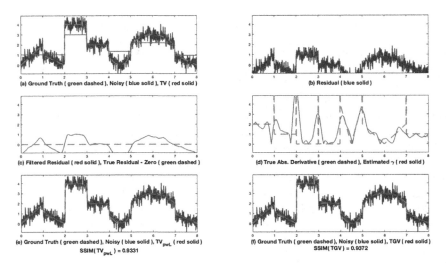

Fig. 1. The pipeline for the reconstruction with TV_{pwL} (a–e). An overregularized TV reconstruction is used to detect and partially eliminate the jumps (a). The residual (b) is filtered (c) and numerically differentiated (d). The absolute value of the obtained derivative is used as the parameter γ (d) for TV_{pwL}^{γ}. The reconstruction using TV_{pwL}^{γ} (e) follows well the structure of the ground truth apart from a small artefact at around 2. TGV also yields a good reconstruction (f), although it tends to approximate the solution with a piecewise-affine function in areas where the ground truth is not affine (e.g., between 4 and 5). Both regularizers yield similar (high) values of SSIM. (Color figure online)

small artefact at the value of the argument of around 2; examining the estimate of the derivative in Fig. 1d and the residual in Fig. 1b, we notice that TV was not able the remove the jump at this location and therefore the estimate of the derivative was too large. This allowed the reconstruction to get too close to the data at this point.

We also notice that the jumps are sometimes reduced, with a characteristic linear cut near the jump (e.g., near $x = 1; 2; 3$ and 4). This can have different reasons. For the jumps near $x = 3$ and 4 we see that the estimate of γ is too large (the true derivative is zero), which allows the regulariser to cut the edges. For the jumps near $x = 1$ and 2 the situation is different. At these positions a negative slope in the ground truth is followed by a positive jump. Since γ only constraints the absolute value of the gradient, even with a correct estimate of γ the regulariser will reduce the jump, going with the maximum slope in the direction of the jump. Functions with a negative slope followed by a positive jump are also problematic for TV, since they do not satisfy the source condition (their subdifferential is empty [7]). In such cases TV will also always reduce the jump.

Figure 1e shows the reconstruction obtained with TGV. The reconstruction is quite good, although it is often piecewise-affine where the ground truth is not, e.g. between 4 and 5 or between 5 and 7. As expected, both regularizers tend to push the reconstructions towards their kernels, but, since TV_{pwL}^{γ} with a good choice of γ contains the ground truth in its kernel (up to the jumps), it yields reconstructions that are more faithful to the structure of the ground truth.

4.2 2D Experiments

In this Section we study the performance of TV_{pwL} in denoising of 2D images. We use two images - "cameraman" (Fig. 2a) and "owl" (Fig. 3a). Both images have the resolution 256×256 pixels and values in the interval $[0, 255]$. The images are corrupted with Gaussian noise with standard deviation $\sigma = 0.1 \cdot 255 = 25.5$ (Figs. 2b and 3b).

(a) Original Image

(b) Noisy Image

(c) Overregularised TV

(d) Residual (Abs. Value)

(e) Filtered Residual

(f) Gamma (Scaled)

(g) TV_{pwL} Reconstruction
SSIM = 0.796

(h) TGV Reconstruction
SSIM = 0.801

Fig. 2. The pipeline for the reconstruction of the "cameraman" image with TV_{pwL} (c–g). An overregularized TV reconstruction is used to detect and partially eliminate the jumps (c). (Note that TV managed to segment the picture into piecewise-constant regions rather well.) The residual (d) is filtered (e) and numerically differentiated. The norm of the obtained gradient is used as the parameter γ for TV_{pwL}^{γ} ((f), γ is scaled to the interval $[0, 255]$ for presentation purposes). The reconstructions using TV_{pwL}^{γ} (g) and TGV (h) are almost identical. Both preserve edges and are rather smooth away from them. Details are rather well preserved (see, e.g., the pillars of the building in the background as well as the face of the cameraman; the texture of the grass is lost in both cases, however). Relatively homogeneous regions in the original image are also relatively homogeneous in the reconstruction, yet they are not piecewise constant. SSIM values differ very little.

The pipeline for the reconstruction using TV_{pwL} is the same as in 1D. We obtain a piecewise constant image by solving an overregularized ROF problem (4)

(a) Original Image   (b) Noisy Image   (c) Overregularised TV   (d) Residual (Abs. Value)

(e) Filtered Residual   (f) Gamma (Scaled)   (g) TV$_{pwL}$ Reconstruction SSIM = 0.676   (h) TGV Reconstruction SSIM = 0.687

Fig. 3. The pipeline for the reconstruction of the "owl" image with TV_{pwL} (c–g). An overregularized TV reconstruction is used to detect and partially eliminate the jumps (c). (This time the segmentation obtained by TV is not perfect – perhaps too detailed – but still rather good.) The residual (d) is filtered (e) and numerically differentiated. The norm of the obtained gradient is used as the parameter γ for TV_{pwL}^{γ} ((f), γ is scaled to the interval $[0, 255]$ for presentation purposes). This time the residual is not as clear as in the "cameraman" example and the estimated γ seems noisier. However, it still mainly follows the structure of the original image. The reconstruction using TV_{pwL}^{γ} (g) preserves the edges and well reconstructs some details in the image, e.g., the feathers of the owl. Other details, however, are lost (the needles of the pine tree in the background). Looking at γ in this region, we notice that it is rather irregular and does not capture the structure of the ground truth. The TGV reconstruction (h) is again very similar to TV_{pwL} and SSIM values are very close.

with $\alpha = 500$ (Figs. 2c and 3c) and compute the residuals (Figs. 2d and 3d). Then we smooth the residuals using a Gauss filter with $\sigma = 2$ (Figs. 2e and 3e) and compute its derivatives in the x- and y-directions using the same approach as in 1D (central differences with a different step size; we used step size 3 in this example). These derivatives are used to estimate γ, which is set equal to the norm of the gradient. Figures 2f and 3f show γ scaled to the interval $[0, 255]$ for better visibility. We use the same parameters (for Gaussian filtering and numerical differentiation) to estimate γ in both images. Reconstructions obtained using TV_{pwL}^{γ} and TGV are shown in Figs. 2g–h and 3g–h. As in the 1D example, the parameter β for TGV was set to 1.25.

Comparing the results for both images, we notice that the residual (as well as its filtered version) captures the details in the "cameraman" image much better than in the "owl" image. The filtered residual in the "owl" image seems to miss some of the structure of the original image and this is reflected in the estimated γ (which looks much noisier in the "owl" image and, in particular, does not capture the structure of the needles of the pine tree in the upper left corner). This might be due to the segmentation achieved by TV, which seems better in the "cameraman" image (the one in the "owl" image seems to a bit too detailed). This effect might be mitigated by using a better segmentation technique.

This difference is reflected in the reconstructions. While in the "cameraman" image the details are well preserved (e.g., the face of the cameraman or his camera, as well as the details of the background), in the "owl" image part of them are lost and replaced by rather blurry (if not constant) regions; however, in other regions, such as the feathers of the owl, the details are preserved much better, which can be also seen from the estimated γ that is much more regular in this area and closer to the structure of the ground truth. We also notice some loss of contrast in the TV_{pwL} reconstruction. Perhaps, it could be dealt with by adopting the concept of debiasing [6,13] in the setting of TV_{pwL}, however, it is not clear yet, what is the structure of the model manifold in this case.

The TGV reconstructions look strikingly similar to those obtained by TV_{pwL}. Structural similarity between these reconstructions (i.e. SSIM computed using on of them as the reference) is 0.98 for the "cameraman" image and 0.97 for the "owl" image. Although the TGV reconstructions depend on the parameter β any may differ more from TV_{pwL} for other values of β, the one we chose here ($\beta = 1.25$) is reasonable and lies within the optimal range reported in [12].

There are two main messages to be taken from these experiments. The first one is that TV_{pwL} is able to almost reproduce the reconstructions obtained using TGV with a reasonable choice of the parameter β, which is a very good performance for a method that does not use higher-order derivatives. The second one is that the performance of TV_{pwL} greatly depends on the quality of the estimate of γ. When we are able to well capture the structure of the original image in this estimate, the structure of the reconstructions is rather close to that of the ground truth. To further illustrate this point, we show in Fig. 4 TV_{pwL}^{γ} reconstructions in the ideal scenario when γ is estimated from the ground truth as the local magnitude of the gradient. The quality of the reconstructions is very good, suggesting that with a better strategy of estimating the gradient TV_{pwL}^{γ} could achieve even better performance.

(a) TV_{pwL}^{γ} with exact γ, SSIM = 0.876   (b) TV_{pwL}^{γ} with exact γ, SSIM = 0.801

Fig. 4. In the ideal scenario when γ is estimated from the ground truth, TV_{pwL}^{γ} is able to reproduce the original image almost perfectly

5   Conclusions

We proposed a new TV-type regularizer that can be used to decompose the image into a jump part and a part with Lipschitz continuous gradient (with a given Lipschitz constant that is also allowed to vary spatially). Functions whose gradient does not exceed this constant lie in the kernel of the regularizer and are not penalized. By smartly choosing this bound we can hope to put the ground truth into the kernel (up to the jumps) and thus not penalize any structure that is present in the ground truth.

In this paper we presented, in the context of denoising, an approach to estimating this bound from the noisy image. The approach is based on segmenting the image (and compensating for the jumps) using overregularized TV and estimating the local bound on the gradient using filtering. Our numerical experiments showed that TV_{pwL} can produce reconstructions that are very similar to TGV, however, the results significantly depend on the quality of the estimation of the local bound on the gradient. Using a more sophisticated estimation technique is expected to further improve the reconstructions. The ideal application for TV_{pwL} would be one where there is some information about the magnitude of the Lipschitz part of the gradient.

Acknowledgments. This work was supported by the European Union's Horizon 2020 research and innovation programme under the Marie Sklodowska-Curie grant agreement No 777826 (NoMADS). MB acknowledges further support by ERC via Grant EU FP7 – ERC Consolidator Grant 615216 LifeInverse. YK acknowledges support of the Royal Society through a Newton International Fellowship. YK also acknowledges support of the Humbold Foundataion through a Humbold Fellowship he held at the University of Münster when this work was initiated. CBS acknowledges support from the Leverhulme Trust project on Breaking the non-convexity barrier, EPSRC grant Nr. EP/M00483X/1, the EPSRC Centre Nr. EP/N014588/1, the RISE projects CHiPS and NoMADS, the Cantab Capital Institute for the Mathematics of Information and the Alan Turing Institute. We gratefully acknowledge the support of NVIDIA Corporation with the donation of a Quadro P6000 and a Titan Xp GPUs used for this research.

References

1. Ambrosio, L., Fusco, N., Pallara, D.: Functions of Bounded Variation and Free Discontinuity Problems. Clarendon Press, Oxford (2000)
2. Borwein, P., Zhu, Q.: Techniques of Variational Analysis. CMS Books in Mathematics. Springer, Heidelberg (2005). https://doi.org/10.1007/0-387-28271-8
3. Bredies, K., Kunisch, K., Pock, T.: Total generalized variation. SIAM J. Imaging Sci. **3**, 492–526 (2011)
4. Bredies, K., Lorenz, D.: Mathematische Bildverarbeitung. Einführung in Grundlagen und moderne Theorie. Springer, Heidelberg (2011). https://doi.org/10.1007/978-3-8348-9814-2
5. Bredies, K., Valkonen, T.: Inverse problems with second-order total generalized variation constraints. In: Proceedings of the 9th International Conference on Sampling Theory and Applications (SampTA) 2011, Singapore (2011)

6. Brinkmann, E.M., Burger, M., Rasch, J., Sutour, C.: Bias reduction in variational regularization. J. Math. Imaging Vis. **59**(3), 534–566 (2017)
7. Bungert, L., Burger, M., Chambolle, A., Novaga, M.: Nonlinear spectral decompositions by gradient flows of one-homogeneous functionals (2019). arXiv:1901.06979
8. Burger, M., Korolev, Y., Rasch, J.: Convergence rates and structure of solutions of inverse problems with imperfect forward models. Inverse Probl. **35**(2), 024006 (2019)
9. Burger, M., Papafitsoros, K., Papoutsellis, E., Schönlieb, C.B.: Infimal convolution regularisation functionals of BV and L^p spaces. Part i: the finite p case. J. Math. Imaging Vis. **55**(3), 343–369 (2016)
10. Burger, M., Papafitsoros, K., Papoutsellis, E., Schönlieb, C.B.: Infimal convolution regularisation functionals of BV and L^p spaces. The case $p = \infty$. In: Bociu, L., Désidéri, J., Habbal, A. (eds.) CSMO 2015. IFIP AICT, vol. 494, pp. 169–179. Springer, Heidelberg (2016). https://doi.org/10.1007/978-3-319-55795-3_15
11. Chambolle, A., Duval, V., Peyré, G., Poon, C.: Geometric properties of solutions to the total variation denoising problem. Inverse Probl. **33**(1), 015002 (2017)
12. De los Reyes, J.C., Schönlieb, C.B., Valkonen, T.: Bilevel parameter learning for higher-order total variation regularisation models. J. Math. Imaging Vis. **57**(1), 1–25 (2017)
13. Deledalle, C.-A., Papadakis, N., Salmon, J.: On debiasing restoration algorithms: applications to total-variation and nonlocal-means. In: Aujol, J.-F., Nikolova, M., Papadakis, N. (eds.) SSVM 2015. LNCS, vol. 9087, pp. 129–141. Springer, Cham (2015). https://doi.org/10.1007/978-3-319-18461-6_11
14. Engl, H., Hanke, M., Neubauer, A.: Regularization of Inverse Problems. Springer, Heidelberg (1996)
15. Grant, M., Boyd, S.: CVX: Matlab software for disciplined convex programming, version 2.1, March 2014. http://cvxr.com/cvx
16. Iglesias, J.A., Mercier, G., Scherzer, O.: A note on convergence of solutions of total variation regularized linear inverse problems. Inverse Probl. **34**, 055011 (2018)
17. Jalalzai, K.: Some remarks on the staircasing phenomenon in total variation-based image denoising. J. Math. Imaging Vis. **54**(2), 256–268 (2016)
18. Luxemburg, W.: Banach function spaces. Ph.D. thesis, T.U. Delft (1955)
19. Ring, W.: Structural properties of solutions to total variation regularization problems. ESAIM: M2AN **34**(4), 799–810 (2000)
20. Rudin, L.I., Osher, S., Fatemi, E.: Nonlinear total variation based noise removal algorithms. Phys. D: Nonlinear Phenom. **60**(1), 259–268 (1992)

A Non-convex Nonseparable Approach to Single-Molecule Localization Microscopy

Raymond H. Chan[1], Damiana Lazzaro[2], Serena Morigi[2(✉)],
and Fiorella Sgallari[2]

[1] Department of Mathematics, City University of Hong Kong,
Kowloon Tong, Hong Kong
rchan.sci@cityu.edu.hk
[2] Department of Mathematics, University of Bologna, Bologna, Italy
{damiana.lazzaro,serena.morigi,fiorella.sgallari}@unibo.it

Abstract. We present a method for high-density super-resolution microscopy which integrates a sparsity-promoting penalty and a blur kernel correction into a nonsmooth, non-convex, nonseparable variational formulation. An efficient majorization minimization strategy is applied to reduce the challenging optimization problem to the solution of a series of easier convex problems.

1 Introduction

Single-molecule localization microscopy (SMLM) is a powerful microscopical technique that is used to detect with high precision molecule localization by sequentially activating and imaging only a random sparse subset of fluorescent molecules in the sample at the same time, localizing these few emitters very precisely, deactivating them and activating another subset. Repeating the process several thousand times ensures that all fluorophores can go through the bright state and are recorded sequentially in frames. A high density map of fluorophore positions is then reconstructed by a sequential imaging process of sparse subsets of fluorophores distributed over thousands of frames. Even when theoretical characteristics on the blur kernel involved in the formation of the images are given, the acquisition process is so complicated that also the slightest difference to the theoretical ideal conditions, results in distortions which affect Point Spread Function (PSF), and, consequently, the image recovering process [12]. Several algorithms have been developed for point source localization in the context of the SMLM challenge. In [5] the variational model is equipped with a sparsity-promoting CEL0 penalty and solved by iterative reweighting. In [10] the blur kernel inaccuracy is addressed with a Taylor approximation of the PSF. For a detailed list of the software proposed to solve the SMLM challenge, and on the physical background of SMLM, we refer the reader to [12]. We formulate the localization problem as a variational sparse image reconstruction problem which integrates a nonseparable structure-preserving penalty. To overcome the problem of inaccurate blur kernel which can cause severe distorsions on the solution,

J. Lellmann et al. (Eds.): SSVM 2019, LNCS 11603, pp. 498–509, 2019.
https://doi.org/10.1007/978-3-030-22368-7_39

we combine our sparsity-promoting formulation with an effective blur kernel model correction. We perform a nonsmooth nonconvex optimization algorithm for the minimization task, based on a majorization minimization strategy. The proposed algorithm is validated on both simulated and experimental datasets and compared with other challenging high density localization softwares.

2   Image Formation Modelling

Let $u \in \mathbb{R}^{N \times N}$ be the unknown high resolution image to be reconstructed, and $g \in \mathbb{R}^{n \times n}$ the acquisition following the molecules activation, with $n = \frac{N}{d}$, and d is the downsampling factor. The linear acquisition process can be formulated as

$$g = \mathcal{P}(M_d(\mathcal{B}_t * u)) + \eta, \tag{1}$$

where $\mathcal{P}$ models the degradation with Poisson noise, η is the zero-mean Gaussian image noise, $\mathcal{B}_t$ is the convolution blurring operator with Gaussian kernel, and $M_d : \mathbb{R}^{N \times N} \rightarrow \mathbb{R}^{n \times n}$ is the downsampling operator which averages pixels by patches of size $d \times d$ in order to map the high resolution image to the coarser one. In the image formation model (1) we assumed that the given blur kernel model of the optical system (microscope) is free of error. The challenge in [12] provided parameters to model a Gaussian PSF model for each experiment data set. However, as assessed in [12] a simple Gaussian PSF model can be sufficiently accurate for low-density data, whereas the quality of high-density imaging depends strongly on the model of the PSF and the PSF model will have an even more significant role in 3D SMLM applications. When an inaccurate blur kernel is used as the input, significant distortions can appear in the recovered image. In this work, we assume an inaccurate blur kernel $\mathcal{B}$, with unknown model error $\delta_{\mathcal{B}}$, that is the true blur kernel $\mathcal{B}_t = \mathcal{B} - \delta_{\mathcal{B}}$ and we neglect the Poisson shot noise contribution, thus (1) becomes

$$g = M_d((\mathcal{B} - \delta_{\mathcal{B}}) * u) + \eta = M_d((\mathcal{B} * u) - (\delta_{\mathcal{B}} * u)) + \eta. \tag{2}$$

In Fig. 1 the high pass nature of the model error $\delta_{\mathcal{B}}$ is shown. Since $\delta_{\mathcal{B}}$ is the difference between two low pass filter, the input blur kernel $\mathcal{B}$ and the unknown true blur kernel $\mathcal{B}_t$, the corrector term $\delta_{\mathcal{B}} * u$ has an enhancing effect of the edges in the image, see [7].

Let us introduce a matrix-vector notation that will be useful in the algorithmic description. In particular, let $B \in \mathbb{R}^{N \times N}$ be the blurring matrix corresponding to the operator $\mathcal{B}$, then

$$\mathcal{B} * u = BuB^T = (B \otimes B)vec(u) = \bar{B}vec(u), \tag{3}$$

where $\otimes$ is the Kronecker product, and $vec(u)$ denotes the vectorization of u. Let $M \in \mathbb{R}^{n \times N}$ be the downsampling matrix such that

$$M_d(u) = MuM^T = (M \otimes M)vec(u) = \bar{M}vec(u). \tag{4}$$

To accurately estimate u in (2) we only need to know the residual term $\delta_{\mathcal{B}} * u$ instead of the perturbation operator $\delta_{\mathcal{B}}$ itself which is hard to estimate due to the lacking of information of the blurring process.

3 Penalty Function

In this section we introduce the sparsity-inducing function used in our variational model, and we highlight some of its properties: be non-convex, parameterized with μ so that we can tune its non-convex behaviour, structure preserving, as required by the high density molecule localization problem. In light of the previous requirements, the proposed penalty function is defined on the local data set consisting of a neighborhood of the pixel (i, j). In particular, we consider a square window centered at u_{ij} containing all the $(2\ell + 1)^2$ neighbors, $\ell \geq 1$, and we denote by $I_{ij} = \{(i + r, j + s) : r, s = -\ell, \ldots, \ell\}$ the neighborhood index set of size ℓ, and by $t := u|_{I_{ij}}$, $t \in \mathbb{R}_+^{(2\ell+1) \times (2\ell+1)}$ the restriction of u to the window I_{ij}. Following [9], to fulfill our goals, we define the non-convex nonseparable penalty function $\psi : \mathbb{R}_+^{(2\ell+1) \times (2\ell+1)} \to \mathbb{R}$ as follows:

$$\psi(t; \mu) = \frac{1}{\log(2)} \log \left(\frac{2}{1 + exp(-\|vec(t)\|_1/\mu)} \right), \tag{5}$$

where $\mu > 0$ represents a parameter which controls the degree of non-convexity of the penalty function and $vec(t) \in \mathbb{R}_+^{(2\ell+1)^2}$. The partial derivatives of $\psi(t; \mu)$ in (5), $\forall (r, s) \in I_{ij}$ are given by

$$\frac{\partial \psi}{\partial |u_{r,s}|}(t; \mu) = \frac{1}{\mu \log(2)} \frac{1}{1 + exp(\|vec(u|_{I_{rs}})\|_1/\mu)}. \tag{6}$$

Simple investigations of the first and second order partial derivatives lead to the following properties for $\psi(t; \mu)$, which characterize a sparsity-promoting function:

- $\psi(t; \mu)$ is concave and non-decreasing;
- $\psi(t; \mu)$ has continuous bounded partial derivatives for $t \neq 0$, and $\psi(0; \mu) = 0$;
- for μ values approaching to zero, $\psi(t; \mu)$ tends to the ℓ_0 quasi-norm.

4 Optimization Model NCNS for SMLM Problem

In the SMLM problem the aim is to recover sparse images with non-zero pixels clustered into elongated structures, whose number, dimension and position are unknown. The problem can also be classified as a blind cluster structured sparse image recovery problem [9]. For its solution we propose to minimize the following nonconvex cost function involving the non-convex nonseparable (NCNS) penalty function introduced in Sect. 3. Let $h := \delta_\mathcal{B} * u$ be the correction term. Then we will denote by NCNS model the following optimization problem

$$\min_{u,h \in \mathbb{R}^{N \times N}} \{ J(u, h; \lambda_1, \lambda_2) := F(u, h) + \lambda_1 R(u; \mu) + \lambda_2 H(h) \} \tag{7}$$

where $\lambda_1, \lambda_2 > 0$ are regularization parameters,

$$F(u, h) = \frac{1}{2} \| M_d((\mathcal{B} * u) - h) - g \|_2^2, \tag{8}$$

is the fidelity term, the penalty function $R(u; \mu)$ reads as

$$R(u; \mu) = \sum_{i=1}^{N} \sum_{j=1}^{N} \psi_{i,j}(u; \mu), \tag{9}$$

where $\psi_{i,j}(u; \mu) = \psi(u|_{I_{ij}}; \mu)$ is defined in (5), and $H(h) = \|h\|_p^p, p = \{1, 2\}$. The aforementioned properties of $\psi_{i,j}$ induce similar properties in the sparsity-promoting function $R(\cdot; \mu)$, which turns out to be both non-convex and non-separable. From (9) we can define the partial derivative with respect to a pixel $(p, q) \in I_{ij}$ as

$$\frac{\partial R(u; \mu)}{\partial |u_{p,q}|} = \sum_{i=1}^{N} \sum_{j=1}^{N} \frac{\partial \psi_{i,j}}{\partial |u_{p,q}|}(u; \mu) = \sum_{(r,s) \in I_{ij}} \frac{\partial \psi_{r,s}}{\partial |u_{p,q}|}(u; \mu). \tag{10}$$

Formula (10) is obtained taking into account that, due to the local support of ψ, the partial derivatives $\frac{\partial \psi_{i,j}}{\partial |u_{p,q}|}$ that are non-zero are those defined on the $(2\ell + 1)^2$ windows containing the pixel $u_{p,q}$ itself.

The effect of the nonseparable penalty $R(u; \mu)$ on a pixel $u_{p,q}$ depends on its neighbors defined in I_{pq}. In particular, the pixel $u_{p,q}$ is considered as belonging to a structure and thus preserved if the ℓ_1 norm of the vector of the pixels in its surrounding window is greater than μ, otherwise, it is forced to be zero, because it could be an isolated artifact. This fulfills the requirements of the SMLM data, where the fluorescent molecules are in general aggregated forming elongated thin structures.

Proposition 1. *For any couple of positive parameters (λ_1, λ_2) the functional $J(u, h; \lambda_1, \lambda_2) : \mathbb{R}^{N \times N} \times \mathbb{R}^{N \times N} \to \mathbb{R}$, defined in (7) is non-convex, proper, continuous, bounded from below by zero but not coercive in u, hence the existence of global minimizers for J is not guaranteed.*

The lack of coercivity not only stems from $R(u; \mu)$, but also from the down-sampling operator $\mathcal{M}_d$ which has a nontrivial kernel, and the non-convexity is due to $R(u; \mu)$. The problem (7) is in general a challenging non-convex nonseparable optimization problem. A minimizer for J in (7) is carried out by applying the Majorization-Minimization (MM) strategy which iteratively minimizes a convexification of J obtained by replacing R with its linearization $\tilde{R}$ around the previous iterate, [8].

In the kth **majorization step**, we generate a tangent majorant of the function (surrogate functional) $J(u, h; \lambda_1, \lambda_2)$ defined as

$$\tilde{J}(u, h; \lambda_1, \lambda_2, u^{(k)}, \mu^{(k)}) = F(u, h) + \lambda_1 \tilde{R}(u; u^{(k)}, \mu^{(k)}) + \lambda_2 H(h), \tag{11}$$

where the linear tangent majorant of $R(u; \mu^{(k)})$ at $u^{(k)}$ is

$$\tilde{R}(u; u^{(k)}, \mu^{(k)}) = R(u^{(k)}; \mu^{(k)}) + \sum_{i=1}^{N} \sum_{j=1}^{N} \left(\frac{\partial R(u; \mu^{(k)})}{\partial |u_{i,j}|} \Big|_{u=u^{(k)}} (|u_{i,j}| - |u_{i,j}^{(k)}|) \right). \tag{12}$$

A suitable reduction of the parameter $\mu^{(k)}$ is carried out at each iteration k, namely $\mu^{(k+1)} = c_\mu \mu^{(k)}$, with $0 < c_\mu < 1$, in such a way that, as the number of iterations increases, the sparsity inducing function gets closer to its limit ℓ_0 quasi-norm.

In the **minimization step**, the following convex nonsmooth minimization problem is solved

$$\{u^{(k+1)}, h^{(k+1)}\} = \arg\min_{u,h} \left\{ \tilde{J}(u, h; \lambda_1, \lambda_2, u^{(k)}, \mu^{(k)}) \right\}. \tag{13}$$

By neglecting the constant terms, problem (13) can be simplified to:

$$\{u^{(k+1)}, h^{(k+1)}\} = \arg\min_{u,h} \{F(u, h) + \lambda_1 \sum_{i=1}^{N} \sum_{j=1}^{N} w_{i,j}^{(k)} |u_{i,j}| + \lambda_2 H(h)\} \tag{14}$$

where, using (10) and (6), the positive weights are defined as

$$w_{i,j}^{(k)} = \frac{\partial R(u; \mu^{(k)})}{\partial |u_{i,j}|}\Big|_{u=u^{(k)}} = \sum_{(r,s) \in I_{ij}} \frac{1}{\mu^{(k)} \log(2)} \frac{1}{1 + exp(\|vec(u|_{I_{ij}})\|_1 / \mu^{(k)})}. \tag{15}$$

Equation (14) can be rewritten in vectorized form as

$$\{u^{(k+1)}, h^{(k+1)}\} = \arg\min_{u,h} \{F(u, h) + \lambda_1 \underbrace{\|W^{(k)} u\|_1}_{G(u)} + \lambda_2 H(h)\}, \tag{16}$$

where $W^{(k)} \in \mathbb{R}^{N^2 \times N^2}$ is a diagonal matrix of weights $w_{ij}^{(k)}$, which assume high values for isolated pixel (i, j) and small values for pixels representing structures. For the sake of simplicity, from now on we will represent the image variables in vectorized form.

4.1   Solving the Minimization Step

In this section we determine an approximate solution of the minimization step (16), which can be rewritten in the form

$$\{u^*, h^*\} = \arg\min_{u,h} \{F(u, h) + \lambda_1 G(u) + \lambda_2 H(h)\}, \tag{17}$$

where we neglected the iteration index (k).

A standard approach for solving (17) is thus to adopt an alternating minimization strategy. However, its convergence is only guaranteed under restrictive assumptions. Therefore, alternative strategies based on proximal tools have been proposed [4]. In particular, in this work, following [1], we propose to adopt the alternating accelerated Forward Backward algorithm which alternates the minimization on the two variable blocks (u, h).

Assuming that $F(u, h)$ is a C^1 coupling function which is required to have only partial Lipschitz continuous gradients $\nabla_u(F(u, h))$ and $\nabla_h(F(u, h))$, and that each of the regularizers $G(u)$ and $H(h)$ is proper, lower semicontinuous with an efficiently computable proximal mapping. In particular, $G(u)$ is convex and nonsmooth, while $H(h)$ is convex and, eventually, nonsmooth. We cannot claim the same for the optimization problem (7).

Proposition 2. *For any fixed h the function $u \rightarrow F(u, h)$ has partial Lipschitz continuous gradient with moduli $L_1 = \rho(A^T A)$, with ρ denoting the spectral radius, that is*

$$\|\nabla_u F(x, h) - \nabla_u F(y, h)\| \leq L_1 \|x - y\|, \quad \forall x, y \in \mathbb{R}^{N^2},$$

where

$$\nabla_u(F(u, h)) = A^T(Au - g + \bar{M}h), \tag{18}$$

with $A = \bar{M}\bar{B}$, $\bar{M}$ defined in (4) and $\bar{B}$ in (3). For any fixed u the function $h \rightarrow F(u, h)$ has partial Lipschitz continuous gradient $\nabla_h F(u, h)$ with moduli $L_2 = \rho(M^T M)$ that is

$$\|\nabla_h F(u, x) - \nabla_h F(u, y)\| \leq L_2 \|x - y\|, \quad \forall x, y \in \mathbb{R}^{N^2},$$

where

$$\nabla_h(F(u, h)) = \bar{M}^T(\bar{M}h - Au + g). \tag{19}$$

Formulas (18) and (19) can be derived from (8), which is rewritten as

$$F(u, h) = \frac{1}{2}\|\bar{M}(\bar{B}u - h) - g\|_2^2,$$
$$= \frac{1}{2}(u^T A^T Au + h^T \bar{M}^T \bar{M}h - 2u^T A^T \bar{M}h + 2g^T \bar{M}h - 2u^T A^T g + g^T g).$$

Let $0 < \beta_1 < \frac{1}{L_1}$ and $0 < \beta_2 < \frac{1}{L_2}$, the approximate solution of the optimization problem (17) is obtained by the iterative procedure sketched below.

– Initialization: start with $u_0 = \tilde{u}_0 = u^{(k)}$, $h_0 = \tilde{h}_0 = h^{(k)}$, $\lambda_1, \lambda_2 > 0$,
– For each $\ell \geq 1$ generate the sequence (u_ℓ, h_ℓ) by iterating
  • Accelerated FB for u

$$v_\ell = u_{\ell-1} - \beta_1 \nabla_u(F(u_{\ell-1}, h_{\ell-1})) \tag{20}$$

$$\tilde{u}_\ell = \arg \min_u \{\frac{1}{2\beta_1}\|u - v_\ell\|_2^2 + \lambda_1 G(u)\} \tag{21}$$

$$u_\ell = \tilde{u}_\ell + \tau_\ell(\tilde{u}_\ell - \tilde{u}_{\ell-1}) \tag{22}$$

  • Accelerated FB for h

$$s_\ell = h_{\ell-1} - \beta_2 \nabla_h(F(u_\ell, h_{\ell-1})) \tag{23}$$

$$\tilde{h}_\ell = \arg \min_h \{\frac{1}{2\beta_2}\|h - s_\ell\|_2^2 + \lambda_2 H(h)\} \tag{24}$$

$$h_\ell = \tilde{h}_\ell + \tau_\ell(\tilde{h}_\ell - \tilde{h}_{\ell-1}). \tag{25}$$

The FB procedure is stopped when the functional (11), evaluated in the current u_ℓ, h_ℓ solutions, drops below 10^{-8}.

The weights τ_ℓ in (22) and (25) used for convergence acceleration are computed as in [2]. The optimization subproblem (21) for u reduces to a weighted soft thresholding with an explicitly given closed-form solution

$$\tilde{u}_\ell = S_{\lambda_1 \beta_1 diag(W)}(v_\ell),$$

where $S_t(\nu)$ is a point-wise soft-thresholding function which, for given vectors t and ν, applies soft thresholding with parameter t_i to the element ν_i of ν, namely $[S_t(\nu)]_i = \text{sign}(\nu_i) \max(0, |\nu_i| - t_i), \quad \forall i$.

The minimization of (24) is easily obtained as follows

$$\begin{aligned} \tilde{h}_\ell &= \frac{1}{1+\lambda_2 \beta_2} s_\ell \text{ for } H(h) = \|h\|_2^2, \\ \tilde{h}_\ell &= S_{\lambda_2 \beta_2}(s_\ell) \text{ for } H(h) = \|h\|_1. \end{aligned} \quad (26)$$

At each Majorization step the parameter λ_1 is decreased following the well-known continuation framework [6], that significantly reduces the number of iterations required. In particular, we adopt the following reduction:

$$\lambda_1^{(k+1)} = c_\lambda \cdot \tilde{J}(u^{(k+1)}, h^{(k+1)}; \lambda_1^{(k)}, \lambda_2^{(k)}, u^{(k)}), \qquad 0 < c_\lambda < 1. \quad (27)$$

The algorithm starts with an initial over-regularized problem and then, at each subsequent majorization step, it reduces the value of parameter λ_1 proportionally to the decreasing of the functional.

For what concerns the parameter choice for λ_2 we consider an a priori fixed value which can be estimated regarding the accuracy of the PSF.

Finally, each nonzero entry $u_{i,j}^*$ in the minimizer of (7) is selected as a fluorescent molecule with localization (X_i, Y_j).

5   Numerical Experiments

We compared the proposed NCNS algorithm, applied with $\ell = 1$ and $\mu^0 = 1$ in (5), with the methods FALCON [10], ThunderSTORM [11], IRL1-CEL0 [5], which are super-resolution localization algorithms currently among the best state-of-the-art methods for high-density molecules estimation according to the 2013/2016 IEEE ISBI Single-Molecule Localisation Microscopy (SMLM) challenge [12]. The algorithms have been provided by the authors. In the experimental results, the methods FALCON and ThunderSTORM are equipped with a post-processing phase, while the proposed NCNS method does not exploit any post-processing. Further improvements will be integrated for removing false positive using a centroid method as suggested in [3].

For all the examples, the reconstructed images $N \times N$ are obtained from the acquired images $n \times n$ where $N = n \times d$, $d = 4$ and $n = 64$, that is $N = 256$.

In the simulated data delivered the xy-Gaussian PSF $\mathcal{B}$ is applied to very high resolution images $(n \times 20)$ and is characterized by a Full Width at Half

Maximum (FWHM) parameter provided with the dataset which is related to the standard deviation σ by the relation $\sigma = FWHM/2.355$ nm. The Gaussian PSF $\mathcal{B}$ applied in the reconstruction algorithms which process high resolution images of size N, is characterized by a standard deviation σ obtained from the relation

$$\sigma = \left(\frac{N}{(n \times 20)} FWHM \right) /2.355.$$

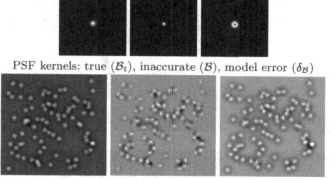

PSF kernels: true $(\mathcal{B}_t)$, inaccurate $(\mathcal{B})$, model error $(\delta_{\mathcal{B}})$

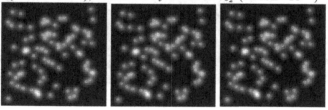

Correction term h: ground-truth ($h = \delta_{\mathcal{B}} * u_{GT}$), estimated by the NCNS$_{\ell_1}$ (RMSE=10^{-3}), estimated by the NCNS$_{\ell_2}$ (RMSE=10^{-4})

Molecule localizations (JAC(%)): NCNS-0 (86.6), NCNS$_{\ell_1}$ (89.6), NCNS$_{\ell_2}$ (90.7).

Fig. 1. Comparisons among different regularization terms for h in (7).

5.1 Performance Evaluation

The performances are evaluated in terms of molecule localizations measured by the **detection rate** via the Jaccard index (JAC), and the **localization accuracy**, measured by the root-mean-square-error (RMSE). The evaluation of both the metrics are performed by the tool in http://bigwww.epfl.ch/smlm/challenge2016/. In particular, let R and T be the two sets of reference (ground truth) molecules and test molecules respectively, the localized molecules successfully paired with some test molecules are classified as true positives (TP), while the remaining localized molecules unpaired are categorized as false positives (FP), and the ground truth molecules not associated with any localized molecules are categorized as false negatives (FN) and related by $FN = |R| - TP$ and $FP = |T| - TP$. A test molecule is paired with a reference one only if the distance between them is lower than a tolerance TOL which should be less than the FWHM of the PSF. The Jaccard index defined by

$$JAC(\%) := \frac{TP}{TP + FP + FN} \times 100 = \frac{|R \cap T|}{|R| + |T| - |R \cap T|}. \qquad (28)$$

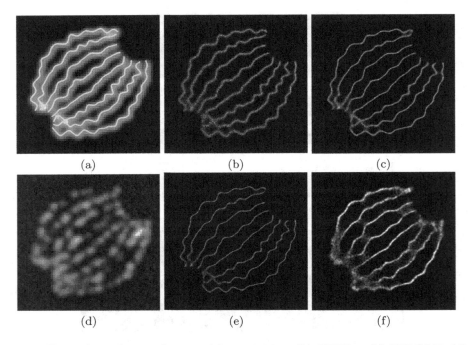

(a)   (b)   (c)

(d)   (e)   (f)

Fig. 2. Example 2: Averaged image (a) acquisition; (b) NCNS_{ℓ_2}; (c) FALCON; (d) single image (frame 58); (e) IRL1-CEL0; (f) ThunderSTORM.

Example 1: Performance of the Blur Correction Term. We first illustrate the benefits introduced by the proposed blur kernel correction in NCNS algorithm applied to a simple synthetic image "Toy" provided by the authors of [5], for which also the ground-truth u_{GT} is given. In particular, we compare the results obtained by the proposed NCNS algorithm without the h regularization term in (7) by optimizing only over u (NCNS–0), with NCNS and $H(h) = \|h\|_2^2$ (NCNS_{ℓ_2}), and with NCNS and $H(h) = \|h\|_1$ (NCNS_{ℓ_1}).

The test image "Toy" of dimension 256×256 has been blurred by the true blur Gaussian kernel $\mathcal{B}_t$ with unknown standard deviation illustrated in Fig. 1

Table 1. Example 2: JAC (and RMSE) for different JAC TOLs.

Method - TOL (nm)	100	150	200	250
$NCNS_{\ell_2}$	55.95 (52.12)	64.55 (60.70)	66.07 (64.01)	66.96 (65.75)
IRL1-CEL0	46.79 (**43.15**)	49.33 (**47.91**)	50.06 (**50.59**)	50.49 (**53.15**)
FALCON	**61.92** (49.75)	**72.58** (59.80)	**76.34** (65.66)	**78.09** (69.76)
ThunderSTORM	14.17 (51.70)	17.43 (68.41)	18.61 (77.41)	18.83 (80.10)

Table 2. Example 3 - Dataset MT0.N1.HD and MT0.N2.HD: JAC and RMSE values for different reconstruction methods using a fixed $TOL = 250$.

	MT0.N1.HD		MT0.N2.HD	
Method	JAC	RMSE	JAC	RMSE
$NCNS_{\ell_2}$	**59.18**	69.20	**49.10**	72.4
NCNS-0	56.75	69.70	48.65	72.4
IRL1-CEL0	37.90	73.04	34.33	71.4
FALCON	44.78	**56.23**	44.55	87.0
ThunderSTORM	52.92	59.61	46.06	**61.7**

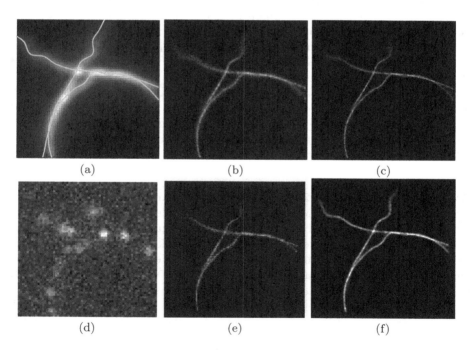

Fig. 3. Example 3 - `MT0.N1.HD` dataset (a)–(f): Averaged image (a) acquisition; (b) $NCNS_{\ell_2}$; (c) IRL1-CEL0; (d) single image (frame 700); (e)FALCON; (f) Thunder-STORM.

(first row, left), and subsampled according to the M_d operator with $d = 4$. All the algorithms have been initialized by an inaccurate Gaussian blur kernel $\mathcal{B}$ with $\sigma = 10^{-4}$, shown in Fig. 1 (first row, center). The model error $\delta_\mathcal{B}$ obtained by the difference between $\mathcal{B} - \mathcal{B}_t$ is shown in Fig. 1 (first row, right). The algorithms minimizing the functional (7) produce the approximate solutions (u^*, h^*). Figure 1 (second row) reports from left to right: the true h computed by $h = \delta_\mathcal{B} * u_{GT}$, the solution h^* of $NCNC_{\ell_1}$ and of $NCNC_{\ell_2}$. In the third row of Fig. 1 we illustrate the acquired blurred image $g \in \mathbb{R}^{64 \times 64}$ with overimposed the ground-truth molecule locations by green circles, and the estimated molecule locations by red crosses. For each method we also reported the Jaccard index obtained and the RMSE computed on h^* results with respect to the ground truth. For what concerns the h reconstructions, qualitative and quantitative results confirm that the use of ℓ_2 norm regularization term in (7) instead of the ℓ_1 norm, provides a more accurate and smooth reconstruction, avoiding the well-known staircase effects. The Jaccard indices highlight the noticeable advantages of the presence of the model error $\delta_\mathcal{B}$ to correct the inaccuracy of the guessed blur kernel during the reconstruction process. More pronounced is the error of the initial blur kernel $\mathcal{B}$ compared to the one that has really corrupted the data $\mathcal{B}_t$, and more significant is the contribution of the correction term.

Example 2: Challenge 2013 Bundled Tubes HD. The Bundled Tubes HD SMLM challenge is part of the Challenge 2013 which represents a set of high density simulated acquisitions of a bundle of 8 simulated tubes of 30 nm diameter. For this simulation, the camera resolution is 64×64 pixels of PixelSize 100 nm, the PSF is modelled by a Gaussian function whose $FWHM = 258.21$ nm, and the stack simulates 81049 emitters activated on 361 different frames. Figure 2 shows the averaged acquisition image with the ground truth in green (Fig. 2(a)), a single image extracted from the stack (Fig. 2(c)), together with the averaged reconstructions of the whole stack, given by the average of the reconstructions of the 361 frames obtained by the compared methods. In Table 1 the Jaccard index results are reported for different tolerances TOL; the best results are shown in bold.

Example 3: Challenge 2016 MT0.N1.HD and MT0.N2.HD. The datasets MT0.N1.HD and MT0.N2.HD in the Challenge 2016 represent three microtubules in the field of view of $6.4 \times 6.4 \times 1.5\,\mu m$. The resolution of the camera is 64 pixels, the pixelsize is 100 nm, the stack simulates 31612 emitters activated on 2500 different frames, the PSF is modelled by a Gaussian function whose $FWHM = 270.21$ nm. The two datasets MT0.N1.HD and MT0.N2.HD differ in the noise corruption, and in the molecule density which are respectively of 2.0 and 0.2. Figure 3 shows the reconstructions of the whole stack MT0.N2.HD, given by the average of the reconstructions of the 2500 frames, processed by the several methods. In Table 2 the Jaccard index (JAC) and RMSE values are reported for the different reconstruction methods for the two different cases.

Results shown in Tables 1, 2 and illustrated in Figs. 2, 3, highlight the good performance of the proposed NCNS algorithm, further improved in Example 3 where the data sparsity is more pronounced with respect to data in Example 2.

6 Conclusion and Future Work

In this paper, we have proposed a non-convex nonseparable optimization algorithm for the 2D molecule localization in high-density super-resolution microscopy which combines a sparsity-promoting formulation with an accurate estimate of the inaccurate blur kernel. The performance results confirm the efficacy of the proposed variational model in the SMLM context.

References

1. Bolte, J., Sabach, S., Teboulle, M.: Proximal alternating linearized minimization for nonconvex and nonsmooth problems. Math. Program. **146**, 459–494 (2014)
2. Chambolle, A., Dossal, C.: On the convergence of the iterates of the "fast iterative shrinkage/thresholding algorithm". J. Optim. Theory Appl. **166**(3), 968–982 (2015)
3. Chan, R., Wang, C., Nikolova, M., Plemmons, R., Prasad, S.: Non-convex optimization for 3D point source localization using a rotating point spread function. SIAM J. Imaging Sci. (2018, in press)
4. Chouzenoux, E., Pesquet, J.C., Repetti, A.: A block coordinate variable metric forward-backward algorithm. J. Glob. Optim. **66**(3), 457–485 (2016)
5. Gazagnes, S., Soubies, E., Blanc-Féraud, L.: High density molecule localization for super-resolution microscopy using CEL0 based sparse approximation. In: IEEE 14th International Symposium on Biomedical Imaging, Melbourne, VIC, pp. 28–31 (2017)
6. Hale, E.T., Yin, W., Zhang, Y.: Fixed-point continuation for ℓ_1-minimization: methodology and convergence. SIAM J. Optim. **19**(3), 1107–1130 (2008)
7. Ji, H., Wang, K.: Robust image deblurring with an inaccurate blur kernel. IEEE Trans. Image Process. **21**(4), 1624–1634 (2012)
8. Lanza, A., Morigi, S., Selesnick, I., Sgallari, F.: Nonconvex nonsmooth optimization via convex-nonconvex majorization-minimization. Numer. Math. **136**(2), 343–381 (2017)
9. Lazzaro, D., Montefusco, L.B., Papi, S.: Blind cluster structured sparse signal recovery: a nonconvex approach. Signal Process. **109**, 212–225 (2015)
10. Min, J., Holden, S.J., Carlini, L., Unser, M., Manley, S., Ye, J.C.: 3D high-density localization microscopy using hybrid astigmatic/ biplane imaging and sparse image reconstruction. Biomed. Opt. Express **5**(11), 3935–3948 (2014)
11. Ovesny, M., Křížek, P., Borkovec, J., Svindrych, Z., Hagen, G.: ThunderSTORM: a comprehensive ImageJ plug-in for PALM and STORM data analysis and super-resolution imaging. Bioinformatics **30**(16), 2389–2390 (2014)
12. Sage, D., et al.: Quantitative evaluation of software packages for single molecule localization microscopy. Nature Methods **12**(8), 717 (2015)

Preservation of Piecewise Constancy under TV Regularization with Rectilinear Anisotropy

Clemens Kirisits[1]([✉]), Otmar Scherzer[1,2], and Eric Setterqvist[1]

[1] Faculty of Mathematics, University of Vienna, Vienna, Austria
{clemens.kirisits,otmar.scherzer,eric.setterqvist}@univie.ac.at
[2] Johann Radon Institute for Computational and Applied Mathematics (RICAM), Austrian Academy of Sciences, Linz, Austria

Abstract. A recent result by Łasica, Moll and Mucha about the ℓ^1-anisotropic Rudin-Osher-Fatemi model in $\mathbb{R}^2$ asserts that the solution is piecewise constant on a rectilinear grid, if the datum is. By means of a new proof we extend this result to $\mathbb{R}^n$. The core of our proof consists in showing that averaging operators associated to certain rectilinear grids map subgradients of the ℓ^1-anisotropic total variation seminorm to subgradients.

Keywords: Rudin-Osher-Fatemi model · ℓ^1-anisotropy · Piecewise constant functions

1 Introduction

This article is concerned with a variant of the Rudin-Osher-Fatemi (ROF) image denoising model [13]. More specifically, we consider minimization of

$$\frac{1}{2}\|u - f\|_{L^2}^2 + \alpha J(u), \tag{1}$$

where $J(u) = \int_\Omega \|\nabla u(x)\|_{\ell^1} dx$ is the total variation with ℓ^1-anisotropy. This model and variations thereof have been used in imaging applications for data exhibiting a rectilinear geometry [2,7,14,15]. Numerical algorithms for minimizing (1) have been studied, for example, in [6,10,12].

The ℓ^1-anisotropic total variation has a special property from a theoretical point of view as well. It has been shown in [3, Thm. 3.4, Rem. 3.5] that approximation of a general $u \in BV \cap L^p$ by functions u_m piecewise constant on rectilinear grids, in the sense that

$$\|u - u_m\|_{L^p} \to 0 \qquad \text{and} \qquad J(u_m) \to J(u),$$

is not possible for $J(u) = \int_\Omega \|\nabla u(x)\|_{\ell^q} dx$, unless $q = 1$.

Let $\Omega \subset \mathbb{R}^n$ be a finite union of hyperrectangles, each aligned with the coordinate axes. Our main result, Theorem 2, states that if the given function

© Springer Nature Switzerland AG 2019
J. Lellmann et al. (Eds.): SSVM 2019, LNCS 11603, pp. 510–521, 2019.
https://doi.org/10.1007/978-3-030-22368-7_40

$f : \Omega \to \mathbb{R}$ is piecewise constant on a rectilinear grid, then the minimizer of (1) is too. This extends a recent result by Łasica, Moll and Mucha about two-dimensional domains [11, Thm. 5]. Their proof is based on constructing the solution by means of its level sets and relies on minimization of an anisotropic Cheeger-type functional over subsets of Ω.

The proof we present below is centred around the averaging operator A_G associated to the grid G on which f is piecewise constant. In addition to being a contraction, it has the crucial property of mapping subgradients of J to subgradients, that is, $A_G(\partial J(0)) \subset \partial J(0)$, see Theorem 1. Combined with the dual formulation of (1) we obtain that the minimizer must be piecewise constant on the same grid as f. While it might be possible to extend the techniques of [11] to higher dimensions, we believe that modifying the so-called "squaring step" in the proof of [11, Lem. 2] could lead to difficulties.

Theorem 2 implies that, if f is piecewise constant on a rectilinear grid, then minimization of functional (1) becomes a finite-dimensional problem. More precisely, in this case the solution can be found by minimizing a discrete energy of the form

$$\sum_i w_i |u_i - f_i|^2 + \alpha \sum_{i,j} w_{ij} |u_i - u_j|,$$

where the weights $w_i, w_{ij} \geq 0$ depend only on the grid. For problems of this sort there are many efficient algorithms, such as graph cuts [4,5,8]. Extending the preservation of piecewise constancy to domains $\Omega \subset \mathbb{R}^n$, $n \geq 3$, means that the discrete reformulation can also be exploited for processing higher dimensional data such as volumetric images or videos.

This article is organized as follows. Section 2 contains the basic concepts that will be required throughout. In Sect. 2.1 we define several spaces of piecewise constant functions, while Sect. 2.2 is devoted to the ℓ^1-anisotropic ROF model. Section 3 is the main part of this paper. It starts with introducing the averaging operator A_G and ends with Theorem 2. The article is concluded in Sect. 4.

2   Mathematical Preliminaries

2.1   PCR Functions

In this section we introduce several notions related to functions which are piecewise constant on rectilinear subsets of $\mathbb{R}^n$. Some of these are n-dimensional analogues of notions from [11, Sect. 2.3].

A bounded set $R \subset \mathbb{R}^n$ which can be written as a Cartesian product of n proper intervals is called an n-*dimensional hyperrectangle*. Recall that an interval is proper, if it is neither empty nor a singleton. Finite unions of n-dimensional hyperrectangles will be referred to as *rectilinear n-polytopes*.

A *rectilinear grid*, or simply *grid*, is a finite family of affine hyperplanes, each being perpendicular to one of the coordinate axes of $\mathbb{R}^n$. For a rectilinear n-polytope P we denote by $G(P)$ the smallest grid with the property that the union of all its affine hyperplanes contains the entire boundary of P.

Throughout this article $\Omega \subset \mathbb{R}^n$ is an *open* rectilinear n-polytope. A finite family of rectilinear n-polytopes $\mathcal{Q} = \{P_1, \ldots, P_N\}$ is called a *partition* of Ω, if they have pairwise disjoint interiors and the union of their closures equals $\overline{\Omega}$. Every grid G defines a partition $\mathcal{Q}(G)$ of Ω into rectilinear n-polytopes in the following way: $P \subset \Omega$ belongs to $\mathcal{Q}(G)$, if and only if its boundary is contained in $\partial \Omega \cup \bigcup G$ while $\operatorname{int} P$ and $\bigcup G$ are disjoint. Note that if G contains $G(\Omega)$, then $\mathcal{Q}(G)$ consists only of hyperrectangles.

We adopt the notation $PCR(\Omega)$, or simply PCR, from [11] for the set of all integrable functions $f : \Omega \to \mathbb{R}$ which can be written as finite linear combinations of indicator functions of rectilinear n-polytopes. That is, $f \in PCR$ if there is an $N \in \mathbb{N}$, $c_i \in \mathbb{R}$ and rectilinear n-polytopes $P_i \subset \Omega$, $1 \le i \le N$, such that

$$f = \sum_{i=1}^{N} c_i \mathbf{1}_{P_i} \qquad (2)$$

almost everywhere. Here, $\mathbf{1}_A$ is the indicator function of the set A, defined by

$$\mathbf{1}_A(x) = \begin{cases} 1, & x \in A, \\ 0, & x \notin A. \end{cases}$$

We can assume, without loss of generality, that the values c_i are pairwise distinct and that the polytopes P_i are pairwise disjoint, which makes the representation (2) unique almost everywhere. To every $f \in PCR$ we associate its *minimal grid*, that is, the unique smallest grid covering the boundaries of all level sets of f, given by

$$G_f = \bigcup_{i=1}^{N} G(P_i).$$

Note that the partition $\mathcal{Q}(G_f)$ always consists of hyperrectangles only. See Fig. 1 for an illustration of G_f and $\mathcal{Q}(G_f)$.

For a given grid G we denote by PCR_G the set of all functions in PCR which are equal almost everywhere to a finite linear combination of indicator functions of $P_i \in \mathcal{Q}(G)$.

The following notions are essential for the proof of Theorem 1. Fix an $i \in \{1, \ldots, n\}$ as well as coordinates $x_1, \ldots, x_{i-1}, x_{i+1}, \ldots, x_n$. The set $\{x_i \in \mathbb{R} : (x_1, \ldots, x_n) \in \Omega\}$ is a union of finitely many disjoint intervals

$$I_i^k = (a_i^k, b_i^k), \quad k = 1, \ldots, m. \qquad (3)$$

Note that the number of intervals m as well as the intervals I_i^k themselves depend on, and are uniquely determined by, the coordinates $x_1, \ldots, x_{i-1}, x_{i+1}, \ldots, x_n$. For an illustration of the intervals I_i^k see Example 1 below. Next, let $\mathcal{G}$ be the set of all rectilinear grids of $\mathbb{R}^n$. For every $G \in \mathcal{G}$ and $i \in \{1, \ldots, n\}$ we define

$$\Gamma_G^i = \left\{ g \in PCR_G : \sup_{s \in (a_i^k, b_i^k)} \left| \int_{a_i^k}^{s} g \, dx_i \right| \le 1, \int_{a_i^k}^{b_i^k} g \, dx_i = 0, 1 \le k \le m \right\}.$$

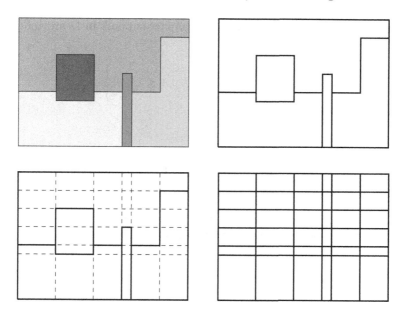

Fig. 1. Upper left: A PCR function f on a planar domain Ω. Each level set of f is visualized using a different grey tone. Upper right: The boundaries of the level sets of f. Lower left: Extension of the boundaries of the level sets of f. Lower right: The minimal grid G_f (lines) and the partition $\mathcal{Q}(G_f)$ of Ω (rectangular cells).

The restrictions on g are to be understood for almost every

$$(x_1, \ldots, x_{i-1}, x_{i+1}, \ldots, x_n) \in \mathbb{R}^{n-1}$$

such that there is an $x_i \in \mathbb{R}$ satisfying $(x_1, \ldots, x_n) \in \Omega$. The sum of the spaces Γ_G^i is denoted by

$$\Gamma_G = \left\{ \sum_{i=1}^{n} g_i : g_i \in \Gamma_G^i, 1 \leq i \leq n \right\},$$

and we further set

$$\Gamma = \bigcup_{G \in \mathcal{G}} \Gamma_G.$$

Remark 1. The set Γ_G consists of divergences of certain piecewise affine vector fields. More precisely, Theorem 1 below implies that $\Gamma_G = \partial J(0) \cap PCR_G$, that is, Γ_G is the set of all subgradients of J which are piecewise constant on G.

Finally, those elements of Γ_G^i which have compact support in Ω are collected in the set $\Gamma_{G,c}^i$, and we define analogously

$$\Gamma_{G,c} = \left\{ \sum_{i=1}^{n} g_i : g_i \in \Gamma_{G,c}^i, 1 \leq i \leq n \right\},$$

$$\Gamma_c = \bigcup_{G \in \mathcal{G}} \Gamma_{G,c}.$$

Example 1. For the rectilinear 2-polytope Ω of Fig. 2 we have the following intervals I_1^k and I_2^k

$$\{x_1 \in \mathbb{R} : (x_1, x_2) \in \Omega\} = \begin{cases} (0,6), & \text{if } x_2 \in (0,1) \cup (2,3), \\ (0,2) \cup (4,6), & \text{if } x_2 \in [1,2], \\ (3,6), & \text{if } x_2 \in [3,4), \end{cases}$$

and

$$\{x_2 \in \mathbb{R} : (x_1, x_2) \in \Omega\} = \begin{cases} (0,3), & \text{if } x_1 \in (0,2), \\ (0,1) \cup (2,3), & \text{if } x_1 \in [2,3], \\ (0,1) \cup (2,4), & \text{if } x_1 \in (3,4], \\ (0,4), & \text{if } x_1 \in (4,6). \end{cases}$$

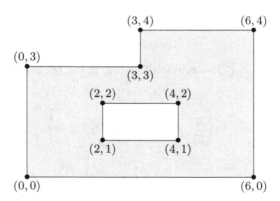

Fig. 2. A rectilinear 2-polytope Ω.

2.2 The ℓ^1-Anisotropic ROF Model

The notion of anisotropic total variation was introduced in [1]. In this article we exclusively consider one particular variant.

For every $\alpha > 0$ we denote by $\mathcal{B}_\alpha$ the set of all smooth compactly supported vector fields on Ω whose components are bounded by α, that is,

$$\mathcal{B}_\alpha = \left\{ H \in C_c^\infty(\Omega, \mathbb{R}^n) : \max_{1 \le i \le n} |H_i(x)| \le \alpha, \forall x \in \Omega \right\}.$$

The ℓ^1-*anisotropic total variation* $J : L^2(\Omega) \to \mathbb{R} \cup \{+\infty\}$ is given by

$$J(u) = \sup_{H \in \mathcal{B}_1} \int_\Omega u \operatorname{div} H \, dx = \sup_{h \in \overline{\operatorname{div} \mathcal{B}_1}} \int_\Omega uh \, dx, \tag{4}$$

where the bar denotes closure in $L^2(\Omega)$. Thus, J is the support function of the closed and convex set $\overline{\operatorname{div} \mathcal{B}_1}$, which implies that $\overline{\operatorname{div} \mathcal{B}_1} = \partial J(0)$, or more generally

$$\overline{\operatorname{div} \mathcal{B}_\alpha} = \alpha \partial J(0) \tag{5}$$

for every $\alpha > 0$. If u is a Sobolev function, then $J(u) = \int_\Omega \|\nabla u(x)\|_{\ell^1} dx$.

The next lemma states that the ℓ^1-anisotropic ROF model is equivalent to constrained L^2-minimization. However, the way it is formulated it actually applies to every support function J of a closed and convex subset of $L^2(\Omega)$.

Lemma 1. *For every $\alpha > 0$ and $f \in L^2(\Omega)$ the minimization problem*

$$\min_{u \in L^2(\Omega)} \frac{1}{2} \|u - f\|_{L^2}^2 + \alpha J(u) \tag{6}$$

is equivalent to

$$\min_{u \in f - \alpha \partial J(0)} \|u\|_{L^2}.$$

Proof. The dual problem associated to (6) is given by

$$\min_{w \in L^2(\Omega)} \frac{1}{2} \|w - f\|_{L^2}^2 + (\alpha J)^*(w), \tag{7}$$

where the asterisk stands for convex conjugation. The two solutions u_α and w_α of (6) and (7), respectively, satisfy the optimality conditions

$$\begin{aligned} u_\alpha &= f - w_\alpha, \\ w_\alpha &\in \partial(\alpha J)(u_\alpha). \end{aligned} \tag{8}$$

Concerning the derivation of (7) and (8) we refer to [9, Chap. III, Rem. 4.2]. Since αJ is the support function of the set $\alpha \partial J(0)$, recall (4), its conjugate is the characteristic function

$$(\alpha J)^*(w) = \begin{cases} 0, & w \in \alpha \partial J(0), \\ +\infty, & w \notin \alpha \partial J(0). \end{cases}$$

Therefore, problem (7) is equivalent to

$$\min_{w \in \alpha \partial J(0)} \|w - f\|_{L^2}.$$

Finally, using the optimality condition (8) we get

$$\|u_\alpha\|_{L^2} = \|w_\alpha - f\|_{L^2} = \min_{w \in \alpha \partial J(0)} \|w - f\|_{L^2} = \min_{u \in f - \alpha \partial J(0)} \|u\|_{L^2}.$$

$\square$

3   The Averaging Operator A_G

Let Ω be a rectilinear n-polytope and G a grid. Define the averaging operator $A_G : L^1(\Omega) \to PCR_G(\Omega)$ by

$$A_G g = \sum_{i=1}^{N} \left(\frac{1}{|P_i|} \int_{P_i} g(s)ds \right) \mathbf{1}_{P_i},$$

where $P_i \in \mathcal{Q}(G)$ and $|P_i|$ is its n-dimensional volume.

Two properties of the operator A_G turn out to be important when establishing the main result of this paper, Theorem 2. The first one is

Lemma 2. *For every $u \in L^1(\Omega)$ and convex $\varphi : \mathbb{R} \to \mathbb{R}$*

$$\int_\Omega \varphi\left((A_G u)(x)\right) dx \leq \int_\Omega \varphi\left(u(x)\right) dx.$$

Proof. By applying Jensen's inequality we obtain

$$\int_\Omega \varphi\left((A_G u)(x)\right) dx = \sum_{i=1}^{N} \int_{P_i} \varphi\left((A_G u)(x)\right) dx = \sum_{i=1}^{N} \varphi\left(\frac{1}{|P_i|} \int_{P_i} u(x)dx\right) |P_i|$$

$$\leq \sum_{i=1}^{N} \int_{P_i} \varphi(u(x))dx = \int_\Omega \varphi(u(x))dx.$$

$\square$

Remark 2. Lemma 2 implies in particular that A_G is a contraction,

$$\|A_G\|_{L^p \to L^p} \leq 1, \quad 1 \leq p < \infty.$$

The second property of A_G is that it maps subgradients of J to subgradients. Recall that $G(\Omega)$ is the smallest grid covering the entire boundary of Ω.

Theorem 1. *Let G be a grid containing $G(\Omega)$. Then $A_G\left(\partial J(0)\right) \subset \partial J(0)$.*

Proof. The proof is divided into three steps, each being proved in a separate lemma

$$A_G\left(\partial J(0)\right) \overset{\text{Lem. 3}}{\subset} \Gamma_G \subset \Gamma \overset{\text{Lem. 4}}{\subset} \overline{\Gamma_c} \overset{\text{Lem. 5}}{\subset} \partial J(0).$$

Note that the inclusion $\Gamma_G \subset \Gamma$ is trivial. $\square$

Lemma 3. *Let G be a grid containing $G(\Omega)$. Then $A_G(\partial J(0)) \subset \Gamma_G$.*

Proof. Throughout this proof we exploit the fact that $\partial J(0) = \overline{\text{div}\,\mathcal{B}_1}$, recall Eq. (5).

First, note that PCR_G is a finite-dimensional subspace of L^2 and that the sets Γ_G^i, $i = 1, ..., n$, are bounded and closed subsets of PCR_G. It follows that Γ_G is a closed subset of PCR_G and in particular of L^2. Therefore, it suffices to show $A_G(\text{div}\,\mathcal{B}_1) \subset \Gamma_G$, as we then have $A_G(\overline{\text{div}\,\mathcal{B}_1}) \subset \overline{A_G(\text{div}\,\mathcal{B}_1)} \subset \overline{\Gamma_G} = \Gamma_G$, because A_G is continuous.

Take $H = (H_1, \ldots, H_n) \in \mathcal{B}_1$. We want to show that $A_G \partial H_i / \partial x_i \in \Gamma_G^i$, that is,

$$\sup_{s \in (a_i^k, b_i^k)} \left| \int_{a_i^k}^s A_G \frac{\partial H_i}{\partial x_i} dx_i \right| \leq 1, \quad \text{and} \quad \int_{a_i^k}^{b_i^k} A_G \frac{\partial H_i}{\partial x_i} dx_i = 0,$$

for $i = 1, \ldots, n$ and each k, where (a_i^k, b_i^k) are the intervals defined in Eq. (3).

Consider the second integral first. From the definition of A_G it follows that (a_i^k, b_i^k) can be divided into a finite number of subintervals in such a way that the integrand is constant on each. In addition the assumption $G \supset G(\Omega)$ implies that the partition $\mathcal{Q}(G)$ consists of hyperrectangles only. Thus, after a potential relabelling of the $R_j \in \mathcal{Q}(G)$, we can write

$$\int_{a_i^k}^{b_i^k} A_G \frac{\partial H_i}{\partial x_i} dx_i = \sum_{j=1}^M \int_{s_{j-1}}^{s_j} \left(\frac{1}{|R_j|} \int_{R_j} \frac{\partial H_i}{\partial x_i} dx \right) dx_i$$

$$= \sum_{j=1}^M \frac{s_j - s_{j-1}}{|R_j|} \int_{R_j} \frac{\partial H_i}{\partial x_i} dx$$

for some $M \in \{1, \ldots, N\}$ and $a_i^k = s_0 < s_1 < \cdots < s_M = b_i^k$. Note that $|R_j|/(s_j - s_{j-1})$ is the $(n-1)$-dimensional volume of $\partial R_j \cap \partial R_{j+1}$, and that this volume is independent of $j \in \{1, \ldots M\}$. In other words, the hyperrectangles R_j only differ in their extent in x_i-direction, compare Fig. 1, bottom right. The reason is that $\mathcal{Q}(G)$ is not an arbitrary partition of Ω into hyperrectangles, but rather formed by a grid. Setting $C = (s_j - s_{j-1})/|R_j|$ we further obtain

$$= C \int_{\bigcup_j R_j} \frac{\partial H_i}{\partial x_i} dx.$$

The remaining integral can be computed by turning it into an iterated one, integrating with respect to x_i first, and recalling that H is compactly supported

$$= C \underbrace{\int \cdots \int}_{n-1} \int_{a_i^k}^{b_i^k} \frac{\partial H_i}{\partial x_i} dx_i = C \int \cdots \int H_i \Big|_{x_i = a_i^k}^{x_i = b_i^k} = 0.$$

Here $F\big|_{x_i = a}^{x_i = b}$ stands for $F(x_i = b) - F(x_i = a)$, where $F(x_i = c)$ denotes the restriction of F to the affine hyperplane defined by $x_i = c$.

Now integrate up to an arbitrary $s \in (a_i^k, b_i^k]$. We can assume $s \in (s_{\ell-1}, s_\ell]$ for some $\ell \in \{1, \ldots, M\}$ and a brief computation similar to the one above shows that

$$\int_{a_i^k}^s A_G \frac{\partial H_i}{\partial x_i} dx_i = C \int \cdots \int H_i \Big|_{x_i=a_i^k}^{x_i=s_{\ell-1}} + \frac{s - s_{\ell-1}}{|R_\ell|} \int \cdots \int H_i \Big|_{x_i=s_{\ell-1}}^{x_i=s}$$

$$= C \int \cdots \int H_i(x_i = s_{\ell-1}) + \frac{s - s_{\ell-1}}{|R_\ell|} \int \cdots \int H_i \Big|_{x_i=s_{\ell-1}}^{x_i=s}.$$

Recalling that we can write $C = (s_\ell - s_{\ell-1})/|R_\ell|$ we rearrange terms

$$= \frac{s_\ell - s}{|R_\ell|} \int \cdots \int H_i(x_i = s_{\ell-1}) + \frac{s - s_{\ell-1}}{|R_\ell|} \int \cdots \int H_i(x_i = s).$$

Finally, we estimate $H_i \leq 1$ and obtain

$$\leq \frac{s_\ell - s}{|R_\ell|} \frac{|R_\ell|}{s_\ell - s_{\ell-1}} + \frac{s - s_{\ell-1}}{|R_\ell|} \frac{|R_\ell|}{s_\ell - s_{\ell-1}} = 1.$$

Similarly, we get $\int_{a_i^k}^s A_G \frac{\partial H_i}{\partial x_i} dx_i \geq -1$. Thus we have $A_G \partial H_i / \partial x_i \in \Gamma_G^i$.   □

Lemma 4. $\Gamma \subset \overline{\Gamma_c}$.

Proof. Let $j \in \mathbb{N}$ and define $\Omega_j \subset \Omega$ by removing strips of width $1/j$ from the boundary of Ω. It is assumed that j is chosen large enough such that the strips are contained in Ω. See Fig. 3 for an example of the construction of Ω_j in the plane. Next, for $i = 1, \ldots, n$ we define $\Omega_j^i \subset \Omega_j$, by removing strips of width $1/j$ from those parts of the boundary of Ω_j which are orthogonal to the x_i-axis. By choosing j large enough, the strips will be contained in Ω_j. For an illustration of the construction of Ω_j^i in the plane, see Fig. 4.

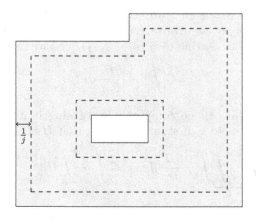

Fig. 3. The construction of Ω_j by removing strips of width $1/j$ from Ω.

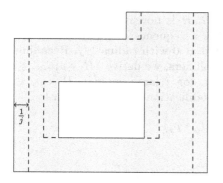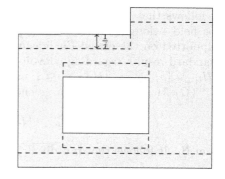

Fig. 4. The construction of Ω_j^1 (left) and Ω_j^2 (right) by removing strips from Ω_j.

Take $h \in \Gamma$. So, $h \in \Gamma_G$ for some $G \in \mathcal{G}$ and in particular $h = \sum_{i=1}^n h_i$ where $h_i \in \Gamma_G^i$. Let $g_j = \sum_{i=1}^n g_{j,i}$ where

$$
g_{j,i}(x) = \begin{cases} 0, & \text{if } x \in \Omega \setminus \Omega_j, \\ 2h_i(x), & \text{if } x \in \Omega_j \setminus \Omega_j^i, \\ h_i(x), & \text{otherwise.} \end{cases}
$$

Note that there is a grid $G_j \supset G$ such that $g_{j,i} \in PCR_{G_j}$ for every $i = 1, \ldots, n$, and that for j large enough

$$
\left| \int_{a_i^k}^s g_{j,i} \, dx_i \right| \leq \left| \int_{a_i^k}^s h_i \, dx_i \right|
$$

for every interval (a_i^k, b_i^k), recall Eq. (3), and $s \in (a_i^k, b_i^k]$. It follows that $g_{j,i} \in \Gamma_{G_j,c}^i$ and therefore $g_j \in \Gamma_{G_j,c}$. Finally, it can be directly verified that

$$
\lim_{j \to \infty} \|g_j - h\|_{L^2} = 0.
$$

As $h \in \Gamma$ was chosen arbitrarily we conclude that $\Gamma \subset \overline{\Gamma_c}$. □

Lemma 5. $\overline{\Gamma_c} \subset \partial J(0)$.

Proof. As in Lemma 3 we use the fact that $\partial J(0) = \overline{\operatorname{div} \mathcal{B}_1}$.

Take $h \in \Gamma_c$. So there is a grid G such that $h = \sum_{i=1}^n h_i \in \Gamma_{G,c}$ where $h_i \in \Gamma_{G,c}^i$. From h we now construct a vector field $H = (H_1, \ldots, H_n)$. For every $i \in \{1, \ldots, n\}$ and $x \in \Omega$ there is a unique interval $I_i^k = (a_i^k, b_i^k)$ containing x_i, recall Eq. (3). Based on this observation we define the components of H by

$$
H_i(x) = \int_{a_i^k}^{x_i} h_i(x_1, \ldots, x_{i-1}, s, x_{i+1} \ldots, x_n) \, ds.
$$

It follows that $\|H_i\|_{L^\infty} \leq 1$ and $\text{supp}(H_i) \subset \Omega$. H is now modified into a vector field belonging to $\mathcal{B}_1$. Let $\{\rho_j\}_{j\in\mathbb{N}}$ denote a sequence of mollifiers on $\mathbb{R}^n$ supported on the closed Euclidean ball centred at 0 with radius $1/j$. Recalling standard results regarding convolution and mollifiers, we derive $\|H_i * \rho_j\|_{L^\infty} \leq \|H_i\|_{L^\infty}\|\rho_j\|_{L^1} = \|H_i\|_{L^\infty} \leq 1$ and moreover, for j large enough, $H_i * \rho_j \in C_c^\infty(\Omega)$. Hence, for $j \in \mathbb{N}$ large enough, the modification H_{ρ_j} of H given by

$$H_{\rho_j} = (H_1 * \rho_j, \ldots, H_n * \rho_j)$$

is in $\mathcal{B}_1$. It follows that $h \in \overline{\text{div}\,\mathcal{B}_1}$, as

$$\left\|\text{div}\,H_{\rho_j} - h\right\|_{L^2} = \left\|\sum_{i=1}^n \left(\frac{\partial}{\partial x_i}(H_i * \rho_j) - h_i\right)\right\|_{L^2} = \left\|\sum_{i=1}^n (h_i * \rho_j - h_i)\right\|_{L^2}$$
$$\leq \sum_{i=1}^n \left\|h_i * \rho_j - h_i\right\|_{L^2} \xrightarrow{j\to\infty} 0.$$

The element $h \in \Gamma_c$ was chosen arbitrarily and $\overline{\text{div}\,\mathcal{B}_1}$ is closed, therefore $\overline{\Gamma_c} \subset \overline{\text{div}\,\mathcal{B}_1}$. □

3.1   Preservation of Piecewise Constancy

We are now ready to prove the following result.

Theorem 2. *Given $f \in PCR$ with minimal grid G_f, the minimizer u_α of the corresponding anisotropic ROF functional*

$$\min_{u\in L^2(\Omega)} \frac{1}{2}\|u - f\|_{L^2}^2 + \alpha J(u)$$

lies in PCR_{G_f}.

Proof. Recall that, according to Lemma 1, u_α is the unique element with minimal L^2-norm in $f - \alpha\partial J(0)$. From Theorem 1 and the fact that $A_{G_f}f = f$ it follows that also $A_{G_f}u_\alpha \in f - \alpha\partial J(0)$. As $\|A_{G_f}u_\alpha\|_{L^2} \leq \|u_\alpha\|_{L^2}$, because of Remark 2, we have $A_{G_f}u_\alpha = u_\alpha$. Therefore, $u_\alpha \in PCR_{G_f}$. □

4   Conclusion

In [11, Thm. 5] the authors have shown that, for Ω being a rectilinear 2-polytope, $f \in PCR$ implies $u_\alpha \in PCR$. We have extended this preservation of piecewise constancy to rectilinear n-polytopes. Our proof can be summarized in the following way

$$\|A_{G_f}u_\alpha\|_{L^2} \overset{\text{Lem. 2}}{\leq} \|u_\alpha\|_{L^2} \overset{\text{Lem. 1}}{=} \min_{u\in f-\alpha\partial J(0)} \|u\|_{L^2} \overset{\text{Thm. 1}}{\leq} \|A_{G_f}u_\alpha\|_{L^2}.$$

The crucial step is Theorem 1, asserting that

$$A_G(\partial J(0)) \subset \partial J(0),$$

which exploits the fact that the anisotropy of J is compatible with the rectilinearity of the grid G.

Acknowledgements. We acknowledge support by the Austrian Science Fund (FWF) within the national research network "Geometry + Simulation," S117, subproject 4.

References

1. Amar, M., Bellettini, G.: A notion of total variation depending on a metric with discontinuous coefficients. Ann. Inst. H. Poincaré Anal. Non Linéaire **11**(1), 91–133 (1994)
2. Berkels, B., Burger, M., Droske, M., Nemitz, O., Rumpf, M.: Cartoon extraction based on anisotropic image classification. In: Vision, Modeling, and Visualization Proceedings, pp. 293–300 (2006)
3. Casas, E., Kunisch, K., Pola, C.: Regularization by functions of bounded variation and applications to image enhancement. Appl. Math. Optim. **40**(2), 229–257 (1999)
4. Chambolle, A.: Total variation minimization and a class of binary MRF models. In: Rangarajan, A., Vemuri, B., Yuille, A.L. (eds.) EMMCVPR 2005. LNCS, vol. 3757, pp. 136–152. Springer, Heidelberg (2005). https://doi.org/10.1007/11585978_10
5. Chambolle, A., Darbon, J.: On total variation minimization and surface evolution using parametric maximum flows. Int. J. Comput. Vis. **84**(3), 288–307 (2009)
6. Chen, H., Wang, C., Song, Y., Li, Z.: Split Bregmanized anisotropic total variation model for image deblurring. J. Vis. Commun. Image Represent. **31**, 282–293 (2015)
7. Choksi, R., van Gennip, Y., Oberman, A.: Anisotropic total variation regularized L^1 approximation and denoising/deblurring of 2D bar codes. Inverse Prob. Imaging **5**(3), 591–617 (2011)
8. Darbon, J., Sigelle, M.: Image restoration with discrete constrained total variation. Part I: fast and exact optimization. J. Math. Imaging Vis. **26**(3), 261–276 (2006)
9. Ekeland, I., Temam, R.: Convex Analysis and Variational Problems. North-Holland, Amsterdam (1976)
10. Goldstein, T., Osher, S.: The split Bregman method for L1-regularized problems. SIAM J. Imaging Sci. **2**, 323–343 (2009)
11. Łasica, M., Moll, S., Mucha, P.B.: Total variation denoising in ℓ^1 anisotropy. SIAM J. Imaging Sci. **10**, 1691–1723 (2017)
12. Li, Y., Santosa, F.: A computational algorithm for minimizing total variation in image restoration. IEEE Trans. Image Process. **5**, 987–995 (1996)
13. Rudin, L., Osher, S., Fatemi, E.: Nonlinear total variation based noise removal algorithms. Physica D **60**, 259–268 (1992)
14. Sanabria, S.J., Ozkan, E., Rominger, M., Goksel, O.: Spatial domain reconstruction for imaging speed-of-sound with pulse-echo ultrasound: simulation and in vivo study. Phys. Med. Biol. **63**(21), 215015 (2018)
15. Setzer, S., Steidl, G., Teuber, T.: Restoration of images with rotated shapes. Numer. Algorithms **48**(1–3), 49–66 (2008)

Total Directional Variation for Video Denoising

Simone Parisotto[1(✉)] and Carola-Bibiane Schönlieb[2]

[1] CCA, University of Cambridge, Wilberforce Road, Cambridge CB3 0WA, UK
sp751@cam.ac.uk
[2] DAMTP, University of Cambridge, Wilberforce Road, Cambridge CB3 0WA, UK
cbs31@cam.ac.uk

Abstract. In this paper we propose a variational approach for video denoising, based on a total directional variation (TDV) regulariser proposed in [20,21] for image denoising and interpolation. In the TDV regulariser, the underlying image structure is encoded by means of weighted derivatives so as to enhance the anisotropic structures in images, e.g. stripes or curves with a dominant local directionality. For the extension of TDV to video denoising, the space-time structure is captured by the volumetric structure tensor guiding the smoothing process. We discuss this and present our whole video denoising workflow. The numerical results are compared with some state-of-the-art video denoising methods.

Keywords: Total directional variation · Video denoising ·
Anisotropy · Structure tensor · Variational methods

1 Introduction

Video denoising refers to the task of removing noise in digital videos. Compared to image denoising, video denoising is usually a more challenging task due to the computational cost in processing large data and the redundancy of information, i.e. the expected similarity between two consecutive frames that should be inherited by the denoised video. A straightforward approach to video denoising is to denoise each frame of the video independently, by using the broad literature on image denoising methods, see e.g. [3,4,6,11,12,15,19,21,23,24,26]. Computational cost is then stratified across image frames by sequentially processing them, which is seen as an advantage. However, a significant disadvantage of this frame-by-frame processing is the appearance of flickering artefacts and post-processing motion compensation step may be required [2,18].

SP acknowledges UK EPSRC grant EP/L016516/1 for the CCA DTC. CBS acknowledges support from Leverhulme Trust project on Breaking the non-convexity barrier, EPSRC grant Nr. EP/M00483X/1, the EPSRC Centre EP/N014588/1, the RISE projects CHiPS and NoMADS, the CCIMI and the Alan Turing Institute.

© Springer Nature Switzerland AG 2019
J. Lellmann et al. (Eds.): SSVM 2019, LNCS 11603, pp. 522–534, 2019.
https://doi.org/10.1007/978-3-030-22368-7_41

In recent years different approaches have been proposed for solving the video denoising problem: we refer to the introduction of [1] for an extensive survey. Notably, patch-based approaches are usually considered among the most promising video denoising methods in that they are able to achieve qualitatively good denoising results. For example, V-BM3D is the 3D extension of the BM3D collaborative filters [10]: without inspecting the motion time-consistency, V-BM3D independently filters 2D patches resulting similar in the 3D spatio-temporal neighbourhood domain. As mentioned in [1], while generally receiving good denoising results, the problem of flickering still occurs in V-BM3D. For this reason, authors of V-BM3D developed an extension, called V-BM4D, where the patch-similarity is explored along space-temporal blocks defined by a motion vector, see [17]. Similarly, in [5] the authors propose to group patches via an optical flow equation based on [28] and implemented in [25]. In these approaches, while the incorporation of motion helps to provide consistency in time, denoising results also suffer from the lack of accuracy in the estimated motion. A possible way to avoid the motion estimation is to consider 3D rectangular patches so as to inherently model the 3D structure and motion in the spatio-temporal video dimensions, based on the fact that rectangular 3D patches are less repeatable than motion-compensated patches. However, such approach is not efficient for uniform motion or homogeneous spatial patterns, cf. the discussion on this topic in [1]. Motivated by this reasoning the authors of [1] introduce a Bayesian patch-based video denoising approach with rectangular 3D patches modelled as independent and identically distributed samples from an unknown *a priori* distribution: then each patch is denoised by minimising the expected mean square error. Other approaches in video denoising are the straightforward extension of the Rudin-Osher-Fatemi (ROF) model [24] to 3D data, by using a spatio-temporal total variation (referred in the next as ROF 2D+t), the joint video denoising with the computation of the flow [7] and CNN approaches [13].

Scope of the Paper. In this paper we propose an extension of the recently introduced *total directional variation* (TDV) regulariser [20,21] for video denoising, via the following variational regularisation model:

$$u^\star \in \arg\min_u \left(\text{TDV}(u, \mathbf{M}) + \frac{\eta}{2} \|u - u^\circ\|_2^2 \right), \tag{1}$$

where $u^\star$ is the denoised video, $\mathbf{M}$ is a weighting field that encodes directional features in two spatial and one temporal dimension, $\eta > 0$ is the regularisation parameter and u° is a given noisy video. The model (1) will be made more precise in the next sections where we mainly focus on its discrete and numerical aspects. In order to accommodate for spatial-temporal data, we consider here a modification of the TDV regulariser given in [20,21] that derives directionality in the temporal dimension. Differently from the patch-based approach, we compute for each voxel the vector field of the motion, to be encoded as a weight in the TDV regulariser. With this voxel-based approach we will reduce the flickering artefact which appears in patch-based approaches due to the patch selection, especially in regions of smooth motion. Results are presented for a variety of videos corrupted with Gaussian white noise.

Organisation of the Paper. This paper is organised as follows: in Sect. 2 we describe the estimation of the vector fields, the TDV regulariser and the variational model to be minimised; in Sect. 3 we describe the optimisation method for solving the TDV video denoising problem and comment on the selection of parameters; in Sect. 4 we show denoising results on a selection of videos corrupted with Gaussian noise of varying strength.

2   Total Directional Variation for Video Denoising

Let $\bar{u} : \Omega \times [1,\dots,T] \rightarrow \mathbb{R}_+^C$ be a clean video and Ω a spatial, rectangular domain indexed by $\boldsymbol{x} = (x,y)$, with number of T frames and C colours. Let $u^\diamond$ be a corrupted version of $\bar{u}$ in each space-time voxel $(\boldsymbol{x},t) \in \Omega \times [1,\dots,T]$ by i.i.d. Gaussian noise n of zero mean and (possibly known) variance $\varsigma^2 > 0$:

$$u^\diamond(\boldsymbol{x},t) = \bar{u}(\boldsymbol{x},t) + n(\boldsymbol{x},t), \quad \forall(\boldsymbol{x},t) \in \Omega \times [1,\dots,T]. \tag{2}$$

In what follows, we propose to compute a denoised video $u^\star \approx \bar{u}$ by solving

$$u^\star \in \arg\min_u \left(\text{TDV}(u,\mathbf{M}) + \frac{\eta}{2} \|u - u^\diamond\|_2^2 \right), \tag{3}$$

where $\text{TDV}(u,\mathbf{M})$ is the proposed total direction regulariser w.r.t. a weighting field $\mathbf{M}$, both specified in the next sections, and $\eta > 0$ a regularisation parameter.

2.1   The Directional Information

In order to capture directional information of u in (3), we eigen-decompose the two-dimensional structure tensor [27] in each coordinate plane.

To do so, we first construct the 3D structure tensor: let $\rho \geq \sigma > 0$ be two smoothing parameters, K_σ, K_ρ be the Gaussian kernels of standard deviation σ and ρ, respectively, and let $u_\sigma = K_\sigma * u$. Then the 3D structure tensor reads as

$$\mathbf{S} := K_\rho * (\boldsymbol{\nabla} u_\sigma \otimes \boldsymbol{\nabla} u_\sigma) = \begin{pmatrix} u_{\sigma,\rho}^{x,x} & u_{\sigma,\rho}^{x,y} & u_{\sigma,\rho}^{x,t} \\ u_{\sigma,\rho}^{y,x} & u_{\sigma,\rho}^{y,y} & u_{\sigma,\rho}^{y,t} \\ u_{\sigma,\rho}^{t,x} & u_{\sigma,\rho}^{t,y} & u_{\sigma,\rho}^{t,t} \end{pmatrix}, \tag{4}$$

where $\boldsymbol{\nabla} u_\sigma \otimes \boldsymbol{\nabla} u_\sigma = \boldsymbol{\nabla} u_\sigma \boldsymbol{\nabla} u_\sigma^T$, $u_{\sigma,\rho}^{p,q} := K_\rho * (\partial_p u_\sigma \otimes \partial_q u_\sigma)$ for each $p,q \in \{x,y,t\}$.

For a straightforward application to the TDV regulariser in [21], we extract the 2D sub-tensors of (4), whose eigen-decomposition encodes structural information in each of the coordinate frames spanned by $\{x,y\}$, $\{x,t\}$ and $\{y,t\}$:

$$\text{on coordinates } \{x,y\}: \quad \mathbf{S}^{x,y} = \begin{pmatrix} u_{\sigma,\rho}^{x,x} & u_{\sigma,\rho}^{x,y} \\ u_{\sigma,\rho}^{y,x} & u_{\sigma,\rho}^{y,y} \end{pmatrix} = \lambda_1(\boldsymbol{e}_1 \otimes \boldsymbol{e}_1) + \lambda_2(\boldsymbol{e}_2 \otimes \boldsymbol{e}_2);$$

$$\text{on coordinates } \{x,t\}: \quad \mathbf{S}^{x,t} = \begin{pmatrix} u_{\sigma,\rho}^{x,x} & u_{\sigma,\rho}^{x,t} \\ u_{\sigma,\rho}^{t,x} & u_{\sigma,\rho}^{t,t} \end{pmatrix} = \lambda_3(\boldsymbol{e}_3 \otimes \boldsymbol{e}_3) + \lambda_4(\boldsymbol{e}_4 \otimes \boldsymbol{e}_4); \tag{5}$$

$$\text{on coordinates } \{y,t\}: \quad \mathbf{S}^{y,t} = \begin{pmatrix} u_{\sigma,\rho}^{y,y} & u_{\sigma,\rho}^{y,t} \\ u_{\sigma,\rho}^{t,y} & u_{\sigma,\rho}^{t,t} \end{pmatrix} = \lambda_5(\boldsymbol{e}_5 \otimes \boldsymbol{e}_5) + \lambda_6(\boldsymbol{e}_6 \otimes \boldsymbol{e}_6).$$

For each $s \in \{1,\ldots,6\}$, the eigenvector $\boldsymbol{e}_s = (e_{s,1}, e_{s,2})$ has eigenvalue λ_s. The tangential directions in the 2D planes $\{x,y\}, \{x,t\}$ and $\{y,t\}$ are $\boldsymbol{e}_2, \boldsymbol{e}_4, \boldsymbol{e}_6$, respectively, with $\boldsymbol{e}_1, \boldsymbol{e}_3, \boldsymbol{e}_5$ the gradient directions, see Fig. 1.

From (5), the ratios between the eigenvalues, called *confidence*, measure the local anisotropy of the gradient on the slices within a certain neighbourhood:

$$a^{x,y} = \frac{\lambda_2}{\lambda_1 + \varepsilon}, \quad a^{x,t} = \frac{\lambda_4}{\lambda_3 + \varepsilon}, \quad a^{y,t} = \frac{\lambda_6}{\lambda_5 + \varepsilon}, \quad \text{with} \quad \varepsilon > 0. \tag{6}$$

Here, $a^{x,y}, a^{x,t}, a^{y,t} \in [0,1]$ and the closer to 0, the higher is the local anisotropy.

Fig. 1. Left: grey-scale video `xylophone.mp4`, corrupted by Gaussian noise ($\varsigma = 20$); right: streamlines of the weighting field with $\boldsymbol{e}_2$ (blue), $\boldsymbol{e}_4$ (red) and $\boldsymbol{e}_6$ (yellow). (Color figure online)

2.2   The Regulariser

The TDV regulariser is composed of a gradient operator weighted by a tensor $\mathbf{M}$, whose purpose is to smooth along selected directions. In view of the spatial-temporal data, we extend the natural gradient operator to the Cartesian planes $\{x,y\}, \{x,t\}$ and $\{y,t\}$. We will denote with $\widetilde{\boldsymbol{\nabla}}$ the concatenation of resulting 2-dimensional gradients. Further, we encode (5) and (6) in $\mathbf{M}$, leading to the *weighted gradient* $\mathbf{M}\widetilde{\boldsymbol{\nabla}}$ for the video function $u = u(x,y,t)$:

$$\mathbf{M}\widetilde{\boldsymbol{\nabla}} \otimes u = \underbrace{\begin{pmatrix} a^{x,y} & 0 & 0 & 0 & 0 & 0 \\ 0 & 1 & 0 & 0 & 0 & 0 \\ 0 & 0 & a^{x,t} & 0 & 0 & 0 \\ 0 & 0 & 0 & 1 & 0 & 0 \\ 0 & 0 & 0 & 0 & a^{y,t} & 0 \\ 0 & 0 & 0 & 0 & 0 & 1 \end{pmatrix}}_{\mathbf{M}} \underbrace{\begin{pmatrix} e_{1,1} & e_{1,2} & 0 & 0 & 0 & 0 \\ e_{2,1} & e_{2,2} & 0 & 0 & 0 & 0 \\ 0 & 0 & e_{3,1} & e_{3,2} & 0 & 0 \\ 0 & 0 & e_{4,1} & e_{4,2} & 0 & 0 \\ 0 & 0 & 0 & 0 & e_{5,1} & e_{5,2} \\ 0 & 0 & 0 & 0 & e_{6,1} & e_{6,2} \end{pmatrix} \begin{pmatrix} \partial_x \\ \partial_y \\ \partial_x \\ \partial_t \\ \partial_y \\ \partial_t \end{pmatrix}}_{\widetilde{\boldsymbol{\nabla}}} \otimes u \tag{7}$$

$$= \left(a^{x,y} \boldsymbol{\nabla}_{\boldsymbol{e}_1}^{x,y} u, \ \boldsymbol{\nabla}_{\boldsymbol{e}_2}^{x,y} u, \ a^{x,t} \boldsymbol{\nabla}_{\boldsymbol{e}_3}^{x,t} u, \ \boldsymbol{\nabla}_{\boldsymbol{e}_4}^{x,t} u, \ a^{y,t} \boldsymbol{\nabla}_{\boldsymbol{e}_5}^{y,t} u, \ \boldsymbol{\nabla}_{\boldsymbol{e}_6}^{y,t} u \right)^{\mathrm{T}}. \tag{8}$$

Note that $\mathbf{M}$ is computed once from the noisy input $u^\diamond$. For a fixed frame $\{p,q\}$ with $p,q \in \{x,y,t\}$ and direction $\boldsymbol{z} = (z_1, z_2)$ the gradient $\boldsymbol{\nabla}_{\boldsymbol{z}}^{p,q} u =$

$\partial_p u \cdot z_1 + \partial_q u \cdot z_2$ is the directional derivative of u along $\boldsymbol{z}$ w.r.t. the frame $\{p, q\}$. See [22, Fig. 3.12] for more details about this choice. With this notation in place, we consider the *total directional variation* (TDV) regulariser,

$$\text{TDV}(u, \mathbf{M}) = \sup_{\boldsymbol{\Psi}} \left\{ \int_{\Omega} (\mathbf{M}\widetilde{\boldsymbol{\nabla}} \otimes u) \cdot \boldsymbol{\Psi} \, d\boldsymbol{x} \, \middle| \, \text{for all suitable test functions } \boldsymbol{\Psi} \right\}. \tag{9}$$

By plugging (8) into (9) we reinterpret (9) as a penalisation of the rate of change along $\boldsymbol{e}_2, \boldsymbol{e}_4, \boldsymbol{e}_6$, with coefficients $a^{x,y}, a^{x,t}, a^{y,t}$ as bias in the gradient estimation. Note, that while in [21] the TDV regulariser has been proposed for a general order of derivatives, we consider here only a TDV regulariser of first differential order.

2.3   Connections to Optical Flow

Let $(x, y, t) \in \Omega \times [1, \dots T]$ be a voxel and $u(x, y, t)$ its intensity in the grey-scale video sequence u. If $u(x, y, t)$ is moved by a small increment $(\delta_x, \delta_y, \delta_t)$ between two frames, then the *brightness constancy* constraint reads

$$u(x, y, t) = u(x + \delta_x, y + \delta_y, t + \delta_t). \tag{10}$$

If u is sufficiently smooth, then the *optical flow* constraint is derived [14,16] as a linearisation of (10) with respect to a velocity field $\boldsymbol{z}$:

$$\boldsymbol{\nabla} u(x, y, t)^T \cdot \boldsymbol{z} = 0, \quad \text{for all } (x, y, t) \in \Omega \times [1, T]. \tag{11}$$

For a specific field $\boldsymbol{z} = (\widetilde{\boldsymbol{z}}, 1)$ with $\widetilde{\boldsymbol{z}} = (z_1(x, y), z_2(x, y))$, Eq. (11) is equivalent to

$$- \partial_t u = \partial_x u \cdot z_1 + \partial_y u \cdot z_2 = \boldsymbol{\nabla}_{\widetilde{\boldsymbol{z}}}^{x,y} u \quad \text{for all } (x, y, t) \in \Omega \times [1, T]. \tag{12}$$

We can now re-write (11) by means of the following velocity vector fields:

$$\begin{array}{ccc} a^{x,y}(e_{1,1}, e_{1,2}, 1), & a^{x,t}(e_{3,1}, 1, e_{3,2}), & a^{y,t}(1, e_{5,1}, e_{5,2}), \\ (e_{2,1}, e_{2,2}, 1), & (e_{4,1}, 1, e_{4,2}), & (1, e_{6,1}, e_{6,2}), \end{array} \tag{13}$$

leading to

$$\begin{pmatrix} -a^{x,y}\partial_t u \\ -\partial_t u \\ -a^{x,t}\partial_y u \\ -\partial_y u \\ -a^{y,t}\partial_x u \\ -\partial_x u \end{pmatrix} = \begin{pmatrix} a^{x,y}\partial_x u \cdot e_{1,1} + a^{x,y}\partial_y u \cdot e_{1,2} \\ \partial_x u \cdot e_{2,1} + \partial_y u \cdot e_{2,2} \\ a^{x,t}\partial_x u \cdot e_{3,1} + a^{x,t}\partial_t u \cdot e_{3,2} \\ \partial_x u \cdot e_{4,1} + \partial_t u \cdot e_{4,2} \\ a^{y,t}\partial_y u \cdot e_{5,1} + a^{y,t}\partial_t u \cdot e_{5,2} \\ \partial_y u \cdot e_{6,1} + \partial_t u \cdot e_{6,2} \end{pmatrix} = \begin{pmatrix} a^{x,y}\boldsymbol{\nabla}_{e_1}^{x,y} u \\ \boldsymbol{\nabla}_{e_2}^{x,y} u \\ a^{x,t}\boldsymbol{\nabla}_{e_3}^{x,t} u \\ \boldsymbol{\nabla}_{e_4}^{x,t} u \\ a^{y,t}\boldsymbol{\nabla}_{e_5}^{y,t} u \\ \boldsymbol{\nabla}_{e_6}^{y,t} u \end{pmatrix}. \tag{14}$$

Here, the right-hand side of (14) encodes the components that we aim to penalise in (8). Thus, the penalisation of (8) is equivalent to the penalisation of the left-hand side of (14), assumed (11) holds with velocity fields in (13). Note that the weights $a^{x,y}, a^{x,t}$ and $a^{y,t}$ add a contribution in the direction of the gradients $\boldsymbol{e}_1, \boldsymbol{e}_3, \boldsymbol{e}_5$, respectively.

2.4   The Minimisation Problem

We aim to find the denoised video $u^\star$ from the noisy input video u° by solving the $\mathrm{TDV} - \mathrm{L}^2$ minimisation problem (3). For the numerical optimisation of (3) we use a primal-dual scheme [9]. For this, we rewrite (3) as a saddle point problem for the operator $\mathcal{K} := \mathbf{M}\tilde{\nabla}$, whose adjoint will be denoted by $\mathcal{K}^*$. In what follows, we denote by u the primal variable, y the dual variable, f^* the Fenchel conjugate of f, by g the fidelity term and by $\sigma, \tau > 0$ the dedicated parameters of the primal-dual algorithm, see [9] for more details on the primal-dual schemes in image processing and [21] for their application to variational problems with TDV regulariser. The resulting saddle-point problem reads

$$u^\star \in \arg\min_u \max_y \left(\langle \mathcal{K}u, y \rangle - \underbrace{\delta_{\{\|\cdot\|_{2,\infty} \leq 1\}}(y)}_{f^*(y)} + \underbrace{\frac{\eta}{2}\|u - u^\circ\|_2^2}_{g(u^\circ)} \right). \tag{15}$$

In the primal-dual algorithm solving (15) we need the proximal operators:

$$\mathbf{prox}_{\sigma f^*}(y) = \frac{y}{\max\{1, \|y\|_2\}}, \quad \mathbf{prox}_{\tau g}(u) = u + (\mathbf{I} + \tau\eta)^{-1}\tau\eta(u^\circ - u), \tag{16}$$

where $\mathbf{I}$ is the identity matrix. Note that g is uniformly convex, with convexity parameter η, so the dual problem is smooth. An accelerated version of the primal-dual algorithm can be used in this case, e.g. [8, Alg. 2], starting with $\tau_0, \sigma_0 > 0$ where $\tau_0\sigma_0 L^2 \leq 1$ and L^2 is the squared operator norm, $L^2 := \|\mathcal{K}\|^2 \leq 24$ (which holds in connection with the discretisation in (17) and stepsize $h = 1$).

3   The Discrete Model

In the discrete model, Ω is a rectangular grid of size $M \times N$ and a video $\boldsymbol{u}$ is a volumetric data of size $M \times N \times C \times T$ (height $\times$ width $\times$ colours $\times$ frames). Here, we consider grey-scale videos ($C = 1$) along the axes $(i, j, k) \in \Omega \times [1, T]$, with $i = 1, \ldots, M$, $j = 1, \ldots, N$ and $k = 1, \ldots, T$. An extension to coloured videos is straightforward by processing each colour channel separately. Here, a fixed $(i, j, k) \in \Omega \times [1, T]$ identifies a voxel in the gridded video domain, i.e. a small cube of size h in each axis direction. Then, $\boldsymbol{u}_{i,j,k} := \boldsymbol{u}(i, j, k)$ is the intensity in the voxel (i, j, k) in the grey-scale video sequence $\boldsymbol{u}$. The noisy input video is denoted by $\boldsymbol{u}^\circ$ as well as the other discrete vectorial quantities, namely $\boldsymbol{a}^{1,2}, \boldsymbol{a}^{1,3}, \boldsymbol{a}^{2,3}$ and $\boldsymbol{\lambda}_s$ for $s = 1, \ldots, 6$.

3.1   Discretisation of Derivative Operators and Vector Fields

We describe a finite difference scheme on the voxels by introducing the discrete gradient operator $\nabla : \mathbb{R}^{M \times N \times T} \to \mathbb{R}^{M \times N \times T \times 3}$, with $\nabla = (\partial_1, \partial_2, \partial_3)$ defined via the central finite differences on half step-size and Neumann conditions as

$$(\boldsymbol{u}^1)_{i,j,k} := (\partial_1 \boldsymbol{u})_{i+0.5,j,k} = \begin{cases} \dfrac{\boldsymbol{u}_{i+1,j,k} - \boldsymbol{u}_{i,j,k}}{h}, & \text{if } i = 1,\ldots,M-1, \\ 0 & \text{if } i = M; \end{cases}$$

$$(\boldsymbol{u}^2)_{i,j,k} := (\partial_2 \boldsymbol{u})_{i,j+0.5,k} = \begin{cases} \dfrac{\boldsymbol{u}_{i,j+1,k} - \boldsymbol{u}_{i,j,k}}{h}, & \text{if } j = 1,\ldots,N-1, \\ 0 & \text{if } j = N; \end{cases} \tag{17}$$

$$(\boldsymbol{u}^3)_{i,j,k} := (\partial_3 \boldsymbol{u})_{i,j,k+0.5} = \begin{cases} \dfrac{\boldsymbol{u}_{i,j,k+1} - \boldsymbol{u}_{i,j,k}}{h}, & \text{if } k = 1,\ldots,T-1, \\ 0 & \text{if } k = T. \end{cases}$$

Remark 1. While $\boldsymbol{u}$ lies at the vertices of the discrete grid, $\nabla \boldsymbol{u}$ lies on its edges. Thus, (17) is advantageous for local anisotropy since it has sub-pixel precision and a more compact stencil radius than the classical forward scheme.

In (7), $\widetilde{\nabla} : \mathbb{R}^{M \times N \times T} \to \mathbb{R}^{M \times N \times T \times 6}$ acts on $\boldsymbol{u}$ as follows:

$$\left(\widetilde{\nabla} \otimes \boldsymbol{u}\right)_{i,j,k} := \left(\boldsymbol{u}^1, \boldsymbol{u}^2, \boldsymbol{u}^1, \boldsymbol{u}^3, \boldsymbol{u}^2, \boldsymbol{u}^3\right)^{\mathrm{T}}_{i,j,k}. \tag{18}$$

Algorithm 1. TDV for video denoising

Input: A grey-scale video $\boldsymbol{u}^\diamond \in [0,255]$ ($M \times N \times 1 \times T$), $\varsigma \in [0,255]$.
Output   : the denoised video $\boldsymbol{u}$;
Parameters: for the primal-dual maxiter, tol; for the variational model: (σ, ρ, η).

Function TDV_video_denoising:

> // Compute operators for the weighted derivative
> $[\partial_1, \partial_2, \partial_3]$ = compute_derivative_operator (M,N,T) ;
> $\mathbf{S}$ = compute_3D_structure_tensor $(\boldsymbol{u}^\diamond, \sigma, \rho)$;
> $[e_1, e_2, e_3, e_4, e_5, e_6, \lambda_1, \lambda_2, \lambda_3, \lambda_4, \lambda_5, \lambda_6]$ = eigendecomposition $(\mathbf{S})$;
> $[a^{1,2}, a^{1,3}, a^{2,3}]$ = compute_anisotropy $(\lambda_1, \lambda_2, \lambda_3, \lambda_4, \lambda_5, \lambda_6)$;
> $\mathbf{M}$ = compute_weights $(a^{1,2}, a^{1,3}, a^{2,3}, e_1, e_2, e_3, e_4, e_5, e_6)$;
>
> // Proximal operators, adjoints and primal-dual from [8][8]
> $[\mathcal{K}, \mathcal{K}^*]$ = compute_K_and_adjoint $(\mathbf{M}, \partial_1, \partial_2, \partial_3)$;
> prox$_{f^*}$ = @$(\boldsymbol{y})$ $\boldsymbol{y}./\max\{1, \|\boldsymbol{y}\|_2\}$;
> prox$_g$ = @$(\boldsymbol{u}, \tau)$ $\boldsymbol{u} + (\mathbf{I} + \tau\eta)^{-1}\tau\eta(\boldsymbol{u}^\diamond - \boldsymbol{u})$;
> $\boldsymbol{u}$ = primal_dual $(\boldsymbol{u}^\diamond, \mathcal{K}, \mathcal{K}^*, \text{prox}_{f^*}, \text{prox}_g, \text{maxiter}, \text{tol})$;

return

Any field $\boldsymbol{e}_s$ with $s = 1,\ldots,6$ and confidence $a^{1,2}$, $a^{1,3}$, $a^{2,3}$ will be discretised in the cell centres $(i+0.5, j+0.5, k+0.5)$ of the discrete grid domain. The weighting multiplication in (7) is performed via an intermediate averaging interpolation operator $\mathcal{W} : \mathbb{R}^{M \times N \times T \times 6} \to \mathbb{R}^{(M-1) \times (N-1) \times (T-1) \times 6}$ that avoids artefacts due to the grid offset: this gives $\mathbf{M}\mathcal{W}\widetilde{\nabla} : \mathbb{R}^{M \times N \times T} \to \mathbb{R}^{(M-1) \times (N-1) \times (T-1) \times 6}$.

3.2   TDV for Video Denoising

The TDV-based workflow consists of two steps, with pseudo-code in Algorithm 1. The first one computes the directions via the eigen-decomposition in (5) while the second one is the primal-dual algorithm [8, Alg. 2], whose stopping criterion is the root mean square difference between two consecutive dual variable iterates.

4   Results

In this section we discuss the numerical results for video denoising obtained with Algorithm 1. Considered videos have been taken from a benchmark video dataset[1,2]. Each video has values in $[0, 255]$ corrupted with Gaussian noise. We tested different noise levels with standard deviation $\varsigma = [10, 20, 35, 50, 70, 90]$ without clipping the videos so as to conform to the observation model.

The quality of the denoised result $\boldsymbol{u}^\star$ is evaluated by the peak signal-to-noise ratio (PSNR) value w.r.t. a ground truth video $\overline{\boldsymbol{u}}$. The model requires the parameters (σ, ρ, η) as input. Once provided, we solve the saddle-point minimisation problem in (15) via the accelerated primal-dual algorithm with $L^2 = \|\mathcal{K}\|^2 = 24$, see [8, Alg. 2]. Here the tolerance for the stopping criterion is fixed to 10^{-4} (on average reached in 300 iterations). However, we experienced faster convergence and similar results with $L^2 \ll 24$ and bigger tolerances, e.g. 10^{-3}.

4.1   Selection of Parameters

In the model, $\boldsymbol{u}^\star$ is sensitive to the choice of both (σ, ρ) for the vector fields, and the regularisation parameter η that is chosen according to the noise level. Choosing those parameters by a trial and error approach is computationally expensive and the best parameters may differ, even for videos with the same noise level. In particular, the parameters (σ, ρ) depend on structure in the data, e.g. flat regions versus motions versus small details. Therefore, a strategy for tuning them is needed.

To estimate appropriate values for (σ, ρ, η) that render good results for a variety of videos we compute optimal parameters via line-search for maximising the PSNR for a small selection of video denoising examples for which the ground truth is available. The result of this optimisation is given in Table 1. For the line-search the parameters for the maximal PSNR values are computed iteratively, by applying Algorithm 1 for two different choices of (σ, ρ, η) at a time, and subsequently adapt this parameter-set for the next iteration towards the ones in the neighbourhood of the one that returns a larger PSNR. In this search we constrain

---

[1] Videos are freely available: *Salesman* and *Miss America* at www.cs.tut.fi/~foi/GCF-BM3D *Xylophone* in MATLAB; *Water* (re-scaled, grey-scaled and clipped, Jay Miller, CC 3.0) at www.videvo.net/video/water-drop/477; *Franke*'s function (a synthetic surface moving on fixed trajectories: the coloured one changes with the `parula` colormap).

[2] Results are available at http://www.simoneparisotto.com/TDV4videodenoising.

$\sigma \le \rho$ [27]. The line-search is stopped when, for the currently best parameters (σ, ρ, η) all the other neighbours in a certain radius of distance report an inferior PSNR value. In Fig. 2 we show the trajectory of the parameters during this line-search for the *Franke* video corrupted with Gaussian noise with $\varsigma = 10$. We observe that there exists a range of parameters in which the PSNR values are almost the same.

By looking at the estimated parameters from the line-search approach in Table 1, we suggest the following rule of thumb for their selection in Algorithm 1:

$$\sigma = \rho = 3.2\eta^{-0.5} \quad \text{and} \quad \eta = 255\varsigma^{-1}. \tag{19}$$

4.2   Numerical Results

For the so-found optimal parameters we compare in Table 1 the PSNR values achieved for our approach (TDV) with patch-based filters (V-BM3D v2.0 and V-BM4D v1.0, default parameters and *normal-complexity profile*).

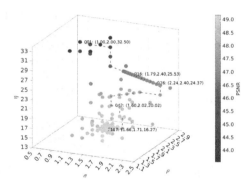

Fig. 2. Line-search (*Franke*, $\varsigma = 10$): optimal trajectory (dashed red line); PSNR values (coloured bullets). Optimal PSNR: 49.16, 117th iteration, $(\sigma, \rho, \eta) = (1.66, 1.71, 16.27)$. (Color figure online)

In Figs. 3 and 4 the visual comparison is shown for selected frames of the *Franke* and *Water* videos (corrupted by noise with $\varsigma = 70$). The time-consistency achieved by our approach is apparent in the frame-by-frame PSNR comparison.

Video denoising results that use the quasi-optimal parameters computed with (19) are reported in Table 2: selected frames of videos corrupted with a high noise level of $\varsigma = 90$ are shown in Figs. 5 and 6, with frame-by-frame PSNR values.

4.3   Discussion of Results

We compared our variational TDV denoising approach with patch-based (V-BM3D/V-BM4D) and variational (ROF 2D+t) methods. Patch-based methods

Table 1. PSNR comparison (best in bold), with TDV parameters from line-search.

Name (M, N, C, T)	ς	Input	V-BM3D	V-BM4D	TDV (σ, ρ, η)	ROF 2D+t (η)
Franke grey-scale (120, 120, 1, 120)	10	28.13	45.99	46.90	**49.16** (1.66, 1.71, 16.27)	42.56 (16.27)
	20	22.11	41.64	42.67	**45.23** (2.00, 2.00, 08.10)	38.18 (08.10)
	35	17.25	38.63	39.34	**41.89** (2.40, 2.40, 04.70)	34.59 (04.70)
	50	14.15	36.37	37.17	**39.64** (2.70, 2.70, 03.30)	32.30 (03.30)
	70	11.23	30.60	35.03	**37.44** (3.00, 3.00, 02.45)	30.45 (02.45)
Franke coloured (120, 120, 3, 120)	10	28.13	47.13	48.21	**50.51** (1.89, 1.92, 16.59)	44.10 (16.59)
	20	22.11	42.96	43.97	**46.46** (2.35, 2.35, 08.35)	39.93 (08.35)
	35	17.25	40.18	40.47	**42.97** (2.79, 2.83, 04.74)	36.36 (04.74)
	50	14.15	38.11	38.15	**40.74** (3.13, 3.17, 03.45)	34.41 (03.45)
	70	11.23	31.72	35.90	**38.62** (3.50, 3.50, 02.45)	32.29 (02.45)
Salesman (288, 352, 1, 050)	10	28.13	**37.30**	37.12	35.24 (0.55, 0.68, 29.25)	31.48 (29.25)
	20	22.11	**34.13**	33.33	31.96 (0.70, 0.75, 13.93)	28.16 (13.93)
	35	17.25	**30.79**	30.20	29.36 (0.89, 0.89, 07.95)	26.01 (07.95)
	50	14.15	28.32	**28.33**	27.78 (1.05, 1.06, 05.45)	24.78 (05.45)
	70	11.23	24.55	**26.68**	26.34 (1.27, 1.32, 03.96)	23.87 (03.96)
Water (180, 320, 1, 120)	10	28.13	43.83	**44.68**	43.13 (0.93, 1.15, 25.75)	39.18 (25.75)
	20	22.11	40.59	**41.02**	39.84 (1.18, 1.35, 12.60)	35.94 (12.60)
	35	17.25	37.75	**37.90**	37.14 (1.40, 1.40, 06.95)	33.36 (06.95)
	50	14.15	35.58	**35.85**	35.41 (1.61, 1.65, 04.80)	31.83 (04.80)
	70	11.23	30.11	**33.87**	33.78 (1.80, 1.85, 03.45)	30.51 (03.45)

Table 2. PSNR comparison (best in bold), with quasi-optimal TDV parameters.

Name (M, N, C, T)	ς	Input	V-BM3D	V-BM4D	TDV (σ, ρ, η)	ROF 2D+t (η)
Miss America (288, 360, 1, 150)	10	28.13	**39.64**	39.93	39.25 (0.63, 0.63, 25.50)	36.93 (25.50)
	20	22.11	**37.95**	37.78	37.28 (0.90, 0.90, 12.75)	34.60 (12.75)
	35	17.25	**36.03**	35.77	35.44 (1.19, 1.19, 07.29)	32.72 (07.29)
	50	14.15	**34.19**	34.26	34.14 (1.42, 1.42, 05.10)	31.47 (05.10)
	70	11.23	28.86	32.64	**32.85** (1.68, 1.68, 03.64)	30.29 (03.64)
	90	09.05	27.42	31.27	**31.87** (1.90, 1.90, 02.83)	29.42 (02.83)
Xylophone coloured (240, 320, 3, 141)	10	28.13	**37.82**	37.49	35.96 (0.63, 0.63, 25.50)	32.70 (25.50)
	20	22.11	**34.70**	34.13	33.06 (0.90, 0.90, 12.75)	29.57 (12.75)
	35	17.25	**32.06**	31.65	30.93 (1.19, 1.19, 07.29)	27.16 (07.29)
	50	14.15	29.98	**30.07**	29.58 (1.42, 1.42, 05.10)	25.72 (05.10)
	70	11.23	25.89	**28.51**	28.32 (1.68, 1.68, 03.64)	24.43 (03.64)
	90	09.05	24.50	27.32	**27.37** (1.90, 1.90, 02.83)	23.57 (02.83)

are usually computationally faster than the variational approaches (including ours) but they tend to suffer from flickering and staircasing artefacts due to their patch-based nature. We experienced that our MATLAB code (not optimised for speed) is approximately 7× slower than V-BM4D (C++ code with MEX interface) with *normal-complexity profile*. Both quantitative (via PSNR) and qualitative results (visual inspection) are relevant indicators for video denoising.

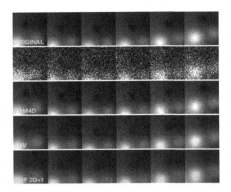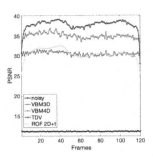

Fig. 3. *Franke*: frames [10, 20, 30, 40, 50, 60] for $\varsigma = 70$ and PSNR comparison.

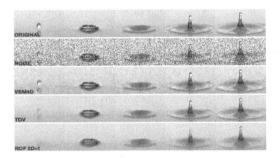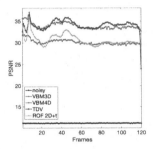

Fig. 4. *Water*: frames [5, 25, 45, 65, 85] for $\varsigma = 70$ and PSNR comparison.

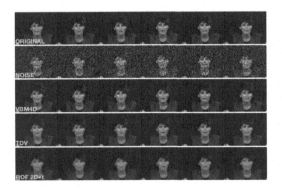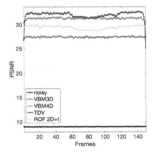

Fig. 5. *Miss America*: frames [5, 10, 15, 20, 25, 30] for $\varsigma = 90$ and PSNR comparison.

From the PSNR values in Tables 1 and 2 the TDV approach is comparable with the patch-based ones, with many single frames achieving higher PSNR value than the patch-based methods did. Also, by changing the noise level, the PSNR values are deteriorating less than with the patch-based methods, demonstrating the consistency of our approach.

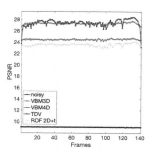

Fig. 6. *Xylophone*: frames $[5, 10, 15, 20, 25, 30]$ for $\varsigma = 90$ and PSNR comparison.

Visual results confirm that the TDV approach improves upon patch-based methods producing less flickering and stair-casing artefacts, especially when the motion is smooth due to the coherence imposed also along the time dimension.

5   Conclusions

In this paper, we proposed a variational approach with the total directional variation (TDV) regulariser for video denoising. We extended the range of applications of TDV regularisation from image processing as demonstrated in [21] to videos. We compared TDV with some state of the art patch-based algorithms for video denoising and obtained comparable results especially for high level noises while reducing artefacts in regions with smooth large motion, where the patch-based approach shows some weakness. We expect to improve further the results by refining the estimation of the anisotropic fields [5] and by using higher-order derivatives in the TDV definition [21]. This is left for future research.

References

1. Arias, P., Morel, J.M.: Video denoising via empirical Bayesian estimation of space-time patches. J. Math. Imaging Vis. **60**(1), 70–93 (2018)
2. Brailean, J.C., et al.: Noise reduction filters for dynamic image sequences: a review. Proc. IEEE **83**(9), 1272–1292 (1995)
3. Bredies, K., Kunisch, K., Pock, T.: Total generalized variation. SIAM J. Imaging Sci. **3**(3), 492–526 (2010)
4. Buades, A., Coll, B., Morel, J.: A non-local algorithm for image denoising. In: IEEE CVPR 2005, vol. 2, pp. 60–65 (2005)
5. Buades, A., Lisani, J., Miladinović, M.: Patch-based video denoising with optical flow estimation. IEEE Trans. Image Process. **25**(6), 2573–2586 (2016)
6. Buades, A., Coll, B., Morel, J.M.: A review of image denoising algorithms, with a new one. SIAM J. Multiscale Model. Simul. **4**(2), 490–530 (2005)
7. Burger, M., Dirks, H., Schönlieb, C.: A variational model for joint motion estimation and image reconstruction. SIAM J. Imaging Sci. **11**(1), 94–128 (2018)

8. Chambolle, A., Pock, T.: A first-order primal-dual algorithm for convex problems with applications to imaging. J. Math. Imaging Vis. **40**(1), 120–145 (2011)
9. Chambolle, A., Pock, T.: An introduction to continuous optimization for imaging. Acta Numerica **25**, 161–319 (2016)
10. Dabov, K., Foi, A., Egiazarian, K.: Video denoising by sparse 3D transform-domain collaborative filtering. In: 15th European Signal Process, pp. 145–149 (2007)
11. Dabov, K., Foi, A., Katkovnik, V., Egiazarian, K.: BM3D image denoising with shape-adaptive principal component analysis. In: SPARS 2009 (2009)
12. Dalgas Kongskov, R., Dong, Y., Knudsen, K.: Directional total generalized variation regularization. ArXiv e-prints (2017)
13. Davy, A., Ehret, T., Facciolo, G., Morel, J., Arias, P.: Non-local video denoising by CNN. ArXiv e-prints (2018)
14. Horn, B.K., Schunck, B.: "determining optical flow": a retrospective. Artif. Intell. **59**(1), 81–87 (1993)
15. Lebrun, M., Buades, A., Morel, J.: A nonlocal Bayesian image denoising algorithm. SIAM J. Imaging Sci. **6**(3), 1665–1688 (2013)
16. Lucas, B.D., Kanade, T.: An iterative image registration technique with an application to stereo vision. In: IJCAI 1981, pp. 674–679. Morgan Kaufmann (1981)
17. Maggioni, M., Boracchi, G., Foi, A., Egiazarian, K.: Video denoising using separable 4D nonlocal spatiotemporal transforms. In: Proceedings of SPIE, vol. 7870 (2011)
18. Ozkan, M.K., Sezan, M.I., Tekalp, A.M.: Adaptive motion-compensated filtering of noisy image sequences. IEEE Trans. Circuits Syst. Video Technol. **3**(4), 277–290 (1993)
19. Papafitsoros, K., Schönlieb, C.B.: A combined first and second order variational approach for image reconstruction. J. Math. Imaging Vis. **48**(2), 308–338 (2014)
20. Parisotto, S., Masnou, S., Schönlieb, C.B.: Higher order total directional variation. Part II: analysis. ArXiv e-prints (2018)
21. Parisotto, S., Masnou, S., Schönlieb, C.B., Lellmann, J.: Higher order total directional variation. Part I: imaging applications. ArXiv e-prints (2018)
22. Parisotto, S.: Anisotropic variational models and PDEs for inverse imaging problems. Ph.D. thesis, University of Cambridge (2019)
23. Portilla, J., et al.: Image denoising using scale mixtures of Gaussians in the wavelet domain. IEEE Trans. Image Process. **12**(11), 1338–1351 (2003)
24. Rudin, L.I., Osher, S., Fatemi, E.: Nonlinear total variation based noise removal algorithms. Physica D **60**(1), 259–268 (1992)
25. Sánchez Prez, J., Meinhardt-Llopis, E., Facciolo, G.: TV-L1 optical flow estimation. IPOL **3**, 137–150 (2013)
26. Tomasi, C., Manduchi, R.: Bilateral filtering for gray and color images. In: Proceedings of IEEE International Conference on Computer Vision, pp. 839–846 (1998)
27. Weickert, J.: Anisotropic diffusion in image processing (1998)
28. Zach, C., Pock, T., Bischof, H.: A duality based approach for realtime TV-L^1 optical flow. In: Hamprecht, F.A., Schnörr, C., Jähne, B. (eds.) DAGM 2007. LNCS, vol. 4713, pp. 214–223. Springer, Heidelberg (2007). https://doi.org/10.1007/978-3-540-74936-3_22

Joint CNN and Variational Model for Fully-Automatic Image Colorization

Thomas Mouzon, Fabien Pierre[✉], and Marie-Odile Berger

Laboratoire Lorrain de Recherche en Informatique et ses Applications, UMR CNRS 7503, Université de Lorraine, INRIA projet Magrit, Vandœuvre-lès-Nancy, France
fabien.pierre@univ-lorraine.fr

Abstract. This paper aims to couple the powerful prediction of the convolutional neural network (CNN) to the accuracy at pixel scale of the variational methods. In this work, the limitations of the CNN-based image colorization approaches are described. We then focus on a CNN which is able to compute a statistical distribution of the colors for each pixel of the image based on a learning stage on a large color image database. After describing its limitation, the variational method of [17] is briefly recalled. This method is able to select a color candidate among a given set while performing regularization of the result. By combining this approach with a CNN, we designed a fully automatic image colorization framework with an improved accuracy in comparison with CNN alone. Some numerical experiments demonstrate the increased accuracy reached by our method.

Keywords: Colorization · Convolutional Neural Networks · Total variation · Optimization

1 Introduction

In video colorization, academic research has reached great improvement since the 1970s. In methodological terms, three types of approaches were proposed.

The first one is based on the diffusion of color points drawn by the user [13,19]. These methods are based onto a diffusion of color with a local assumption: if the contours, so the gradients, are high for the luminance channel, they should be also high for the chrominance ones. The two drawbacks of this kind of methods are the tedious work needed by the user in case of complex images (for instance with textures), and the failure of diffusion method in case when there is not enough contrast to stop the diffusion of colors at the contours. The second category of approaches is based on a reference colored image [9] which is used as an example. In addition, the difficult problem of colorization based on example faces has been solved by calculating diffeomorphisms between images [15]. Segmentation and patch methods have been introduced to take into account example images. Some variational models for colorization that

© Springer Nature Switzerland AG 2019
J. Lellmann et al. (Eds.): SSVM 2019, LNCS 11603, pp. 535–546, 2019.
https://doi.org/10.1007/978-3-030-22368-7_42

combine several results under the assumption of spatial and temporal regularity have been proposed [16, 17]. Nevertheless, whereas theses methods are fully automatic when the reference image is given, its choice may be critical.

Some models are based on the minimization of a cost function. They have been developed in order to regularize the results of colorization [14, 17], both for example and manual methods.

The third colorization approach uses some large image databases [23]. Neural networks (Convolutional Neural Networks, Generative Adversarial Networks, Autoencoder, Recursive Neural Networks) have also been used successfully leading to a significant number of recent contributions. This literature can be divided into two categories of methods. The first evaluates the statistical distribution of colours for each pixel [3, 18, 23]. The network computes, for each pixel of the gray-scale image, the probability distribution of the possible colors. The second takes a grayscale image as input and provides a color image as output, mostly in the form of chrominance channels [1, 5–8, 10, 12, 21]. Some methods use a hybrid of both (*e.g.*, [24]).

Both techniques require image resizing, that is either done by deconvolution layers or performed *a posteriori* with standard interpolation techniques.

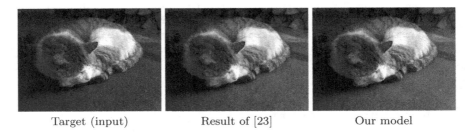

Target (input)   Result of [23]   Our model

Fig. 1. Example of halo effects produced by the method of [23]. Based on a variational model, our method is able to remove such artifacts.

In the case of [23], the network computes a probability distribution of the color on a down-sampled version of the original image. The choice of a color in each pixel at high resolution is made by linear interpolation without taking into account the grayscale image. Hence, the contours of chrominance and luminance may be not aligned, producing halo effects. Figure 1 shows some grey halo effects at the bottom of the cat that are visible on the red part, near the tail. In the other hand, in comparison to the others approaches of the state-of-the-art, the method of [23], produces images which are shiniest. We also show in this paper that we can make it a little bit shinier. Visually, the results of the competitive methods [8, 12] look drabber. In the following, the method of [23] is integrated in our system to predict colors.

In image colorization, convolutional neural networks can be used to compute in each pixel a set of possible colors and their associated probabilities [23].

However, since the final choice is made without taking into account the regularity of the image, this leads to halo effects. To improve this, we first propose to adapt the functional of [17] to the regularization of such results within the framework of colorization. The method of [17] being able to choose between several color candidates in each pixel, it will be quite easy to use on the color distribution provided by the CNN described in [23]. In addition, the numerical results of [17] demonstrate the ability to remove halos, which is relevant to the limitations of [23]. This functional will have to face two main problems: on the one hand, the transition from a low to a high resolution, and on the other hand, the maintenance of a higher saturation than current methods.

In this paper, the CNN described in [23] is presented in the first section. In the second one, the functional of [17] is recalled. The next section describes the way to couple the methods of [17,23]. Finally, the last section shows some numerical comparisons with some state-of-the art methods.

2   A CNN to Compute a Statistical Distribution of Color

The method of [23] is based on a discretization of the CIE Lab color space into C = 313 colors. This number of reference colors comes from the intersection gamut of the RGB color space and the discretization of the Lab space. The authors designed a CNN based on a VGG network [20] in order to compute a statistical distribution of the C colors in each pixel. The input of the network is the L lightness channel of the Lab transform of an image of size 256×256. The output is a distribution of probability over a set of 313 couples of a, b chrominance values for each pixel of a 64×64 size image. The quantification of the color space in 313 colors is computed from two assumptions. First, the colors are regularly spaced onto the CIE Lab color space. On this color space two colors are close with respect to the Euclidean norm when the human visual system feels them close. The second assumption that rules the set of colors is the respect of the RGB gamut. The colors have to be displayable onto a standard screen.

To train this CNN, the database ImageNet [4] is used without the gray-scale images. The images are resized at size 256×256 and then transformed into the CIE Lab colorspace. The images are then resized at size 64×64 to compute the a and b channels. The loss-function used is the cross-entropy between the luminance (a, b) of the training image and the distribution over the 313 original colors. Let us denote by Δ the probability simplex in C = 313 dimensions.

Denoting by $(\hat{w}_i(x))_{i=1..C} \in \Delta^N$ the probability distribution of dimension C in the N pixels of the 64×64 image (over a domain Ω), and denoting by $(w_i(x))$ the ground truth distribution computed with a soft-encoding scheme (see [23] for details), the loss-function is given by:

$$L(\hat{w}, w) = - \sum_{x \in \Omega} \sum_{i=1}^{C} w_i(x) \log(\hat{w}_i(x)). \tag{1}$$

The forward propagation in the network provides a probability distribution over the C colors. In order to compute a colorization result, a choice among all these colors has to be performed. Basically, the authors of [23] proposed an annealed-mean in each pixel, independently. After that, a resizing of the (a, b) channels at original size is done and recombined with brightness channel to obtain the color image.

Nevertheless, this recombination is done without taking into account any spatial consideration. In the next section we recall how the functional of [17] works to adapt it.

3   A Variational Model for Image Colorization with Channels Coupling

In [17], the authors have proposed a functional that selects a color among candidates extracted from a patch-based method. Assuming that C candidates are available in each pixel of a domain Ω and assuming that two chrominance channels are available for each candidate. Let us denote for each pixel at position x the i-th candidate by $c_i(x)$, $u(x) = (U(x), V(x))$ stands for chrominances to compute, and $w(x) = \{w_i(x)\}$ with $i = 1, \ldots, C$ for the candidate weights. Let us minimize the following functional with respect to (u, w):

$$F(u, w) := TV_{\mathfrak{C}}(u) + \frac{\lambda}{2} \int_\Omega \sum_{i=1}^{C} w_i(x) \|u(x) - c_i(x)\|_2^2 \, dx + \chi_\mathcal{R}(u(x)) + \chi_\Delta(w(x)).$$
(2)

The central part of this model is based on the term

$$\int_\Omega \sum_{i=1}^{C} w_i(x) \|u(x) - c_i(x)\|_2^2 \, dx.$$
(3)

This term is a weighted average of some L2 norms with respect to the candidates c_i. The weights w_i can be seen as a probability distribution of the c_i. For instance, if $w_1 = 1$ and $w_i = 0$ for $2 \le i \le C$, the minimum of F with respect to u is equal to the minimization of

$$TV_{\mathfrak{C}}(u) + \frac{\lambda}{2} \int_\Omega \|u(x) - c_1(x)\|_2^2 \, dx + \chi_\mathcal{R}(u(x)).$$
(4)

To simplify the notations, the dependence of each value to the position x of the current pixel will be removed in the following. For instance, the second term of (2) will be denoted by $\int_\Omega \sum_{i=1}^{C} w_i(x) \|u(x) - c_i(x)\|_2^2 \, dx$.

This model is a classical one with a fidelity-data term $\int_\Omega \sum_{i=1}^{C} w_i \|u - c_i\|_2^2$ and a regularization term $TV_{\mathfrak{C}}(u)$. Since the first step of the method extracts many candidates, we propose averaging the fidelity-data term issued from each candidate. This average is weighted by w_i. Thus, the term

$$\int_\Omega \sum_{i=1}^C w_i \|u - c_i\|_2^2 \tag{5}$$

connects the candidate color c_i to the color u that will be retained. The minimum of this term with respect to u is reached when u is equal to the weighted average of candidates c_i.

Since the average is weighted by w_i, these weights are constrained to be onto the probability simplex. This constraint is formalized by $\chi_\Delta(w)$ whose value is 0 if $w \in \Delta$ and $+\infty$ otherwise, with Δ defined as:

$$\Delta := \left\{ (w_1, \cdots, w_C) \text{ s.t. } 0 \leq w_i \leq 1 \text{ and } \sum_{i=1}^C w_i = 1 \right\}. \tag{6}$$

Let $TV_\mathfrak{C}$ be a *coupled* total variation defined as

$$TV_\mathfrak{C}(u) = \int_\Omega \sqrt{\gamma \partial_x Y^2 + \gamma \partial_y Y^2 + \partial_x U^2 + \partial_y U^2 + \partial_x V^2 + \partial_y V^2}, \tag{7}$$

where Y, U and V are the luminance and chrominance channels. γ is a parameter which enforces the coupling of the channels. Some others total variation formulations have been proposed to couple the channels, see for instance [11] or [2].

In order to compute a suitable solution for problem (2), authors of [17] propose a primal-dual algorithm with alternating minimization of the terms depending of w. They also proposed numerical experiments showing the convergence of their algorithm. Let us note that this recent reference shows that the convergence of such numerical schemes can be demonstrated after smoothing of the total variation term. Among all the numerical schemes proposed in the references [17,22], we choose the methodology having the best convergence rate as well as a convergence proof. This scheme is given in Algorithm 2 in [22]. This algorithm is a block coordinate forward backward algorithm. To increase the speed-up of the convergence, Algorithm 2 of [22] is initialized with the result of 500 iterations of the primal-dual algorithm of [17]. Whereas this algorithm has no guaranty of convergence, the authors of [22] have experimentally observed that it numerically converges faster.

Unfortunately, the functional (2) is highly non-convex and contains many critical points. More precisely, the functional is convex with respect to u with fixed w and reversely, it is convex with respect to w for fixed u. Nevertheless, the functional is not convex with respect to the joint variables (u, w). Thus, even if the numerical scheme would converge to a local minimum, the solution of the problem highly depends on the initialization.

In the next section, we will show how the powerful prediction of CNN can be used to tackle this last problem.

4   Joining Total Variation Model with CNN

In this section, a method to couple the prediction power of CNN with the precision of variational methods is described. To this aim, let us remark that the variable w of the functional (2) represents the ratio of each color candidate which is represented in the final result. This comes from the fact that, for a given vector $w \in \mathbb{R}^C$, the minimum of

$$\sum_{i=1}^{C} w_i \|u - c_i\| \tag{8}$$

with respect to u is given by

$$\sum_{i=1}^{C} w_i c_i. \tag{9}$$

Thus, it can bee seen as a probability distribution of the colors in the desired color image, which is exactly the same purpose of the CNN in [23].

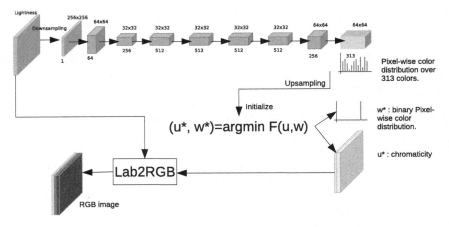

Fig. 2. Overview of our method. A CNN computes color distribution on each pixel. A variational method selects then a color for each pixel based on a regularity hypothesis.

Figure 2 shows an overview of our method. First, the gray-scale image, considered as the luminance L is given as an input to the CNN. The output of the CNN is a probability distribution over 313 possible chromaticity at low resolution (64×64). In order to initialize the minimization algorithm, the output weights of the CNN can be used. The CNN provides a coarse scale output, that needs an up-sampling before producing a suitable output at original definition. Two ways can be considered. For the first one, the variational method can be used at coarse scale (low definition), and then an interpolation can be performed to recover a result at fine scale (high definition). For the second one, the probability distributions can be interpolated to get a high definition array. In the

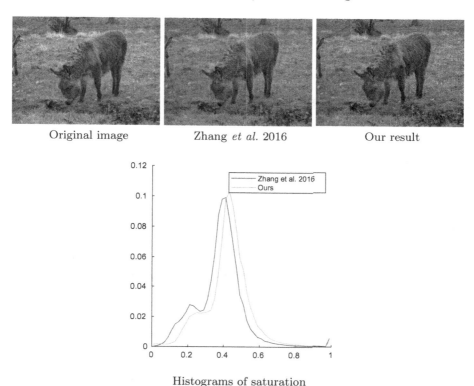

Histograms of saturation

Fig. 3. Results of Zhang *et al.* [23] compared with our result. The histogram of the saturation shows that our result is shinier than the original method. Indeed, the average value of the saturation is higher for our model (0.4228) than the one of [23] (0.3802).

following, the second approach will be preferred. Indeed, the interpolation of a color image produces a decrease of the saturation, that makes images drabber. By interpolating the probability distributions instead of the color images, the variational method will be able to compute a color for each pixel based on a coupling of the channels at high resolution. The given probability distribution is then used as initialization value for the numerical scheme. As it was still proposed in [17], the variable u is initialized with $\sum_{i=1}^{C} w_i c_i$. After the iterations of the functional, the result, denoted by (u^*, w^*), provides some binary weights (see, *eg*, [17], Section 2.3.2) and a regularized result u^* that gives two chromaticity channels, a and b, at initial definition. Recombined with the luminance L and transformed into the RGB space, that produces a color image.

Let us remark that the authors of [23] proposed to first produce the color image and then to resize it with bi-cubic interpolation. Unfortunately, upsampling or down-sampling images with bi-linear or bi-cubic interpolations reduce the saturation of the colors and make them drabber than the original. To avoid that, we propose here the opposite approach: we first up-sample the color distribution, and then we compute a color image at full definition by using it. Since the numerical scheme is used at full definition, the required memory of the algorithm for all the weights and the colors is a limitation to process high resolution images on a standard PC. To tackle this issue, we propose to select some of the 313 colors. This selection is done with respect to the probability distribution of the colors, by choosing the 10 highest modes.

This choice of 10 has been done experimentally. For most images, 8 or 9 candidates are enough and taking more of them does not improve the result, but it increases the computational time. On the other hand, taking less candidates decreases the quality of the result on a significant number of images. Finally, the number of 10 is a fair trade-off.

The training step of the CNN is done as in [23]. The variational step is not taken into account during the training process. Indeed, the relation between the initialization of the weights and the result is not analytically described and the gradient back-propagation algorithms is not suitable for this problem. Thus, the training is done by feeding the CNN with a gray-scale image as input and a color distribution as output. The variational step remains independent of the full framework during the training step. Its integration will be the purpose of future works.

In the next section, the numerical results are presented.

5   Numerical Results

In this section we show a qualitative comparison between [23] and our framework. A lot of results provided by [23] are accurate and reliable. We will show on these examples that our method does not reduce the quality of the images. We then propose some comparisons with erroneous results of [23], which shows that our method is reliable to fully automatically colorize images without artifacts and halo effects. A time comparison between the CNN inference computation and the variational step will be proposed to show that the regularization of the result is not a burden on the CNN approach. Finally, to show the limitation of CNN in image colorization, we will show some results where neither the approach of [23] nor our framework are able to produce some reliable results.

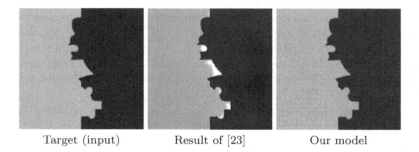

Target (input)   Result of [23]   Our model

Fig. 4. Comparison of our method with [23]. This example provides a proof of concept. Our method is able to remove the halo effects on the colorization result of [23].

Target (input)   Result of [23]   Our model

Fig. 5. Comparison of our method with [23].

Figure 3 shows the colorization results of the method of Zhang *et al.* [23]. Whereas it is hard to see that our method produces a shinier result that the result of [23] unless being a calibration expert, the histogram of the saturation is able to show the improvement. Indeed, since the histogram is right-shifted, it means that globally, the saturation is higher on our result. Quantitatively, the average of the saturation is equal to 0.4228 for our method, while it is equal to 0.3802 for the method of [23]. This improvement comes from the fact that our method selects one color among the ones given by the results of the CNN, whereas the method of [23] computes the annealed mean of them. The mean of the colors of the chrominances produces a decrease of the saturation and makes the colors drabber. By using a selection algorithm based on the image regularization, our method is able to avoid this drawback.

The result in Fig. 4 is a proof of concept for the proposed framework. We can see a toy example which is automatically colorized by the method of [23]. The result given by the method of [23] produces some halo effect near the only contour of the image, which is unnatural. The regularization of the result is able to remove this halo effect and to recover an image looking less artificial. This toy example contains only two constant parts. The aim of the variational

Target (input)   Result of [23]   Our model

Fig. 6. Additional comparisons of our method with [23].

method is to couple the contours of the chrominance channels and the ones of the luminance. The result produced with our method contains no halo effect, showing the benefits of our framework.

In Fig. 5, we show some results and we compare them to the method of [23]. For the lion, (first line), a misalignment of the colors with the gray scale image is visible (a part of the lion is colorized in blue and a part of the sky is brown beige). This is a typical case of halo effect where our framework is able to remove the artifacts. For the image of mountaineer, on the result of [23] some pink stains appear. With our method, the minimization of the total variation ensures the regularity of the image, thus it removes these strains.

Figure 6 shows additional results. The first line is an old port-card. Its colorization is reliable with the CNN and, in addition, the variational approach makes it a little bit shinier. This example shows the ability of our approach to colorize historical images. In the second example, most of the image is well colorized by the original method of [23]. Nevertheless, the lighthouse as well as the right-side building contain some orange halos that are not reliable. With the variational method, the colors are convincing. Additional results are available on http://www.fabienpierre.fr/ssvm2019.

The computational time of the CNN forward pass is about 1.5 s in GPU, whereas the minimization of the variational model (2) is about 15 s in Matlab in CPU. In [16], the authors provide a computation time almost equal to 1 s with unoptimized GPU implementation. Since the minimization scheme of [22] is about the same, the computational time would be almost equal. Thus, the computational time of our approach is not a burden in comparison with the method of [23].

In Fig. 7, a failure case is shown. In this case, since the minimization of the variational model strongly depends on its initialization, our method is not able to recover realistic colors. Actually, fully automatic colorization remains an open problem.

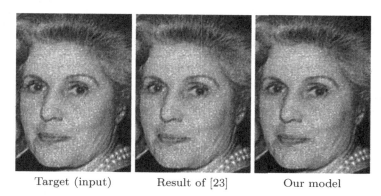

Target (input)   Result of [23]   Our model

Fig. 7. Fail case. The prediction of the CNN is not able to recover a reliable color.

6   Conclusion

In this paper, we propose a novel approach to couple the power of the CNN with the precision of the variational models. This coupling is done with a transfer of information based on probability distributions. The computation of the two parts of the framework is based on standard techniques issued from the literature. The numerical results show the improvement of colorization results performed with the two methods considered together. Some results where neither the approach of [23] nor our framework are able to produce some reliable results are presented. Thus, image colorization remains an open issue despite the huge number of CNN-based approaches proposed in state-of-the-art.

References

1. Cao, Y., Zhou, Z., Zhang, W., Yu, Y.: Unsupervised diverse colorization via generative adversarial networks. In: Ceci, M., Hollmén, J., Todorovski, L., Vens, C., Džeroski, S. (eds.) ECML PKDD 2017. LNCS (LNAI), vol. 10534, pp. 151–166. Springer, Cham (2017). https://doi.org/10.1007/978-3-319-71249-9_10
2. Caselles, V., Facciolo, G., Meinhardt, E.: Anisotropic Cheeger sets and applications. SIAM J. Imaging Sci. **2**(4), 1211–1254 (2009)
3. Chen, Y., Luo, Y., Ding, Y., Yu, B.: Automatic colorization of images from Chinese black and white films based on CNN. In: IEEE International Conference on Audio, Language and Image Processing, pp. 97–102 (2018)
4. Deng, J., Dong, W., Socher, R., Li, L.J., Li, K., Fei-Fei, L.: Imagenet: a large-scale hierarchical image database. In: IEEE Conference on Computer Vision and Pattern Recognition, pp. 248–255 (2009)

5. Deshpande, A., Lu, J., Yeh, M.C., Chong, M.J., Forsyth, D.A.: Learning diverse image colorization. In: IEEE Conference on Computer Vision and Pattern Recognition, pp. 2877–2885 (2017)
6. Guadarrama, S., Dahl, R., Bieber, D., Shlens, J., Norouzi, M., Murphy, K.: Pixcolor: pixel recursive colorization. In: British Machine Vision Conference (2017)
7. He, M., Chen, D., Liao, J., Sander, P.V., Yuan, L.: Deep exemplar-based colorization. ACM Trans. Graph. **37**(4), 47:1–47:16 (2018)
8. Iizuka, S., Simo-Serra, E., Ishikawa, H.: Let there be color! Joint end-to-end learning of global and local image priors for automatic image colorization with simultaneous classification. ACM Trans. Graph. **35**(4), 110 (2016)
9. Irony, R., Cohen-Or, D., Lischinski, D.: Colorization by example. In: Eurographics Symposium on Rendering, vol. 2. Citeseer (2005)
10. Isola, P., Zhu, J.Y., Zhou, T., Efros, A.A.: Image-to-image translation with conditional adversarial networks. In: IEEE Conference on Computer Vision and Pattern Recognition (2017)
11. Kang, S.H., March, R.: Variational models for image colorization via chromaticity and brightness decomposition. IEEE Trans. Image Process. **16**(9), 2251–2261 (2007)
12. Larsson, G., Maire, M., Shakhnarovich, G.: Learning representations for automatic colorization. In: Leibe, B., Matas, J., Sebe, N., Welling, M. (eds.) ECCV 2016. LNCS, vol. 9908, pp. 577–593. Springer, Cham (2016). https://doi.org/10.1007/978-3-319-46493-0_35
13. Levin, A., Lischinski, D., Weiss, Y.: Colorization using optimization. ACM Trans. Graph. **23**(3), 689–694 (2004)
14. Lézoray, O., Ta, V.T., Elmoataz, A.: Nonlocal graph regularization for image colorization. In: IEEE International Conference on Pattern Recognition, pp. 1–4 (2008)
15. Persch, J., Pierre, F., Steidl, G.: Exemplar-based face colorization using image morphing. J. Imaging **3**(4), 48 (2017)
16. Pierre, F., Aujol, J.F., Bugeau, A., Ta, V.T.: Interactive video colorization within a variational framework. SIAM J. Imaging Sci. **10**(4), 2293–2325 (2017)
17. Pierre, F., Aujol, J.F., Bugeau, A., Papadakis, N., Ta, V.T.: Luminance-chrominance model for image colorization. SIAM J. Imaging Sci. **8**(1), 536–563 (2015)
18. Royer, A., Kolesnikov, A., Lampert, C.H.: Probabilistic image colorization. In: British Machine Vision Conference (2017)
19. Sapiro, G.: Inpainting the colors. In: IEEE International Conference on Image Processing, vol. 2, p. II-698 (2005)
20. Simonyan, K., Zisserman, A.: Very deep convolutional networks for large-scale image recognition. In: International Conference on Learning Representations (2015)
21. Su, Z., Liang, X., Guo, J., Gao, C., Luo, X.: An edge-refined vectorized deep colorization model for grayscale-to-color images. Neurocomputing **311**, 305–315 (2018)
22. Tan, P., Pierre, F., Nikolova, M.: Inertial alternating generalized forward-backward splitting for image colorization. J. Math. Imaging Vis. (2019, to appear)
23. Zhang, R., Isola, P., Efros, A.A.: Colorful image colorization. In: Leibe, B., Matas, J., Sebe, N., Welling, M. (eds.) ECCV 2016. LNCS, vol. 9907, pp. 649–666. Springer, Cham (2016). https://doi.org/10.1007/978-3-319-46487-9_40
24. Zhang, R., et al.: Real-time user-guided image colorization with learned deep priors. ACM Trans. Graph. **36**(4), 119:1–119:11 (2017). https://doi.org/10.1145/3072959.3073703

A Variational Perspective
on the Assignment Flow

Fabrizio Savarino$^{(\boxtimes)}$ and Christoph Schnörr

Image and Pattern Analysis Group, Heidelberg University, Heidelberg, Germany
fabrizio.savarino@iwr.uni-heidelberg.de,
schnoerr@math.uni-heidelberg.de

Abstract. The image labeling problem can be described as assigning to each pixel a single element from a finite set of predefined labels. Recently, a smooth geometric approach for inferring such label assignments was proposed by following the Riemannian gradient flow of a given objective function on the so-called assignment manifold. Due to the specific Riemannian structure, this results in a coupled replicator dynamic incorporating local spatial geometric averages of lifted data-dependent distances. However, in this framework an approximation of the flow is necessary in order to arrive at explicit formulas. We propose an alternative variational model, where lifting and averaging are decoupled in the objective function so as to stay closer to established approaches and at the same time preserve the main ingredients of the original approach: the overall smooth geometric setting and regularization through geometric local averages. As a consequence the resulting flow is explicitly given, without the need for any approximation. Furthermore, there exists an interesting connection to graphical models.

Keywords: Image labeling · Assignment manifold · Assignment flow ·
Geometric optimization · Riemannian gradient flow ·
Replicator equation · Multiplicative updates

1 Introduction

Overview, Motivation. Let $\mathcal{G} = (\mathcal{V}, \mathcal{E})$ be a graph representing a certain spatial structure and denote by $f \colon \mathcal{V} \to \mathcal{F}$ some given data on that graph with values in a feature space $\mathcal{F}$. A *labeling* of f on $\mathcal{V}$ with predefined labels $\mathcal{L} = \{l_1, \dots, l_n\}$ is a map $A \colon \mathcal{V} \to \mathcal{L}$ assigning to every vertex $i \in \mathcal{V}$ a label $A_i \in \mathcal{L}$. By identifying the nodes $\mathcal{V}$ with the numbers $\{1, \dots, m\}$, for $m := |\mathcal{V}|$, a labeling A corresponds to a vector $A \in \mathcal{L}^m$. In the case of *image labeling*, the graph $\mathcal{G}$ might be a grid graph embedded into the image domain $\Omega \subset \mathbb{R}^2$ and f represents some observed raw image data $\mathcal{F} = [0,1]^3$ or some features extracted from the image by standard methods. Depending on the domain of application, one is usually interested in finding an optimal labeling with respect to a quality measure, called *objective function*. In general the task of computing globally

© Springer Nature Switzerland AG 2019
J. Lellmann et al. (Eds.): SSVM 2019, LNCS 11603, pp. 547–558, 2019.
https://doi.org/10.1007/978-3-030-22368-7_43

optimal labels results in an NP-hard problem and therefore several relaxations are used to arrive at a computationally feasible formulation [7].

In [2] a new smooth geometric approach was suggested, which is modeled on the manifold of row-stochastic matrices with full support, called the *assignment manifold* and denoted by $\mathcal{W} \subset \mathbb{R}^{m \times n}$ (for details see Sect. 2). By choosing the Fisher-Rao (information) metric, $\mathcal{W}$ is turned into a Riemannian manifold. Their basic idea is to encode labelings as points on the assignment manifold, exploit the Riemannian setting for constructing an objective function $E : \mathcal{W} \to \mathbb{R}$ using Riemannian means and optimizing it by following the Riemannian gradient flow. After a simplifying assumption and approximating the Riemannian mean to first order by the geometric mean, they arrive at the following dynamical system, called *assignment flow*

$$\dot{W}(t) = \Pi_{W(t)} S(W(t)) \tag{1.1}$$

where S consists of certain geometric means and Π_W is a linear map (see Sect. 2).

While the overall geometric model constitutes an interesting new approach to the labeling problem and performs very well, there are some mathematical points to address: The relation to classical approaches with objective function $E = E_{\text{data}} + E_{\text{reg}}$ has not been worked out, where E_{data} is a data dependent and E_{reg} a regularization term. It can be shown, that there exists no potential of the vector field (1.1) in the Fisher-Rao geometry, which implies that the flow is not variational. Furthermore, the above mentioned simplifying assumption and approximation are unavoidable, since otherwise there is no closed form solution for the Riemannian mean.

Contribution. We propose a variational model where the lifting and averaging is decoupled in a way similar to more classical approaches of the form $E = E_{\text{data}} + E_{\text{reg}}$ mentioned above. In this alternative model we are able to completely avoid the need for any approximation and simplifying assumptions, while still exploiting the Riemannian structure of the setting. Additionally, there is an interesting connection to graphical models, a well established formulation of image labeling.

2 Preliminaries

Basic Notation. We assume $\mathcal{G} = (\mathcal{V}, \mathcal{E})$ is an *undirected* graph. If two nodes i and j are connected by an undirected edge $ij \in \mathcal{E}$ then we call i and j adjacent and denote this relation by $i \sim j$. The neighborhood of node i is the set $\mathcal{N}(i) := \{j \in \mathcal{V} : i \sim j\}$. The number of nodes will be denoted by $m := |\mathcal{V}|$ and the number of labels by $n := |\mathcal{L}|$. We use the abbreviation $[k] = \{1, 2, \dots, k\}$ for $k \in \mathbb{N}$ and identify $[m]$ with $\mathcal{V}$ as well as $[n]$ with $\mathcal{L}$. For a matrix $M \in \mathbb{R}^{m \times n}$ we denote the i-th row of M as M_i. For any two vectors $x \in \mathbb{R}^n$ and $y \in \mathbb{R}^n_{>0}$, we denote the componentwise product and division by $xy = (x_1 y_1, \dots, x_n y_n)^\top$ and $\frac{x}{y} = (\frac{x_1}{y_1}, \dots, \frac{x_n}{y_n})^\top$ respectively. If $g : \mathbb{R} \to \mathbb{R}$ is a scalar function, then $g(x)$ denotes the componentwise application of g, i.e. $e^x = (e^{x_1}, \dots, e^{x_n})^\top$. The standard basis of $\mathbb{R}^n$ is denoted by $\{e_1, \dots, e_n\}$ and the standard inner product on

$\mathbb{R}^n$ and $\mathbb{R}^{m \times n}$ respectively by $\langle \cdot, \cdot \rangle$. We set $\mathbb{1}_n = (1, 1, \ldots, 1)^\top \in \mathbb{R}^n$.

The Assignment Manifold. We briefly introduce the necessary geometric setting of the assignment manifold from [2]. Let $\Delta_n = \{p \in \mathbb{R}^n : p_i \geq 0 \text{ for } i = 1, \ldots, n , \langle p, \mathbb{1} \rangle = 1\}$ denote the probability simplex and $c := \frac{1}{n} \mathbb{1}_n$ the barycenter. The relative interior of Δ_n is given by

$$\mathcal{S} := \text{rint}(\Delta_n) = \{p \in \Delta_n : p_i > 0 \text{ for } i = 1, \ldots, n\} \qquad (2.1)$$

and is a smooth manifold of dimension $n - 1$ with a global chart and an $n - 1$ dimensional constant tangent space

$$T_p \mathcal{S} = \{v \in \mathbb{R}^n : \langle v, \mathbb{1} \rangle = 0\} =: T \subset \mathbb{R}^n \quad \text{for all} \quad p \in \mathcal{S}. \qquad (2.2)$$

The orthogonal projection of $\mathbb{R}^n$ to T with respect to the standard inner product is given by

$$P_T[x] := \left(I - \frac{1}{n} \mathbb{1} \mathbb{1}^\top\right) x \qquad (2.3)$$

The *lifting map* $\exp : T\mathcal{S} = \mathcal{S} \times T \to \mathcal{S}$ is defined as

$$(p, u) \mapsto \exp_p(u) := \frac{p e^u}{\langle p, e^u \rangle}. \qquad (2.4)$$

The map $\exp_p : T \to \mathcal{S}$ is a diffeomorphism for every $p \in \mathcal{S}$ with inverse $\exp_p^{-1}(q) = P_T \log(\frac{q}{p})$ and since $T \subset \mathbb{R}^n$ is a linear space, it can be used as a chart for $\mathcal{S}$. The lifting map can also be viewed as $\exp_p : \mathbb{R}^n \to \mathcal{S}$ with $\exp_p \circ P_T = \exp_p$, however, this is not an invertible map anymore.

The Fisher-Rao metric endows $\mathcal{S}$ with a Riemannian structure given by

$$g_p : T \times T \to \mathbb{R}, \quad g_p(u, v) = \langle u, \text{Diag}(\tfrac{1}{p}) v \rangle, \qquad (2.5)$$

for $p \in \mathcal{S}$ and $u, v \in T$. Denote by $2\mathbb{S}^{n-1} \subset \mathbb{R}^n$ the sphere of radius 2 with Riemannian metric induced by the Euclidean inner product of $\mathbb{R}^n$. There is an isomorphism, called *sphere map* $\psi : \mathcal{S} \to 2\mathbb{S}^{n-1} \cap \mathbb{R}_{>0}^n$, given by $p \mapsto 2\sqrt{p}$ (cf. [2, Sect. 2.1]). Due to the form of ψ, the geometry can be continuously extended to Δ_n. As a consequence, the Riemannian distance between $p, q \in \mathcal{S}$ is given by

$$d_{\mathcal{S}}(p, q) = 2 \arccos(\langle \sqrt{p}, \sqrt{q} \rangle) \leq \pi. \qquad (2.6)$$

There is also an explicit formula for the exponential map of the Riemannian manifold Exp_p and its inverse (cf. [2, Prop 2 and Eq. (7.16b)]), where the latter is given by

$$\text{Exp}_p^{-1}(q) = \frac{d_{\mathcal{S}}(p, q)}{\sqrt{1 - \langle \sqrt{p}, \sqrt{q} \rangle^2}} \left(\sqrt{pq} - \langle \sqrt{p}, \sqrt{q} \rangle p\right). \qquad (2.7)$$

For a scalar valued function $f : \mathcal{S} \to \mathbb{R}$, the Riemannian gradient (cf. [1]) at $p \in \mathcal{S}$ is the vector $\nabla_{\mathcal{S}} f(p) \in T$ uniquely characterized via the differential of f by $Df(p)[v] = g_p(\nabla_{\mathcal{S}} f(p), v)$ for all $v \in T$. Define the *replicator operator* as

$$\Pi_p : \mathbb{R}^n \to T \quad x \mapsto \Pi_p[x] := \left(\text{Diag}(p) - p p^\top\right) x \quad \text{for} \quad p \in \mathcal{S}. \qquad (2.8)$$

An elementary calculation shows that $\Pi_p \circ P_T = \Pi_p$. Viewed as a linear map $\Pi_p \colon T \to T$ an inverse exists and is given by $(\Pi_p)^{-1} = P_T \operatorname{Diag}(\frac{1}{p})$.

Denote by $\nabla f(p) \in T$ the Riemannian gradient of $\mathcal{S}$ with the standard inner product as Riemannian metric, then $\nabla_{\mathcal{S}} f$ and ∇f are connected by (cf. [2, Prop. 1])

$$\nabla_{\mathcal{S}} f(p) = p \nabla f(p) - \langle p, \nabla f(p) \rangle p = \Pi_p[\nabla f(p)]. \tag{2.9}$$

The Riemannian gradient flow on $\mathcal{S}$ can be transformed onto T by $\exp_c$. According to [9, Cor. 1 and Lem. 4] we have $t \mapsto p(t) \in \mathcal{S}$ with $p(0) = c$ solves gradient flow (2.10)(a) if and only if $t \mapsto v(t) \in T$ with $v(0) = 0$ and $p(t) = \exp_c(v(t))$ solves the gradient flow (2.10)(b)

$$\text{(a)} \quad \dot{p}(t) = \nabla_{\mathcal{S}} f(p(t)) \qquad \text{(b)} \quad \dot{v}(t) = \nabla f\big(\exp_c(v(t))\big) \tag{2.10}$$

The *assignment manifold* $\mathcal{W}$ is defined to be the product manifold $\mathcal{W} := \mathcal{S}^m$ with tangent space given by $T_W \mathcal{W} = T^m =: \mathcal{T}$ for $W \in \mathcal{W}$. The Fisher Rao metric on $\mathcal{S}$ induces a Riemannian metric on $\mathcal{W}$ via the product metric, thus the above formulas carry over to the product manifold setting by applying them componentwise. In the following we use the description of $\mathcal{W}$ as a set of matrices $\mathcal{W} = \{W \in \mathbb{R}_{>0}^{m \times n} \colon W\mathbb{1} = \mathbb{1}\}$ together with $\mathcal{T} = \{V \in \mathbb{R}^{m \times n} \colon V\mathbb{1} = 0\}$.

3   Model

Motivation: The Assignment Flow. In order to better motivate the form of our variational approach below and the parallels to [2], we review the core concept of the assignment flow in a bit more detail. The basic idea of modeling the labeling problem on the assignment manifold is to encode label $l_j \in \mathcal{L}$ by the j-th standard basis vector e_j of $\mathbb{R}^n$, which is a corner of Δ_n. With this, a labeling $A \in \mathcal{L}^m$ corresponds to an assignment matrix $W \in \Delta_n$ with i-th row $W_i = e_j$ if $A_i = l_j$. These integral labelings are relaxed to the assignment manifold $\mathcal{W}$ by allowing them to be fully probabilistic. Let $f \colon \mathcal{V} \to \mathcal{F}$ be some given data on the graph $\mathcal{G}$ with values in a feature space $\mathcal{F}$. Suppose some labels $\mathcal{L} = \{l_1, \ldots, l_n\}$ and a distance function $d \colon \mathcal{F} \times \mathcal{L} \to \mathbb{R}$ are given. Then the data dependent distance matrix $D \in \mathbb{R}^{m \times n}$ measuring the fit of labels to the data is defined as

$$D_{ij} := d(f_i, l_j) \quad \text{for} \quad i \in [m], j \in [n]. \tag{3.1}$$

These distances are lifted to the assignment manifold by the lifting map

$$L_i = L_i(W_i, D_i) := \exp_{W_i}(-\tfrac{1}{\rho} D_i), \tag{3.2}$$

where D_i denotes the i-th row of D. These vectors are used to build the similarity matrix $S \in \mathcal{W}$, where each row $S_i = S_i(W, D)$ is the Riemannian mean given by the unique minimizer of

$$\frac{1}{2} \sum_{j \in \mathcal{N}(i)} \omega_{ij} d_{\mathcal{S}}^2\big((p, L_j(W_j, D_j))\big) \tag{3.3}$$

for some chosen weights $\omega_{ij} > 0$ with $\sum_{j \in \mathcal{N}(i)} \omega_{ij} = 1$ for every $i \in \mathcal{V}$. A label assignment is inferred by maximizing the correlation between the current assignment state W and the similarity matrix S and has the form

$$\sup_{W \in \mathcal{W}} E(W) := \langle W, S(W, D) \rangle. \tag{3.4}$$

Adopting the simplifying assumption that averaging changes slowly and approximating the Riemannian mean by the geometric mean, one arrives at the assignment flow formulation (1.1).

Variational Model. Intuitively, the above described assignment flow also contains the two classical components of labeling, using the data for label decisions while locally regularizing these decisions in a spatial neighborhood. The goal for our variational approach is to disentangle these two components in the assignment flow, while keeping the basic building blocks: the overall smooth geometric setting and regularization through geometric local averages.

Using the definition of the distance matrix D from (3.1), we propose the following quality measure $J(W)$ evaluating an assignment matrix $W \in \mathcal{W}$ by

$$J(W) := J_{\text{data}} + \rho J_{\text{reg}} := \langle W, D \rangle + \frac{\rho}{2} \sum_{ij \in \mathcal{E}} \omega_{ij} d_{\mathcal{S}}^2(W_i, W_j) \tag{3.5a}$$

$$= \sum_{i \in \mathcal{V}} \left(\langle W_i, D_i \rangle + \frac{\rho}{2} \sum_{j \in \mathcal{N}(i)} \omega_{ij} d_{\mathcal{S}}^2(W_i, W_j) \right), \tag{3.5b}$$

with weights $\omega_{ij} > 0$ and $\sum_{j \in \mathcal{N}(i)} \omega_{ij} = 1$ for every $i \in \mathcal{V}$ and a parameter $\rho > 0$ regulating the amount of regularization.

The purpose of the data term is to choose the best fit to the data, i.e. the smallest distance at every node $i \in \mathcal{V}$. Regularization is induced by comparing the assignments W_i at every node $i \in \mathcal{V}$ with their neighboring assignments W_j with $j \in \mathcal{N}(i)$, similar to [11] and [4]. The main **take-home message of this paper** is: In contrast to the assignment flow, we do not directly compare the assignments W_i to the Riemannian mean itself (cf. (3.4)), but rather use the Riemannian mean defining objective functions (3.3) as a measure for similarity in a spatial neighborhood. This way we avoid the need for an explicit expression of the Riemannian mean and its derivative while at the same time favoring those assignments similar to the Riemannian mean of neighboring assignments.

A label assignment is inferred by minimizing the obj. function: $\inf_{W \in \mathcal{W}} J(W)$. Since $\langle W, D \rangle$ as well as the Riemannian distance are defined and continuous on the whole space Δ^m, so is the objective function J. Yet, this extension is not smooth due to the square root in the sphere map ψ, which isomorphically embeds $\mathcal{S}$ into the Euclidean sphere. Because of $J \colon \Delta^m \to \mathbb{R}$ being continuous and Δ^m compact, the existence of minimizers is assured and

$$\inf_{W \in \mathcal{W}} J(W) = \min_{W \in \Delta^m} J(W). \tag{3.6}$$

As a consequence, the minimizer W^* might not be an element of the assignment manifold $\mathcal{W} = \mathrm{rint}(\Delta^m)$ anymore.

Parameter Influence of ρ. In the following we analyse the set of minimizers for the two extreme cases of the parameter $\rho > 0$ being close to 0 and very large.

Case $\rho \to 0$: For ρ tending to 0, we loose the regularization term in the limit and obtain a minimization problem which separates over the nodes

$$\min_{W \in \Delta^m} \sum_{i \in \mathcal{V}} \langle W_i, D_i \rangle = \sum_{i \in \mathcal{V}} \min_{W_i \in \Delta} \langle W_i, D_i \rangle. \tag{3.7}$$

Assume that every D_i has a unique minimal entry denoted by $D_{im_i} < D_{ij}$ for $j \neq m_i$. Then the unique minimizer W_i^* of $\langle W_i, D_i \rangle$ is given by $W_i^* = e_{m_i} \in \partial\Delta$ for every $i \in \mathcal{V}$. Thus, the objective function J has a unique minimum at a corner point of Δ^m, i.e. W^* is an integral labeling.

Case $\rho \to \infty$: Equivalently characterizing the set of minimizer W^* by

$$\mathrm{argmin}_{W \in \Delta^m} J_{\mathrm{data}}(W) + \rho J_{\mathrm{reg}}(W) = \mathrm{argmin}_{W \in \Delta^m} \tfrac{1}{\rho} J_{\mathrm{data}}(W) + J_{\mathrm{reg}}(W), \tag{3.8}$$

shows, that for ρ tending towards infinity the influence of the data term vanishes in the limit and we are only minimizing with respect to the regularizer, i.e.

$$\mathrm{argmin}_{W \in \Delta^m} \frac{1}{2} \sum_{ij \in \mathcal{E}} \omega_{ij} d_\mathcal{S}^2(W_i, W_j). \tag{3.9}$$

In this case the set of minimizers is given by all $W^* \in \Delta^m$ having identical rows, $W_i^* = W_j^*$ for all $i, j \in \mathcal{V}$, showing the existence of interior optima.

Due to the behaviour of the model in these two extreme cases, it is expected that for larger ρ there exist local optima in the interior constituting fully probabilistic assignments while for decreasing ρ closer to 0 local optima tend to lie closer to the corners, i.e. result in integral assignments. Experimentally, this intuition is indeed confirmed (see Experiments below).

Enforcing Integrality and Connection to Graphical Models. In order to enforce (approximate) integral solutions of the model, also called *rounding*, we add an additional cost term punishing large deviations from integral assignments similar to [6]. This can be done by adding the entropy of W defined by

$$H(W) := \sum_{i \in \mathcal{V}} H(W_i) = -\langle W, \log(W) \rangle \in [0, m\log(n)], \tag{3.10}$$

thus obtaining the extended model with integrality enforcing parameter $\alpha > 0$

$$J_\alpha(W) := J(W) + \alpha H(W) = \langle W, D \rangle + \frac{\rho}{2} \sum_{ij \in \mathcal{E}} \omega_{ij} d_\mathcal{S}^2(W_i, W_j) + \alpha H(W). \tag{3.11}$$

For $\alpha \to 0$ we obtain the the previous model. For $\alpha \to \infty$ we end up with the following minimization problem

$$\min_{W \in \Delta^m} J(W), \quad \text{s.t.} \quad W_i \in \{0,1\}^n \quad \text{for all} \quad i \in \mathcal{V}. \tag{3.12}$$

Under these constraints, every assignment vector equals a standard basis vector $W_i \in \{e_1, \dots, e_n\} \subset \mathbb{R}^n$ and the Riemannian distance in the regularizer only takes on two different values

$$d_{\mathcal{S}}^2(W_i, W_j) = \begin{cases} \pi^2 & \text{, for } W_i \neq W_j \\ 0 & \text{, for } W_i = W_j. \end{cases} \tag{3.13}$$

Therefore, this minimization problem has the form of a graphical model with Potts prior (cf. [10, Sect. 3.3]). It is more appropriate however, to consider our approach as a geometric alternative to the continuous cut variational formulation of the image segmentation problem [5, 8]. For an in-depth discussion of evaluating discrete graphical models using the assignment flow we refer to [6].

4 Optimization Approach

Riemannian Gradient Flow. Our optimization strategy is to follow the Riemannian gradient descend flow of J on the manifold $\mathcal{W}$ with a natural unbiased initialization given by the barycenter at every node

$$\dot{W}(t) = -\nabla_{\mathcal{W}} J_\alpha(W(t)), \quad W(0) = \mathbb{1}_m c^\top \tag{4.1}$$

or due to (2.10) equivalently the transformed flow on $\mathcal{T}$

$$\dot{V}(t) = -\nabla J_\alpha(W(t)), \quad V(0) = 0, \tag{4.2}$$

with $W(t) = \exp_C(V(t))$.

Proposition 1. *Setting* $\bar{\omega}_{ij} = \frac{1}{2}(\omega_{ij} + \omega_{ji})$, *the* i-*th row of the Riemannian gradient and the Euclidean gradient of* J_α *at* $W \in \mathcal{W}$ *are given by*

$$\left(\nabla_{\mathcal{W}} J_\alpha(W)\right)_i = \Pi_{W_i} D_i - \rho \sum_{j \in \mathcal{N}(i)} \bar{\omega}_{ij} \operatorname{Exp}_{W_i}^{-1}(W_j) - \alpha \Pi_{W_i} \log(W_i) \tag{4.3a}$$

$$\left(\nabla J_\alpha(W)\right)_i = P_T D_i - \rho \sum_{j \in \mathcal{N}(i)} \bar{\omega}_{ij} P_T \left[\tfrac{1}{W_i} \operatorname{Exp}_{W_i}^{-1}(W_j)\right] - \alpha P_T \log(W_i). \tag{4.3b}$$

Proof. A standard calculation shows $\nabla\left(J_{\text{data}} + \alpha H\right)(W) = P_T D - \alpha P_T \log(W)$ and therefore $\left(\nabla_{\mathcal{W}}(J_{\text{data}} + \alpha H)(W)\right)_i = \Pi_{W_i} D_i - \alpha \Pi_{W_i} \log(W_i)$ by (2.9) and $\Pi_{W_i} P_T = \Pi_{W_i}$. As for the Riemannian gradient of J_{reg}, we note that the Riemannian gradient of $d_{\mathcal{S}}^2(p,q)$ with respect to p has the form $\nabla_{\mathcal{S},p} d_{\mathcal{S}}^2(p,q) = -2 \operatorname{Exp}_p^{-1}(q)$. The Euclidean gradient then follows by applying $\Pi_{W_i}^{-1}$ to (2.9). $\square$

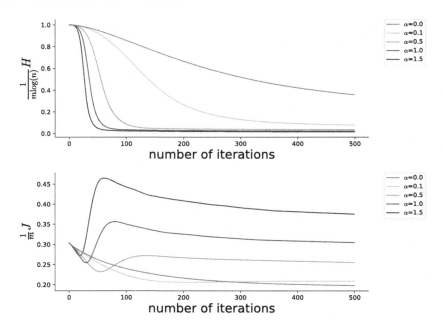

Fig. 1. Convergence and rounding. The normalized average entropy $\frac{1}{m \log(n)} H$ and average energy $\frac{1}{m} J$ for $\rho = 1.5$ and varying α are shown for the first 500 iterations of the algorithm. TOP: Larger values of α causes a faster decay in entropy which forces the flow to converge more rapidly to integral assignments. BOTTOM: Two phases can be identified depending on the parameter α. In the first phase, the flow moves towards a local optima of J up to the point where the dynamics of the added entropy term takes over in the second phase and pushes the flow towards an integral assignment, leading to higher J values. For smaller values of α, the algorithm spends more time minimizing J while for larger α a faster convergence towards integral solutions is favoured.

Numerical Integration of the Flow. For simplicity of notation we denote the i-th row of the Euclidean gradient by $\nabla_i J_\alpha(W)$. We follow [9] for discretizing the gradient flow (4.1) on $\mathcal{W}$. This is done by choosing the common *explicit Euler method* for the transformed flow (4.2) on the linear space $\mathcal{T}$, which reads

$$V_i^{(k+1)} = V_i^{(k)} - h^{(k)} \nabla_i J_\alpha(W^{(k)}), \quad V_i^{(0)} = 0 \quad \text{for all} \quad i \in \mathcal{V}, \quad (4.4)$$

where $h^{(k)} > 0$ denotes the step-size. Transforming this update scheme back onto $\mathcal{W}$ by $W_i^{(k)} = \exp_c(V_i^{(k)})$ with initial condition $W_i^{(0)} = \frac{1}{n} \mathbb{1}_n$ and using the fact that $\nabla_i J_\alpha = \nabla_i J(W) - \alpha P_T \log(W_i)$, we obtain a *multiplicative update*

$$W_i^{(k+1)} = \frac{1}{Z_i} W_i^{(k)} e^{-h^{(k)} \nabla_i J_\alpha(W^{(k)})} = \frac{1}{Z_i'} \left(W_i^{(k)}\right)^{1+h^{(k)}\alpha} e^{-h^{(k)} \nabla_i J(W^{(k)})} \quad (4.5)$$

where Z_i and Z_i' are normalizing constants ensuring $\langle W_i^{(k)}, \mathbb{1} \rangle = 1$. This update formula clearly illustrates the influence of the integrality parameter α as some sort of built in rounding mechanism of the flow.

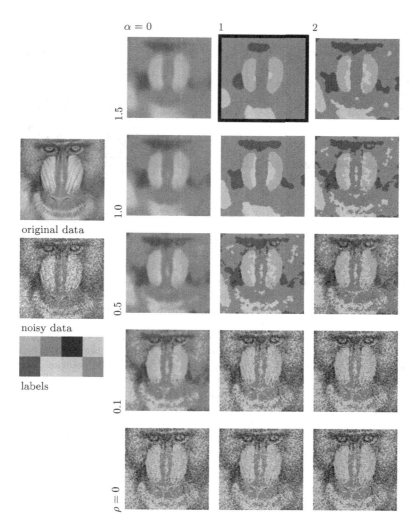

Fig. 2. Parameter influence. Influence of the parameter ρ controlling the amount of regularization and the parameter α enforcing integrality of the solution. The images depict the expected value of the labels given the assignment W after numerical optimization, i.e. the value of pixel i is given by $\mathbb{E}_{W_i}[\mathcal{L}] = \sum_{j \in [n]} W_{ij} l_j$. The flow corresponding to the image in the top row framed in black reached the maximum number of 2000 iteration without $r(W)$ dropping below 10^{-4}. For $\alpha = 0$, even relatively small values $\rho = 0.1$ and larger ones lead to non integral solutions as indicated by the blurred images in the first column, while for $\rho = 0$ we obtain an integral solution without regularization, as expected from (3.7). The other parameter values for $\alpha > 0$ clearly illustrate the mechanism explained in Fig. 1. For smaller α more time is spent to minimize J during the algorithm resulting in more regularized label assignments. Larger α values cause a faster integral decision in an earlier stage of the algorithm leading to less regularized results.

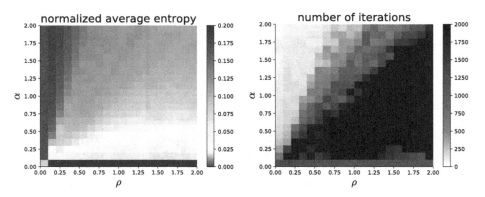

Fig. 3. Interior vs. integral minima. Evaluation of model (3.11) for all parameter combinations (ρ, α) with values $\rho, \alpha \in \{0.0, 0.1, 0.2, \ldots, 1.9, 2.0\}$. LEFT: The normalized average entropy values $\frac{1}{m \log(n)} H$ after numerical optimization are displayed. As expected from the model, larger α values lead to a decrease in entropy favouring integral solutions as discussed in Fig. 1. If ρ increases relative to α then the entropy term weakens and local optima tend towards fractional assignments in the interior, leading to higher entropy values. Two special cases are clearly visible. For $\alpha = 0$ and $\rho > 0$ we are in the regime of (3.5) where local optima tend to lie in the interior causing large entropy values. For $\alpha = \rho = 0$ the optimization problem seperates over the nodes and the flow theoretically converges towards an integral solution as discussed after (3.7) and illustrated in Fig. 2 bottom left. However, the numerical integration scheme (4.6) slows down for $\alpha = 0$ as the flow converges towards an integral solution, preventing it from reaching a low entropy state before the termination criterion of the algorithm is fulfilled. RIGHT: The corresponding number of iterations are shown. For $\alpha = 0$ the termination criterion is reached after about 1000 iterations, while for $\rho > \alpha > 0$ the averaging and integrality enforcing effects compensate each other and result in a slowdown of the algorithm, thus reaching the maximum number of 2000 iterations as indicated by the black region. For $\alpha = \rho$ and increasing α the entropy term dominates and accelerates the integration of the flow, leading to faster convergence.

Assignment Normalization. The flow $W(t)$ solving (4.1) evolves on the manifold $\mathcal{W}$ and therefore all entries $W_{ij}(t) > 0$ are positive all the time. If $W_i(t)$ approaches an integral label, then all but one entries approach 0, for $t \to \infty$. However, the multiplicative update (4.5) is only valid on the manifold $\mathcal{W}$. Since there is a difference between mathematical and numerical positivity we adopt the strategy of [2, Sect. 3.3] for ensuring numerical positivity of the discretized flow. The basic idea is to do an ε-*normalization* every time an entry of W_i drops below $\varepsilon > 0$ given by

$$W_i' = W_i + (\varepsilon - \min_j W_{ij}) \mathbb{1}_n \quad \text{and} \quad W_i \leftarrow \frac{1}{\langle W_i', \mathbb{1} \rangle} W_i'. \tag{4.6}$$

As shown in [3], the ε-normalization has an influence for the discrete flow of [2]. Since the model (3.5) is continuous on Δ^m, applying this normalization strategy only has a negligible effect in our situation for ε close to 0.

5 Experiments

We assess the parameter influence of ρ and α by applying our geometric variational approach to the following image labeling problem. For this, we take a grid graph $\mathcal{G} = (\mathcal{V}, \mathcal{E})$ with neighborhood size $|\mathcal{N}(i)| = 3 \times 3$ for every $i \in \mathcal{V}$ representing the spatial structure of a noisy RGB-image $f : \mathcal{V} \to [0,1]^3$, depicted in Fig. 2. Eight prototypical colors $\{l_1, \ldots, l_8\} \subset [0,1]^3$ (Fig. 2) were used as labels. The distance function $d(f_i, l_j) = \|f_i - l_j\|_1/3$ is used for constructing the distance matrix (3.1) and the weights in the regularizer are set to $\omega_{ij} = 1/|\mathcal{N}(i)|$. For obtaining a fine discretization of the flow, a constant step-size with value $h = 0.1$ is used for numerical integration by applying (4.5) together with $\varepsilon = 10^{-10}$ for the assignment normalization (4.6). As convergence criterion we use the normalized relative change of the objective function J_α, defined by $r(W^{(k)}) := |J_\alpha(W^{(k)}) - J_\alpha(W^{(k-1)})|/|hJ_\alpha(W^{(k-1)})|$. Due to the small (save) step-size and in order to give the dynamics enough time to trade off regularization against rounding in our model, we set the maximum number of iterations to 2000 and stop the algorithm if either $r(W^{(k)})$ drops below a threshold of 10^{-4} or if the maximum number of iterations is reached. The figures provide quantitative illustrations of all aspects of the variational model introduced in Sect. 3. We refer to the figure captions for a detailed discussion.

6 Conclusion

This work clarifies some mathematically open points in connection with the assignment flow. Due to the way in which we utilized the smooth geometric setting and regularization through geometric local averages, the proposed variational model (3.5) avoids the need for an explicit expression of the Riemannian mean and its derivative. To enforce integral assignments, an extended version of the model was introduced in (3.11), by adding an entropy term. The interaction between regularization and entropy minimization results in an interesting dynamics of the Riemannian gradient flow, illustrated by preliminary experiments. The question of convergence properties and a closer investigation of the interplay between entropy and regularization in the optimization process provides an opportunity for further research.

Acknowledgments. We gratefully acknowledge support by the German Science Foundation, grant GRK 1653.

References

1. Absil, P.-A., Mathony, R., Sepulchre, R.: Optimization Algorithms on Matrix Manifolds. Princeton University Press, Princeton (2008)
2. Åström, F., Petra, S., Schmitzer, B., Schnörr, C.: Image labeling by assignment. J. Math. Imag. Vis. **58**(2), 211–238 (2017)
3. Bergmann, R., Fitschen, J.H., Persch, J., Steidl, G.: Iterative multiplicative filters for data labeling. Int. J. Comput. Vis. **123**(3), 435–453 (2017)

4. Bergmann, R., Tenbrinck, D.: A graph framework for manifold-valued data. SIAM J. Imag. Sci. **11**(1), 325–360 (2018)
5. Chan, T.F., Esedoglu, S., Nikolova, M.: Algorithms for finding global minimizers of image segmentation and denoising models. SIAM J. Appl. Math. **66**(5), 1632–1648 (2006)
6. Hühnerbein, R., Savarino, F., Åström, F., Schnörr, C.: Image labeling based on graphical models using wasserstein messages and geometric assignment. SIAM J. Imag. Sci. **11**(2), 1317–1362 (2018)
7. Kappes, J.H., et al.: A comparative study of modern inference techniques for structured discrete energy minimization problems. In: IJCV, vol. 115, no. 2, pp. 155–184 (2015)
8. Lellmann, J., Schnörr, C.: Continuous multiclass labeling approaches and algorithms. SIAM J. Imag. Sci. **4**(4), 1049–1096 (2011)
9. Savarino, F., Hühnerbein, R., Åström, F., Recknagel, J., Schnörr, C.: Numerical integration of riemannian gradient flows for image labeling. In: Lauze, F., Dong, Y., Dahl, A.B. (eds.) SSVM 2017. LNCS, vol. 10302, pp. 361–372. Springer, Cham (2017). https://doi.org/10.1007/978-3-319-58771-4_29
10. Wainwright, M.J., Jordan, M.I.: Graphical models, exponential families, and variational inference. Found. Trends Mach. Learn. **1**(1–2), 1–305 (2008)
11. Weinmann, A., Demaret, L., Storath, M.: Total variation regularization for manifold-valued data. SIAM J. Imag. Sci. **7**(4), 2226–2257 (2014)

Functional Liftings of Vectorial Variational Problems with Laplacian Regularization

Thomas Vogt$^{(\boxtimes)}$ and Jan Lellmann

Institute of Mathematics and Image Computing (MIC), University of Lübeck,
Maria-Goeppert-Str. 3, 23562 Lübeck, Germany
{vogt,lellmann}@mic.uni-luebeck.de

Abstract. We propose a functional lifting-based convex relaxation of variational problems with Laplacian-based second-order regularization. The approach rests on ideas from the calibration method as well as from sublabel-accurate continuous multilabeling approaches, and makes these approaches amenable for variational problems with vectorial data and higher-order regularization, as is common in image processing applications. We motivate the approach in the function space setting and prove that, in the special case of absolute Laplacian regularization, it encompasses the discretization-first sublabel-accurate continuous multilabeling approach as a special case. We present a mathematical connection between the lifted and original functional and discuss possible interpretations of minimizers in the lifted function space. Finally, we exemplarily apply the proposed approach to 2D image registration problems.

Keywords: Variational methods · Curvature regularization · Convex relaxation · Functional lifting · Measure-based regularization

1 Introduction

Let $\Omega \subset \mathbb{R}^d$ and $\Gamma \subset \mathbb{R}^s$ both be bounded sets. In the following, we consider the variational problem of minimizing the functional

$$F(u) = \int_\Omega f(x, u(x), \Delta u(x))dx, \tag{1}$$

that acts on vector-valued functions $u \in C^2(\Omega; \Gamma)$. Convexity of the integrand $f\colon \Omega \times \Gamma \times \mathbb{R}^s \to \mathbb{R}$ is only assumed in the last entry, so that $u \mapsto F(u)$ is generally *non-convex*. The Laplacian Δu is understood component-wise and reduces to u'' if the domain Ω is one-dimensional.

Variational problems of this form occur in a wide variety of image processing tasks, including image reconstruction, restoration, and interpolation. Commonly, the integrand is split into data term and regularizer:

$$f(x, z, p) = \rho(x, z) + \eta(p). \tag{2}$$

© Springer Nature Switzerland AG 2019
J. Lellmann et al. (Eds.): SSVM 2019, LNCS 11603, pp. 559–571, 2019.
https://doi.org/10.1007/978-3-030-22368-7_44

As an example, in *image registration* (sometimes referred to as large-displacement optical flow), the data term $\rho(x,z) = d(R(x), T(x+z))$ encodes the pointwise distance of a reference image $R \colon \mathbb{R}^d \to \mathbb{R}^k$ to a deformed template image $T \colon \mathbb{R}^d \to \mathbb{R}^k$ according to a given distance measure $d(\cdot, \cdot)$, such as the squared Euclidean distance $d(a,b) = \frac{1}{2}\|a-b\|_2^2$. While often a suitable convex regularizer η can be found, the highly non-convex nature of ρ renders the search for global minimizers of (1) a difficult problem.

Instead of directly minimizing F using gradient descent or other local solvers, we will aim to *replace* it by a convex functional $\mathcal{F}$ that acts on a higher-dimensional (*lifted*) function space. If the lifting is chosen in such a way that we can construct global minimizers of F from global minimizers of $\mathcal{F}$, we can find a global solution of the original problem by applying convex solvers to $\mathcal{F}$. While we cannot claim this property for our choice of lifting, we believe that the mathematical motivation and some of the experimental results show that this approach can be a good basis for future work on global solutions of variational models with higher-order regularization.

Calibrations in Variational Calculus. The lifted functional $\mathcal{F}$ proposed in this work is motivated by previous lifting approaches for *first-order* variational problems of the form

$$\min_u F(u) = \int_\Omega f(x, u(x), \nabla u(x))dx, \tag{3}$$

where F acts on functions $u \colon \Omega \to \Gamma$ with $\Omega \subset \mathbb{R}^d$ and scalar range $\Gamma \subset \mathbb{R}$.

The *calibration method* as introduced in [1] gives a globally sufficient optimality condition for functionals of the form (3) with $\Gamma = \mathbb{R}$. Importantly, $f(x, z, p)$ is not required to be convex in (x, z), but only in p. The method states that u minimizes F if there exists a divergence-free vector field $\phi \colon \Omega \times \mathbb{R} \to \mathbb{R}^{d+1}$ (a *calibration*) in a certain admissible set X of vector fields on $\Omega \times \mathbb{R}$ (see below for details), such that

$$F(u) = \int_{\Omega \times \mathbb{R}} \phi \cdot D\mathbf{1}_u, \tag{4}$$

where $\mathbf{1}_u$ is the characteristic function of the subgraph of u in $\Omega \times \mathbb{R}$, $\mathbf{1}_u(x,z) = 1$ if $u(x) > z$ and 0 otherwise, and $D\mathbf{1}_u$ is its distributional derivative. The duality between subgraphs and certain vector fields is also the subject of the broader theory of Cartesian currents [8].

A convex relaxation of the original minimization problem then can be formulated in a higher-dimensional space by considering the functional [5,19]

$$\mathcal{F}(v) := \sup_{\phi \in X} \int_{\Omega \times \mathbb{R}} \phi \cdot Dv, \tag{5}$$

acting on functions v from the *convex* set

$$\mathcal{C} = \{v \colon \Omega \times \mathbb{R} \to [0,1] : \lim_{z \to -\infty} v(x,z) = 1, \lim_{z \to \infty} v(x,z) = 0\}. \tag{6}$$

In both formulations, the set of admissible test functions is

$$X = \{\phi \colon \Omega \times \mathbb{R} \to \mathbb{R}^{d+1} \colon \ \phi^t(x, z) \geq f^*(x, z, \phi^x(x, z)) \tag{7}$$
$$\text{for every } (x, z) \in \Omega \times \mathbb{R}\}, \tag{8}$$

where $f^*(x, z, p) := \sup_q \langle p, q \rangle - f(x, z, q)$ is the convex conjugate of f with respect to the last variable. In fact, the equality

$$F(u) = \mathcal{F}(\mathbf{1}_u) \tag{9}$$

has been argued to hold for $u \in W^{1,1}(\Omega)$ under suitable assumptions on f [19]. A rigorous proof of the case of $u \in BV(\Omega)$ and $f(x, z, p) = f(z, p)$ (f independent of x), but not necessarily continuous in z, can be found in the recent work [2].

In [17], it is discussed how the choice of discretization influences the results of numerical implementations of this approach. More precisely, motivated by the work [18] from continuous multilabeling techniques, the choice of piecewise linear finite elements on Γ was shown to exhibit so-called *sublabel-accuracy*, which is known to significantly reduce memory requirements.

Vectorial Data. The application of the calibration method to *vectorial* data $\Gamma \subset \mathbb{R}^s$, $s > 1$, is not straightforward, as the concept of subgraphs, which is central to the idea, does not translate easily to higher-dimensional range. While the original sufficient minimization criterion has been successfully translated [16], functional lifting approaches have not been based on this generalization so far. In [20], this approach is considered to be intractable in terms of memory and computational performance.

There are functional lifting approaches for vectorial data with first-order regularization that consider the subgraphs of the components of u [9,21]. It is not clear how to generalize this approach to nonlinear data $\Gamma \subset \mathcal{M}$, such as a manifold $\mathcal{M}$, where other functional lifting approaches exist at least for the case of total variation regularization [14].

An approach along the lines of [18] for vectorial data with total variation regularization was proposed in [12]. Even though [17] demonstrated how [18] can be interpreted as a discretized version of the calibration-based lifting, the equivalent approach [12] for vectorial data lacks a fully-continuous formulation as well as a generalization to arbitrary integrands that would demonstrate the exact connection to the calibration method.

Higher-Order Regularization. Another limitation of the calibration method is its limitation to first-order derivatives of u, which leaves out higher-order regularizers such as the Laplacian-based curvature regularizer in image registration [7]. Recently, a functional lifting approach has been successfully applied to second-order regularized image registration problems [15], but the approach was limited to a single regularizer, namely the integral over the 1-norm of the Laplacian (*absolute Laplacian regularization*).

Projection of Lifted Solutions. In the scalar-valued case with first-order regularization, the calibration-based lifting is known to generate minimizers that can be *projected* to minimizers of the original problem by thresholding [19, Theorem 3.1]. This method is also used for vectorial data with component-wise lifting as in [21]. In the continuous multi-labeling approaches [12,14,18], simple averaging is demonstrated to produce useful results even though no theoretical proof is given addressing the accuracy in general. In convex LP relaxation methods, projection (or *rounding*) strategies with provable optimality bounds exist [11] and can be extended to the continuous setting [13]. We demonstrate that rounding is non-trivial in our case, but will leave a thorough investigation to future work.

Contribution. In Sect. 2, we propose a calibration method-like functional lifting approach in the fully-continuous vector-valued setting for functionals that depend in a convex way on Δu. We show that the lifted functional satisfies $\mathcal{F}(\delta_u) \leq F(u)$, where δ_u is the lifted version of a function u and discuss the question of whether the inequality is actually an equality. For the case of absolute Laplacian regularization, we show that our model is a generalization of [15]. In Sect. 2.3, we clarify how convex saddle-point solvers can be applied to our discretized model. Section 3 is concerned with experimental results. We discuss the problem of projection and demonstrate that the model can be applied to image registration problems.

2   A Calibration Method with Vectorial Second-Order Terms

2.1   Continuous Formulation

We propose the following lifted substitute for F:

$$\mathcal{F}(\mathbf{u}) := \sup_{(p,q)\in X} \int_\Omega \int_\Gamma (\Delta_x p(x,z) + q(x,z))\, d\mathbf{u}_x(z) dx, \tag{10}$$

acting on functions $\mathbf{u}\colon \Omega \to \mathcal{P}(\Gamma)$ with values in the space $\mathcal{P}(\Gamma)$ of Borel probability measures on Γ. This means that, for each $x \in \Omega$ and any measurable set $U \subset \Gamma$, the expression $\mathbf{u}_x(U) \in \mathbb{R}$ can be interpreted as the "confidence" of an assumed underlying function on Ω to take a value inside of U at point x. A function $u\colon \Omega \to \Gamma$ can be *lifted* to a function $\mathbf{u}\colon \Omega \to \mathcal{P}(\Gamma)$ by defining $\mathbf{u}_x := \delta_{u(x)}$, the Dirac mass at $u(x) \in \Gamma$, for each $x \in \Omega$.

We propose the following set of test functions in the definition of $\mathcal{F}$:

$$X = \{(p,q):\ p \in C_c^2(\Omega \times \Gamma), q \in L^1(\Omega \times \Gamma), \tag{11}$$

$$z \mapsto p(x,z)\ \text{concave} \tag{12}$$

$$\text{and}\ q(x,z) + f^*(x,z,\nabla_z p(x,z)) \leq 0 \tag{13}$$

$$\text{for every}\ (x,z) \in \Omega \times \Gamma\}, \tag{14}$$

where $f^*(x,z,q) := \sup_{p\in\mathbb{R}^s} \langle q,p \rangle - f(x,z,p)$ is the convex conjugate of f with respect to the last argument.

A thorough analysis of $\mathcal{F}$ requires a careful choice of function spaces in the definition of X as well as a precise definition of the properties of the integrand f and the admissible functions $\mathbf{u}\colon \Omega \to \mathcal{P}(\Gamma)$, which we leave to future work. Here, we present a proof that the lifted functional $\mathcal{F}$ bounds the original functional F from below.

Proposition 1. *Let $f\colon \Omega \times \Gamma \times \mathbb{R}^s \to \mathbb{R}$ be measurable in the first two, and convex in the third entry, and let $u \in C^2(\Omega; \Gamma)$ be given. Then, for $\mathbf{u}\colon \Omega \to \mathcal{P}(\Gamma)$ defined by $\mathbf{u}_x := \delta_{u(x)}$, it holds that*

$$F(u) \geq \mathcal{F}(\mathbf{u}). \tag{15}$$

Proof. Let p, q be any pair of functions satisfying the properties from the definition of X. By the chain rule, we compute

$$\Delta_x p(x, u(x)) = \Delta\left[p(x, u(x))\right] - \sum_{i=1}^{d} \langle \partial_i u(x), D_z^2 p(x, u(x)) \partial_i u(x) \rangle \tag{16}$$
$$- 2\langle \nabla_x \nabla_z p(x, u(x)), \nabla u(x) \rangle - \langle \nabla_z p(x, u(x)), \Delta u(x) \rangle.$$

Furthermore, the divergence theorem ensures

$$-\int_\Omega \langle \nabla_x \nabla_z p(x, u(x)), \nabla u(x) \rangle dx = \int_\Omega \langle \nabla_z p(x, u(x)), \Delta u(x) \rangle dx \tag{17}$$
$$+ \int_\Omega \sum_{i=1}^{d} \langle \partial_i u(x), D_z^2 p(x, u(x)) \partial_i u(x) \rangle dx,$$

as well as $\int_\Omega \Delta\left[p(x, u(x))\right] dx = 0$ by the compact support of p. As $p \in C_c^2(\Omega \times \Gamma)$, concavity of $z \mapsto p(x, z)$ implies a negative semi-definite Hessian $D_z^2 p(x, z)$, so that, together with (16)–(17),

$$\int_\Omega \Delta_x p(x, u(x)) \, dx \leq \int_\Omega \langle \nabla_z p(x, u(x)), \Delta u(x) \rangle \, dx. \tag{18}$$

We conclude

$$\mathcal{F}(\mathbf{u}) = \int_\Omega \int_\Gamma \left(\Delta_x p(x, z) + q(x, z)\right) d\mathbf{u}_x(z) dx \tag{19}$$

$$= \int_\Omega \Delta_x p(x, u(x)) + q(x, u(x)) \, dx \tag{20}$$

$$\overset{(13)}{\leq} \int_\Omega \Delta_x p(x, u(x)) - f^*(x, u(x), \nabla_z p(x, u(x))) \, dx \tag{21}$$

$$\overset{(18)}{\leq} \int_\Omega \langle \nabla_z p(x, u(x)), \Delta u(x) \rangle - f^*(x, u(x), \nabla_z p(x, u(x))) \, dx \tag{22}$$

$$\leq \int_\Omega f(x, u(x), \Delta u(x)) \, dx, \tag{23}$$

where we used the definition of f^* in the last inequality.   □

By a standard result from convex analysis, $\langle p, g \rangle - f^*(x, z, g) = f(x, z, p)$ whenever $g \in \partial_p f(x, z, p)$, the subdifferential of f with respect to p. Hence, for equality to hold in (15), we would need to find a function $p \in C_c^2(\Omega \times \Gamma)$ with

$$\nabla_z p(x, u(x)) \in \partial_p f(x, u(x), \Delta u(x)) \tag{24}$$

and associated $q(x, z) := -f^*(x, z, \Delta u(x))$, such that $(p, q) \in X$ or (p, q) can be approximated by functions from X.

Separate Data Term and Regularizer. If the integrand can be decomposed into $f(x, z, p) = \rho(x, z) + \eta(p)$ as in (2), with $\eta \in C^1(\mathbb{R}^s)$ and u sufficiently smooth, the optimal pair (p, q) in the sense of (24) can be explicitly given as

$$p(x, z) := \langle z, \nabla \eta(\Delta u(x)) \rangle, \tag{25}$$
$$q(x, z) := \rho(x, z) - \eta^*(\nabla \eta(\Delta u(x))). \tag{26}$$

A rigorous argument that such p, q exist for any given u could be made by approximating them by compactly supported functions from the admissible set X using suitable cut-off functions on $\Omega \times \Gamma$.

2.2   Connection to the Discretization-First Approach [15]

In [15], data term ρ and regularizer η are lifted independently from each other for the case $\eta = \| \cdot \|_1$. Following the continuous multilabeling approaches in [6,12,18], the setting is fully discretized in $\Omega \times \Gamma$ in a first step. Then the lifted data term and regularizer are defined to be the convex hull of a constraint function, which enforces the lifted terms to agree on the Dirac measures δ_u with the original functional applied to the corresponding function u. The data term is taken from [12], while the main contribution concerns the regularizer that now depends on the Laplacian of u.

In this section, we show that our fully-continuous lifting is a generalization of the result from [15] after discretization.

Discretization. In order to formulate the discretization-first lifting approach given in [15], we have to clarify the used discretization.

For the image domain $\Omega \subset \mathbb{R}^d$, discretized using points $X^1, \ldots, X^N \in \Omega$ on a rectangular grid, we employ a finite-differences scheme: We assume that, on each grid point X^{i_0}, the discrete Laplacian of $u \in \mathbb{R}^{N,s}$, $u^i \approx u(X^i) \in \mathbb{R}^s$, is defined using the values of u on $m + 1$ grid points $X^{i_0}, \ldots, X^{i_m}$ such that

$$(\Delta u)^{i_0} = \sum_{l=1}^{m} (u^{i_l} - u^{i_0}) \in \mathbb{R}^s. \tag{27}$$

For example, in the case $d = 2$, the popular five-point stencil means $m = 4$ and the X^{i_l} are the neighboring points of X^{i_0} in the rectangular grid. More precisely,

$$\sum_{l=1}^{4} (u^{i_l} - u^{i_0}) = [u^{i_1} - 2u^{i_0} + u^{i_2}] + [u^{i_3} - 2u^{i_0} + u^{i_4}]. \tag{28}$$

The range $\Gamma \subset \mathbb{R}^s$ is triangulated into simplices $\Delta_1, \ldots, \Delta_M$ with altogether L vertices (or *labels*) $Z^1, \ldots, Z^L \in \Gamma$. We write $T := (Z^1 | \ldots | Z^L)^T \in \mathbb{R}^{L,s}$, and define the sparse indexing matrices $P^j \in \mathbb{R}^{s+1,L}$ in such a way that the rows of $T_j := P^j T \in \mathbb{R}^{s+1,s}$ are the labels that make up Δ_j.

There exist piecewise linear finite elements $\Phi_k \colon \Gamma \to \mathbb{R}$, $k = 1, \ldots, L$ satisfying $\Phi_k(t_l) = 1$ if $k = l$, and $\Phi_k(t_l) = 0$ otherwise. In particular, the Φ_k form a partition of unity for Γ, i.e., $\sum_k \Phi_k(z) = 1$ for any $z \in \Gamma$. For a function $p \colon \Gamma \to \mathbb{R}$ in the function space spanned by the Φ_k, with a slight abuse of notation, we write $p = (p_1, \ldots, p_L)$, where $p_k = p(Z^k)$ so that $p(z) = \sum_k p_k \Phi_k(z)$.

Functional Lifting of the Discretized Absolute Laplacian. Along the lines of classical continuous multilabeling approaches, the absolute Laplacian regularizer is lifted to become the convex hull of the constraint function $\phi :$ $\mathbb{R}^L \to \mathbb{R} \cup \{+\infty\}$,

$$
\phi(p) := \begin{cases} \mu \left\| \sum_{l=1}^m (T_{j_l} \alpha^l - T_{j_0} \alpha^0) \right\|, & \text{if } p = \mu \sum_{l=1}^m (P^{j_l} \alpha^l - P^{j_l} \alpha^0), \\ +\infty, & \text{otherwise,} \end{cases} \tag{29}
$$

where $\mu \geq 0$, $\alpha^l \in \Delta_{s+1}^U$ (for Δ_{s+1}^U the unit simplex) and $1 \leq j_l \leq M$ for each $l = 0, \ldots, m$. The parameter $\mu \geq 0$ is enforcing positive homogeneity of ϕ which makes sure that the convex conjugate ϕ^* of ϕ is given by the characteristic function $\delta_\mathcal{K}$ of a set $\mathcal{K} \subset \mathbb{R}^L$. Namely,

$$
\mathcal{K} = \bigcap_{1 \leq j_l \leq M} \{ f \in \mathbb{R}^L : \sum_{l=1}^m (f(t^l) - f(t^0)) \leq \left\| \sum_{l=1}^m (t^l - t^0) \right\|, \tag{30}
$$

$$
\text{for any } \alpha^l \in \Delta_{s+1}^U, l = 0, 1, \ldots, m \}, \tag{31}
$$

where $t^l := T_{j_l} \alpha^l$ and $f(t^l)$ is the evaluation of the piecewise linear function f defined by the coefficients $(f_1, \ldots, f_L)$ (cf. above). The formulation of $\mathcal{K}$ comes with infinitely many constraints so far.

We now show two propositions which give a meaning to this set of constraints for arbitrary dimensions s of the labeling space and an arbitrary choice of norm in the definition of $\eta = \| \cdot \|$. They extend the component-wise (anisotropic) absolute Laplacian result in [15] to the vector-valued case.

Proposition 2. *The set $\mathcal{K}$ can be written as*

$$
\mathcal{K} = \{ f \in \mathbb{R}^L : f \colon \Gamma \to \mathbb{R} \text{ is concave and 1-Lipschitz continuous} \}.
$$

Proof. If the piecewise linear function induced by $f \in \mathbb{R}^L$ is concave and 1-Lipschitz continuous, then

$$
\frac{1}{m} \sum_{l=1}^m (f(t^l) - f(t^0)) = \left(\frac{1}{m} \sum_{l=1}^m f(t^l) \right) - f(t^0) \leq f \left(\frac{1}{m} \sum_{l=1}^m t^l \right) - f(t^0) \tag{32}
$$

$$
\leq \left\| \left(\frac{1}{m} \sum_{l=1}^m t^l \right) - t^0 \right\| = \frac{1}{m} \left\| \sum_{l=1}^m (t^l - t^0) \right\|. \tag{33}
$$

Hence, $f \in \mathcal{K}$. On the other hand, if $f \in \mathcal{K}$, then we recover Lipschitz continuity by choosing $t^l = t^1$, for any l in (30). For concavity, we first prove mid-point concavity. That is, for any $t^1, t^2 \in \Gamma$, we have

$$\frac{f(t^1)+f(t^2)}{2} \leq f\left(\frac{t^1+t^2}{2}\right) \tag{34}$$

or, equivalently, $[f(t^1) - f(t^0)] + [f(t^2) - f(t^0)] \leq 0$, where $t^0 = \frac{1}{2}(t^1 + t^2)$. This follows from (30) by choosing $t^0 = \frac{1}{2}(t^1 + t^2)$ and $t^l = t^0$ for $l > 2$. With this choice, the right-hand side of the inequality in (30) vanishes and the left-hand side reduces to the desired statement. Now, f is continuous by definition and, for these functions, mid-point concavity is equivalent to concavity.   □

The following theorem is an extension of [15, Theorem 1] to the vector-valued case and is crucial for numerical performance, as it shows that the constraints in Proposition 2 can be reduced to a finite number:

Proposition 3. *The set $\mathcal{K}$ can be expressed using not more than $|\mathcal{E}|$ (nonlinear) constraints, where $\mathcal{E}$ is the set of faces (or edges in the 2D-case) in the triangulation.*

Proof. Usually, Lipschitz continuity of a piecewise linear function requires one constraint on each of the simplices in the triangulation, and thus as many constraints as there are gradients. However, together with concavity, it suffices to enforce a gradient constraint on each of the boundary simplices, of which there are fewer than the number of outer faces in the triangulation. This can be seen by considering the one-dimensional case where Lipschitz constraints on the two outermost pieces of a concave function enforce Lipschitz continuity on the whole domain. Concavity of a function $f \colon \Gamma \to \mathbb{R}$ expressed in the basis (Φ_k) is equivalent to its gradient being monotonously decreasing across the common boundary between any neighboring simplices. Together, we need one gradient constraint for each inner, and at most one for each outer face in the triangulation.   □

2.3   Numerical Aspects

For the numerical experiments, we restrict to the special case of integrands $f(x, z, p) = \rho(x, z) + \eta(p)$ as motivated in Sect. 2.1.

Discretization. We base our discretization on the setting in Sect. 2.2. For a function $p \colon \Gamma \to \mathbb{R}$ in the function space spanned by the Φ_k, we note that

$$p(z) = \textstyle\sum_{k=1}^{L} p_k \Phi_k(z) = \langle A^j z - b^j, P^j p \rangle \quad \text{whenever } z \in \Delta_j, \tag{35}$$

where A^j and b^j are such that $\alpha = A^j z - b^j \in \Delta_{s+1}^U$ contains the barycentric coordinates of z with respect to Δ_j. More precisely, for $\bar{T}^j := (P^j T| - e)^{-1} \in \mathbb{R}^{s+1,s+1}$ with $e = (1, \ldots, 1) \in \mathbb{R}^{s+1}$, we set

$$A^j := \bar{T}^j (1{:}\mathsf{s}, :) \in \mathbb{R}^{s,s+1}, \quad b^j := \bar{T}^j (\mathsf{s+1}, :) \in \mathbb{R}^{s+1}. \tag{36}$$

The functions $\mathbf{u}\colon \Omega \to \mathcal{P}(\Gamma)$ are discretized as $u^{ik} := \int_\Gamma \Phi_k(z)d\mathbf{u}_{X^i}(z)$, hence $u \in \mathbb{R}^{N,L}$. Furthermore, whenever $\mathbf{u}_x = \delta_{u(x)}$, the discretization u^i contains the barycentric coordinates of $u(X^i)$ relative to Δ_j. In the context of first-order models, this property is described as sublabel-accuracy in [12,17].

Dual Admissibility Constraints. The admissible set X of dual variables is realized by discretizing the conditions (12) and (13).

Concavity (12) of a function $p\colon \Gamma \to \mathbb{R}$ expressed in the basis (Φ_k) is equivalent to its gradient being monotonously decreasing across the common boundary between any neighboring simplices. This amounts to

$$\langle g^{j_2} - g^{j_1}, n_{j_1,j_2} \rangle \leq 0, \tag{37}$$

where g^{j_1}, g^{j_2} are the (piecewise constant) gradients $\nabla p(z)$ on two neighboring simplices $\Delta_{j_1}, \Delta_{j_2}$, and $n_{j_1,j_2} \in \mathbb{R}^s$ is the normal of their common boundary pointing from Δ_{j_1} to Δ_{j_2}.

The inequality (13) is discretized using (35) similar to the one-dimensional setting presented in [17]. We denote the dependence of p and q on $X^i \in \Omega$ by a superscript i as in q^i and p^i. Then, for any $j = 1, \ldots, M$, we require

$$\sup_{z \in \Delta_j} \langle A^j z - b^j, P^j q^i \rangle - \rho(X^i, z) + \eta^*(g^{ij}) \leq 0 \tag{38}$$

which, for $\rho_j := \rho + \delta_{\Delta_j}$, can be formulated equivalently as

$$\rho_j^*(X^i, (A^j)^T P^j q^i) + \eta^*(g^{ij}) \leq \langle b^j, P^j q \rangle. \tag{39}$$

The fully discretized problem can be expressed in convex-concave saddle point form to which we apply the primal-dual hybrid gradient (PDHG) algorithm [4] with adaptive step sizes from [10]. The epigraph projections for ρ_j^* and η are implemented along the lines of [18,19].

3   Numerical Results

We implemented the proposed model in Python 3 with NumPy and PyCUDA. The examples were computed on an Intel Core i7 4.00 GHz with 16 GB of memory and an NVIDIA GeForce GTX 1080 Ti with 12 GB of dedicated video memory. The iteration was stopped when the Euclidean norms of the primal and dual residuals [10] fell below $10^{-6} \cdot \sqrt{n}$ where n is the respective number of variables.

Image Registration. We show that the proposed model can be applied to two-dimensional image registration problems (Figs. 1 and 2). We used the sum of squared distances (SSD) data term $\rho(x, z) := \frac{1}{2}\|R(x) - T(x+z)\|_2^2$ and squared Laplacian (curvature) regularization $\eta(p) := \frac{1}{2}\|\cdot\|^2$. The image values $T(x + z)$ were calculated using bilinear interpolation with Neumann boundary conditions. After minimizing the lifted functional, we projected the solution by taking averages over Γ in each image pixel.

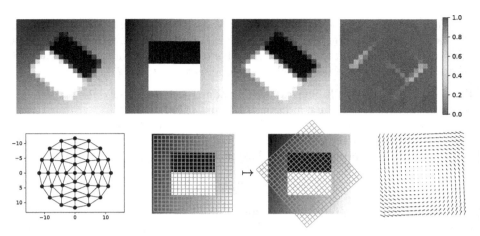

Fig. 1. Application of the proposed higher-order lifting to image registration with SSD data term and squared Laplacian regularization. The method accurately finds a deformation (bottom row, middle and right) that maps the template image (top row, second from left) to the reference image (top row, left), as also visible from the difference image (top row, right). The result (top row, second from right) is almost pixel-accurate, although the range Γ of possible deformation vectors at each point is discretized using only 25 points (second row, left).

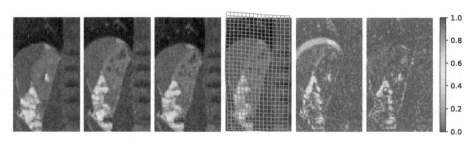

Fig. 2. DCE-MRI data of a human kidney; data courtesy of Jarle Rørvik, Haukeland University Hospital Bergen, Norway; taken from [3]. The deformation (from the left: third and fourth picture) mapping the template (second) to the reference (first) image, computed using our proposed model, is able to significantly reduce the misfit in the left half while fixing the spinal cord at the right edge as can be observed in the difference images from before (fifth) and after (last) registration.

In the first experiment (Fig. 1), the reference image R was synthesized by numerically rotating the template T by 40°. The grid plot of the computed deformation as well as the deformed template are visually very close to the rigid ground-truth deformation (a rotation by 40°). Note that the method obtains almost pixel-accurate results although the range Γ of the deformation is discretized on a disk around the origin, triangulated using only 25 vertices, which is far less than the image resolution.

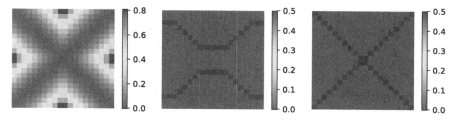

Fig. 3. Minimizers of the lifted functional for the non-convex data term $\rho(x, z) = (|x| - |z|)^2$ (left). With classical first-order total variation-regularized lifting (middle), the result is a composition of two solutions, which can be easily discriminated using thresholding. For the new second-order squared-Laplacian regularized lifting (right), this simple approach fails to separate the two possible (straight line) solutions.

The second experiment (Fig. 2) consists of two coronal slices from a DCE-MRI dataset of a human kidney (data courtesy of Jarle Rørvik, Haukeland University Hospital Bergen, Norway; taken from [3]). The deformation computed using our proposed model is able to significantly reduce the misfit in liver and kidney in the left half while accurately fixing the spinal cord at the right edge.

Projecting the Lifted Solution. In the scalar-valued case with first-order regularization, the minimizers of the calibration-based lifting can be projected to minimizers of the original problem [19, Theorem 3.1]. In our notation, the thresholding technique used there corresponds to mapping $\mathbf{u}$ to

$$u(x) := \inf\{t : \mathbf{u}_x((-\infty, t] \cap \Gamma) > s\}, \tag{40}$$

which is (provably) a global minimizer of the original problem for any $s \in [0, 1)$.

To investigate whether a similar property can hold in our higher-order case, we applied our model with Laplacian regularization $\eta(p) = \frac{1}{2}\|p\|^2$ as well as the calibration method approach with total variation regularization to the data term $\rho(x, z) = (|x| - |z|)^2$ with one-dimensional domain $\Omega = [-1, 1]$ and scalar data $\Gamma = [-1, 1]$ using 20 regularly-spaced discretization points (Fig. 3).

The result from the first-order approach is easily interpretable as a composition of two solutions to the original problem, each of which can be obtained by thresholding (40). In contrast, thresholding applied to the result from the second-order approach yields the two hat functions $v_1(x) = |x|$ and $v_2(x) = -|x|$, neither of which minimizes the original functional. Instead, the solution turns out to be of the form $\mathbf{u} = \frac{1}{2}\delta_{u_1} + \frac{1}{2}\delta_{u_2}$, where u_1 and u_2 are in fact global minimizers of the original problem: namely, the straight lines $u_1(x) = x$ and $u_2(x) = -x$.

4   Conclusion

In this work we presented a novel fully-continuous functional lifting approach for non-convex variational problems that involve Laplacian second-order terms and vectorial data, with the aim to ultimately provide sufficient optimality conditions and find global solutions despite the non-convexity. First experiments indicate that the method can produce subpixel-accurate solutions for the non-convex image registration problem. We argued that more involved projection strategies than in the classical calibration approach will be needed for obtaining a good (approximate) solution of the original problem from a solution of the lifted problem. Another interesting direction for future work is the generalization to functionals that involve arbitrary second- or higher-order terms.

Acknowledgments. The authors acknowledge support through DFG grant LE 4064/1-1 "Functional Lifting 2.0: Efficient Convexifications for Imaging and Vision" and NVIDIA Corporation.

References

1. Alberti, G., Bouchitté, G., Dal Maso, G.: The calibration method for the mumford-shah functional and free-discontinuity problems. Calc. Var. Partial Differ. Equ. **16**(3), 299–333 (2003)
2. Bouchitté, G., Fragalà, I.: A duality theory for non-convex problems in the calculus of variations. Arch. Rational. Mech. Anal. **229**(1), 361–415 (2018)
3. Brehmer, K., Wacker, B., Modersitzki, J.: A novel similarity measure for image sequences. In: Klein, S., Staring, M., Durrleman, S., Sommer, S. (eds.) WBIR 2018. LNCS, vol. 10883, pp. 47–56. Springer, Cham (2018). https://doi.org/10.1007/978-3-319-92258-4_5
4. Chambolle, A., Pock, T.: A first-order primal-dual algorithm for convex problems with applications to imaging. J. Math. Imaging Vis. **40**(1), 120–145 (2011)
5. Chambolle, A.: Convex representation for lower semicontinuous envelopes of functionals in L^1. J. Convex Anal. **8**(1), 149–170 (2001)
6. Chambolle, A., Cremers, D., Pock, T.: A convex approach to minimal partitions. SIAM J. Imaging Sci. **5**(4), 1113–1158 (2012)
7. Fischer, B., Modersitzki, J.: Curvature based image registration. J. Math. Imaging Vis. **18**(1), 81–85 (2003)
8. Giaquinta, M., Modica, G., Souček, J.: Cartesian currents in the calculus of variations I. Cartesian Currents, vol. 37. Springer, Heidelberg (1998)
9. Goldluecke, B., Strekalovskiy, E., Cremers, D.: Tight convex relaxations for vector-valued labeling. SIAM J. Imaging Sci. **6**(3), 1626–1664 (2013)
10. Goldstein, T., Esser, E., Baraniuk, R.: Adaptive primal dual optimization for image processing and learning. In: Proceedings 6th NIPS Workshop Optimize Machine Learning (2013)
11. Kleinberg, J., Tardos, E.: Approximation algorithms for classification problems with pairwise relationships: metric labeling and markov random fields. J. ACM (JACM) **49**(5), 616–639 (2002)
12. Laude, E., Möllenhoff, T., Moeller, M., Lellmann, J., Cremers, D.: Sublabel-accurate convex relaxation of vectorial multilabel energies. In: Leibe, B., Matas, J., Sebe, N., Welling, M. (eds.) ECCV 2016. LNCS, vol. 9905, pp. 614–627. Springer, Cham (2016). https://doi.org/10.1007/978-3-319-46448-0_37
13. Lellmann, J., Lenzen, F., Schnörr, C.: Optimality bounds for a variational relaxation of the image partitioning problem. J. Math. Imaging Vis. **47**(3), 239–257 (2013)

14. Lellmann, J., Strekalovskiy, E., Koetter, S., Cremers, D.: Total variation regularization for functions with values in a manifold. In: Proceedings of the IEEE International Conference on Computer Vision, pp. 2944–2951 (2013)
15. Loewenhauser, B., Lellmann, J.: Functional lifting for variational problems with higher-order regularization. In: Tai, X.-C., Bae, E., Lysaker, M. (eds.) IVLOPDE 2016. MV, pp. 101–120. Springer, Cham (2018). https://doi.org/10.1007/978-3-319-91274-5_5
16. Mora, M.G.: The calibration method for free-discontinuity problems on vector-valued maps. J. Convex Anal. **9**(1), 1–29 (2002)
17. Möllenhoff, T., Cremers, D.: Sublabel-accurate discretization of nonconvex free-discontinuity problems. In: Proceedings of the IEEE International Conference on Computer Vision (ICCV), October 2017
18. Möllenhoff, T., Laude, E., Moeller, M., Lellmann, J., Cremers, D.: Sublabel-accurate relaxation of nonconvex energies. In: Proceedings of the IEEE Conference on Computer Vision and Pattern Recognition, pp. 3948–3956 (2016)
19. Pock, T., Cremers, D., Bischof, H., Chambolle, A.: Global solutions of variational models with convex regularization. SIAM J. Imaging Sci. **3**(4), 1122–1145 (2010)
20. Strekalovskiy, E., Chambolle, A., Cremers, D.: A convex representation for the vectorial mumford-shah functional. In: IEEE Conference on Computer Vision and Pattern Recognition (CVPR), pp. 1712–1719. IEEE (2012)
21. Strekalovskiy, E., Chambolle, A., Cremers, D.: Convex relaxation of vectorial problems with coupled regularization. SIAM J. Imaging Sci. **7**(1), 294–336 (2014)

Author Index

Adams, Matthew   3
Aggrawal, Hari Om   251
Alvarez, Luis   447
Andris, Sarah   67
Augustin, Matthias   67
Aujol, Jean-François   104
Aviles-Rivero, Angelica I.   263

Bačák, Miroslav   183
Batard, Thomas   459
Berger, Marie-Odile   535
Bertalmío, Marcelo   459, 472
Brehmer, Kai   251
Breuß, Michael   79, 199
Bungert, Leon   291
Burger, Martin   291, 485

Calatroni, Luca   472
Cárdenas, Marcelo   303
Carreira, José M.   447
Castan, Fabien   51
Chan, Raymond H.   498
Chung, Julianne   119
Chung, Matthias   119
Cohen, Ido   315
Contelly, Jan   92
Corona, Veronica   263

Dahl, Anders Bjorholm   357
Dahl, Vedrana Andersen   357
De Bortoli, Valentin   13
Debroux, Noémie   263
Deledalle, Charles-Alban   131
Desolneux, Agnès   13
Deutsch, Shay   25
Dong, Yiqiu   144, 156
Duits, Remco   211
Durou, Jean-Denis   51, 104

Effland, Alexander   171

Falik, Adi   315
Franceschi, Valentina   472
Franceschiello, Benedetta   472

Galerne, Bruno   13
Gilboa, Guy   315
Graves, Martin   263

Hansen, Jacob Daniel Kirstejn   369
He, Yuchen   381
Heldmann, Stefan   251
Hertrich, Johannes   183
Hoeltgen, Laurent   79
Horvath, Peter   419
Hosoya, Kento   275
Hühnerbein, Ruben   393
Huo, Limei   144

Imiya, Atsushi   275

Kang, Sung Ha   381
Kiefer, Lukas   406
Kimmel, Ron   38
Kirisits, Clemens   510
Kobler, Erich   171
Korolev, Yury   485

Lauze, François   104, 369
Lazzaro, Damiana   498
Le Guyader, Carole   263
Leclaire, Arthur   13, 341
Lellmann, Jan   559
Li, Lingfeng   224
Lichtenstein, Moshe   38
Lindeberg, Tony   328
Luo, Shousheng   144, 224

Maldonado, Eduard Ramon   459
Masi, Iacopo   25
Mélou, Jean   51
Modersitzki, Jan   251
Molnar, Jozsef   419
Morigi, Serena   498
Mouzon, Thomas   535

Neumayer, Sebastian   183

Pai, Gautam   38
Papadakis, Nicolas   131

Parisotto, Simone   522
Peter, Pascal   92, 303
Petra, Stefania   393, 406, 432
Pierre, Fabien   104, 535
Pock, Thomas   171
Portegies, Jim   211
Prandi, Dario   472

Quéau, Yvain   51

Rabin, Julien   341
Radow, Georg   79
Renaudeau, Arthur   104
Riis, Nicolai André Brogaard   156
Rumpf, Martin   171

Salmon, Joseph   131
Santana-Cedrés, Daniel   447
Savarino, Fabrizio   393, 547
Scherzer, Otmar   510
Schnörr, Christoph   393, 432, 547
Schönlieb, Carola-Bibiane   263, 485, 522
Setterqvist, Eric   510
Sgallari, Fiorella   498
Slagel, J. Tanner   119

Smets, Bart   211
Soatto, Stefano   25
Steidl, Gabriele   183, 459
Stollenwerk, Christiane   485
St-Onge, Etienne   211
Storath, Martin   406

Tahoces, Pablo G.   447
Tai, Xue-Cheng   144, 224
Tasnadi, Ervin   419
Tenbrinck, Daniel   291

Vaiter, Samuel   131
Vogt, Thomas   559

Wang, Yang   144
Weickert, Joachim   67, 92, 236, 303
Weinmann, Andreas   406
Welk, Martin   199, 236
Williams, Guy   263

Yang, Jiang   224

Zern, Artjom   432
Zisler, Matthias   432

Printed in the United States
By Bookmasters